Springer Texts in Statistics

Springer
New York
Berlin
Heidelberg
Hong Kong
London
Milan
Paris
Tokyo

Springer Texts in Statistics

(continued after index)

Yuan Shih Chow   Henry Teicher

Probability Theory

Independence, Interchangeability, Martingales

Third Edition

Springer

Yuan Shih Chow
Department of Statistics
Columbia University
New York, NY 10027
USA

Henry Teicher
Department of Statistics
Rutgers University
New Brunswick, NJ 08903
USA

Library of Congress Cataloging-in-Publication Data
Chow, Yuan Shih, 1924–
Probability theory : independence, interchangeability, martingales / Yuan Shih Chow, Henry Teicher. – 3rd ed.
p. cm. – (Springer texts in statistics)
Includes bibliographical references and index.
1. Probabilities. 2. Martingales (Mathematics) I. Teicher, Henry. II. Title. III. Series.
QA273.C573 1997
519.2 – dc21 97-9299

ISBN 0-387-40607-7 Printed on acid-free paper.

First softcover printing, 2003.

Printed in the United States of America.

9 8 7 6 5 4 3 2 1 SPIN 10947210

www.springer-ny.com

Springer-Verlag New York Berlin Heidelberg
A member of BertelsmannSpringer Science+Business Media GmbH

To our teachers
J. L. Doob and J. Wolfowitz

Preface to the Third Edition

Apart from some additional theorems and examples, simplification of proofs, and correction of typographical errors, the main change is the addition of section 7.5, dealing with U-statistics. In the first two editions (1978, 1988) Lemma 5.4.5 was contained in the proof of Theorem 10.4.1 but seems worth highlighting here.

Preface to the Second Edition

Apart from new examples and exercises, some simplifications of proofs, minor improvements, and correction of typographical errors, the principal change from the first edition is the addition of section 9.5, dealing with the central limit theorem for martingales and more general stochastic arrays.

Preface to the First Edition

Probability theory is a branch of mathematics dealing with chance phenomena and has clearly discernible links with the real world. The origins of the subject, generally attributed to investigations by the renowned French mathematician Fermat of problems posed by a gambling contemporary to Pascal, have been pushed back a century earlier to the Italian mathematicians Cardano and Tartaglia about 1570 (Ore, 1953). Results as significant as the Bernoulli weak law of large numbers appeared as early as 1713, although its counterpart, the Borel strong law of large numbers, did not emerge until 1909. Central limit theorems and conditional probabilities were already being investigated in the eighteenth century, but the first serious attempts to grapple with the logical foundations of probability seem to be Keynes (1921), von Mises (1928; 1931), and Kolmogorov (1933).

An axiomatic mold and measure-theoretic framework for probability theory was furnished by Kolmogorov. In this so-called objective or measure-theoretic approach, definitions and axioms are so chosen that the empirical realization of an event is the outcome of a not completely determined physical experiment—an experiment which is at least conceptually capable of indefinite repetition (this notion is due to von Mises). The concrete or intuitive counterpart of the probability of an event is a long run or limiting frequency of the corresponding outcome.

In contradistinction to the objective approach—where typical realizations of events might be: a coin falls heads, more than 50 cars reach a busy intersection during a specified period, a continuously burning light bulb fails within 1000 hours—the subjective approach to probability advocated by Keynes is designed to encompass realizations such as: it will rain tomorrow, life exists on the planet Saturn, the Iliad and the Odyssey were written by the same author—despite the fact that the experiments in question are clearly unrepeatable. Here the empirical counterpart of probability is degree or intensity of belief.

It is tempting to try to define probability as a limit of frequencies (as advocated by von Mises) rather than as a real number between zero and one satisfying certain postulates (as in the objective approach). Unfortunately, incorporation of repeatability as a postulate (von Mises' "randomness axiom") complicates matters while simultaneously circumscribing the notion of an event. Thus, the probability of the occurrence infinitely often of some particular event in an infinite sequence of repetitions of an experiment—which is of considerable interest in the Kolmogorov schema—is proscribed in (the 1964 rendition of) the von Mises approach (1931). Possibly for these reasons, the frequency approach appears to have lost out to the measure-theoretic. It should be pointed out, however, that justification of the measure-theoretic approach via the Borel strong law of large numbers is circular in that the convergence of the observed frequency of an event to its theoretically defined probability (as the number of repetitions increases) is not pointwise but can only be defined in terms of the concept being justified, viz., probability. If, however, one is willing to ascribe an intuitive meaning to the notion of probability one (hence also, probability zero) then the probability p of any intermediate value can be interpreted in this fashion.

A number of axiomatizations for subjective probability have appeared since Keynes with no single approach dominating. Perhaps the greatest influence of subjective probability is outside the realm of probability theory proper and rather in the recent emergence of the Bayesian school of statistics.

The concern of this book is with the measure-theoretic foundations of probability theory and (a portion of) the body of laws and theorems that emerge therefrom. In the 45 years since the appearance of von Mises' and Kolmogorov's works on the foundations of probability, the theory itself has expanded at an explosive pace. Despite this burgeoning, or perhaps because of the very extent thereof, only the topics of independence, interchangeability, and martingales will be treated here. Thus, such important concepts as Markov and stationary processes will not even be defined, although the special cases of sums of independent random variables and interchangeable random variables will be dealt with extensively. Likewise, continuous parameter stochastic processes, although alluded to, will not be discussed. Indeed, the time seems propitious for the appearance of a book devoted solely to such processes and presupposing familiarity with a significant portion of the material contained here.

Particular emphasis is placed in this book on stopping times—on the one one hand, as tools in proving theorems, and on the other, as objects of interest in themselves. Apropos of the latter, randomly stopped sums, optimal stopping problems, and limit distributions of sequences of stopping rules (i.e., finite stopping times) are of special interest. Wald's equation and its second-moment analogue, in turn, show the usefulness of such stopped sums in renewal theory and elsewhere in probability. Martingales provide a natural vehicle for stopping times, but a formal treatment of the latter cannot await development of the former. Thus, stopping times and, in particular, a sequence of copies of a fixed stopping rule appear as early as Chapter 5, thereby facilitating discussion of the limiting behavior of random walks.

Many of the proofs given and a few of the results are new. Occasionally, a classical notion is looked at through new lenses (e.g., reformulation of the Lindeberg condition). Examples, sprinkled throughout, are used in various guises; to extend theory, to illustrate a theorem that has just appeared, to obtain a classical result from one recently proven.

A novel feature is the attempt to intertwine measure and probability rather than, as is customary, set up between them a sharp demarcation. It is surprising how much probability can be developed (Chapters 2, 3) without even a mention of integration. A number of topics treated later in generality are foreshadowed in the very tractable binomial case of Chapter 2.

This book is intended to serve as a graduate text in probability theory. No knowledge of measure or probability is presupposed, although it is recognized that most students will have been exposed to at least an elementary treatment of the latter. The former is confined for the most part to Chapters 1, 4, 6, with convergence appearing in Section 3.3 (i.e., Section 3 of Chapter 3).[1] Readers familiar with measure theory can plunge into Chapter 5 after reading Section 3.2 and portions of Sections 3.1, 3.3, 4.2, 4.3. In any case, Chapter 2 and also Section 3.4 can be omitted without affecting subsequent developments.

Martingales are introduced in Section 7.4, where the upward case is treated and then developed more generally in Chapter 11. Interchangeable random variables are discussed primarily in Sections 7.3 and 9.2. Apropos of terminology, "interchangeable" is far more indicative of the underlying property than the current "exchangeable," which seems to be a too literal rendition of the french word "échangeable."

A one-year course presupposing measure theory can be built around Chapters 5, 7, 8, 9, 10, 11, 12.

Our warm thanks and appreciation go to Mary Daughaday, Beatrice Williams, and Pat Wolf for their expert typing of the manuscript.

References

J. M. Keynes, *A Treatise on Probability*, 1921; MacMillan, London, 1943.

A. Kolmogorov, *Foundations of the Theory of Probability*, 1933; Chelsea, New York, 1950.

R. von Mises, *Probability, Statistics and Truth*, 1928; Wm. Hodge, London, 1939.

R. von Mises, *Mathematical Theory of Probability and Statistics*, 1931 (H. Geiringer, editor), Academic Press, N.Y., 1964.

O. Ore, "Appendix," *Cardano, The Gambling Scholar*, Princeton University Press, 1953; Holt, New York, 1961.

I. Todhunter, *A History of the Mathematical Theory of Probability*, 1865; Chelsea, New York, 1949.

[1] In the same notational vein, Theorem 3.4.2 signifies Theorem 2 of Section 4 of Chapter 3.

Contents

List of Abbreviations

r.v.	random variable
r.v.s	random variables
d.f.	distribution function
c.f.	characteristic function
p.d.f.	probability density function
u.i.	uniform integrability or uniformly integrable
i.o.	infinitely often
a.c.	almost certainly
a.s.	almost surely
a.e.	almost everywhere
i.d.	infinitely divisible
i.i.d.	independent, identically distributed
iff	if and only if
CLT	Central Limit Theorem
WLLN	Weak Law of Large Numbers
SLLN	Strong Law of Large Numbers
LIL	Law of the Iterated Logarithm
m.g.f.	moment generating function
Cov	Covariance

List of Symbols and Conventions

$\sigma(\mathscr{G})$	σ-algebra generated by the class $\mathscr{G}$
$\sigma(X)$	σ-algebra generated by the random variable X
$\mathrm{E}\,X$	expectation of the random variable X
$\int X$	abbreviated form of the integral $\int X\,dP$
$\mathrm{E}^p\,X$	abbreviated form of $(\mathrm{E}\,X)^p$
$\Vert X \Vert_p$	p-norm of X, that is, $(\mathrm{E}\,\vert X\vert^p)^{1/p}$
$\mathrm{C}(F)$	continuity set of the function F
$\xrightarrow{\text{a.c. or a.s. or a.e.}}$	convergence almost certainly or almost surely or almost everywhere
$\xrightarrow{P \text{ or } d \text{ or } \mu}$	convergence in probability or in distribution or in μ-measure
$\xrightarrow{\mathscr{L}_p}$	convergence in mean of order p
$\xrightarrow{\text{w or c}}$	weak or complete convergence
$\mathscr{B}^n$ or $\mathscr{B}^\infty$	class of n-dimensional or infinite-dimensional Borel sets
$\mathscr{R}\{\ \}$	real part of
$\mathscr{I}\{\ \}$	imaginary part of
$\wedge$	minimum of
$\vee$	maximum of
$a \le \overline{\underline{\lim}}\, Y_n \le b$	simultaneous statement that $a \le \underline{\lim}_{n\to\infty} Y_n \le \overline{\lim}_{n\to\infty} Y_n \le b$
$Z \le z_2$ $\ge z_1$	simultaneous statement that $Z \le z_2$ and $Z \ge z_1$

X_F	fictitious r.v. with d.f. F
$m(X)$	median of X
N_{μ,σ^2}	normal r.v. with mean μ, variance σ^2

1 Classes of Sets, Measures, and Probability Spaces

1.1 Sets and Set Operations

A **set**, in the words of Georg Cantor, the founder of modern set theory, is a collection into a whole of definite, well-distinguished objects of our perception or thought, The objects are called elements and the set is the aggregate of these elements. It is very convenient to extend this notion and also envisage a set devoid of elements, a so-called empty set, and this will be denoted by $\varnothing$. Each element of a set appears only once therein and its order of appearance within the set is irrelevant. A set whose elements are themselves sets will be called a class.

Examples of sets are (i) the set of positive integers denoted by either $\{1, 2, \ldots\}$ or $\{\omega: \omega \text{ is a positive integer}\}$ and (ii) the closed interval with end points a and b denoted by either $\{\omega: a \le \omega \le b\}$ or $[a, b]$. Analogously, the open interval with end points a and b is denoted by $\{\omega: a < \omega < b\}$ or (a, b), while $(a, b]$ and $[a, b)$ are designations for $\{\omega: a < \omega \le b\}$ and $\{\omega: a \le \omega < b\}$ respectively.

The statement that $\omega \in A$ means that ω is an element of the set A and analogously the assertion $\omega \notin A$ means that ω is not an element of the set A or alternatively that ω does not belong to A. If A and B are sets and every element of A is likewise an element of B, this situation is depicted by writing $A \subset B$ or $B \supset A$, and in such a case the set A is said to be a subset of B or contained in B. If both $A \subset B$ and $B \subset A$, then A and B contain exactly the same elements and are said to be equal, denoted by $A = B$. Note that for every set A, $\varnothing \subset A \subset A$.

A set A is termed **countable** if there exists a one-to-one correspondence between (the elements of) A and (those of) some subset B of the set of all positive integers. If, in this correspondence, $B = \{1, 2, \ldots, n\}$, then A is called a **finite** set (with n elements). It is natural to consider $\varnothing$ as a finite set (with zero elements). A set A which is not countable is called **uncountable** or **nondenumerable**.

If A and B are two sets, the difference $A - B$ is the set of all elements of A which do not belong to B; the intersection $A \cap B$ or $A \cdot B$ or simply AB is the set of all elements belonging to both A and B; the union $A \cup B$ is the set of all elements belonging to either A or B (or both); and the symmetric difference $A \,\Delta\, B$ is the set of all elements which belong to A or B but not both. Note that

$$A \cup A = A, \qquad A \cap A = A, \qquad A - A = \varnothing, \qquad A - B = A - (AB) \subset A,$$
$$A \cup B = B \cup A \supset A \supset AB = BA, \qquad A \,\Delta\, B = (A - B) \cup (B - A).$$

Union, intersection, difference, and symmetric difference are termed **set operations**.

If A, B, C are sets and several set operations are indicated, it is, strictly speaking, necessary to indicate via parentheses which operations are to be performed first. However, such specification is frequently unnecessary. For instance, $(A \cup B) \cup C = A \cup (B \cup C)$ and so this double union is independent of order and may be designated simply by $A \cup B \cup C$. Analogously,

$$(AB)C = A(BC) = ABC, \qquad (A \,\Delta\, B) \,\Delta\, C = A \,\Delta\, (B \,\Delta\, C) = A \,\Delta\, B \,\Delta\, C,$$
$$A(B \cup C) = AB \cup AC, \qquad A(B \,\Delta\, C) = AB \,\Delta\, AC.$$

If Λ is a nonempty set whose elements λ may be envisaged as tags or labels, $\{A_\lambda : \lambda \in \Lambda\}$ is a nonempty class of sets. The intersection $\bigcap_{\lambda \in \Lambda} A_\lambda$ (resp. union $\bigcup_{\lambda \in \Lambda} A_\lambda$) is defined to be the set of all elements which belong to A_λ for all $\lambda \in \Lambda$ (resp. for some $\lambda \in \Lambda$). Apropos of order of carrying out set operations, if $*$ denotes any one of $\cup$, $\cap$, $-$, Δ, for any set A it follows from the definitions that

$$\bigcup_{\lambda \in \Lambda} A_\lambda * A = \left(\bigcup_{\lambda \in \Lambda} A_\lambda\right) * A, \qquad A * \bigcup_{\lambda \in \Lambda} A_\lambda = A * \left(\bigcup_{\lambda \in \Lambda} A_\lambda\right),$$
$$\bigcap_{\lambda \in \Lambda} A_\lambda * A = \left(\bigcap_{\lambda \in \Lambda} A_\lambda\right) * A, \qquad A * \bigcap_{\lambda \in \Lambda} A_\lambda = A * \left(\bigcap_{\lambda \in \Lambda} A_\lambda\right).$$

Then

$$A - \bigcup_{\lambda \in \Lambda} A_\lambda = \bigcap_{\lambda \in \Lambda} (A - A_\lambda), \qquad A - \bigcap_{\lambda \in \Lambda} A_\lambda = \bigcup_{\lambda \in \Lambda} (A - A_\lambda).$$

For any sequence $\{A_n, n \ge 1\}$ of sets, define

$$\varliminf_{n \to \infty} A_n = \bigcup_{n=1}^{\infty} \bigcap_{k=n}^{\infty} A_k, \qquad \varlimsup_{n \to \infty} A_n = \bigcap_{n=1}^{\infty} \bigcup_{k=n}^{\infty} A_k$$

and note that, employing the abbreviation i.o. to designate "infinitely often,"

$$\begin{aligned} \varlimsup_{n \to \infty} A_n &= \{\omega : \omega \in A_n \text{ for infinitely many } n\} = \{\omega : \omega \in A_n, \text{ i.o.}\} \\ \varliminf_{n \to \infty} A_n &= \{\omega : \omega \in A_n \text{ for all but a finite number of indices } n\}. \end{aligned} \tag{1}$$

To prove, for example, the first relation, let $A = \{\omega: \omega \in A_n, \text{ i.o.}\}$. Then $\omega \in A$ iff for every positive integer m, there exists $n \geq m$ such that $\omega \in A_n$, that is, iff for every positive integer m, $\omega \in \bigcup_{n=m}^{\infty} A_n$ i.e., iff $\omega \in \bigcap_{m=1}^{\infty} \bigcup_{n=m}^{\infty} A_n$.

In view of (1), $\underline{\lim}\, A_n \subset \overline{\lim}\, A_n$, but these two sets need not be equal (Exercise 3). If $\underline{\lim}\, A_n = \overline{\lim}\, A_n = A$ (say), A is called the limit of the sequence A_n; this situation is depicted by writing $\lim A_n = A$ or $A_n \to A$. If $A_1 \subset A_2 \subset \cdots$ (resp. $A_1 \supset A_2 \supset \cdots$) the sequence A_n is said to be **increasing** (resp. **decreasing**). In either case, $\{A_n, n \geq 1\}$ is called **monotone**.

Palpably, for every monotone sequence A_n, $\lim_{n\to\infty} A_n$ exists; in fact, if $\{A_n\}$ is increasing, $\lim_{n\to\infty} A_n = \bigcup_{n=1}^{\infty} A_n$, while if $\{A_n\}$ is decreasing, $\lim_{n\to\infty} A_n = \bigcap_{n=1}^{\infty} A_n$. Consequently, for any sequence of sets A_n,

$$\overline{\lim_{n\to\infty}}\, A_n = \lim_{n\to\infty} \bigcup_{k=n}^{\infty} A_k,$$

$$\underline{\lim_{n\to\infty}}\, A_n = \lim_{n\to\infty} \bigcap_{k=n}^{\infty} A_k.$$

Exercises 1.1

1. Prove (i) if A_n is countable, $n \geq 1$, so is $\bigcup_{n=1}^{\infty} A_n$; (ii) if A is uncountable and $B \supset A$, then B is uncountable.
2. Show that $\bigcup_{n=1}^{\infty} [0, n/(n+1)) = [0, 1)$, $\bigcap_{n=1}^{\infty} (0, 1/n) = \emptyset$.
3. Prove that $\underline{\lim}_{n\to\infty} A_n \subset \overline{\lim}_{n\to\infty} A_n$. Specify $\overline{\lim}\, A_n$ and $\underline{\lim}\, A_n$ when $A_{2j} = B$, $A_{2j-1} = C$, $j = 1, 2, \ldots$.
4. Verify that $\bigcup_{n=1}^{\infty} A_n = \lim_{n\to\infty} \bigcup_{1}^{n} A_j$ and $\bigcap_{n=1}^{\infty} A_n = \lim_{n\to\infty} \bigcap_{j=1}^{n} A_j$. Moreover, if $\{A_n, n \geq 1\}$ is a sequence of disjoint sets, i.e., $A_i A_j = \emptyset$, $i \neq j$, then
$$\lim_{n\to\infty} \bigcup_{j=n}^{\infty} A_j = \emptyset.$$
5. Prove that $\overline{\lim}_n (A_n \cup B_n) = \overline{\lim}_n A_n \cup \overline{\lim}_n B_n$ and $\underline{\lim}_n A_n \cdot B_n = \underline{\lim}_n A_n \cdot \underline{\lim}_n B_n$. Moreover, $\lim A_n = A$ and $\lim B_n = B$ imply $\lim_n (A_n \cup B_n) = A \cup B$ and
$$\lim A_n B_n = AB.$$
6. Demonstrate that if B is a countable set and $B_n = \{(b_1, \ldots, b_n): b_i \in B \text{ for } 1 \leq i \leq n\}$, then B_n is countable, $n \geq 1$.
7. Prove that the set S consisting of all infinite sequences with entries 0 or 1 is nondenumerable and conclude that the set of real numbers in $[0, 1]$ or any nondegenerate interval is nondenumerable. *Hint*: If S were countable, i.e., $S = \{s_n, n \geq 1\}$ where $s_n = (x_{n1}, x_{n2}, \ldots)$, then $(1 - x_{11}, 1 - x_{22}, \ldots, 1 - x_{nn}, \ldots)$ would be an infinite sequence of zeros and ones not in S.
8. If a_n is a sequence of real numbers, $0 \leq a_n \leq \infty$, prove that
$$\bigcup_{n=1}^{\infty} [0, a_n) = \left[0, \sup_{n\geq 1} a_n\right), \qquad \bigcup_{n=1}^{\infty} \left[0, \left(\frac{n+1}{n}\right)^n\right] \neq \left[0, \sup_{n\geq 1} \left(\frac{n+1}{n}\right)^n\right].$$
9. For any sequence of sets $\{A_n, n \geq 1\}$, define $B_1 = A_1$, $B_{n+1} = B_n \,\Delta\, A_{n+1}$, $n \geq 1$. Prove that $\lim_n B_n$ exists iff $\lim A_n$ exists and is empty.

1.2 Spaces and Indicators

A **space** Ω is an arbitrary, nonempty set and is usually postulated as a reference or point of departure for further discussion and investigation. Its elements are referred to as points (of the space) and will be denoted generically by ω. Thus $\Omega = \{\omega: \omega \in \Omega\}$.

For any reference space Ω, the **complement** A^c of a subset A of Ω is defined by $A^c = \Omega - A$ and the **indicator** I_A of $A \subset \Omega$ is a function defined on Ω by

$$I_A(\omega) = 1 \quad \text{for } \omega \in A, \qquad I_A(\omega) = 0 \quad \text{for } \omega \in A^c.$$

Similarly, for any real function f on Ω and real constants a, b, $I_{[a \le f \le b]}$ signifies the indicator of the set $\{\omega: a \le f(\omega) \le b\}$.

For any subsets A, B of Ω

$$A \subset B \quad \text{iff } A^c \supset B^c,$$

$$(A^c)^c = A, \qquad A \cup A^c = \Omega, \qquad A \cdot A^c = \emptyset, \qquad I_{A^c} = 1 - I_A,$$

$$A - B = AB^c, \qquad I_A \le I_B \quad \text{iff } A \subset B, \qquad I_{A \cup B} \le I_A + I_B,$$

with the last inequality becoming an equality for all ω iff $AB = \emptyset$. Let Λ be an arbitrary set and $\{A_\lambda, \lambda \in \Lambda\}$ a class of subsets of Ω. It is convenient to adopt the conventions

$$\bigcup_{\lambda \in \emptyset} A_\lambda = \emptyset, \qquad \bigcap_{\lambda \in \emptyset} A_\lambda = \Omega.$$

Moreover,

$$\left(\bigcup_{\lambda \in \Lambda} A_\lambda\right)^c = \bigcap_{\lambda \in \Lambda} A_\lambda^c, \qquad \left(\bigcap_{\lambda \in \Lambda} A_\lambda\right)^c = \bigcup_{\lambda \in \Lambda} A_\lambda^c,$$

$$I_{\bigcap_{\lambda \in \Lambda} A_\lambda} = \inf_{\lambda \in \Lambda} I_{A_\lambda}, \qquad I_{\bigcup_{\lambda \in \Lambda} A_\lambda} = \sup_{\lambda \in \Lambda} I_{A_\lambda}.$$

If $A_\lambda \cdot A_{\lambda'} = \emptyset$ for $\lambda, \lambda' \in \Lambda$ and $\lambda \ne \lambda'$, the sets A_λ are called **disjoint**. A class of disjoint sets will be referred to as a **disjoint class**.

If $\{A_n, n \ge 1\}$ is a sequence of subsets of Ω, then $\{I_{A_n}, n \ge 1\}$ is a sequence of functions on Ω with values 0 or 1 and

$$I_{\underline{\lim} A_n} = \underline{\lim}_{n \to \infty} I_{A_n}, \qquad I_{\overline{\lim} A_n} = \overline{\lim}_{n \to \infty} I_{A_n}.$$

Moreover,

$$I_{\bigcup_1^\infty A_n} \le \sum_1^\infty I_{A_n}. \tag{1}$$

Equality holds in (1) iff $\{A_n, n \geq 1\}$ is a disjoint class. The following identity (2) is a refinement of the finite counterpart of (1): For $A_i \subset \Omega, 1 \leq i \leq n$, set

$$s_1 = \sum_1^n I_{A_j}, \qquad s_2 = \sum_{1 \leq j_1 < j_2 \leq n} I_{A_{j_1}A_{j_2}}, \ldots,$$

$$s_n = \sum_{1 \leq j_1 < \cdots < j_n \leq n} I_{A_{j_1}A_{j_2}\cdots A_{j_n}} = I_{A_1A_2\cdots A_n}.$$

Then

$$I_{\cup_1^n A_j} = s_1 - s_2 + s_3 - \cdots + (-1)^{n-1}s_n. \tag{2}$$

In proof of (2), if for some $\omega \in \Omega$, $I_{\cup_1^n A_j}(\omega) = 0$, clearly $s_k(\omega) = 0, 1 \leq k \leq n$, whence (2) obtains. On the other hand, if $I_{\cup_1^n A_j}(\omega) = 1$, then $\omega \in A_j$ for at least one j, $1 \leq j \leq n$. Suppose that ω belongs to exactly m of the sets $A_1, \ldots, A_n$. Then $s_1(\omega) = m$, $s_2(\omega) = \binom{m}{2}, \ldots, s_m(\omega) = 1$, $s_{m+1}(\omega) = \cdots = s_n(\omega) = 0$, whence

$$s_1 - s_2 + \cdots + (-1)^{n-1}s_n = m - \binom{m}{2} + \cdots + (-1)^{m-1}\binom{m}{m} = 1 = I_{\cup_1^n A_j}.$$

Exercises 1.2

1. Verify that

$$A \,\Delta\, B = A^c \,\Delta\, B^c, \qquad C = A \,\Delta\, B \quad \text{iff } A = B \,\Delta\, C,$$

$$\bigcup_1^\infty A_n \,\Delta \bigcup_1^\infty B_n \subset \bigcup_1^\infty (A_n \,\Delta\, B_n),$$

$$\bigcap_1^\infty A_n \,\Delta \bigcap_1^\infty B_n \subset \bigcup_1^\infty (A_n \,\Delta\, B_n).$$

2. Prove that $(\underline{\lim}_{n\to\infty} A_n)^c = \overline{\lim}_{n\to\infty} A_n^c$ and $(\overline{\lim}_{n\to\infty} A_n)^c = \underline{\lim}_{n\to\infty} A_n^c$; also that $\lim B_n = B$ implies $\lim B_n^c = B^c$ and $\lim A \,\Delta\, B_n = A \,\Delta\, B$.

3. Prove that $I_{\overline{\lim}\, A_n} = \overline{\lim}\, I_{A_n}$ and that $I_{\lim A_n} = \lim I_{A_n}$ whenever either side exists.

4. If $A_n \subset \Omega$, $n \geq 1$, show that

$$I_{\cup_{n=1}^\infty A_n} = \max_{n\geq 1} I_{A_n}, \qquad I_{\cap_{n=1}^\infty A_n} = \min_{n\geq 1} I_{A_n}.$$

5. If f is a real function on Ω, then $f^2 = f$ iff f is an indicator of some subset of Ω.

6. Apropos of (2), prove that if B_m is the set of points belonging to exactly $m (1 \leq m \leq n)$ of $A_1, \ldots, A_n$, then

$$I_{B_m} = s_m - \binom{m+1}{m}s_{m+1} + \binom{m+2}{m}s_{m+2} - \cdots + (-1)^{n-m}\binom{n}{m}s_n. \tag{3}$$

7. If $\{f_n, n \geq 0\}$ is a sequence of real functions on Ω with $f_n \uparrow f_0$ and $A_n = \{\omega: f_n(\omega) > c\}$, then $A_n \subset A_{n+1}$ and $\lim A_n = A_0$.

8. If $\{f_n, n \geq 0\}$ is a sequence of real functions with $f_n \uparrow f_0$ and $g_n = f_n I_{[a \leq f_n \leq b]}$ for some constants $-\infty \leq a < b \leq \infty$, then $\{g_n, n \geq 1\}$ is not necessarily increasing. However, if for $n \geq 0$

$$f'_n = f_n I_{[a \leq f_n \leq b]} + a I_{[f_n < a]} + b I_{[f_n > b]},$$

then $f'_n \uparrow f'_0$.

9. If f_1 and f_2 are real functions on Ω, prove that for all real x and rational r

$$\{\omega: f_1(\omega) + f_2(\omega) < x\} = \bigcup_{\text{all } r} \{\omega: f_1(\omega) < r\} \cdot \{\omega: f_2(\omega) < x - r\}.$$

1.3 σ-Algebras, Measurable Spaces, and Product Spaces

Let Ω be a space.

Definition. A nonempty class $\mathscr{A}$ of subsets of Ω is an **algebra** if

i. $A^c \in \mathscr{A}$ whenever $A \in \mathscr{A}$,
ii. $A_1 \cup A_2 \in \mathscr{A}$ whenever $A_j \in \mathscr{A}, j = 1, 2$.

Moreover, $\mathscr{A}$ is a **σ-algebra** if, in addition,

iii. $\bigcup_{n=1}^{\infty} A_n \in \mathscr{A}$ whenever $A_n \in \mathscr{A}, n \geq 1$.

Evidently, (ii) implies that for every positive integer n, $\bigcup_1^n A_j \in \mathscr{A}$ whenever $A_j \in \mathscr{A}$, $1 \leq j \leq n$, while both (i) and (ii) entail $A_1 A_2 \in \mathscr{A}$, if $A_j \in \mathscr{A}$, $j = 1, 2,$; also, since $\mathscr{A}$ is nonempty, $\Omega \in \mathscr{A}$, $\varnothing \in \mathscr{A}$. Clearly, (iii) implies (ii) by taking $A_n = A_2$, $n \geq 2$. Note that a σ-algebra is closed under countable intersections.

Definition. A nonempty class $\mathscr{A}$ of subsets of Ω is a **monotone class** if $\lim A_n \in \mathscr{A}$ for every monotone sequence $A_n \in \mathscr{A}, n \geq 1$.

Obviously, a σ-algebra is a monotone class. Conversely, a monotone algebra $\mathscr{A}$ (i.e., a monotone class which is simultaneously an algebra) is a σ-algebra. For if $A_n \in \mathscr{A}$, $n \geq 1$, then $B_n = \bigcup_{j=1}^n A_j \in \mathscr{A}$, $n \geq 1$, whence $\bigcup_{j=1}^{\infty} A_j = \lim_n B_n \in \mathscr{A}$.

Let S_Ω be the class of all subsets of Ω and $T_\Omega = \{\varnothing, \Omega\}$. Then S_Ω and T_Ω are σ-algebras and for any σ-algebra U_Ω of subsets of Ω, $T_\Omega \subset U_\Omega \subset S_\Omega$.

Definition. The **minimal algebra** $\mathscr{E}'$ (resp. σ-algebra, monotone class) containing a nonempty class $\mathscr{E}$ of subsets of Ω, is an algebra (resp. σ-algebra, monotone class) such that

i. $\mathscr{E}' \supset \mathscr{E}$,

ii. $\mathscr{E}'' \supset \mathscr{E}'$ whenever $\mathscr{E}'' \supset \mathscr{E}$ and $\mathscr{E}''$ is an algebra (resp. σ-algebra, monotone class).

Such a minimal algebra $\mathscr{E}'$ (resp. σ-algebra, monotone class) containing $\mathscr{E}$ is also called the algebra (resp. σ-algebra, monotone class) **generated by** $\mathscr{E}$ and is denoted by $\mathscr{A}(\mathscr{E})$ (resp. $\sigma(\mathscr{E})$, $m(\mathscr{E})$).

To demonstrate the existence of $\mathscr{A}(\mathscr{E})$ (resp. $\sigma(\mathscr{E})$, $m(\mathscr{E})$), let

$$Q_{\mathscr{E}} = \{\mathscr{B} : \mathscr{B} \supset \mathscr{E}, \mathscr{B} \text{ is an algebra (resp. } \sigma\text{-algebra, monotone class)}\}$$

Then $Q_{\mathscr{E}}$ is nonempty since $S_\Omega \in Q_{\mathscr{E}}$. Since an arbitrary intersection of algebras (resp. σ-algebras, monotone classes) is itself an algebra (resp. σ-algebra, monotone class), if $\mathscr{D} = \bigcap_{\mathscr{B} \in Q_{\mathscr{E}}} \mathscr{B}$, then $\mathscr{D}$ is an algebra (resp. σ-algebra, monotone class) and $\mathscr{D} \supset \mathscr{E}$. Obviously, $\mathscr{D}$ is minimal.

For an arbitrary class $\mathscr{E}$ of sets, it is impossible to give a constructive procedure for obtaining $\sigma(\mathscr{E})$ (see Hausdorff (1957)). However, some indication of its structure is given by

Theorem 1. *If $\mathscr{A}$ is an algebra, $m(\mathscr{A}) = \sigma(\mathscr{A})$.*

PROOF. By a prior observation, $\mathscr{A} \subset m(\mathscr{A}) \subset \sigma(\mathscr{A})$. To prove that $m(\mathscr{A}) \supset \sigma(\mathscr{A})$ it suffices to show that $m(\mathscr{A})$ is a σ-algebra, and indeed merely an algebra, again via a prior comment. To this end, for any $B \subset \Omega$, set $\mathscr{D}_B = \{A : A \cdot B^c \in m(\mathscr{A}),\ B \cdot A^c \in m(\mathscr{A}),\ A \cup B \in m(\mathscr{A})\}$. If $A_n \in \mathscr{D}_B$, $A_{n+1} \subset A_n$, $n \geq 1$, and $A = \bigcap_{n=1}^\infty A_n$, then $\{A_n \cdot B^c\}$, $\{B \cdot A_n^c\}$, and $\{A_n \cup B\}$ are monotone sequences in $m(\mathscr{A})$. Thus, $A \cdot B^c$, $B \cdot A^c$, and $A \cup B$ belong to $m(\mathscr{A})$, whence $A \in \mathscr{D}_B$. Similarly, for increasing A_n in $\mathscr{D}_B$, $n \geq 1$, $\lim A_n \in \mathscr{D}_B$. Hence $\mathscr{D}_B$ is a monotone class for all $B \subset \Omega$. If $B \in \mathscr{A}$, $\mathscr{D}_B \supset \mathscr{A}$ since $\mathscr{A}$ is an algebra. Therefore, $\mathscr{D}_B \supset m(\mathscr{A})$ for $B \in \mathscr{A}$. Since $A \in \mathscr{D}_B$ iff $B \in \mathscr{D}_A$, necessarily $\mathscr{D}_B \supset m(\mathscr{A})$ for $B \in m(\mathscr{A})$. This latter means that if $B \in m(\mathscr{A})$ and $E \in m(\mathscr{A})$, then $B\,E^c$, $E\,B^c$, and $B \cup E$ all are in $m(\mathscr{A})$. In other words, $m(\mathscr{A})$ is an algebra. (Recall that $\Omega \in \mathscr{A} \subset m(\mathscr{A})$.) □

Definitions. A nonempty class $\mathscr{A}$ of subsets of Ω is a **π-class** if $AB \in \mathscr{A}$ whenever both $A, B \in \mathscr{A}$, whereas it is a **λ-class** if (i) $\Omega \in \mathscr{A}$, (ii) $A - B \in \mathscr{A}$ for $A, B \in \mathscr{A}$ and $A \supset B$, (iii) $\lim A_n \in \mathscr{A}$ for every increasing sequence $A_n \in \mathscr{A}$, $n \geq 1$.

Note that a λ-class is closed under complementation and hence (iv) $A \cup B \in \mathscr{A}$ if $A, B \in \mathscr{A}$ and $AB = \varnothing$ (since $(A \cup B)^c = B^c - A \in \mathscr{A}$ via (ii)). In view of $A \cup B = A \cup BA^c$, a λ-class which is simultaneously a π-class is a σ-algebra via (iv) and (iii). Furthermore (same proof as for $\mathscr{A}(\mathscr{E})$), there exists a unique minimal λ-class $\lambda(\mathscr{E})$ (resp. π-class) containing a given class $\mathscr{E}$ of sets.

Theorem 2. *If a λ-class $\mathscr{A}$ contains a π-class $\mathscr{D}$, then $\mathscr{A} \supset \sigma(\mathscr{D})$, the σ-algebra generated by $\mathscr{D}$. In particular, if $\mathscr{D}$ is a π-class, $\lambda(\mathscr{D}) = \sigma(\mathscr{D})$.*

PROOF. If $\mathcal{G}$ is the minimal λ-class containing $\mathcal{D}$, it suffices to show that $\mathcal{G}$ is a π-class since for any other λ-class $\mathcal{G}'$ containing $\mathcal{D}$ necessarily $\mathcal{G}' \supset \mathcal{G} \supset \sigma(\mathcal{D})$ because $\mathcal{G}$ is a σ-algebra. To this end, set $\mathcal{G}_1 = \{A: A \subset \Omega, AD \in \mathcal{G}$ for all $D \in \mathcal{D}\}$. Clearly, $\mathcal{G}_1$ is a λ-class containing $\mathcal{D}$, whence $\mathcal{G}_1 \supset \mathcal{G}$. Hence, $AD \in \mathcal{G}$ whenever $D \in \mathcal{D}$, $A \in \mathcal{G}$. Thus, if $\mathcal{G}_2 = \{B: B \subset \Omega, AB \in \mathcal{G}$ for all $A \in \mathcal{G}\}$, then $\mathcal{G}_2$ is a λ-class containing $\mathcal{D}$. Consequently, $\mathcal{G}_2 \supset \mathcal{G}$ and so $AB \in \mathcal{G}$ for both $A, B \in \mathcal{G}$. □

If A is a fixed subset of Ω and $\mathcal{G}$ is any class of subsets of Ω, the class of all sets of the form BA with $B \in \mathcal{G}$ will be denoted by $A \cdot \mathcal{G}$ or $\mathcal{G} \cdot A$ or $\mathcal{G} \cap A$. If A (rather than Ω) is considered as the reference space (relative to which complementation occurs) then $\mathcal{G} \cap A$ is a σ-algebra (resp. π-class, algebra) relative to A if $\mathcal{G}$ is a σ-algebra (resp. π-class, algebra) relative to Ω. When more than one reference space crops up, this will be appended to the class of sets under consideration. Thus, $\sigma_A(\mathcal{D})$ denotes the σ-algebra generated by the class $\mathcal{D}$ but relative to the reference space A.

Theorem 3. *For every non-empty class $\mathcal{D}$ of subsets of Ω and every non-empty set $A \subset \Omega$, $\sigma_\Omega(\mathcal{D}) \cap A = \sigma_A(\mathcal{D} \cap A)$.*

PROOF. Let $\mathcal{G} = \{B: B \subset \Omega, B \cdot A \in \sigma_A(\mathcal{D} \cap A)\}$. Then $\mathcal{G}$ is a σ-algebra relative to Ω and $\mathcal{G} \supset \mathcal{D}$, whence $\mathcal{G} \supset \sigma_\Omega(\mathcal{D})$. In other words, $\sigma_\Omega(\mathcal{D}) \cap A \subset \sigma_A(\mathcal{D} \cap A)$. On the other hand, $\sigma_\Omega(\mathcal{D}) \cap A$ is a σ-algebra containing $\mathcal{D} \cap A$, whence $\sigma_\Omega(\mathcal{D}) \cap A \supset \sigma_A(\mathcal{D} \cap A)$. □

Definition. If $\mathcal{A}$ is a σ-algebra relative to Ω, then the pair $(\Omega, \mathcal{A})$ is called a **measurable space**. The sets of $\mathcal{A}$ are called **measurable sets**.

It is interesting to contrast the definition of a measurable space with that of a topological space. Both measurable and topological spaces engender natural product spaces.

For any measurable spaces $(\Omega_i, \mathcal{A}_i)$, $i = 1, 2, \ldots$ define for $n \geq 2$

$$\bigtimes_{i=1}^{n} A_i = \{(\omega_1, \omega_2, \ldots, \omega_n): \omega_i \in A_i \subset \Omega_i, 1 \leq i \leq n\},$$

$$A_1 \times A_2 \times \cdots \times A_n = \bigtimes_{i=1}^{n} A_i,$$

$$\bigtimes_{i=1}^{n} \mathcal{A}_i = \sigma\left(\left\{\bigtimes_{i=1}^{n} A_i: A_i \in \mathcal{A}_i, 1 \leq i \leq n\right\}\right),$$

$$\bigtimes_{i=1}^{n} (\Omega_i, \mathcal{A}_i) = \left(\bigtimes_{i=1}^{n} \Omega_i, \bigtimes_{i=1}^{n} \mathcal{A}_i\right).$$

Then $\bigtimes_{i=1}^{n} \Omega_i$ is called the **product space** (with components Ω_i, $1 \leq i \leq n$). Moreover, the measurable space $\bigtimes_{i=1}^{n} (\Omega_i, \mathcal{A}_i)$ is termed the n-dimensional **product-measurable space** and $\bigtimes_{i=1}^{n} \mathcal{A}_i$ is the **product σ-algebra**.

Sets of the form $\times_{i=1}^n A_i$ with $A_i \subset \Omega_i$, $1 \le i \le n$, are called (n-dimensional) **rectangles** of the product space $\times_{i=1}^n \Omega_i$, and, moreover, if $A_i \in \mathscr{A}_i$, $1 \le i \le n$, they are dubbed **measurable rectangles** or **rectangles with measurable sides**. Clearly, the intersection (but not the union) of any two measurable rectangles of a given product space is a measurable rectangle in that space. In other words, the class of measurable rectangles of $\times_{i=1}^n \Omega_i$ is a π-class.

Theorem 4. *If $(\Omega_i, \mathscr{A}_i)$, $1 \le i \le n$ are measurable spaces, the class $\mathscr{A}$ of all finite unions of disjoint rectangles $\times_{i=1}^n A_i$ with $A_i \in \mathscr{A}_i$, $1 \le i \le n$, is the algebra generated by the class of all measurable rectangles of the product space $\times_{i=1}^n \Omega_i$.*

PROOF. Let $\mathscr{G}$ denote the π-class of measurable rectangles of $\times_{i=1}^n \Omega_i$. Now $\mathscr{A}$ is also a π-class since if $A_i = \bigcup_{j=1}^{n_i} E_{ij} \in \mathscr{A}$, $i = 1, 2$, with $E_{ij} \in \mathscr{G}$, then $A_1 \cdot A_2 = \bigcup_{h,k} E_{1h} E_{2k} \in \mathscr{A}$. Moreover, if $E = E_1 \times \cdots \times E_n \in \mathscr{G}$,

$$E^c = E_1 \times \cdots \times E_{n-1} \times E_n^c \cup E_1 \times \cdots \times E_{n-2} \times E_{n-1}^c \times \Omega_n$$

$$\cup \cdots \cup E_1^c \times \Omega_2 \times \cdots \times \Omega_n = \bigcup_1^n D_i \text{ (say)} \in \mathscr{A}$$

and so, if $A = \bigcup_{j=1}^r E_j \in \mathscr{A}$ with $E_j \in \mathscr{G}$,

$$A^c = \bigcap_{j=1}^r E_j^c = \bigcap_{j=1}^r \bigcup_{i=1}^n D_i^{(j)} = \bigcup D_{i_1}^{(1)} D_{i_2}^{(2)} \cdots D_{i_r}^{(r)} \in \mathscr{A}.$$

Consequently, if $A, B \in \mathscr{A}$, the preceding ensures that

$$A \cup B = A \cup B \cdot A^c \in \mathscr{A}.$$

Hence $\mathscr{A}$ is an algebra, and since $\mathscr{A} \supset \mathscr{G}$, necessarily $\mathscr{A} \supset \mathscr{A}(\mathscr{G})$. On the other hand, every finite union of disjoint rectangles with measurable sides $\in \mathscr{A}(\mathscr{G})$ and so $\mathscr{A} \subset \mathscr{A}(\mathscr{G})$, whence $\mathscr{A} = \mathscr{A}(\mathscr{G})$. □

Clearly, the σ-algebra $\times_{i=1}^n \mathscr{A}_i$, generated by the rectangles with measurable sides, is also the σ-algebra generated by the algebra $\mathscr{A}$ of Theorem 4.

The points of $\times_{i=1}^3 \Omega_i$, $(\Omega_1 \times \Omega_2) \times \Omega_3$, and $\Omega_1 \times (\Omega_2 \times \Omega_3)$ are respectively of the form $(\omega_1, \omega_2, \omega_3)$, $((\omega_1, \omega_2), \omega_3)$, and $(\omega_1, (\omega_2, \omega_3))$, where $\omega_i \in \Omega_i$ for $i = 1, 2, 3$ and formally they are different. However, between any two of the three points there is a natural one-to-one correspondence and so each one of the points will be identified with $(\omega_1, \omega_2, \omega_3)$. Similarly, $((\omega_1, \omega_2), (\omega_3, \omega_4))$ will be identified with $(\omega_1, \omega_2, \omega_3, \omega_4)$, etc. Under this convention it is easily seen that for $1 \le m < n$

$$\mathop{\times}_{i=1}^n \Omega_i = \left(\mathop{\times}_{i=1}^m \Omega_i\right) \times \left(\mathop{\times}_{i=m+1}^n \Omega_i\right)$$

and not so easily that

$$\mathop{\times}_{i=1}^n \mathscr{A}_i = \left(\mathop{\times}_{i=1}^m \mathscr{A}_i\right) \times \left(\mathop{\times}_{i=m+1}^n \mathscr{A}_i\right).$$

To verify the latter, let

$$X = \bigtimes_{i=1}^{n} \Omega_i, \qquad Y = \bigtimes_{i=1}^{m} \Omega_i, \qquad Z = \bigtimes_{i=m+1}^{n} \Omega_i,$$

$$\mathscr{F} = \bigtimes_{i=1}^{n} \mathscr{A}_i, \qquad \mathscr{G} = \bigtimes_{i=1}^{m} \mathscr{A}_i, \qquad \mathscr{H} = \bigtimes_{i=m+1}^{n} \mathscr{A}_i,$$

$$\mathscr{F}' = \left\{ \bigtimes_{i=1}^{n} A_i : A_i \in \mathscr{A}_i, 1 \le i \le n \right\}, \qquad \mathscr{G}' = \left\{ \bigtimes_{i=1}^{m} A_i : A_i \in \mathscr{A}_i, 1 \le i \le m \right\},$$

$$\mathscr{H}' = \left\{ \bigtimes_{i=m+1}^{n} A_i : A_i \in \mathscr{A}_i, m + 1 \le i \le n \right\}.$$

By definition, $\mathscr{F} = \sigma_X(\mathscr{F}')$, $\mathscr{G} = \sigma_Y(\mathscr{G}')$, $\mathscr{H} = \sigma_Z(\mathscr{H}')$, and $\mathscr{G} \times \mathscr{H} = \sigma(\{A \times B : A \in \mathscr{G}, B \in \mathscr{H}\})$, where σ_Q is the σ-algebra relative to the space Q. Now if $\bigtimes_{i=1}^{n} A_i \in \mathscr{F}'$, then $\bigtimes_{i=1}^{m} A_i \in \mathscr{G}'$ and $\bigtimes_{i=m+1}^{n} A_i \in \mathscr{H}'$, whence $\bigtimes_{i=1}^{n} A_i = (\bigtimes_{i=1}^{m} A_i) \times (\bigtimes_{i=m+1}^{n} A_i) \in \mathscr{G} \times \mathscr{H}$, implying $\mathscr{F} \subset \mathscr{G} \times \mathscr{H}$. On the other hand, if $A \in \mathscr{G}$ and $B \in \mathscr{H}$, then $A \times B = (A \times \Omega_{m+1} \times \cdots \times \Omega_n) \cap (\Omega_1 \times \cdots \times \Omega_m \times B) \in \mathscr{F}$ and so $\mathscr{G} \times \mathscr{H} \subset \mathscr{F}$. □

In the product measurable space $\bigtimes_{i=1}^{n}(\Omega_i, \mathscr{A}_i)$, $n = 1, 2, \ldots$, sets of the form $A \times (\bigtimes_{i=m+1}^{n} \Omega_i)$ with $A \in \bigtimes_{i=1}^{m} \mathscr{A}_i$ for $1 \le m < n$ are called **cylinders** with m-dimensional base A. The quintessence of a **cylinder set** is that all but a finite number of coordinates are unrestricted. The notion of a cylinder is important in the case of an infinite-dimensional product space.

Let $(\Omega_i, \mathscr{A}_i)$, $i = 1, 2, \ldots$, be measurable spaces and define

$$\bigtimes_{i=1}^{\infty} A_i = \{(\omega_1, \omega_2, \ldots) : \omega_i \in A_i \subset \Omega_i, i = 1, 2, \ldots\},$$

$$\mathscr{G} = \bigcup_{m=1}^{\infty} \left\{ \bigtimes_{i=1}^{\infty} A_i : A_i \in \mathscr{A}_i, 1 \le i \le m, \text{and } A_i = \Omega_i, i > m \right\},$$

$$\bigtimes_{i=1}^{\infty} \mathscr{A}_i = \sigma(\mathscr{G}),$$

$$\bigtimes_{i=1}^{\infty} (\Omega_i, \mathscr{A}_i) = \left(\bigtimes_{i=1}^{\infty} \Omega_i, \bigtimes_{i=1}^{\infty} \mathscr{A}_i \right).$$

Then $\bigtimes_{i=1}^{\infty}(\Omega_i, \mathscr{A}_i)$ is the **infinite-dimensional product-measurable space** with components $(\Omega_i, \mathscr{A}_i)$. The σ-algebra in question, $\bigtimes_{i=1}^{\infty} \mathscr{A}_i$, is that generated by cylinders with finite-dimensional (measurable) bases (see Exercise 1.3.6).

In the special case where $(\Omega_i, \mathscr{A}_i) = (\Omega, \mathscr{A})$ for all i, a convenient notation is

$$\Omega^n = \bigtimes_{i=1}^{n} \Omega_i, \qquad \mathscr{A}^n = \bigtimes_{i=1}^{n} \mathscr{A}_i, \qquad (\Omega^n, \mathscr{A}^n) = \bigtimes_{i=1}^{n} (\Omega_i, \mathscr{A}_i), \qquad 1 \le n \le \infty.$$

A prominent example of the preceding considerations arises when $\Omega = R = [-\infty, \infty] = \{\omega: -\infty \le \omega \le \infty\}$. The set R will be called the **real line** and $\pm\infty$ will be considered as real numbers. In contradistinction, $(-\infty, \infty) = \{\omega: -\infty < \omega < \infty\}$ will be termed the **finite real line,** whose elements are finite real numbers. The relationships between $\pm\infty$ and the finite real numbers are such that for every **finite** real number x

$$-\infty < x < \infty, \qquad x + \infty = \infty + x = \infty,$$

$$x + (-\infty) = -\infty + x = -\infty,$$

$$\frac{x}{\infty} = \frac{x}{-\infty} = 0,$$

$$\infty + \infty = (\infty)\cdot(\infty) = (-\infty)\cdot(-\infty) = -(-\infty) = \infty,$$

$$-\infty + (-\infty) = \infty\cdot(-\infty) = (-\infty)\cdot\infty = -(\infty) = -\infty,$$

$$x\cdot\infty = \infty\cdot x = (-x)\cdot(-\infty) = (-\infty)\cdot(-x) = \infty \quad \text{or} \quad 0 \quad \text{or} \quad -\infty,$$

according as $x > 0, x = 0, x < 0$. Note that ∞/∞ and $\infty - \infty$ are not defined. For $x \in R$, the sets $[-\infty, x]$, $(-\infty, x)$, $(-\infty, x]$, $[-\infty, x)$, $[x, \infty]$, (x, ∞), $[x, \infty)$, $(x, \infty]$ are termed infinite intervals.

The elements of the σ-algebra $\mathscr{B}$ generated by the class of infinite intervals of the form $[-\infty, x)$, $-\infty < x < \infty$, are known as the **Borel sets** (of the line) or **linear Borel sets** or Borel sets in R. The measurable space $(R, \mathscr{B})$ is called the **Borel line** or 1-dimensional Borel space.

The product measurable spaces $(R^n, \mathscr{B}^n)$ and $(R^\infty, \mathscr{B}^\infty)$, emanating from the Borel line $(R, \mathscr{B})$, are called the ***n*-dimensional Borel space** and **infinite-dimensional Borel space** respectively. Moreover, the sets of $\mathscr{B}^n$ are termed ***n*-dimensional Borel sets.** $1 \le n \le \infty$. Since every point $\omega \in R^\infty$ is an infinite sequence of real numbers, the infinite dimensional Borel space $(R^\infty, \mathscr{B}^\infty)$ is also alluded to as a sequence space.

For any interval $J \subset R$ the σ-algebra generated by the class of all subintervals of J coincides with $\mathscr{B} \cap J$ according to Theorem 3. Thus, it is natural to describe the σ-algebra in question as the class of Borel subsets of J. Similar comments apply to any n-dimensional interval (rectangle) $J^n \subset R^n$.

EXERCISES 1.3

1. Prove that $\mathscr{A}$ is a σ-algebra if and only if it is both a λ-class and a π-class. If a class $\mathscr{A} \subset \mathscr{D}$, then $\sigma(\mathscr{A}) \subset \sigma(\mathscr{D})$.

2. If $\mathscr{F}$ is an algebra such that for every disjoint sequence $\{F_n, n \ge 1\}$ of sets in $\mathscr{F}$ $\bigcup_{n=1}^\infty F_n \in \mathscr{F}$, then $\mathscr{F}$ is a σ-algebra.

3. Let $(\Omega_i, \mathscr{A}_i)$ be measurable spaces and $\mathscr{A}_i = \sigma(\mathscr{G}_i)$ with $\Omega_i \in \mathscr{G}_i$, $1 \le i \le n$. If $\mathscr{G} = \{A_1 \times A_2 \times \cdots \times A_n: A_i \in \mathscr{G}_i, 1 \le i \le n\}$, then $\sigma(\mathscr{G}) = \mathscr{A}_1 \times \mathscr{A}_2 \times \cdots \times \mathscr{A}_n$ on $\Omega_1 \times \Omega_2 \times \cdots \times \Omega_n$.

4. If $\mathscr{A}_n$, $n \geq 1$ is an increasing sequence of σ-algebras, then $\mathscr{A} = \bigcup_{n=1}^{\infty} \mathscr{A}_n$ is merely an algebra.

5. The σ-algebra generated by a countable class of disjoint, nonempty sets whose union $= \Omega$ is the class of all unions of these sets.

6. If $(\Omega_i, \mathscr{A}_i)$, $i \geq 1$, are measurable spaces, the class $\mathscr{E}$ of all cylinder sets of $\times_1^\infty \Omega_i$ with bases in $\times_1^m \mathscr{A}_i$ for some $m \geq 1$ is an algebra, but not a σ-algebra. Moreover, setting $\mathscr{D} = \{\times_{i=1}^\infty A_i : A_i \in \mathscr{A}_i\}$ verify that $\times_{i=1}^\infty \mathscr{A}_i = \sigma(\mathscr{D}) = \sigma(\mathscr{E})$.

7. Let $\mathscr{D}$ be a π-class of subsets of Ω and $\mathscr{G}$ the class of all finite unions of disjoint sets of $\mathscr{D}$ with $\varnothing \in \mathscr{D}$. If $D^c \in \mathscr{G}$ for every $D \in \mathscr{D}$, prove that $\mathscr{G}$ is the algebra generated by $\mathscr{D}$.

8. Show that the class of Borel sets $\mathscr{B}$ may be generated by $\{(x, \infty], -\infty < x < \infty\}$ or by $\mathscr{S} = \{\{+\infty\}, [-\infty, \infty], [a, b), -\infty \leq a \leq b \leq \infty\}$.

9. Prove that $A = \{(x, y): x^2 + y^2 < r^2\}$ is a Borel set of R^2. *Hint*: A is a countable union of open (classical) rectangles. Utilize this to prove that $\{(x, y): x^2 + y^2 \leq r^2\}$ is likewise a Borel set and hence also the circumference of a circle.

10. If $\mathscr{A}$ is the class of open sets of $Q^n = (-\infty, \infty)^n$, then $\sigma(\mathscr{A}) = \mathscr{B}^n \cap Q^n$.

1.4 Measurable Transformations

Let Ω_1 and Ω_2 be reference sets and X a function defined on Ω_1, with values in Ω_2, the latter being denoted by $X: \Omega_1 \to \Omega_2$. For every subset A of Ω_2 and class $\mathscr{G}$ of subsets of Ω_2, define

$$X^{-1}(A) = \{\omega: \omega \in \Omega_1, X(\omega) \in A\}, \qquad X^{-1}(\mathscr{G}) = \{X^{-1}(A): A \in \mathscr{G}\}.$$

The set $X^{-1}(A)$ and the class $X^{-1}(\mathscr{G})$ are called respectively the **inverse images** of the set A and of the class $\mathscr{G}$. Clearly, if for each $\lambda \in \Lambda$ the set $A_\lambda \subset \Omega_2$, then

$$X^{-1}\left(\bigcup_{\lambda \in \Lambda} A_\lambda\right) = \bigcup_{\lambda \in \Lambda} X^{-1}(A_\lambda), \qquad X^{-1}(A_\lambda - A_{\lambda'}) = X^{-1}(A_\lambda) - X^{-1}(A_{\lambda'}),$$

$$X^{-1}\left(\bigcap_{\lambda \in \Lambda} A_\lambda\right) = \bigcap_{\lambda \in \Lambda} X^{-1}(A_\lambda),$$

and hence the inverse image $X^{-1}(\mathscr{G})$ of a σ-algebra $\mathscr{G}$ on Ω_2 is a σ-algebra on Ω_1 and the class $\{B: B \subset \Omega_2, X^{-1}(B) \in \mathscr{H}\}$ is a σ-algebra on Ω_2 if $\mathscr{H}$ is a σ-algebra on Ω_1.

Lemma 1. *For any mapping $X: \Omega_1 \to \Omega_2$ and class $\mathscr{A}$ of subsets of Ω_2,*

$$X^{-1}(\sigma_{\Omega_2}(\mathscr{A})) = \sigma_{\Omega_1}(X^{-1}(\mathscr{A})).$$

PROOF. Since $X^{-1}(\sigma_{\Omega_2}(\mathscr{A}))$ is a σ-algebra containing $X^{-1}(\mathscr{A})$, it must be shown for any σ-algebra $\mathscr{D}$ of subsets of Ω_1 with $\mathscr{D} \supset X^{-1}(\mathscr{A})$ that $\mathscr{D} \supset X^{-1}(\sigma_{\Omega_2}(\mathscr{A}))$. Since $\mathscr{D}$ is a σ-algebra, the class $\mathscr{B} = \{B: B \subset \Omega_2, X^{-1}(B) \in \mathscr{D}\}$ is likewise a

σ-algebra and this together with the relation $\mathscr{B} \supset \mathscr{A}$ implies $\mathscr{B} \supset \sigma_{\Omega_2}(\mathscr{A})$. Thus

$$\mathscr{D} \supset X^{-1}(\mathscr{B}) \supset X^{-1}(\sigma_{\Omega_2}(\mathscr{A})). \qquad \square$$

Suppose that $X: \Omega_1 \to \Omega_2$ and $Y: \Omega_2 \to \Omega_3$. If $Y(X)$ is defined by the usual composition, that is,

$$Y(X)(\omega) = Y(X(\omega)), \qquad \omega \in \Omega,$$

then $Y(X): \Omega_1 \to \Omega_3$ and $(Y(X))^{-1}(A) = X^{-1}(Y^{-1}(A))$ for every $A \subset \Omega_3$.

If $(\Omega_1, \mathscr{A}_1)$ and $(\Omega_2, \mathscr{A}_2)$ are measurable spaces and $X: \Omega_1 \to \Omega_2$ then X is said to be a **measurable transformation** from $(\Omega_1, \mathscr{A}_1)$ to $(\Omega_2, \mathscr{A}_2)$ or an $\mathscr{A}_1$-**measurable function** from Ω_1 to $(\Omega_2, \mathscr{A}_2)$ provided $X^{-1}(\mathscr{A}_2) \subset \mathscr{A}_1$.

Suppose that X is a measurable transformation from $(\Omega_1, \mathscr{A}_1)$ to $(\Omega_2, \mathscr{A}_2)$ and Y is a measurable transformation from $(\Omega_2, \mathscr{A}_2)$ to $(\Omega_3, \mathscr{A}_3)$. It follows immediately from the definition that $Y(X)$ is a measurable transformation from $(\Omega_1, \mathscr{A}_1)$ to $(\Omega_3, \mathscr{A}_3)$.

Theorem 1. *If X_i is a measurable transformation from $(\Omega, \mathscr{A})$ to $(\Omega_i, \mathscr{A}_i)$, $1 \leq i \leq n$, where n may be infinite, then $X(\omega) = (X_1(\omega), \ldots, X_n(\omega))$ is a measurable transformation from $(\Omega, \mathscr{A})$ to the product measurable space $\times_{i=1}^{n}(\Omega_i, \mathscr{A}_i)$.*

PROOF. If $n < \infty$, let

$$\mathscr{G} = \left\{ \underset{i=1}{\overset{n}{\times}} A_i : A_i \in \mathscr{A}_i,\ 1 \leq i \leq n \right\},$$

while if $n = \infty$, take

$$\mathscr{G} = \bigcup_{m=1}^{\infty} \left\{ \underset{i=1}{\overset{\infty}{\times}} A_i : A_i \in \mathscr{A}_i,\ 1 \leq i \leq m, \text{ and } A_i = \Omega_i,\ i > m \right\}.$$

Then $\sigma(\mathscr{G}) = \times_{i=1}^{n} \mathscr{A}_i$ and by the prior lemma

$$\sigma(X^{-1}(\mathscr{G})) = X^{-1}(\sigma(\mathscr{G})) = X^{-1}\left(\underset{i=1}{\overset{n}{\times}} \mathscr{A}_i\right).$$

Since $X_i^{-1}(\mathscr{A}_i) \subset \mathscr{A}$ for each i, $X^{-1}(\mathscr{G}) \subset \mathscr{A}$, whence

$$\mathscr{A} \supset \sigma(X^{-1}(\mathscr{G})) = X^{-1}\left(\underset{i=1}{\overset{n}{\times}} \mathscr{A}_i\right). \qquad \square$$

Next, if $\times_{i=1}^{2}(\Omega_i, \mathscr{A}_i)$ is a product-measurable space, define for $A \subset \Omega_1 \times \Omega_2$

$$A^{(1)}(\omega_1) = \{\omega_2 : (\omega_1, \omega_2) \in A\} \quad \text{for } \omega_1 \in \Omega_1,$$
$$A^{(2)}(\omega_2) = \{\omega_1 : (\omega_1, \omega_2) \in A\} \quad \text{for } \omega_2 \in \Omega_2.$$

The sets $A^{(1)}(\omega_1)$ and $A^{(2)}(\omega_2)$ are called **sections** of A (at ω_1 and ω_2 respectively).

Theorem 2. *Let* $\times_{i=1}^{2}(\Omega_i, \mathscr{A}_i)$ *be a product measurable space.*

i. *For* $A \in \mathscr{A}_1 \times \mathscr{A}_2$ *and* $\omega_j \in \Omega_j$, *the sections* $A^{(j)}(\omega_j) \in \mathscr{A}_{3-j}$, $j = 1, 2$.
ii. *If* T *is a measurable transformation from* $\times_{i=1}^{2}(\Omega_i, \mathscr{A}_i)$ *to a measurable space* $(\Omega, \mathscr{A})$, *then for every* $\omega_2 \in \Omega_2$, $T(\omega_1, \omega_2)$ *defines a measurable transformation from* $(\Omega_1, \mathscr{A}_1)$ *to* $(\Omega, \mathscr{A})$.

PROOF. Let

$$\mathscr{G} = \{A: A \in \mathscr{A}_1 \times \mathscr{A}_2, A^{(1)}(\omega_1) \in \mathscr{A}_2 \text{ for each } \omega_1 \in \Omega_1\},$$
$$\mathscr{D} = \{D: D = D_1 \times D_2 \text{ where } D_i \in \mathscr{A}_i, i = 1, 2\}.$$

Then $\mathscr{A}_1 \times \mathscr{A}_2 = \sigma(\mathscr{D})$ by definition. For $D = D_1 \times D_2 \in \mathscr{D}$, $D^{(1)}(\omega_1) = D_2$ or $\varnothing$ according as $\omega_1 \in D_1$ or not. Hence $\mathscr{G} \supset \mathscr{D}$. It is easy to verify that the section of a union (resp. difference) is the union (resp. difference) of the sections, i.e., for $\bigcup_1^\infty A_n \subset \Omega_1 \times \Omega_2$ and all $\omega_1 \in \Omega_1$

$$(A_1 - A_2)^{(1)}(\omega_1) = A_1^{(1)}(\omega_1) - A_2^{(1)}(\omega_1), \quad \left(\bigcup_1^\infty A_n\right)^{(1)}(\omega_1) = \bigcup_1^\infty A_n^{(1)}(\omega_1).$$

Therefore, $\mathscr{G}$ is a σ-algebra and so $\mathscr{G} \supset \sigma(\mathscr{D}) = \mathscr{A}_1 \times \mathscr{A}_2$, proving (i) for $j = 1$. Similarly, for $A^{(2)}(\omega_2)$.

To prove (ii), let $B \in \mathscr{A}$. Then $T^{-1}(B) \in \mathscr{A}_1 \times \mathscr{A}_2$ and for every $\omega_2 \in \Omega_2$, by (i) $\{\omega_1: T(\omega_1, \omega_2) \in B\} = (T^{-1}(B))^{(2)}(\omega_2) \in \mathscr{A}_1$. Therefore, $T(\omega_1, \omega_2)$ is measurable from $(\Omega_1, \mathscr{A}_1)$ to $(\Omega, \mathscr{A})$ for every $\omega_2 \in \Omega_2$. □

Since outcomes of interest in a chance experiment can usually be quantized, numerical-valued functions play an important role in probability theory. A measurable transformation X from a measurable space $(\Omega, \mathscr{A})$ to the Borel line $(R, \mathscr{B})$ is called a **(real) measurable function on** $(\Omega, \mathscr{A})$. Since $\mathscr{B}$ is the σ-algebra generated by the class of intervals $[-\infty, x)$, $-\infty < x < \infty$, or by $(x, \infty]$, $-\infty < x < \infty$, it follows from Lemma 1 that a real-valued function X on Ω is measurable iff for every finite x, $\{X(\omega) < x\} \in \mathscr{A}$ or

$$\{X(\omega) > x\} \in \mathscr{A}, x \in (-\infty, \infty).$$

In the special case of a measurable transformation f from the Borel line $(R, \mathscr{B})$ to itself, f is termed a **Borel** or **Borel-measurable function**. Since every open set in $(-\infty, \infty)$ is a countable union of open intervals and every open interval is a Borel set, every continuous function on $(-\infty, \infty)$ is a Borel function. Hence a real function on $[-\infty, \infty]$ with range $[-\infty, \infty]$ which is continuous on $(-\infty, \infty)$ is always a Borel function.

Similarly, for any finite or infinite interval $J \subset R$ measurable transformations from $(J, \mathscr{B} \cdot J)$ to $(R, \mathscr{B})$ are termed Borel functions or **Borel functions on** J. Analogously, for any finite or infinite rectangle $J^n \subset R^n$, measurable functions from $(J^n, \mathscr{B}^n \cdot J^n)$ to $(R, \mathscr{B})$ are called **Borel functions of** n **variables** or Borel functions on J^n. Since every open set of $(-\infty, \infty)^n$ is a countable union of open rectangles, continuous functions on $(-\infty, \infty)^n$ are likewise Borel functions.

Let $(\Omega, \mathscr{A})$ be a measurable space and X, Y real functions on Ω. A **complex-valued function** $Z = X + iY$ on Ω is called **$\mathscr{A}$-measurable** or simply measurable whenever both X and Y are $\mathscr{A}$-measurable.

If $\{X_n, n \geq 1\}$ are real measurable functions on $(\Omega, \mathscr{A})$, the set equality $\{\omega: \sup_{n\geq 1} X_n(\omega) > x\} = \bigcup_{n=1}^{\infty}\{\omega: X_n(\omega) > x\}$ reveals that $\sup_{n\geq 1} X_n$ is a measurable function. Analogously, $\inf_{n\geq 1} X_n$, $\overline{\lim}_{n\to\infty} X_n = \inf_{k\geq 1}\sup_{n\geq k} X_n$, and $\underline{\lim}\, X_n$ are all measurable. In particular, $\max(X_1, X_2)$ and $\min(X_1, X_2)$ are measurable. Since an identically constant function is trivially measurable, it follows that if X is a real measurable function, so are its **positive** and **negative parts** defined by

$$X^+ = \max(0, X) \quad \text{and} \quad X^- = \max(0, -X).$$

Measurability of the sum (if defined) of two measurable functions is a simple consequence of the relation (Exercise 1.2.9)

$$\{\omega: X_1(\omega) + X_2(\omega) < x\} = \bigcup_{\text{rational } r} \{\omega: X_1(\omega) < r\}\cdot\{\omega: r < x - X_2(\omega)\}.$$

Thus, $X = X^+ - X^-$ is measurable iff X^+ and X^- are measurable.

Clearly, if n is a positive integer and c is any constant, measurability of X implies that of X^n and cX. Measurability of the product of two measurable functions is then a simple consequence of the identity

$$X_1 X_2 = \tfrac{1}{4}[(X_1 + X_2)^2 - (X_1 - X_2)^2].$$

Likewise, the ratio of two measurable functions (if defined) is measurable.

If $\{X_n, n \geq 1\}$ are real measurable functions, measurability of $\underline{\lim}_{n\to\infty} X_n$ and $\overline{\lim}_{n\to\infty} X_n$ and Exercise 1.4.5 ensure that $\lim_{n\to\infty} X_n$ (if it exists) is measurable.

As an aid in the discussion of integration in later chapters, it is helpful to define classes of functions with properties mirroring those of a λ-class of sets.

Definition. A family $\mathscr{H}$ of nonnegative functions is called a **λ-system** if

i. $1 \in \mathscr{H}$,
ii. $X_n \in \mathscr{H}, n \geq 1, X_n \uparrow X \Rightarrow X \in \mathscr{H}$,
iii. $X_i \in \mathscr{H}$, c_i real, finite, $i = 1, 2$, and $c_1 X_1 + c_2 X_2 \geq 0$ $\Rightarrow c_1 X_1 + c_2 X_2 \in \mathscr{H}$.

Definition. A nonempty family $\mathscr{M}$ of nonnegative functions is called a **monotone system** if the following conditions are satisfied:

i. If $X_i \in \mathscr{M}$ and c_i is a finite nonnegative number, $i = 1, 2$, then $c_1 X_1 + c_2 X_2 \in \mathscr{M}$,
ii. if $X_n \in \mathscr{M}$ and $X_n \uparrow X$, then $X \in \mathscr{M}$.

Note that a λ-system is always a montone system but not conversely. A connection between classes of functions and classes of sets beyond that furnished by indicator functions is given next. In the course of the proof, an arbitrary nonnegative measurable function is shown to be an increasing limit of simple (see Exercise 1.4.3) nonnegative measurable functions, a phenomenon that will reappear frequently in Chapters 4 and 6.

Theorem 3. *Let $\mathscr{H}$ be a family of nonnegative functions on Ω which contains all indicators of sets of some class $\mathscr{D}$ of subsets of Ω. If either* (i) *$\mathscr{D}$ is a π-class and $\mathscr{H}$ is a λ-system, or* (ii) *$\mathscr{D}$ is a σ-algebra and $\mathscr{H}$ is a monotone system, then $\mathscr{H}$ contains all nonnegative $\sigma(\mathscr{D})$-measurable functions.*

PROOF. Set $\mathscr{G} = \{A: I_A \in \mathscr{H}\}$. (i) By hypothesis $\mathscr{G} \supset \mathscr{D}$ and, according to (i), (ii), (iii) of the definition of a λ-system, $\mathscr{G}$ is a λ-class, whence $\mathscr{G} \supset \sigma(\mathscr{D})$ by Theorem 1.3.2. If X is a nonnegative $\sigma(\mathscr{D})$-measurable function, then $J_{k,n} = \{k/2^n \le X < (k+1)/2^n\} \in \mathscr{G}$, $0 \le k \le 4^n - 1$, as does $J_{4^n,n} = \{X \ge 2^n\}$. Hence, if

$$X_n = \sum_{k=1}^{4^n} \frac{k}{2^n} I_{J_{k,n}},$$

$X_n \in \mathscr{H}$ by (iii). Clearly $X_n \uparrow X$ and so $X \in \mathscr{H}$ by (ii). In case (ii), where $\mathscr{H}$ is a monotone system and $\mathscr{D}$ is a σ-algebra, $\mathscr{G} \supset \mathscr{D} = \sigma(\mathscr{D})$ and the rest follows as before. □

If X is a real $\mathscr{A}$-measurable function so that $X^{-1}(\mathscr{B}) \subset \mathscr{A}$, it is natural to speak of $X^{-1}(\mathscr{B})$ as the σ-algebra generated by X and hence to denote it by $\sigma(X)$. Then X is also a measurable function from $(\Omega, \sigma(X))$ to $(R, \mathscr{B})$ and indeed $\sigma(X)$ is the smallest σ-algebra of subsets of Ω for which that statement is true. Moreover, if g is a Borel function on R, then $g(X)$ is a real $\mathscr{A}$-measurable function and $\sigma(g(X)) \subset \sigma(X)$.

Definition. The σ-algebra $\sigma(X_\lambda, \lambda \in \Lambda)$ generated by a nonempty family $\{X_\lambda, \lambda \in \Lambda\}$ of real measurable functions on $(\Omega, \mathscr{A})$ is defined to be the σ-algebra generated by the class of sets $\{\omega: X_\lambda(\omega) \in B\}$, $B \in \mathscr{B}$, $\lambda \in \Lambda$, or equivalently by $\{X_\lambda < x\}$, $x \in (-\infty, \infty)$, $\lambda \in \Lambda$.

Clearly, $\sigma(X_\lambda, \lambda \in \Lambda) \subset \mathscr{A}$ and $\sigma(X_\lambda, \lambda \in \Lambda)$ is the smallest σ-algebra relative to which all X_λ, $\lambda \in \Lambda$, are measurable.

In particular, for $\Lambda = \{1, 2, \ldots, n\}$ the σ-algebra generated by $X_1, \ldots, X_n$, namely $\sigma(X_1, \ldots, X_n)$, is the σ-algebra generated by the class of sets $\{X_i \in B_i\}$, $B_i \in \mathscr{B}$, $1 \le i \le n$, or equivalently by the sets of $\sigma(X_i)$, $1 \le i \le n$. Thus, $\sigma(X_1, \ldots, X_n) = \sigma(\bigcup_1^n \sigma(X_i))$.

When $\Lambda = \{1, 2, \ldots\}$, it follows analogously that $\sigma(X_1, X_2, \ldots) = \sigma(\bigcup_1^\infty \sigma(X_i))$. Consequently, for $n \ge 2$

$$\sigma(X_1) \subset \sigma(X_1, X_2) \subset \cdots \subset \sigma(X_1, \ldots, X_n),$$

$$\sigma(X_1, X_2, \ldots) \supset \sigma(X_2, X_3, \ldots) \supset \cdots \supset \sigma(X_n, X_{n+1}, \ldots).$$

The σ-algebra $\bigcap_{n=1}^\infty \sigma(X_n, X_{n+1}, \ldots)$, which is called the tail σ-algebra of $\{X_n, n \ge 1\}$, will be encountered in Chapter 3.

If X_i is a measurable function from $(\Omega, \mathscr{A})$ to $(R, \mathscr{B})$, $1 \le i \le n$, then X_i is likewise a (real) measurable function from $(\Omega, \sigma(X_i))$ to $(R, \mathscr{B})$, $1 \le i \le n$, and, moreover, each X_i is a measurable function from $(\Omega, \sigma(X_1 \cdots X_n))$ to $(R, \mathscr{B})$, $1 \le i \le n$. The next theorem characterizes functions which are $\sigma(X_1, \ldots, X_n)$-measurable.

Theorem 4. *Let $X_1, \ldots, X_n$ be real measurable functions on $(\Omega, \mathscr{A})$. A (real) function on Ω is $\sigma(X_1, \ldots, X_n)$-measurable iff it has the form $f(X_1, \ldots, X_n)$, where f is a Borel function on R^n.*

PROOF. Let $\mathscr{G}$ denote the class of Borel functions on R^n and define

$$\mathscr{H} = \{f(X_1, \ldots, X_n): f \in \mathscr{G}, f(X_1, \ldots, X_n) \ge 0\},$$
$$\mathscr{D} = \{\{(\omega: X_i(\omega) \le x_i, 1 \le i \le n\}: -\infty < x_i < \infty, 1 \le i \le n\}$$

Then $\mathscr{D}$ is a π-class of subsets of Ω, $\sigma(\mathscr{D}) = \sigma(X_1, \ldots, X_n)$, and $I_D \in \mathscr{H}$ for $D \in \mathscr{D}$. Moreover, $\mathscr{H}$ is a λ-system. In fact, $1 \in \mathscr{H}$, and if $f_i(X_1, \ldots, X_n) \in \mathscr{H}$, c_i is a finite real number for $i = 1, 2$, and

$$c_1 f_1(X_1, \ldots, X_n) + c_2 f_2(X_1, \ldots, X_n) \ge 0,$$

then $c_1 f_1(X_1, \ldots, X_n) + c_2 f_2(X_1, \ldots, X_n) = f(X_1, \ldots, X_n) \in \mathscr{H}$ where

$$f = (c_1 f_1 + c_2 f_2) I_{[c_1 f_1 = -c_2 f_2 = \pm\infty]^c}.$$

Now let $f_m(X_1, \ldots, X_n) \in \mathscr{H}$, $m \ge 1$, and $f_m(X_1, \ldots, X_n) \uparrow f(X_1, \ldots, X_n)$. Set $g_m = \max(f_1, \ldots, f_m)$. Then $g_m \uparrow g \in \mathscr{G}$ and $0 \le f_m(X_1, \ldots, X_n) = g_m(X_1, \ldots, X_n) \le f_{m+1}(X_1, \ldots, X_n)$. Therefore,

$$f(X_1, \ldots, X_n) = g(X_1, \ldots, X_n) \in \mathscr{H}$$

and $\mathscr{H}$ is a λ-system. By Theorem 3, $\mathscr{H}$ contains all nonnegative $\sigma(\mathscr{D})$-measurable functions. Hence, if Y is $\sigma(X_1, \ldots, X_n)$-measurable, both Y^+ and Y^- are in $\mathscr{H}$ so that $Y = Y^+ - Y^- = f(X_1, \ldots, X_n)$ for some Borel function f on R^n.

Conversely, let f be a Borel function on R^n and

$$X(\omega) = (X_1(\omega), \ldots, X_n(\omega)), \qquad \omega \in \Omega.$$

Since X_i is measurable from $(\Omega, \sigma(X_1, \ldots, X_n))$ to $(R, \mathscr{B})$ for $1 \le i \le n$, X is measurable from $(\Omega, \sigma(X_1, X_2, \ldots, X_n))$ to $(R^n, \mathscr{B}^n)$ by Theorem 1. By hypothesis, f is measurable from $(R^n, \mathscr{B}^n)$ to $(R, \mathscr{B})$, whence $f(X_1, \ldots, X_n) = f(X)$ is measurable from $(\Omega, \sigma(X_1, \ldots, X_n))$ to $(R, \mathscr{B})$, that is, $f(X_1, \ldots, X_n)$ is $\sigma(X_1, \ldots, X_n)$-measurable. □

In similar fashion one can prove

Theorem 5. *If $\{X_\lambda, \lambda \in \Lambda\}$ is an infinite family of real measurable functions on $(\Omega, \mathscr{A})$, then a real measurable function on Ω is $\sigma(X_\lambda, \lambda \in \Lambda)$-measurable iff it is of the form $f(X_{\lambda_1}, X_{\lambda_2}, \ldots, X_{\lambda_n}, \ldots)$, where $\lambda_i \in \Lambda$, $i = 1, 2, \ldots$, and f is an infinite-dimensional Borel function.*

EXERCISES 1.4

1. Let $T: \Omega_1 \to \Omega_2$ and $A_\lambda \subset \Omega_1$ for $\lambda \in \Lambda$. Show that $T(\bigcup_\lambda A_\lambda) = \bigcup_\lambda T(A_\lambda)$ but that $T(A_\lambda - A_{\lambda'})$ need not equal $T(A_\lambda) - T(A_{\lambda'})$, where $T(A) = \{T(\omega_1): \omega_1 \in A\}$, $T(\mathscr{A}) = \{T(A): A \in \mathscr{A}\}$. If T is-one-to-one and onto, then $\mathscr{A}$ a σ-algebra on Ω_1 entails $T(\mathscr{A})$ is a σ-algebra on Ω_2.

2. In the notation of Exercise 1, prove that $A \subset T^{-1}(T(A))$ with set equality holding if T is one-to-one. Also, $T(T^{-1}(B)) \subset B \subset \Omega_2$ with equality if T is onto.

3. A function X on $(\Omega, \mathscr{F}, P)$ is called **simple** if for some finite integer n and real numbers $x_1, \ldots, x_n$ it is representable as $X = \sum_{i=1}^n x_i I_{A_i}$ for $\{A_i, 1 \le i \le n\}$ a disjoint subclass of $\mathscr{F}$. Prove that any non-negative measurable function is an increasing limit of non-negative simple functions.

4. Prove that $(X + Y)^+ \le X^+ + Y^+$, $(X + Y)^- \le X^- + Y^-$, $X^+ \le (X + Y)^+ + Y^-$. Also if $B_n \uparrow B$, then $X^{-1}(B_n) \uparrow X^{-1}(B)$.

5. Let X and Y be real measurable functions on $(\Omega, \mathscr{A})$ and c any real constant. Show that $\{\omega: X(\omega) < Y(\omega) + c\}$, $\{\omega: X(\omega) \le Y(\omega) + c\}$, $\{\omega: X(\omega) = Y(\omega)\} \in \mathscr{A}$.

6. Prove that if X is a real measurable function on $(\Omega, \mathscr{A})$, so is $|X|$. Is the converse true?

7. If X and Y are real functions on $(\Omega, \mathscr{A})$ then $X + iY$ is measurable iff (X, Y) is measurable from $(\Omega, \mathscr{A})$ to $(R^2, \mathscr{B}^2)$.

8. If $(\Omega_i, \mathscr{A}_i)$, $i = 1, 2$ are measurable spaces and $X: \Omega_1 \to \Omega_2$ then X is a measurable transformation from $(\Omega_1, \mathscr{A}_1)$ to $(\Omega_2, \mathscr{A}_2)$ provided $X^{-1}(\mathscr{D}) \subset \mathscr{A}_1$ for some class $\mathscr{D}$ of sets for which $\sigma(\mathscr{D}) = \mathscr{A}_2$.

9. Prove that any real monotone function on R is Borel measurable and has at most a countable number of discontinuities.

10. For any linear Borel set B and r.v. X prove that $\sigma(X I_{[X \in B]}) = X^{-1}(B) \cap \sigma(X)$.

1.5 Additive Set Functions, Measures, and Probability Spaces

Let Ω be a space and $\mathscr{A}$ be a nonempty class of subsets of Ω. A **set function** μ on $\mathscr{A}$ is a real-valued function defined on $\mathscr{A}$. If $\mu(A)$ is finite for each $A \in \mathscr{A}$, μ is said to be **finite**, and if there exists $\{A_n, n \ge 1\} \subset \mathscr{A}$ such that $\mu(A_n)$ is finite for each n and $\bigcup_{n=1}^\infty A_n = \Omega$, μ is said to be **σ-finite** on $\mathscr{A}$.

If $A \subset \Omega$ and $A = \bigcup_{n=1}^m A_n$, where $\{A_n, n = 1, \ldots, m\}$ is a disjoint subclass of $\mathscr{A}$, the latter subclass is called a **finite partition of A in $\mathscr{A}$**. If $\{A_n, n = 1, 2, \ldots\}$ is a disjoint subclass of $\mathscr{A}$ and $\bigcup_{n=1}^\infty A_n = A$, it is called a **σ-partition of A in $\mathscr{A}$**.

Definition. A set function μ on $\mathscr{A}$, denoted $\mu(A)$ or $\mu\{A\}$ for $A \in \mathscr{A}$ is **additive** (or more precisely, **finitely additive**), if for every $A \in \mathscr{A}$ and every finite partition $\{A_n, n = 1, \ldots, m\}$ of A in $\mathscr{A}$, $\sum_1^m \mu(A_n)$ is defined and

$$\mu(A) = \sum_1^m \mu(A_n);$$

moreover, μ is **σ-additive** or **countably additive** or **completely additive** if for every $A \in \mathscr{A}$ and every σ-partition $\{A_n, n = 1, 2, \ldots\}$ of A in $\mathscr{A}$, $\sum_1^\infty \mu(A_n)$ is defined and

$$\mu(A) = \sum_1^\infty \mu(A_n).$$

Note in the σ-additive case that the definition precludes conditional convergence of the series. Clearly, if $\varnothing \in \mathscr{A}$, countable additivity implies finite additivity.

If an additive set function is finite on some set A of an algebra $\mathscr{A}$, it is necessarily finite on all $B \in \mathscr{A}$ with $B \subset A$. Examples of set functions that are additive but not σ-additive appear in Exercises 1.5.6 and 1.5.7.

Definition. A nonnegative σ-additive set function μ on a class $\mathscr{A}$ containing $\varnothing$ with $\mu\{\varnothing\} = 0$ is called a **measure**. If μ is a measure on a σ-algebra $\mathscr{F}$ of subsets of Ω, the triplet $(\Omega, \mathscr{F}, \mu)$ is called a **measure space**. A measure space $(\Omega, \mathscr{F}, \mathrm{P})$ is a **probability space** if $\mathrm{P}\{\Omega\} = 1$.

From a purely formal vantage point the prior definition relegates probability theory to a special case of measure theory. However, such important notions as independence (Chapter 3) and other links between probability and the real world have nurtured probability theory and given it a life and direction of its own.

In addition to the basic property of σ-additivity, a measure μ on an algebra $\mathscr{A}$ is **monotone**, that is, $\mu\{A_1\} \leq \mu\{A_2\}$ whenever $A_1 \subset A_2$, $A_i \in \mathscr{A}$, $i = 1, 2$ (Exercise 1), and, moreover, **subadditive**, that is, if $\bigcup_1^\infty A_j \in \mathscr{A}$, $A_j \in \mathscr{A}$, $j \geq 1$, then

$$\mu\left\{\bigcup_1^\infty A_j\right\} \leq \sum_{j=1}^\infty \mu\{A_j\}, \tag{1}$$

as noted in (iii) of the forthcoming Theorem 2. A prominent example of a measure is ordinary "length" on the class $\mathscr{A}$ of all finite half-open intervals $[a, b)$, i.e., $\mu\{[a, b)\} = b - a$. The extension of the definition of length to $\sigma(\mathscr{A})$ (= class of all Borel sets) is known as Lebesgue measure and will be encountered in Chapter 6.

Probability spaces will underlie most if not all of this book. In a probability space $(\Omega, \mathscr{F}, \mathrm{P})$, the sets $A \in \mathscr{F}$ are called **events** and the nonnegative, real number $\mathrm{P}\{A\}$ is referred to as the **probability** of the event A. The monotone property of a measure ensures that for every event A

$$0 = \mathrm{P}\{\varnothing\} \leq \mathrm{P}\{A\} \leq \mathrm{P}\{\Omega\} = 1. \tag{2}$$

Thus, in probability theory "event" is simply a name attached to an element of a σ-algebra of subsets of a basic reference set Ω, frequently called the **sample space**. From an intuitive or applicational standpoint an event is the abstract counterpart of an observable outcome of a (not completely determined) physical experiment. The numerical-valued probability of an event is in some sense an idealization of the intuitive notion of a long-run frequency as attested to by (2) and the additivity property.

Events of probability zero are called **null events**. A real-valued measurable function X on a probability space $(\Omega, \mathscr{F}, P)$ is called a **random variable** (abbreviated r.v.) if $\{\omega: |X(\omega)| = \infty\}$ is a null event. If some property obtains except on a null event, this property is said to obtain **almost surely** (abbreviated a.s.) or **almost certainly** (abbreviated a.c.) or with probability one. Hence, a r.v. on a probability space $(\Omega, \mathscr{F}, P)$ is just an a.c. finite $\mathscr{F}$-measurable function on Ω.

It is an extremely useful state of affairs that a probability or more generally a measure defined on an algebra $\mathscr{A}$ or even a semi-algebra (definition forthcoming) may be uniquely extended to $\sigma(\mathscr{A})$. The proof will be deferred to Chapter 6 but a first step in this direction is Theorem 1.

Definition. A **semi-algebra** $\mathscr{S}$ is a π-class of subsets of Ω such that $\Omega \in \mathscr{S}$, $\emptyset \in \mathscr{S}$ and for each $A \in \mathscr{S}$ there is a finite partition of A^c in $\mathscr{S}$.

Let $\mathscr{S}$ be a semi-algebra. It follows easily that

i. the class $\mathscr{G}$ of all finite unions of disjoint sets of $\mathscr{S}$ is the algebra generated by $\mathscr{S}$ (clearly, $\mathscr{G} \subset \mathscr{A}(\mathscr{S})$ and $\mathscr{G}$ is an algebra containing $\mathscr{S}$ and hence $\mathscr{A}(\mathscr{S})$);
ii. for $\{A_n, n = 1, \ldots, m\} \subset \mathscr{S}$, there is a finite partition of $A_1 A_2^c \cdots A_m^c$ in $\mathscr{S}$;
iii. for each $A \in \mathscr{S}$ and each countable class $\{A_n, n = 1, 2, \ldots\} \subset \mathscr{S}$ such that $\bigcup_1^\infty A_n \supset A$, there is a σ-partition $\{B_n, n = 1, 2, \ldots\}$ of A in $\mathscr{S}$ for which each B_n is a subset of some A_m (write $A = \bigcup_n AA_n$ as a disjoint union and utilize (ii)).

Let $\mathscr{G}$ and $\mathscr{H}$ be two classes of subsets of Ω with $\mathscr{G} \subset \mathscr{H}$. If μ and ν are set functions defined on $\mathscr{G}$ and $\mathscr{H}$ respectively such that $\mu(A) = \nu(A)$ for $A \in \mathscr{G}$, ν is said to be an **extension of μ to $\mathscr{H}$**, and μ the **restriction of ν to $\mathscr{G}$**, denoted by $\mu = \nu|_{\mathscr{G}}$.

Theorem 1. *If μ is a nonnegative additive set function on a semi-algebra $\mathscr{S}$, then there exists a unique extension ν of μ to the algebra $\mathscr{A}$ generated by $\mathscr{S}$ such that ν is additive. Moreover, if μ is σ-additive on $\mathscr{S}$, so is ν on $\mathscr{A}$.*

PROOF. Since $\mathscr{A}$ is the class of all finite unions of disjoint sets in $\mathscr{S}$, every $A \in \mathscr{A}$ has a finite partition $\{A_n, 1 \le n \le m\}$ in $\mathscr{S}$. For such an A, define

$$\nu(A) = \sum_1^m \mu(A_n).$$

Then ν is consistently defined on $\mathscr{A}$, since if $A \in \mathscr{A}$ has two distinct finite partitions $\{A_n, 1 \le n \le m\}$ and $\{B_j, 1 \le j \le k\}$ in $\mathscr{S}$,

$$\nu(A) = \sum_{n=1}^{m} \mu(A_n) = \sum_{n=1}^{m} \mu(AA_n) = \sum_{n=1}^{m} \sum_{j=1}^{k} \mu(A_n B_j)$$

$$= \sum_{j=1}^{k} \sum_{n=1}^{m} \mu(A_n B_j) = \sum_{j=1}^{k} \mu(AB_j) = \sum_{j=1}^{k} \mu(B_j).$$

It is easy to see that ν is additive on $\mathscr{A}$. The uniqueness of ν follows from the fact that if ν^* is additive on $\mathscr{A}$ and $\nu^*|_{\mathscr{S}} = \mu$, then for any finite partition $\{A_n, 1 \le n \le m\}$ in $\mathscr{S}$ of $A \in \mathscr{A}$,

$$\nu^*(A) = \sum_{n=1}^{m} \nu^*(A_n) = \sum_{n=1}^{m} \mu(A_n) = \nu(A).$$

Suppose next that μ is σ-additive on $\mathscr{S}$, $\{A_n, n \ge 1\}$ is a σ-partition in $\mathscr{A}$ of $A \in \mathscr{A}$ and $\{C_n, 1 \le n \le m\}$ is a partition of A in $\mathscr{S}$. For each n, let $\{B_j, j_{n-1} < j \le j_n\}$ be a finite partition of A_n in $\mathscr{S}$, where $j_0 = 0$. Then $\{B_j, j \ge 1\}$ is a σ-partition of A in $\mathscr{S}$ and

$$\sum_1^{\infty} \nu(A_n) = \sum_{n=1}^{\infty} \sum_{j=j_{n-1}+1}^{j_n} \mu(B_j) = \sum_1^{\infty} \mu(B_j) = \sum_1^{\infty} \mu(AB_j)$$

$$= \sum_{j=1}^{\infty} \sum_{n=1}^{m} \mu(C_n B_j) = \sum_{n=1}^{m} \sum_{j=1}^{\infty} \mu(C_n B_j)$$

$$= \sum_1^{m} \mu(C_n) = \nu(A). \qquad \square$$

The subadditive property of a measure and alternative characterizations of countable additivity appear in

Theorem 2. *Let μ be a nonnegative additive set function on an algebra $\mathscr{A}$ and $\{A_n, n \ge 0\} \subset \mathscr{A}$.*

i. *if $\{A_n, n \ge 1\}$ is a disjoint class and $\bigcup_1^{\infty} A_n \subset A_0$, then*

$$\mu(A_0) \ge \sum_1^{\infty} \mu(A_n).$$

ii. *if $A_0 \subset \bigcup_1^m A_n$ for some $m = 1, 2, \ldots$, then*

$$\mu(A_0) \le \sum_1^{m} \mu(A_n).$$

iii. *μ is σ-additive* iff *for every increasing sequence $\{A_n, n \ge 1\} \subset \mathscr{A}$ with* $\lim_n A_n = A \in \mathscr{A}$,

$$\lim_n \mu(A_n) = \mu(A). \tag{3}$$

In this case μ *is subadditive on* $\mathscr{A}$, *i.e., if* $\bigcup_1^\infty A_j \in \mathscr{A}$, $A_j \in \mathscr{A}$, $j \geq 1$, *then*

$$\mu\left(\bigcup_1^\infty A_j\right) \leq \sum_1^\infty \mu(A_j).$$

iv. *if* μ *is* σ*-additive, then for every decreasing sequence* $A_n \in \mathscr{A}$ *with* $\mu(A_1) < \infty$ *and* $\lim_n A_n = A \in \mathscr{A}$,

$$\lim_n \mu(A_n) = \mu(A). \tag{4}$$

Conversely, if for every decreasing sequence $A_n \in \mathscr{A}$ *with* $\lim_n A_n = \varnothing$,

$$\lim_n \mu(A_n) = 0, \tag{5}$$

then μ *is* σ*-additive.*

PROOF. (i) Since (i) is obvious if $\mu(A_0) = \infty$, let $\mu(A_0) < \infty$. Now $A_0 - \bigcup_1^m A_n \in \mathscr{A}$ $(m = 1, 2, \ldots)$, whence by subtractivity (see Exercise 1.5.1)

$$0 \leq \mu\left(A_0 - \bigcup_1^m A_n\right) = \mu(A_0) - \mu\left(\bigcup_1^m A_n\right) = \mu(A_0) - \sum_1^m \mu(A_n),$$

$$\mu(A_0) \geq \sum_1^m \mu(A_n) \xrightarrow{m \to \infty} \sum_1^\infty \mu(A_n).$$

(ii) Since $A_0 = \bigcup_1^m A_0 A_n = A_0 A_1 \cup A_0 A_2 A_1^c \cup \cdots \cup A_0 A_m A_{m-1}^c \cdots A_1^c$, by additivity and monotonicity of μ

$$\begin{aligned}\mu(A_0) = \mu\left(\bigcup_1^m A_0 A_n\right) &= \mu(A_0 A_1) + \mu(A_0 A_2 A_1^c) + \cdots \\ &\quad + \mu(A_0 A_m A_{m-1}^c \cdots A_1^c) \\ &\leq \mu(A_1) + \mu(A_2) + \cdots + \mu(A_m).\end{aligned}$$

(iii) If (3) holds for every increasing sequence $\{A_n, n \geq 1\} \subset \mathscr{A}$ with $\lim A_n = A \in \mathscr{A}$, then for any σ-partition $\{B_n, n \geq 1\}$ in $\mathscr{A}$ of $B \in \mathscr{A}$,

$$\begin{aligned}\mu(B) = \mu\left(\bigcup_1^\infty B_n\right) &= \mu\left(\lim_m \bigcup_1^m B_n\right) = \lim_m \mu\left(\bigcup_1^m B_n\right) \\ &= \lim_m \sum_1^m \mu(B_n) = \sum_1^\infty \mu(B_n).\end{aligned}$$

Hence, μ is σ-additive. Conversely, let μ be σ-additive and A_n, $n \geq 1$, an increasing sequence of $\mathscr{A}$ sets with $\lim_n A_n = A \in \mathscr{A}$. Then

$$\begin{aligned}\mu(A) = \mu\left(\bigcup_1^\infty A_n\right) &= \mu(A_1 \cup (A_2 - A_1) \cup (A_3 - A_2) \cup \cdots) \\ &= \mu(A_1) + \mu(A_2 - A_1) + \mu(A_3 - A_2) + \cdots \\ &= \lim_n(\mu(A_1) + \mu(A_2 - A_1) + \cdots + \mu(A_n - A_{n-1})) \\ &= \lim_n \mu(A_n).\end{aligned}$$

The second part of (iii) follows from the first part and (ii).

(iv) If μ is σ-additive, and $A_n, n \geq 1$, is a decreasing sequence of $\mathscr{A}$ sets with $\mu(A_1) < \infty$ and $\lim_n A_n = A \in \mathscr{A}$, then $A_1 - A_n$ is an increasing sequence in $\mathscr{A}$, whence by (iii)

$$\mu(A_1) - \mu(A) = \mu(A_1 - A) = \mu\left(\lim_n (A_1 - A_n)\right)$$
$$= \lim_n \mu(A_1 - A_n) = \mu(A_1) - \lim_n \mu(A_n).$$

Since $\mu(A_1) < \infty$, (4) holds. Conversely, if $\{A_n\}$ is any σ-partition in $\mathscr{A}$ of some set $A \in \mathscr{A}$, then $\bigcup_{j=n}^{\infty} A_j \in \mathscr{A}$, $n \geq 1$, and $\bigcup_{j=n}^{\infty} A_j \downarrow \overline{\lim}\, A_n = \emptyset$, whence by hypothesis $\lim \mu\{\bigcup_{j=n}^{\infty} A_j\} = 0$. By finite additivity,

$$\mu\{A\} = \sum_{1}^{n-1} \mu\{A_j\} + \mu\left\{\bigcup_{j=n}^{\infty} A_j\right\},$$

and so countable additivity follows upon letting $n \to \infty$. □

Corollary 1. *A nonnegative additive set function μ on a semi-algebra $\mathscr{S}$ satisfies* (i) *and* (ii) *on $\mathscr{S}$. Moreover, a measure μ on $\mathscr{S}$ is subadditive on $\mathscr{S}$.*

PROOF. Extend μ to the algebra $\mathscr{A}(\mathscr{S})$ generated by $\mathscr{S}$ via Theorem 1. By Theorem 2, the set function (resp. measure) μ satisfies (i) and (ii) (resp. σ-additivity, hence subadditivity) on $\mathscr{A}(\mathscr{S})$ and *a fortiori* on $\mathscr{S}$. □

A **finite measure space** is a measure space $(\Omega, \mathscr{F}, \mu)$ with μ finite. In such a case, the finiteness proviso in Theorem 2 (iv) is superfluous.

Corollary 2. *If $(\Omega, \mathscr{F}, \mathrm{P})$ is a probability space, $\mathrm{P}\{\lim_n A_n\} = \lim_n \mathrm{P}\{A_n\}$ for every sequence of sets $A_n \in \mathscr{F}$ whose limit exists.*

Theorem 3. *Let $(\Omega, \mathscr{F}, \mathrm{P})$ be a probability space with $\mathscr{F} = \sigma(\mathscr{G})$, where $\mathscr{G}$ is an algebra of subsets of Ω. Then for $A \in \mathscr{F}$ and all $\varepsilon > 0$, there exists a set $B_\varepsilon \in \mathscr{G}$ such that $\mathrm{P}\{A \,\Delta\, B_\varepsilon\} < \varepsilon$ and so $|\mathrm{P}\{A\} - \mathrm{P}\{B_\varepsilon\}| < \varepsilon$.*

PROOF. Let $\mathscr{A} = \{A: A \in \mathscr{F}$ and for every $\varepsilon > 0$ there exists $B_\varepsilon \in \mathscr{G}$ with $\mathrm{P}\{A \,\Delta\, B_\varepsilon\} < \varepsilon\}$. Then $\mathscr{A} \supset \mathscr{G}$, and moreover $\mathscr{A}$ is a σ-algebra since, setting $B = B_\varepsilon$,

$$A \,\Delta\, B = A^c \,\Delta\, B^c \Rightarrow A^c \in \mathscr{A} \quad \text{if } A \in \mathscr{A},$$

$$\bigcup_1^\infty A_n \,\Delta\, \bigcup_1^\infty B_n \subset \bigcup_1^\infty (A_n \,\Delta\, B_n) \Rightarrow \bigcup_1^\infty A_n \in \mathscr{A} \quad \text{if } A_j \in \mathscr{A}, \quad j \geq 1,$$

(recall Exercise 1.2.1), noting that

$$\lim_{n \to \infty} \mathrm{P}\left\{\bigcup_1^\infty A_j \,\Delta\, \bigcup_{j=1}^n B_j\right\} = \mathrm{P}\left\{\bigcup_1^\infty A_j \,\Delta\, \bigcup_{j=1}^\infty B_j\right\}.$$

Hence $\mathscr{A} \supset \sigma(\mathscr{G}) = \mathscr{F}$. □

EXERCISES 1.5

1. Show that a finite measure μ on a class $\mathscr{A}$ is (i) **subtractive**, i.e., $A_i \in \mathscr{A}$, $i = 1, 2$, $A_1 \subset A_2$ and $A_2 - A_1 \in \mathscr{A}$ implies $\mu(A_2 - A_1) = \mu(A_2) - \mu(A_1)$, and (ii) **monotone**, i.e., $\mu(A_1) \le \mu(A_2)$ with A_i as in (i). In view of (i), if there is one set in $\mathscr{A}$ with finite measure, the proviso $\mu(\varnothing) = 0$ is automatically satisfied.

2. If $(\Omega, \mathscr{A}, \mu)$ is a measure space, prove that $\mu(\underline{\lim}\, A_n) \le \underline{\lim}\, \mu(A_n)$ for $A_n \in \mathscr{A}$. Analogously, if $\mu(\bigcup_{i=n}^{\infty} A_i) < \infty$, some $n \ge 1$, then $\mu(\overline{\lim}\, A_n) \ge \overline{\lim}\, \mu(A_n)$.

3. If $\Omega = R = [-\infty, \infty]$, then $\mathscr{S} = \{\{+\infty\}, R, [a, b), -\infty \le a \le b \le \infty\}$ is a semi-algebra, not an algebra. Also, if $(\Omega_i, \mathscr{A}_i)$, $i = 1, 2$, are measurable spaces,
$$\{A_1 \times A_2 : A_i \in \mathscr{A}_i, i = 1, 2\}$$
is a semi-algebra. Also, if $n \ge 2$ and $\mathscr{S}_i$ is a semi-algebra of subsets of Ω_i, $1 \le i \le n$, then $\mathscr{S} = \{\times_{i=1}^{n} S_i : S_i \in \mathscr{S}_i, 1 \le i \le n$ is a semi-algebra of subsets of $\times_{i=1}^{n} \Omega_i$.

4. Prove that the class of all finite unions of disjoint sets of a semi-algebra $\mathscr{S}$ is an algebra.

5. If $(\Omega, \mathscr{A}, P)$ is a probability space and $\{A_n, n \ge 1\}$ is a sequence of events,
   (i) verify that
$$P\Big\{\underline{\lim}_{n\to\infty} A_n\Big\} = \lim_{n\to\infty} P\Big\{\bigcap_{j=n}^{\infty} A_j\Big\}, \quad P\Big\{\overline{\lim}_{n\to\infty} A_n\Big\} = \lim_{n\to\infty} P\Big\{\bigcup_{j=n}^{\infty} A_j\Big\}$$
   (ii) show that $P\{\bigcap_{n=1}^{\infty} A_n\} = 1$ if $P\{A_n\} = 1, n \ge 1$.

6. Let $\Omega = \{\text{positive integers}\}$, $\mathscr{A} = \{A : A \subset \Omega\}$. Define $\mu_1\{A\} =$ number of elements in A and $\mu_2\{A\} = 0$ or ∞ according as A is a finite or infinite set. Note that if $A_n = \{n\}$, then $\bigcup_{j=n}^{\infty} A_j \downarrow \lim A_n = \varnothing$ and $\mu_1\{\bigcup_n^{\infty} A_j\} = \infty$. Since μ_1 is a bona fide measure on $\mathscr{A}$, called **counting measure**, the finite hypothesis in the first part of (iv) of Theorem 2 is indispensable. The set function μ_2 is additive but not σ-additive. Thus, if in the second part of (iv), condition (5), were stipulated only when $A_n \downarrow \varnothing$ and $\mu\{A_n\} < \infty$, some $n \ge 1$, then a finiteness requirement for μ would also be necessary.

7. For $\Omega, \mathscr{A}, \mu_1$ as in Exercise 6, define $N_n(A) = \mu_1\{A[1, n]\}$ and
$$\mathscr{D} = \left\{A \in \mathscr{A} : \mu\{A\} = \lim_{n\to\infty} \frac{N_n(A)}{n} \text{ exists}\right\}.$$
Prove that $\mathscr{D}$ is closed under complementation and finite disjoint unions but that nonetheless $\mathscr{D}$ is not an algebra. Also $\mu\{A\}$, called the asymptotic density of A, is additive but not σ-additive. *Hint*: Let $B_k = \{\text{odd integers in } [2^{2k}, 2^{2k+1})\}$ and $B'_k = \{\text{even integers in } [2^{2k-1}, 2^{2k})\}$. If $B = \bigcup_{k=1}^{\infty}(B_k \cup B'_k)$ and $A = \{\text{odd integers of } \Omega\}$, then $A \in \mathscr{D}$, $B \in \mathscr{D}$, but $AB \notin \mathscr{D}$.

8. Let f be a monotone increasing function on $\Omega = [0, 1)$ such that $0 \le f \le 1$, and $\mathscr{G}$ the class of finite unions of disjoint intervals $[a, b) \subset \Omega$.

   i. $\mathscr{G}$ is an algebra.

ii. Put $\mu(A) = \sum_{j=1}^{n} (f(b_j) - f(a_j))$ for $A = \bigcup_{j=1}^{n} [a_j, b_j) \in \mathscr{G}$, where $0 \le a_1 < b_1 \le a_2 < b_2 \le \cdots \le a_n < b_n \le 1$. Then μ is additive on $\mathscr{G}$.

iii. If μ is σ-additive on $\mathscr{G}$, then f is **left continuous**, i.e., $f(t) = \lim_{0<\delta\to 0} f(t-\delta)$ for every $t \in \Omega$.

iv. If f is left continuous, then μ is σ-additive on $\mathscr{G}$. *Hint*: For part (iv), see the proof of Lemma 6.1.1.

9. Let $(\Omega, \mathscr{F}, \mu)$ be a measure space and $\mathscr{G}$ a sub-σ-algebra of $\mathscr{F}$. Prove that

$$\mathscr{G}^* = \{G \,\Delta\, N : G \in \mathscr{G}, N \in \mathscr{F}, \mu\{N\} = 0\}$$

is a sub-σ-algebra of $\mathscr{F}$. Then $\mathscr{G}^*$ is called the **completion** of $\mathscr{G}$. When $\mathscr{G} = \mathscr{F}$ then $\mathscr{G}^* = \mathscr{F}$ and so the completion of $\mathscr{F}$ is rather defined by $\mathscr{F}^* = \{A \,\Delta\, M : A \in \mathscr{F}, M \subset N \in \mathscr{F}, \mu\{N\} = 0\}$. If μ^* on $\mathscr{F}^*$ is defined by $\mu^*\{A \,\Delta\, M\} = \mu\{A\}$, check that $(\Omega, \mathscr{F}^*, \mu^*)$ is a measure space. It is called the completion of $(\Omega, \mathscr{F}, \mu)$.

10. (Cantor ternary set.) Let $I_1 = I_{11} = (\frac{1}{3}, \frac{2}{3})$ be the "open middle third" of $I = [0, 1]$. Analogously, for $j \ge 2$, let $I_{j,k}$, $k = 1, \ldots, 2^{j-1}$, be the open middle third of each of the 2^{j-1} intervals of $I - \bigcup_{i=1}^{j-1} I_i$, where $I_j = \bigcup_{k=1}^{2^{j-1}} I_{j,k}$, $j \ge 1$. Show that the Cantor set $C = I - \bigcup_{j=1}^{\infty} I_j$ satisfies: (i) $x \in C$ iff the digit 1 is not needed in the triadic expansion $x = \sum_{i=1}^{\infty} x_i 3^{-i}$, where, in general, $x_i = 0, 1, 2$; (ii) C is uncountable; (iii) C is perfect (all its points are limit points); (iv) C is nowhere dense (the closure of C contains no nondegenerate interval); (v) if μ is a measure on the Borel sets of $[0, 1]$ whose restriction to intervals coincides with "length" (such a "Lebesgue measure" appears in Section 6.1), then $\mu\{C\} = 0$.

1.6 Induced Measures and Distribution Functions

Consider a measurable transformation X from a measure space $(\Omega, \mathscr{A}, \mu)$ to a measurable space $(\tilde{\Omega}, \tilde{\mathscr{A}})$, i.e., $X: \Omega \to \tilde{\Omega}$ and $X^{-1}(\tilde{\mathscr{A}}) \subset \mathscr{A}$. The measure μ on $\mathscr{A}$ induces in a very natural fashion a measure $\tilde{\mu}$ on $\tilde{\mathscr{A}}$, referred to as the **induced measure**, denoted by μ_X and defined for all $\tilde{A} \in \tilde{\mathscr{A}}$ by

$$\mu_X\{\tilde{A}\} = \tilde{\mu}\{\tilde{A}\} = \mu\{X^{-1}(\tilde{A})\}.$$

It is readily verified that μ_X is a measure on $\tilde{\mathscr{A}}$ and consequently that $(\tilde{\Omega}, \tilde{\mathscr{A}}, \mu_X)$ is a measure space.

In particular, a random variable X on a probability space $(\Omega, \mathscr{F}, \mathrm{P})$ engenders a new (Borel) probability space $(R, \mathscr{B}, \mathrm{P}_X)$, where $\mathscr{B}$ is the class of Borel sets of $R = [-\infty, \infty]$. To focus attention on the latter, i.e., to consider X of primary interest, is tantamount to restricting the former to $(\Omega, \sigma(X), \mathrm{P})$. Associated with any random variable X on a probability space $(\Omega, \mathscr{F}, \mathrm{P})$ is a real function F_X called the distribution function of the random variable X and defined for all $x \in [-\infty, \infty]$ by

$$F_X(x) = \mathrm{P}\{\omega: X(\omega) < x\} = \mathrm{P}\{X^{-1}((-\infty, x))\} \tag{1}$$

Definition. A real-valued function $G(x)$ on $R = [-\infty, \infty]$ is called a **distribution function** (abbreviated d.f.) if

i. G is nondecreasing,
ii. G is left continuous, i.e.,

$$\lim_{\substack{y \to x \\ y < x}} G(y) = G(x), \quad \text{all } x \in R,$$

iii. $G(-\infty) = \lim_{x \to -\infty} G(x) = 0, \qquad G(\infty) = \lim_{x \to \infty} G(x) = 1.$

It follows from Theorem 1.5.2 and the fact that a r.v. is finite a.s., that the "distribution function of any r.v." is a distribution function in the sense of the prior definition. Conversely, for every distribution function G there exists a r.v. X on some probability space $(\Omega, \mathscr{F}, P)$ such that $F_X = G$. The proof for a special case is given in this section but a general treatment will be deferred to Section 6.3.

Analogously to the preceding, n random variables $X_1, \ldots, X_n$ on a probability space $(\Omega, \mathscr{F}, P)$ induce via the map $X = (X_1, \ldots, X_n)$, the new (Borel) probability space $(R^n, \mathscr{B}^n, P_X)$, where $\mathscr{B}^n$ is the class of n-dimensional Borel sets of R^n and P_X is the induced measure. Rather than shift from the basic probability space, one may restrict attention to $(\Omega, \sigma(X_1, \ldots, X_n), P)$. Associated with the **random vector** X i.e., with the n-tuple of random variables $(X_1, \ldots, X_n)$, is a real function on R^n called the (joint) distribution function of $(X_1, \ldots, X_n)$, denoted by $F_{X_1, \ldots, X_n}$ and defined by

$$F_{X_1, \ldots, X_n}(x_1, \ldots, x_n) = P\{\omega: X_1 < x_1, \ldots, X_n < x_n\},$$
$$x = (x_1, \ldots, x_n) \in R^n. \qquad (2)$$

The joint d.f. $F_{X_1, \ldots, X_n}$ satisfies properties akin to (i), (ii), (iii) (see Section 6.3), but monotonicity in each argument is insufficient and the analogue of (i) involves a mixed difference. Nonetheless, if a d.f. on R^n is properly defined, it is again true that the "d.f. of $(X_1, \ldots, X_n)$" is a d.f. on R^n and that, conversely, given a d.f. $G(x_1, \ldots, x_n)$ on R^n, there always exist random variables $X_1, \ldots, X_n$ on some probability space whose joint d.f. $F_{X_1, \ldots, X_n} = G$.

In the same vein, a sequence $\{X_n, n \geq 1\}$ of r.v.s on a probability space $(\Omega, \mathscr{F}, P)$ induces a Borel probability space $(R^\infty, \mathscr{B}^\infty, P_X)$ via the map $X = (X_1, X_2, \ldots)$ or alternatively places in relief $(\Omega, \sigma(X_n, n \geq 1), P)$. Note that the d.f.s of all finite n-tuples of r.v.s $(X_{i_1}, \ldots, X_{i_n})$ are determined via

$$F_{X_{i_1}, \ldots, X_{i_n}}(x_1, \ldots, x_n) = P\{\omega: X_{i_1} < x_1, \ldots, X_{i_n} < x_n\}.$$

It will be proved in Section 6.4 that if the d.f.s $G_{i_1, \ldots, i_n}$ are prescribed in a consistent manner for all choices of indices $1 \leq i_1 < \cdots < i_n$ and all $n \geq 1$, there exists a sequence $\{X_n, n \geq 1\}$ of r.v.s on a probability space such that

$$F_{X_{i_1}, \ldots, X_{i_n}} = G_{i_1, \ldots, i_n}.$$

A d.f. G on R is called **discrete** if

$$G(x) = \sum_{j: x_j < x} p_j, \qquad x \in R, \tag{3}$$

where $p_j > 0$ for all j, $\sum_{\text{all } j} p_j = 1$, and $S = \{x_j: 1 \le j < n + 1 \le \infty\}$ is a subset of $(-\infty, \infty)$. The associated function

$$f(x)\begin{cases} = p_j & \text{for } x = x_j, \quad 1 \le j < n + 1 \le \infty \\ = 0 & \text{for } x \ne x_j \end{cases} \tag{4}$$

is termed a **probability density function** (abbreviated p.d.f.) on

$$S = \{x_j: 1 \le j < n + 1 \le \infty\}.$$

Clearly, a probability density function is completely determined by $\{x_j, p_j, 1 \le j < n + 1 \le \infty\}$. Typically, S is the set of positive or nonnegative integers or some finite subset thereof. This will be the case with the binomial and Poisson d.f.s occurring in Chapter 2.

To construct a probability space $(\Omega, \mathscr{F}, \mathrm{P})$ and a r.v. X on it whose d.f. F_X is equal to a preassigned discrete d.f. G say (as in (3)), it suffices to choose $\Omega = S$, $\mathscr{F}$ = class of all subsets of Ω, $\mathrm{P}\{A\} = \sum_{j: x_j \in A} p_j$, and $X(\omega) = \omega$. Note that then $\mathrm{P}\{\omega: X(\omega) = x_j\} = p_j$, $1 \le j < n + 1 \le \infty$, where $\sum_1^n p_j = 1$.

A d.f. G is called **absolutely continuous** if there exists a Borel function g on $R = [-\infty, \infty]$ such that

$$G(x) = \int_{-\infty}^{x} g(t)dt, \qquad x \in R. \tag{5}$$

The associated function g is termed a **density function** and necessarily satisfies

$$g \ge 0, \quad \text{a.e.,} \qquad \int_{-\infty}^{\infty} g(t)dt = 1. \tag{6}$$

Here, a.e. abbreviates "almost everywhere" (the analogue of "almost certainly" when probability is replaced by Lebesgue measure) and both integrals will be interpreted in the Lebesgue sense. The Lebesgue integral, Lebesgue measure, and related questions will be discussed in Chapter 6.

A third type of d.f., called singular, is defined in Chapter 8. The most general d.f. on R occurring in probability theory is a (convex) linear combination of the three types mentioned (see Section 8.1).

Distribution functions occupy a preeminent position in probability theory, which is primarily concerned with properties of r.v.s verifiable via their d.f.s.

Random variables with discrete d.f.s as in (3) will be called **discrete r.v.s** (with values x_j) while r.v.s with absolutely continuous d.f.s as in (5) will be called **absolutely continuous r.v.s** (with density g). The next chapter deals with some important special cases thereof.

Theorem 1. *If $(X_1, \dots, X_n)$ and $(Y_1, \dots, Y_n)$ are random vectors with identical distribution functions, that is,*

$$F_{X_1, \dots, X_n} = F_{Y_1, \dots, Y_n},$$

then $g(X_1, \dots, X_n)$ and $g(Y_1, \dots, Y_n)$ have identical d.f.s *for any finite Borel function g on R^n.*

PROOF. It follows from Theorem 1.4.4 that $g(X_1, \dots, X_n)$ and $g(Y_1, \dots, Y_n)$ are r.v.s. Set

$$\mathscr{G} = \{B: B \in \mathscr{B}^n, \mathrm{P}\{(X_1, \dots, X_n) \in B\} = \mathrm{P}\{(Y_1, \dots, Y_n) \in B\}\},$$

$$\mathscr{D} = \left\{D: D = \bigtimes_{j=1}^{n} [-\infty, c_j), c_j \text{ real}\right\}.$$

Then $\mathscr{G} \supset \mathscr{D}$ by hypothesis, and, moreover, it is easy to verify that $\mathscr{G}$ is a λ-class and $\mathscr{D}$ is a π-class. By Theorem 1.3.2, $\mathscr{G} \supset \sigma(\mathscr{D}) = \mathscr{B}^n$. Hence, since for $\lambda \in (-\infty, \infty)$, $A \equiv \{(x_1, \dots, x_n) \in R^n: g(x_1, \dots, x_n) < \lambda\} \in \mathscr{B}^n$,

$$\begin{aligned} \mathrm{P}\{g(X_1, \dots, X_n) < \lambda\} &= \mathrm{P}\{(X_1, \dots, X_n) \in A\} = \mathrm{P}\{(Y_1, \dots, Y_n) \in A\} \\ &= \mathrm{P}\{g(Y_1, \dots, Y_n) < \lambda\}. \end{aligned}$$ □

Corollary 1. *If $(X_1, \dots, X_n)$ and $(Y_1, \dots, Y_n)$ are random vectors with identical distribution functions, then for any Borel function g on R^n and linear Borel set B*

$$\mathrm{P}\{g(X_1, \dots, X_n) \in B\} = \mathrm{P}\{g(Y_1, \dots, Y_n) \in B\}.$$

EXERCISES 1.6

1. Prove that G as defined in (3) is a d.f.; verify that $(\Omega, \mathscr{F}, \mathrm{P})$ as defined thereafter is a probability space and that $F_X = G$.

2. If $p_{\rho,t} \geq 0$, $\rho \in S$, $t \in T$, and $\sum_{t \in T} \sum_{\rho \in S} p_{\rho,t} = 1$, where S and T are countable subsets of $(-\infty, \infty)$, define a probability space and random variables X, Y on it with

$$F_X(x) = \sum_{\rho < x} \sum_{t \in T} p_{\rho,t}, \qquad F_Y(y) = \sum_{t < y} \sum_{\rho \in S} p_{\rho,t}.$$

Hint: Take $\Omega = S \times T$.

3. Prove that if $H(x) = \mathrm{P}\{X \leq x\}$, where X is a r.v. on $(\Omega, \mathscr{F}, P)$, then H is a nondecreasing right-continuous function with $H(-\infty) = 0$, $H(+\infty) = 1$. Some authors define such functions to be d.f.s.

4. For a d.f. F, the set

$$S(F) = \{x: F(x + \varepsilon) - F(x - \varepsilon) > 0 \text{ for all } \varepsilon > 0\}$$

is called the **support** of F. Show that each jump point of F belongs to the support and that each isolated point of the support is a jump point. Prove that $S(F)$ is a closed set and give an example of a discrete d.f. whose support is $(-\infty, \infty)$. Any point $x \in S(F)$ is called a **point of increase** of F.

Comments

The notion of a π-class and λ-class (Section 3) seems to have originated with Dynkin (1961).

References

J. L. Doob, "Supplement," *Stochastic Processes*, Wiley, New York, 1953.

E. B. Dynkin, *Theory of Markov Processes* (D. E. Brown, translator), Prentice-Hall, Englewood Cliffs, New Jersey, 1961.

Paul R. Halmos, *Measure Theory*, Van Nostrand, Princeton, 1950; Springer-Verlag, Berlin and New York, 1974.

Felix Hausdorff, *Set Theory* (J. Aumman *et al.*, translators), Chelsea, New York, 1957.

Stanislaw Saks, *Theory of the Integral*, (L. C. Young, translator), Stechert–Hafner, New York, 1937.

2 Binomial Random Variables

The major theorems of probability theory fall into a natural dichotomy—those which are analytic in character and those which are measure-theoretic. In the latter category are zero–one laws, the Borel–Cantelli lemma, strong laws of large numbers, and indeed any result which requires the apparatus of a probability space.

On the other hand, findings such as central limit theorems, weak laws of large numbers, etc., may be stated entirely in terms of distribution functions and hence are intrinsically analytic. The fact that these distribution functions (d.f.s) are frequently attached to random variables (r.v.s) does not alter the fact that the underlying probability space (on which the r.v.s are defined) is of no consequence in the statement of the analytic result. Indeed, it would be possible, although in many cases unnatural, to express the essential finding without any mention of r.v.s (and *a fortiori* of a probability space).

In presenting theorems, distributions will generally be attached to r.v.s. For analytic results, the r.v.s are inessential but provide a more colorful and intuitive background. In the case of measure-theoretic results, it suffices to recognize that a probability space and r.v.s on it with the given distributions can always be constructed.

2.1 Poisson Theorem, Interchangeable Events, and Their Limiting Probabilities

The term **distribution** will be used in lieu of either d.f. or p.d.f. The binomial distribution is not only one of the most frequently encountered in practice but plays a considerable theoretical role due in part to its elemental simplicity and in part to its leading (in the limit) to even more important distributions, such as the Poisson and the normal.

As is customary, the combinatorial symbol $\binom{n}{j}$ abbreviates $n!/j!(n-j)!$ for integers $n \geq j \geq 0$ and is defined as zero when $j > n$ or $j < 0$.

The **binomial d.f.** is a discrete d.f. with jumps of sizes $\binom{n}{j}p^j(1-p)^{n-j}$ at the points $j = 0, 1, \ldots, n$. In other words, its p.d.f. is completely characterized (Section 1.6) by $S = \{0, 1, 2, \ldots, n\}$ and $\{\binom{n}{j}p^j(1-p)^{n-j}, 0 \leq j \leq n\}$. Here, $p \in (0, 1)$ and n is a positive integer.

The construction of Section 1.6 shows that it is a simple matter to define a probability space and a r.v. X on this space whose associated d.f. is binomial. Thus, a r.v. X will be alluded to as a **binomial r.v.** with p.d.f. $b(k; n, p)$ if for some positive integer n and $p \in (0, 1)$

$$P\{X = k\} = \binom{n}{k}p^k q^{n-k} \equiv b(k; n, p), \qquad k = 0, 1, \ldots, n, \quad q = 1 - p. \tag{1}$$

The **Poisson d.f.** is a discrete d.f. with jumps of sizes $\lambda^j e^{-\lambda}/j!$ at the points $j = 0, 1, 2, \ldots$, and so X will be called a **Poisson r.v.** with p.d.f. $p(k; \lambda)$ if for some $\lambda \in (0, \infty)$

$$P\{X = k\} = \frac{\lambda^k e^{-\lambda}}{k!} \equiv p(k; \lambda), \qquad k = 0, 1, 2, \ldots. \tag{2}$$

The quantities n and p of the binomial distribution and λ of the Poisson distribution are referred to as **parameters.** If $\lambda = 0$ in (2) or $p = 0$ in (1), the r.v.s (or their distributions) are called **degenerate** since then $P\{X = 0\} = 1$ (i.e., its d.f. has a single jump of size one). Degeneracy likewise occurs if $p = 1$ in (1) since then $P\{X = n\} = 1$.

The normal d.f. is an absolutely continuous d.f. (Section 1.6) G with

$$G(x) = \int_{-\infty}^{x} \frac{1}{\sigma\sqrt{2\pi}} e^{-(u-\theta)^2/2\sigma^2}\, du, \qquad x \in R = [-\infty, \infty]. \tag{3}$$

The parameters of the distribution are $\theta \in (-\infty, \infty)$ and $\sigma^2 \in (0, \infty)$. Here, $\sigma^2 = 0$ may be regarded as a degenerate distribution. It is customary to denote the **standard normal distribution** corresponding to the case $\theta = 0$, $\sigma^2 = 1$ by

$$\Phi(x) = \int_{-\infty}^{x} \frac{1}{\sqrt{2\pi}} e^{-u^2/2}\, du, \qquad \varphi(x) = \frac{1}{\sqrt{2\pi}} e^{-x^2/2}. \tag{4}$$

Here $\varphi(x)$ is called the (standard) normal density function. A r.v. X is normally distributed with parameters θ and σ^2, written X is $\boldsymbol{N(\theta, \sigma^2)}$ if

$$P\{X < x\} = G(x) = \Phi\left(\frac{x-\theta}{\sigma}\right), \qquad x \in R. \tag{5}$$

Commencing here and sporadically throughout the remaining chapters, a simple but incredibly expedient device for asymptotic calculations, introduced by the English mathematician G. H. Hardy, will be utilized. Thus, if $r_n > 0$, $n \geq 1$, a sequence $\{b_n, n \geq 1\}$ of real numbers is said to be **little o of r_n** (resp. **capital O of r_n**), denoted $b_n = o(r_n)$ (resp. $b_n = O(r_n)$) if $\lim_{n\to\infty}(b_n/r_n) = 0$ (resp. $|b_n|/r_n < C < \infty, n \geq 1$). It is a simple matter to check that the sum of two quantities that are $o(r_n)$ (resp. $O(r_n)$) is likewise $o(r_n)$ (resp. $O(r_n)$). Thus, a veritable calculus can be established involving one or the other or both of o and O. In a similar vein, $b_n \sim r_n$ if $\lim_{n\to\infty}(b_n/r_n) = 1$. The same notations and calculus apply to real functions $f(x)$ where either $x \to \infty$ or $x \to 0$. Thus, for example, $\log(1 + x) = x + O(x^2) = O(|x|)$ as $x \to 0$.

The first theorem, purely analytic in character, states that the limit of binomial p.d.f.s may be a Poisson p.d.f. and hence implies the analogous statement for d.f.s.

Theorem 1 (Poisson, 1832). *If S_n is a binomial* r.v. *with* p.d.f. $b(k; n, p_n)$, $n \geq 1$, *and as $n \to \infty$, $np_n = \lambda + o(1)$ for some $\lambda \in [0, \infty)$, then for $k = 0, 1, 2, \ldots$*

$$\lim_{n\to\infty} P\{S_n = k\} = \frac{\lambda^k e^{-\lambda}}{k!}.$$

PROOF. Set $q_n = 1 - p_n$. Since $n \log(1 - (\lambda/n) + o(1/n)) = n((-\lambda/n) + o(1/n)) \to -\lambda$,

$$\begin{aligned} P\{S_n = k\} &= \binom{n}{k} p_n^k q_n^{n-k} \\ &= \frac{n(n-1)\cdots(n-k+1)}{k!}\left[\frac{\lambda}{n} + o\left(\frac{1}{n}\right)\right]^k \left[1 - \frac{\lambda}{n} + o\left(\frac{1}{n}\right)\right]^{n-k} \\ &= \frac{n}{n}\left(\frac{n-1}{n}\right)\cdots\left(\frac{n-k+1}{n}\right)\frac{1}{k!}[\lambda + o(1)]^k \\ &\quad \times \left[1 - \frac{\lambda}{n} + o\left(\frac{1}{n}\right)\right]^n \left[1 - \frac{\lambda}{n} + o\left(\frac{1}{n}\right)\right]^{-k} \to \frac{\lambda^k}{k!} e^{-\lambda}. \end{aligned}$$ □

By refining the arguments of the preceding proof, a more general result is obtainable. To pave the way for the introduction of a new p.d.f. subsuming the binomial p.d.f., two lemmas which are probabilistic or measure analogues of (2) and (3) of Section 1.2 will be proved.

Lemma 1. *Let φ be an additive set function on an algebra $\mathscr{A}$ and $A_j \in \mathscr{A}$, $1 \leq j \leq n$. If $|\varphi(\bigcup_1^n A_j)| < \infty$, then*

$$\varphi\left(\bigcup_1^n A_j\right) = \sum_{j=1}^n \varphi(A_j) - \sum_{1 \leq j_1 < j_2 \leq n} \varphi(A_{j_1}A_{j_2}) + \cdots + (-1)^{n-1}\varphi(A_1 A_2 \cdots A_n). \tag{6}$$

PROOF. Set $\varphi_j = \varphi(A_j), \varphi_{j_1, j_2 \cdots j_r} = \varphi(A_{j_1} A_{j_2} \cdots A_{j_r}), r \geq 2$. Clearly, (6) holds for $n = 1$. Suppose, inductively, that it holds for a fixed but arbitrary positive integer m in $[1, n)$, whence $|\varphi(\bigcup_1^{m+1} A_j)| < \infty$. Now,

$$\varphi\left(\bigcup_1^{m+1} A_j\right) = \varphi\left(\bigcup_1^{m} A_j\right) + \varphi\left(A_{m+1} - \bigcup_1^{m} A_j\right),$$
$$\varphi(A_{m+1}) = \varphi\left(\bigcup_1^{m} A_j A_{m+1}\right) + \varphi\left(A_{m+1} - \bigcup_1^{m} A_j\right), \tag{7}$$

where $|\varphi(\bigcup_1^m A_j)| < \infty$, $|\varphi(A_{m+1})| < \infty$, $|\varphi(\bigcup_1^m A_j A_m)| < \infty$. By the induction hypothesis

$$\varphi\left(\bigcup_1^{m} A_j\right) = \sum_{j=1}^{m} \varphi_j - \sum_{1 \leq j_1 < j_2 \leq m} \varphi_{j_1, j_2} + \cdots + (-1)^{m-1} \varphi_{1, 2, \ldots, m},$$
$$\varphi\left(\bigcup_1^{m} A_j A_{m+1}\right) = \sum_{j=1}^{m} \varphi_{j, m+1} - \sum_{1 \leq j_1 < j_2 \leq m} \varphi_{j_1, j_2, m+1} + \cdots \tag{8}$$
$$+ (-1)^{m-1} \varphi_{1, 2, \ldots, m, m+1},$$

and so employing (8) in (7) the conclusion (6) with $n = m + 1$ follows. □

Lemma 1 yields immediately

Corollary 1 (Poincaré Formula). *If $A_1, \ldots, A_n$ are events in a probability space $(\Omega, \mathscr{F}, \mathrm{P})$ and $T_k = \sum_{1 \leq j_1 < j_2 < \cdots < j_k \leq n} \mathrm{P}\{A_{j_1} A_{j_2} \cdots A_{j_k}\}$, then*

$$\mathrm{P}\left\{\bigcup_1^n A_j\right\} = \sum_1^n (-1)^{k-1} T_k. \tag{9}$$

Definition. Events $A_1, \ldots, A_n$ of a probability space $(\Omega, \mathscr{F}, \mathrm{P})$ are called **interchangeable** (exchangeable) if for all choices of $1 \leq i_1 < \cdots < i_j \leq n$ and all $1 \leq j \leq n$,

$$\mathrm{P}\{A_{i_1} \cdot A_{i_2} \cdots A_{i_j}\} = p_j. \tag{10}$$

Interchangeable events seem to have been introduced by Haag (1924–1928) but the basic results involving these are due to de Finetti (1937).

Corollary 2. *If $\{A_1, \ldots, A_n\}$ are interchangeable events of a probability space $(\Omega, \mathscr{F}, \mathrm{P})$ with p_j defined by* (10), *then*

$$\mathrm{P}\left\{\bigcup_{j=1}^n A_j\right\} = np_1 - \binom{n}{2} p_2 + \binom{n}{3} p_3 - \cdots + (-1)^{n-1} p_n. \tag{11}$$

The novel part of the next lemma is (13), since for $A = \Omega$, (12) is merely the complementary version of the Poincaré formula (9).

Lemma 2. *Let φ be an additive set function on an algebra $\mathscr{A}$ and $A_j \in \mathscr{A}$, $1 \le j \le n$. If $|\varphi(\bigcup_1^n A_j)| < \infty$ and*

$$B_m = \{\omega : \omega \in \textit{exactly } m \textit{ of the events } A_1, \ldots, A_n\}, \qquad 0 \le m \le n,$$

then for any $A \in \mathscr{A}$

$$\varphi(B_0 A) = \varphi(A) - \sum_{j=1}^{n} \varphi(A_j A) + \sum_{1 \le j_1 < j_2 \le n} \varphi(A_{j_1} A_{j_2} A) - \cdots + (-1)^n \varphi(A_1 \cdots A_n A), \tag{12}$$

$$\varphi(B_m) = \sum_{1 \le j_1 < \cdots < j_m \le n} \varphi(A_{j_1} \cdots A_{j_m}) - \binom{m+1}{m} \sum_{1 \le j_1 < \cdots < j_{m+1} \le n} \times \varphi(A_{j_1} \cdots A_{j_{m+1}}) + \cdots + (-1)^{n-m} \binom{n}{m} \varphi(A_1 \cdots A_n). \tag{13}$$

PROOF. By Lemma 1

$$\varphi(B_0 A) = \varphi(A) - \varphi\left(\bigcup_1^n A_j A\right)$$
$$= \varphi(A) - \sum_{j=1}^{n} \varphi(A_j A) + \sum_{1 \le j_1 < j_2 \le n} \varphi(A_{j_1} A_{j_2} A) - \cdots + (-1)^n \varphi(A_1 \cdots A_n A),$$

proving (12). For any choice of j_i, $1 \le i \le m$ with $1 \le j_1 < \cdots < j_m \le n$, let $1 \le i_1 < \cdots < i_{n-m} \le n$ signify the remaining integers in $[1, n]$. Then taking $A = A_{j_1} \cdots A_{j_m}$ in (12)

$$\varphi(A_{j_1} \cdots A_{j_m} A_{i_1}^c \cdots A_{i_{n-m}}^c) = \varphi(A \cdot A_{i_1}^c \cdots A_{i_{n-m}}^c)$$
$$= \varphi(A) - \sum_{h=1}^{n-m} \varphi(A_{i_h} A) + \sum_{1 \le h_1 < h_2 \le n-m} \varphi(A_{i_{h_1}} A_{i_{h_2}} A) - \cdots + (-1)^{n-m} \varphi(A_{i_1} \cdots A_{i_{n-m}} A). \tag{14}$$

Since for any $1 \le m \le n$

$$B_m = \bigcup_{1 \le j_1 < \cdots < j_m \le n} A_{j_1} \cdots A_{j_m} \cdot A_{i_1}^c \cdots A_{i_{n-m}}^c$$

represents B_m as a disjoint union, (13) follows by summing over $j_1, \ldots, j_m$ in (14). □

The following corollaries are direct consequences of Lemma 2.

Corollary 3. *If $A_1, \ldots, A_n$ are events of a probability space $(\Omega, \mathscr{F}, P)$ and $T_k = \sum_{1 \le j_1 < \cdots < j_k \le n} P\{A_{j_1} A_{j_2} \cdots A_{j_k}\}$ for $k = 1, 2, \ldots, n$, then*

$$P\{\textit{exactly } m \textit{ of } A_1, \ldots, A_n \textit{ occur}\} = T_m - \binom{m+1}{m} T_{m+1} + \cdots + (-1)^{n-m} \binom{n}{m} T_n. \tag{15}$$

In the case of interchangeable events, (15) simplifies considerably and may be expressed in terms of the classical difference operator:

$$\Delta p_n = \Delta^1 p_n = p_{n+1} - p_n, \qquad \Delta^k p_n = \Delta(\Delta^{k-1} p_n), \qquad k \geq 2.$$

Corollary 4. *If $A_1, \ldots, A_n$ are interchangeable events of a probability space $(\Omega, \mathscr{F}, \mathrm{P})$ with $p_j = \mathrm{P}\{A_1 A_2 \cdots A_j\}$, $1 \leq j \leq n$, then, setting $p_0 = 1$, for any integer m in $[0, n]$*

$$\begin{aligned} \mathrm{P}\{\textit{exactly } m \textit{ of } A_1, \ldots, A_n \textit{ occur}\} &= \sum_{i=m}^{n} (-1)^{i-m} \binom{n}{i}\binom{i}{m} p_i \\ &= \binom{n}{m} \sum_{j=0}^{n-m} (-1)^j \binom{n-m}{j} p_{m+j} \\ &= \binom{n}{m} (-1)^{n-m} \Delta^{n-m} p_m. \end{aligned} \tag{16}$$

Since the events B_m, $0 \leq m \leq n$, of (16) constitute a finite partition of Ω, their probabilities are nonnegative and add up to one. This shows that the probabilities p_j in (10) cannot be arbitrary positive numbers (see Exercise 9).

In view of the correspondence between events A_j and their indicators I_{A_j}, it should not be surprising, setting $S_n = \sum_{j=1}^n I_{A_j}$, that the event B_m of (16) is equivalent to $\{\omega: S_n = m\}$. Indeed, discussion of the limiting behavior of the probabilities in (16) will be couched in terms of the r.v.s S_n.

Theorem 2. *If for each $n \geq 1$, S_n is a random variable with*

$$\mathrm{P}\{S_n = m\} = \sum_{j=0}^{n-m} (-1)^j \binom{n}{m+j}\binom{m+j}{m} p_{m+j}^{(n)}, \qquad 0 \leq m \leq n, \tag{17}$$

such that for some $\lambda_0 \in (0, \infty)$ and $\lambda_n \in [0, \infty)$

$$0 \leq n(n-1)\cdots(n-k+1)p_k^{(n)} \leq \lambda_0^k, \qquad 1 \leq k \leq n, \tag{18}$$

$$n^k p_k^{(n)} = \lambda_n^k + o(1), \qquad k \geq 1, \quad \textit{as } n \to \infty, \tag{19}$$

then

$$\mathrm{P}\{S_n = m\} - \frac{\lambda_n^m e^{-\lambda_n}}{m!} \to 0, \qquad m \geq 0.$$

PROOF. For $m \geq 0$, fixed n_0 and all large n

$$\begin{aligned} \mathrm{P}\{S_n = m\} - \frac{\lambda_n^m e^{-\lambda_n}}{m!} &= \sum_{j=0}^{n_0-1} \frac{(-1)^j}{m!\,j!}\left[\frac{n!}{(n-m-j)!} p_{m+j}^{(n)} - \lambda_n^{m+j}\right] \\ &\quad + \sum_{j=n_0}^{n-m} (-1)^j \frac{n!}{m!\,j!\,(n-m-j)!} p_{m+j}^{(n)} \\ &\quad - \sum_{j=n_0}^{\infty} (-1)^j \frac{\lambda_n^{m+j}}{m!\,j!} = I_1 + I_2 + I_3 \text{ (say)}. \end{aligned}$$

Since $\lambda_n + o(1) = np_1^{(n)} \le \lambda_0$, there exists $\lambda > \lambda_0$ such that $\lambda_n < \lambda$, all n. For any $\varepsilon > 0$ and fixed m, choose n_0 to satisfy

$$\sum_{j=n_0}^{\infty} \frac{\lambda^{m+j}}{m!j!} < \varepsilon,$$

whence $|I_3| < \varepsilon$ and via (18)

$$|I_2| \le \sum_{j=n_0}^{n-m} \frac{n(n-1)\cdots(n-m-j+1)}{m!j!} p_{m+j}^{(n)} \le \sum_{j=n_0}^{n-m} \frac{\lambda_0^{m+j}}{m!j!} < \varepsilon.$$

Now, for $1 \le m + j \le m + n_0$, by (19)

$$\begin{aligned} \frac{n!}{(n-m-j)!} p_{m+j}^{(n)} - \lambda_n^{m+j} &\le n^{m+j} p_{m+j}^{(n)} - \lambda_n^{m+j} \\ &= \lambda_n^{m+j} + o(1) - \lambda_n^{m+j} = o(1), \\ \frac{n!}{(n-m-j)!} p_{m+j}^{(n)} - \lambda_n^{m+j} &\ge (n-m-j)^{m+j} p_{m+j}^{(n)} - \lambda_n^{m+j} \\ &= \left(1 - \frac{m+j}{n}\right)^{m+j} n^{m+j} p_{m+j}^{(n)} - \lambda_n^{m+j} \\ &= \left(1 - \frac{m+j}{n}\right)^{m+j} (\lambda_n^{m+j} + o(1)) - \lambda_n^{m+j} = o(1), \end{aligned}$$

implying $I_1 = o(1)$, and the theorem follows. □

This result will be utilized in Theorem 3.1.2 to obtain (under suitable conditions) a limiting Poisson distribution for the number of empty cells in the casting of balls into cells at random. Note that this implies Theorem 1.

It follows immediately from Theorem 1 that the sequence $F_n(x) = \sum_{k<x} b(k; n, p)$ of binomial d.f.s tends to the Poisson d.f. $F(x) = \sum_{k<x} p(k; \lambda)$ as $n \to \infty$, $np \to \lambda \in (0, \infty)$. In this case, not only does

$$b(k; n, p) \to p(k; \lambda), \tag{20}$$

for every positive integer k, but also

$$\sum_{k=0}^{\infty} |b(k; n, p) - p(k; \lambda)| \to 0 \tag{21}$$

as $n \to \infty$, $np \to \lambda \in (0, \infty)$, where $b(k; n, p)$ is defined as zero for $k > n$. This is an instance of the more general phenomenon of the following

Example 1. Let $\Omega = \{\omega_1, \omega_2, \ldots\}$, $\mathscr{F}$ = class of all subsets of Ω, and let $\{\mathrm{P}, \mathrm{P}_n, n \ge 1\}$ be a sequence of probability measures on $(\Omega, \mathscr{F})$ such that $\mathrm{P}_n \to \mathrm{P}$ as $n \to \infty$, that is,

$$p_{n,j} \equiv \mathrm{P}_n\{\omega_j\} \to \mathrm{P}\{\omega_j\} \equiv p_j, \qquad j = 1, 2, \ldots. \tag{22}$$

Then, as $n \to \infty$

$$\sum_{j=1}^{\infty} |p_{n,j} - p_j| \to 0. \tag{23}$$

PROOF. For any $\varepsilon > 0$, choose $N = N_\varepsilon$ such that

$$\sum_{j>N} p_j < \varepsilon. \tag{24}$$

Now (22) ensures that as $n \to \infty$

$$\sum_{j=1}^{N} |p_{n,j} - p_j| \to 0, \tag{25}$$

whence as $n \to \infty$

$$\left| \sum_{j>N} p_{n,j} - \sum_{j>N} p_j \right| = \left| \sum_{j=1}^{N} p_{n,j} - \sum_{j=1}^{N} p_j \right| \to 0. \tag{26}$$

Then (24) and (26) entail

$$\overline{\lim_{n\to\infty}} \sum_{j>N} p_{n,j} \le \varepsilon, \tag{27}$$

so that via (24), (25), (26), and (27)

$$\overline{\lim_{n\to\infty}} \sum_{j=1}^{\infty} |p_{n,j} - p_j| = \overline{\lim_{n\to\infty}} \sum_{j>N} |p_{n,j} - p_j| \le \overline{\lim_{n\to\infty}} \sum_{j>N} (p_{n,j} + p_j) \le 2\varepsilon,$$

which, in view of the arbitrariness of ε, is tantamount to (23). □

EXERCISES 2.1

1. Prove that for $n = 1, 2, \ldots$

$$\int_{-\infty}^{\infty} \varphi(x)dx = 1, \qquad \int_{-\infty}^{\infty} x^{2n-1}\varphi(x)dx = 0, \qquad \int_{-\infty}^{\infty} x^{2n}\varphi(x)dx = 1 \cdot 3 \cdots (2n-1).$$

2. Verify that

$$\sum_{k=0}^{\infty} kp(k; \lambda) = \lambda = \sum_{k=0}^{\infty} k^2 p(k; \lambda) - \lambda^2,$$

$$\sum_{k=0}^{n} p(k; \lambda) = \frac{1}{n!} \int_{\lambda}^{\infty} e^{-x} x^n \, dx,$$

$$\sum_{k=0}^{\infty} k^3 p(k; \lambda) = \lambda^3 + 3\lambda^2 + \lambda,$$

$$\sum_{k=0}^{n} k^j b(k; n, p) = np \quad \text{or} \quad npq + (np)^2 \quad \text{as} \quad j = 1 \text{ or } 2.$$

3. Verify for all positive integers n_1, n_2 and nonnegative integers k that

$$\sum_{j=0}^{n_1} \binom{n_1}{j}\binom{n_2}{k-j} = \binom{n_1+n_2}{k}.$$

4. Find the d.f. of X^2 when

   i. X is a Poisson r.v. with parameter λ,

   ii. X is $N(0, 1)$.

   Prove under (i) that $\lambda^n e^{-\lambda}/n! \leq \mathrm{P}\{X \geq n\} \leq \lambda^n/n!$ for $n \geq 0$.

5. A deck of N cards numbered $1, 2, \ldots, N$ is shuffled. A "match" occurs at position j if the card numbered j occupies the jth place in the deck. If $p_m^{(N)}$ denotes the probability of exactly m matches, $0 \leq m \leq N$, prove that

$$p_m^{(N)} = \frac{1}{m!}\left(1 - 1 + \frac{1}{2!} - \frac{1}{3!} + \cdots + \frac{(-1)^{N-m}}{(N-m)!}\right), \qquad \text{(i)}$$

$$\left|p_m^{(N)} - \frac{e^{-1}}{m!}\right| < \frac{1}{m!(N-m+1)!}, \qquad \text{(ii)}$$

$$q_2^{(N)} \to e^{-1}\left(\frac{1}{2!} + \frac{1}{3!} + \cdots\right), \qquad \text{(iii)}$$

where $q_m^{(N)}$ is the probability of at least m matches.

6. For $k = 0, 1, 2, \ldots, r = 1, 2, \ldots, 0 < p < 1$, and $q = 1 - p$, put $f(k; r, p) = \binom{r+k-1}{k} p^r q^k$. Prove that $\sum_{k=0}^{\infty} f(k; r, p) = 1$ for every r and p and that if $q = q(r)$ and $rq \to \lambda \in (0, 1)$ as $r \to \infty$, then

$$f(k; r, p) \to e^{-\lambda}\frac{\lambda^k}{k!}.$$

($f(k; r, p)$ is called a **negative binomial** p.d.f. with parameters r and p.)

7. **Bonferroni Inequalities.** If $A_1, \ldots, A_n$ are events and

$$T_k = \sum_{1 \leq j_1 < \cdots < j_k \leq n} \mathrm{P}\{A_{j_1} A_{j_2} \cdots A_{j_k}\},$$

then

$$\sum_{1}^{2m} (-1)^{k-1} T_k \leq \mathrm{P}\left\{\bigcup_{1}^{n} A_j\right\} \leq \sum_{1}^{2m-1} (-1)^{j-1} T_j, \qquad 2m \leq n.$$

8. Find the formula analogous to (15) for the probability that at least m of the events $A_1, \ldots, A_n$ occur.

9. Let $p_0 = 1$ and p_m, $1 \leq m \leq n$, be real numbers such that

$$q_m = \binom{n}{m} \sum_{j=0}^{n-m} (-1)^j \binom{n-m}{j} p_{m+j} \geq 0$$

for $0 \leq m \leq n$. Then $\sum_{m=0}^{n} q_m = p_0 = 1$ and there exists a probability space $(\Omega, \mathscr{F}, \mathrm{P})$ and interchangeable events $A_1, \ldots, A_n$ with $p_j = \mathrm{P}\{A_1 A_2 \cdots A_j\}$, $1 \leq j \leq n$. *Hint*: Take $\Omega = \{\omega = (\omega_1, \ldots, \omega_n)\colon \omega_j = 0 \text{ or } 1, 1 \leq j \leq n\}$, $\mathscr{F}$ = class of all subsets of Ω, and $\mathrm{P}(\{\omega\}) = q_m/\binom{n}{m}$ whenever $\sum_{j=1}^{n} \omega_j = m$. Set

$$A_j = \{\omega = (\omega_1, \ldots, \omega_n)\colon \omega_j = 1\}, 1 \leq j \leq n.$$

10. Verify that if $p_j = p^j$ in (16) i.e., in Exercise 9, the probabilities q_m coalesce to the binomial and $P\{A_{i_1} A_{i_2} \cdots A_{i_k}\} = \prod_{j=1}^{k} P\{A_{i_j}\}, 1 \le i_1 < \cdots < i_k \le n$.

11. Prove that if $p_j = \binom{N-j}{r-j}/\binom{N}{r}, j = 0, \ldots, r, r \le N$, the p.d.f. in (16) is **hypergeometric**, i.e., $q_m = \binom{n}{m}\binom{N-n}{r-m}/\binom{N}{r}$; also if $p_j = 1/(j+1)$, $\{q_m\}$ is the discrete **uniform** distribution, i.e., $q_m = 1/(n+1), 0 \le m \le n$.

12. Prove that if $r = r_n = \lambda_n N/n \to \infty$ and $\lambda_n \to \lambda \in [0, \infty)$ in the hypergeometric case of Exercise 11, then $p_j = p_j^{(n)}$ satisfies (18) and (19) of Theorem 2, so that the hypergeometric p.d.f. tends to the Poisson p.d.f. under the stated condition.

13. Let $\{A_n, n \ge 1\}$ be an infinite sequence of interchangeable events on a probability space $(\Omega, \mathscr{F}, P)$, that is, for all $n \ge 1$ and indices

$$1 \le i_1 < \cdots < i_n \qquad P\{A_{i_1} A_{i_2} \cdots A_{i_n}\} = p_n.$$

Prove that

$$P\{A_n, \text{i.o.}\} = P\left\{\bigcup_1^\infty A_j\right\} = 1 - \lim_{k\to\infty} (-1)^k \Delta^k p_0,$$

$$P\{\underline{\lim}\, A_n\} = P\left\{\bigcap_{n=1}^\infty A_n\right\} = \lim_{k\to\infty} p_k.$$

Hint: Recall Exercise 1.5.5.

2.2 Bernoulli, Borel Theorems

Due to the ease of computations involving the binomial distribution, many important notions such as weak and strong laws of large numbers and the central limit theorem applicable to wide classes of random variables may be adumbrated here.

Theorem 1 (Bernoulli Weak Law of Large Numbers, 1713). *Let $\{S_n\}$ be a sequence of binomial random variables with* p.d.f.s $b(k; n, p)$, $n \ge 1$. *Then for all* $\varepsilon > 0$,

$$\lim_{n\to\infty} P\left\{\left|\frac{S_n}{n} - p\right| \ge \varepsilon\right\} = 0.$$

PROOF.

$$P\left\{\left|\frac{S_n}{n} - p\right| \ge \varepsilon\right\} = P\{|S_n - np| \ge n\varepsilon\} = \sum_{|k-np|\ge n\varepsilon} P\{S_n = k\}$$

$$\le \sum_{|k-np|\ge n\varepsilon} \frac{(k-np)^2}{n^2\varepsilon^2} P\{S_n = k\}$$

$$\le \frac{1}{n^2\varepsilon^2} \sum_{k=0}^{n} (k-np)^2 P\{S_n = k\}.$$

The prior inequality is a special case of the simple but very useful Tchebychev inequality discussed in Chapter 4. Moreover, the last sum which will be identified in Chapter 4 as the variance of S_n (or its d.f.) equals

$$\begin{aligned}
&\sum_{k=0}^{n} (k - np)^2 \binom{n}{k} p^k q^{n-k} \\
&= \sum_{k=0}^{n} [k(k-1) - (2np-1)k + n^2p^2] \binom{n}{k} p^k q^{n-k} \\
&= n(n-1)p^2 \sum_{k=2}^{n} \frac{(n-2)!}{(k-2)!(n-k)!} p^{k-2} q^{n-k} \\
&\quad - (2np-1)np \sum_{k=1}^{n} \frac{(n-1)!}{(k-1)!(n-k)!} p^{k-1} q^{n-k} + n^2p^2 \\
&= n(n-1)p^2 - np(2np-1) + n^2p^2 = npq. \qquad (1)
\end{aligned}$$

Therefore,

$$P\left\{\left|\frac{S_n}{n} - p\right| \geq \varepsilon\right\} \leq \frac{pq}{n\varepsilon^2} = o(1). \qquad \square \quad (2)$$

The prior theorem may be strengthened by the simple device of replacing $(k - np)^2$ by $(k - np)^4$.

Theorem 2. *Let* S_n, $n \geq 1$, *be binomial* r.v.s *with* p.d.f.s $b(k; n, p)$, $0 < p < 1$, $n \geq 1$. *Then for every* $\varepsilon > 0$

$$\sum_{n=1}^{\infty} P\left\{\left|\frac{S_n}{n} - p\right| \geq \varepsilon\right\} < \infty.$$

PROOF. Set $k_j = k(k-1)\cdots(k-j+1)$, $1 \leq j \leq k$, whence

$$k = k_1, \qquad k^2 = k_2 + k_1, \qquad k^3 = k_3 + 3k_2 + k_1,$$

$$k^4 = k_4 + 6k_3 + 7k_2 + k_1.$$

Since

$$\sum_{k=0}^{n} k_j \binom{n}{k} p^k q^{n-k} = n_j p^j, \qquad 0 \leq j \leq n,$$

it follows that

$$\begin{aligned}
&\sum_{k=0}^{n} (k - np)^4 \binom{n}{k} p^k q^{n-k} \\
&= \sum_{k=0}^{n} (k^4 - 4k^3np + 6k^2n^2p^2 - 4kn^3p^3 + n^4p^4) \binom{n}{k} p^k q^{n-k} \\
&= (n_4 p^4 + 6n_3 p^3 + 7n_2 p^2 + n_1 p) - 4np(n_3 p^3 + 3n_2 p^2 + n_1 p) \\
&\quad + 6n^2p^2(n_2 p^2 + n_1 p) - 4n^3p^3(n_1 p) + n^4p^4 \\
&= p^4(n_4 - 4nn_3 + 6n^2n_2 - 4n^3n_1 + n^4) + p^3(6n_3 - 12nn_2 + 6n^2n_1) \\
&\quad + p^2(7n_2 - 4nn_1) + n_1 p \\
&= (3n^2 - 6n)p^4 - (6n^2 - 12n)p^3 + (3n^2 - 7n)p^2 + np \\
&= 3n(n - 2)p^2(p - 1)^2 + npq = npq(3npq - 6pq + 1).
\end{aligned}$$

Consequently, proceeding as at the outset of Theorem 1,

$$\begin{aligned}
P\left\{\left|\frac{S_n}{n} - p\right| \geq \varepsilon\right\} &= \sum_{|k-np| \geq n\varepsilon} P\{S_n = k\} \leq \frac{1}{n^4\varepsilon^4} \sum_{k=0}^{n} (k - np)^4 \binom{n}{k} p^k q^{n-k}, \\
&= \frac{pq(3npq - 6pq + 1)}{n^3\varepsilon^4} = O\left(\frac{1}{n^2}\right),
\end{aligned}$$

and therefore the series in question converges. □

The strong law of large numbers involves the pointwise convergence of a sequence of random variables on a probability space. A discussion of this will be facilitated by

Lemma 1. *If* $\{Y_n, n \geq 1\}$ *is any sequence of random variables on a probability space* $(\Omega, \mathscr{F}, P)$, *then* $P\{\lim_{n\to\infty} Y_n = 0\} = 1$ iff

$$P\left\{|Y_n| > \frac{1}{k}, \text{i.o.}\right\} = 0, \qquad k = 1, 2, \ldots.$$

Proof. Let $A = \bigcup_{k=1}^{\infty} A_k$, where $A_k = \{|Y_n| > 1/k, \text{i.o.}\}$. If $\omega \notin A$, then $|Y_n(\omega)| > 1/k$ for only finitely many n for every positive integer k, implying $\lim Y_n(\omega) = 0$. Conversely, if $\lim Y_n(\omega) = 0$, then $\omega \notin A_k$ for $k = 1, 2, \ldots$, and so $\omega \notin A$. Thus $A^c = \{\lim Y_n = 0\}$, whence $P\{\lim Y_n = 0\} = 1$ iff $P\{A\} = 0$ or equivalently

$$P\{A_k\} = 0, \qquad k \geq 1.$$ □

The lemma which follows plays a pivotal role in probability theory in establishing the existence of limits and constitutes one-half the Borel–Cantelli theorem of Section 3.2.

Lemma 2 (Borel–Cantelli Lemma). *If $\{A_n, n \geq 1\}$ is a sequence of events for which $\sum_1^\infty \mathrm{P}\{A_n\} < \infty$, then $\mathrm{P}\{A_n, \text{i.o.}\} = 0$.*

PROOF. Since $\{A_n, \text{i.o.}\} = \bigcap_{k=1}^\infty \bigcup_{n=k}^\infty A_n \subset \bigcup_{n=k}^\infty A_n$, all $k \geq 1$, by Theorem 1.5.2

$$\mathrm{P}\{A_n, \text{i.o.}\} \leq \mathrm{P}\left\{\bigcup_{n=k}^{\infty} A_n\right\} \leq \sum_{n=k}^{\infty} \mathrm{P}\{A_n\},$$

and so

$$0 \leq \mathrm{P}\{A_n, \text{i.o.}\} \leq \lim_{k \to \infty} \sum_{n=k}^{\infty} \mathrm{P}\{A_n\} = 0. \qquad \square$$

The last two lemmas in conjunction with Theorem 2 yield

Theorem 3 (Borel Strong Law of Large Numbers, 1909). *Let S_n constitute a sequence of binomial* r.v.s *on some probability space* $(\Omega, \mathscr{F}, \mathrm{P})$ *with* p.d.f.s *$b(k; n, p)$, $n \geq 1$. Then*

$$\mathrm{P}\left\{\lim_{n \to \infty} \frac{S_n}{n} = p\right\} = 1.$$

PROOF. According to Theorem 2, for every $\varepsilon > 0$

$$\sum_{n=1}^{\infty} \mathrm{P}\left\{\left|\frac{S}{n} - p\right| > \varepsilon\right\} < \infty,$$

whence the Borel–Cantelli lemma guarantees that

$$\mathrm{P}\left\{\left|\frac{S}{n} - p\right| > \varepsilon, \text{i.o.}\right\} = 0, \qquad \varepsilon > 0.$$

Thus, by Lemma 1

$$\mathrm{P}\left\{\lim_{n \to \infty}\left(\frac{S_n}{n} - p\right) = 0\right\} = 1,$$

which is tantamount to that which was to be proved. $\square$

The existence of such a probability space will follow from Theorems 3.1.1 and 6.4.3.

S. Bernstein ingeniously exploited the binomial distribution and Theorem 1 to prove Weierstrass' approximation theorem, which asserts that every continuous function on [0, 1] can be uniformly approximated by polynomials.

EXAMPLE 1. If f is a continuous function on [0, 1] and the Bernstein polynomials are defined by

$$B_n(p) = \sum_{j=0}^{n} f\left(\frac{j}{n}\right)\binom{n}{j} p^j (1 - p)^{n-j}, \qquad p \in [0, 1], \tag{3}$$

then

$$\lim_{n\to\infty} B_n(p) = f(p) \quad \text{uniformly for } p \in [0, 1]. \tag{4}$$

PROOF. Let S_n be a binomial r.v. with p.d.f. $b(k; n, p)$. Since every continuous function on $[0, 1]$ is bounded and uniformly continuous thereon, $|f(p)| \le M < \infty$ for $p \in [0, 1]$, and for every $\varepsilon > 0$ there exists $\delta > 0$ such that $|f(p) - f(p')| < \varepsilon$ if $|p - p'| < \delta$ and $0 \le p, p' \le 1$. Then, setting $q = 1 - p$ and $A_n = \{j: |j/n - p| < \delta\}$,

$$\begin{aligned} |B_n(p) - f(p)| &= \left| \sum_{j=0}^{n} \binom{n}{j} p^j q^{n-j} \left[f\left(\frac{j}{n}\right) - f(p) \right] \right| \\ &\le \sum_{j=0}^{n} \binom{n}{j} p^j q^{n-j} \left| f\left(\frac{j}{n}\right) - f(p) \right| \\ &\le \varepsilon \sum_{A_n} b(j; n, p) + 2M \sum_{A_n^c} b(j; n, p) \\ &\le \varepsilon + 2M\, \mathrm{P}\left\{ \left| \frac{S_n}{n} - p \right| \ge \delta \right\}. \end{aligned}$$

By (2),

$$\mathrm{P}\left\{ \left| \frac{S_n}{n} - p \right| \ge \delta \right\} \le \frac{p(1-p)}{n\delta^2} \le \frac{1}{4n\delta^2},$$

and so if $n \ge M(\varepsilon\delta^2)^{-1}$,

$$|B_n(p) - f(p)| \le \varepsilon + \varepsilon = 2\varepsilon, \qquad 0 \le p \le 1,$$

yielding (4). □

If $\{Y_n, n \ge 0\}$ are r.v.s on a probability space $(\Omega, \mathscr{F}, \mathrm{P})$, then $\{Y_n, n \ge 1\}$ is said to **converge in probability** to Y_0, denoted $Y_n \xrightarrow{\mathrm{P}} Y_0$, if

$$\lim_{n\to\infty} \mathrm{P}\{|Y_n - Y_0| \le \varepsilon\} = 1,$$

all $\varepsilon > 0$. Alternatively, $\{Y_n, n \ge 1\}$ **converges almost certainly** (a.c.) to Y_0, denoted $Y_n \xrightarrow{\text{a.c.}} Y_0$, if $\mathrm{P}\{\lim_{n\to\infty} Y_n = Y_0\} = 1$. Theorems 1 and 3 of this section assert that if $\{S_n, n \ge 1\}$ are binomial r.v.s on $(\Omega, \mathscr{F}, \mathrm{P})$, then both types of convergence hold with $Y_n = S_n/n$ and $Y_0 = p$. A detailed discussion of these concepts is given in Section 3.3, where it is shown that $Y_n \xrightarrow{\text{a.c.}} Y_0$ implies $Y_n \xrightarrow{\mathrm{P}} Y_0$, the converse being untrue in general. However, the case of a countable sample space Ω is exceptional according to

EXAMPLE 2. If $\{Y_n, n \ge 0\}$ is a sequence of r.v.s on a countable probability space $(\Omega, \mathscr{F}, \mathrm{P})$ where $\mathscr{F} = \{\text{all subsets of } \Omega\}$ and $Y_n \xrightarrow{\mathrm{P}} Y_0$, then $Y_n \xrightarrow{\text{a.c.}} Y_0$.

PROOF. Set $A = \{\omega: \lim_{n\to\infty} Y_n = Y_0\}$ and suppose that $P\{A\} < 1$ or equivalently $P\{A^c\} > 0$. Since A^c is countable and, moreover, a countable union of null events is a null event, there exists $\omega_0 \in A^c$ with $P\{\omega_0\} = \delta > 0$. Moreover, $\omega_0 \in A^c$ implies that for some $\varepsilon > 0$ and subsequence n_j of the positive integers, $|Y_{n_j}(\omega_0) - Y_0(\omega_0)| > \varepsilon, j \geq 1$, whence

$$P\{|Y_{n_j} - Y_0| > \varepsilon\} \geq P\{\omega_0\} \geq \delta,$$

contradicting $Y_n \xrightarrow{P} Y_0$. □

EXERCISES 2.2

1. Verify that if $np \to \lambda \in (0, \infty)$, then $\sum_0^n k^j b(k; n, p) \to \sum_0^\infty k^j p(k; \lambda)$, $j = 1, 2$. *Hint*: Recall Exercise 2.1.2.

2. If $g_m(\lambda) = \sum_{k=0}^\infty (k - \lambda)^m p(k; \lambda)$, show that $g_1(\lambda) = 0$, $g_2(\lambda) = g_3(\lambda) = \lambda$, $g_4(\lambda) = 3\lambda^2 + \lambda$, $g_6(\lambda) = 15\lambda^3 + 25\lambda^2 + \lambda$.

3. (i) Prove a weak law of large numbers where S_n has p.d.f. $p(k; n\lambda)$, $n \geq 1$, that is, $P\{|(S_n/n) - \lambda| > \varepsilon\} = o(1)$, $\varepsilon > 0$. (ii) If $\{X, X_n, n \geq 1\}$ are r.v.s with a common d.f. and $n\,P\{|X| > n\} = o(1)$, then $(1/n)\max_{1\leq i\leq n}|X_i| \xrightarrow{P} 0$.

4. Prove a strong law of large numbers where S_n has p.d.f. $p(k; n\lambda)$, $n \geq 1$, that is, $\lim_{n\to\infty}(S_n/n) = \lambda$, a.c. *Hint*: Consider $P\{|S_n - n\lambda|^4 > n^4\varepsilon^4\}$.

5. Show that the Borel–Cantelli lemma is valid on any measure space $(\Omega, \mathscr{A}, \mu)$, that is, $\sum_1^\infty \mu\{A_n\} < \infty$ implies $\mu\{\overline{\lim}_n A_n\} = 0$ for $A_n \in \mathscr{A}$, $n \geq 1$.

6. Let $\{X_{n,j}, j \geq 1, n \geq 1\}$ be a sequence of r.v.s such that $\sum_{j=1}^\infty P\{|X_{n,j}| > \varepsilon\} \xrightarrow{n\to\infty} 0$, all $\varepsilon > 0$. Prove that $\sup_{j\geq 1}|X_{n,j}| \xrightarrow{P} 0$.

7. Prove that for $0 < p < 1$

$$\lim_{n\to\infty} \sum_{k=0}^n \left(\frac{k - np}{\sqrt{npq}}\right)^4 b(k; n, p) = \frac{1}{\sqrt{2\pi}} \int_{-\infty}^\infty t^4 e^{-t^2/2}\, dt = \lim_{\lambda\to\infty} \sum_{k=0}^\infty \left(\frac{k - \lambda}{\sqrt{\lambda}}\right)^4 p(k; \lambda).$$

8. A sequence $\{S_n, n \geq 1\}$ of r.v.s is said to **converge completely** (Hsu–Robbins) if for every $\varepsilon > 0$, $\sum P\{|S_n| > \varepsilon\} < \infty$. Prove that if $\{S_n, n \geq 1\}$ converges completely, then $\lim S_n = 0$, a.c.

9. (i) For any r.v. X and constants $a_n \to 0$, verify that $a_n X \xrightarrow{\text{a.c.}} 0$. *Hint*: It suffices to prove $c_n|X| \xrightarrow{\text{a.c.}} 0$, where $c_n = \sup_{j\geq n}|a_j|$. (ii) If $\{X_n, n \geq 1\}$ are r.v.s with identical d.f.s and $\{a_n\}$ are as in (i), then $a_n X_n \xrightarrow{P} 0$.

2.3 Central Limit Theorem for Binomial Random Variables, Large Deviations

To show that the limit of binomial p.d.f.s may be normal an asymptotic estimate of the combinatorial occurring therein is essential and this follows readily from

Lemma 1 (Stirling's Formula). *For every positive integer $n \geq 2$*

$$n! = n^{n+(1/2)}e^{-n+\varepsilon_n}\sqrt{2\pi} \quad \textit{where } \frac{1}{12n+1} < \varepsilon_n < \frac{1}{12n}. \tag{1}$$

PROOF. Define

$$a_n = \left(1 + \frac{1}{n}\right)^{n+(1/2)}, \qquad b_n = \frac{1}{2n+1}, \qquad n \geq 1.$$

Then

$$\log a_n = (n + \tfrac{1}{2})\log\frac{n+1}{n} = \frac{1}{2b_n}\log\frac{1+b_n}{1-b_n} = 1 + \frac{b_n^2}{3} + \frac{b_n^4}{5} + \cdots$$

$= 1 + \delta_n$ (say), where

$$\frac{1}{12n+1} - \frac{1}{12(n+1)+1} < \frac{b_n^2}{3} < \delta_n < \frac{b_n^2}{3}(1 + b_n^2 + b_n^4 + \cdots)$$

$$= \frac{b_n^2}{3(1-b_n^2)} = \frac{1}{12n(n+1)}, \tag{2}$$

so that

$$0 < \sum_{n=1}^{\infty} \delta_n = C < \infty.$$

Therefore,

$$\log\frac{(n+1)^{n+(1/2)}}{n!} = \sum_1^n \log a_j = n + \sum_{j=1}^{n} \delta_j = n + C - \sum_{j=n+1}^{\infty} \delta_j$$

$= n + C - \varepsilon_{n+1}$ (say), where via (2)

$$\frac{1}{12n+1} < \varepsilon_n < \frac{1}{12n}.$$

Hence, for $n \geq 1$

$$n! = (n+1)^{n+(1/2)}\exp(-n - C + \varepsilon_{n+1}),$$

implying

$$(n+1)! = (n+1)^{n+(3/2)}\exp(-n - C + \varepsilon_{n+1}). \tag{3}$$

Set $K = e^{1-C} > 0$. Then, replacing n by $n - 1$ in (3),

$$n! = n^{n+(1/2)}e^{-n+\varepsilon_n}K, \qquad n \geq 2. \tag{4}$$

The identification of K as $\sqrt{2\pi}$ will be made at the end of this section via probabilistic reasoning, thereby completing the proof. □

Lemma 2 (DeMoivre–Laplace, 1730). *For $n = 1, 2, \ldots$ let $k = k_n$ be a non-negative integer and set $x = x_k = (k - np)(npq)^{-1/2}$, where $q = 1 - p$, $0 < p < 1$. If $x = o(n^{1/6})$ and $\varphi(x)$ is the standard normal density, there exist positive constants A, B, C such that*

$$\left| \frac{b(k; n, p)}{(npq)^{-1/2}\varphi(x)} - 1 \right| < \frac{A}{n} + \frac{B|x|^3}{\sqrt{n}} + \frac{C|x|}{\sqrt{n}}. \tag{5}$$

Proof. Since $x = o(n^{1/6})$, necessarily $(k/n) \to p$. By Stirling's formula

$$b(k; n, p) = \binom{n}{k} p^k q^{n-k} = \frac{n^{n+(1/2)} \exp(-n + \varepsilon_n)(2\pi)^{-1/2} p^k q^{n-k}}{k^{k+(1/2)}(n-k)^{n-k+(1/2)} \exp(-n + \varepsilon_k + \varepsilon_{n-k})}$$

$$= \frac{e^{\varepsilon}}{\sqrt{2\pi}} \left(\frac{k}{np}\right)^{-k-(1/2)} \left(\frac{n-k}{nq}\right)^{-n+k-(1/2)} (npq)^{-1/2}$$

where $\varepsilon = \varepsilon_n - \varepsilon_k - \varepsilon_{n-k}$. Since $k/n \to p$, $\varepsilon = O(n^{-1})$. Now

$$\log\{(2\pi npq)^{1/2} b(k; n, p)\} = \varepsilon - (k + \tfrac{1}{2})\log \frac{k}{np} - (n - k + \tfrac{1}{2})\log \frac{n-k}{nq}$$

$$= \varepsilon - (np + x\sqrt{npq} + \tfrac{1}{2})\log\left(1 + x\sqrt{\frac{q}{np}}\right)$$

$$- (nq - x\sqrt{npq} + \tfrac{1}{2})\log\left(1 - x\sqrt{\frac{p}{nq}}\right)$$

$$= \varepsilon - (np + x\sqrt{npq} + \tfrac{1}{2})\left[x\sqrt{\frac{q}{np}} - \frac{x^2 q}{2np} + O\left(\frac{|x|^3}{n^{3/2}}\right)\right]$$

$$- (nq - x\sqrt{npq} + \tfrac{1}{2})\left[-x\sqrt{\frac{p}{nq}} - \frac{x^2 p}{2nq} + O\left(\frac{|x|^3}{n^{3/2}}\right)\right]$$

$$= \varepsilon - \left[x\sqrt{npq} + x^2 q - \frac{x^2 q}{2} + \frac{x}{2}\sqrt{\frac{q}{np}} + O\left(\frac{|x|^3}{n^{1/2}}\right) + O\left(\frac{x^2}{n}\right)\right]$$

$$- \left[-x\sqrt{npq} + x^2 p - \frac{x^2 p}{2} - \frac{x}{2}\sqrt{\frac{p}{nq}} + O\left(\frac{|x|^3}{n^{1/2}}\right) + O\left(\frac{x^2}{n}\right)\right]$$

$$= -\frac{x^2}{2} + O\left(\frac{|x|^3}{\sqrt{n}}\right) + O\left(\frac{|x|}{\sqrt{n}}\right) + O\left(\frac{1}{n}\right),$$

whence

$$(npq)^{1/2}b(k; n, p) = \frac{1}{\sqrt{2\pi}} \exp\left(\frac{-x^2}{2} + O\left(\frac{|x|^3}{\sqrt{n}}\right) + O\left(\frac{|x|}{\sqrt{n}}\right) + O\left(\frac{1}{n}\right)\right)$$

$$= \varphi(x)\left[1 + O\left(\frac{|x|^3}{\sqrt{n}}\right) + O\left(\frac{|x|}{\sqrt{n}}\right) + O\left(\frac{1}{n}\right)\right]$$

yielding (5). □

The preceding lemma is an example of a **local limit theorem** since it concerns densities, while the forthcoming corollary provides a **global limit** theorem involving d.f.s.

Theorem 1 (DeMoivre–Laplace (Central Limit) Theorem). *If S_n is a sequence of binomial* r.v.s *with* p.d.f.s $b(k; n, p)$, $n \geq 1$, *and* $\{m_n\}$, $\{M_n\}$ *are sequences of nonnegative integers satisfying* $m_n \leq M_n$, $n \geq 1$, *for which*

$$\frac{(m_n - np)^3}{n^2} \to 0, \qquad \frac{(M_n - np)^3}{n^2} \to 0 \tag{6}$$

as $n \to \infty$, *then*

$$P\{m_n \leq S_n \leq M_n\} \sim \Phi\left(\frac{M_n - np + \frac{1}{2}}{(npq)^{1/2}}\right) - \Phi\left(\frac{m_n - np - \frac{1}{2}}{(npq)^{1/2}}\right). \tag{7}$$

PROOF. For $k = m_n, m_n + 1, \ldots, M_n$ let $x_k = (k - np)h_n$, where $h_n = (npq)^{-1/2}$. By Lemma 2

$$P\{m_n \leq S_n \leq M_n\} = \sum_{k=m_n}^{M_n} b(k; n, p) = \sum_{k=m_n}^{M_n} \varphi(x_k) \cdot h_n(1 + o_k(1)), \tag{8}$$

where $o_k(1) \to 0$ uniformly for $m_n \leq k \leq M_n$ (via (6)). Now

$$\int_{(k-(1/2)-np)h_n}^{(k+(1/2)-np)h_n} \varphi(u)du = h_n\varphi(\zeta_k) = h_n \exp(\tfrac{1}{2}(x_k^2 - \zeta_k^2))\varphi(x_k),$$

where $x_k - (h_n/2) < \zeta_k < x_k + (h_n/2)$. Hence,

$$h_n\varphi(x_k) = \exp(\tfrac{1}{2}(\zeta_k^2 - x_k^2))[\Phi(x_k + \tfrac{1}{2}h_n) - \Phi(x_k - \tfrac{1}{2}h_n)]. \tag{9}$$

Since $|\zeta_k^2 - x_k^2| = |\zeta_k - x_k| \cdot |\zeta_k + x_k| \leq h_n[|x_k| + (h_n/4)]$, hypothesis (6) ensures that $\exp\{\frac{1}{2}(\zeta_k^2 - x_k^2)\} \to 1$ uniformly for $m_n \leq k \leq M_n$.

Consequently, from (8) and (9)

$$P\{m_n \leq S_n \leq M_n\} = (1 + o(1)) \sum_{k=m_n}^{M_n} [\Phi(x_k + \tfrac{1}{2}h_n) - \Phi(x_k - \tfrac{1}{2}h_n)]$$

$$= (1 + o(1))[\Phi(x_{M_n} + \tfrac{1}{2}h_n) - \Phi(x_{m_n} - \tfrac{1}{2}h_n)]$$

$$\sim \Phi\left(\frac{M_n - np + \frac{1}{2}}{\sqrt{npq}}\right) - \Phi\left(\frac{m_n - np - \frac{1}{2}}{\sqrt{npq}}\right). \qquad \square$$

Note that the second statement in (11) of Corollary 1 asserts that the d.f. of the normalized binomial r.v. S_n^* as defined in (10) tends to the normal d.f. Such a tendency toward normality is a widespread phenomenon and exemplifies a class of theorems known as central limit theorems (Chapter 9).

Corollary 1. *If* $\{S_n, n \geq 1\}$ *is a sequence of binomial* r.v.s *with* p.d.f.s $b(k; n, p)$, $0 < p < 1$, *and*

$$S_n^* = \frac{S_n - np}{\sqrt{npq}}, \tag{10}$$

then for every pair of finite constants $a < b$ *and all* x

$$\lim_{n\to\infty} P\{a \leq S_n^* \leq b\} = \Phi(b) - \Phi(a), \qquad \lim_{n\to\infty} P\{S_n^* < x\} = \Phi(x). \tag{11}$$

PROOF. Let M_n be the largest integer $\leq np + b\sqrt{npq}$ and m_n the smallest integer $\geq np + a\sqrt{npq}$. Then $M_n \geq m_n \geq 0$ for all large n and (6) obtains. Thus, by the DeMoivre–Laplace theorem

$$\begin{aligned} P\{a \leq S_n^* \leq b\} &= P\{m_n \leq S_n \leq M_n\} \\ &\sim \Phi\left(\frac{M_n - np + \frac{1}{2}}{\sqrt{npq}}\right) - \Phi\left(\frac{m_n - np - \frac{1}{2}}{\sqrt{npq}}\right). \end{aligned} \tag{12}$$

Since the right side of (12) approaches $\Phi(b) - \Phi(a)$ as $n \to \infty$ via the continuity of $\Phi(\cdot)$, the first part of (11) follows. As for the second, for

$$-\infty < a < b < \infty$$

$$\varliminf_{n\to\infty} P\{S_n^* \leq b\} \geq \lim_{n\to\infty} P\{a \leq S_n^* \leq b\} = \Phi(b) - \Phi(a),$$

whence, letting $a \to -\infty$,

$$\varliminf_{n\to\infty} P\{S_n^* \leq b\} \geq \Phi(b).$$

Thus, for $a < b$,

$$\varliminf_{n\to\infty} P\{S_n^* < b\} \geq \varliminf_{n\to\infty} P\{S_n^* \leq a\} \geq \Phi(a),$$

yielding as $a \to b$

$$\varliminf_{n\to\infty} P\{S_n^* < b\} \geq \Phi(b).$$

Similarly,

$$1 - \varlimsup_{n\to\infty} P\{S_n^* < a\} = \varliminf_{n\to\infty} P\{S_n^* \geq a\} \geq \varliminf_{n\to\infty} P\{a \leq S_n^* \leq b\} = \Phi(b) - \Phi(a),$$

yielding as $b \to \infty$

$$\varlimsup_{n\to\infty} P\{S_n^* < a\} \leq \Phi(a),$$

and the second part of (11) follows. □

As an application of Corollary 1, the constant K occurring in the proof of Stirling's formula will be shown to equal $(2\pi)^{1/2}$: if (4) rather than (1) is utilized in the proof of the prior theorem, then setting $a = -b$, (11) becomes

$$\lim_{n\to\infty} P\{|S_n^*| \le b\} = \frac{1}{K}\int_{-b}^{b} e^{-x^2/2}\,dx, \tag{13}$$

which readily implies $K \ge (2\pi)^{1/2}$. On the other hand,

$$P\{|S_n^*| > b\} \le \frac{1}{b^2}\sum_{|k-np|>b\sqrt{npq}}\left(\frac{k-np}{\sqrt{npq}}\right)^2\binom{n}{k}p^k q^{n-k}$$

$$\le \frac{1}{b^2}\sum_{k=0}^{n}\frac{(k-np)^2}{npq}\binom{n}{k}p^k q^{n-k} = \frac{1}{b^2}$$

via (1) of Section 2. Hence,

$$1 = P\{|S_n^*| \le b\} + P\{|S_n^*| > b\} \le P\{|S_n^*| \le b\} + \frac{1}{b^2}$$

$$\xrightarrow{n\to\infty} \frac{1}{K}\int_{-b}^{b} e^{-x^2/2}\,dx + \frac{1}{b^2} \xrightarrow{b\to\infty} \frac{(2\pi)^{1/2}}{K}$$

yielding $K \le (2\pi)^{1/2}$. Consequently, $K = (2\pi)^{1/2}$. □

As will become increasingly apparent, an important characteristic of any d.f. $F(x)$ is the manner in which it increases to one and decreases to zero—in other words, the order of magnitude of its **tails** $1 - F(x)$ and $F(-x)$ as $x \to +\infty$. In the case of a normal r.v., symmetry of the distribution entails identical behavior of the **upper and lower tails**

Lemma 3. *For all* $x > 0$,

$$\frac{x}{1+x^2}e^{-x^2/2} < \int_x^\infty e^{-u^2/2}\,du < \frac{1}{x}e^{-x^2/2}.$$

PROOF. For $x > 0$

$$\frac{1}{x^2}\int_x^\infty e^{-u^2/2}\,du > \int_x^\infty \frac{1}{u^2}e^{-u^2/2}\,du = \frac{1}{x}e^{-x^2/2} - \int_x^\infty e^{-u^2/2}\,du$$

whence

$$\frac{1}{x}e^{-x^2/2} = \int_x^\infty\left(1+\frac{1}{u^2}\right)e^{-u^2/2}\,du > \int_x^\infty e^{-u^2/2}\,du > \frac{x}{1+x^2}e^{-x^2/2}. \quad \square$$

Corollary 2. *For the standard normal distribution* $\Phi(x)$,

$$1 - \Phi(x) \sim \frac{1}{x(2\pi)^{1/2}}e^{-x^2/2}, \qquad \textit{as } x \to \infty,$$

Corollary 1 ensures that the tails of the distribution of the normalized binomial r.v. $S_n^* = (S_n - np)/\sqrt{npq}$, say $1 - F_n^*(x)$ and $F_n^*(-x)$, tend to $1 - \Phi(x)$ and $\Phi(-x)$ respectively as $n \to \infty$. The next assertion is that this remains true for $x = x_n \to \infty$ provided the approach of x_n to ∞ is sufficiently slow. This is the prototype of what is called a large deviation theorem.

Theorem 2. *If S_n is a sequence of binomial* r.v.s *with* p.d.f.s $b(k; n, p), n \geq 1$, *and* $\{a_n, n \geq 1)$ *is a sequence of real numbers with* $a_n \to \infty$, $a_n = o(n^{1/6})$ *as* $n \to \infty$, *then setting* $S_n^* = (S_n - np)/(npq)^{1/2}$ *as earlier,*

$$P\{S_n^* \geq a_n\} \sim 1 - \Phi(a_n) \sim \frac{1}{a_n}\varphi(a_n). \tag{14}$$

PROOF. For simplicity suppose $a_n > 1, n \geq 1$, and define M_n to be the largest integer $\leq np + (a_n + \log a_n)(npq)^{1/2}$ and m_n the largest integer $\leq np + a_n(npq)^{1/2}$. Then

$$x_{M_n} = \frac{M_n - np}{(npq)^{1/2}} = a_n + \log a_n + O(n^{-1/2}), \qquad x_{m_n} = a_n + O(n^{-1/2})$$

whence the DeMoivre–Laplace theorem is applicable, yielding

$$P\{m_n \leq S_n \leq M_n\} \sim \Phi\left(\frac{M_n + \frac{1}{2} - np}{(npq)^{1/2}}\right) - \Phi\left(\frac{m_n - \frac{1}{2} - np}{(npq)^{1/2}}\right)$$

$$= \Phi(a_n + \log a_n + O(n^{-1/2})) - \Phi(a_n + O(n^{-1/2})).$$

By Lemma 3 and the fact that $a_n^3 = o(n^{1/2})$,

$$1 - \Phi(a_n + O(n^{-1/2})) \sim \frac{(2\pi)^{-1/2}}{(a_n + O(n^{-1/2}))}\exp\left\{-\frac{1}{2}(a_n + O(n^{-1/2}))^2\right\} \sim \frac{1}{a_n}\varphi(a_n)$$

$$1 - \Phi(a_n + \log a_n + O(n^{-1/2})) \sim \frac{(2\pi)^{-1/2}}{a_n + \log a_n + O(n^{-1/2})}$$

$$\times \exp\{-\tfrac{1}{2}(a_n + \log a_n + O(n^{-1/2}))^2\} = o\left(\frac{\varphi(a_n)}{a_n}\right).$$

Hence,

$$P\{m_n \leq S_n \leq M_n\} \sim \frac{1}{a_n}\varphi(a_n).$$

Now, for $k = M_n + 1, \ldots, n$

$$\frac{P\{S_n = k\}}{P\{S_n = k - 1\}} = \frac{n - k + 1}{k}\frac{p}{q} \leq \frac{n - M_n}{M_n + 1}\frac{p}{q},$$

whence

$$P\{S_n \geq M_n\} = \sum_{j=0}^{n-M_n} P\{S_n = M_n + j\} \leq \sum_{j=0}^{n-M_n} \left(\frac{n - M_n}{M_n + 1}\frac{p}{q}\right)^j P\{S_n = M_n\}$$

$$\leq \frac{(M_n + 1)q}{M_n + q - np} P\{S_n = M_n\} \leq \frac{n}{M_n - np} P\{S_n = M_n\}.$$

Since $M_n - np = x_{M_n}(npq)^{1/2}$, by Lemma 2

$$\frac{n}{M_n - np} P\{S_n = M_n\} \sim \frac{\varphi(x_{M_n})}{p \cdot q x_{M_n}}$$

$$= \frac{(2\pi p^2 q^2)^{-1/2}}{[a_n + \log a_n + O(n^{-1/2})]} \exp(-\tfrac{1}{2}[a_n + \log a_n + O(n^{-1/2})]^2)$$

$$= o\left(\frac{\varphi(a_n)}{a_n}\right).$$

Analogously,

$$P\{S_n = m_n\} \sim \frac{\varphi(a_n)}{(npq)^{1/2}} = o\left(\frac{\varphi(a_n)}{a_n}\right). \tag{15}$$

Thus,

$$P\{S_n \geq M_n\} = o\left(\frac{\varphi(a_n)}{a_n}\right)$$

and

$$P\{S_n \geq m_n\} = P\{m_n \leq S_n \leq M_n\} + P\{S_n > M_n\}$$

$$\sim \frac{\varphi(a_n)}{a_n} + o\left(\frac{\varphi(a_n)}{a_n}\right) \sim \frac{1}{a_n}\varphi(a_n). \tag{16}$$

Consequently, (14) follows from (15), (16) and

$$P\{S_n > m_n\} \leq P\{S_n^* \geq a_n\} \leq P\{S_n \geq m_n\}. \qquad \square$$

EXERCISES 2.3

1. Show that the Bernoulli weak law of large numbers is consequence of the DeMoivre–Laplace theorem.

2. Prove that as $x \to \infty$.

$$\int_x^\infty u\varphi(u)du \sim x \int_x^\infty \varphi(u)du,$$

$$\frac{1 - \Phi(x + (a/x))}{1 - \Phi(x)} \to e^{-a}.$$

3. Use Lemma 2 to show that $b(n; 2n, \frac{1}{2}) \sim (n\pi)^{-1/2}$.

4. (Normal approximation to the Poisson distribution.) If X_λ is a Poisson r.v. with parameter λ, prove that for any $-\infty < a < b < \infty$

$$\lim_{\lambda\to\infty} \mathrm{P}\left[a < \frac{X_\lambda - \lambda}{\sqrt{\lambda}} < b \right] = \Phi(b) - \Phi(a)$$

by first showing for

$$z = \frac{k-\lambda}{\sqrt{\lambda}} = o(\lambda^{1/6}) \qquad \text{that} \qquad \left| \frac{p(k;\lambda)}{\lambda^{-1/2}\phi(z)} - 1 \right| \le \frac{A}{\lambda} + \frac{B|z|^3}{\sqrt{\lambda}} + \frac{C|z|}{\sqrt{\lambda}}.$$

Obtain a large deviation result akin to Theorem 2.

5. Let S_n be binomial r.v.s with p.d.f. $b(k; n, p)$, $0 < p < 1$, and $S_n^* = (S_n - np)/\sqrt{npq}$. Prove that for every $\alpha > 1$,

$$\mathrm{P}\{|S_n^*| > (2\alpha \log n)^{1/2}, \text{ i.o.}\} = 0.$$

6. Utilize Exercise 5 to prove that for S_n as defined there and for every $\beta > \frac{1}{2}$,

$$\lim_{n\to\infty} \frac{S_n - np}{n^\beta} = 0 \text{ a.c.}$$

(If $\beta = 1$, this yields the Borel strong law of large numbers.)

7. Prove for X_λ as in Exercise 4 that $\mathrm{P}\{X_\lambda \le \lambda\} > c_\lambda$, $\lambda > 0$, where $c_\lambda = \frac{1}{2}$ for integer λ and $c_\lambda = e^{-1}$ otherwise. *Hint*: $A_{n,\lambda} \equiv \sum_{j=0}^{n} (\lambda^j/j!)e^{-\lambda} = \mathrm{P}\{X_\lambda \le \lambda\}$ for $n \le \lambda < n+1$, and $A_{n,n} - A_{n,n+1} = \int_n^{n+1} (\lambda^n e^{-\lambda}/n!)\,d\lambda > A_{n+1,n+1} - A_{n,n+1}$, implying $A_{n,n} > A_{n+1,n+1}$. Also $A_{n,n} - A_{n,n+1} < A_{n,n} - A_{n-1,n}$ and by Exercise 4 $\mathrm{P}\{X_\lambda \le \lambda\} \to \frac{1}{2}$ as $\lambda \to \infty$.

8. For $q > 0$ and $k = 0, 1, \ldots, n$, let

$$e_k = b\left(k; n, \frac{1}{q+1}\right) = \binom{n}{k}(q+1)^{-n} q^{n-k}, \qquad c = \frac{q+1}{2q},$$

$$N = \left[\frac{n+1}{q+1}\right], \qquad h = k - N.$$

Prove that if $\frac{1}{2} < \alpha < \frac{2}{3}$, then

$$\sum_{|h| > n^\alpha} e_k = O(\exp(-n^\eta)) \tag{i}$$

where $\eta < 2\alpha - 1$, and that for $|h| \le n^\alpha$,

$$e_k = \left(\frac{c}{\pi N}\right)^{1/2} e^{-ch^2/N}\left\{1 + O\left(\frac{|h|+1}{n}\right) + O\left(\frac{|h|^3}{n^2}\right)\right\}. \tag{ii}$$

(*Hint*: For (i) apply Theorem 2, and for (ii) apply Lemma 2.)

9. (Renyi) Let S_n be a binomial r.v. with p.d.f. $b(k; n, p)$, where $0 < p < 1$. If $[np]$ is the largest integer $\le np$ and j_n, k_n are integers such that $j_n = O(n^\alpha)$, $k_n = O(n^\alpha)$ for some α in $(0, \frac{2}{3})$, and $k_n^2 - j_n^2 = o(n)$, prove that

$$\lim_{n\to\infty} \frac{\mathrm{P}\{S_n = k_n + [np]\}}{\mathrm{P}\{S_n = j_n + [np]\}} = 1.$$

Hint: Apply Exercise 8(ii).

10. If $0 < \sigma \leq 1$ and $-\infty < a < \infty$ show that $\sup_x |\Phi(x\sigma^{-1} - a) - \Phi(x)| \leq (2\pi)^{-1/2}$. $[\sigma^{-1} - 1 + |a|]$. *Hint*: Add and subtract $\Phi(x\sigma^{-1})$, and use the mean value theorem.

References

J. Bernoulli, *Ars Conjectandi*, Basel, 1713.

S. Bernstein, "Demonstration du théorème de Weierstrass fondée sur le calcul des probabilités," *Soob. Charkov. Mat. Obs.* **13** (1912), 1–2.

É. Borel, "Sur les probabilités démombrables et leurs applications arithmétiques," *Rend. Circ. Mat. Palermo* **27** (1909), 247–271.

W. Feller, *An Introduction to Probability Theory and Its Applications*, Vol. 1, 3rd ed., Wiley, New York, 1950.

B. de Finetti, "La prévision, ses lois logiques, ses sources subjectives," *Annales de l'Institut Henri Poincaré* **7** (1937), 1–68.

J. Haag, "Sur un problème général de probabilités et ses diverses applications," *Proc. Inst. Congr. Math., Toronto, 1924, 1928*, 629–674.

G. H. Hardy, *Divergent Series*, Clarendon Press, Oxford, 1949.

P. L. Hsu and H. Robbins, "Complete convergence and the law of large numbers," *Proc. Nat. Acad. Sci. U.S.A.* **33** (1947), 25–31.

P. S. Laplace, *Théorie analytique de probabilités*, 1812 [Vol. 7 in *Oeuvres complètes de Laplace*, Gauthier–Villars, Paris, 1886].

A. de Moivre, *The Doctrine of Chances*, 1718; 3rd ed., London, 1756.

S. D. Poisson, *Recherches sur la probabilité des judgements*, Paris, 1837.

A. Renyi, *Foundations of Probability*, Holden–Day, San Francisco, 1970.

H. Robbins, "A remark on Stirling's formula," *Amer. Math. Montlhy* **62** (1955), 26–29.

J. Stirling, *Methodus Differentialis*, London, 1730.

H. Teicher, "An inequality on Poisson probabilities," *Ann. Math. Stat.* **26** (1955), 147–149.

3 Independence

Independence may be considered the single most important concept in probability theory, demarcating the latter from measure theory and fostering an independent development. In the course of this evolution, probability theory has been fortified by its links with the real world, and indeed the definition of independence is the abstract counterpart of a highly intuitive and empirical notion. Independence of random variables $\{X_i\}$, the definition of which involves the events of $\sigma(X_i)$, will be shown in Section 2 to concern only the joint distribution functions.

3.1 Independence, Random Allocation of Balls into Cells

Definition. If $(\Omega, \mathscr{F}, \mathrm{P})$ is a probability space and T a nonempty index set, classes $\mathscr{G}_t$ of events, $t \in T$, are termed **independent** if for each $m = 2, 3, \ldots$, each choice of distinct $t_j \in T$, and events $A_j \in \mathscr{G}_{t_j}$, $1 \leq j \leq m$,

$$\mathrm{P}\{A_1 \cdot A_2 \cdots A_m\} = \mathrm{P}\{A_1\} \cdot \mathrm{P}\{A_2\} \cdots \mathrm{P}\{A_m\}. \tag{1}$$

Events A_t, $t \in T$, are called independent if the one-element classes $\mathscr{G}_t = \{A_t\}$, $t \in T$, are independent.

Clearly, nonempty subclasses of independent classes are likewise independent classes. Conversely, if for every nonempty finite subset $T_1 \subset T$ the classes $\mathscr{G}_t$, $t \in T_1$, are independent, then so are the classes $\mathscr{G}_t$, $t \in T$. It may be noted that the validity of (1) for some fixed integer $m > 2$ is not sufficient to guarantee independence of the events $A_1, A_2, \ldots, A_m$ (Exercise 4).

On the other hand, it is easily verified via (14) of Lemma 2.1.2 that A_t, $t \in T$, are independent events iff the classes $\mathscr{G}_t = \{\varnothing, \Omega, A_t, A_t^c\}$ are independent.

Definition. $\{X_n, n \geq 1\}$ are termed **independent random variables** if the classes $\mathscr{G}_n = \sigma(X_n)$, $n \geq 1$, are independent. More generally, **stochastic processes,** i.e., families of r.v.s $\{X_t^{(n)}, t \in T_n\}$, $n \geq 1$, are independent (of one another) if the classes $\mathscr{G}_n = \sigma(X_t^{(n)}, t \in T_n)$, $n \geq 1$, are independent.

Random variables which are not independent are generally referred to as **dependent**. Clearly, subsets of independent families or of independent random variables are themselves independent. Of course, independence of the families (or random vectors) (X_1, X_2) and (X_3, X_4) postulates nothing about the independence or dependence of X_1 and X_2. Note that random variables X and Y are independent if and only if for all $A \in \sigma(X)$, $B \in \sigma(Y)$

$$\mathrm{P}\{A \cdot B\} = \mathrm{P}\{A\} \cdot \mathrm{P}\{B\}.$$

A sequence $\{X_n, n \geq 1\}$ (or the random variables comprising this sequence) is called independent, identically distributed (abbreviated i.i.d.) if X_n, $n \geq 1$, are independent and their distribution functions are identical.

If $\{X_n, n \geq 1\}$ are random variables with $\mathrm{P}\{X_n \in A_n\} = 1$, where $A_n \subset A = \{a_1, a_2, \ldots\}$, $n \geq 1$, define

$$p_{1,\ldots,n}(a_1, \ldots, a_n) = \mathrm{P}\{X_1 = a_1, \ldots, X_n = a_n\}. \tag{2}$$

Then $p_{1,\ldots,n}(a_1, \ldots, a_n)$ is called the **joint probability density function** of $X_1, \ldots, X_n$ and the latter are termed discrete r.v.s. It is not difficult to ascertain that the discrete random variables $X_1, \ldots, X_n$ are independent iff for every choice of $(a_1, \ldots, a_n)$ in $A \times A \times \cdots \times A$

$$p_{1,\ldots,n}(a_1, \ldots, a_n) = p_1(a_1) \cdots p_n(a_n), \tag{3}$$

where p_j is the (one-dimensional) probability density function of X_j, $1 \leq j \leq n$. It follows from (3) that if the discrete random variables $X_1, \ldots, X_n$ are independent then for all real $\lambda_1, \ldots, \lambda_n$

$$\mathrm{P}\{X_1 < \lambda_1, \ldots, X_n < \lambda_n\} = \mathrm{P}\{X_1 < \lambda_1\} \cdots \mathrm{P}\{X_n < \lambda_n\}, \tag{4}$$

and the converse is also true. Even without the proviso of discreteness (4) is still equivalent to independence, but the proof is more involved and hence deferred to Section 2. This condition may be rephrased in terms of the joint and one-dimensional d.f.s, namely, for all real λ_i, $1 \leq i \leq n$,

$$F_{X_1,\ldots,X_n}(\lambda_1, \ldots, \lambda_n) = \prod_{i=1}^{n} F_{X_i}(\lambda_i). \tag{5}$$

If r.v.s X_1 and X_2 on $(\Omega, \mathscr{F}, \mathrm{P})$ are finite everywhere or even if merely $A_1 = \{X_1 = \infty, X_2 = -\infty\} = \varnothing$, $A_2 = \{X_1 = -\infty, X_2 = \infty\} = \varnothing$, then the definition of their sum is standard, i.e., $(X_1 + X_2)(\omega) = X_1(\omega) + X_2(\omega)$.

However, if $A = A_1 \cup A_2 \neq \varnothing$, then $X_1(\omega) + X_2(\omega)$ is undefined on A. For definiteness, set $(X_1 + X_2)(\omega) = 0, \omega \in A$. Since $P\{|X_i| = \infty\} = 0, i = 1, 2$, $P\{A\} = 0$ and so $(X_1 + X_2)(\omega) = X_1(\omega) + X_2(\omega)$, a.c. Hence, for any r.v. X_3 on $(\Omega, \mathcal{F}, P)$,

$$X_1 + X_2 + X_3 = (X_1 + X_2) + X_3 = X_1 + (X_2 + X_3), \qquad \text{a.c.}$$

Unless the contrary is stated, any subsequent relationship among r.v.s will be interpreted in the a.c. sense. Since $f(t_1, t_2) = t_1 + t_2$ is a Borel function on $(R^2, \mathscr{B}^2)$ (with the convention that $\infty + (-\infty) = (-\infty) + \infty = 0$), the sum $X_1 + X_2$ is a r.v. by Theorem 1.4.4. By induction, the sum $\sum_1^n X_i$ of n r.v.s is a r.v. for every $n = 1, 2, \ldots$.

Definition. Random variables X_j, $1 \le j \le n \le \infty$, are called **Bernoulli trials with success probability** p if X_j, $1 \le j \le n \le \infty$, are i.i.d. with $P\{X_j = 1\} = p \in (0, 1)$ and $P\{X_j = 0\} = q = 1 - p$.

The event $\{X_j = 1\}$ is frequently interpreted as a success on the jth trial (and concomitantly $\{X_j = 0\}$ as a failure on the jth trial) so that it is natural to refer to p as the **success probability**. Then $S_n = \sum_1^n X_j$ is the number of successes in the first n trials.

Theorem 1. *If* $\{X_n, n \ge 1\}$ *are* i.i.d. r.v.s *on a probability space* $(\Omega, \mathcal{F}, P)$ *with* $P\{X_1 = 1\} = p \in (0, 1)$ *and* $P\{X_1 = 0\} = q = 1 - p$, *then* $S_n = \sum_1^n X_i$, $n \ge 1$, *is a sequence of binomial* r.v.s *on* $(\Omega, \mathcal{F}, P)$ *with* p.d.f.s $b(k; n, p)$, $n \ge 1$, *and*

$$\lim_{n\to\infty} P\left\{\frac{S_n - np}{(npq)^{1/2}} < x\right\} = \frac{1}{\sqrt{2\pi}} \int_{-\infty}^{x} e^{-u^2/2}\, du, \qquad x \in (-\infty, \infty), \tag{6}$$

$$\frac{S_n}{n} \xrightarrow{\text{a.c.}} p. \tag{7}$$

PROOF. Clearly, $\{S_n, n \ge 1\}$ are r.v.s, and since $P\{X_n = x\} = p^x q^{1-x}$ for $x = 0$ or 1, by independence for $k = 0, 1, \ldots, n$

$$P\{S_n = k\} = \sum P\{X_1 = x_1, \ldots, X_n = x_n\} = \sum p^{\sum_1^n x_i} q^{n - \sum_1^n x_i},$$

where the summation is over all $\binom{n}{k}$ choices of $x_1, \ldots, x_n$ such that $\sum_1^n x_i = k$ and $x_i = 0$ or 1 for $i = 1, \ldots, n$. Hence,

$$P\{S_n = k\} = \sum p^k q^{n-k} = \binom{n}{k} p^k q^{n-k}.$$

Now, (6) and (7) follow from Corollary 2.3.1 and Theorem 2.2.3. □

To say that r balls have been cast **at random** into n cells will mean that every possible arrangement of these r balls into the n cells has the same probability, i.e., is "equally likely." A mathematical model or probability space describing such an experiment is concocted by taking

$$\Omega = \{\omega: \omega = (\omega_1, \ldots, \omega_r), \omega_j = 1, 2, \ldots, n, 1 \le j \le r\}$$
$$= (1, \ldots, n) \times \cdots \times (1, \ldots, n),$$

$$\mathscr{F} = \text{class of all subsets of } \Omega,$$

$$\mathrm{P}\{A\} = \sum_{\omega \in A} n^{-r} = mn^{-r}, \qquad A \in \mathscr{F},$$

where m = number of elements in A. Define $X_1, \ldots, X_r$ to be the **coordinate random variables** of Ω, i.e., $X_j(\omega) = \omega_j$, $1 \le j \le r$. In words, X_j is the number of the cell containing the jth ball. Then $X_1, \ldots, X_r$ are discrete r.v.s with joint p.d.f.

$$p_{1,\ldots,r}(k_1, \ldots, k_r) = \mathrm{P}\{X_1 = k_1, \ldots, X_r = k_r\} = n^{-r},$$
$$k_i = 1, \ldots, n, 1 \le i \le r.$$

Moreover, if p_i denotes the p.d.f. of X_i, $1 \le i \le r$, for $k_i = 1, \ldots, n$, $1 \le i \le r$,

$$p_i(k_i) = \mathrm{P}\{X_i = k_i\} = \sum_{k_j=1, j \ne i}^{n} p_{1,\ldots,r}(k_1, \ldots, k_r) = \frac{n^{r-1}}{n^r} = \frac{1}{n}, \tag{8}$$

whence for all choices of $k_1, \ldots, k_r$

$$p_{1,\ldots,r}(k_1, \ldots, k_r) = \prod_{i=1}^{r} p_i(k_i).$$

According to (3), $X_1, \ldots, X_r$ are independent and, taking cognizance of (8), X_j, $1 \le j \le r$, are i.i.d. random variables. Thus, a random allocation of r balls into n cells is tantamount to considering i.i.d. r.v.s $X_1, \ldots, X_r$ with $\mathrm{P}\{X_1 = k\} = 1/n$, $k = 1, \ldots, n$. Let

$$A_i = \{X_1 \ne i, \ldots, X_r \ne i\} = \{\omega: \text{cell } i \text{ is empty}\}.$$

Then in view of independence and (8), for $1 \le i \le n$

$$\mathrm{P}\{A_i\} = \prod_{j=1}^{r} \mathrm{P}\{X_j \ne i\} = \left(1 - \frac{1}{n}\right)^r;$$

for $1 \le i_1 < i_2 \le n$

$$\mathrm{P}\{A_{i_1}A_{i_2}\} = \mathrm{P}\{X_j \ne i_1 \text{ or } i_2, 1 \le j \le r\}$$
$$= (1 - \mathrm{P}\{X_j = i_1\} - \mathrm{P}\{X_j = i_2\})^r = \left(1 - \frac{2}{n}\right)^r,$$

and, in general, for $1 \le m \le n$ and $1 \le i_1 < \cdots < i_m \le n$

$$\mathrm{P}\{A_{i_1} \cdots A_{i_m}\} = \mathrm{P}\{X_j \ne i_1 \text{ or } \cdots \text{ or } i_m, 1 \le j \le r\} = \left(1 - \frac{m}{n}\right)^r.$$

In other words, $A_1, \ldots, A_n$ are interchangeable events and consequently if

$$p_m(r, n) = P\{\text{exactly } m \text{ cells are vacant}\}, \qquad 0 \le m \le n,$$

it follows from Corollary 2.1.4 that

$$p_m(r, n) = \sum_{j=m}^{n} (-1)^{j-m} \binom{j}{m} \binom{n}{j} \left(1 - \frac{j}{n}\right)^r, \qquad 0 \le m \le n. \tag{9}$$

Theorem 2 (von Mises). *If r_n balls are cast at random into n cells so that each arrangement has probability n^{-r_n}, where*

$$ne^{-r_n/n} = \lambda_n \le \lambda_0 < \infty, \qquad n \ge 1, \tag{10}$$

then as $n \to \infty$ the probability $p_m(r_n, n)$ of exactly m vacant cells satisfies

$$p_m(r_n, n) - \frac{\lambda_n^m e^{-\lambda_n}}{m!} = o(1) \quad \textit{for } m = 0, 1, \ldots. \tag{11}$$

In particular, if $\lambda_n = \lambda + o(1)$, then $\{p_m(r_n, n), m \ge 0\}$ tends as $n \to \infty$ to the Poisson p.d.f. *with parameter λ.*

PROOF. Set $p_j^{(n)} = (1 - (j/n))^{r_n}$ and rewrite (9) as

$$p_m(r_n, n) = \sum_{j=0}^{n-m} (-1)^j \binom{n}{m+j} \binom{m+j}{m} p_{m+j}^{(n)}.$$

Since $1 - x \le e^{-x}$ for $0 < x < 1$,

$$n(n-1)\cdots(n-k+1)p_k^{(n)} \le n^k e^{-kr_n/n} = \lambda_n^k \le \lambda_0^k$$

by (10). Moreover, for any fixed $k \ge 1$, if $r_n \le n^{3/2}$, then

$$\begin{aligned} n^k p_k^{(n)} &= n^k \left(1 - \frac{k}{n}\right)^{r_n} = n^k[e^{-k/n} + O(n^{-2})]^{r_n} \\ &= n^k e^{-kr_n/n}[1 + O(n^{-2}e^{k/n})]^{r_n} = \lambda_n^k[1 + o(1)] = \lambda_n^k + o(1) \end{aligned}$$

by (10), while if $r_n > n^{3/2}$, then

$$\begin{aligned} n^k p_k^{(n)} &= n^k \left(1 - \frac{k}{n}\right)^{r_n} \le n^k \left[\left(1 - \frac{k}{n}\right)^n\right]^{\sqrt{n}} \le n^k e^{-k\sqrt{n}} = o(1), \\ \lambda_n^k &= n^k e^{-kr_n/n} \le n^k e^{-k\sqrt{n}} = o(1), \\ n^k p_k^{(n)} &= \lambda_n^k + o(1), \end{aligned}$$

and so the desired conclusion (11) follows from Theorem 2.1.2. □

In Chapter 9 it will be shown that the d.f. of the number of empty cells tends to the normal distribution when $r_n/n \to \alpha > 0$ and even in certain cases if $r_n/n \to 0$ or ∞.

Definition. The **conditional probability** of an event A given an event B of positive probability, denoted $P\{A|B\}$, is defined by

$$P\{A|B\} = \frac{P\{AB\}}{P\{B\}}. \tag{12}$$

Clearly, when the events A and B are independent, $P\{A|B\} = P\{A\}$.

Theorem 3. *If $\{A_n, n \geq 1\}$ is a sequence of events with $\sum_{n=1}^{\infty} P\{A_n\} = \infty$, then*

$$\limsup_{n\to\infty} P\left\{\bigcup_{j=n+1}^{\infty} A_j | A_n\right\} = 1. \tag{13}$$

PROOF. Evidently, $P\{A_n\} > 0$ for infinitely many values of n, and it may be supposed without loss of generality that $P\{A_n\} > 0$, $n \geq 1$. Since

$$1 \geq P\left\{\bigcup_{n=1}^{\infty} A_n\right\} \geq \sum_{n=1}^{\infty} P\left\{A_n \bigcap_{j=n+1}^{\infty} A_j^c\right\} = \sum_{n=1}^{\infty} P\{A_n\}\left[1 - P\left\{\bigcup_{j=n+1}^{\infty} A_j | A_n\right\}\right],$$

divergence of the series requires

$$\liminf_{n\to\infty}\left[1 - P\left\{\bigcup_{j=n+1}^{\infty} A_j | A_n\right\}\right] = 0,$$

which is tantamount to (13).

EXERCISES 3.1

1. Events A_n, $n \geq 1$, are independent iff their indicator functions I_{A_n}, $n \geq 1$, are independent r.v.s iff $\mathscr{G}_n = \{\varnothing, \Omega, A_n, A_n^c\}$, $n \geq 1$, are independent classes.
2. If the classes of events $\mathscr{A}_n$ and $\mathscr{D}$ are independent, $n \geq 1$, so are $\bigcup_{n=1}^{\infty} \mathscr{A}_n$ and $\mathscr{D}$.
3. (i) Any r.v. X is independent of a degenerate r.v. Y. (ii) Two disjoint events are independent iff one of them has probability zero. (iii) If $P\{X = \pm 1, Y = \pm 1\} = \frac{1}{4}$ for all four pairs of signs, then X and Y are independent r.v.s.
4. Let $\Omega = \{\omega_0, \omega_1, \omega_2, \omega_3\}$, $\mathscr{F} = \{A: A \subset \Omega\}$, $P\{\omega_i\} = \frac{1}{4}$, $0 \leq i \leq 3$, and otherwise P is defined by additivity. If $A_i = \{\omega_0, \omega_i\}$, $1 \leq i \leq 3$, then each of $\{A_1, A_2\}$, $\{A_1, A_3\}$, $\{A_2, A_3\}$ is a pair of independent events but the events A_1, A_2, A_3 are not independent. On the other hand, if $B_1 = \{\omega_0\}$, $B_2 = A_2$, $B_3 = \varnothing$, then (1) obtains for $m = 3$ but the events B_1, B_2 are not independent.
5. (i) If $X_1, \ldots, X_n$ are independent r.v.s and g_i, $1 \leq i \leq n$, are finite Borel functions on $(-\infty, \infty)$, then $Y_i = g_i(X_i)$, $1 \leq i \leq n$ are independent r.v.s. In particular, $-X_1, \ldots, -X_n$ are independent r.v.s. (ii) If B_i, $1 \leq i \leq n$, are linear Borel sets and $Y_i = X_i I_{[X_i \in B_i]}$, where I_A is the indicator function of the set A, show that $\{Y_i, 1 \leq i \leq n\}$ are independent r.v.s.
6. If $\{X_n, n \geq 1\}$ are i.i.d. r.v.s with $P\{X_1 = 0\} < 1$ and $S_n = \sum_1^n X_i$, $n \geq 1$, then for every $c > 0$ there exists an integer $n = n_c$ such that $P\{|S_n| > c\} > 0$.

7. If X, Y, Z are r.v.s on $(\Omega, \mathscr{F}, P)$ and $\circ$ signifies the relation of independence, prove or give counter examples for:

   i. $X \circ Y$ iff $X^2 \circ Y^2$.
   ii. $X \circ Y, X \circ Z$ iff $X \circ (Y + Z)$.
   iii. $X \circ Y, Y \circ Z$ imply $X \circ Z$.
   iv. $X \circ (Y, Z), Y \circ Z$ imply X, Y, Z independent r.v.s.

8. If X and Y are independent r.v.s and $X + Y$ is degenerate, then both X and Y are degenerate.

9. Let $\{X_n, n \geq 1\}$ be i.i.d. r.v.s with $P\{X_i = X_j\} = 0$ for $i \neq j$. If $Y_i = \sum_{j=1}^{i} I_{[X_j \leq X_i]}$, $i \geq 1$, prove that $\{Y_i, i \geq 1\}$ are independent r.v.s with $P\{Y_i = j\} = 1/i, j = 1, 2, \ldots, i$, and $i \geq 1$.

10. In Bernoulli trials with success probability p, let Y be the waiting time beyond the 1st trial until the first success occurs, i.e., $Y = j$ iff $X_{j+1} = 1$, $X_i = 0$, $i \leq j$, where $\{X_1, \ldots, X_j\}$ are i.i.d. with $P\{X_i = 1\} = p = 1 - P\{X_i = 0\}$. Find $P\{Y = j\}$, $j = 0, 1, \ldots$, which for obvious reasons is called the **geometric distribution.** If $Y_1, \ldots, Y_r$ are i.i.d. r.v.s with a geometric distribution, find the p.d.f. of $S_r = \sum_{i=1}^{r} Y_i$, known as the **negative binomial distribution** (with parameter r). *Hint*: S_r may be envisaged as the waiting time beyond the rth trial until the rth success.

11. If Y and Z are independent Binomial (resp. Poisson, negative binomial) r.v.s with parameters (n, p) and (m, p) (resp. λ_1 and λ_2, r_1 and r_2), then $Y + Z$ is a binomial (resp. Poisson, negative binomial) r.v. with parameter $(m + n, p)$ (resp. $\lambda_1 + \lambda_2$, $r_1 + r_2$). Thus, if $\{X_n, n \geq 1\}$ are i.i.d. Poisson r.v.s with parameter λ, the sum $S_n = \sum_i^n X_i$ is a Poisson r.v. with parameter $n\lambda$.

12. In the random casting of r balls into n cells, let $Y_i = 1$ if the ith cell is empty and $=0$ otherwise. For any $k < n$, show that the p.d.f. of $Y_{i_1}, \ldots, Y_{i_k}$ depends only on k and not on $i_1, \ldots, i_k$. *Hint*: It suffices to consider $P\{Y_{i_1} = 1, \ldots, Y_{i_k} = 1\}$ for all $k \leq n$.

13. If $\{A_n\}$ is a sequence of independent events with $P\{A_n\} < 1, n \geq 1$, and $P\{\bigcup_1^\infty A_n\} = 1$, then $P\{A_n, \text{ i.o.}\} = 1$.

14. Let $\Omega = [0, 1]$ and $\mathscr{A} =$ Borel subsets of Ω, and let P be a probability measure on $\mathscr{A}$ such that $P\{[a, b)\} = b = a$ for $0 \leq a \leq b \leq 1$. Such a "Lebesgue measure" exists and is unique (Section 6.1). For $\omega \in \Omega$, let $\omega = \omega_1\omega_2\ldots$ be the decimal expansion of ω (for definiteness, no "finite expansion" is permitted). Prove that $\{\omega_n, n \geq 1\}$ are independent r.v.s with $P\{\omega_n = j\} = \frac{1}{10}, j = 0, 1, \ldots, 9$.

15. If N_p is a geometric r.v. with parameter p, that is, $P\{N_p = k\} = pq^k, k \geq 0$, prove that $\lim_{p \to 0} P\{pN_p < x\} = 1 - e^{-x}, x > 0$, and check that $F(x; \lambda) = (1 - e^{-\lambda x})I_{[x>0]}$ is a d.f. for any $\lambda > 0$. A r.v. with d.f. $F(x; \lambda)$ is said to have an **exponential distribution** with parameter λ.

16. If $\{A_n, n \geq 1\}$ is a sequence of events with $\sum_{n=1}^\infty P\{A_n\} = \infty$, show that $\limsup_{n \to \infty} P\{\bigcup_{j=m}^{n-1} A_j | A_n\} = 1$ for all $m \geq 1$.

3.2 Borel–Cantelli Theorem, Characterization of Independence, Kolmogorov Zero–One Law

The Borel–Cantelli theorem is a *sine qua non* of probability theory and is instrumental in proving strong laws of large numbers, the law of the iterated logarithm (Chapter 10), etc. The portion of Theorem 1 that is complementary to Lemma 2.2.2 postulates independent events and while this proviso can be weakened (Lemma 4.2.4) some such restriction is necessary (see Example 1).

Theorem 1 (Borel–Cantelli). *If* $\{A_n, n \geq 1\}$ *is a sequence of events with* $\sum_1^\infty P\{A_n\} < \infty$, *then* $P\{A_n, \text{i.o.}\} = 0$. *Conversely, if the events* $\{A_n, n \geq 1\}$ *are independent and* $\sum_1^\infty P\{A_n\} = \infty$, *then* $P\{A_n, \text{i.o.}\} = 1$.

PROOF. The first part is just the Borel–Cantelli lemma of the prior chapter. If the events are independent, Theorem 3.1.3 ensures that $\limsup_{n\to\infty} P\{\bigcup_{j=n+1}^\infty A_j\} = 1$; therefore, since $P\{\bigcup_{j=n}^\infty A_j\}$ is a monotone sequence,

$$1 = \lim_{n\to\infty} P\left\{\bigcup_{j=n}^\infty A_j\right\} = P\{A_n, \text{i.o.}\}. \qquad \square$$

Corollary 1 (Borel Zero–One Criterion). *If* $\{A_n, n \geq 1\}$ *are independent events,* $P\{A_n, \text{i.o.}\} = 0$ *or* 1 *according as* $\sum_1^\infty P\{A_n\} < \infty$ *or* $\sum_1^\infty P\{A_n\} = \infty$.

EXAMPLE 1 (D. J. Newman).[1] In a sequence of Bernoulli trials $\{X_n\}$ with success probability $p \in (0, 1)$, define N_n to be the length of the maximal success run commencing at trial n, that is,

$$\{\omega: N_n = j\} = \{\omega: X_{n+j} = 0, X_i = 1, n \leq i < n + j\}, \qquad j \geq 0.$$

If Log n denotes logarithm to the base $1/p$, then

$$P\left\{\overline{\lim_{n\to\infty}} \frac{N_n}{\text{Log } n} = 1\right\} = 1. \tag{1}$$

PROOF. Since (Exercise 3.1.10) N_n has a geometric distribution

$$P\{N_n > a \text{ Log } n\} = \sum_{j > a \text{ Log } n} qp^j \leq p^{a \text{ Log } n} = \frac{1}{n^a},$$

and so by the Borel–Cantelli theorem

$$P\{N_n > a \text{ Log } n, \text{i.o.}\} = 0, \qquad a > 1,$$

implying

$$P\left\{\overline{\lim_{n\to\infty}} \frac{N_n}{\text{Log } n} \leq 1\right\} = 1. \tag{2}$$

[1] See (Feller, 1950, p. 210).

To circumvent the dependence of the events $\{N_n > a \operatorname{Log} n\}, n \geq 1$, define $k_n = [n \operatorname{Log} n]$ = greatest integer equal to or less than $n \operatorname{Log} n$. If $\operatorname{Log}_2 n$ denotes Log Log n and $0 < a < 1$.

$$\begin{aligned} k_n + [a \operatorname{Log} k_n] &\leq (n+1)\operatorname{Log} n - \operatorname{Log} n + a(\operatorname{Log} n + \operatorname{Log}_2 n) \\ &\leq k_{n+1} - (1-a)\operatorname{Log} n + a \operatorname{Log}_2 n + 1, \end{aligned}$$

whence

$$k_{n+1} - (k_n + [a \operatorname{Log} k_n]) \geq (1-a)\operatorname{Log} n - a \operatorname{Log}_2 n - 1 > 1$$

for $n \geq n_0$. Consequently, the events

$$A_n = \{N_{k_n} \geq [a \operatorname{Log} k_n] + 1\}, \qquad n \geq n_0,$$

are independent and, moreover,

$$\begin{aligned} P\{A_n\} &= P\{X_{k_n} = 1, \ldots, X_{k_n + [a \operatorname{Log} k_n]} = 1\} = p^{[a \operatorname{Log} k_n] + 1} \\ &\geq p^{a \operatorname{Log} k_n + 1} \geq \frac{p}{(n \operatorname{Log} n)^a}, \end{aligned}$$

implying $\sum_{n=n_0}^{\infty} P\{A_n\} = \infty$. Thus, by the Borel–Cantelli theorem, for $a \in (0, 1)$

$$P\{N_{k_n} \geq a \operatorname{Log} k_n, \text{ i.o.}\} \geq P\{N_{k_n} \geq [a \operatorname{Log} k_n] + 1, \text{ i.o.}\} = 1,$$

yielding

$$P\left\{\overline{\lim_{n\to\infty}} \frac{N_n}{\operatorname{Log} n} \geq 1\right\} \geq P\left\{\overline{\lim_{n\to\infty}} \frac{N_{k_n}}{\operatorname{Log} k_n} \geq 1\right\} = 1,$$

which, in conjunction with (2), proves (1). □

Next, it will be shown that independence of a finite set of r.v.s depends only upon their joint d.f. Some preliminary lemmas permitting enlargement of independent π-classes are needed.

Lemma 1. *If $\mathscr{G}$ and $\mathscr{D}$ are independent classes of events and $\mathscr{D}$ is a π-class, then $\mathscr{G}$ and $\sigma(\mathscr{D})$ are independent.*

PROOF. For any $B \in \mathscr{G}$, define

$$\mathscr{D}^* = \{A : A \in \sigma(\mathscr{D}) \text{ and } P\{A \cdot B\} = P\{A\} \cdot P\{B\}\}.$$

Then: $\Omega \in \mathscr{D}^*$; $A_1 - A_2 \in \mathscr{D}^*$ if $A_1, A_2 \in \mathscr{D}^*$ and $A_1 \supset A_2$; $A \in \mathscr{D}^*$ if $A = \lim A_n$, where $A_n \in \mathscr{D}^*$ and $A_n \subset A_{n+1}$, $n \geq 1$. Thus, $\mathscr{D}^*$ is a λ-class which contains $\mathscr{D}$ by hypothesis. Consequently, $\mathscr{D}^* \supset \sigma(\mathscr{D})$ by Theorem 1.3.2. In other words, $P\{A \cdot B\} = P\{A\} \cdot P\{B\}$ for any $B \in \mathscr{G}$ and every $A \in \sigma(\mathscr{D})$. □

Lemma 2. *Let $\{X_t, t \in T\}$ be a stochastic process on $(\Omega, \mathscr{F}, P)$ and T_1, T_2 nonempty disjoint subsets of T. For $i = 1, 2$ define $\mathscr{D}_i$ to be the class of all sets*

$D_m^i = \{X_{t_1^{(i)}} < \lambda_1, \ldots, X_{t_m^{(i)}} < \lambda_m\}$, *where m is a positive integer,* λ_j *a real number, and* $t_j^{(i)} \in T_i$, $1 \le j \le m$. *If* $\mathscr{D}_1$ *and* $\mathscr{D}_2$ *are independent classes, then so are* $\sigma(X_t, t \in T_1)$ *and* $\sigma(X_t, t \in T_2)$.

PROOF. Since $\mathscr{D}_1$ and $\mathscr{D}_2$ are π-classes and $\mathscr{D}_1$ is independent of $\mathscr{D}_2$, Lemma 1 ensures that $\mathscr{D}_1$ and $\sigma(\mathscr{D}_2)$ are independent. A second application of Lemma 1 yields independence of $\sigma(\mathscr{D}_1)$ and $\sigma(\mathscr{D}_2)$. Since $\sigma(\mathscr{D}_i) = \sigma(X_t, t \in T_i)$, $i = 1, 2$, Lemma 2 is proved.

Corollary 2. *If the* r.v.s X_t, $t \in T$, *are independent and* T_1, T_2 *are nonempty disjoint subsets of* T, *then* $\sigma(X_t, t \in T_1)$ *and* $\sigma(X_t, t \in T_2)$ *are independent.*

The joint d.f. of any finite set of r.v.s $X_1, \ldots, X_n$ has been defined in (2) of Section 1.6.by

$$F_{X_1, \ldots, X_n}(x_1, \ldots, x_n) = \mathrm{P}\{X_1 < x_1, \ldots, X_n < x_n\}. \tag{3}$$

Theorem 2. *Random variables* $X_1, \ldots, X_n$ *on a probability space* $(\Omega, \mathscr{F}, \mathrm{P})$ *are independent if and only if*

$$F_{X_1, \ldots, X_n} = \prod_{j=1}^{n} F_{X_j}. \tag{4}$$

PROOF. Necessity is trivial since $\{X_j < x_j\}$ is an event in $\sigma(X_j)$, $1 \le j \le n$. To prove sufficiency, let $\mathscr{D}_1$ be the class of sets of the type $\{X_n < \lambda_n\}$ while $\mathscr{D}_2$ is the class of sets of the form $\{X_1 < \lambda_1, \ldots, X_{n-1} < \lambda_{n-1}\}$. Then $\mathscr{D}_1$ and $\mathscr{D}_2$ are π-classes and independent. It follows from Lemma 2 that $\sigma(X_n)$ and $\sigma(X_1, \ldots, X_{n-1})$ are independent. Consequently, if $A_n \in \sigma(X_n)$ and $A_i \in \sigma(X_i) \subset \sigma(X_1, \ldots, X_{n-1})$, $1 \le i < n$,

$$\mathrm{P}\{A_1 \cdots A_{n-1} A_n\} = \mathrm{P}\{A_1 \cdots A_{n-1}\}\mathrm{P}\{A_n\}. \tag{5}$$

If $n = 2$, the proof is complete. Otherwise, since (4) holds with n replaced by $n - 1$ so does (5), and repeating the argument a finite number of times

$$\mathrm{P}\{A_1 A_2 \cdots A_n\} = \prod_{i=1}^{n} \mathrm{P}\{A_i\}$$

for all $A_i \in \sigma(X_i)$, $1 \le i \le n$. Thus, the classes $\sigma(X_i)$, $1 \le i \le n$, and hence also the r.v.s X_i, $1 \le i \le n$, are independent. □

Corollary 3. (i) *The random variables comprising a stochastic process* $\{X_t, t \in T\}$ *are independent iff for all finite subsets of indices* $\{t_j \in T, 1 \le j \le m\}$ *the joint distribution of* $X_{t_1}, \ldots, X_{t_m}$ *coincides with* $\prod_{j=1}^{m} F_{X_{t_j}}$.

(ii) *If* $\{X_n, n \ge 1\}$ *are independent* r.v.s, $\{Y_n, n \ge 1\}$ *are independent* r.v.s *and* X_n *and* Y_n *are identically distributed for* $n \ge 1$, *then* $F_{X_1, \ldots, X_n} = F_{Y_1, \ldots, Y_n}$, $n \ge 1$.

It is a remarkable fact that probabilities of sets of a certain class defined in terms of independent r.v.s cannot be other than zero or one. The sets in question are encompassed by the following

Definition. The **tail σ-algebra** of a sequence $\{X_n, n \geq 1\}$ of r.v.s on a probability space $(\Omega, \mathscr{F}, P)$ is $\bigcap_{n=1}^{\infty} \sigma(X_j, j \geq n)$. The sets of the tail σ-algebra are called **tail events** and functions measurable relative to the tail σ-algebra are dubbed **tail functions.**

A typical example of a tail event of $\{X_n, n \geq 1\}$ is $\{\omega: \sum_{n=1}^{\infty} X_n(\omega)$ converges$\}$ since convergence depends entirely upon the "tail of the series."

If $A_n, n \geq 1$, are independent events, then $X_n = I_{A_n}$ are independent r.v.s. Moreover, $\bigcup_{j=n}^{\infty} A_j \in \sigma(X_j, j \geq m)$ for $n \geq m \geq 1$ whence

$$\{A_n, \text{ i.o.}\} = \bigcap_{n=m}^{\infty} \bigcup_{j=n}^{\infty} A_j \in \sigma(X_m, X_{m+1}, \ldots), \quad \text{all } m \geq 1.$$

The Kolmogorov zero–one law (below) confines the value of $P\{A_n, \text{ i.o.}\}$ to zero or one while the Borel zero–one criterion provides a touchstone. The zero–one law, however, applies to independent r.v.s other than indicator functions.

Theorem 3 (Kolmogorov Zero–One Law). *Tail events of a sequence $\{X_n, n \geq 1\}$ of independent random variables have probabilities zero or one.*

PROOF. By Corollary 2, for each $n \geq 1$ the classes $\sigma(X_i, 1 \leq i \leq n)$ and $\sigma(X_j, j > n)$ are independent and so *a fortiori* are $\sigma(X_i, 1 \leq i \leq n)$ and $\bigcap_{n=0}^{\infty} \sigma(X_j, j > n) = \mathscr{D}$ (say) for every $n \geq 1$. The latter implies that $\mathscr{A} = \bigcup_{n=1}^{\infty} \sigma(X_j, 1 \leq j \leq n)$ is independent of $\mathscr{D}$. Since $\mathscr{A}$ is an algebra (Exercise 1.3.4) and hence a π-class, Lemma 1 ensures that $\sigma(\mathscr{A})$ and $\mathscr{D}$ are independent. But $\mathscr{D} \subset \sigma(X_n, n \geq 1) = \sigma(\mathscr{A})$, whence the tail σ-algebra $\mathscr{D}$ is independent of itself! In other words, for every $B \in \mathscr{D}$, $P\{B\} = P\{B \cdot B\} = P^2\{B\}$, implying $P\{B\} = 0$ or 1. □

Corollary 4. *Tail functions of a sequence of independent r.v.s are* **degenerate,** *that is,* a.c. *constant.*

PROOF. For any tail function Y, by the zero–one law $P\{Y < c\} = 0$ or 1 for every c in $(-\infty, \infty)$. If $P\{Y < c\} = 0$ for all c, then $P\{Y = \infty\} = 1$, whereas if $P\{Y < c\} = 1$ for all c, then $P\{Y = -\infty\} = 1$. Otherwise $c_0 = \inf\{c\colon P\{Y < c\} = 1\}$ is finite, whence $P\{Y = c_0\} = 1$ via the definition of c_0. □

Corollary 5. *If $\{X_n, n \geq 1\}$ is a sequence of independent* r.v.s *then $\overline{\lim}_{n\to\infty} X_n$ and $\underline{\lim}_{n\to\infty} X_n$ are degenerate.*

PROOF. Since for each $n \ge k \ge 1$, X_n is $\sigma(X_j, j \ge k)$-measurable, $Y_k = \sup_{n \ge k} X_n$ is $\sigma(X_j, j \ge k)$-measurable (Section 1.4), whence Y_n is $\sigma(X_j, j \ge k)$-measurable for $n \ge k \ge 1$, implying $\overline{\lim}_{n\to\infty} X_n = \lim_{n\to\infty} \sup_{j \ge n} X_j = \lim_{n\to\infty} Y_n$ is $\sigma(X_j, j \ge k)$-measurable for all $k \ge 1$. Thus, $\overline{\lim}_{n\to\infty} X_n$ is $\bigcap_{k=1}^{\infty} \sigma(X_j, j \ge k)$-measurable, i.e., a tail function, and similarly for $\underline{\lim}\, X_n$, so that the conclusion follows from Corollary 4. □

EXERCISES 3.2

1. Find a trivial example of dependent events A_n with $\sum P\{A_n\}$ divergent but $P\{A_n, \text{i.o.}\} < 1$.
2. If X_n, $n \ge 1$, are independent r.v.s, prove that $P\{\lim_{n\to\infty} X_n = 0\} = 1$ iff
$$\sum_{n=1}^{\infty} P\{|X_n| \ge \varepsilon\} < \infty,$$
all $\varepsilon > 0$.
3. Prove that if (i) $\{X_n, n \ge 1\}$ are i.i.d. $N(0, \sigma^2)$ random variables, then
$$P\left\{\overline{\lim_{n\to\infty}} \frac{X_n}{\sqrt{2 \log n}} = \sigma\right\} = 1.$$
If, rather, $\{X_n, n \ge 1\}$ are i.i.d. exponential r.v.s with parameter λ (Exercise 3.1.15) then $P\{\overline{\lim}_{n\to\infty} X_n/\log n = \lambda^{-1}\} = 1$. (ii) If $\{X_n, n \ge 1\}$ are i.i.d. Poisson r.v.s with parameter λ, then $P\{\overline{\lim}_{n\to\infty} X_n(\log \log n)/\log n = 1\} = 1$ irrespective of λ. *Hint*: Recall Exercise 2.1.4.
4. Show that random vectors $X = (X_1, \ldots, X_m)$ and $Y = (Y_1, \ldots, Y_n)$ are independent (of one another) iff the joint d.f. of X and Y is the product of the d.f.s of X and of Y.
5. If $\{X_n, n \ge 1\}$ is a sequence of finite-valued r.v.s and $S_n = \sum_1^n X_i$, determine which of the following are tail events of $\{X_n, n \ge 1\}$: $\{S_n \text{ converges}\}$; $\{\overline{\lim}\, S_n > \underline{\lim}\, S_n\}$; $\{\overline{\lim}\, S_n = \infty\}$; $\{X_n > c, \text{i.o.}\}$; $\{\underline{\lim}\, X_n = 0\}$; $\{S_n > c_n, \text{i.o.}\}$.
6. If $\{X_n, n \ge 1\}$ is a sequence of independent finite-valued r.v.s and $\{a_n, n \ge 1\}$ is a sequence of finite constants with $0 < a_n \to \infty$, show that $\overline{\lim}\, S_n/a_n$ and $\underline{\lim}\, S_n/a_n$ are degenerate.
7. (Barndorff-Nielsen,) If $\{A_n, n \ge 1\}$ are events satisfying $\sum_{n=1}^{\infty} P\{A_n A_{n+1}^c\} < \infty$ and $P\{A_n\} = o(1)$, then $P\{A_n, \text{i.o.}\} = 0$.
8. Let $\{X_n, n \ge 1\}$ be i.i.d. r.v.s with $P\{X_1 = 1\} = P\{X_1 = -1\} = \frac{1}{2}$ and set $S_n = \sum_{i=1}^n X_i$. Prove that
   i. $\lim_{n\to\infty} P\{S_n/\sqrt{n} < x\} = \Phi(x)$, $-\infty < x < \infty$,
   ii. $P\{\inf_{n\ge1} S_n = -\infty\} = P\{\sup_{n\ge1} S_n = \infty\} = 1$.
9. If X_n, $n \ge 1$, are i.i.d. demonstrate equivalence of the following relations:
   i. $\overline{\lim}_{n\to\infty} |X_n|/n \le 1$, a.s.;
   ii. $\sum_{n=1}^{\infty} P\{|X_n| > n\} < \infty$;
   iii. $\sum_{n=1}^{\infty} n\, P\{n - 1 \le |X_1| < n\} < \infty$.
10. Verify for i.i.d. r.v.s X_n, $n \ge 1$, that $P\{\overline{\lim}_{n\to\infty} X_n = \infty\} = 1$ if and only if X_1 is unbounded above, i.e., $P\{X_1 < C\} < 1$, all $C < \infty$.

11. For each $n \geq 1$, let $\{X_{n,j}, j \geq 1\}$ be a sequence of independent r.v.s. Then $\sup|X_{n,j}| \xrightarrow{P} 0$ iff $\sum_{j=1}^{\infty} P\{|X_{n,j}| > \varepsilon\} = o(1)$, all $\varepsilon > 0$.

12. If $\{X_n, n \geq 1\}$ are independent r.v.s with $E X_n = 0$, $E X_n^2 = 1$ which obey a central limit theorem, i.e., $\lim_{n\to\infty} P\{S_n \geq xn^{1/2}\} = 1 - \Phi(x)$, $x \in (-\infty, \infty)$, where $S_n = \sum_i^n X_i$, $n \geq 1$, prove that $\overline{\lim}_{n\to\infty} S_n/n^{1/2} = \infty$, a.c. *Hint*: Utilize the zero–one law.

3.3 Convergence in Probability, Almost Certain Convergence, and Their Equivalence for Sums of Independent Random Variables

If $\{X_n, n \geq 1\}$ is a sequence of r.v.s on $(\Omega, \mathscr{F}, P)$, the set C, where $X_n, n \geq 1$, converges (i.e., $\lim X_n(\omega)$ exists and is finite) is a measurable set since

$$C = \bigcap_{k=1}^{\infty} \bigcup_{n=1}^{\infty} \bigcap_{i=1}^{\infty} \left\{ |X_{n+i}(\omega) - X_n(\omega)| < \frac{1}{k} \right\}.$$

Moreover, if $P\{C\} = 1$, there exists a r.v. X on $(\Omega, \mathscr{F}, P)$ such that

$$P\left\{ \lim_{n\to\infty} X_n = X \right\} = 1. \tag{1}$$

It suffices to define

$$\begin{aligned} X(\omega) &= \lim_{n\to\infty} X_n(\omega), \qquad \omega \in C, \\ &= \overline{\lim_{n\to\infty}}\, X_n(\omega), \qquad \omega \in C^c, \end{aligned}$$

and clearly X is a r.v. In such a situation, the r.v.s $X_n, n \geq 1$, are said to **converge almost certainly** (or almost surely) to the r.v. X, denoted symbolically by $X_n \xrightarrow{\text{a.c.}} X$. Even if $X(\omega)$ is not a r.v. it is still possible that $\underline{\lim}_{n\to\infty} X_n(\omega) = \overline{\lim}_{n\to\infty} X_n(\omega) = X(\omega)$ with probability one, likewise denoted by $X_n \xrightarrow{\text{a.c.}} X$. The former will be distinguished from the latter on occasion by writing $X_n \xrightarrow{\text{a.c.}}$ a r.v. X or $X_n \xrightarrow{\text{a.c.}} X$, finite.

Definition. A sequence of r.v.s $\{X_n, n \geq 1\}$ on a probability space $(\Omega, \mathscr{F}, P)$ is said to **converge in probability** (to X) if there exists a r.v. X on $(\Omega, \mathscr{F}, P)$ such that

$$P\{|X_n - X| > \varepsilon\} = o(1), \quad \text{all } \varepsilon > 0.$$

This will be denoted symbolically by $X_n \xrightarrow{P} X$.

If r.v.s $X_n \xrightarrow{P} X$ and $X_n \xrightarrow{P} Y$, then $X = Y$, a.c. (Exercise 1). Moreover, the calculus of convergence of real numbers carries over to convergence in probability (Exercise 1(ii), 1(iv)).

The next lemmas analyze a.c. convergence and convergence in probability.

Lemma 1. *Random variables* $X_n \xrightarrow{\text{a.c.}}$ *a* r.v. X iff $\sup_{j\ge n}|X_j - X| \xrightarrow{P} 0$ iff *for all* $\varepsilon, \delta > 0$

$$P\left\{\bigcap_{j=n}^{\infty}[|X_j - X| \le \varepsilon]\right\} \ge 1 - \delta \quad \textit{for } n \ge N(\varepsilon, \delta).$$

PROOF. By Lemma 2.2.1, $X_n \xrightarrow{\text{a.c.}} X$, a r.v., or equivalently $X_n - X \xrightarrow{\text{a.c.}} 0$, iff

$$P\{|X_n - X| > \varepsilon, \text{i.o.}\} = 0, \qquad \varepsilon > 0.$$

Hence, $X_n \xrightarrow{\text{a.c.}} X$ iff for all $\varepsilon > 0$

$$\lim_{n\to\infty} P\left\{\sup_{j\ge n}|X_j - X| > \varepsilon\right\} = \lim_{n\to\infty} P\left\{\bigcup_{j=n}^{\infty}[|X_j - X| > \varepsilon]\right\}$$
$$= P\left\{\bigcap_{n=1}^{\infty}\bigcup_{j=n}^{\infty}[|X_j - X| > \varepsilon]\right\} = 0, \qquad (2)$$

that is, iff $\sup_{j\ge n}|X_j - X| \xrightarrow{P} 0$. The final condition of the lemma is simply a restatement in finite terms of the following alternative form of (2):

$$\lim_{n\to\infty} P\left\{\bigcap_{j=n}^{\infty}[|X_j - X| \le \varepsilon]\right\} = P\left\{\bigcup_{n=1}^{\infty}\bigcap_{j=n}^{\infty}[|X_j - X| \le \varepsilon]\right\} = 1, \qquad \varepsilon > 0.$$

□

If $X, X_n, n \ge 1$ are r.v.s with $\sum_{n=1}^{\infty} P\{|X_n - X| > \varepsilon\} < \infty$, all $\varepsilon > 0$ then the Borel-Cantelli lemma ensures that $X_n \xrightarrow{\text{a.c.}} X$.

Corollary 1. *If random variables* $X_n \xrightarrow{\text{a.c.}}$ *a* r.v. X, *then* $X_n \xrightarrow{P} X$.

Although a.c. convergence of r.v.s X_n, $n \ge 1$, is, in general, stronger than convergence in probability, the latter does circumscribe $\overline{\lim}\, X_n$ and $\underline{\lim}\, X_n$ (Exercise 5).

Corollary 2. *If random variables* $X_n \xrightarrow{\text{a.c.}}$ *a* r.v. X, *then* $g(X_n) \xrightarrow{\text{a.c.}} g(X)$ *for* g *continuous.*

Strictly speaking, the latter is not a corollary of the lemma but rather an immediate consequence of the definition and continuity.

Lemma 2. *Random variables* $X_n \xrightarrow{P}$ *a* r.v. X iff (i) $\sup_{m>n} P\{|X_m - X_n| > \varepsilon\} = o(1)$, $\varepsilon > 0$, iff (ii) *every subsequence of* $\{X_n, n \ge 1\}$ *has itself a subsequence converging* a.c. *to the same* r.v. X.

PROOF. If $X_n \xrightarrow{P} X$, then for all $\varepsilon > 0$, $n \ge n_0(\varepsilon)$ implies $P\{|X_n - X| > \varepsilon\} < \varepsilon$. Hence, for $m > n \ge n_0(\varepsilon/2)$

$$P\{|X_m - X_n| > \varepsilon\} \le P\left\{|X_m - X| > \frac{\varepsilon}{2}\right\} + P\left\{|X - X_n| > \frac{\varepsilon}{2}\right\} < \varepsilon,$$

which implies (i). Conversely, if (i) obtains, then for any integer $k \geq 1$, $P\{|X_n - X_m| > 2^{-k}\} < 2^{-k}$ provided $n > m \geq m_k$. Set $n_1 = m_1$, $n_{i+1} = \max(n_i + 1, m_{i+1})$, and $X'_k = X_{n_k}$ and $A_k = \{|X'_{k+1} - X'_k| > 2^{-k}\}$. Then $\sum_{k=1}^{\infty} P\{A_k\} < \infty$, whence the Borel–Cantelli theorem ensures that, apart from an ω-set A of measure zero, $|X'_{k+1}(\omega) - X'_k(\omega)| \leq 2^{-k}$ provided $k \geq$ some integer $k_0(\omega)$. Hence, for $\omega \in A^c$, as $n \to \infty$

$$\sup_{m>n} |X'_m - X'_n| \leq \sum_{k=n}^{\infty} |X'_{k+1} - X'_k| \leq \sum_{k=n}^{\infty} 2^{-k} = \frac{1}{2^{n-1}} = o(1),$$

and so

$$P\left\{\overline{\lim_{k\to\infty}}\, X'_k = \underline{\lim_{k\to\infty}}\, X'_k, \text{ finite}\right\} = 1. \tag{3}$$

If $X = \underline{\lim}\, X'_k$, then X is a r.v. and, according to (3), $X_{n_k} \xrightarrow{\text{a.c.}} X$ as $k \to \infty$. By Corollary 1, $X_{n_k} \xrightarrow{P} X$. Since for any $\varepsilon > 0$

$$P\{|X_k - X| > \varepsilon\} \leq P\left\{|X_k - X_{n_k}| > \frac{\varepsilon}{2}\right\} + P\left\{|X_{n_k} - X| > \frac{\varepsilon}{2}\right\} = o(1)$$

as $k \to \infty$, $X_k \xrightarrow{P} X$.

Next, if $X_n \xrightarrow{P} X$, any subsequence of $\{X_n\}$, say $X''_n \xrightarrow{P} X$, whence, as already shown, there exists a subsequence of $\{X''_n\}$, say $X'_n \xrightarrow{\text{a.c.}}$ some r.v. Y. Then $X'_n \xrightarrow{P} Y$ but also $X'_n \xrightarrow{P} X$, necessitating $X = Y$, a.c. (Exercise 1). Thus, the subsequence X''_n has a further subsequence $X'_n \xrightarrow{\text{a.c.}} X$.

Finally, if X_n does not converge in probability to X, there exists an $\varepsilon > 0$ and a subsequence X_{n_k} with $P\{|X_{n_k} - X| > \varepsilon\} > \varepsilon$. But then **no** subsequence of X_{n_k} converges in probability to X (*a fortiori* almost certainly to X) in violation of (ii).

Corollary 3. *If random variables $X_n \xrightarrow{P} X$, then $g(X_n) \xrightarrow{P} g(X)$ for every continuous function g.*

PROOF. Every subsequence of $Y_n = g(X_n)$ has a further subsequence $Y_{n_k} = g(X_{n_k})$ with $X_{n_k} \xrightarrow{\text{a.c.}} X$. By Corollary 2, $Y_{n_k} = g(X_{n_k}) \xrightarrow{\text{a.c.}} g(X)$, whence, by Lemma 2, $g(X_n) = Y_n \xrightarrow{P} g(X)$.

Corollary 4. *Random variables $X_n \xrightarrow{\text{a.c.}}$ a r.v. X iff (iii) $\sup_{m>n} |X_m - X_n| \xrightarrow{P} 0$.*

PROOF. If $X_n \xrightarrow{\text{a.c.}} X$, both $\sup_{m>n} |X_m - X| \xrightarrow{P} 0$ and $|X_n - X| \xrightarrow{P} 0$ via Lemma 1, whence for any $\varepsilon > 0$

$$P\left\{\sup_{m\geq n} |X_m - X_n| > \varepsilon\right\} \leq P\left\{\sup_{m\geq n} |X_m - X| > \frac{\varepsilon}{2}\right\} + P\left\{|X - X_n| > \frac{\varepsilon}{2}\right\} = o(1).$$

Conversely, (iii) entails $\sup_{m>n} P\{|X_m - X_n| > \varepsilon\} = o(1)$, all $\varepsilon > 0$, and so by Lemma 2 there exists a r.v. X with $X_n \xrightarrow{P} X$. Thus, for all $\varepsilon > 0$

$$P\left\{\sup_{m\geq n}|X_m - X| > \varepsilon\right\} \leq P\left\{\sup_{m\geq n}|X_m - X_n| > \frac{\varepsilon}{2}\right\} + P\left\{|X_n - X| > \frac{\varepsilon}{2}\right\} = o(1)$$

as $n \to \infty$, implying $X_n \xrightarrow{\text{a.c.}} X$ by Lemma 1. □

The question of a.c. convergence of a sequence of r.v.s $\{X_n, n \geq 1\}$ depends only upon the corresponding sequence of joint d.f.s $\{F_{X_1,\ldots,X_n}, n \geq 1\}$. In other words, if $\{X_n, n \geq 1\}$ and $\{Y_n, n \geq 1\}$ have the same finite-dimensional joint d.f.s, then $X_n \xrightarrow{\text{a.c.}} X$ iff $Y_n \xrightarrow{\text{a.c.}} Y$. In fact, setting $A_m = \{|X_m - X_n| > \varepsilon\}$,

$$\begin{aligned} P\left\{\sup_{m>n}|X_m - X_n| > \varepsilon\right\} &= P\left\{\bigcup_{m=n+1}^{\infty} A_m\right\} \\ &= P\{A_{n+1}\} + P\{A_{n+1}^c A_{n+2}\} \\ &\quad + P\{A_{n+1}^c A_{n+2}^c A_{n+3}\} + \cdots \\ &= P\left\{\sup_{m>n}|Y_m - Y_n| > \varepsilon\right\} \end{aligned}$$

by Theorem 1.6.1, and so the equivalence follows from Lemma 1.

Suppose that for some constants $0 < b_n \uparrow \infty$, $b_{n+1}/b_n \to 1$, i.i.d. r.v.s $\{X_n, n \geq 1\}$ and partial sums $S_n = \sum_1^n X_i$, $n \geq 1$,

$$\frac{S_n}{b_n} \xrightarrow{\text{a.c.}} S, \qquad \text{finite.} \tag{4}$$

Then, clearly, it is necessary that

$$\frac{X_n}{b_n} = \frac{S_n}{b_n} - \frac{b_{n-1}}{b_n}\frac{S_{n-1}}{b_{n-1}} \xrightarrow{\text{a.c.}} 0, \tag{5}$$

and via the Borel–Cantelli theorem

$$\sum_{n=1}^{\infty} P\{|X_1| > \varepsilon b_n\} = \sum_{n=1}^{\infty} P\{|X_n| > \varepsilon b_n\} < \infty, \quad \text{all } \varepsilon > 0. \tag{6}$$

This is a restriction on the d.f. of X_1 (see Corollary 4.1.3) and thereby provides a necessary condition for (4). On the other hand, if (4) is replaced by

$$\frac{1}{b_n}\sum_{j=1}^{n} \frac{X_{n+1-j}}{a_j} \xrightarrow{\text{a.c.}} 0, \tag{7}$$

the d.f. of X_1 should likewise be constrained, but the simple subtraction of (5) leads nowhere. However, (6) is still necessary for (7) if the r.v.s $\{X_n, n \geq 1\}$ are i.i.d. as follows from the second of the ensuing lemmas.

Lemma 3 (Feller–Chung). *Let $\{A_n, n \geq 1\}$ and $\{B_n, n \geq 1\}$ be sequences of events on $(\Omega, \mathscr{F}, \mathrm{P})$ and set $A_0 = \varnothing$. If either* (i) *B_n and $A_n A_{n-1}^c \cdots A_0^c$ are independent for all $n \geq 1$ or* (ii) *the classes $\{B_n\}$ and*

$$\{A_n, A_n A_{n+1}^c, A_n A_{n+1}^c A_{n+2}^c, \ldots\}$$

are independent for all $n \geq 1$, then

$$\mathrm{P}\left\{\bigcup_{n=1}^{\infty} A_n B_n\right\} \geq \mathrm{P}\left\{\bigcup_{n=1}^{\infty} A_n\right\} \inf_{n \geq 1} \mathrm{P}\{B_n\}.$$

PROOF. In case (i)

$$\mathrm{P}\left\{\bigcup_{n=1}^{\infty} A_n B_n\right\} = \mathrm{P}\left\{\bigcup_{n=1}^{\infty} B_n A_n \bigcap_{j=0}^{n-1} (B_j A_j)^c\right\} = \sum_{n=1}^{\infty} \mathrm{P}\left\{B_n A_n \bigcap_{j=0}^{n-1} (B_j A_j)^c\right\}$$

$$\geq \sum_{n=1}^{\infty} \mathrm{P}\left\{B_n A_n \bigcap_{j=0}^{n-1} A_j^c\right\} = \sum_{n=1}^{\infty} \mathrm{P}\{B_n\} \mathrm{P}\left\{A_n \bigcap_{j=0}^{n-1} A_j^c\right\}$$

$$\geq \left(\inf_{n \geq 1} \mathrm{P}\{B_n\}\right) \mathrm{P}\left\{\bigcup_{n=1}^{\infty} A_n\right\},$$

and in case (ii)

$$\mathrm{P}\left\{\bigcup_{1}^{n} A_j B_j\right\} = \sum_{j=1}^{n} \mathrm{P}\left\{A_j B_j \bigcap_{j+1}^{n} (A_i B_i)^c\right\} \geq \sum_{j=1}^{n} \mathrm{P}\left\{A_j B_j \bigcap_{j+1}^{n} A_i^c\right\}$$

$$= \sum_{j=1}^{n} \mathrm{P}\{B_j\} \mathrm{P}\left\{A_j \bigcap_{j+1}^{n} A_i^c\right\} \geq \inf_{1 \leq j \leq n} \mathrm{P}\{B_j\} \mathrm{P}\left\{\bigcup_{1}^{n} A_j\right\},$$

whence the conclusion follows as $n \to \infty$. □

Lemma 4. *Let $\{Y_n, n \geq 1\}$ and $\{Z_n, n \geq 1\}$ be sequences of* r.v.s *such that either* (i) *Y_n and $(Z_1, \ldots, Z_n)$ are independent for all $n \geq 1$ or* (ii) *Y_n and $(Z_n, Z_{n+1}, \ldots)$ are independent for all $n \geq 1$. Then for any constants ε_n, δ_n, ε, and δ,*

$$\mathrm{P}\left\{\bigcup_{n=1}^{\infty} [Z_n + Y_n > \varepsilon_n]\right\} \geq \mathrm{P}\left\{\bigcup_{n=1}^{\infty} [Z_n > \varepsilon_n + \delta_n]\right\} \inf_{n \geq 1} \mathrm{P}\{Y_n \geq -\delta_n\}, \quad (8)$$

$$\mathrm{P}\left\{\overline{\lim_{n \to \infty}} (Z_n + Y_n) \geq \varepsilon\right\} \geq \mathrm{P}\left\{\overline{\lim_{n \to \infty}} Z_n > \varepsilon + \delta\right\} \cdot \underline{\lim}_{n \to \infty} \mathrm{P}\{Y_n \geq -\delta\}. \quad (9)$$

Moreover, if $\underline{\lim}_{n \to \infty} \mathrm{P}\{Y_n \geq -\delta\} > 0$, all $\delta > 0$, then $\overline{\lim}_{n \to \infty}(Z_n + Y_n) < \varepsilon$, a.c., *entails* $\overline{\lim}_{n\to\infty} Z_n \le \varepsilon$, a.c. *Furthermore, if* $\underline{\lim}_{n\to\infty} P\{Y_n > -\delta\} \cdot \underline{\lim}_{n\to\infty} P\{Y_n < \delta\} > 0$ *for all* $\delta > 0$ (*a fortiori, if* $Y_n \xrightarrow{P} 0$), *then* $Z_n + Y_n \xrightarrow{\text{a.c.}} 0$ *implies* $Z_n \xrightarrow{\text{a.c.}} 0$.

PROOF. Set $A_n = \{Z_n > \varepsilon_n + \delta_n\}$, $B_n = \{Y_n \ge -\delta_n\}$. By Lemma 3, for $m \ge 1$

$$P\left\{\bigcup_{n=m}^{\infty} [Y_n + Z_n > \varepsilon_n]\right\} \ge P\left\{\bigcup_{n=m}^{\infty} A_n B_n\right\} \ge P\left\{\bigcup_{n=m}^{\infty} A_n\right\} \inf_{n\ge m} P\{B_n\},$$

yielding (8) for $m = 1$ and (9) via $\varepsilon_n \equiv \varepsilon$, $\delta_n \equiv \delta$, $m \to \infty$. The penultimate statement follows easily from (9), and since both (8) and (9) also hold for $\{-Y_n\}$, $\{-Z_n\}$ the final assertion likewise obtains. □

EXAMPLE 1 (Chung). Let $\{X_n, n \ge 1\}$ be independent r.v.s with partial sums $S_n = \sum_1^n X_i$, $n \ge 1$. If $S_n/n \xrightarrow{P} 0$ and $S_{2^n}/2^n \xrightarrow{\text{a.c.}} 0$, then $S_n/n \xrightarrow{\text{a.c.}} 0$.

PROOF. For $k = 1, 2, \ldots$ there is a unique integer $n(k)$ such that $2^{n(k)-1} \le k < 2^{n(k)}$. Take $\varepsilon > 0$ and set

$$A_k = \{|S_k| > 2^{n(k)+1}\varepsilon\}, \qquad B_k = \{|S_{2^{n(k)}} - S_k| \le 2^{n(k)}\varepsilon\},$$
$$C_n = \{|S_{2^n}| > 2^n\varepsilon\}.$$

Since B_k and $A_k A_{k-1}^c \cdots A_1^c A_0^c$ ($A_0 = \emptyset$) are independent and $C_n \supset \bigcup_{2^{n-1}}^{2^n-1} A_k B_k$, by Lemma 3

$$P\left\{\bigcup_{m+1}^{\infty} C_k\right\} \ge P\left\{\bigcup_{2^m}^{\infty} A_k B_k\right\} \ge P\left\{\bigcup_{2^m}^{\infty} A_k\right\} \inf_{k\ge 2^m} P\{B_k\}. \tag{10}$$

By hypothesis and Lemma 1 the left side of (10) is $o(1)$ as $m \to \infty$, and moreover,

$$P\{B_k^c\} \le P\{|S_{2^{n(k)}}| > 2^{n(k)-1}\varepsilon\} + P\{|S_k| > 2^{n(k)-1}\varepsilon\} = o(1)$$

as $k \to \infty$. Consequently, $P\{\bigcup_{2^m}^{\infty} A_k\} = o(1)$ as $m \to \infty$, and so $S_n/n \xrightarrow{\text{a.c.}} 0$. □

Although, in general, a.c. convergence is much stronger than convergence in probability, in the special case of sums S_n of independent random variables the two are equivalent. A basic tool in demonstrating this equivalence for sums of independent random variables is an inequality due to P. Lévy. This in turn necessitates the

Definition. For any r.v. X a real number $m(X)$ is called a **median** of X if

$$P\{X \le m(X)\} \ge \tfrac{1}{2} \le P\{X \ge m(X)\}.$$

In fact, if $a = \inf\{\lambda: P\{X \le \lambda\} \ge \frac{1}{2}\}$, then $|a| < \infty$ and, since (Exercise 1.6.3) $P\{X \le \lambda\}$ is right continuous, $P\{X \le a\} \ge \frac{1}{2}$. By definition,

$$P\{X \le a - \varepsilon\} < \tfrac{1}{2}$$

for all $\varepsilon > 0$, and so letting $\varepsilon \to 0$, $P\{X < a\} \le \frac{1}{2}$ or equivalently $P\{X \ge a\} \ge \frac{1}{2}$. Thus, a is a median of X.

A pertinent observation concerning medians is that if for some constant c $P\{|X| \ge c\} < \varepsilon \le \frac{1}{2}$, then $|m(X)| \le c$. Moreover, if c is any finite constant, $cm(X)$ and $m(X) + c$ are medians of cX and $X + c$ respectively.

Lemma 5 (Lévy Inequalities). *If* $\{X_j, 1 \le j \le n\}$ *are independent* r.v.s, $S_j = \sum_{i=1}^{j} X_i$, *and* $m(Y)$ *denotes a median of* Y, *then for every* $\varepsilon > 0$

$$P\left\{\max_{1 \le j \le n} [S_j - m(S_j - S_n)] \ge \varepsilon\right\} \le 2\,P\{S_n \ge \varepsilon\}, \tag{11}$$

$$P\left\{\max_{1 \le j \le n} |S_j - m(S_j - S_n)| \ge \varepsilon\right\} \le 2\,P\{|S_n| \ge \varepsilon\}. \tag{12}$$

PROOF. Set $S_0 = 0$ and define T to be the smallest integer j in $[1, n]$ for which $S_j - m(S_j - S_n) \ge \varepsilon$ (if such an integer exists) and $T = n + 1$ otherwise. If

$$B_j = \{m(S_j - S_n) \ge S_j - S_n\}, \qquad 1 \le j \le n,$$

then $P\{B_j\} \ge \frac{1}{2}$. Since $\{\omega: T = j\} \in \sigma(X_1, \ldots, X_j)$, $B_j \in \sigma(X_{j+1}, \ldots, X_n)$, and $\{S_n \ge \varepsilon\} \supset \bigcup_{j=1}^{n} B_j\{T = j\}$,

$$P\{S_n \ge \varepsilon\} \ge \sum_{j=1}^{n} P\{B_j[T = j]\} = \sum_{j=1}^{n} P\{B_j\} \cdot P\{T = j\} \ge \tfrac{1}{2} P\{1 \le T \le n\},$$

which is tantamount to (11). Rewrite (11) with X_j replaced by $-X_j$, $1 \le j \le n$, recalling that $m(-Y) = -m(Y)$, and add this to (11) to obtain (12). □

Definition. A r.v. X is called **symmetric** or said to have a symmetric d.f. if X and $-X$ have the same distribution function.

It is easy to verify that X is symmetric iff $P\{X < x\} = P\{X > -x\}$ for every real x and also that zero is a median of a symmetric r.v. It follows from Corollary 3.2.3 that sums of independent symmetric r.v.s are themselves symmetric r.v.s. This leads directly to

Corollary 5. *If* $\{X_j, 1 \le j \le n\}$ *are independent, symmetric* r.v.s *with partial sums* $S_n = \sum_1^n X_j$, *then for every* $\varepsilon > 0$

$$P\left\{\max_{1 \le j \le n} S_j \ge \varepsilon\right\} \le 2\,P\{S_n \ge \varepsilon\}, \tag{13}$$

$$P\left\{\max_{1 \le j \le n} |S_j| \ge \varepsilon\right\} \le 2\,P\{|S_n| \ge \varepsilon\}. \tag{14}$$

Theorem 1 (Lévy). *If* $\{X_n, n \ge 1\}$ *is a sequence of independent* r.v.s, *then* $S_n = \sum_1^n X_i$ *converges* a.c. iff *it converges in probability.*

PROOF. It suffices to verify sufficiency. By Lemma 2, for any ε in $(0, \frac{1}{4})$, there exists an integer h_0 such that $n > h \geq h_0$ implies, setting $S_{h,n} = S_n - S_h$, that $P\{|S_{h,n}| > \varepsilon\} < \varepsilon$. In view of an earlier comment, this entails $|m(S_{h,n})| \leq \varepsilon$ for $n > h \geq h_0$. By Lévy's inequality (12), for $k > h \geq h_0$

$$P\left\{\max_{h<n\leq k} |S_{h,n}| > 2\varepsilon\right\} = P\left\{\max_{h<n\leq k} |S_{h,n}| > 2\varepsilon, \max_{h<n\leq k} |m(S_{n,k})| \leq \varepsilon\right\}$$

$$\leq P\left\{\max_{h<n\leq k} |S_{h,n} - m(S_{h,n} - S_{h,k})| > \varepsilon\right\}$$

$$\leq 2\,P\{|S_{h,k}| > \varepsilon\} < 2\varepsilon.$$

Hence, letting $k \to \infty$, if $h \geq h_0$,

$$P\left\{\sup_{n>h} |S_{h,n}| > 2\varepsilon\right\} \leq 2\varepsilon,$$

and so $S_n \xrightarrow{\text{a.c.}}$ some r.v. S by Corollary 4. □

Lemmas 4 and 2 may be exploited to give an alternative proof of Lévy's Theorem. Since $S_n \xrightarrow{P} S$, the latter ensures the existence of a subsequence k_n with $S_{k_n} \xrightarrow{\text{a.c.}} S$. Now for every integer $m > 0$, there is an integer $n = n(m)$ such that $k_n \leq m < k_{n+1}$. Clearly, $m \to \infty$ entails $k_n \to \infty$.

Set $Y_m = S - S_m$ and $Z_m = S_m - S_{k_n}$. By hypothesis, $Y_m \xrightarrow{P} 0$ and moreover $Y_m + Z_m \xrightarrow{\text{a.c.}} 0$ by the choice of k_n. Clearly, $(Z_1, \ldots, Z_m)$ is $\sigma(X_1, \ldots, X_m)$-measurable and as noted in the proof of Corollary 3.2.5, Y_m is $\sigma(X_j, j > m)$-measurable. Since the two σ-algebras are independent via Corollary 3.2.2, so are $(Z_1, \ldots, Z_m)$ and Y_m for every $m \geq 1$. Hence, by Lemma 4, $Z_m \xrightarrow{\text{a.c.}} 0$ implying $Y_m \xrightarrow{\text{a.c.}} 0$. □

EXERCISES 3.3

1. i. If $X_n \xrightarrow{P} X$ and $X_n \xrightarrow{P} Y$, then $P\{X = Y\} = 1$.
   ii. $X_n \xrightarrow{P} X$ and $Y_n \xrightarrow{P} Y$ imply $X_n + Y_n \xrightarrow{P} X + Y$.
   iii. $X_n \xrightarrow{P} 0$ implies $m(X_n) \to 0$.
   iv. If $X_n \xrightarrow{P} X$, $Y_n \xrightarrow{P} Y$, and g is a continuous function on R^2, then $g(X_n, Y_n) \xrightarrow{P} g(X, Y)$.

2. Let $\{X_n, n \geq 1\}$ and $\{Y_n, n \geq 1\}$ be two sequences of r.v.s with $F_{X_1,\ldots,X_n} = F_{Y_1,\ldots,Y_n}$ for $n \geq 1$. If $X_n \xrightarrow{P} X$, prove that $Y_n \xrightarrow{P} Y$ and that X and Y are identically distributed. *Hint*: Apply Lemma 3.3.2.

3. i. What is wrong with the following "proof" of Corollary 3?

$$P\{|g(X_n) - g(X)| > \varepsilon\} \leq P\{|X_n - X| > \delta\} = o(1).$$

ii. A r.v. X is symmetric iff X^+ and X^- have identical d.f.s.
iii. If 0 is a median of a r.v. X, it is also a median of $XI_{[|X|<c]}$ for any $c > 0$.

4. If independent r.v.s $X_n \xrightarrow{P} X$, then X is degenerate. Prove for nondegenerate i.i.d. r.v.s $\{X_n\}$ that $P\{X_n \text{ converges}\} = 0$.

5. For any sequence of r.v.s $\{X_n, n \geq 0\}$ with $X_n \xrightarrow{P} X_0$,
$$P\left\{\varliminf_{n\to\infty} X_n \leq X_0 \leq \varlimsup_{n\to\infty} X_n\right\} = 1.$$
Conversely, if $\varlimsup_{n\to\infty} X_n = X_0$ (resp. $\varliminf_{n\to\infty} X_n = X_0$), a.c., then for any $\varepsilon > 0$ $P\{X_n > X_0 + \varepsilon\} = o(1)$ (resp. $P\{X_n < X_0 - \varepsilon\} = o(1)$).

6. If $\{X_n, n \geq 1\}$ are independent, symmetric r.v.s such that $(1/b_n)\sum_1^n X_i \xrightarrow{P} 0$ for some positive constants b_n, then $(1/b_n)\max_{1\leq i\leq n} X_i \xrightarrow{P} 0$. *Hint*:
$$P\left\{\max_{1\leq j\leq n} |X_j| > \varepsilon b_n\right\} \leq P\left\{\max_{1\leq j\leq n} \left|\sum_1^j X_i\right| > \frac{\varepsilon}{2} b_n\right\}.$$

7. If the r.v.s $X_n/b_n \xrightarrow{P} 0$, where the constants b_n satisfy $0 < b_n \uparrow \infty$, then
$$\max_{1\leq j\leq n} |m(X_j - X_n)| = o(b_n).$$

8. Prove for any r.v.s $\{X_n\}$ and constants $\{b_n\}$ with $0 < b_n \uparrow \infty$ that (i) $(1/b_n)\sum_1^n X_i \xrightarrow[\text{a.c.}]{P} 0$ implies $X_n/b_n \xrightarrow[\text{a.c.}]{P} 0$, (ii) if for identically distributed $\{X_n\}$, some nonzero constant c and $0 < b_n \to \infty$, $(1/b_n)\sum_1^n X_i \xrightarrow{P} c$, then $b_n \sim b_{n-1}$.

9. Let the stochastic processes $\{X_n, 1 \leq n \leq k\}$ and $\{X_n', 1 \leq n \leq k\}$ be independent of one another and have the same joint distributions. If m_n is a median of X_n, $1 \leq n \leq k$, then for $\varepsilon > 0$ (see Lemma 10.1.1)
$$P\left\{\max_{1\leq n\leq k} |X_n - m_n| \geq \varepsilon\right\} \leq 2P\left\{\max_{1\leq n\leq k} |X_n - X_n'| \geq \varepsilon\right\}.$$

10. If r.v.s $X_n \xrightarrow{\text{a.c.}} X$ as $n \to \infty$ and $\{N_n, n \geq 1\}$ are positive-integer valued r.v.s with (i) $N_n \xrightarrow{\text{a.c.}} \infty$, then $X_{N_n} \xrightarrow{\text{a.c.}} X$. If, rather, (ii) $N_n \xrightarrow{P} \infty$, that is, $P\{N_n < C\} = o(1)$ all $C > 0$ and X is a r.v., then $X_{N_n} \xrightarrow{P} X$.

11. If the r.v. W_n on $(\Omega, \mathscr{F}, P)$ is $\mathscr{F}_n$-measurable, $n \geq 1$, where $\mathscr{F}_n$ is a decreasing sequence of sub-σ-algebras of $\mathscr{F}$ and $W_n \xrightarrow{\text{a.c.}} W$, then W is $\bigcap_{n=1}^\infty \mathscr{F}_n$-measurable.

12. If X_1, X_2, X_3 are independent, symmetric r.v.s with $P\{|X_1 + X_2 + X_3| \leq M\} = 1$, then $P\{\sum_{i=1}^3 |X_i| \leq M\} = 1$.

13. If $\{X, X_n, n \geq 1\}$ are finite measurable functions on a measure space (S, Σ, μ) with $\mu\{|X_n - X| > \varepsilon\} = o(1)$, $\varepsilon > 0$, then $\sup_{m>n} \mu\{|X_m - X_n| > \varepsilon\} = o(1)$, $\varepsilon > 0$, and there exists a subsequence X_{n_j} with $\mu\{\lim_{j\to\infty} X_{n_j} \neq X\} = 0$.

14. Let $\{X_n, n \geq 1\}$ be r.v.s such that $P\{|X_n| \geq c > 0\} \geq \delta > 0$, $n \geq 1$. If $\{a_n\}$ are finite constants for which $a_n X_n \xrightarrow{P} 0$, then $a_n \to 0$.

15. If r.v.s $X_n \xrightarrow{\text{a.c.}} X$, finite, prove that for every $\varepsilon > 0$ there is a set A_ε with $P\{A_\varepsilon\} < \varepsilon$ such that $\lim X_n(\omega) = X(\omega)$ uniformly on A_ε^c. This is known as **Egorov's theorem**. (*Hint*: Verify that if $A_{n,k} = \bigcap_{j=n}^{\infty} \{|X_j - X| < 2^{-k}\}$ and $A = \{\lim X_n = X\}$, then $\lim_{n\to\infty} P\{A_{n,k}\} = P\{\lim_{n\to\infty} A_{n,k}\} \ge P\{A\} = 1$, whence for some integers n_k $P\{A_{n_k,k}^c\} < \varepsilon/2^k$. Take $A_\varepsilon = \bigcup_{k=1}^{\infty} A_{n_k,k}^c$). State and prove the converse.

16. If $\{X_n, n \ge 1\}$ are independent r.v.s, $S_{m,n} = \sum_{j=m+1}^{n} X_j$, $S_n = S_{0,n}$, then for any $\varepsilon > 0$

$$P\left\{\max_{1\le j\le n} |S_j| > 2\varepsilon\right\} \le \frac{P\{|S_n| > \varepsilon\}}{\min_{1\le j\le n} P\{|S_{j,n}| \le \varepsilon\}}.$$

This is **Ottaviani's inequality**. *Hint*: If $T = \inf\{j \ge 1 : |S_j| > 2\varepsilon\}$, then

$$\bigcup_{j=1}^{n} \{T = j, |S_{j,n}| \le \varepsilon\} \subset \{|S_n| > \varepsilon\}.$$

17. If $\{X, X_n, n \ge 1\}$ are i.i.d. symmetric r.v.s and $S_n = \sum_1^n X_i$, then (i)

$$P\{S_n > x\} \ge \frac{n}{2} P\{X > 2x\} P^{n-1}\{X \le 2x\}$$

for $x > 0$, and (ii) $P\{S_n > x\} \ge (n/2)P\{X > x\}[1 - (n-1)P\{X > x\}]$. Part (i) is untrue if **all** "twos" are deleted (Take $n = 2$, $x = \frac{1}{2}$, and $P\{X = \pm 1\} = \frac{1}{2}$). *Hint*: Apropos of (ii), define $T = \inf\{j \ge 1: X_j > x\}$.

18. Let $S_n = \sum_1^n X_i$ where $\{X_n, n \ge 1\}$ are independent r.v.s and suppose that $\underline{\lim}_{n\to\infty} P\{S_{n-1} \ge -\delta n\} > 0$, all $\delta > 0$. Then $\overline{\lim}\, S_n/n \le C < \infty$, a.c. implies $\sum_{n=1}^{\infty} P\{X_n > \varepsilon n\} < \infty$, all $\varepsilon > C$.

3.4 Bernoulli Trials

A sequence $\{X_n, n \ge 1\}$ of i.i.d. r.v.s with $P\{X_n = 1\} = p \in (0, 1)$ and $P\{X_n = -1\} = q = 1 - p$ constitutes a **sequence of Bernoulli trials with parameter *p***. Define $S_n = \sum_1^n X_i, n \ge 1$. If $Y_n = (X_n + 1)/2$, clearly $\{Y_n, n \ge 1\}$ is a sequence of Bernoulli trials with success probability p (Section 1) and so $\{(S_n + n)/2, n \ge 1\}$ is a sequence of binomial r.v.s. Thus, the DeMoivre–Laplace, Bernoulli, and Borel theorems all pertain to Bernoulli trials with parameter p.

According to the intuitive notion of fairness, a sequence of tosses of a fair coin should at any stage n assign equal probabilities to each of the 2^n n-tuples of outcomes. If a gambler bets one dollar on correctly guessing each individual outcome and X_i denotes his gain (or loss) at the ith toss, this is tantamount to requiring that $P\{X_1 = \pm 1, \ldots, X_n = \pm 1\} = 2^{-n}$ for each of the 2^n choices of sign, $n \ge 1$. Thus, his cumulative gain (or loss) after n tosses is $S_n = \sum_1^n X_j$, where $\{X_n, n \ge 1\}$ are Bernoulli trials with parameter $\frac{1}{2}$. The graph of S_n, $n \ge 1$, shows the random cumulative fortunes (gains) of the gambler as a function of n (which may be envisaged as time), and the fortunes S_n, $n \ge 1$, are said to undergo or constitute a random walk.

The distribution of the length of time T_k to achieve a gain of k dollars is implicit in Theorem 1 (Exercise 5) while the limit distribution of the "first passage time" T_k as well as that of $\max_{1 \le j \le n} S_j$ appear in Theorem 2.

The same limit distributions are shown in Chapter 9 to hold for a large class of **random walks**, i.e., sequence of partial sums S_n, $n \ge 1$, of i.i.d. random variables.

Clearly the r.v.s $\{X_n, n \ge 1\}$ constituting Bernoulli trials with parameter $p = \frac{1}{2}$ are independent symmetric r.v.s, and so by Corollary 3.2.3 the joint d.f.s of $(X_1, \ldots, X_n)$ and $(-X_1, \ldots, -X_n)$ are identical and Theorem 1.6.1 guarantees

$$P\{(X_1, \ldots, X_n) \in B\} = P\{(-X_1, \ldots, -X_n) \in B\}$$

for any Borel set B of R^n.

Theorem 1. *If $\{X_n, n \ge 1\}$ are* i.i.d. *with* $P\{X_1 = 1\} = P\{X_1 = -1\} = \frac{1}{2}$ *and $S_n = \sum_{i=1}^n X_i$, then for every positive integer N:*

$$P\Big\{\max_{1 \le j \le n} S_j \ge N, S_n < N\Big\} = P\{S_n > N\}; \tag{1}$$

$$P\Big\{\max_{1 \le j \le n} S_j \ge N\Big\} = 2\,P\{S_n \ge N\} - P\{S_n = N\}; \tag{2}$$

$$P\Big\{\max_{1 \le j \le n} S_j = N\Big\} = P\{S_n = N\} + P\{S_n = N + 1\}$$

$$= 2^{-n}\binom{n}{[(n + N + 1)/2]}, \tag{3}$$

where $[\lambda]$ is the integral part of λ if $\lambda \ge 0$, $[-\lambda] = -[\lambda]$ if $\lambda \ge 0$;

$$P\{S_j \ne 0, 1 \le j \le n + 1\} = P\Big\{\max_{1 \le j \le n} S_j \le 0\Big\}$$

$$= P\{S_n = 0\} + P\{S_n = 1\} = 2^{-n}\binom{n}{[n/2]}; \tag{4}$$

$$P\{S_1 \ne 0, \ldots, S_n \ne 0, S_{n+1} = 0\} = P\Big\{\max_{1 \le j \le n-1} S_j \le 0, S_n > 0\Big\}. \tag{5}$$

Proof. Define T to be the smallest integer j in $[1, n]$ for which $S_j = N$ if such j exists and $T = n + 1$ otherwise. Then $[T = k] \in \sigma(X_1, \ldots, X_k)$ and $S_T = N$ on $[T \le n]$ since $N \ge 1$. Hence, in view of independence,

$$P\Big\{\max_{1 \le j \le n} S_j \ge N, S_n < N\Big\} = P\{T \le n, S_n < N\} = P\{T < n, S_n < N\}$$

$$= \sum_{k=1}^{n-1} P\Big\{T = k, \sum_{i=k+1}^{n} X_i < 0\Big\}$$

$$= \sum_{k=1}^{n-1} P\{T = k\} P\left\{ \sum_{i=k+1}^{n} X_i < 0 \right\}$$

$$= \sum_{k=1}^{n-1} P\{T = k\} P\left\{ \sum_{i=k+1}^{n} X_i > 0 \right\}$$

$$= \sum_{k=1}^{n-1} P\{T = k, S_n > N\}$$

$$= P\{T < n, S_n > N\} = P\{S_n > N\},$$

yielding (1). To obtain (2), note that via (1)

$$P\left\{ \max_{1 \le j \le n} S_j \ge N \right\} = P\left\{ \max_{1 \le j \le n} S_j \ge N, S_n < N \right\} + P\left\{ \max_{1 \le j \le n} S_j \ge N, S_n \ge N \right\}$$

$$= P\{S_n > N\} + P\{S_n \ge N\} = 2\,P\{S_n \ge N\} - P\{S_n = N\}.$$

The first equality of (3) follows via (2) since

$$P\left\{ \max_{1 \le j \le n} S_j = N \right\} = P\left\{ \max_{1 \le j \le n} S_j \ge N \right\} - P\left\{ \max_{1 \le j \le n} S_j \ge N + 1 \right\}$$

$$= 2\,P\{S_n \ge N\} - 2\,P\{S_n \ge N + 1\} - P\{S_n = N\}$$

$$+ P\{S_n = N + 1\}$$

$$= P\{S_n = N\} + P\{S_n = N + 1\}.$$

In proving the second equality of (3), if $n + N = 2m$ for some $m = 0, 1, \ldots,$

$$P\{S_n = N\} = P\left\{ \frac{S_n + n}{2} = \frac{n + N}{2} \right\} = \binom{n}{m} 2^{-n} = \binom{n}{[(n + N)/2]} 2^{-n}, \quad (6)$$

and clearly $P\{S_n = N + 1\} = 0$. Similarly, if $n + N + 1 = 2m$ for some $m = 0,$ $1, \ldots$ then (6) holds with N replaced by $N + 1$.

Apropos of (4),

$$P\{S_1 \ne 0, \ldots, S_n \ne 0\} = P\{S_1 = -1, S_2 \le -1, \ldots, S_n \le -1\}$$

$$+ P\{S_1 = 1, S_2 \ge 1, \ldots, S_n \ge 1\}$$

$$= 2\,P\{S_1 = -1, S_2 \le -1, \ldots, S_n \le -1\}$$

$$= 2\,P\{X_1 = -1, X_2 \le 0, \ldots, X_2 + \cdots + X_n \le 0\}$$

$$= P\{X_2 \le 0, \ldots, X_2 + \cdots + X_n \le 0\}$$

$$= P\{X_1 \le 0, \ldots, X_1 + \cdots + X_{n-1} \le 0\}$$

$$= P\left\{ \max_{1 \le j \le n-1} S_j \le 0 \right\},$$

which is tantamount to the first equality of (4). To obtain the second, note that via (2)

$$\mathrm{P}\left\{\max_{1\le j\le n} S_j \le 0\right\} = 1 - \mathrm{P}\left\{\max_{1\le j\le n} S_j \ge 1\right\} = 1 - 2\,\mathrm{P}\{S_n \ge 1\} + \mathrm{P}\{S_n = 1\}$$

$$= \mathrm{P}\{S_n = 0\} + \mathrm{P}\{S_n = 1\}.$$

The last equality of (4) derives from

$$\mathrm{P}\{S_{2m} = 0\} = \binom{2m}{m}2^{-2m}, \qquad \mathrm{P}\{S_{2m+1} = 1\} = \binom{2m+1}{m}2^{-2m-1}.$$

Finally, to obtain (5), note that via (4)

$$\begin{aligned}\mathrm{P}\{S_1 \ne 0, \ldots, S_n \ne 0, S_{n+1} = 0\} &= \mathrm{P}\{S_1 \ne 0, \ldots, S_n \ne 0\} \\ &\quad - \mathrm{P}\{S_1 \ne 0, \ldots, S_{n+1} \ne 0\} \\ &= \mathrm{P}\left\{\max_{1\le j\le n-1} S_j \le 0\right\} \\ &\quad - \mathrm{P}\left\{\max_{1\le j\le n} S_j \le 0\right\} \\ &= \mathrm{P}\left\{\max_{1\le j\le n-1} S_j \le 0, S_n > 0\right\}.\end{aligned}$$

□

Next, a local limit theorem (7) and global limit (8) will be obtained for $\bar{S}_n = \max_{1\le j\le n} S_j$. According to (8), the d.f. of $n^{-1/2}\bar{S}_n$ tends as $n \to \infty$ to the **positive normal** d.f. The name arises from the fact that if X is $N(0, 1)$ then $|X|$ has the d.f. $2\Phi(x) - 1$, $x > 0$.

Theorem 2. *Let* $\{X_n, n \ge 1\}$ *be* i.i.d. r.v.s *with* $P\{X_1 = 1\} = \mathrm{P}\{X_1 = -1\} = \frac{1}{2}$ *and partial sums* $S_n = \sum_1^n X_i$, $n \ge 1$, $S_0 = 0$. *If* $\bar{S}_n = \max_{1\le j\le n} S_j$ and $T_k = \inf\{n \ge 1\colon S_n = k\}$, $k \ge 1$ *then for any sequence of positive integers* N_n *with* $N_n = o(n^{2/3})$

$$\mathrm{P}\{\bar{S}_n = N_n\} \sim \left(\frac{2}{\pi n}\right)^{1/2} e^{-N_n^2/2n}, \tag{7}$$

$$\lim_{n\to\infty} \mathrm{P}\{\bar{S}_n < xn^{1/2}\} = 2\Phi(x) - 1 = \left(\frac{2}{\pi}\right)^{1/2}\int_0^x e^{-u^2/2}\,du, \quad x > 0, \tag{8}$$

$$\lim_{k\to\infty} \mathrm{P}\{T_k < xk^2\} = 2[1 - \Phi(x^{-1/2})], \qquad x > 0. \tag{9}$$

Proof. By (3) of Theorem 1

$$\mathrm{P}\{\bar{S}_n = N\} = \mathrm{P}\{S_n = N\} + \mathrm{P}\{S_n = N + 1\},$$

and so, by DeMoivre–Laplace local central limit theorem (Lemma 2.3.2), if $N_n + n$ is even,

$$P\{\bar{S}_n = N_n\} = P\{S_n = N_n\} = b\left(\frac{N_n + n}{2}; n, \tfrac{1}{2}\right) \sim \frac{2}{(2\pi n)^{1/2}} e^{-N_n^2/2n},$$

and similarly when $N_n + n$ is odd.

Next, if $x > 0$, setting $N = [xn^{1/2}]$ and employing (2) and Lemma 2.3.2

$$\begin{aligned} P(\bar{S}_n \le xn^{1/2}\} &= P\{\bar{S}_n \le N\} = 1 - P\{\bar{S}_n \ge N + 1\} \\ &= 1 - 2\,P\{S_n \ge N + 1\} + P\{S_n = N + 1\} \\ &= 2\,P\{S_n \le N\} - 1 + o(1). \end{aligned} \tag{10}$$

By the DeMoivre–Laplace global central limit theorem (Corollary 2.3.1)

$$P\{S_n \le N\} = P\left\{\frac{S_n}{n^{1/2}} \le x(1 + o(1))\right\} \to \Phi(x),$$

whence (8) follows from (10) and (7).

Finally, if $x > 0$, setting $n = [xk^2]$, via (8) and continuity of the normal d.f.

$$\begin{aligned} P\{T_k \le xk^2\} &= P\{T_k \le n\} = P\{\bar{S}_n \ge k\} \\ &\to 1 - [2\Phi(x^{-1/2}) - 1] = 2[1 - \Phi(x^{-1/2})], \end{aligned}$$

and since

$$P\{T_k = xk^2\} \le P\{T_k = n\} \le P\{S_n = k\} = o(1)$$

by the local central limit theorem, (9) follows. □

Theorem 3. *Let* $\{X_n, n \ge 1\}$ *be* i.i.d. *with* $P\{X_1 = 1\} = P\{X_1 = -1\} = \frac{1}{2}$ *and set* $S_n = \sum_{i=1}^n X_i$, $n \ge 1$, $S_0 = 0$. *Then*

$$P\left\{\overline{\lim_{n\to\infty}}\, S_n = \infty\right\} = 1 = P\left\{\underline{\lim}_{n\to\infty}\, S_n = -\infty\right\}, \tag{11}$$

$$P\{S_n \textit{ assumes every integer value, i.o.}\} = 1, \tag{12}$$

$P\{S_n, n \ge 0$ *reaches* k *before* $-j\} = j/(j + k)$ *for every pair of positive integers* j, k. (13)

Proof. Since, probabilistically speaking, the sequence $\{X_n, n \ge 2\}$ does not differ from $\{X_n, n \ge 1\}$, defining $q_j = P\{\sup_{n\ge 0} S_n \ge j\}$, $j \ge 0$, it follows that for $j \ge 1$

$$q_j = P\left\{\sup_{n\ge 1} S_n \ge j\right\}$$

$$= \tfrac{1}{2} P\left\{\sup_{n\geq 1} \sum_{i=2}^{n} X_i \geq j - 1\right\} + \tfrac{1}{2} P\left\{\sup_{n\geq 1} \sum_{i=2}^{n} X_i \geq j + 1\right\}$$

$$= \tfrac{1}{2} P\left\{\sup_{n\geq 0} S_n \geq j - 1\right\} + \tfrac{1}{2} P\left\{\sup_{n\geq 1} S_n \geq j + 1\right\} = \tfrac{1}{2}(q_{j-1} + q_{j+1}).$$

Hence, for $j \geq 1, q_j - q_{j-1} = q_{j+1} - q_j =$ constant $= c$ (say). Therefore, for $j \geq 1, q_j = cj + q_0$. Since $0 \leq q_j \leq 1$, necessarily $c = 0$, whence $q_j = q_0 = 1$ for $j \geq 1$. That is, $P\{\sup_{n\geq 1} S_n \geq j\} = 1$ for every $j \geq 1$, whence

$$P\left\{\sup_{n\geq 1} S_n = \infty\right\} = P\left\{\overline{\lim_{n\to\infty}}\, S_n = \infty\right\} = 1.$$

By symmetry, $P\{\underline{\lim}_{n\to\infty} S_n = -\infty\} = 1$ and (11) is proved.

Next, if $A = \{\sup_{n\geq 1} S_n = \infty, \inf_{n\geq 1} S_n = -\infty\}$, then by (11), $P\{A\} = 1$, necessitating (12).

To prove (13), set $r = j + k$ and

$$y_i = P\{\{S_n, n \geq 0\} \text{ reaches } i \text{ before } i - r\}, \qquad i \geq 0.$$

Then $y_0 = 1$, $y_r = 0$. For $0 < i < r$

$$\begin{aligned} y_i &= P\{X_1 = 1, \{S_n - X_1, n \geq 1\} \text{ reaches } i - 1 \text{ before } i - r - 1\} \\ &\quad + P\{X_1 = -1, \{S_n - X_1, n \geq 1\} \text{ reaches } i + 1 \text{ before } i - r + 1\} \\ &= p y_{i-1} + q y_{i+1}, \end{aligned} \tag{14}$$

where $p = q = \frac{1}{2}$.

As earlier, $y_i = c + y_{i-1} = ci + y_0, 1 \leq i \leq r$. Since $y_0 = 1, y_r = 0$, necessarily $y_i = 1 - (i/r) = (r - i)/r, 0 \leq i \leq r$, and $y_k = (r - k)/r = j/(j + k)$.

□

Theorem 4. *Let $\{X_n, n \geq 1\}$ be a sequence of Bernoulli trials with parameter $p \neq \frac{1}{2}$. If the partial sums are defined by $S_n = \sum_1^n X_i, n \geq 1$, and $S_0 = 0$, then for any positive integers j, k*

$$P\{\textit{partial sums } \{S_n, n \geq 0\} \textit{ reach } k \textit{ before } -j\} = \frac{(p/q)^k - (p/q)^{k+j}}{1 - (p/q)^{k+j}}. \tag{15}$$

PROOF. Set $r = j + k$ and $s = p/q$, where, as usual, $q = 1 - p$, and define for any integer $i \geq 0$

$$y_i = P\{\{S_n, n \geq 0\} \text{ reaches } i \text{ before } i - r\}.$$

They $y_0 = 1$, $y_r = 0$, and (14) obtains but with $p \neq \frac{1}{2}$. Hence, for $0 < i < r$

$$p(y_i - y_{i-1}) = q(y_{i+1} - y_i) \quad \text{or} \quad y_{i+1} - y_i = s(y_i - y_{i-1}).$$

Thus, for $0 < i < r$

$$y_{i+1} - y_i = s^2(y_{i-1} - y_{i-2}) = \cdots = s^i(y_1 - y_0), \tag{16}$$

and clearly (16) holds for $i = 0$. Since $s \neq 1$, for $0 < i \leq r$

$$y_i - y_0 = \sum_{m=1}^{i} (y_m - y_{m-1}) = \sum_{m=1}^{i} s^{m-1}(y_1 - y_0) = \frac{1 - s^i}{1 - s}(y_1 - y_0). \quad (17)$$

Taking $i = r$ in (17) reveals that $-(1 - s) = (1 - s^r)(y_1 - y_0)$, and hence

$$y_i - y_0 = \frac{-(1 - s^i)}{(1 - s^r)} \quad \text{or} \quad y_i = \frac{s^i - s^r}{1 - s^r} = \frac{s^i - s^{j+k}}{1 - s^{j+k}},$$

yielding (15) for $i = k$. □

When $p = q$, the right side of (15) becomes 0/0 and by l'Hospital's rule

$$\frac{s^k - s^{j+k}}{1 - s^{j+k}} \to \frac{j}{j + k} \quad \text{as } s \to 1.$$

If it were known that the left side of (15) was a continuous function of p, then (15) would imply (13).

EXAMPLE 1. Suppose that a gambler A with a capital of j dollars and his adversary B whose capital is k dollars play the following game. A coin is tossed repeatedly (sequentially) and at each toss A wins one dollar if the outcome is a head while B wins a dollar if the outcome is a tail. The game terminates when one of the players is ruined, i.e., when either A loses j dollars or B loses k dollars. If the probability of heads is $p \in (0, 1)$, then by Theorems 3 and 4

$$\begin{aligned} P\{A \text{ ultimately wins}\} &= \frac{j}{j + k} && \text{if } p = q \\ &= \frac{s^k - s^{j+k}}{1 - s^{j+k}} && \text{if } s = \frac{p}{q} \neq 1. \end{aligned} \quad (18)$$

Interchanging p with q and j with k,

$$\begin{aligned} P\{B \text{ ultimately wins}\} &= \frac{k}{j + k} && \text{if } p = q \\ &= \frac{s^{-j} - s^{-(j+k)}}{1 - s^{-(j+k)}} = \frac{1 - s^k}{1 - s^{j+k}} && \text{if } s = \frac{p}{q} \neq 1, \end{aligned} \quad (19)$$

and so for all $p \in (0, 1)$

$$P\{A \text{ ultimately wins}\} + P\{B \text{ ultimately wins}\} = 1,$$

that is, the game terminates (in a finite number of tosses) with probability one.

If $p \leq q$, that is, $s \leq 1$ and B has infinite capital ($k = \infty$), then letting $k \to \infty$ in (18) and (19)

$$P\{A \text{ ultimately wins}\} = 0,$$

$$P\{B \text{ ultimately wins}\} = 1,$$

whereas if $p > q$, that is, $s > 1$ and B has infinite capital,

$$\mathrm{P}\{\text{games terminates}\} = \mathrm{P}\{B \text{ ultimately wins}\} = s^{-j}.$$

The next result demonstrates that when the gamblers A and B have the same initial capital r, the duration of the game and the final capital of A are independent random variables. The duration of the game foreshadows the notion of a finite stopping time (Chapter 5.3).

EXAMPLE 2 (S. Samuels). Let $S_n = \sum_{j=1}^n X_j$ where $\{X_j, j \geq 1\}$ are Bernoulli trials with parameter $p \in (0, 1)$ and define $T = \inf\{n \geq 1: |S_n| = r\}, r > 0$ and $T = \infty$ otherwise. If $U = r + S_T$ where $S_T = \sum_{n=1}^\infty S_n \cdot I_{[T=n]}$, then U and T are independent random variables.

PROOF. Note that according to Example 1, $\mathrm{P}\{T < \infty\} = 1$. Let $C(n, r) =$ number of n-tuplets $(X_1, \ldots, X_n)$ with $S_n = r$, $|S_j| < r$, $1 \leq j < n$. Then

$$\mathrm{P}\{T = n, U = 2r\} = \mathrm{P}\{T = n, S_n = r\} = C(n, r)p^{(n+r)/2}q^{(n-r)/2}, \quad n \geq r > 0.$$

and by symmetry

$$\mathrm{P}\{T = n, U = 0\} = \mathrm{P}\{T = n, S_n = -r\} = C(n, r)p^{(n-r)/2}q^{(n+r)/2}.$$

Hence,

$$\mathrm{P}\{T = n\} = [1 + (p/q)^r] \cdot \mathrm{P}\{T = n, U = 0\}, \quad n \geq r > 0 \tag{20}$$

and so

$$\mathrm{P}\{U = 0\} = \sum_{n=r}^{\infty} \mathrm{P}\{T = n, U = 0\} = [1 + (p/q)^r]^{-1}.$$

Consequently, according to (20)

$$\mathrm{P}\{T = n, U = 0\} = \mathrm{P}\{T = n\} \cdot \mathrm{P}\{U = 0\}, \quad n \geq r > 0$$

which, in turn, implies

$$\mathrm{P}\{T = n, U = 2r\} = \mathrm{P}\{T = n\} \cdot \mathrm{P}\{U = 2r\}, \quad n \geq r > 0. \qquad \square$$

EXERCISES 3.4

1. Let $S_n = \sum_1^n X_i, n \geq 1$, where $\{X_n, n \geq 1\}$ are i.i.d. with $\mathrm{P}\{X_1 = 1\} = \mathrm{P}\{X_1 = -1\} = \frac{1}{2}$. If $T_i = \inf\{n \geq 1: S_n = i\}$, $i = 1, 2$, then

   a. $\mathrm{P}\{T_1 < \infty\} = 1, \sum_{n=1}^\infty \mathrm{P}\{T_1 > n\} = \infty$,
   b. T_1 and $T_2 - T_1$ are i.i.d.

2. Show that (a) and (b) of Exercise 1 hold if rather

$$T_1 = \inf\{n \geq 1: S_n = 0\} \quad \text{and} \quad T_2 = \inf\{n > T_1: S_n = 0\}.$$

3. Let $S_n = \sum_1^n a_i X_i$, where $\{a_i, i \geq 1\}$ are constants and $\{X_n, n \geq 1\}$ are as in Exercise 1. If $S_n \xrightarrow{\text{a.c.}} S$, where $|S| \leq M < \infty$, a.c., prove via Lévy's inequality that (i) $\sup_{n \geq 1} |S_n| \leq M$, a.c., and, moreover, by an extension of Exercise 3.3.12 that (ii) $\sum_1^n |a_n| \leq M$.

4. Let $\{X_n, n \geq 1\}$ be i.i.d. with $P\{X_1 = 1\} = P\{X_1 = -1\} = \frac{1}{2}$ and $S_n = \sum_1^n X_i$, $n \geq 1$. For positive integers i and j, let T be the smallest positive integer for which $S_n = i$ or $-j$. Then

$$i\,P\{S_T = i\} - j\,P\{S_T = j\} = 0, \qquad i^2\,P\{S_T = i\} + j^2\,P\{S_T = j\} = i \cdot j,$$

where $S_T = S_m$ on the set $\{T = m\}$, $m \geq 1$.

5. If in Bernoulli trials with parameter $\frac{1}{2}$, $T_k = \inf\{n \geq 1 : S_n = k\}$, prove that

$$P\{T_k = n\} = 2^{-n}\left\{\binom{n-1}{[(n-k)/2]} - \binom{n-1}{[(n-k-1)/2]}\right\} \quad \text{for } n \geq k.$$

Hint: Apply (2) to $P\{T_k \leq n\} = P\{\max_{1 \leq j \leq n} S_j \geq k\}$.

6. If S_n is as in Exercise 1, prove for $N = 0, 1, \ldots$ that

$$P\left\{\max_{j \leq n} S_j \leq N, S_{n+1} > N\right\} = 2P\{S_{n+1} \geq N + 1\} - P\{S_{n+1} = N + 1\}$$
$$- 2P\{S_n \geq N + 1\} + P\{S_n = N + 1\}.$$

7. If $S_n = \sum_1^n X_j$, $n \geq 1$ where $\{X_n, n \geq 1\}$ are Bernoulli trials with parameter p, verify that $P\{\max_{j \leq n}(S_j - jp) > x\} \leq c\,P\{S_n - np > x\}$ where $c > 0$.

8. Let $\{A_n, n \geq 1\}$ and $\{B_n, n \geq 1\}$ be events such that for all large m and $n > m$, A_n is independent of $B_n(\bigcup_{j=m}^{n-1} A_j B_j)^c$. Then $\sum_{n=1}^{\infty} P\{A_n\} = \infty$ implies $P\{A_n B_n, \text{ i.o.}\} \geq \underline{\lim}_{n\to\infty} P\{B_n\}$. *Hint*: Use the techniques of Theorem 3.2.1 and Lemma 3.3.3.

References

O. Barndorff-Nielsen, "On the rate of growth of the partial maxima of a sequence of independent, identically distributed random variables," *Math. Scand.* **9** (1961), 383–394.

L. Baum, M. Katz, and H. Stratton, "Strong Laws for Ruled Sums," *Ann. Math. Stat.* **42** (1971), 625–629.

F. Cantelli, "Su due applicazioni di un teorema di G. Boole," *Rend. Accad. Naz. Lincei* **26** (1917).

K. L. Chung, "The strong law of large numbers," *Proc. 2nd Berkeley Symp. Stat. and Prob.* (1951), 341–352.

K. L. Chung, *Elementary Probability Theory with Stochastic Processes*, Springer-Verlag, Berlin, New York, 1974.

J. L. Doob, *Stochastic Processes*, Wiley, New York, 1953.

W. Feller, *An Introduction to Probability Theory and Its Applications*, Vol. 1, 3rd ed., Wiley, New York, 1950.

A. Kolmogorov, *Foundations of Probability*, (Nathan Morrison, translator), Chelsea, New York, 1950.

P. Lévy, *Théorie de l'addition des variables aléatoires*, Gauthier-Villars, Paris, 1937; 2nd ed., 1954.

M. Loève, *Probability Theory*, 3rd ed. Van Nostrand, Princeton, 1963; 4th ed., Springer-Verlag, Berlin and New York, 1977-1978.

R. von Mises, "Uber aufteilungs und Besetzungs Wahrscheinlichkeiten," *Revue de la Faculté des Sciences de l'Université d'Istanbul*, N.S. **4** (1939), 145–163.

A. Renyi, *Foundations of Probability*, Holden-Day, San Francisco, 1970.

4 Integration in a Probability Space

4.1 Definition, Properties of the Integral, Monotone Convergence Theorem

There are two basic avenues to integration. In the modern approach the integral is introduced first for simple functions—as a weighted average of the values of the function—and then defined for any nonnegative measurable function f as a limit of the integrals of simple nonnegative functions increasing to f. Conceptually this is extremely simple, but a certain price is paid in terms of proofs. The alternative classical approach, while employing a less intuitive definition, achieves considerable simplicity in proofs of elementary properties.

If X is a (real) measurable function on a probability space $(\Omega, \mathscr{F}, \mathrm{P})$, the **integral** of X over Ω with respect to P is denoted by $\int_\Omega X\,d\mathrm{P}$, abbreviated by $\mathbf{E}\,X$ or $\mathrm{E}[X]$, referred to as the **expectation** or **mean** of X and defined (when possible) by:

i. If $X \geq 0$, then $\mathrm{E}\,X = \infty$ when $\mathrm{P}\{X = \infty\} > 0$, while if $\mathrm{P}\{X = \infty\} = 0$,

$$\mathrm{E}\,X = \lim_{n\to\infty} \sum_{i=1}^{\infty} \frac{i}{2^n}\mathrm{P}\left\{\frac{i}{2^n} < X \leq \frac{i+1}{2^n}\right\}. \tag{1}$$

ii. For general X, if either $\mathrm{E}\,X^+ < \infty$ or $\mathrm{E}\,X^- < \infty$, then

$$\mathrm{E}\,X = \mathrm{E}\,X^+ - \mathrm{E}\,X^-, \tag{2}$$

where $X^+ = \max(X, 0)$, $X^- = \max(-X, 0)$. In this case, the expectation of X is said to **exist** and $\mathrm{E}\,X \in [-\infty, \infty]$, denoted by $|\mathrm{E}\,X| \leq \infty$. If $|\mathrm{E}\,X| < \infty$, X is called **integrable**.

iii. If $\mathrm{E}\,X^+ = \mathrm{E}\,X^- = \infty$, $\mathrm{E}\,X$ is undefined.

It is not difficult to see that the limit in (1) always exists, since setting

$$p_{n,i} = \mathrm{P}\left\{\frac{i}{2^n} < X \le \frac{i+1}{2^n}\right\}, \qquad s_n = \sum_{i=1}^{\infty} \frac{i}{2^n} p_{n,i}, \tag{3}$$

additivity of P guarantees $p_{n,i} = p_{n+1,2i} + p_{n+1,2i+1}$, whence

$$s_n = \sum_{i=1}^{\infty} \frac{2i}{2^{n+1}} (p_{n+1,2i} + p_{n+1,2i+1}) \le s_{n+1},$$

and so $\lim_{n\to\infty} s_n$ exists. Furthermore, $\mathrm{E}\, X \ge s_n$, $n \ge 1$, for $X \ge 0$.

It is trivial to check that

$$\begin{gathered} \mathrm{E}[1] = 1, \qquad 0 \le \mathrm{E}\, X \le \infty \quad \text{when } X \ge 0, \\ X \overset{\text{a.c.}}{=\!=} 0 \quad \text{if } \mathrm{E}|X| = 0, \qquad \mathrm{P}\{|X| < \infty\} = 1 \quad \text{if } E|X| < \infty, \\ \mathrm{E}\, X = \mathrm{E}\, Y \text{ if } \mathrm{P}\{X = Y\} = 1, \qquad \mathrm{E}[-X] = -\mathrm{E}\, X \quad \text{if } |\mathrm{E}\, X| \le \infty, \end{gathered} \tag{4}$$

and easy to verify that if $X \ge 0$ and $\mathrm{P}\{X = \infty\} = 0$,

$$\begin{aligned} \mathrm{E}\, X &= \lim_{n\to\infty} \sum_{i=0}^{\infty} \frac{i+1}{2^n} \mathrm{P}\left\{\frac{i}{2^n} < X \le \frac{i+1}{2^n}\right\} \\ &= \lim_{n\to\infty} \sum_{i=0}^{\infty} \frac{i+1}{2^n} \mathrm{P}\left\{\frac{i}{2^n} \le X < \frac{i+1}{2^n}\right\} \\ &= \lim_{n\to\infty} \sum_{i=1}^{\infty} \frac{i}{2^n} \mathrm{P}\left\{\frac{i}{2^n} \le X < \frac{i+1}{2^n}\right\}. \end{aligned} \tag{5}$$

For example, the last line of (5) equals

$$\begin{aligned} &\lim_{n\to\infty}\left[\sum_{i=1}^{\infty} \frac{i}{2^n} \mathrm{P}\left\{\frac{i}{2^n} < X < \frac{i+1}{2^n}\right\} + \sum_{i=1}^{\infty} \frac{i-1}{2^n} \mathrm{P}\left\{X = \frac{i}{2^n}\right\}\right] \\ &= \lim_{n\to\infty}\left[\sum_{i=1}^{\infty} \frac{i}{2^n} \mathrm{P}\left\{\frac{i}{2^n} < X < \frac{i+1}{2^n}\right\} + \sum_{i=0}^{\infty} \frac{i}{2^n} \mathrm{P}\left\{X = \frac{i+1}{2^n}\right\}\right], \end{aligned}$$

which, in turn, coincides with the definition of $\mathrm{E}\, X$.

Every integrable, symmetric r.v. has mean zero as follows from (2) and Exercise 3.3.3(ii), whereas the expectation of a nonintegrable symmetric r.v. is undefined.

A measurable function X is called **elementary** if for some sequence $\{A_n, n \ge 1\}$ of disjoint sets of $\mathscr{F}$

$$X = \sum_{n=1}^{\infty} x_n I_{A_n}, \tag{6}$$

where $-\infty \le x_n \le \infty$. Clearly, simple functions (Exercise 1.4.3) are elementary. An elementary function X always engenders a partition $\{B_n\}$ of Ω in $\mathscr{F}$ (i.e., $B_n \in \mathscr{F}$, B_n disjoint, and $\bigcup_1^\infty B_n = \Omega$). It suffices to note that

$$X = \sum_{n=1}^{\infty} y_n I_{B_n}, \tag{6'}$$

where $B_1 = (\bigcup_1^\infty A_n)^c$, $y_1 = 0$, $y_{n+1} = x_n$, $B_{n+1} = A_n$, $n \geq 1$.

The basic properties of the integral—linearity, order preservation, and monotone convergence—are embodied in

Theorem 1. *Let X, Y, X_n, Y_n, $n \geq 1$, be measurable functions on a probability space $(\Omega, \mathscr{F}, \mathrm{P})$.*

i. *If $X = \sum_{n=1}^\infty x_n I_{A_n}$ is a nonnegative elementary function, then*

$$\mathrm{E}\,X = \sum_1^\infty x_n \,\mathrm{P}\{A_n\}. \tag{7}$$

ii. *If $X \geq 0$, a.c., there exist elementary functions X_n with $0 \leq X_n \uparrow X$ a.c. and $\mathrm{E}\,X_n \uparrow \mathrm{E}\,X$; moreover, if $\mathrm{P}\{X = \infty\} = 0$, then X_n may be chosen so that $\mathrm{P}\{X - X_n \leq 2^{-n}, n \geq 1\} = 1$.*

iii. *If $X \geq Y \geq 0$, a.c., then $\mathrm{E}\,X \geq \mathrm{E}\,Y \geq 0$ and for any a in $(0, \infty)$,*

$$\mathrm{P}\{X \geq a\} \leq \frac{1}{a}\,\mathrm{E}\,X \qquad (\textit{Markov inequality}). \tag{8}$$

iv. *If $0 \leq X_n \uparrow X$ a.c., then $\mathrm{E}\,X_n \uparrow \mathrm{E}\,X$ (monotone convergence theorem).*

v. *If $\mathrm{E}\,X, \mathrm{E}\,Y, \mathrm{E}\,X + \mathrm{E}\,Y$ are all defined, then $\mathrm{E}[X + Y]$ exists and*

$$\mathrm{E}[X + Y] = \mathrm{E}\,X + \mathrm{E}\,Y \qquad (\textit{additivity}). \tag{9}$$

vi. *If $X \geq Y$, a.c., and $\mathrm{E}\,X, \mathrm{E}\,Y$ exist, then*

$$\mathrm{E}\,X \geq \mathrm{E}\,Y \qquad (\textit{order preservation}). \tag{10}$$

vii. *If $\mathrm{E}\,X$ exists and Y is integrable, then for all finite real a, b, $\mathrm{E}[aX + bY]$ exists and*

$$\mathrm{E}[aX + bY] = a\,\mathrm{E}\,X + b\,\mathrm{E}\,Y \qquad (\textit{linearity}). \tag{11}$$

PROOF. (i) If $\mathrm{P}\{X = \infty\} > 0$, then $x_n = \infty$ for some n for which $\mathrm{P}\{A_n\} > 0$ so that (7) obtains. If, rather $\mathrm{P}\{X = \infty\} = 0$, then $x_m = \infty$ for some m requires that $\mathrm{P}\{A_m\} = 0$, whence $x_m \,\mathrm{P}\{A_m\} = 0$. By setting $x_n' = 0$ on any $A_n \subset \{X = \infty\}$ and $x_n' = x_n$ otherwise, and defining $X' \equiv X \cdot I_{[X < \infty]} = \sum_1^\infty x_n' I_{A_n}$, it is evident that $\mathrm{P}\{X' = X\} = 1$, $\mathrm{E}\,X' = \mathrm{E}\,X$, and $X'(\omega) < \infty$ for all ω. Let

$$I_{ni} = \left(\frac{i}{2^n}, \frac{i+1}{2^n}\right], \qquad p_{ni} = \mathrm{P}\{X' \in I_{ni}\} = \sum_{j: x_j' \in I_{ni}} \mathrm{P}\{A_j\}$$

and note that for all $n \geq 1$

$$\sum_{i=1}^\infty \frac{i}{2^n} p_{ni} \leq \sum_{i=1}^\infty \sum_{j: x_j' \in I_{ni}} x_j' \,\mathrm{P}\{A_j\} \leq \sum_{j=1}^\infty x_j' \,\mathrm{P}\{A_j\}, \tag{12}$$

$$\sum_{i=0}^\infty \frac{i+1}{2^n} p_{ni} \geq \sum_{i=0}^\infty \sum_{j: x_j' \in I_{ni}} x_j' \,\mathrm{P}\{A_j\} = \sum_{j=1}^\infty x_j' \mathrm{P}\{A_j\}, \tag{13}$$

so that by prior observations

$$\mathrm{E}\,X = \mathrm{E}\,X' = \sum_1^\infty x_j' \,\mathrm{P}\{A_j\} = \sum_1^\infty x_j \mathrm{P}\{A_j\}.$$

To prove (ii), set

$$\varphi_{ni} = \frac{i}{2^n} I_{[i/2^n < X \le (i+1)/2^n]}$$

and define the simple functions $X_n^{(1)}$ and elementary functions $X_n^{(2)}$ by

$$X_n^{(1)} = nI_{[X>n]} + \sum_{i=1}^{n\cdot 2^n - 1} \varphi_{ni}, \qquad X_n^{(2)} = \sum_{i=1}^{\infty} \varphi_{ni}. \tag{14}$$

Then $X_n^{(1)} \uparrow X$, $X_n^{(2)} \uparrow X'$, and $X_n^{(j)} \le X_{n+1}^{(j)}$, $j = 1, 2$; furthermore, via (i) $\mathrm{E}\,X_n^{(2)} = s_n \le s_{n+1} = \mathrm{E}\,X_{n+1}^{(2)}$. Consequently, if $\mathrm{P}\{X = \infty\} = 0$, $\mathrm{E}\,X_n^{(2)} \uparrow \mathrm{E}\,X$ according to the definition (1). Moreover, under the same proviso $\mathrm{P}\{X = \infty\} = 0$, subadditivity guarantees

$$\mathrm{P}\left\{\bigcup_{n=1}^{\infty}[X - X_n^{(2)} > 2^{-n}]\right\} \le \sum_{n=1}^{\infty} \mathrm{P}\{X - X_n^{(2)} > 2^{-n}\} = 0,$$

which is tantamount to the last assertion in (ii). On the other hand, if $\mathrm{P}\{X = \infty\} > 0$, then $\mathrm{E}\,X_n^{(1)} \ge n\,\mathrm{P}\{X = \infty\} \to \infty = \mathrm{E}\,X$ and it will follow from the initial portion of the proof of (iii) that $\mathrm{E}\,X_n^{(1)}$ is monotone.

Apropos of (iii), suppose first that $X = \sum_{i=1}^\infty x_i I_{A_i}$, $Y = \sum_{j=1}^\infty y_j I_{B_j}$, where $\{A_i\}$ and $\{B_j\}$ are each partitions of Ω in $\mathscr{F}$. Then

$$X = \sum_{i,j} x_i I_{A_i B_j}, \qquad Y = \sum_{i,j} y_j I_{A_i B_j},$$

and $\mathrm{P}\{X \ge Y \ge 0\} = 1$ entails $x_i \ge y_j \ge 0$ if $\mathrm{P}\{A_i B_j\} > 0$. By (i)

$$\mathrm{E}\,X = \sum_{i,j}^{\infty} x_i \,\mathrm{P}\{A_i B_j\} \ge \sum_{i,j}^{\infty} y_j \,\mathrm{P}\{A_i B_j\} = \mathrm{E}\,Y.$$

For the general case it may be supposed that $\mathrm{P}\{X = \infty\} = 0$, whence also $\mathrm{P}\{Y = \infty\} = 0$. If $X_n = X_n^{(2)}$ as in (14) and Y_n is defined analogously via Y, then these elementary functions satisfy $\mathrm{P}\{X_n \ge Y_n \ge 0\} = 1$, so that by the part already proved $\mathrm{E}\,X_n \ge \mathrm{E}\,Y_n$. Hence, by (ii), $\mathrm{E}\,X \ge \mathrm{E}\,Y$.

Consequently, the Markov inequality follows directly from $X \ge aI_{[X \ge a]}$ and (i).

To prove the monotone convergence theorem, note first in case $\mathrm{P}\{X = \infty\} > 0$ that by the Markov inequality and Corollary 1.5.2

$$\mathrm{E}\,X_n \ge a\,\mathrm{P}\{X_n > a\} \xrightarrow{n\to\infty} a\,\mathrm{P}\{X > a\} \ge a\,\mathrm{P}\{X = \infty\} \xrightarrow{a\to\infty} \infty,$$

which is fitting since $\mathrm{E}\,X = \infty$.

If rather $P\{X = \infty\} = 0$, note that by this same corollary as $m \to \infty$

$$P\left\{\frac{i}{2^n} < X_m \le \frac{i+1}{2^n}\right\} = P\left\{X_m > \frac{i}{2^n}\right\} - P\left\{X_m > \frac{i+1}{2^n}\right\}$$
$$\to P\left\{X > \frac{i}{2^n}\right\} - P\left\{X > \frac{i+1}{2^n}\right\}$$
$$= P\left\{\frac{i}{2^n} < X \le \frac{i+1}{2^n}\right\}. \tag{15}$$

For any $a < E\,X$, the definition of $E\,X$ ensures the existence of positive integers n, k such that

$$\sum_{i=1}^{k} \frac{i}{2^n} P\left\{\frac{i}{2^n} < X \le \frac{i+1}{2^n}\right\} > a,$$

whence, via (15), for all large m

$$E\,X_m \ge \sum_{i=1}^{k} \frac{i}{2^n} P\left\{\frac{i}{2^n} < X_m \le \frac{i+1}{2^n}\right\} > a. \tag{16}$$

By (iii), $E\,X_m \le E\,X_{m+1} \le E\,X$, which in conjunction with (16) yields $E\,X_m \uparrow E\,X$.

Apropos of (v), if $X = \sum_1^\infty x_i I_{A_i} \ge 0$, $Y = \sum_1^\infty y_j I_{B_j} \ge 0$, where $\{A_i\}$ and $\{B_j\}$ are partitions of Ω in $\mathscr{F}$, then $X + Y = \sum_{i,j} (x_i + y_j) I_{A_i B_j}$, yielding via (7) and σ-additivity

$$E\,[X + Y] = \sum_{i,j} (x_i + y_j) P\{A_i B_j\}$$
$$= \sum_i x_i\, P\{A_i\} + \sum_j y_j\, P\{B_j\} = E\,X + E\,Y. \tag{17}$$

In conjunction with (ii), (17) yields additivity for $X \ge 0$, $Y \ge 0$.

In general, if $E\,X = \infty$, then $E\,X^+ = \infty$, $E\,X^- < \infty$, $E\,Y > -\infty$, $E\,Y^- < \infty$, so that $P\{X > -\infty, Y > -\infty\} = 1$. Now (Exercise 1.4.4), $(X + Y)^- \le X^- + Y^-$ and $X^+ \le (X + Y)^+ + Y^-$, whence by (iii) and the part already proved

$$E(X + Y)^- \le E(X^- + Y^-) = E\,X^- + E\,Y^- < \infty,$$

$$E(X + Y)^+ + E\,Y^- = E\,[(X + Y)^+ + Y^-] \ge E\,X^+ = \infty,$$

implying $E[X + Y] = \infty$ and hence (9). Similarly (9) holds if $E\,X = -\infty$ or $E\,Y = -\infty$ or $E\,Y = \infty$.

Lastly, if $|E\,X| + |E\,Y| < \infty$, then $E\,X^+ < \infty$, $E\,X^- < \infty$, implying $E|X| = E\,X^+ + E\,X^- < \infty$ by the portion already proved. Similarly, $E|Y| < \infty$, whence $P\{|X| < \infty, |Y| < \infty\} = 1$ and

$$E|X + Y| \le E|X| + E|Y| < \infty.$$

Thus, from

$$X^+ + Y^+ = (X + Y)^+ + [X^- + Y^- - (X + Y)^-],$$

by the part already proved

$$\mathrm{E}\,X^+ + \mathrm{E}\,Y^+ = \mathrm{E}[X^+ + Y^+] = \mathrm{E}(X + Y)^+ + \mathrm{E}[X^- + Y^- - (X + Y)^-],$$

yielding again via the additivity or subtractivity already proved

$$\mathrm{E}\,X + \mathrm{E}\,Y = \mathrm{E}(X + Y).$$

To dispatch (vi) in the nontrivial case $\mathrm{E}\,X < \infty$, $\mathrm{E}\,Y > -\infty$, note that via (v) and (iii)

$$\mathrm{E}\,X = \mathrm{E}\,Y + \mathrm{E}[X - Y] \geq \mathrm{E}\,Y.$$

Finally, apropos of (vii), let $0 < a < \infty$ and $X \geq 0$. If $\mathrm{P}\{X = \infty\} > 0$, then $\mathrm{E}[aX] = \infty = a\,\mathrm{E}\,X$. If $\mathrm{P}\{X = \infty\} = 0$, then, recalling (ii), $X_n^{(2)} = \sum_{i=1}^{\infty} \varphi_{ni} \uparrow X$, a.c., whence $0 \leq aX_n^{(2)} \uparrow aX$, a.c., and by (i) and (iv)

$$\mathrm{E}[aX] = \lim_n \mathrm{E}[aX_n^{(2)}] = \lim_n \mathrm{E}\left[\sum_{i=1}^{\infty} a\varphi_{ni}\right] = a\,\mathrm{E}\,X.$$

In general, for $0 < a < \infty$ and $|\mathrm{E}\,X| \leq \infty$

$$\mathrm{E}[aX] = \mathrm{E}(aX)^+ - \mathrm{E}(aX^-) = a\,\mathrm{E}\,X^+ - a\,\mathrm{E}\,X^- = a\,\mathrm{E}\,X.$$

Since $\mathrm{E}[0 \cdot X] = 0 \cdot \mathrm{E}\,X = 0$ and $\mathrm{E}[-X] = -\mathrm{E}\,X$,

$$\mathrm{E}\,aX = a\,\mathrm{E}\,X \tag{18}$$

for any finite a whenever $\mathrm{E}\,X$ exists. Finally, (11) follows via (9) and (18). □

Note that if X, Z are measurable functions with $X \leq Z$ (respectively, $X \geq Z$) where $\mathrm{E}Z^+ < \infty$ (respectively, $\mathrm{E}Z^- < \infty$) then $\mathrm{E}X$ exists. A fortiori if X is bounded above or below by an integrable random variable, then $\mathrm{E}X$ exists.

Corollary 1. *A measurable function X on $(\Omega, \mathscr{F}, \mathrm{P})$ is integrable* iff $|X|$ *is integrable, and in such a case*

$$\mathrm{P}\{|X| \geq a\} \leq \frac{1}{a}\,\mathrm{E}\,|X|, \qquad a > 0 \qquad (\textit{Markov inequality}) \tag{19}$$

and

$$|\mathrm{E}\,X| \leq \mathrm{E}\,|X|. \tag{20}$$

If X is a discrete r.v. with values $\{x_j, 1 \leq j \leq n\}$ and p.d.f. $\{p_j, 1 \leq j \leq n\}$, where $1 \leq n \leq \infty$, Theorem 1(i) ensures that

$$\mathrm{E}\,X = \sum_{j=1}^{n} x_j p_j \tag{21}$$

when $x_j \geq 0$. Consequently, (21) holds for any discrete r.v. provided that $\sum_{j=1}^{\infty} x_j^+ p_j$ or $\sum_{j=1}^{\infty} x_j^- p_j$ is finite when $n = \infty$. On the other hand, if X is an absolutely continuous r.v. with density g and $\mathrm{E}\,X$ exists,

$$\mathrm{E}\,X = \int_{-\infty}^{\infty} tg(t)dt \tag{22}$$

according to Corollary 6.5.4.

In the case of an infinite series of nonnegative measurable functions, the operations of expectation and summation may be interchanged as follows directly from Lebesgue's monotone convergence theorem (Theorem 1(iv)):

Corollary 2. *If $\{X_n, n \geq 1\}$ are nonnegative measurable functions on $(\Omega, \mathscr{F}, \mathrm{P})$, then*

$$\mathrm{E}\sum_{n=1}^{\infty} X_n = \sum_{n=1}^{\infty} \mathrm{E}\,X_n. \tag{23}$$

Without the nonnegativity proviso, (23) is, in general, false even when $\sum_{i=1}^{\infty} X_i$ converges, a.c.

EXAMPLE 1. Let $\{Y_n, n \geq 1\}$ be i.i.d. random variables with $\mathrm{P}\{Y_n = \pm 1\} = \frac{1}{2}$ and define $T = \inf\{n \geq 1: \sum_{i=1}^{n} Y_i = 1\}$ where $\inf\{\emptyset\} = \infty$. Then $T < \infty$, a.c. (Exercise 3.4.1), and, setting $X_n = Y_n I_{[T \geq n]}$,

$$\sum_{n=1}^{\infty} X_n = \sum_{n=1}^{\infty} Y_n I_{[T \geq n]} = \sum_{n=1}^{T} Y_n = 1$$

by definition of T, and so $\mathrm{E}\sum_{n=1}^{\infty} X_n = 1$. However, since the event $\{T \geq n\} \in \sigma(Y_1, \ldots, Y_{n-1})$, the r.v.s Y_n and $I_{[T \geq n]}$ are independent by Corollary 3.2.2, whence it follows from Theorem 4.3.3 that

$$\mathrm{E}\,X_n = \mathrm{E}\,Y_n\,\mathrm{E}\,I_{[T \geq n]} = 0, \qquad n \geq 1,$$

so that (23) fails.

Definition. Given any nonnegative constants $\{b_n, n \geq 0\}$, a continuous function $b(\cdot)$ on $[0, \infty)$ is called an **extension** of $\{b_n\}$ to $[0, \infty)$ if $b(n) = b_n$, $n \geq 0$. Moreover, when $\{b_n\}$ is strictly monotone, $b(\cdot)$ is a **strictly monotone extension** of $\{b_n\}$ if it is both strictly monotone and an extension of $\{b_n\}$.

Corollary 3. *Let $\{b_n, n \geq 0\}$ be a strictly increasing sequence with $0 \leq b_n \uparrow \infty$ and let $b(\cdot)$ be a strictly monotone extension of $\{b_n\}$ to $[0, \infty)$. Then for any* r.v. $X \geq 0$, a.c.

$$\sum_{n=1}^{\infty} \mathrm{P}\{X \geq b_n\} \leq \mathrm{E}\,b^{-1}(X) \leq \sum_{n=0}^{\infty} \mathrm{P}\{X > b_n\}. \tag{24}$$

In particular, for any $r > 0$ and any r.v. X,

$$\sum_{n=1}^{\infty} \mathrm{P}\{|X| \geq n^{1/r}\} \leq \mathrm{E}|X|^r \leq \sum_{n=0}^{\infty} \mathrm{P}\{|X| > n^{1/r}\}. \tag{25}$$

$$E|X|^r = r\int_0^\infty t^{r-1}P\{|X| > t\}\,dt. \tag{26}$$

PROOF. Set $\varphi(x) = b^{-1}(x)$,

$$Y = \sum_1^\infty jI_{[j\le\varphi(X)<j+1]}, \qquad Z = \sum_0^\infty (j+1)I_{[j<\varphi(X)\le j+1]},$$

whence $Y \le \varphi(X) \le Z$, a.c., since $P\{X < \infty\} = 1 = P\{\varphi(X) < \infty\}$. Thus,

$$E\,Y \le E\,\varphi(X) \le E\,Z.$$

But

$$\sum_1^\infty P\{X \ge b_n\} = \sum_1^\infty P\{\varphi(X) \ge n\} = \sum_{n=1}^\infty \sum_{j=n}^\infty P\{j \le \varphi(X) < j+1\}$$

$$= \sum_{j=1}^\infty jP\{j \le \varphi(X) < j+1\} = E\,Y$$

and

$$\sum_0^\infty P\{X > b_n\} = \sum_{n=0}^\infty \sum_{j=n}^\infty P\{j < \varphi(X) \le j+1\}$$

$$= \sum_{j=0}^\infty (j+1)P\{j < \varphi(X) \le j+1\} = E\,Z,$$

completing the proof of (24).

Clearly, (25) follows from (24) with $b(x) = x^r$, $r > 0$. Finally, it suffices to verify (26) when $r = 1$ since the statement for $r > 0$ then follows by a change of variable. Now, for any $a > 0$, (25) ensures

$$\int_1^\infty P\{|aX| \ge u\}\,du \le \sum_{n=1}^\infty P\{|aX| \ge n\} \le a\cdot E|X| \le 1 + \sum_{n=1}^\infty P\{|aX| > n\}$$

$$\le 1 + \int_0^\infty P\{|aX| > u\}\,du$$

whence setting $t = u/a$

$$\int_{a^{-1}}^\infty P\{|X| \ge t\}\,dt \le E|X| \le a^{-1} + \int_0^\infty P\{|X| > t\}\,dt$$

and the conclusion follows as $a \to \infty$. □

EXERCISES 4.1

1. If X is a geometric r.v. with parameter p, that is, $P\{X = k\} = pq^k$, $k = 0, 1, \ldots$, $0 < p < 1$ verify that $EX = q/p$, $EX(X-1) = 2(q/p)^2$, $EX(X-1)(X-2) = 6(q/p)^3$, $EX^3 = 6(q/p)^3 + 6(q/p)^2 + q/p$ and $EX^2 - (EX)^2 = (q/p)^2 + q/p$.

2. If X is an integrable r.v., for every $\varepsilon > 0$ there is a simple function X_ε with $\mathrm{E}|X - X_\varepsilon| < \varepsilon$.

3. Utilize Exercise 1.2.6 to give an alternative proof of Corollary 2.1.3.

4. (i) If $\mathrm{P}\{X = 1\} = p$, $\mathrm{P}\{X = 0\} = q = 1 - p$, prove that $\mathrm{E}(X - \mathrm{E}\,X)^k = pq[q^{k-1} - (-p)^{k-1}]$, $k = 1, 2, \ldots$. (ii) if X is a r.v. with $\sum_{j=0}^{\infty} \mathrm{P}\{X = j\} = 1$, then $\mathrm{E}\,X = \sum_{j=1}^{\infty} \mathrm{P}\{X \geq j\}$.

5. If X is an integrable r.v., $n\,\mathrm{P}\{|X| \geq n\} = o(1)$, but the converse does not hold.

6. Construct a sequence of discrete r.v.s X_n such that $X_n \xrightarrow{\mathrm{P}} 0$ but $\mathrm{E}\,X_n \not\to 0$.

7. If $\{X, X_n, n \geq 1\}$ is a sequence of r.v.s on some probability space such that

$$\sum_{n=1}^{\infty} \mathrm{E}|X_n - X|^r < \infty,$$

some $r > 0$, then $X_n \xrightarrow{\text{a.c.}} X$.

8. If (i) $\{X_n, n \geq 1\}$ is a sequence of nonnegative, integrable r.v.s with $S_n = X_1 + \cdots + X_n$ and (ii) $\sum_{n=1}^{\infty} \mathrm{E}\,X_n < \infty$, then S_n converges a.c. Hence, if (i) obtains, $\mathrm{E}\,S_n > 0$, $n \geq 1$, and $\sum_{n=1}^{\infty} \mathrm{E}\,X_n/\mathrm{E}\,S_n < \infty$, then $S_n/\mathrm{E}\,S_n$ converges a.c.

9. If X and Y are measurable functions on $(\Omega, \mathscr{F}, \mathrm{P})$ with $\mathrm{E}\,XI_A = \mathrm{E}\,YI_A$ for all $A \in \mathscr{F}$, then $X = Y$, a.c. *Hint*: Consider $A_{r,s} = \{\omega\colon \infty \geq X(\omega) \geq r > s \geq Y(\omega) \geq -\infty\}$.

10. If $\{X_n, n \geq 1\}$ are i.i.d. random variables then (i) $1/n \max_{1 \leq j \leq n} |X_j| \xrightarrow{\text{a.c.}} 0$ iff $\mathrm{E}|X_1| < \infty$. (ii) $1/n \max_{1 \leq j \leq n} |X_j| \xrightarrow{\mathrm{P}} 0$ iff $n \cdot \mathrm{P}\{|X_1| > n\} \to 0$.

4.2 Indefinite Integrals, Uniform Integrability, Mean Convergence

For any measurable function X on a probability space $(\Omega, \mathscr{F}, \mathrm{P})$ whose expectation $\mathrm{E}\,X$ exists, define

$$\nu\{A\} = \mathrm{E}\,X \cdot I_A = \int_A X\,d\mathrm{P}, \qquad A \in \mathscr{F}. \tag{1}$$

The set function $\nu\{A\}$ defined on $\mathscr{F}$ by (1) is called the **indefinite integral of** X.

Moreover, any nonnegative measurable function X on $(\Omega, \mathscr{F}, \mathrm{P})$ generates a new measure ν on $\mathscr{F}$ via (1), and if $\mathrm{E}\,X = 1$, this will be a probability measure.

Lemma 1. *The indefinite integral ν of any measurable function X on $(\Omega, \mathscr{F}, \mathrm{P})$ whose expectation exists is a σ-additive set function on $\mathscr{F}$, that is, for disjoint $A_j \in \mathscr{F}, j \geq 1$,*

$$\nu\left\{\bigcup_1^{\infty} A_j\right\} = \sum_1^{\infty} \nu\{A_j\}.$$

PROOF. By Corollary 4.1.2,

$$\sum_{j=1}^{\infty} \int_{A_j} X^+ \, dP = \sum_{j=1}^{\infty} E\, X^+ I_{A_j} = E \sum_{j=1}^{\infty} X^+ I_{A_j}$$

$$= E\, X^+ I_{\cup_1^\infty A_j} = \int_{\cup_1^\infty A_j} X^+ \, dP$$

and, similarly,

$$\sum_{j=1}^{\infty} \int_{A_j} X^- \, dP = \int_{\cup_1^\infty A_j} X^- \, dP.$$

Thus, since E X exists,

$$\sum_{j=1}^{\infty} \int_{A_j} X \, dP = \sum_{1}^{\infty} \left(\int_{A_j} X^+ dP - \int_{A_j} X^- dP \right)$$

$$= \int_{\cup A_j} X^+ \, dP - \int_{\cup A_j} X^- \, dP = \int_{\cup_1^\infty A_j} X \, dP. \qquad \square$$

Corollary 1. *If X is a nonnegative measurable function on $(\Omega, \mathscr{F}, P)$, its indefinite integral is a measure on $\mathscr{F}$.*

The integrability of a measurable function X can be characterized via its indefinite integral according to

Lemma 2. *A measurable function X is integrable* iff *for every $\varepsilon > 0$ there corresponds a $\delta > 0$ such that $A \in \mathscr{F}$, $P\{A\} < \delta$ implies*

$$\int_A |X| \, dP < \varepsilon, \qquad E|X| \leq \frac{1}{\delta}. \tag{2}$$

PROOF. If X is integrable and $X_k = |X| I_{[|X| \leq k]}$, then $X_k \uparrow |X|$, a.c., whence by Theorem 4.1.1(iv), E $X_k \uparrow E|X|$, which entails $E|X| I_{[|X| > k]} \to 0$ as $k \to \infty$. If K is a positive integer for which $E|X| I_{[|X| > K]} < \varepsilon/2$, set $\delta = \min(\varepsilon/2K, 1/E|X|)$. Then for $A \in \mathscr{F}$ with $P\{A\} < \delta$,

$$\int_A |X| \, dP = \int_A |X| I_{[|X| > K]} \, dP + \int_A |X| I_{[|X| \leq K]} \, dP \leq \frac{\varepsilon}{2} + \frac{\varepsilon}{2} = \varepsilon,$$

and so (2) holds.

Conversely, (2) implies $|X|$ and therefore X integrable. $\square$

This suggests the following

Definition. A sequence of r.v.s $\{X_n, n \geq 1\}$ is called **uniformly integrable** (abbreviated u.i.) if for every $\varepsilon > 0$ there corresponds a $\delta > 0$ such that

$$\sup_{n \geq 1} \int_A |X_n| \, dP < \varepsilon \tag{3}$$

whenever $P\{A\} < \delta$ and, in addition,

$$\sup_{n \geq 1} E|X_n| < \infty. \tag{4}$$

Furthermore, $\{X_n\}$ is said to be **u.i. from above** or **below** according as $\{X_n^+\}$ or $\{X_n^-\}$ is u.i.

Some immediate consequences of the definition are:

i. $\{X_n\}$ is u.i. iff $\{|X_n|\}$ is u.i.
ii. If $\{X_n\}$ and $\{Y_n\}$ are each u.i., so is $\{X_n + Y_n\}$.
iii. If $\{X_n, n \geq 1\}$ is u.i., so is any subsequence of $\{X_n\}$.
iv. $\{X_n\}$ is u.i. iff it is u.i. from above and from below.
v. If $|X_n| \leq Y$ with $E\,Y < \infty$, then $\{X_n\}$ is u.i.

An alternative characterization of uniform integrability appears in

Theorem 1 (u.i. criterion). *A sequence of* r.v.s $\{X_n, n \geq 1\}$ *is* u.i. iff

$$\lim_{a \to \infty} \sup_{n \geq 1} \int_{[|X_n| > a]} |X_n| \, dP = 0. \tag{5}$$

PROOF. If $\{X_n\}$ is u.i., then $\sup E|X_n| \leq C < \infty$, whence for any $\varepsilon > 0$, choosing δ as in (3), the Markov inequality ensures that

$$P\{|X_n| > a\} \leq a^{-1} E|X_n| \leq \frac{C}{a} < \delta, \qquad n \geq 1,$$

provided $a > C/\delta$. Consequently, from (3) for $a > C/\delta$,

$$\sup_{n \geq 1} \int_{[|X_n| > a]} |X_n| \, dP < \varepsilon, \tag{6}$$

which is tantamount to (5). Conversely, for any $\varepsilon > 0$, choosing a sufficiently large to activate (6),

$$E|X_n| \leq a + \int_{[|X_n| > a]} |X_n| \, dP \leq a + \varepsilon, \qquad n \geq 1,$$

yielding (4). Moreover, selecting $\delta = \varepsilon/a$, for any A with $P\{A\} < \delta$ and all $n \geq 1$

$$\begin{aligned} \int_A |X_n| \, dP &= \int_{A[|X_n| \leq a]} |X_n| \, dP + \int_{A[|X_n| > a]} |X_n| \, dP \\ &\leq a\,P\{A\} + \int_{[|X_n| > a]} |X_n| \, dP \leq \varepsilon + \varepsilon = 2\varepsilon, \end{aligned}$$

so that (3) holds. □

The importance of uniform integrability will become apparent from Theorem 3.

Associated with any probability space $(\Omega, \mathscr{F}, P)$ are the $\mathscr{L}_p$ **spaces** of all measurable functions X (necessarily r.v.s) for which $E|X|^p < \infty$, denoted by $\mathscr{L}_p$ or $\mathscr{L}_p(\Omega, \mathscr{F}, P)$, $p > 0$. Random variables in $\mathscr{L}_p$ will be dubbed $\mathscr{L}_p$ **r.v.s.**

The inequalities of Section 1 show that a r.v. $X \in \mathscr{L}_p$ iff the series $\sum_1^\infty P\{|X| > n^{1/p}\}$ converges.

Definition. A sequence $\{X_n, n \geq 1\}$ of $\mathscr{L}_p$ r.v.s is said to **converge in mean of order** p (to a r.v. X) if $E|X_n - X|^p \to 0$ as $n \to \infty$. This will be denoted by $X_n \xrightarrow{\mathscr{L}_p} X$. Convergence in mean of orders one and two are called convergence in mean and convergence in mean square (or quadratic mean) respectively.

Convergence in mean of order p for any $p > 0$ implies convergence in probability as follows from the Markov inequality via

$$P\{|X_n - X| \geq \varepsilon\} = P\{|X_n - X|^p \geq \varepsilon^p\} \leq \varepsilon^{-p} E|X_n - X|^p.$$

Moreover, convergence of X_n to X in mean of order p entails $X \in \mathscr{L}_p$ in view of the inequality

$$E|X|^p \leq 2^p[E|X_n|^p + E|X_n - X|^p],$$

the latter being an immediate consequence of

Lemma 3. *If X and Y are nonnegative* r.v.s *and $p > 0$, then*

$$E(X + Y)^p \leq 2^p[E\, X^p + E\, Y^p]. \tag{7}$$

PROOF. For $a > 0$, $b > 0$, $(a + b)^p \leq [2 \max(a, b)]^p \leq 2^p[a^p + b^p]$ and (7) follows.

In particular, $X \in \mathscr{L}_p$, $Y \in \mathscr{L}_p$ imply $X + Y \in \mathscr{L}_p$. □

Among the most important and frequently used results of integration are Lebesgue's dominated convergence theorem (Corollary 3) and monotone convergence theorem, and Fatou's lemma. The second will be obtained by first verifying the latter.

Theorem 2 (i) (Monotone convergence theorem). *If the* r.v.s $\{X_n, n \geq 1\}$ *are* u.i. *from below and* $X_n \uparrow X$, a.c., *then* $E\, X^- < \infty$ *and* $E\, X_n \uparrow E\, X$.

(ii) (Fatou's lemma). *If the* r.v.s $\{X_n, n \geq 1\}$ *are* u.i. *from below and* $E \varliminf X_n$ *exists, then*

$$E \varliminf_{n\to\infty} X_n \leq \varliminf_{n\to\infty} E\, X_n. \tag{8}$$

The most typical usage of Fatou's lemma and (i) would be as in

Corollary 2. *If the* r.v.s $\{X_n, n \geq 0\}$ *satisfy* $X_n \geq X_0$, a.c., $n \geq 1$, *where* X_0 *is integrable (a fortiori, if* $X_n \geq 0$, a.c., $n \geq 1$) *then* $E \varliminf_{n\to\infty} X_n$ *exists,* (8) *holds, and, moreover, if* $X_n \uparrow X$, a.c., *then* $E\, X_n \uparrow E\, X$.

PROOF OF THEOREM 2. (i) $X_n \uparrow X$ a.c. entails $X_n^- \downarrow X^-$, a.c., whence $\sup \mathrm{E}\, X_n^- < \infty$ implies $\mathrm{E}\, X^- < \infty$. Then, $0 \le X_n + X_1^- \uparrow X + X_1^-$ whence, by Theorem 4.1 (iv), $\mathrm{E}(X_n + X_1^-) \uparrow \mathrm{E}(X + X_1^-)$ implying that $\mathrm{E}\, X_n \uparrow \mathrm{E}\, X$.

To prove (ii), define, for $K > 0$,

$$X_n' = X_n \wedge (-K), \qquad X_n'' = X_n - X_n' \tag{9}$$

and note that, for any $\varepsilon > 0$ and sufficiently large K,

$$\inf_{n\ge 1} \mathrm{E}\, X_n'' = \inf_{n\ge 1} \mathrm{E}(X_n + K) I_{[X_n \le -K]} \ge \inf_{n\ge 1} \mathrm{E}(-X_n^- I_{[X_n^- \ge K]}) \ge -\varepsilon. \tag{10}$$

Hence, setting $Y_n = \inf_{i \ge n} X_i'$, it follows from $-K \le Y_n \uparrow \underline{\lim}_{m\to\infty} X_m'$ and (10) that

$$\mathrm{E} \varliminf_{m\to\infty} X_m' = \varliminf_{n\to\infty} \mathrm{E}\, Y_n \le \varliminf_{n\to\infty} \mathrm{E}\, X_n'. \tag{11}$$

Now $X_n \le X_n'$ and $\mathrm{E}\, \underline{\lim}_{n\to\infty} X_n$ exists by hypothesis, whence $\mathrm{E}\, \underline{\lim}_{n\to\infty} X_n \le \mathrm{E}\, \underline{\lim}_{n\to\infty} X_n'$. Thus, by (9), (10), and (11)

$$\begin{aligned} \varliminf \mathrm{E}\, X_n = \varliminf \mathrm{E}(X_n' + X_n'') &\ge \varliminf \mathrm{E}\, X_n' - \varepsilon \\ &\ge \mathrm{E} \varliminf X_n' - \varepsilon \ge \mathrm{E} \varliminf X_n - \varepsilon, \end{aligned}$$

and (8) obtains as $\varepsilon \to 0$. □

The next lemma, especially when $A_n = B_n$, may be regarded as an extension of the Borel–Cantelli theorem which relaxes the independence requirement therein.

Lemma 4. *If $\{A_n, n \ge 1\}$ and $\{B_n, n \ge 1\}$ are sequences of events satisfying*

$$\mathrm{P}\left\{A_i \cap \bigcup_{j=1}^{\infty} A_{i+jk}\right\} \le \mathrm{P}\{A_i\}\mathrm{P}\left\{\bigcup_{j=k}^{\infty} B_j\right\} \tag{12}$$

for all large i and some positive integer k and

$$\sum_{i=1}^{\infty} \mathrm{P}\{A_i\} = \infty, \tag{13}$$

then

$$\mathrm{P}\left\{\bigcup_{j=k}^{\infty} B_j\right\} = 1. \tag{14}$$

Moreover, if (12) *holds for infinitely many integers k and* (13) *obtains, then*

$$\mathrm{P}\{B_n, \text{i.o.}\} = 1. \tag{15}$$

PROOF. Replacing i by $i + nk$ and then setting $A_n^* = A_{n,i}^* = A_{i+nk}$ in (12),

$$P\left\{\bigcup_{j=k}^{\infty} B_j\right\} \ge P\left\{\bigcup_{j=1}^{\infty} A_{i+(n+j)k} \middle| A_{i+nk}\right\} = P\left\{\bigcup_{h=n+1}^{\infty} A_h^* \middle| A_n^*\right\}$$

for all large n, whence (14) follows from Theorem 3.1.3, noting that

$$\sum_{i=1}^{k} \sum_{n=1}^{\infty} P\{A_{i+nk}\} = \infty = \sum_{n=1}^{\infty} P\{A_{n,i}^*\}$$

for some i. The final statement is a consequence of the monotonicity in the left side of (14).

EXAMPLE 1. If $\{S_n = \sum_{i=1}^n X_i, n \geq 1\}$ where $\{X_n, n \geq 1\}$ are i.i.d. random variables, then for any $\varepsilon \geq 0$, $\mathrm{P}\{|S_n| \leq \varepsilon, \text{i.o.}\} = 1$ or 0 according as

$$\sum_{n=1}^{\infty} \mathrm{P}\{|S_n| \leq \varepsilon\}$$

diverges or converges.

PROOF. In view of the Borel–Cantelli lemma, it suffices to consider the case of divergence and here it may be supposed that $\sum_{n=1}^{\infty} \mathrm{P}\{0 \leq S_n \leq \varepsilon\} = \infty$ since otherwise $\sum_{n=1}^{\infty} \mathrm{P}\{0 \geq S_n \geq -\varepsilon\} = \infty$ whence X_n may be replaced by $-X_n$. Setting $A_i = \{0 \leq S_i \leq \varepsilon\}$, $B_j = \{|S_j| \leq \varepsilon\}$, $C_{ij} = \{|S_j - S_i| \leq \varepsilon\}$, clearly $A_i \bigcup_{j=i+k}^{\infty} A_j \subset A_i \bigcup_{j=i+k}^{\infty} C_{ij}$, implying via independence and Theorem 1.6.1 that for all $k \geq 1$

$$\mathrm{P}\left\{A_i \bigcup_{j=i+k}^{\infty} A_j\right\} \leq \mathrm{P}\{A_i\}\mathrm{P}\left\{\bigcup_{j=i+k}^{\infty} C_{ij}\right\} = \mathrm{P}\{A_i\}\mathrm{P}\left\{\bigcup_{j=i+k}^{\infty} B_{j-i}\right\}.$$

Thus, $\mathrm{P}\{B_n, \text{i.o.}\} = 1$ by Lemma 4. □

Lemma 5. *If $\{S_n = \sum_{i=1}^n X_i, n \geq 1, S_0 = 0\}$, where $\{X_n, n \geq 1\}$ are* i.i.d. *random variables, then for any $\delta > 0$ and positive integers k, N*

$$\sum_{i=0}^{N} \mathrm{P}\{|S_i| < k\delta\} \leq 2k \sum_{i=0}^{N} \mathrm{P}\{|S_i| < \delta\}. \tag{16}$$

PROOF. If $T_j = \inf\{n \geq 0 : j\delta \leq S_n < (j+1)\delta\}$, where $\inf\{\emptyset\} = \infty$, then

$$\sum_{i=0}^{N} \mathrm{P}\{j\delta \leq S_i < (j+1)\delta\} = \sum_{i=0}^{N} \sum_{n=0}^{i} \mathrm{P}\{T_j = n, j\delta \leq S_i < (j+1)\delta\}$$

$$\leq \sum_{n=0}^{N} \sum_{i=n}^{N} \mathrm{P}\{T_j = n\}\mathrm{P}\{|S_i - S_n| < \delta\} \leq \sum_{n=0}^{N} \mathrm{P}\{|S_n| < \delta\},$$

and so (16) follows by summing on j from $-k$ to $k-1$. □

A sequence of partial sums $\{S_n, n \geq 1\}$ of i.i.d. random variables $\{X_n, n \geq 1\}$ is called a **random walk** and it is customary to set $S_0 = 0$. The origin and the random walk itself are said to be **recurrent** if $\mathrm{P}\{|S_n| < \varepsilon, \text{i.o.}\} = 1$, for all $\varepsilon > 0$. Otherwise, the random walk is **nonrecurrent** or transient. Thus, Example 1 furnishes a criterion for a random walk to be recurrent. However, here as in virtually all questions concerning random walks, criteria involving the underlying distribution F of X are far preferable to those involving S_n for the simple reason that the former are much more readily verifiable (see Example 5.2.1).

The definition of a recurrent random walk applies also to sums S_n of i.i.d. random vectors where $|S_n|$ signifies Euclidean distance from the origin. According to Example 2 and Exercise 4.2.10, there is an abrupt change in the behavior of simple random walks in going from the plane to three dimensions. This is likewise true for general random walks (Chung and Fuchs, 1951).

EXAMPLE 2 (Polya). Let $\{S_n, n \geq 0\}$ be a **simple random walk** on $(-\infty, \infty)^m$ whose initial position is at the origin, that is, $S_n = \sum_1^n X_i$, where $\{X_n, n \geq 1\}$ are i.i.d. random vectors with

$$P\{X_n = (e_1, \ldots, e_m)\} = \frac{1}{2m} \quad \text{where } e_i = 0, 1, -1 \quad \text{and} \quad \sum_{i=1}^m e_i^2 = 1.$$

When $m = 1$, $\{X_n, n \geq 1\}$ constitute Bernoulli trials with parameter $p = \frac{1}{2}$ and it follows immediately from Theorem 3.4.3 that $\{S_n, n \geq 0\}$ is recurrent in this case. When $m = 2$, setting $A_n = \{S_{2n} = (0, 0)\}$ and recalling Exercises 2.1.3, 2.3.3,

$$P\{A_n\} = 4^{-2n} \sum_{j=0}^n \frac{(2n)!}{[j!(n-j)!]^2} = \binom{2n}{n} 4^{-2n} \sum_{j=0}^n \binom{n}{j}^2 = \left[\binom{2n}{n} 2^{-2n}\right]^2 \sim \frac{1}{\pi n},$$

and so $\sum_1^\infty P\{A_n\} = \infty$. Moreover, $A_i A_j = A_i C_{ij}$ for $i < j$ where $C_{ij} = \{S_{2j} - S_{2i} = (0, 0)\}$, implying

$$P\left\{A_i \bigcup_{j=i+k}^\infty A_j\right\} = P\{A_i\} P\left\{\bigcup_{j=i+k}^\infty C_{ij}\right\} = P\{A_i\} P\left\{\bigcup_{j=i+k}^\infty A_{j-i}\right\}.$$

Thus, Lemma 4 applies with $B_i = A_i$, revealing that a simple random walk in the plane is recurrent. On the other hand, for $m \geq 3$ the origin and random walk are nonrecurrent (Exercise 4.2.10).

Lemma 6. *Let $\{X_n, n \geq 1\}$ be independent,* symmetric r.v.s *and $\{a_n, n \geq 1\}$, $\{c_n, n \geq 1\}$ sequences of positive numbers with $a_n \to \infty$. If $X'_n = X_n I_{[|X_n| \leq c_n]}$, $S'_n = \sum_{j=1}^n X'_j$, $S_n = \sum_{j=1}^n X_j$, then*

$$P\left\{\varlimsup_{n\to\infty} \frac{S'_n}{a_n} > 1\right\} = 1 \quad \textit{implies} \quad P\left\{\varlimsup_{n\to\infty} \frac{S_n}{a_n} \geq 1\right\} = 1. \tag{17}$$

PROOF. The hypothesis ensures that $N_m = \inf\{j \geq m: S'_j > a_j\}$ ($= \infty$ otherwise) is a bona fide r.v. for all $m \geq 1$. If whenever $n \geq m$

$$P\{S_n \geq S'_n, N_m = n\} = P\{S_n \leq S'_n, N_m = n\}, \tag{18}$$

then

$$\mathrm{P}\left\{\bigcup_{j=m}^{\infty}[S_j > a_j]\right\} \geq \mathrm{P}\left\{\bigcup_{j=m}^{\infty}[S_j' > a_j,\, S_j \geq S_j']\right\} \geq \mathrm{P}\{S_{N_m} \geq S_{N_m}',\, N_m < \infty\}$$

$$= \sum_{n=m}^{\infty} \mathrm{P}\{S_n \geq S_n',\, N_m = n\} \geq \tfrac{1}{2},$$

implying

$$\mathrm{P}\left\{\varlimsup_{n\to\infty} \frac{S_n}{a_n} \geq 1\right\} \geq \mathrm{P}\{S_n > a_n, \text{ i.o.}\} \geq \tfrac{1}{2}$$

and hence also the conclusion of (17) by the Kolmogorov zero-one law.

To verify (18), set $X_j^* = X_j I_{[|X_j| \leq c_j]} - X_j I_{[|X_j| > c_j]}$ and note that by symmetry and independence the joint distributions of $(X_1, \ldots, X_n)$ and $(X_1^*, \ldots, X_n^*)$ are identical for all n. Hence, if $n > m$,

$$\mathrm{P}\{S_n \geq S_n',\, N_m = n\}$$

$$= \mathrm{P}\left\{\sum_1^n X_j I_{[|X_j| > c_j]} \geq 0,\, \sum_1^n X_j I_{[|X_j| \leq c_j]} > a_n,\, S_i' \leq a_i,\, m \leq i < n\right\}$$

$$= \mathrm{P}\left\{\sum_1^n X_j^* I_{[|X_j^*| > c_j]} \geq 0,\, \sum_1^n X_j^* I_{[|X_j^*| \leq c_j]} > a_n,\, \sum_1^i X_j^* I_{[|X_j^*| \leq c_j]} \leq a_i,\, m \leq i < n\right\}$$

$$= \mathrm{P}\{S_n \leq S_n',\, N_m = n\},$$

and equality also holds for $n = m$, *mutatis mutandis*. □

The next result reveals the extent to which mean convergence is stronger than convergence in probability and provides a Cauchy criterion for the former.

Theorem 3 (i) (Mean Convergence Criterion). *If the* r.v.s $\{|X_n|^p, n \geq 1\}$ *are* u.i. *for some* $p > 0$ *and* $X_n \xrightarrow{\mathrm{P}} X$, *then* $X \in \mathscr{L}_p$ *and* $X_n \xrightarrow{\mathscr{L}_p} X$.

Conversely, if X_n, $n \geq 1$, *are* $\mathscr{L}_p$ r.v.s *with* $X_n \xrightarrow{\mathscr{L}_p} X$, *then* $X \in \mathscr{L}_p$, $X_n \xrightarrow{\mathrm{P}} X$, *and* $\{|X_n|^p, n \geq 1\}$ *are* u.i.

(ii) (Cauchy Convergence Criterion). *If* $\{X_n, n \geq 1\}$ *are* $\mathscr{L}_p$ r.v.s *with* $sup_{m>n}\, \mathrm{E}|X_m - X_n|^p = o(1)$ *as* $n \to \infty$, *there exists a* r.v. $X \in \mathscr{L}_p$ *such that* $X_n \xrightarrow{\mathscr{L}_p} X$ *and conversely.*

PROOF. If $X_n \xrightarrow{\mathrm{P}} X$, by Lemma 3.3.2 there exists a subsequence X_{n_k} with $X_{n_k} \xrightarrow{\text{a.c.}} X$. By Fatou's lemma (Corollary 2)

$$\mathrm{E}|X|^p = \mathrm{E}\left[\varliminf_{k\to\infty} |X_{n_k}|^p\right] \leq \varliminf_{k\to\infty} \mathrm{E}|X_{n_k}|^p \leq \sup_{m\geq 1} \mathrm{E}|X_m|^p < \infty$$

since $\{|X_n|^p\}$ is u.i. Again employing the latter, for any $\varepsilon > 0$, $\delta > 0$ may be chosen such that $\mathrm{P}\{A\} < \delta$ implies

$$\sup_{n\geq 1} \mathrm{E}\{|X_n|^p I_A] < \varepsilon, \qquad \mathrm{E}\{|X|^p I_A] < \varepsilon, \tag{19}$$

the second part holding by Lemma 2. Moreover, $X_n \xrightarrow{P} X$ ensures that $n \geq N$ entails

$$P\{|X_n - X| > \varepsilon\} < \delta. \tag{20}$$

Consequently, for $n \geq N$, by (19), (20) and Lemma 3

$$\begin{aligned} E|X_n - X|^p &= E[|X_n - X|^p(I_{[|X_n - X| \leq \varepsilon]} + I_{[|X_n - X| > \varepsilon]})] \\ &\leq \varepsilon^p + 2^p E[I_{[|X_n - X| > \varepsilon]}(|X_n|^p + |X|^p)] < 2^{p+1}(\varepsilon + \varepsilon^p), \end{aligned}$$

so that $X_n \xrightarrow{\mathscr{L}_p} X$.

Conversely, if $X_n \xrightarrow{\mathscr{L}_p} X$, then, as noted earlier, $X_n \xrightarrow{P} X \in \mathscr{L}_p$ and by Lemma 3

$$\sup_{n \geq 1} E|X_n|^p I_A \leq 2^p \sup_{n \geq 1} E[|X_n - X|^p I_A + |X|^p I_A] < \infty \tag{21}$$

for all $A \in \mathscr{F}$ and, in particular, for $A = \Omega$. Since

$$\begin{aligned} \sup E|X_n - X|^p I_{[|X_n - X| > K]} &\leq \max_{n \leq m} E|X_n - X|^p I_{[|X_n - X| > K]} + \sup_{n > m} E|X_n - X|^p \\ &\xrightarrow{K \to \infty} \sup_{n > m} E|X_n - X|^p \xrightarrow{m \to \infty} 0, \end{aligned}$$

$\{|X_n - X|^p, n \geq 1\}$ is u.i. whence (21) guarantees that $\{|X_n|^p, n \geq 1\}$ is u.i. Apropos of (ii), if the Cauchy criterion holds, the Markov inequality ensures that for any $\varepsilon > 0$

$$\sup_{m > n} P\{|X_m - X_n| > \varepsilon\} \leq \varepsilon^{-p} \sup_{m > n} E|X_m - X_n|^p = o(1)$$

and Lemma 3.3.2 guarantees a subsequence n_k with $X_{n_k} \xrightarrow{\text{a.c.}}$ some r.v. X. By Corollary 2 (Fatou's lemma), $E|X_m - X|^p \leq \underline{\lim}_{k \to \infty} E|X_m - X_{n_k}|^p$, implying

$$0 \leq \lim_{m \to \infty} E|X_m - X|^p \leq \lim_{m \to \infty} \underline{\lim}_{k \to \infty} E|X_m - X_{n_k}|^p = 0,$$

so that $X_n \xrightarrow{\mathscr{L}_p} X$. Conversely, if $X_n \xrightarrow{\mathscr{L}_p} X$, by Lemma 3

$$\sup_{m > n} E|X_m - X_n|^p \leq 2^p \left[\sup_{m > n} E|X_m - X|^p + E|X - X_n|^p\right] = o(1)$$

and the Cauchy criterion obtains. □

Corollary 3 (Lebesgue Dominated Convergence Theorem). *Let* $\{X, X_n, n \geq 1\}$ *be a sequence of* r.v.s *with* $X_n \xrightarrow{P} X$. *If* $E[\sup_{n \geq 1} |X_n|] < \infty$, *then* $E|X_n - X| \to 0$ *and a fortiori* $E X_n \to E X$.

PROOF. The hypothesis ensures that $\{X_n, n \geq 1\}$ is u.i., whence the conclusion follows from Theorem 3(i) with $p = 1$. □

Corollary 4. *If* $\{X, X_n, n \geq 1\}$ *is a sequence of nonnegative* $\mathscr{L}_1$ r.v.s *with* $X_n \xrightarrow{P} X$, *then* $E X_n \to E X$ iff $E|X_n - X| \to 0$ iff $\{X_n, n \geq 1\}$ *is* u.i.

PROOF. Sufficiency is immediate from $|E X_n - E X| = |E(X_n - X)| \le E|X_n - X|$. Apropos of necessity, since $0 \le (X - X_n)^+ \le X$ and $(X - X_n)^+ \xrightarrow{P} 0$ by Corollary 3.3.3, dominated convergence (Corollary 3) guarantees $E(X - X_n)^+ \to 0$. By hypothesis $E(X - X_n) \to 0$, whence $E(X - X_n)^- \to 0$, and so $E|X - X_n| \to 0$. □

Among other things, Corollary 4 underlines the importance of the concept of u.i.

Corollary 5. *If* $X_n, n \ge 1$, *are* $\mathscr{L}_p$ r.v.s *with* $X_n \xrightarrow{\mathscr{L}_p} X$ *for some* $p > 0$, *then* $X \in \mathscr{L}_p$ *and* $E|X_n|^p \to E|X|^p$.

PROOF. The hypothesis implies that $X_n \xrightarrow{P} X$ and Corollary 3.3.3 ensures that $|X_n|^p \xrightarrow{P} |X|^p$, whence $E|X_n|^p \to E|X|^p$ by Corollary 4. □

In the proof of Theorem 2.2.2 it was shown for binomial r.v.s S_n that $E|S_n - np|^4 = O(n^2)$ and so $E|(S_n/n) - p|^4 = o(1)$. Thus, it follows directly from Corollary 5 that $E(S_n/n)^4 \to p^4$. More generally, since $S_n/n \xrightarrow{P} p$ and $|S_n/n| \le 1$ Corollaries 3 and 5 ensure that $E|S_n/n|^\beta \to p^\beta$ for every $\beta > 0$.

EXERCISES 4.2

1. Improve the inequality of Lemma 3 by showing that
$$(a + b)^p \le (a^p + b^p) \cdot \max(1, 2^{p-1}) \quad \text{for } a > 0, b > 0, p > 0;$$
also, if $a_i \ge 0$, then $(\sum_{i=1}^n a_i)^p \ge \sum_{i=1}^n a_i^p$ (resp. $\le$) for $p \ge 1$ (resp. $\le$).

2. Prove that if r.v.s $X_n \xrightarrow{\text{a.c.}} X$ and $X_n \xrightarrow{\mathscr{L}_p} Y$, then $X = Y$, a.c. Construct r.v.s X, $X_n, n \ge 1$, such that (i) $X_n \xrightarrow{\text{a.c.}} X$ but $X_n \not\xrightarrow{\mathscr{L}_p} X$ for any $p > 0$, (ii) $X_n \xrightarrow{\mathscr{L}_p} X$ for all all $p > 0$ but $X_n \not\xrightarrow{\text{a.c.}} X$.

3. Let $\{A_n, n \ge 1\}$ be events with $\sum_{n=1}^\infty P\{A_n\} = \infty$. If there exist events $\{B_i, i \ge 1\}$ and $\{D_{ij}, j > i, i \ge 1\}$ such that for all large i and some positive integer k (resp. infinitely many $k > 0$)

   i. $A_i A_j \subset D_{ij}, i < j$,
   ii. $P\{\bigcup_{j=k+i}^\infty D_{ij}\} = P\{\bigcup_{j=k+i}^\infty B_{j-i}\}$,
   iii. the classes $\{A_i\}$ and $\{D_{i,i+k}, D_{i,i+k+1}D_{i,i+k}^c, D_{i,i+k+2}D_{i,i+k+1}^c D_{i,i+k}^c, \ldots\}$ are independent,

   then $P\{\bigcup_{j=k}^\infty B_j\} = 1$ (resp. $P\{B_n, \text{i.o.}\} = 1$).

4. If X_1 and X_2 are independent r.v.s and $X_1 + X_2 \in \mathscr{L}_p$ for some $p \in (0, \infty)$, then $X_i \in \mathscr{L}_p, i = 1, 2$. *Hint*: For all large $\lambda > 0$,
$$P\{|X_1| > \lambda\} \le 2\,P\left\{|X_1| > \lambda, |X_2| < \frac{\lambda}{2}\right\} \le 2\,P\left\{|X_1 + X_2| > \frac{\lambda}{2}\right\}.$$

5. Let $P\{X_n = a_n > 0\} = 1/n = 1 - P\{X_n = 0\}, n \ge 1$. Is $\{X_n, n \ge 1\}$ u.i. if

   i. $a_n = o(n)$,
   ii. $a_n = cn > 0$?

6. If $\{X_n, n \geq 1\}$ are r.v.s with $\sup_{n \geq 1} E|X_n|^\beta < \infty$ for some $\beta > 0$, then $\{|X_n|^\alpha, n \geq 1\}$ is u.i. for $0 < \alpha < \beta$.
7. If the r.v.s $X_n, n \geq 1$, are u.i., so are $S_n/n, n \geq 1$, where $S_n = \sum_{i=1}^n X_i$; in particular, if $X_n, n \geq 1$, are identically distributed $\mathscr{L}_1$ random variables, then $\{S_n/n, n \geq 1\}$ is u.i.
8. Show that (i) if the Poisson r.v.s S_n have p.d.f. $p(k; n\lambda)$, $n \geq 1$, then $E|(S_n/n) - \lambda| \to 0$. *Hint*: Recall Exercise 2.2.2. (ii) Any sequence of r.v.s $Y_n \xrightarrow{P} 0$ iff $E(|Y_n|/(1 + |Y_n|)) = o(1)$.
9. (i) Construct a sequence of r.v.s that is u.i. from below and for which the expectation of $\underline{\lim}\, X_n$ does not exist. (ii) Show that the hypothesis $\sup_{n \geq 1} |X_n| \in \mathscr{L}_1$ of Corollary 3 is equivalent to $|X_n| \leq Y \in \mathscr{L}_1, n \geq 1$.
10. Let $\{X_n, n \geq 1\}$ be a simple symmetric random walk in R^k, that is, $\{X_n\}$ are i.i.d. random vectors such that $P\{X_n = (e_1, \ldots, e_k)\} = 1/2k$, where $e_j = 0, 1$, or -1 and $\sum_1^k e_j^2 = 1$. Prove that
$$P\left\{\sum_1^n X_i \text{ returns to its origin, i.o.}\right\} = 0 \quad \text{for } k = 3.$$
11. If $\{X, X_n, n \geq 1\}$ are r.v.s on $(\Omega, \mathscr{F}, P)$, show that the indefinite integrals $\int_A X_n \, dP \to \int_A X \, dP$, finite, uniformly for all $A \in \mathscr{F}$ iff X_n converges to X in mean.
12. If the two sequences of integrable r.v.s $\{X_n\}$, $\{Y_n\}$ satisfy $P\{X_n \geq Y_n \geq 0\} = 1$, $X_n \xrightarrow{P} X$, $Y_n \xrightarrow{P} Y$, and $E\, X_n \to E\, X$, finite, then $E|Y_n - Y| \to 0$.
13. Let $\{X_n, n \geq 1\}$ be a sequence of $\mathscr{L}_p$ r.v.s, $p > 0$ with $\sup[\int_A |X_n|^p \, dP\colon n \geq 1$ and $P\{A\} < \delta] = o(1)$ as $\delta \to 0$. Then $X_n \xrightarrow{P} X$ iff $X_n \xrightarrow{\mathscr{L}_p} X$.
14. If $X \in \mathscr{L}_1(\Omega, \mathscr{F}, P)$ and $\sigma(X)$, $\mathscr{G}$ are independent classes of events, prove that $E\, XI_A = E\, X \cdot P\{A\}$, all $A \in \mathscr{G}$.
15. (Kochen–Stone) If $\{Z_n, n \geq 1\}$ is a sequence of r.v.s with $0 < E\, Z_n^2 < \infty$, $E\, Z_n \neq 0$, and $\overline{\lim}_{n \to \infty} (E\, Z_n)^2 / E\, Z_n^2 > 0$, then $P\{\overline{\lim}_{n \to \infty} Z_n / E\, Z_n \geq 1\} > 0$. *Hint*: If $Y_n = Z_n / E\, Z_n$, there is a subsequence $\{n'\}$ with $E\, Y_{n'}^2 < K < \infty$. Replacing $\{n'\}$ by $\{n\}$ for notational simplicity, $E\, \underline{\lim}\, Y_n^2 \leq K$ by Corollary 2. Since $-Y_n \leq 1 + Y_n^2$, necessarily $E\, \underline{\lim}(-Y_n)$ exists. Then Theorem 2(ii) ensures (since $\{Y_n\}$ is u.i.) that $E\, \overline{\lim}\, Y_n \geq \overline{\lim}\, E\, Y_n = 1$.
16. (Kochen–Stone) If $\{A_n, n \geq 1\}$ is a sequence of events such that for some $c > 0$

    i. $P\{A_i A_j\} \leq c\, P\{A_i\}[P\{A_{j-i}\} + P\{A_j\}]$, $i < j$
    ii. $\sum_{n=1}^\infty P\{A_n\} = \infty$,

    then $P\{A_n, \text{i.o.}\} > 0$.
    Hint: If $Z_n = \sum_{i=1}^n I_{A_i}$, note that $E\, Z_n^2 \leq E\, Z_n + 4c(\sum_1^n P\{A_i\})^2 \leq (1 + 4c)(E\, Z_n)^2$ for all large n since $E\, Z_n \to \infty$ and, via Exercise 15,
$$P\{A_n, \text{i.o.}\} \geq P\{\overline{\lim}\, Z_n / E\, Z_n \geq 1\} > 0.$$

4.3 Jensen, Hölder, Schwarz Inequalities

A finite real function g on an interval $J \subset (-\infty, \infty)$ is called **convex** on J if whenever $x_1, x_2 \in J$ and $\lambda \in [0, 1]$

$$g(\lambda x_1 + (1 - \lambda)x_2) \le \lambda g(x_1) + (1 - \lambda)g(x_2). \tag{1}$$

Geometrically speaking, the value of a convex function at any point on the line segment joining x_1 to x_2 lies on or below the line segment joining $g(x_1)$ and $g(x_2)$. Since $t = u((t - s)/(u - s)) + s((u - t)/(u - s))$,

$$g(t) \le \frac{t - s}{u - s} g(u) + \frac{u - t}{u - s} g(s), \qquad s < t < u, \tag{2}$$

or equivalently

$$\frac{g(t) - g(s)}{t - s} \le \frac{g(u) - g(t)}{u - t}, \qquad s < t < u. \tag{3}$$

If g is convex on an open interval J_0, it follows from (2) that $\overline{\lim}_{s<t\to s} g(t) \le g(s)$, $\underline{\lim}_{t>s\to t} g(s) \ge g(t)$, $\underline{\lim}_{t<u\to t} g(u) \ge g(t)$, and $\overline{\lim}_{u>t\to u} g(t) \le g(u)$, whence g is continuous on J_0. Furthermore, as a consequence of (3), a differentiable function g is convex on J_0 iff g' is nondecreasing on J_0. Thus, if g is convex on J_0 and twice differentiable, $g'' \ge 0$. Conversely, if $g'' \ge 0$ on J_0, a two-term Taylor expansion yields, setting $q_1 = \lambda$, $q_2 = 1 - \lambda$,

$$q_i g(x_i) \ge q_i[g(q_1 x_1 + q_2 x_2) + q_{3-i}(x_i - x_{3-i})g'(q_1 x_1 + q_2 x_2)], \qquad i = 1, 2,$$

and summing, (1) holds, that is, g is convex on J_0. Moreover, it is shown in Hardy *et al.* (1934, pp. 91–95) that

i. If g is convex on an open interval J_0, it has left and right derivatives g_l' and g_r' at every point of J_0, with $g_l' \le g_r'$, each derivative being nondecreasing,
ii. If g is convex on an interval J, at each interior point $\xi \in J$,

$$g(t) \ge g(\xi) + (t - \xi)g_r'(\xi), \qquad t \in J. \tag{4}$$

Theorem 1. *If X is an integrable* r.v., *c is a finite constant, and g is a convex function on $(-\infty, \infty)$, then*

$$\mathrm{E}\, g^-(X - \mathrm{E}\, X + c) < \infty. \tag{5}$$

Moreover, if $\alpha(t)$ and $t - \alpha(t)$ are nondecreasing on $(-\infty, \infty)$, then $\mathrm{E}|\alpha(X)| < \infty$, $\mathrm{E}\, g^-(\alpha(X) - \mathrm{E}\,\alpha(X) + c) < \infty$, *and*

$$\mathrm{E}\, g(X - E\, X + c) \ge \mathrm{E}\, g(\alpha(X) - \mathrm{E}\alpha(X) + c). \tag{6}$$

PROOF. By (4), $g(t) \ge g(0) + tg_r'(0)$ for $t \in (-\infty, \infty)$, whence (5) holds. Since monotonicity ensures $t^+ + \alpha(0) \ge \alpha(t^+) \ge \alpha(t) \ge \alpha(-t^-) \ge -t^- + \alpha(0)$, the hypothesis implies $|\alpha(X)| \le |X| + |\alpha(0)|$ and so $\alpha(X)$ is integrable. Consequently, (4) yields $\mathrm{E}\, g^-(\alpha(X) - \mathrm{E}\,\alpha(X) + c) < \infty$. Set

$$\beta(t) = t - \alpha(t) - \mathrm{E}\, X + \mathrm{E}\,\alpha(X), \qquad t \in (-\infty, \infty).$$

Then $\mathrm{E}|\beta(X)| < \infty$ and $\mathrm{E}\,\beta(X) = 0$. If $\mathrm{P}\{\beta(X) = 0\} = 1$, (6) holds trivially. Otherwise, $\beta(t_1) < 0$, $\beta(t_2) > 0$ for some $t_1, t_2 \in (-\infty, \infty)$. If $t_0 = \inf\{t: \beta(t) > 0\}$, then $t_1 \le t_0 \le t_2$ by monotonicity of $t - \alpha(t)$, and

$$t \ge t_0 \quad \text{if } \beta(t) > 0, \qquad t \le t_0 \quad \text{if } \beta(t) < 0. \tag{7}$$

Again employing (4),

$$g(X - \mathrm{E}\,X + c) \geq g(\alpha(X) - \mathrm{E}\,\alpha(X) + c) + \beta(X)g_r'(\alpha(X) - \mathrm{E}\,\alpha(X) + c). \tag{8}$$

By (7), $X \geq t_0$ when $\beta(X) > 0$ and $X \leq t_0$ for $\beta(X) < 0$. Since both g_r' and α are nondecreasing, necessarily

$$\beta(X)g_r'(\alpha(X) - \mathrm{E}\,\alpha(X) + c) \geq \beta(X)g_r'(\alpha(t_0) - \mathrm{E}\,\alpha(X) + c). \tag{9}$$

Taking expectations in (8) and (9), the conclusion (6) follows by recalling that $\mathrm{E}\,\beta(X) = 0$. □

Corollary 1. *If g is a convex function on $(-\infty, \infty)$, for any $\mathscr{L}_1$ r.v. X and any finite constant c*

$$\mathrm{E}\,g(X - \mathrm{E}\,X + c) \geq g(c) \tag{10}$$

and, in particular,

$$\mathrm{E}\,g(X) \geq g(\mathrm{E}\,X) \qquad (\textit{Jensen's inequality}). \tag{11}$$

Corollary 2. *If X is an $\mathscr{L}_1$ r.v., then for $1 \leq p < \infty$*

$$\mathrm{E}|X - \mathrm{E}\,X|^p \geq \mathrm{E}|Y - \mathrm{E}\,Y|^p, \tag{12}$$

where for some choice of $-\infty \leq a < b \leq \infty$

$$Y = XI_{[a \leq X \leq b]} + aI_{[X < a]} + bI_{[X > b]}.$$

Proof. Take $c = 0$, $\alpha(t) = \max[a, \min(t, b)]$, $g(t) = |t|^p$, $p \geq 1$ in (6). □

In particular, abbreviating $(\mathrm{E}|X|)^p$ by $\mathrm{E}^p|X|$, Jensen's inequality (11) yields for $1 \leq p < \infty$

$$\mathrm{E}|X|^p \geq \mathrm{E}^p|X| \quad \text{or} \quad \mathrm{E}^{1/p}|X|^p \geq \mathrm{E}|X|. \tag{13}$$

Replacing p and $|X|$ respectively by p'/p and $|X|^p$ in (13),

$$\mathrm{E}^{1/p}|X|^p \leq \mathrm{E}^{1/p'}|X|^{p'}, \qquad 0 < p < p' < \infty \tag{14}$$

and so convergence in mean of order p implies convergence in mean of any order less than p. A convenient, widespread notation is to set

$$\|X\|_p = \mathrm{E}^{1/p}|X|^p, \qquad p > 0, \tag{15}$$

and it is customary to refer to $\|X\|_p$ as the ***p*-norm** of X. According to (14), $\|X\|_p \leq \|X\|_{p'}$ for $0 < p < p'$. Moreover, $\|X\|_p$ satisfies the triangle inequality for $p \geq 1$ as noted in Exercise 10 and $\|cX\|_p = |c| \cdot \|X\|_p$.

Theorem 2 (Hölder Inequality). *If X, Y are measurable functions on a probability space $(\Omega, \mathscr{F}, \mathrm{P})$, then for $p > 1$, $p' > 1$ with $(1/p) + (1/p') = 1$*

$$\mathrm{E}|XY| \leq \|X\|_p \cdot \|Y\|_{p'}. \tag{16}$$

PROOF. In proving (16), it may be supposed that $0 < \|X\|_p \|Y\|_{p'} < \infty$ since (16) is trivial otherwise. Set

$$U = \frac{|X|}{\|X\|_p}, \qquad V = \frac{|Y|}{\|Y\|_{p'}},$$

entailing $\|U\|_p = 1 = \|V\|_{p'}$. Now, $-\log t$ is a convex function on $(0, \infty)$, whence, via (1), for $a, b > 0$

$$-\log\left(\frac{a^p}{p} + \frac{b^{p'}}{p'}\right) \le -\frac{1}{p}\log a^p - \frac{1}{p'}\log b^{p'} = -\log ab,$$

or equivalently

$$ab \le \frac{a^p}{p} + \frac{b^{p'}}{p'}, \qquad 0 \le a, b \le \infty.$$

Thus,

$$\mathrm{E}\, UV \le \frac{1}{p}\mathrm{E}\, U^p + \frac{1}{p'}\mathrm{E}\, V^{p'} = \frac{1}{p} + \frac{1}{p'} = 1,$$

which is tantamount to (16). □

Corollary 3 (Schwarz Inequality). *For any $\mathscr{L}_2$ random variables X and Y,*

$$|\mathrm{E}\, XY| \le \mathrm{E}|XY| \le (\mathrm{E}\, X^2)^{1/2}(\mathrm{E}\, Y^2)^{1/2}. \tag{17}$$

Corollary 4 (Liapounov). *If X is a non-negative $\mathscr{L}_p$* r.v., *all* $p > 0$, *and*

$$g(p) = \log \mathrm{E}\, X^p, \qquad 0 \le p < \infty, \tag{18}$$

then g is convex on $[0, \infty)$.

PROOF. For $0 \le p_1, p_2 < \infty$ and $q_1, q_2 > 0$ with $q_1 + q_2 = 1$, noting that $1/q_i > 1$, $i = 1, 2$, Hölder's inequality yields

$$\begin{aligned} g(q_1 p_1 + q_2 p_2) &= \log \mathrm{E}(X^{q_1 p_1} \cdot X^{q_2 p_2}) \le \log(\|X^{q_1 p_1}\|_{1/q_1} \cdot \|X^{q_2 p_2}\|_{1/q_2}) \\ &= \log(\mathrm{E}^{q_1} X^{p_1} \cdot \mathrm{E}^{q_2} X^{p_2}) = q_1 g(p_1) + q_2 g(p_2). \end{aligned}$$ □

If X is a r.v. on a probability space, $\mathrm{E}|X|^p$, $p > 0$ is called the ***p*th absolute moment** of X(or its distribution), while for any positive integer k, $\mathrm{E}\, X^k$ (if it exists) is termed the ***k*th moment** of X.

For any $\mathscr{L}_1$ r.v. X, the **variance** of X is defined by

$$\sigma_X^2 = \sigma^2(X) = \mathrm{E}(X - \mathrm{E}\, X)^2, \tag{19}$$

while its positive square root is the **standard deviation** of X. Clearly, for every finite constant c, $\sigma^2(X + c) = \sigma^2(X)$ and $\sigma^2(cX) = c^2\sigma^2(X)$. The variance or standard deviation of X provides information about the extent to which the distribution of X clusters about its mean and this is reflected in the simple but extremely useful **Tchebychev inequality**

$$P\{|X - E X| \geq a\} \leq \frac{\sigma_X^2}{a^2}, \qquad a > 0, \tag{20}$$

which follows from a direct application of the Markov inequality to the r.v. $(X - E X)^2$.

A basic tool of probability theory is **truncation**. Two alternative methods of truncating a r.v. X are

$$Y' = XI_{[a \leq X \leq c]}, \tag{i}$$

$$Y = XI_{[a \leq X \leq c]} + aI_{[X < a]} + cI_{[X > c]}, \tag{ii}$$

where a, c are constants such that $-\infty \leq a < c \leq \infty$. One or both equality signs in the set of the indicator function of (i) may be deleted. Whenever both a and c are finite, Y and Y' are bounded r.v.s and hence have moments of all orders. For $X \in \mathscr{L}_1$, Corollary 2 reveals that $\sigma_Y^2 \leq \sigma_X^2$, whereas no comparable inequality between $\sigma_{Y'}^2$ and σ_X^2 exists (Exercise 1).

If X, Y are r.v.s with $0 < \sigma(X), \sigma(Y) < \infty$, the **correlation coefficient** between X and Y or simply the correlation of X and Y is given by

$$\rho_{X,Y} = \rho(X, Y) = \frac{E(X - E X)(Y - E Y)}{\sigma(X) \cdot \sigma(Y)}. \tag{21}$$

If $\rho(X, Y) = 0$, the r.v.s X and Y are said to be **uncorrelated**.

It follows directly from the Schwarz inequality (17) that $|\rho(X, Y)| \leq 1$. The correlation coefficient $\rho_{X,Y}$ indicates the extent to which there is a linear relationship between the r.v.s X and Y (Exercise 7).

$\mathscr{L}_2$ r.v.s X_n, $n \geq 1$, are called **uncorrelated** if X_n and X_m are uncorrelated for each pair of distinct indices n, m. Independent $\mathscr{L}_2$ r.v.s are necessarily uncorrelated as follows from

Theorem 3. *If X and Y are independent $\mathscr{L}_1$ r.v.s, then $X \cdot Y \in \mathscr{L}_1$ and*

$$E XY = E X \cdot E Y. \tag{22}$$

PROOF. To prove (22) suppose first that $X \geq 0$, $Y \geq 0$. For $m \geq 1$ and $j = 1, 2, \ldots$, set $m_j = j/2^m$, $Y_{m,j} = I_{[m_j < Y \leq m_{j+1}]}$, and $Y_m = \sum_{j=1}^{\infty} m_j Y_{m,j}$. Then $0 \leq Y_m \uparrow Y$, a.c., and $0 \leq XY_m \uparrow XY$, a.c. By the definition of expectation and independence, taking n_i analogous to m_j

$$
\begin{aligned}
\mathrm{E}\, XY_{m,j} &= \lim_{n} \sum_{i=1}^{\infty} n_i \,\mathrm{P}\{n_i < XY_{m,j} \le n_{i+1}\} \\
&= \lim_{n} \sum_{i=1}^{\infty} n_i \,\mathrm{P}\{n_i < X \le n_{i+1}, m_j < Y \le m_{j+1}\} \\
&= \lim_{n\to\infty} \sum_{i=1}^{\infty} n_i \,\mathrm{P}\{n_i < X \le n_{i+1}\} \cdot \mathrm{P}\{m_j < Y \le m_{j+1}\} \\
&= \mathrm{P}\{m_j < Y \le m_{j+1}\} \cdot \mathrm{E}\, X.
\end{aligned}
$$

By the monotone convergence theorem

$$
\begin{aligned}
\mathrm{E}\, XY = \lim_m \mathrm{E}\, XY_m &= \lim_m \mathrm{E} \sum_{j=1}^{\infty} m_j XY_{m,j} = \lim_m \sum_{j=1}^{\infty} m_j \,\mathrm{E}\, XY_{m,j} \\
&= \lim_m \sum_{j=1}^{\infty} m_j \,\mathrm{P}\{m_j < Y \le m_{j+1}\} \mathrm{E}\, X = \mathrm{E}\, Y \cdot \mathrm{E}\, X.
\end{aligned}
$$

Thus, in the general case, recalling Exercise 3.1.5,

$$
\mathrm{E}|XY| = \mathrm{E}|X| \cdot \mathrm{E}|Y| \quad \text{and} \quad \mathrm{E}\, X^{\pm} Y^{\pm} = \mathrm{E}\, X^{\pm} \cdot \mathrm{E}\, Y^{\pm}
$$

for all four pairings of positive and negative parts. Consequently,

$$
\begin{aligned}
\mathrm{E}\, X \cdot \mathrm{E}\, Y &= (\mathrm{E}\, X^+ - \mathrm{E}\, X^-)(\mathrm{E}\, Y^+ - \mathrm{E}\, Y^-) \\
&= \mathrm{E}(X^+ - X^-)(Y^+ - Y^-) = \mathrm{E}\, XY. \qquad \square
\end{aligned}
$$

Corollary 5. *If* $\{X_j, 1 \le j \le n\}$ *are uncorrelated* $\mathscr{L}_2$ r.v.s, (*a fortiori, independent* $\mathscr{L}_2$ r.v.s), *then*

$$
\sigma^2\left(\sum_{j=1}^{n} X_j\right) = \sum_{j=1}^{n} \sigma^2(X_j). \tag{23}
$$

PROOF. Since $\mathrm{E} \sum_{j=1}^{n} X_j = \sum_{j=1}^{n} \mathrm{E}\, X_j$ and $\sigma^2(X + c) = \sigma^2(X)$ for every finite constant c, it may be supposed that $\mathrm{E}\, X_j = 0$, $1 \le j \le n$. Then

$$
\sigma^2\left(\sum_{j=1}^{n} X_j\right) = \mathrm{E}\left(\sum_{j=1}^{n} X_j\right)^2 = \sum_{j=1}^{n} \mathrm{E}\, X_j^2 + \sum_{i \ne j} \mathrm{E}\, X_i X_j = \sum_{j=1}^{n} \sigma^2(X_j). \qquad \square
$$

Even though Tchebychev's inequality is an easy consequence of the order-preserving property of expectations, it is quite useful, especially in verifying convergence in probability (Exercise 4).

Let $\{X_n, n \ge 1\}$ be a sequence of r.v.s with $\mathrm{E}|X_n|^\beta \le C < \infty$ for some $\beta \ge 1$ and $S_n = \sum_1^n X_i$. By Hölder's inequality (Exercise 4.3.5) or Jensen's inequality

$$
\mathrm{E}\left|\frac{S_n}{n}\right|^\beta \le \frac{1}{n} \sum_{j=1}^{n} \mathrm{E}|X_j|^\beta \le C,
$$

and so according to Exercise 4.2.6 $\{|S_n/n|^\alpha, n \geq 1\}$ is u.i. for $0 < \alpha < \beta$. If, moreover, $\{|X_n|^\beta, n \geq 1\}$ is u.i., so is $\{|S_n/n|^\beta, n \geq 1\}$, generalizing Exercise 4.2.7.

EXAMPLE 1. If $\{|X_n|^\beta, n \geq 1\}$ is u.i. for some $\beta \geq 1$ and $S_n = \sum_1^n X_i$, then $|S_n/n|^\beta$ is u.i. In particular, if $\{X_n, n > 1\}$ are identically distributed $\mathscr{L}_\beta$ r.v.s for $\beta \geq 1$, then $\{|S_n/n|^\beta, n \geq 1\}$ is u.i.

PROOF. Since the case $\beta = 1$ is obvious (Exercise 4.2.7), suppose that $\beta > 1$. By Hölder's inequality, $|S_n/n|^\beta \leq \frac{1}{n}\sum_{i=1}^n |X_i|^\beta$, whence uniform integrability follows easily. □

EXAMPLE 2. Let $S_n = \sum_{j=1}^n X_j$, $n \geq 1$ where $\{X_n, n \geq 1\}$ are independent r.v.s with $\mathrm{E}\, X_n = 0$, $\mathrm{E}\, X_n^2 = 1$. If $\{X_n^2, n \geq 1\}$ is u.i., then $\{S_n^2/n, n \geq 1\}$ is u.i.

PROOF. Define $Y_n = X_n I_{[|X_n| \leq K]} - \mathrm{E}\, X_n I_{[|X_n| \leq K]}$, $K > 0$ and $Z_n = X_n - Y_n$, $n \geq 1$. Then $\{Y_n, n \geq 1\}$ are independent r.v.s with $\mathrm{E}\, Y_n = 0$ and $\mathrm{E}\, Y_n^2 \leq 1$. The same statement likewise holds for $\{Z_n, n \geq 1\}$. Hence, if $T_n = \sum_{j=1}^n Y_j$, $W_n = \sum_{j=1}^n Z_j$,

$$\mathrm{E}(T_n/n^{1/2})^4 = \frac{1}{n^2}\left(\sum_{j=1}^n \mathrm{E}\, Y_j^4 + 2\sum_{i<j} \mathrm{E}\, Y_i^2 \cdot \mathrm{E}\, Y_j^2\right)$$

$$\leq \frac{1}{n^2}\left[nK^4 + 2\binom{n}{2}\right] \leq K^4 + 1$$

and so $\{(T_n/n^{1/2})^2, n \geq 1\}$ is u.i. by Exercise 4.2.6. Hence, for any $\varepsilon > 0$, there is a $\delta > 0$ such that $\mathrm{P}\{A\} < \delta$ entails

$$\sup_{n \geq 1} \mathrm{E}\left(\frac{T_n^2}{n} I_A\right) < \frac{\varepsilon}{4}.$$

On the other hand, via u.i. of $\{X_n^2, n \geq 1\}$, $K = K_\varepsilon > 0$ may be chosen so that $\sup_{n \geq 1} \mathrm{E}\, X_n^2 I_{[|X_n| > K]} < \varepsilon/4$ whence

$$\mathrm{E}(W_n^2/n) = \frac{1}{n}\sum_{j=1}^n \mathrm{E}\, Z_j^2 \leq \frac{1}{n}\sum_{j=1}^n \mathrm{E}\, X_j^2 I_{[|X_j| > K]} < \frac{\varepsilon}{4}.$$

Consequently, for $\mathrm{P}\{A\} < \delta$ and all $n \geq 1$,

$$\mathrm{E}\left(\frac{S_n^2}{n} I_A\right) \leq 2\,\mathrm{E}\left(\frac{T_n^2 + W_n^2}{n} I_A\right) < \varepsilon.$$

Clearly, $\mathrm{E}(S_n^2/n) = 1$, $n \geq 1$ and so $\{S_n^2/n, n \geq 1\}$ and a fortiori $\{S_n/n^{1/2}, n \geq 1\}$ is u.i. □

A generalization of Example 2 appears in Corollary 11.3.2.

EXAMPLE 3. If $\{Y_n, n \geq 1\}$ are non-negative r.v.s satisfying $\lim_{n\to\infty} \mathrm{E}\, Y_n = 1 = \lim_{n\to\infty} \mathrm{E}\, Y_n^p$ for some $p > 1$ then $Y_n \xrightarrow{\mathscr{L}_p} 1$.

Proof. Let $r = \min(1, p/2)$. If $p < 2$, then $r < 1 < p$ and so by Exercise 4.3.8

$$\left(\varliminf_{n\to\infty} \mathrm{E}\, Y_n^r\right)^{(p-1)/(p-r)} \geq \varliminf_{n\to\infty} \frac{\mathrm{E}\, Y_n}{(\mathrm{E}\, Y_n^p)^{(1-r)/(p-r)}} = 1$$

implying

$$0 \leq \varlimsup_{n\to\infty} \sigma^2_{Y_n^r} = \varlimsup_{n\to\infty} \mathrm{E}\,(Y_n^{2r} - \mathrm{E}^2\, Y_n^r) \leq 0. \tag{24}$$

Thus, $\sigma^2_{Y_n^r} \to 0$ and $\mathrm{E}\, Y_n^r \to 1$ so that $Y_n^r \xrightarrow{\mathrm{P}} 1$ and hence $Y_n^p \xrightarrow{\mathrm{P}} 1$. Since $\mathrm{E}\, Y_n^p \to 1$, $\{Y_n^p, n \geq 1\}$ is u.i. which, in turn, ensures $\mathrm{E}\,|Y_n - 1|^p \to 0$.

If rather $p \geq 2$, then $r = 1$ and $\mathrm{E}\, Y_n^{2r} = \mathrm{E}\, Y_n^2 \to 1$ since $(\mathrm{E}\, Y_n^q)^{1/q}$ is an increasing function of q. Thus, (24) holds with equality throughout and the conclusion follows as before. □

Exercises 4.3

1. (i) Show for any $\mathscr{L}_1$ r.v. X that $\sigma_X^2 = \mathrm{E}\, X^2 - (\mathrm{E}\, X)^2$. (ii) If $\mathrm{P}\{X = 1\} = p \in (0, 1)$ and $\mathrm{P}\{X = 0\} = \mathrm{P}\{X = 2\} = (1 - p)/2$, then, setting $Z = XI_{[X>1]}$, $W = XI_{[X\leq 1]}$, necessarily $\sigma_W^2 < \sigma_X^2 < \sigma_Z^2$.

2. Calculate the mean and variance of U_n, the number of empty cells in the random assignment of r balls into n cells.

3. Prove (i) for any r.v.s $\{S_n, n \geq 1\}$ with $\sigma^2(S_n) = o(n^2)$ that $n^{-1}(S_n - \mathrm{E}\, S_n) \xrightarrow{\mathrm{P}} 0$, (ii) for any r.v. X and positive numbers a, t that necessarily $\mathrm{P}\{X \geq a\} \leq e^{-at}\, \mathrm{E}\, e^{tX}$ (iii) if m is a median of $X \in \mathscr{L}_1$, then $|m - \mathrm{E}\, X| \leq \mathrm{E}|X - \mathrm{E}\, X|$ and hence $|m - \mathrm{E}\, X| \leq \sigma$ for $X \in \mathscr{L}_2$ with variance σ^2.
   (iv) for $\mathscr{L}_1$ r.v.s. X and Y with $\mathrm{E}\, XY$ finite, their **covariance**, denoted $\mathrm{Cov}[X, Y]$, is defined by $\mathrm{Cov}[X, Y] = \mathrm{E}[X - \mathrm{E}\, X][Y - \mathrm{E}\, Y]$. Prove that if f and g are non-decreasing functions and X is an r.v. with $\mathrm{E}\, f(X)$, $\mathrm{E}\, g(X)$, $\mathrm{E}\, f(X)\cdot g(X)$ finite, then $\mathrm{Cov}[f(X), g(X)] \geq 0$.
   (v) for any $\mathscr{L}_2$ r.v.s. X_j, $1 \leq j \leq n$, check that

$$\sigma^2\left(\sum_{j=1}^n X_j\right) = \sum_{j=1}^n \sigma^2(X_j) + 2 \sum_{1\leq i<j\leq n} \mathrm{Cov}(X_i, X_j).$$

4. For arbitrary real numbers a_i, b_i, $1 \leq i \leq n$, prove that

$$\sum_1^n |a_i b_i| \leq \left(\sum_1^n |a_i|^p\right)^{1/p} \left(\sum_1^n |b_i|^q\right)^{1/q},$$

   provided $(1/p) + (1/q) = 1$, $p > 0$, $q > 0$. *Hint*: Apply Hölder's inequality to suitable r.v.s X, Y.

5. If $S_n = \sum_{j=1}^n X_j$ where $\{X_j, j \geq 1\}$ are independent r.v.s with $\mathrm{E}\, X_j = 0$, $\mathrm{E}\, X_j^2 = \sigma_j^2$, $s_n^2 = \sum_{j=1}^n \sigma_j^2$, then

$$\mathrm{P}\{S_n \geq x\} \geq \left(1 - \frac{2s_n^2}{x^2}\right) \sum_{j=1}^n \mathrm{P}\{X_j \geq 2x\}, \quad x > 0.$$

   Hint: $\mathrm{P}\{S_n \geq x,\ X_j \geq x,\ \max_{i<j} X_i < x\} \geq \mathrm{P}\{X_j \geq 2x\}\cdot \mathrm{P}\{S_n - X_j \geq -x, \max_{i<j} X_i < x\}$.

6. (i) If X is an $\mathscr{L}_1$ r.v., $Y = Y(a, b)$ is as in Corollary 2, and $Z = Y(a', b')$, where $a \le a' < b' \le b$, then $\mathrm{E}|Y - \mathrm{E}\,Y|^p \ge \mathrm{E}|Z - \mathrm{E}\,Z|^p$ for any $p \ge 1$. (ii) For $g(t), \alpha(t)$ as in Theorem 1, $\mathrm{E}\,g(X) \ge \mathrm{E}\,g(\alpha(X) - \mathrm{E}\,\alpha(X) + \mathrm{E}\,X) \ge g(\mathrm{E}\,X)$ and $\mathrm{E}\,g(X) \ge \mathrm{E}\,g(cX + (1 - c)E\,X) \ge g(\mathrm{E}\,X), 0 \le c \le 1$.

7. (i) If $X' = aX + b$, $Y' = cX + d$, verify that $\rho(X', Y') = \pm\rho(X, Y)$ according as $ac > 0$ or $ac < 0$. (ii) if X, Y are r.v.s with $0 < \sigma(X), \sigma(Y) < \infty$, then $\rho(X, Y) = 1$ iff $(X - \mathrm{E}\,X)/\sigma(X) = (Y - \mathrm{E}\,Y)/\sigma(Y)$, a.c. (iii) if $X = \sin Z$, $Y = \cos Z$, where $\mathrm{P}\{Z = \pm 1\} = \frac{1}{4} = \mathrm{P}\{Z = \pm 2\}$, then $\rho(X, Y) = 0$ despite X, Y being functionally related.

8. Verify for $0 < a < b < d$ and any nonnegative r.v. Y that

$$\mathrm{E}\,Y^b \le (\mathrm{E}\,Y^a)^{(d-b)/(d-a)}(\mathrm{E}\,Y^d)^{(b-a)/(d-a)}.$$

Utilize this to conclude for any positive r.v.s $\{Y_n, n \ge 1\}$ and positive constants $\{c_n, n \ge 1\}$ that if $\sum_{n=1}^{\infty} c_n^\alpha\,\mathrm{E}\,Y_n^\alpha < \infty$ for $\alpha = \alpha_1$ and α_2, where $\alpha_i > 0, i = 1, 2$, then the series converges for all α in $[\alpha_1, \alpha_2]$.

9. The **moment-generating function** (m.g.f.) of a r.v. X is the function $\varphi(h) = \mathrm{E}\,e^{hX}$. If $\varphi(h)$ is finite for $h = h_0 > 0$ and $h = -h_0$, verify that (i) $\mathrm{E}\,e^{h|X|} \le \varphi(h) + \varphi(-h) < \infty, 0 < h \le h_0$ (ii) $\log \varphi(h)$ is convex in $[-h_0, h_0]$ and strictly convex if X is non-degenerate (iii) if $\mathrm{E}\,X = 0$, then $\varphi(h) \ge 1$ for $|h| \le h_0$. (iv) The kth moment of X is finite for every positive integer k and equals $\varphi^{(k)}(0)$, (v) if X_1 and X_2 are independent r.v.s with (finite) m.g.f.s $\varphi_1(h), \varphi_2(h)$ (say) for $|h| < h_0$, then the m.g.f. of $X_1 + X_2$ is $\varphi_1(h) \cdot \varphi_2(h)$ in this interval, (vi) if $\varphi(h)$ is finite in $[0, h_0]$, all moments of X^+ are finite.

10. Prove **Minkowski's inequality**, that is, if $X_1 \in \mathscr{L}_p$, $X_2 \in \mathscr{L}_p$, then

$$\|X_1 + X_2\|_p \le \|X_1\|_p + \|X_2\|_p, \qquad p \ge 1.$$

Hint: Apply Hölder's inequality to $\mathrm{E}|X_i||X_1 + X_2|^{p-1}$ when $p > 1$.

11. Let $\{X_n, n \ge 1\}$ be r.v.s with (i) $\mathrm{E}|X_n|^2 \le 1$ and (ii) $\mathrm{E}((X_1 + \cdots + X_n)/n)^2 = O(n^{-\alpha})$ for some $\alpha > 0$. Then $(1/n)\sum_{j=1}^n X_j \xrightarrow{\text{a.c.}} 0$. Note that (i) $\Rightarrow$ (ii) if $\mathrm{E}\,X_i X_j = 0, i \ne j$. *Hint*: Choose n_m = smallest integer $\ge m^{2/\alpha}$, $m = 1, 2, \ldots$ Then $(1/n_m)\sum_1^{n_m} X_j \xrightarrow{\text{a.c.}} 0$ as $m \to \infty$ and

$$\mathrm{E}\left\{\max_{1 \le k < n_{m+1} - n_m}\left|\frac{1}{n_m + k}\sum_1^k X_{n_m + j}\right|^2\right\} \le \mathrm{E}\,\frac{n_{m+1} - n_m}{n_m^2}\sum_1^{n_{m+1} - n_m} X_{n_m + j}^2 = O(m^{-2}).$$

12. For any sequence $\{A_n, n \ge 1\}$ of events, define

$$Y_n = \frac{1}{n}\sum_1^n (I_{A_j} - \mathrm{P}\{A_j\}), \qquad p_1(n) = \frac{1}{n}\sum_1^n \mathrm{P}\{A_j\},$$

$$p_2(n) = \frac{2}{n(n-1)}\sum_{1 \le j < k \le n} \mathrm{P}\{A_j A_k\}, \qquad d_n = p_2(n) - p_1^2(n).$$

Prove that (i) $\mathrm{E}\,Y_n^2 = d_n + n^{-1}[p_1(n) - p_2(n)]$, (ii) $Y_n \xrightarrow{P} 0$ iff $\mathrm{E}\,Y_n^2 \to 0$ iff $d_n \to 0$, (iii) $Y_n \xrightarrow{\text{a.c.}} O$ if $d_n = 0(n^{-\alpha})$ for some $\alpha > 0$. *Hint*: Utilize Exercise 11.

13. For events $\{A_n, n \ge 1\}$ and $\varepsilon > 0$, there exist distinct indices j, k in $[1, n]$, where $n > 1/\varepsilon$, such that $\mathrm{P}\{A_j A_k\} \ge p_1^2(n) - \varepsilon$, where $p_1(n)$ is as in Exercise 12. *Hint*: $p_2(n) \ge p_1^2(n) - 1/n$.

14. **Bernstein's Inequality.** If $S_n = \sum_{j=1}^n X_j$ where $\{X_j, 1 \le j \le n\}$ are independent r.v.s with $E\, X_j = 0$, $E\, X_j^2 = \sigma_j^2$, $s_n^2 = \sum_{j=1}^n \sigma_j^2 > 0$ which satisfy (i) $E|X_j|^k \le (k!/2)\sigma_j^2 c^{k-2}$ for $k > 2$, $1 \le j \le n$, $0 < c < \infty$ (a fortiori if (i)' $P\{|X_j| \le c\} = 1$, $1 \le j \le n$), then

$$P\{S_n > x\} \le \exp\left\{\frac{-x^2}{2(s_n^2 + cx)}\right\}, \qquad x > 0.$$

Hint: $\exp\{tX_j\} \le 1 + tX_j + \sum_{k=2}^\infty t^k |X_j|^k/k!$ valid for $t > 0$ implies $E\, e^{tX_j} \le \exp\cdot \{\sigma_j^2 t^2/2(1 - tc)\}$, $0 < tc < 1$ and choosing $t = x/(s_n^2 + cx)$ yields $E \exp\{xS_n/(s_n^2 + cx)\} \le \exp\{x^2/2(s_n^2 + cx)\}$, $x > 0$. Now apply Exercise 4.3.3(ii).

15. If $S_n = \sum_{j=1}^n X_j$ where $\{X_j, 1 \le j \le n\}$ are independent r.v.s with $E\, X_j = 0$, $E\, X_j^2 = \sigma_j^2$, $s_n^2 = \sum_{j=1}^n \sigma_j^2 > 0$, then for $\lambda > \gamma > 1$,

$$P\left\{\max_{1 \le j \le n} |S_j| \ge \lambda s_n\right\} \le \frac{\gamma^2}{\gamma^2 - 1} P\{|S_n| \ge (\lambda - \gamma)s_n\}.$$

Hint: If $T = \inf\{1 \le j \le n: |S_j| \ge \lambda s_n\}$ and $T = n + 1$ otherwise, then for $1 < \gamma < \lambda$

$$P\left\{\max_{1 \le j \le n} |S_j| \ge \lambda s_n\right\} \le P\{|S_n| \ge (\lambda - \gamma)s_n\} + \sum_{j=1}^{n-1} P\{T = j\} \cdot P\{|S_n - S_j| \ge \gamma s_n\}.$$

Now apply Tchebychev's inequality.

16. If $\{S_n, X_n, n \ge 1\}$ are as in exercise 15 and $c_n \to \infty$, then $(1/c_n s_n)\max_{1 \le j \le n} |S_j| \xrightarrow{P} 0$.

17. For $X, Y \in \mathscr{L}_p$, $p > 1$, define $d(X, Y) = \|X - Y\|_p$ and show that d has all the attributes of a metric except one. For $X, Y \in \mathscr{L}_p$ write $X \sim Y$ if $X = Y$, a.c., and let $\mathscr{L}_p^*$ be the space of equivalence classes of $\mathscr{L}_p$. Show that $\mathscr{L}_p^*$ is a Banach space, i.e., a normed, linear, complete space. When $p = 2$, $\mathscr{L}_p^*$ is a Hilbert space under the inner product $(X, Y) = E\, XY$.

18. Prove that if $S_n = \sum_{i=1}^n X_i$, where $\{X_i, i \ge 1\}$ are independent r.v.'s with $E\, X_i = 0$, $1 \le i \le n$, then for any $\varepsilon > 0$,

$$P\left\{\max_{1 \le j \le n} S_j > \varepsilon\right\} \le 2P\{S_n \ge \varepsilon - E|S_n|\}. \tag{24}$$

Hint. Use the Feller–Chung lemma with $A_j = \{S_j > \varepsilon\}$, $B_j = \{S_n - S_j > -E|S_n|\}$.

References

Y. S. Chow and W. J. Studden, "Monotonicity of the variance under truncation and variations of Jensen's inequality," *Ann. Math. Stat.* **40** (1969), 1106–1108.

K. L. Chung and W. H. J. Fuchs, "On the distribution of values of sums of random variables," *Mem. Amer. Soc.* **6** (1951).

J. L. Doob, *Stochastic Processes*, Wiley, New York, 1953.

P. R. Halmos, *Measure Theory*, Van Nostrand, Princeton, 1950; Springer-Verlag, Berlin and New York, 1974.

P. Hall, "On the $\mathscr{L}_p$ convergence of random variables," *Proc. Cambridge Philos. Soc.* **82** (1977). 439–446.

G. H. Hardy, J. E. Littlewood, and G. Polya, *Inequalities*, Cambridge Univ. Press, London, 1934.

S. B. Kochen and C. J. Stone, "A note on the Borel–Cantelli lemma," *Ill. Jour. Math.* **8** (1964), 248–251.
A. Liapounov, "Nouvelle forme du théoreme sur la limite de probabilité," *Mem. Acad. Sc. St. Petersbourg* **12** (1905), No. 5.
M. Loève, *Probability Theory*, 3rd ed., Van Nostrand, Princeton, 1963; 4th ed., Springer-Verlag, Berlin and New York, 1977–1978.
G. Polya, "Uber eine Aufgabe der Wahrscheinlichkeitsrechnung betreffend die Irrfahrt im Strassennetz," *Math. Ann.* **84** (1921), 149–160.
S. Saks, *Theory of the Integral* (L. C. Young, translation), Stechert–Hafner, New York, 1937.
H. Teicher, "On the law of the iterated logarithm," *Ann. Prob.* **2** (1974), 714–728.

5 Sums of Independent Random Variables

Of paramount concern in probability theory is the behavior of sums $\{S_n, n \geq 1\}$ of independent random variables $\{X_i, i \geq 1\}$. The case where the $\{X_i\}$ are i.i.d. is of especial interest and frequently lends itself to more incisive results. The sequence of sums $\{S_n, n \geq 1\}$ of i.i.d. r.v.s $\{X_n\}$ is alluded to as a random walk; in the particular case when the component r.v.s $\{X_n\}$ are nonnegative, the random walk is referred to as a renewal process.

5.1 Three Series Theorem

The first question to be dealt with apropos of sums of independent r.v.s is when such sums converge a.c. A partial answer is given next, and the ensuing lemmas culminate in the Kolmogorov three series theorem.

Theorem 1 (Khintchine–Kolmogorov Convergence Theorem). *Let* $\{X_n, n \geq 1\}$ *be independent* $\mathscr{L}_2$ r.v.s *with* $\mathrm{E}\, X_n = 0$, $n \geq 1$. *If* $\sum_{j=1}^{\infty} \mathrm{E}\, X_j^2 < \infty$, *then* $\sum_1^{\infty} X_j$ *converges* a.c. *and in quadratic mean and, moreover,* $\mathrm{E}(\sum_1^{\infty} X_j)^2 = \sum_{j=1}^{\infty} \mathrm{E}\, X_j^2$.

PROOF. If $S_n = \sum_1^n X_j$, by Corollary 4.3.5

$$\mathrm{E}(S_m - S_n)^2 = \mathrm{E}\left(\sum_{n+1}^{m} X_j\right)^2 = \sum_{n+1}^{m} \mathrm{E}\, X_j^2 \to 0 \tag{1}$$

as $m > n \to \infty$, whence, according to Theorem 4.2.3, $S_n \xrightarrow{\mathscr{L}_2}$ some r.v. S, denoted by $S = \sum_1^{\infty} X_j$. *A fortiori*, $S_n \xrightarrow{\mathrm{P}} \sum_1^{\infty} X_j$ and so by Lévy's theorem

(Theorem 3.3.1) $S_n \xrightarrow{\text{a.c.}} \sum_1^\infty X_j$. The remainder follows from

$$\mathrm{E}\left(\sum_1^\infty X_j\right)^2 = \lim_{n\to\infty} \mathrm{E}\, S_n^2 = \lim_{n\to\infty} \sum_{j=1}^n \mathrm{E}\, X_j^2 = \sum_1^\infty \mathrm{E}\, X_j^2 \tag{2}$$

via Corollaries 4.2.5 and 4.3.5. □

The first lemma involves "summation by parts" and is as useful as its integral counterpart.

Lemma 1 (Abel). *If $\{a_n\}$, $\{b_n\}$ are sequences of real numbers and $A_n = \sum_{j=0}^n a_j$, $n \ge 0$, then for $n \ge 1$*

$$\sum_{j=1}^n a_j b_j = A_n b_n - A_0 b_1 - \sum_{j=1}^{n-1} A_j(b_{j+1} - b_j); \tag{3}$$

if $\sum_{j=1}^\infty a_j$ converges and $A_n^* = \sum_{j=n+1}^\infty a_j$, then for $n \ge 1$

$$\sum_{j=1}^n a_j b_j = A_0^* b_1 - A_n^* b_n + \sum_{j=1}^{n-1} A_j^*(b_{j+1} - b_j); \tag{4}$$

moreover, if $a_n \ge 0$, $b_{n+1} \ge b_n \ge 0$, $A_n^* = \sum_{j=n+1}^\infty a_j < \infty$, then

$$\sum_{j=1}^\infty a_j b_j = A_0^* b_1 + \sum_{j=1}^\infty A_j^*(b_{j+1} - b_j). \tag{5}$$

PROOF.

$$\sum_1^n a_j b_j = \sum_1^n (A_j - A_{j-1}) b_j = \sum_1^n A_j b_j - \sum_1^n A_{j-1} b_j,$$

yielding (3). Take $a_0 = -\sum_{j=1}^\infty a_j = -A_0^*$ in (3) to obtain (4). Next, assuming $a_n \ge 0$, $b_{n+1} \ge b_n \ge 0$, if $\overline{\lim}_n A_n^* b_n > 0$, then $\sum_{n+1}^\infty a_j b_j \ge A_n^* b_n$ implies $\sum_1^\infty a_j b_j = \infty$. By (4), $A_0^* b_1 + \sum_1^\infty A_j^*(b_{j+1} - b_j) \ge \sum_1^\infty a_j b_j = \infty$, so that (5) holds. If, rather, $\overline{\lim}\, A_n^* b_n = 0$, then (5) obtains by letting $n \to \infty$ in (4). □

The following **Kronecker lemma** is a *sine qua non* for probability theory as will become apparent in the next section.

Lemma 2 (Kronecker). *If $\{a_n\}$, $\{b_n\}$ are sequences of real numbers with $0 < b_n \uparrow \infty$, $\sum_1^\infty (a_j/b_j)$ converging, then*

$$\frac{1}{b_n} \sum_{j=1}^n a_j \to 0. \tag{6}$$

PROOF. This will be demonstrated in the alternative equivalent form that convergence of $\sum_1^\infty a_j$ entails $\sum_1^n a_j b_j = o(b_n)$. By (4),

$$\frac{1}{b_n} \sum_{j=1}^n a_j b_j = -A_n^* + A_0^* \frac{b_1}{b_n} + \frac{1}{b_n} \sum_1^{n-1} A_j^*(b_{j+1} - b_j). \tag{7}$$

For any $\varepsilon > 0$, choose the integer m so that $|A_j^*| < \varepsilon$ for $j \geq m$. Then

$$-\varepsilon \leq \varliminf_n \frac{1}{b_n} \sum_m^{n-1} A_j^*(b_{j+1} - b_j) \leq \varlimsup_n \frac{1}{b_n} \sum_m^{n-1} A_j^*(b_{j+1} - b_j) \leq \varepsilon,$$

whence from (7)

$$-\varepsilon \leq \varliminf_n \frac{1}{b_n} \sum_{j=1}^n a_j b_j \leq \varlimsup_n \frac{1}{b_n} \sum_{j=1}^n a_j b_j \leq \varepsilon$$

and $(1/b_n) \sum_{j=1}^n a_j b_j \to 0$. □

Lemma 3. *Let* $\{X_n\}$ *be independent* r.v.s *with* $\mathrm{E}\, X_n = 0$, $S_n = \sum_1^n X_i$, $n \geq 1$, *and* $\mathrm{E} \sup_n X_n^2 < \infty$. *If* $\mathrm{P}\{\sup_n |S_n| < \infty\} > 0$, *then* S_n *converges* a.c. *and in quadratic mean and*

$$\sum_1^\infty \mathrm{E}\, X_n^2 < \infty. \tag{8}$$

Proof. It suffices to prove (8) since Theorem 1 guarantees the rest. To this end, set $Z^2 = \sup X_n^2$ and choose $K > 0$ sufficiently large so that

$$\mathrm{P}\left\{\sup_n |S_n| < K\right\} > 0.$$

Define $T = \inf\{n: |S_n| \geq K\}$ and note that T is positive integer valued with $\mathrm{P}\{T = \infty\} > 0$. Since $\{T \geq n\} = \{|S_j| < K, 1 \leq j < n\} \in \sigma(X_1, \ldots, X_{n-1})$ for $n \geq 2$, the r.v.s X_n and $I_{[T \geq n]}$ are independent for $n \geq 1$. Let $U_n = \sum_{j=1}^n X_j I_{[T \geq j]}$ and observe that U_n is $\sigma(X_1 \cdots X_n)$-measurable and

$$\begin{aligned} U_n &= S_{\min[T,n]}, \\ U_n^2 &= |S_{\min(T-1,n-1)} + X_{\min(T,n)}|^2 \leq 2(K^2 + Z^2), \\ \mathrm{E}\, U_n^2 &\leq 2(K^2 + \mathrm{E}\, Z^2) = C < \infty. \end{aligned} \tag{9}$$

Now, setting $U_0 = 0$, for $j \geq 1$

$$U_j^2 = U_{j-1}^2 + 2U_{j-1} X_j I_{[T \geq j]} + X_j^2 I_{[T \geq j]},$$

and so by independence and Theorem 4.3.3

$$\mathrm{E}\, U_j^2 - \mathrm{E}\, U_{j-1}^2 = \mathrm{P}\{T \geq j\} \mathrm{E}\, X_j^2.$$

Summing over $1 \leq j \leq n$,

$$C \geq \mathrm{E}\, U_n^2 = \sum_{j=1}^n \mathrm{P}\{T \geq j\} \mathrm{E}\, X_j^2 \geq \mathrm{P}\{T = \infty\} \sum_{j=1}^n \mathrm{E}\, X_j^2,$$

which yields (8) when $n \to \infty$. □

Lemma 4. *If* $\{X_n\}$ *are independent* r.v.s *with* $\mathrm{E} \sup_n |X_n| < \infty$ *and* $S_n = \sum_1^n X_i$ *converges* a.c., *then* $\sum_1^\infty \mathrm{E}\, X_n$ *converges.*

PROOF. Define K, T, U_n as in Lemma 3, which is permissible since $S_n \xrightarrow{\text{a.c.}}$ some r.v. S. Now, $\min[T, n] \to T$ and so (9) ensures that

$$U_n = S_{\min[T,n]} \xrightarrow{\text{a.c.}} S_T, \tag{10}$$

where $S_T = S_n$ on $\{T = n\}, n \geq 1$, and $S_T = S$ on $\{T = \infty\}$. A more extensive discussion of S_T occurs in Section 3 of this chapter, where it is pointed out that S_T is a bona fide r.v. As in (9),

$$\mathrm{E}\sup|U_n| \leq K + \mathrm{E}\sup_n|X_n| < \infty,$$

which, in conjunction with (10) and the Lebesgue dominated convergence theorem, ensures

$$\lim_{n\to\infty} \mathrm{E}\,U_n = \mathrm{E}\,S_T, \qquad \text{finite.} \tag{11}$$

By the independence observed in Lemma 3,

$$\mathrm{E}\,U_n = \mathrm{E}\sum_{j=1}^{n} X_j I_{[T\geq j]} = \sum_{1}^{n} \mathrm{P}\{T \geq j\}\mathrm{E}\,X_j,$$

whence

$$\mathrm{E}\,X_n = \frac{\mathrm{E}\,U_n - \mathrm{E}\,U_{n-1}}{\mathrm{P}\{T \geq n\}}.$$

Employing Lemma 1 with $b_j = 1/\mathrm{P}\{T \geq j\}$, $a_j = \mathrm{E}\,U_j - \mathrm{E}\,U_{j-1}$, $j \geq 1$, $a_0 = A_0 = 0$,

$$\sum_{j=1}^{n} \mathrm{E}\,X_j = \frac{\mathrm{E}\,U_n}{\mathrm{P}\{T \geq n\}} - \sum_{j=1}^{n-1}\left[\frac{1}{\mathrm{P}\{T > j\}} - \frac{1}{\mathrm{P}\{T \geq j\}}\right]\mathrm{E}\,U_j$$

and so, recalling (11) and $\mathrm{P}\{T = \infty\} > 0$, $\sum_1^\infty \mathrm{E}\,X_j$ converges. □

Corollary 1. *If* $\{X_n, n \geq 1\}$ *are independent* r.v.s *which are* **uniformly bounded**, *i.e.*, $\mathrm{P}\{|X_n| \leq C < \infty, n \geq 1\} = 1$, *and moreover*, $S_n = \sum_1^n X_i \xrightarrow{\text{a.c.}} S$ *finite, then* $\sum_1^\infty \mathrm{E}\,X_j$ *and* $\sum_{j=1}^\infty \sigma^2(X_j)$ *converge.*

PROOF. The series of means converges by Lemma 4, whence $\sum_1^\infty (X_i - \mathrm{E}\,X_i)$ converges a.c. and Lemma 3 applies. □

Definition. Two sequences of r.v.s $\{X_n\}$ and $\{Y_n\}$ will be called **equivalent** if $\sum_1^\infty \mathrm{P}\{X_n \neq Y_n\} < \infty$.

If $\{X_n\}$, $\{Y_n\}$ are equivalent sequences of r.v.s, the Borel–Cantelli lemma ensures that $\mathrm{P}\{X_n \neq Y_n, \text{ i.o.}\} = 0$. Hence, $\mathrm{P}\{\sum X_i \text{ converges}\} = 1$ iff $\mathrm{P}\{\sum Y_i \text{ converges}\} = 1$.

The way is now paved for presentation of

Theorem 2 (Kolmogorov Three Series Theorem). *If* $\{X_n\}$ *are independent* r.v.s, *then* $\sum_1^\infty X_i$ *converges* a.c. iff

i. $\sum_1^\infty P\{|X_n| > 1\} < \infty$,
ii. $\sum_1^\infty E\,X'_n$ *converges,*
iii. $\sum_1^\infty \sigma^2_{X'_n} < \infty$,

where $X'_n = X_n I_{[|X_n| \le 1]}$, $n \ge 1$.

PROOF. Sufficiency: If the three series converge, then $\sum_1^\infty (X'_n - E\,X'_n)$ converges a.c. by Theorem 1, whence (ii) implies that $\sum_1^\infty X'_n$ converges a.c. According to (i), $\{X_n\}$, $\{X'_n\}$ are equivalent sequences of r.v.s and so $\sum X_i$ converges a.c.

Conversely, if $\sum X_n$ converges a.c., then $X_n \xrightarrow{\text{a.c.}} 0$, implying $P\{|X_n| > 1$, i.o.$\} = 0$, whence (i) holds by the Borel–Cantelli theorem (Theorem 3.1.1). Also, $\{X_n\}$, $\{X'_n\}$ are equivalent sequences, so that necessarily $\sum_1^\infty X'_n$ converges a.c. The remaining series, (ii) and (iii), now converge by Corollary 1. □

Corollary 2. *If* $\{X_n\}$ *are independent* r.v.s *satisfying* $E\,X_n = 0$, $n \ge 1$, *and*

$$\sum_1^\infty E[X_n^2 I_{[|X_n| \le 1]} + |X_n| I_{[|X_n| > 1]}] < \infty, \tag{12}$$

then $\sum_1^\infty X_n$ *converges* a.c.

PROOF. Since $E\,X_n = 0$,

$$\sum_1^\infty |E\,X_n I_{[|X_n| \le 1]}| = \sum_1^\infty |E\,X_n I_{[|X_n| > 1]}| \le \sum_1^\infty E|X_n| I_{[|X_n| > 1]} < \infty.$$

Moreover, by the Markov inequality (Theorem 4.1.1(iii))

$$P\{|X_n| > 1\} = P\{|X_n| I_{[|X_n| > 1]} > 1\} \le E|X_n| I_{[|X_n| > 1]},$$

whence the corollary flows from Theorem 2. □

Corollary 3 (Loève). *If* $\{X_n\}$ *are independent* r.v.s *and for some constants* $0 < \alpha_n \le 2$, $\sum_1^\infty E|X_n|^{\alpha_n} < \infty$, *where* $E\,X_n = 0$ *when* $1 \le \alpha_n \le 2$, *then* $\sum_1^\infty X_n$ *converges* a.c.

PROOF. It suffices to consider separately the cases $1 \le \alpha_n \le 2$, $n \ge 1$, and $0 < \alpha_n < 1$, $n \ge 1$. In the former instance, $X_n^2 I_{[|X_n| \le 1]} + |X_n| I_{[|X_n| > 1]} \le |X_n|^{\alpha_n}$, whence (12) obtains. In the latter,

$$\sum_1^\infty E(X_n^2 + |X_n|) I_{[|X_n| \le 1]} \le 2 \sum_1^\infty E|X_n|^{\alpha_n} < \infty,$$

and in both cases

$$\sum_1^\infty P[|X_n| \ge 1] \le \sum_1^\infty E|X_n|^{\alpha_n} < \infty,$$

so that the three series of Theorem 2 converge. □

In much of probability theory, integration is with respect to a single probability measure P, and so it seems natural to write $\int_A X$ as an abbreviation for $\int_A X\, dP$. Abundant use will be made of this concise notation.

Turning to the i.i.d. case:

Theorem 3 (Marcinkiewicz–Zygmund). *If* $\{X_n\}$ *are* i.i.d. *with* $E|X_1|^p < \infty$ *for some p in* (0, 2), *then*

$$\sum_1^\infty \left(\frac{X_n}{n^{1/p}} - E\, Y_n\right)$$

converges a.c., *where*

$$Y_n = n^{-1/p} X_n I_{[|X_n| \le n^{1/p}]}.$$

Furthermore, if either (i) $0 < p < 1$ *or* (ii) $1 < p < 2$ *and* $E\, X_1 = 0$, *then* $\sum_1^\infty X_n/n^{1/p}$ *converges* a.c.

Proof. Set $A_j = \{(j-1)^{1/p} < |X_1| \le j^{1/p}\}, j \ge 1$. Then for $\alpha > p > 0$

$$\sum_1^\infty E|Y_n|^\alpha = \sum_{n=1}^\infty \sum_{j=1}^n n^{-\alpha/p} \int_{A_j} |X_1|^\alpha = \sum_{j=1}^\infty \sum_{n=j}^\infty n^{-\alpha/p} \int_{A_j} |X_1|^\alpha$$

$$\le \sum_{j=1}^\infty \left(j^{-\alpha/p} + \frac{p}{\alpha - p} j^{(p-\alpha)/p}\right) \int_{A_j} |X_1|^\alpha$$

$$\le \sum_{j=1}^\infty \left(\frac{1}{j} + \frac{p}{\alpha - p}\right) \int_{A_j} |X_1|^p \le \frac{\alpha}{\alpha - p} E|X_1|^p < \infty, \tag{13}$$

whence ($\alpha = 2$) $\sum_1^\infty (Y_n - E\, Y_n)$ converges a.c. by Theorem 1. Since, recalling Corollary 4.1.3,

$$\sum_1^\infty P\left\{\frac{X_n}{n^{1/p}} \ne Y_n\right\} = \sum_1^\infty P\{|X_1| > n^{1/p}\} \le E|X_1|^p < \infty,$$

the sequences $\{X_n/n^{1/p}\}$, $\{Y_n\}$ are equivalent, and so $\sum_1^\infty (X_n/n^{1/p} - E\, Y_n)$ converges a.c.

In case (i), where $0 < p < 1$, $\sum_{n=1}^\infty |E Y_n| < \infty$ via (13) with $\alpha = 1$, and this same series converges in case (ii), $1 < p < 2$ and $E\, X_1 = 0$, since

$$\sum_1^\infty |E\, Y_n| \le \sum_1^\infty n^{-1/p} \int_{[|X_n| > n^{1/p}]} |X_n| = \sum_{n=1}^\infty \sum_{j=n+1}^\infty n^{-1/p} \int_{A_j} |X_1|$$

$$= \sum_{j=2}^\infty \sum_{n=1}^{j-1} n^{-1/p} \int_{A_j} |X_1| \le \frac{p}{p-1} \sum_{j=1}^\infty (j-1)^{(p-1)/p} \int_{A_j} |X_1|$$

$$\le \frac{p}{p-1} \sum_1^\infty \int_{A_j} |X_1|^p = \frac{p}{p-1} E|X_1|^p < \infty.$$

Thus, the second part of Theorem 3 follows from the first. □

EXAMPLE 1. Let $\{X_n, n \geq 1\}$ be i.i.d. r.v.s with $E|X_1| < \infty$. If $\{a_n, n \geq 1\}$ are real numbers such that $a_n = O(1/n)$ and $\sum_1^\infty a_n$ converges, then $\sum_{n=1}^\infty a_n X_n$ converges a.c.

PROOF. By considering X_n^+ and X_n^- separately, it may and will be supposed that $X_1 \geq 0$. Set

$$X'_n = X_n I_{[X_n \leq n]}, \qquad Y_n = X'_n - E X'_n.$$

By the Borel–Cantelli theorem, $\{X'_n\}$ and $\{X_n\}$ are equivalent sequences, and from (4) it follows that a convergent series remains convergent when its terms are multiplied by any convergent, monotone sequence (Abel's test). Thus, $E X'_n \uparrow E X_1$ implies $\sum a_n E X'_n$ converges. Consequently, to prove $\sum a_n X_n$ converges a.c., it suffices to prove $\sum a_n Y_n$ converges a.c. By hypothesis, $n^2 a_n^2 \leq A < \infty$, and so, setting $A_j = \{j - 1 < X_1 \leq j\}$,

$$\sum_1^\infty a_n^2 E Y_n^2 \leq \sum_1^\infty a_n^2 E X_n'^2 \leq A \sum_{n=1}^\infty n^{-2} \int_{[X_1 \leq n]} X_1^2$$

$$\leq A \sum_{n=1}^\infty \sum_{j=1}^n n^{-2} \int_{A_j} X_1^2 \leq 2A E X_1,$$

where the last inequality follows via (13) with $\alpha = 2$, $p = 1$. Hence, $\sum a_n Y_n$ converges a.c. by the Khintchine–Kolmogorov convergence theorem. □

EXAMPLE 2. If $\{b_n, n \geq 1\}$ are constants satisfying $0 < b_n \uparrow \infty$ and

$$b_n^2 \sum_{j=n}^\infty b_j^{-2} = O(n) \tag{14}$$

and $\{X, X_n, n \geq 1\}$ are i.i.d. r.v.s with $\sum_{n=1}^\infty P\{|X| > b_n\} < \infty$, then

$$\sum_{n=1}^\infty b_n^{-1}(X_n - E X I_{[|X| \leq b_n]}) \tag{15}$$

converges a.c. Moreover, if $E X = 0$ and

$$b_n \sum_{j=1}^n b_j^{-1} = O(n), \tag{16}$$

then

$$\sum_{n=1}^\infty \frac{X_n}{b_n} \tag{17}$$

converges a.c.

Remark. If $b_n/n \uparrow$ or $b_j/b_n \geq A(j/n)^\delta$ for $j \geq n$, where $\delta > \frac{1}{2}$, $A > 0$, then (14) holds; if $b_j/b_n \geq A(j/n)^\delta$ for $j \leq n$, where $0 < \delta < 1$, $A > 0$, then (16) obtains.

PROOF. Set $b_0 = 0$, $A_n = \{b_{n-1} < |X| \le b_n\}$, and $Y_n = X_n I_{[|X_n| \le b_n]}$. Then

$$\infty > \sum_{n=0}^{\infty} P\{|X_n| > b_n\} = \sum_{n=0}^{\infty} \sum_{j=n+1}^{\infty} P\{A_j\} = \sum_{j=1}^{\infty} j\, P\{A_j\} \tag{18}$$

whence for some C in $(0, \infty)$,

$$\sum_{j=1}^{\infty} b_j^{-2} \sigma^2(Y_j) \le \sum_1^{\infty} b_j^{-2}\, E\, Y_j^2 \le \sum_{j=1}^{\infty} b_j^{-2} \sum_{n=1}^{j} \int_{A_n} X_j^2$$

$$= \sum_{n=1}^{\infty} \sum_{j=n}^{\infty} b_j^{-2} \int_{A_n} X^2 \le \sum_{n=1}^{\infty} b_n^2 \sum_{j=n}^{\infty} b_j^{-2}\, P\{A_n\}$$

$$\le C \sum_1^{\infty} n\, P\{A_n\} < \infty$$

via (14) and (18). By Theorem 5.1.1, $\sum_1^{\infty} b_j^{-1}(Y_j - E\, Y_j)$ converges a.c., and so, again employing (18), $\sum_1^{\infty} b_j^{-1}(X_j - E\, Y_j)$ converges a.c., yielding (15).

Moreover, if $E\, X = 0$,

$$\sum_1^{\infty} b_j^{-1} |E\, Y_j| \le \sum_{j=1}^{\infty} b_j^{-1}\, E|X| I_{[|X| > b_j]} = \sum_{j=1}^{\infty} b_j^{-1} \sum_{n=j}^{\infty} \int_{A_{n+1}} |X|$$

$$= \sum_{n=1}^{\infty} \sum_{j=1}^{n} b_j^{-1} \int_{A_{n+1}} |X| \le \sum_{n=1}^{\infty} b_{n+1} \sum_{j=1}^{n} b_j^{-1}\, P\{A_{n+1}\}$$

$$\le C \sum_{n=1}^{\infty} (n+1) P\{A_{n+1}\} < \infty$$

via (16) and (18). Consequently, (17) follows via (15). □

Theorem 4. *Let $\{X_n\}$ be independent* r.v.s *such that for some positive ε, δ*

$$\inf_{n \ge 1} P\{|X_n| > \varepsilon\} = \delta, \tag{19}$$

and suppose that $\sum_1^{\infty} a_n X_n$ converges a.c. *for some sequence $\{a_n\}$ of real numbers. If either* (i) *X_n, $n \ge 1$, are* i.i.d. *and nondegenerate or* (ii) $E\, X_n = 0$, $n \ge 1$, *and $\{X_n\}$ is* u.i., *then*

$$\sum_1^{\infty} a_n^2 < \infty. \tag{20}$$

PROOF. Since $\sum_1^{\infty} a_n X_n$ converges a.c., necessarily $a_n X_n \xrightarrow{P} 0$, implying via (19) that $a_n \to 0$. In proving (20), it may be supposed that $a_n \ne 0$, $n \ge 1$. If $Y_n = X_n I_{[|a_n X_n| \le 1]}$, the three series theorem requires that

$$\sum_1^{\infty} a_n^2 \sigma_{Y_n}^2 < \infty. \tag{21}$$

In case (i), $\sigma_{Y_n}^2 = E(X_1 I_{[|a_n X_1| \le 1]} - E\, Y_n)^2$. If $\sigma_{Y_{n_j}}^2 \to 0$ for some subsequence n_j, then $X_1 I_{[|a_{n_j} X_1| \le 1]} - E\, Y_{n_j} \xrightarrow{P} 0$, implying X_1 degenerate, contrary to the hypothesis of (i). Thus $\underline{\lim}_n\, \sigma_{Y_n}^2 > 0$, whence (21) implies (20).

Under (ii), since $\mathrm{E}\, X_n = 0$,

$$|\mathrm{E}\, Y_n| = \left| \int_{[|X_n| > |a_n|^{-1}]} X_n \right| \le \int_{[|X_n| > |a_n|^{-1}]} |X_n| = o(1)$$

by uniform integrability and $a_n \to 0$. Hence,

$$\underline{\lim}_n\, \sigma_{Y_n}^2 = \underline{\lim}(\mathrm{E}\, Y_n^2 - \mathrm{E}^2\, Y_n) = \underline{\lim}\, \mathrm{E}\, Y_n^2 \ge \underline{\lim}\, \mathrm{E}^2 |Y_n|$$

$$= \underline{\lim}\left(\mathrm{E}|X_n| - \int_{[|a_n X_n| > 1]} |X_n|\right)^2 = \underline{\lim}\, \mathrm{E}^2|X_n| \ge \varepsilon^2\delta^2$$

by (19), and once more (20) flows from (21). □

Corollary 4 (Marcinkiewicz–Zygmund). *If $\{X_n\}$ are independent* r.v.s *with* $\mathrm{E}\, X_n = 0$, $\mathrm{E}\, X_n^2 = 1$, $n \ge 1$, *and* $\inf_n \mathrm{E}|X_n| > 0$, *then* a.c. *convergence of* $\sum a_n X_n$ *for some real numbers a_n entails convergence of* $\sum a_n^2$.

PROOF. Uniform integrability of $\{X_n\}$ is implied by $\mathrm{E}\, X_n^2 = 1$, $n \ge 1$, while $\inf \mathrm{E}|X_n| > 0$ and $\mathrm{E}\, X_n^2 = 1$ ensure (19) (Exercise 5). □

Definition. A series $\sum_{n=1}^{\infty} X_n$ of r.v.s will be said to **converge a.c. unconditionally** if $\sum_{k=1}^{\infty} X_{n_k}$ converges a.c. for every rearrangement $(n_1, n_2, \ldots)$ of $(1, 2, \ldots)$. More specifically, a rearrangement is a one-to-one map of the positive integers onto the positive integers.

In the case of degenerate r.v.s X_n, $n \ge 1$, the series $\sum_1^{\infty} X_n$ converges unconditionally iff $\sum |X_n|$ converges (Weierstrass). However, the analogue for nondegenerate independent random variables is invalid. In fact, if X_n, $n \ge 1$, are independent r.v.s with $P\{X_n = \pm 1/n\} = \frac{1}{2}$, then $\sum X_n$ converges a.c. unconditionally by Theorem 1, but $\sum |X_n| = \infty$, a.c.

Lemma 5. *If $\{X_n\}$ are independent* r.v.s *with* $\mathrm{E}\, X_n = 0$, $n \ge 1$, *and* $\sum_1^{\infty} \mathrm{E}\, X_n^2 < \infty$, *then* $\sum X_n$ *converges* a.c. *unconditionally and* $\sum_{j=1}^{\infty} X_{n_j} = \sum_1^{\infty} X_j$, a.c., *for every rearrangement $\{n_j\}$ of $\{j\}$.*

PROOF. Theorem 1 ensures that $\sum X_n$ converges a.c. unconditionally. Moreover, for any fixed rearrangement $\{n_j\}$, define $S'_m = \sum_{j=1}^{m} X_{n_j}$, $S_m = \sum_1^m X_j$. Then, setting $Q = \{n_1, \ldots, n_m\}\Delta\{1, 2, \ldots m\}$,

$$\mathrm{E}(S'_m - S_m)^2 = \sum_{k \in Q} \mathrm{E}\, X_k^2.$$

Now, if $\{n_1, \ldots, n_m\} \supset \{1, 2, \ldots, j\}$, then

$$\mathrm{E}(S'_m - S_m)^2 \le \sum_{j+1}^{\infty} \mathrm{E}\, X_k^2 = o(1) \quad \text{as } j \to \infty.$$

Hence, $S'_m - S_m \xrightarrow{P} 0$ as $m \to \infty$, implying $\sum_{j=1}^{\infty} X_{n_j} = \sum_1^{\infty} X_j$ a.c. □

Theorem 5. *A series* $\sum_1^\infty X_n$ *of independent* r.v.s X_n *converges* a.c. *unconditionally* iff *for* $Y_n = X_n I_{[|X_n| \le 1]}$

i′. $\sum_1^\infty P\{|X_n| > 1\} < \infty$,
ii′. $\sum_1^\infty |E\, Y_n| < \infty$,
iii′. $\sum_1^\infty E\, Y_n^2 < \infty$,

and if so, $\sum_1^\infty X_{n_j} = \sum_1^\infty X_j$ a.c. *for every rearrangement* $\{n_j\}$ *of* $\{j\}$. *Moreover, a series of independent* r.v.s $\{X_n\}$ *converges absolutely* a.c. iff (i′), (iii′), *and*

ii″. $\sum_1^\infty E|Y_n| < \infty$

hold.

Proof. Since the series appearing in (i′), (ii′), (iii′) are independent of the order of summation, the three series theorem (Theorem 2) guarantees that $\sum_1^\infty X_n$ converges a.c. unconditionally, and by Lemma 5

$$\sum_1^\infty (Y_{n_j} - E\, Y_{n_j}) = \sum_1^\infty (Y_j - E\, Y_j), \qquad \text{a.c.}$$

Then, in view of (ii′), $\sum_1^\infty Y_{n_j} = \sum_1^\infty Y_j$ a.c., whence (i′) ensures that $\sum_1^\infty X_{n_j} = \sum_1^\infty X_j$ a.c.

Conversely, if $\sum X_n$ converges a.c. unconditionally, (i), (ii), (iii) of Theorem 2 are valid for every rearrangement $\{n_j\}$ of $\{j\}$. By the Weierstrass theorem (ii′) holds, and hence also $\sum_1^\infty E^2\, Y_n < \infty$. But this latter and (iii) entail (iii′).

The proof of the final statement is similar. □

Corollary 5. *If the series* $\sum_1^\infty X_n$ *of independent* r.v.s X_n *converges* a.c., *then* $\sum (X_n - c_n)$ *converges* a.c. *unconditionally, where*

$$c_n = E\, X_n I_{[|X_n| \le 1]}. \tag{22}$$

Proof. Set $Y_n = X_n I_{[|X_n| \le 1]}$. By the three series theorem

$$\sum_1^\infty \sigma_{Y_n}^2 < \infty, \qquad \sum_1^\infty P\{X_n \ne Y_n\} < \infty, \tag{23}$$

whence Lemma 5 guarantees that $\sum_1^\infty (Y_n - E\, Y_n)$ converges a.c. unconditionally. Then by (23), $\sum_1^\infty (X_n - E\, Y_n)$ converges a.c. unconditionally. □

Exercises 5.1

1. Let $S_n = \sum_1^n X_i$, where $\{X_n, n \ge 1\}$ are independent r.v.s. (i) If $\sum_{n=1}^\infty P\{|X_n| > c\} = \infty$, all $c > 0$, then $\overline{\lim}_{n\to\infty} |S_n| = \infty$, a.c. (ii) If $\{b_n, n \ge 1\}$ are positive constants for which $\underline{\lim}_{n\to\infty} P\{S_{n-1} > -\delta b_n\} > 0$ for all $\delta > 0$, then $\overline{\lim}\, S_n/b_n \le C < \infty$, a.c., implies $\sum_{n=1}^\infty P\{X_n > \varepsilon b_n\} < \infty$ for all $\varepsilon > C$. *Hint*: Recall Lemma 3.3.4.
2. Any series $\sum_{i=1}^\infty X_i$ of r.v.s converges absolutely a.c. if $\sum_{n=1}^\infty E|X_n|^{r_n} < \infty$ for some sequence r_n in $(0, 1]$.

3. If X_n, $n \geq 1$, are independent r.v.s and $Y_n^c = X_n I_{[|X_n| \leq c]}$, then a.c. convergence of $\sum X_n$ ensures convergence of $\sum_1^\infty P\{|X_n| > c\}$, $\sum_1^\infty E\, Y_n^c$, $\sum_1^\infty \sigma_{Y_n^c}^2$ for every $c > 0$. Conversely, convergence of these series for some $c > 0$ guarantees a.c. convergence of $\sum X_n$.

4. If $\{X_n\}$, $\{Y_n\}$ are equivalent sequences of r.v.s, prove that
   i. $P\{\sum_1^\infty X_n \text{ converges}\} = 1$ iff $P\{\sum_1^\infty Y_n \text{ converges}\} = 1$,
   ii. if $0 < b_n \uparrow \infty$, then $P\{\sum_1^n X_i = o(b_n)\} = 1$ iff $P\{\sum_1^n Y_i = o(b_n)\} = 1$.

5. Let $\{X_n\}$ be u.i.r.v.s. Then $\inf P[|X_n| > \varepsilon] > 0$, for some $\varepsilon > 0$, iff $\inf_n E|X_n| > 0$.

6. If X_n, $n \geq 1$, are i.i.d. $\mathscr{L}_1$ r.v.s, then $\sum (X_n/n)$ converges a.c. if either (i) X_1 is symmetric or (ii) $E|X_1|\log^+|X_1| < \infty$ and $E\, X_1 = 0$.

7. Let $\{a_n\}$ be a sequence of positive numbers with $\sum_1^\infty a_n = \infty$, Then, if $p > 0$, there exist independent r.v.s X_n with $E\, X_n = 0$, $E|X_n|^p = a_n$ such that $\sum_1^\infty X_n$ diverges a.c., thereby furnishing a partial converse to Corollary 3 when $0 < p \leq 2$.

8. If X_n, $n \geq 1$, are independent r.v.s with $P\{X_n = 1\} = P\{X_n = -1\} = \frac{1}{2}$, then $\sum_1^\infty X_n/\sqrt{n}$ diverges a.c. although $\sum_1^\infty E|X_n/\sqrt{n}|^p < \infty$, all $p > 2$. Thus, the restriction of exponents in Corollary 2 is essential.

9. If $\{A_n, n \geq 1\}$ are independent events with $P\{A_n\} > 0$, $n \geq 1$ and $\sum_{n=1}^\infty P\{A_n\} = \infty$ then $\sum_{j=1}^n I_{A_j}/\sum_{j=1}^n P\{A_j\} \xrightarrow{\text{a.c.}} 1$. *Hint*: If $a_n = \sum_{j=1}^n P\{A_j\}$ and $X_j = (I_{A_j} - P\{A_j\})/a_j$ then $\{X_j, j \geq 1\}$ are independent with $E\, X_j = 0$, $E\, X_j^2 \leq P\{A_j\}/a_j^2 \leq 1/a_{j-1} - 1/a_j$, $j > 1$.

10. If X_n, $n \geq 1$, are independent r.v.s with $p_n = P\{X_n = a_n\} = 1 - P(X_n = -a_n\}$, characterize the sequences (a_n, p_n) for which $\sum_1^\infty X_i$ converges a.c.; specialize to $p_n \equiv \frac{1}{2}$, to $a_n \equiv a$, and to $a_n = n^{-\alpha}$ $(\alpha > 0)$.

11. If, in the three series theorem, the alternative truncation $Z_n = \min[1, \max(X_n, -1)]$ is employed, then convergence of the two series $\sum_1^\infty E\, Z_n$, $\sum_1^\infty \sigma_{Z_n}^2$ is equivalent to the a.c. convergence of $\sum_1^\infty X_n$.

12. For any sequence of r.v.s $\{S_n\}$, it is always possible to find constants $0 < a_n \uparrow \infty$ for which $S_n/a_n \xrightarrow{\text{a.c.}} 0$. *Hint*: If $0 < \varepsilon_n \downarrow 0$, choose $a_n > a_{n-1}$ such that $P\{|S_n| > a_n\varepsilon_n\} < 2^{-n}$.

13. (Chung) Let Ψ be a positive, even function with $x^{-2}\Psi(x)\downarrow$, $|x|^{-1}\Psi(x)\uparrow$ as $|x|\uparrow$. If $0 < b_n \uparrow \infty$, $\{X_n\}$ are independent with $E\, X_n = 0$, $\sum (E\,\Psi(X_n)/\Psi(b_n)) < \infty$, then $\sum (X_n/b_n)$ converges a.c. *Hint*: Apply Corollary 2 with X_n replaced by X_n/b_n.

14. If $\{X_n, n \geq 1\}$ are i.i.d. r.v.s with $E|X_1| < \infty$, prove that $\sum_{n=1}^\infty X_n(\sin nt)/n$ converges a.c. for every $t \in (-\infty, \infty)$. Conversely, a.c. convergence of this series for some $t \neq k\pi$, k an integer and i.i.d. $\{X_n, n \geq 1\}$ implies $E|X_1| < \infty$. *Hint*: For $m = 1, 2, \ldots$, choose integers n_m so that $n_m t \in (2m\pi + (\pi/4), 2m\pi + (\pi/2)]$ for $t = \pi/4$.

15. If $\{b_n, n \geq 1\}$ are finite constants with $0 < b_n \uparrow \infty$ and (i) $b_n^2 \sum_{j=n}^\infty b_j^{-2} = O(d(b_n))$, where d is a nondecreasing mapping of $[0, \infty)$ into $[0, \infty)$ for which (ii) $d(b_n) \geq cn > 0$, $n \geq 1$, and (iii) $x^2/d(|x|)\uparrow$ as $|x|\uparrow$, then for any i.i.d. r.v.s $\{X, X_n, n \geq 1\}$ with (iv) $E\, d(|X|) < \infty$, the conclusion (15) obtains. Moreover, if $E\, X = 0$ and (v) $|x|/d(|x|)\downarrow$ as $|x|\uparrow$ and (vi) $b_n \sum_{j=1}^n b_j^{-1} = O(d(b_n))$, then (17) holds.

16. If $\{X, X_n, n \geq 1\}$ are i.i.d. r.v.s with $E\, X = 0$, $E\, X^2(1 + \log^+|X|)^{-2\delta} < \infty$ for some $\delta > 0$, then $\sum n^{-1/2}(\log n)^{-(1/2)-\delta} X_n$ converges a.c.

5.2 Laws of Large Numbers

In a sense, a.c. convergence of a series of independent r.v.s X_n is atypical and in the nondegenerate i.i.d. case it is nonexistent. Thus, the issue rather becomes one of the magnitude of the partial sums $S_n = \sum_1^n X_i$. When $P\{X_n = 1\} = p = 1 - P\{X_n = 0\}$, so that S_n is a binomial r.v. with parameters n and p, it was proved in Theorem 2.2.3 that

$$\frac{S_n - np}{n} = \frac{1}{n} \sum_{i=1}^{n} (X_i - \mathrm{E}\, X_i) \xrightarrow{\text{a.c.}} 0.$$

Definition. A sequence $\{X_n\}$ of $\mathscr{L}_1$ r.v.s is said to obey the **classical strong law of large numbers** (SLLN) if

(a) $$\frac{1}{n} \sum_{i=1}^{n} (X_i - \mathrm{E}\, X_i) \xrightarrow{\text{a.c.}} 0.$$

If, merely,

(b) $$\frac{1}{n} \sum_{i=1}^{n} (X_i - \mathrm{E}\, X_i) \xrightarrow{\text{P}} 0,$$

the sequence $\{X_n\}$ satisfies the **classical weak law of large numbers (WLLN).**

From a wider vista, n may not reflect the real magnitude and the expectations need not exist. Thus, there is occasion to consider the more general **strong and weak laws of large numbers**

$$\frac{1}{a_n} \sum_{i=1}^{n} (X_i - b_i) \xrightarrow[\text{P}]{\text{a.c.}} 0,$$

where $0 < a_n \uparrow \infty$. Here, the smaller the order of magnitude of a_n, the more precise the SLLN becomes; the fuzzy notion of an optimal choice of a_n impinges on the law of the iterated logarithm (Chapter 10). Note, in this context, Exercise 5.1.12.

The first SLLN may be reaped as a direct application of Kronecker's lemma to Corollary 3 of the three series theorem (Theorem 5.1.2), thereby obtaining Loève's generalization of a result of Kolmogorov ($\alpha_n \equiv 2$).

Theorem 1. *If* $\{X_n\}$ *are independent* r.v.s *satisfying*

$$\sum_{1}^{\infty} \frac{\mathrm{E}|X_n|^{\alpha_n}}{n^{\alpha_n}} < \infty \tag{1}$$

for some choice of α_n *in* (0, 2], *where* $\mathrm{E}\, X_n = 0$ *whenever* $1 \le \alpha_n \le 2$, *then*

$$\frac{1}{n} \sum_{j=1}^{n} X_j \xrightarrow{\text{a.c.}} 0. \tag{2}$$

Corollary 1. *Let $\{X_n\}$ be independent $\mathscr{L}_2$ r.v.s with* $\mathrm{E}\,X_n = 0$, $\sum_1^\infty \mathrm{E}\,X_n^2/n^2 < \infty$. *Then* (2) *holds.*

According to Corollary 1, independent r.v.s with means zero and variances $n(\log n)^{-\delta}$ obey the classical SLLN if $\delta > 1$. This remains true for any $\delta > 0$ by Corollary 10.1.4. provided (as is necessary) $\sum_{n=1}^\infty \mathrm{P}\{|X_n| \geq n\varepsilon\} < \infty$, $\varepsilon > 0$.

In the i.i.d. case, the next theorem gives a generalization due to Marcinkiewicz and Zygmund of a classical SLLN ($p = 1$) of Kolmogorov.

Theorem 2 (Marcinkiewicz–Zygmund). *If $\{X_n\}$ are* i.i.d. r.v.s *and* $S_n = \sum_1^n X_i$, *then for any p in* $(0, 2)$

$$\frac{S_n - nc}{n^{1/p}} \xrightarrow{\text{a.c.}} 0 \tag{3}$$

for some finite constant c iff $\mathrm{E}|X_1|^p < \infty$, *and if so,* $c = \mathrm{E}\,X_1$ *when* $1 \leq p < 2$ *while c is arbitrary (and hence may be taken as zero) for* $0 < p < 1$.

PROOF. If (3) holds, then

$$\frac{X_n}{n^{1/p}} = \frac{S_n - nc}{n^{1/p}} - \left(\frac{n-1}{n}\right)^{1/p} \frac{S_{n-1} - nc}{(n-1)^{1/p}} \xrightarrow{\text{a.c.}} 0,$$

whence by the Borel–Cantelli theorem $\sum_1^\infty \mathrm{P}\{|X_1| \geq n^{1/p}\} < \infty$. Thus, $\mathrm{E}|X_1|^p < \infty$ by Corollary 4.1.3. Conversely, if $\mathrm{E}|X_1|^p < \infty$, by Theorem 5.1.3 the following series converge a.c.:

$$\sum_1^\infty \frac{X_n - \mathrm{E}\,X_n}{n^{1/p}}, \qquad 1 < p < 2;$$

$$\sum_1^\infty \frac{X_n - \mathrm{E}\,X_n I_{[|X_n| \leq n]}}{n^{1/p}}, \qquad p = 1; \tag{4}$$

$$\sum_1^\infty \frac{X_n}{n^{1/p}}, \qquad 0 < p < 1.$$

For $p \neq 1$, (3) obtains (with $c = \mathrm{E}\,X_1$ or 0) by Kronecker's lemma. When $p = 1$, since $\mathrm{E}\,X_n I_{[|X_n| \leq n]} = \mathrm{E}\,X_1 I_{[|X_1| \leq n]} \to \mathrm{E}\,X_1$ as $n \to \infty$, Kronecker's lemma, in conjunction with (4), yields

$$\lim_n \frac{S_n}{n} - \mathrm{E}\,X_1 = \lim_n \frac{1}{n}\left(S_n - \sum_{j=1}^n \mathrm{E}\,X_j I_{[|X_j| \leq J]}\right) = 0, \qquad \text{a.c.} \qquad \square$$

Corollary 2 (Kolmogorov). *If $\{X_n\}$ are* i.i.d. r.v.s, $S_n = \sum_1^n X_i$ *and* $\mathrm{E}\,X_1$ *exists, then* $S_n/n \xrightarrow{\text{a.c.}} \mathrm{E}\,X_1$.

PROOF. If $|\mathrm{E}\, X_1| < \infty$, the conclusion is contained in Theorem 2. If $\mathrm{E}\, X_1 = -\infty$, define $Y_n = \max(X_n, -K)$, where $K > 0$. Then $\{Y_n\}$ are i.i.d. with $\mathrm{E}|Y_1| < \infty$. By Theorem 2

$$\overline{\lim}\, \frac{S_n}{n} \le \overline{\lim}\, \frac{1}{n} \sum_1^n Y_i \overset{\text{a.c.}}{=\!=} \mathrm{E}\, Y_1 \xrightarrow{K \to \infty} \mathrm{E}\, X_1.$$

Hence, $S_n/n \xrightarrow{\text{a.c.}} -\infty$. Analogously, $S_n/n \xrightarrow{\text{a.c.}} \infty$ when $\mathrm{E}\, X_1 = \infty$. □

EXAMPLE 1. If $\{X_n, n \ge 1\}$ are i.i.d. $\mathscr{L}_p$ random varibles for some p in $(0, 2)$, then (i) $\sum_{n=1}^{\infty} n^{-2/p} X_n^2 < \infty$, a.c. Moreover, if $\mathrm{E}\, X = 0$ whenever $1 \le p < 2$, (ii) $n^{-2/p} \sum_{j=2}^{n} X_j S_{j-1} \xrightarrow{\text{a.c.}} 0$.

PROOF. According to the proof of Theorem 5.1.3, $Y_n = n^{-1/p} X_n I_{[|X_n| \le n^{1/p}]}$, $n \ge 1$ and $n^{-1/p} X_n$, $n \ge 1$ are equivalent sequences and $\sum_{n=1}^{\infty} \mathrm{E}\, Y_n^2 < \infty$ implying (i). Then, by Kronecker's lemma, $n^{-2/p} \sum_{j=1}^{n} X_j^2 \xrightarrow{\text{a.c.}} 0$, and so Theorem 2 ensures that

$$n^{-2/p} \sum_{j=2}^{n} X_j S_{j-1} = \frac{1}{2} n^{-2/p} \left[S_n^2 - \sum_{j=1}^{n} X_j^2 \right] \xrightarrow{\text{a.c.}} 0. \qquad \square$$

If $\{X_n, n \ge 1\}$ are i.i.d. r.v.s, $S_n = \sum_1^n X_i$ and $\{b_n, n \ge 1\}$ are constants with $0 < b_n \uparrow \infty$, the relationship between $X_n/b_n \xrightarrow{\text{a.c.}} 0$ and $(S_n - C_n)/b_n \xrightarrow{\text{a.c.}} 0$ for some sequence of constants $\{C_n, n \ge 1\}$ is described by

Theorem 3. *Let $\{X, X_n, n \ge 1\}$ be i.i.d. r.v.s with partial sums $\{S_n, n \ge 1\}$ and $\{b_n, n \ge 1\}$ constants such that $0 < b_n \uparrow \infty$ and*

$$\sum_{n=1}^{\infty} \mathrm{P}\{|X_n| > b_n\} < \infty. \tag{5}$$

(i) (Feller) *If* (α) $b_n/n \uparrow \infty$ *or* (β) $b_n/n \downarrow$, *the first half of* (6) *holds and* $\mathrm{E}\, X = 0$ *or* (γ) $\mathrm{E}\, X = 0$ *and*

$$b_n^2 \sum_{j=n}^{\infty} b_j^{-2} = O(n), \qquad b_n \sum_{j=1}^{n} b_j^{-1} = O(n), \tag{6}$$

then

$$\frac{S_n}{b_n} \xrightarrow{\text{a.c.}} 0. \tag{7}$$

(ii) *If $\{a_n, n \ge 1\}$ are positive constants with $A_n = \sum_1^n a_j \to \infty$ such that $b_n = A_n/a_n \uparrow \infty$ and satisfies the first half of* (6), *then*

$$\frac{1}{A_n}\sum_{j=1}^{n} a_j(X_j - \mathrm{E}\, XI_{[|X| \le A_j/a_j]}) \xrightarrow{\text{a.c.}} 0 \tag{8}$$

and, moreover, if E $X = 0$,

$$\frac{1}{A_n}\sum_{1}^{n} a_j X_j \xrightarrow{\text{a.c.}} 0. \tag{9}$$

PROOF. According to Example 5.1.2, condition (5), the initial portion of (6) and Kronecker's lemma ensure

$$\frac{1}{b_n}\sum_{j=1}^{n}(X_j - \mathrm{E}\, XI_{[|X| \le b_j]}) \xrightarrow{\text{a.c.}} 0. \tag{10}$$

The conclusion (8) in case (ii) follows likewise.

Now $b_n/n \uparrow \infty$ implies the first half of (6), hence (10) and also

$$\begin{aligned}\left|\frac{1}{b_n}\sum_{1}^{n}\mathrm{E}\, XI_{[|X| \le b_j]}\right| &\le \frac{1}{b_n}\sum_{1}^{n}\mathrm{E}|X|(I_{[|X| \le b_N]} + I_{[b_N < |X| \le b_j]})\\ &\le \frac{nb_N}{b_n} + \frac{n}{b_n}\int_{[b_N < |X| \le b_n]}|X|\\ &\le \frac{nb_N}{b_n} + \sum_{j=N}^{n}\frac{j}{b_j}\int_{[b_{j-1} < |X| \le b_j]}|X|\\ &\le \frac{nb_N}{b_n} + \sum_{j=N}^{\infty} j\, \mathrm{P}\{b_{j-1} < |X| \le b_j\}\\ &\xrightarrow{n\to\infty} \sum_{j=N}^{\infty} j\, \mathrm{P}\{b_{j-1} < |X| \le b_j\} \xrightarrow{N\to\infty} 0\end{aligned}$$

by Example 5.1.2 (18), and so (7) follows from (10), thereby proving (α) of (i). In case (β) since E $X = 0$,

$$\begin{aligned}\left|\frac{1}{b_n}\sum_{1}^{n}\mathrm{E}\, XI_{[|X| \le b_j]}\right| &\le \frac{1}{b_n}\sum_{j=1}^{N-1}\mathrm{E}|X|I_{[|X| \le b_j]} + \frac{1}{b_n}\left|\sum_{j=N}^{n}\mathrm{E}\, XI_{[|X| > b_j]}\right|\\ \le O\left(\frac{N}{b_n}\right) &+ \frac{1}{b_n}\sum_{j=N}^{n}\int_{[b_j < |X| \le b_n]}|X| + \frac{n}{b_n}\int_{[|X| > b_n]}|X|\\ \le O\left(\frac{N}{b_n}\right) &+ \sum_{j=N}^{\infty}\mathrm{P}\{|X| > b_j\} + \frac{n}{b_n}\sum_{j=n+1}^{\infty} b_j\, \mathrm{P}\{b_{j-1} < |X| \le b_j\}\end{aligned}$$

$$= O\left(\frac{N}{b_n}\right) + \sum_{j=N}^{\infty} \mathrm{P}\{|X| > b_j\} + \sum_{n+1}^{\infty} j\,\mathrm{P}\{b_{j-1} < |X| \le b_j\}$$

$$= o(1)$$

as n, and then $N \to \infty$ by Example 5.1.2 (18).

In case (γ), (7) is a direct consequence of Kronecker's lemma applied to (17) of Example 5.1.2.

To prove the remaining portion of (ii) it suffices, via (8), to note that $\sum_{j=1}^{n} a_j \,\mathrm{E}\, X I_{[|X| \le A_j/a_j]} = o(A_n)$ in view of $\mathrm{E}\, X = 0$ and

$$\left| \frac{1}{A_n} \sum_{1}^{n} a_j \,\mathrm{E}\, X I_{[|X| > A_j/a_j]} \right| \le \frac{1}{A_n} \sum_{1}^{n} a_j \,\mathrm{E}|X| I_{[|X| > A_j/a_j]} = o(1). \qquad \square$$

Corollary 3. *Let* $\{a_n, n \ge 1\}$ *be positive constants with* $A_n = \sum_1^n a_j \to \infty$ *and*

$$\frac{a_n}{A_n} \downarrow 0, \qquad \left(\frac{A_n}{a_n}\right)^2 \sum_{j=n}^{\infty} \left(\frac{a_j}{A_j}\right)^2 = O(n). \tag{11}$$

For any i.i.d. r.v.s $\{X_n, n \ge 1\}$, *there exist constants* C_n *for which*

$$\frac{1}{A_n} \sum_{1}^{n} a_j (X_j - C_j) \xrightarrow{\text{a.c.}} 0 \tag{12}$$

iff

$$\frac{a_n X_n}{A_n} \xrightarrow{\text{a.c.}} 0. \tag{13}$$

PROOF. If (13) obtains, so does (12) with $C_j = \mathrm{E}\, X I_{[|X| \le A_j/a_j]}$ by Theorem 3. Conversely, since $a_n = o(A_n)$ guarantees $A_n/A_{n-1} \to 1$, (12) ensures

$$a_n(X_n - C_n)/A_n \xrightarrow{\text{a.c.}} 0.$$

Moreover, $a_n X_n/A_n \xrightarrow{\mathrm{P}} 0$ via $a_n = o(A_n)$. Then $a_n C_n/A_n \to 0$, which, in turn, ensures (13). $\square$

Although necessary and sufficient conditions for the WLLN are available in the case of independent r.v.s, a complete discussion (see Section 10.1) requires the notion of symmetrization. The i.i.d. situation, however, is amenable to methods akin to those already employed.

Theorem 4 (Feller). *If* $\{X_n\}$ *are* i.i.d. r.v.s *and* $S_n = \sum_{i=1}^n X_i$, *then*

$$\frac{S_n - C_n}{n} \xrightarrow{\mathrm{P}} 0 \tag{14}$$

for some choice of real numbers C_n iff

$$n\,\mathrm{P}\{|X_1| > n\} \to 0, \tag{15}$$

and if so, $C_n/n = \mathrm{E}\,X_1 I_{[|X_1| \le n]} + o(1)$.

PROOF. Sufficiency: Set $X'_j = X_j I_{[|X_j| \le n]}$ for $1 \le j \le n$ and $S'_n = \sum_{j=1}^n X'_j$. Then, for each $n \ge 2$, $\{X'_j, 1 \le j \le n\}$ are i.i.d. and for $\varepsilon > 0$, $\mathrm{P}\{|(S_n/n) - (S'_n/n)| > \varepsilon\} \le \mathrm{P}\{S_n \ne S'_n\} = \mathrm{P}\{\bigcup_1^n [X_j \ne X'_j]\} \le n\,\mathrm{P}\{|X_1| > n]$, so that (15) entails $(S'_n/n) - (S_n/n) \xrightarrow{\mathrm{P}} 0$. Thus, to prove (14) it suffices to verify that

$$\frac{S'_n - \mathrm{E}\,S'_n}{n} = \frac{S'_n}{n} - \mathrm{E}\,X_1 I_{[|X_1| \le n]} \xrightarrow{\mathrm{P}} 0. \tag{16}$$

By Lemma 5.1.1 (4), Corollary 4.3.5, and (15),

$$\begin{aligned}
\mathrm{E}(S'_n - \mathrm{E}\,S'_n)^2 &= \sum_1^n \sigma^2(X'_j) \le \sum_1^n \mathrm{E}(X'_j)^2 = n\,\mathrm{E}(X'_1)^2 \\
&= n \sum_{j=1}^n \int_{[j-1 < |X_1| \le j]} X_1^2 \\
&\le n \sum_{j=1}^n j^2[\mathrm{P}\{|X_1| > j-1\} - \mathrm{P}\{|X_1| > j\}] \\
&= n\Bigg[\mathrm{P}\{|X_1| > 0\} - n^2\,\mathrm{P}\{|X_1| > n\} \\
&\qquad + \sum_{j=1}^{n-1}((j+1)^2 - j^2)\mathrm{P}\{|X_1| > j\}\Bigg] \\
&\le 3n\left[1 + \sum_{j=1}^{n-1} j\,\mathrm{P}\{|X_1| > j\}\right] = o(n^2),
\end{aligned} \tag{17}$$

which implies (16) and hence (14) with $C_n = n\,\mathrm{E}X_1 I_{[|X_1| \le n]}$.

Conversely, if (14) holds, setting $c_n = C_n - C_{n-1}$, $n \ge 1$, $C_0 = 0$,

$$\frac{X_n - c_n}{n} = \frac{S_n - C_n}{n} - \frac{n-1}{n}\left(\frac{S_{n-1} - C_{n-1}}{n-1}\right) \xrightarrow{\mathrm{P}} 0,$$

whence $(X_1 - c_n)/n \xrightarrow{\mathrm{P}} 0$, necessitating $c_n = o(n)$. By Lévy's inequality (Lemma 3.3.5), for any $\varepsilon > 0$

$$\begin{aligned}
&\mathrm{P}\left\{\max_{1 \le j \le n} |S_j - C_j - m(S_j - C_j - S_n + C_n)| \ge \frac{n\varepsilon}{2}\right\} \\
&\le 2\,\mathrm{P}\left\{|S_n - C_n| \ge \frac{n\varepsilon}{2}\right\} = o(1);
\end{aligned} \tag{18}$$

but, taking $X_j = S_j - C_j$ in Exercise 3.3.7,

$$\max_{1 \le j \le n} |m(S_j - C_j - S_n + C_n)| = o(n). \tag{19}$$

Thus, from (18) and (19), for all $\varepsilon > 0$

$$\lim_n \mathrm{P}\left\{\max_{1 \le j \le n} |S_j - C_j| < n\varepsilon\right\} = 1. \tag{20}$$

Moreover, for $\max_{1 \le j \le n}|c_j| < n\varepsilon$, and hence for all large n,

$$\mathrm{P}\left\{\max_{1 \le j \le n} |S_j - C_j| < n\varepsilon\right\} \le \mathrm{P}\left\{\max_{1 \le j \le n} |X_j - c_j| < 2n\varepsilon\right\}$$
$$\le \mathrm{P}\left\{\max_{1 \le j \le n} |X_j| < 3n\varepsilon\right\},$$

which, in conjunction with (20), yields

$$\mathrm{P}^n\{|X_1| < 3n\varepsilon\} = \mathrm{P}\left\{\max_{1 \le j \le n} |X_j| < 3n\varepsilon\right\} \to 1$$

or, equivalently, for all $\varepsilon > 0$

$$n \log[1 - \mathrm{P}\{|X_1| \ge 3n\varepsilon\}] \to 0 \tag{21}$$

as $n \to \infty$. Since $\log(1 - x) = -x + o(x)$ as $x \to 0$, (21) entails (15).

The final characterization of C_n/n results from the fact that (14) entails (15), which, in turn, implies (14) with $C_n/n = \mathrm{E}\, X_1 I_{[|X_1| \le n]}$. □

EXAMPLE 2 (Chung–Ornstein). If $\{X, X_n, n \ge 1\}$ are i.i.d. with $n\,\mathrm{P}\{|X| > n\} = o(1)$ and $\mathrm{E}\, X I_{[|X| \le n]} = o(1)$ (*a fortiori* if $\mathrm{E}\, X = 0$) then the random walk $\{S_n = \sum_1^n X_i, n \ge 1\}$ is recurrent. If, rather, $\mathrm{E}\, X > 0$ or $\mathrm{E}\, X < 0$, the random walk is nonrecurrent.

PROOF. Take $N = k \cdot m$ in Lemma 4.2.5, where m is an arbitrary integer. Then for any $\varepsilon > 0$

$$\sum_{n=0}^{km} \mathrm{P}\{|S_n| < \varepsilon\} \ge \frac{1}{2k} \sum_{n=0}^{km} \mathrm{P}\{|S_n| < k\varepsilon\} \ge \frac{1}{2k} \sum_{n=0}^{km} \mathrm{P}\left\{|S_n| < \frac{n\varepsilon}{m}\right\}.$$

According to Theorem 4, $S_n/n \xrightarrow{\mathrm{P}} 0$, and so

$$\sum_{n=0}^{\infty} \mathrm{P}\{|S_n| < \varepsilon\} \ge \frac{m}{2} \varliminf_{k \to \infty} \frac{1}{km} \sum_{n=0}^{km} \mathrm{P}\left\{\left|\frac{S_n}{n}\right| < \frac{\varepsilon}{m}\right\} = \frac{m}{2}.$$

Since m is arbitrary, the series on the left diverges for all $\varepsilon > 0$ and the conclusion follows from Example 4.2.1. The final remark stems from the strong law of large numbers. □

In Kolmogorov's classical SLLN (Corollary 2) involving identically distributed $\mathscr{L}_1$ random variables $\{X_n, n \geq 1\}$, independence can be weakened to *pairwise independence* (X_i independent of X_j for all $i \neq j$). One way of demonstrating this is via strong laws for non-negative random variables.

Lemma 1 (Etemadi). *If $S_n = \sum_{j=1}^n X_j$, $n \geq 1$, where $\{X_n, n \geq 1\}$ are non-negative $\mathscr{L}_2$ random variables satisfying*

$$\sup_{n \geq 1} \mathrm{E}\, X_n = B < \infty \tag{22}$$

$$\sum_{n=1}^{\infty} \sigma_{X_n}^2 / n^2 < \infty \tag{23}$$

$$\mathrm{E}\, X_m X_n \leq \mathrm{E}\, X_m \cdot \mathrm{E}\, X_n, \qquad n > m \geq 1, \tag{24}$$

then

$$\frac{S_n - \mathrm{E}\, S_n}{n} \xrightarrow{\text{a.s.}} 0. \tag{25}$$

PROOF. Let $k_n = [\alpha^n]$, $\alpha > 1$. Since all pairs of r.v.'s have non-positive covariance, for all $\varepsilon > 0$ and some C in $(0, \infty)$

$$\begin{aligned}\varepsilon^2 \sum_{n=1}^{\infty} \mathrm{P}\{|S_{k_n} - \mathrm{E}\, S_{k_n}| > \varepsilon k_n\} &\leq \sum_{n=1}^{\infty} \sigma_{S_{k_n}}^2 / k_n^2 \leq \sum_{n=1}^{\infty} \sum_{j=1}^{k_n} \sigma_{X_j}^2 / k_n^2 \\ &= \sum_{j=1}^{\infty} \sum_{n=(\log j)/\log \alpha}^{\infty} k_n^{-2} \sigma_{X_j}^2 \leq C \sum_{j=1}^{\infty} \sigma_{X_j}^2 / j^2 < \infty,\end{aligned}$$

whence the Borel–Cantelli lemma ensures

$$(S_{k_n} - \mathrm{E}\, S_{k_n})/k_n \xrightarrow{\text{a.s.}} 0.$$

Via monotonicity of S_k, for any $k \in [k_n, k_{n+1})$,

$$\frac{|S_k - \mathrm{E}\, S_k|}{k} \leq \frac{k_{n+1}}{k_n} \left| \frac{S_{k_{n+1}} - \mathrm{E}\, S_{k_{n+1}}}{k_{n+1}} \right| + \left| \frac{S_{k_n} - \mathrm{E}\, S_{k_n}}{k_n} \right| + \frac{\mathrm{E}(S_{k_{n+1}} - S_{k_n})}{k_n},$$

and so, since $k_n^{-1} \mathrm{E}(S_{k_{n+1}} - S_{k_n}) \leq k_n^{-1}(k_{n+1} - k_n) B \leq (\alpha - 1) B$,

$$\overline{\lim}_{k \to \infty} \left| \frac{S_k - \mathrm{E}\, S_k}{k} \right| \leq (\alpha - 1) B \xrightarrow{\alpha \downarrow 1} 0, \qquad \text{a.s.}$$

yielding (25). □

Corollary 4. *If $S_n = \sum_{j=1}^n X_j$, $n \geq 1$, where $\{X_n, n \geq 1\}$ are non-negative, pairwise independent random variables such that*

$$\sup_{n \geq 1} \mathrm{E}\, X_n < \infty \tag{26}$$

$$\sum_{n=1}^{\infty} n^{-2} \mathrm{E}\, X_n^2 I_{[X_n \leq n]} < \infty \tag{27}$$

$$\sum_{n=1}^{\infty} \mathrm{P}\{X_n > n\} < \infty, \tag{28}$$

then

$$\frac{S_n - \mathrm{E}\, S_n}{n} \xrightarrow{\text{a.s.}} 0. \tag{29}$$

PROOF. The non-negative, pairwise independent, $\mathscr{L}_2$ r.v.'s $Y_n = X_n I_{[X_n \le n]}$, $n \ge 1$ clearly satisfy (24) and also (22) and (23) in view of (26) and (27). Thus, by Lemma 1,

$$\frac{1}{n}\left(\sum_{i=1}^{n} Y_i - \sum_{i=1}^{n} \mathrm{E}\, Y_i\right) \xrightarrow{\text{a.s.}} 0. \tag{30}$$

Now, via (26) and (27),

$$0 \le \frac{1}{n} \mathrm{E}\, X_n I_{[X_n > n]} = \frac{1}{n}(\mathrm{E}\, X_n - \mathrm{E}\, X_n I_{[X_n \le n]}) \le \frac{B}{n} - \frac{1}{n^2} \mathrm{E}\, X_n^2 I_{[X_n \le n]} = o(1)$$

implying as $n \to \infty$ that

$$\frac{1}{n}\left(\mathrm{E}\, S_n - \sum_{i=1}^{n} \mathrm{E}\, Y_i\right) = \frac{1}{n} \sum_{i=1}^{n} \mathrm{E}\, X_i I_{[X_i > i]} \to 0. \tag{31}$$

Since (28) ensures that $\{X_n, n \ge 1\}$ and $\{Y_n, n \ge 1\}$ are equivalent sequences, (30) holds with $\sum_{i=1}^{n} Y_i$ replaced by S_n, and this together with (31) yields (29). □

Theorem 5 (Etemadi). *Let* $S_n = \sum_{i=1}^{n} X_i$, $n \ge 1$, *where* $\{X_n, n \ge 1\}$ *are pairwise independent, identically distributed* $\mathscr{L}_1$ *random variables. Then* $S_n/n \xrightarrow{\text{a.s.}} \mathrm{E}\, X_1$.

PROOF. Evidently, $\{X_n^+, n \ge 1\}$ (likewise $\{X_n^-, n \ge 1\}$) satisfies (26), (28), and also (27) since

$$\begin{aligned}
\sum_{n=1}^{\infty} n^{-2} \mathrm{E}(X_n^+)^2 I_{[X_n^+ \le n]} &= \sum_{n=1}^{\infty} n^{-2} \sum_{j=1}^{n} \mathrm{E}(X^+)^2 I_{[j-1 < X^+ \le j]} \\
&= \sum_{j=1}^{\infty} \sum_{n=j}^{\infty} n^{-2} \mathrm{E}(X^+)^2 I_{[j-1 < X^+ \le j]} \\
&\le C \sum_{j=1}^{\infty} j^{-1} \mathrm{E}(X^+)^2 I_{[j-1 < X^+ \le j]} \\
&\le C\, \mathrm{E} \sum_{j=1}^{\infty} X^+ I_{[j-1 < X^+ \le j]} \\
&= C\, \mathrm{E}\, X^+ < \infty,
\end{aligned}$$

whence Corollary 4 guarantees

$$S_n/n = \frac{1}{n} \sum_{j=1}^{n} (X_i^+ - X_i^-) \xrightarrow{\text{a.s.}} \mathrm{E}\, X_1^+ - \mathrm{E}\, X_1^- = \mathrm{E}\, X_1. \qquad \square$$

An extremely useful tool in probability theory is the Kolmogorov inequality, which, as will be seen in Section 7.4, holds in a much broader context.

Theorem 6 (Kolmogorov inequality). *If* $\{X_j, 1 \le j \le n\}$ *are independent* $\mathscr{L}_2$ r.v.s *with* $\mathrm{E}\, X_j = 0$, $S_j = \sum_1^j X_i$, $1 \le j \le n$, *then for* $\varepsilon > 0$

$$P\left\{\max_{1 \le j \le n} |S_j| \ge \varepsilon\right\} \le \frac{1}{\varepsilon^2} \sum_{j=1}^n \mathrm{E}\, X_j^2. \tag{32}$$

PROOF. Define T = smallest integer j in $[1, n]$ for which $|S_j| \ge \varepsilon$ if such exists and $T = n + 1$ otherwise. Then $\{T \ge j\} \in \sigma(X_1, \ldots, X_{j-1})$ for $1 \le j \le n + 1$, where the σ-algebra in question is $\{\emptyset, \Omega\}$ when $j = 1$. Thus, X_j and $I_{[T \ge j]}$ are independent r.v.s for $1 \le j \le n$. Moreover, since $S_T = S_j$ on $\{T = j\}$, $1 \le j \le n$, and $S_{\min[T,n]} = \sum_{j=1}^n X_j I_{[T \ge j]}$,

$$\begin{aligned}
P\left\{\max_{1 \le j \le n} |S_j| \ge \varepsilon\right\} &= P\{T \le n\} = \sum_{j=1}^n P\{T = j\} \\
&\le \frac{1}{\varepsilon^2} \sum_{j=1}^n \int_{[T=j]} S_j^2 = \frac{1}{\varepsilon^2} \int_{[T \le n]} S_{\min[T,n]}^2 \\
&\le \frac{1}{\varepsilon^2} \mathrm{E}\left(\sum_1^n X_j I_{[T \ge j]}\right)^2 \\
&= \frac{1}{\varepsilon^2} \mathrm{E}\left(\sum_1^n X_j^2 I_{[T \ge j]} + 2\sum_{1 \le i < j \le n} X_i X_j I_{[T \ge j]}\right) \\
&= \frac{1}{\varepsilon^2}\left(\sum_1^n P\{T \ge j\} \mathrm{E}\, X_j^2 + 2\sum_{1 \le i < j \le n} \mathrm{E}(X_i I_{[T \ge j]}) \cdot \mathrm{E}\, X_j\right) \\
&\le \frac{1}{\varepsilon^2} \sum_1^n \mathrm{E}\, X_j^2.
\end{aligned}$$

□

The versatility of the Kolmogorov inequality will be exemplified in extending Theorem 4 to the case of random indices.

Theorem 7. *If* $\{X_n, n \ge 1\}$ *are* i.i.d. r.v.s *obeying*

$$n\, P\{|X_1| > n\} = o(1) \tag{33}$$

and $\{T_n, n \ge 1\}$ *are positive integer-valued* r.v.s *satisfying*

$$\frac{T_n}{n} \xrightarrow{P} c, \qquad \text{where } 0 < c < \infty, \tag{34}$$

then, setting $S_n = \sum_1^n X_i$,

$$(S_{T_n}/T_n) - \mathrm{E}\, X_1 I_{[|X_1| \le n]} \xrightarrow{P} 0. \tag{35}$$

PROOF. Define $X'_j = X_j I_{[|X_j| \le n]}$, $S'_j = \sum_{i=1}^{j} X'_i$, $M'_j = \mathrm{E}\, S'_j$, and $c_n = [c \cdot n] =$ largest integer $\le c \cdot n$. Then,

$$\mathrm{P}\{S_{T_n} \ne S'_{T_n}, T_n \le 2c_n\} \le \mathrm{P}\left\{\bigcup_{1}^{2c_n} [|X_j| > n]\right\}$$

$$\le \sum_{j=1}^{2c_n} \mathrm{P}\{|X_j| > n\} \le 2c_n \mathrm{P}\{|X_1| > n\} = o(1)$$

by (33), whence, taking cognizance of (34),

$$\mathrm{P}\{S_{T_n} \ne S'_{T_n}\} \le \mathrm{P}\{S_{T_n} \ne S'_{T_n}, T_n \le 2c_n\} + \mathrm{P}\{T_n > 2c_n\} = o(1),$$

that is,

$$S_{T_n} - S'_{T_n} \xrightarrow{\mathrm{P}} 0. \tag{36}$$

Now the prior calculation of (17) shows that $\mathrm{E}\, X_1'^2 = o(n)$, whence, defining

$$B_j = \{|S'_j - M'_j| > n\varepsilon\}, \qquad D_n = \bigcup_{1}^{2c_n} B_j,$$

by Kolmogorov's inequality (Theorem 6)

$$\mathrm{P}\{D_n\} \le \frac{1}{n^2\varepsilon^2} \sum_{j=1}^{2c_n} \mathrm{E}\, X_j'^2 \le \frac{2c}{n\varepsilon^2} \mathrm{E}\, X_1'^2 = o(1).$$

Thus,

$$\mathrm{P}\{B_{T_n}\} \le \mathrm{P}\{T_n \le 2c_n, B_{T_n}\} + \mathrm{P}\{T_n > 2c_n\}$$

$$\le \mathrm{P}\{D_n\} + \mathrm{P}\{T_n > 2c_n\} = o(1),$$

so that

$$\frac{S'_{T_n} - M'_{T_n}}{n} \xrightarrow{\mathrm{P}} 0.$$

Hence, by (36) and (34)

$$\frac{S_{T_n} - M'_{T_n}}{T_n} = \frac{n}{T_n}\left(\frac{S_{T_n} - M'_{T_n}}{n}\right) \xrightarrow{\mathrm{P}} 0.$$

But $M'_{T_n} = \sum_{j=1}^{T_n} \mathrm{E}\, X'_j = T_n \mathrm{E}\, X'_1 = T_n \mathrm{E}\, X_1 I_{[|X_1| \le n]}$, whence

$$\frac{S_{T_n}}{T_n} - \mathrm{E}\, X_1 I_{[|X_1| \le n]} \xrightarrow{\mathrm{P}} 0. \qquad \square$$

For any sequence of r.v.s $\{X_n, n \ge 1\}$ and nonnegative numbers u, v define

$$S_{u,v} = \sum_{1 \le j \le v} X_{[u]+j}, \qquad S_v = S_{0,v}, \qquad S_0 = 0. \tag{37}$$

The r.v.s $S_{u,v}$ are called **delayed sums** (of the sequence X_n). If $\sum_{n=1}^{\infty} P\{|S_n| > n\varepsilon\} < \infty$ for every $\varepsilon > 0$, then by the Borel–Cantelli lemma, $S_n/n \xrightarrow{\text{a.c.}} 0$. The converse is false even when the $\{X_n\}$ are i.i.d. since then by the classical SLLN (Theorem 5.2.2, $p = 1$), $S_n/n \xrightarrow{\text{a.c.}} 0$ iff $E X_1 = 0$, whereas according to a theorem of Hsu–Robbins–Erdos (Chapter 10) $\sum_{n=1}^{\infty} P\{|S_n| > n\varepsilon\} < \infty$, all $\varepsilon > 0$ iff $E X_1 = 0$, $E X_1^2 < \infty$. However, in the i.i.d. case $S_n/n \xrightarrow{\text{a.c.}} 0$ does imply that $\sum_{n=1}^{\infty} (1/n)P\{|S_n| > n\varepsilon\} < \infty$ for every $\varepsilon > 0$ by a result of Spitzer, and the next theorem due to Baum–Katz extends this.

For any sequence of r.v.s $\{X_n, n \geq 1\}$ define via (37).

$$S_{u,v}^* = \max_{1 \leq j \leq v} |S_{u,j}|, \qquad S_v^* = \max_{1 \leq j \leq v} |S_j|, \qquad u \geq 0, v \geq 0. \tag{38}$$

Theorem 8 (Baum–Katz–Spitzer). *Let $\{X_n, n \geq 1\}$ be* i.i.d. r.v.s *with* $E|X_1|^p < \infty$ *for some p in $(0, 2)$ and* $E X_1 = 0$ *if $1 \leq p < 2$. If $\alpha p \geq 1$ (hence $\alpha > \frac{1}{2}$). then for every $\varepsilon > 0$*

$$\sum_{n=1}^{\infty} n^{\alpha p - 2} P\{S_n^* > n^{\alpha}\varepsilon\} < \infty. \tag{39}$$

PROOF. By Theorem 5.2.2, $S_n/n^{1/p} \xrightarrow{\text{a.c.}} 0$, implying $S_n^*/n^{1/p} \xrightarrow{\text{a.c.}} 0$ and hence

$$\frac{S_{n,n}^*}{n^{1/p}} \leq \frac{S_n^* + S_{2n}^*}{n^{1/p}} \xrightarrow{\text{a.c.}} 0. \tag{40}$$

Suppose first that $\alpha p = 1$. Since $\{S_{2^n,2^n}^*, n \geq 1\}$ are independent r.v.s, (40) and the Borel–Cantelli theorem (Theorem 3.2.1) imply that for every $\varepsilon > 0$ and $c = (\log 2)^{-1}$

$$\infty > \sum_{n=1}^{\infty} P\{S_{2^n,2^n}^* \geq 2^{\alpha n}\varepsilon\} = \sum_{n=1}^{\infty} P\{S_{2^n}^* \geq 2^{\alpha n}\varepsilon\}$$

$$\geq \int_0^{\infty} P\{S_{2^t}^* \geq 2^{\alpha(t+1)}\varepsilon\}dt > c\int_1^{\infty} x^{-1} P\{S_x^* \geq 2^{\alpha}\varepsilon x^{\alpha}\}dx,$$

and so for all $\varepsilon' > 0$

$$\sum_{n=1}^{\infty} \frac{1}{n} P\{S_n^* \geq n^{\alpha}\varepsilon'\} < \infty.$$

Suppose next that $\alpha p > 1$. Since for $m \geq 1$

$$(m+1)^{\alpha p/(\alpha p - 1)} \geq m^{\alpha p/(\alpha p - 1)} + \frac{\alpha p}{\alpha p - 1} m^{1/(\alpha p - 1)} \geq m^{\alpha p/(\alpha p - 1)} + m^{1/(\alpha p - 1)},$$

the r.v.s

$$\{S_{m^{\alpha p/(\alpha p-1)}, m^{1/(\alpha p - 1)}}^*, m \geq 1\}$$

are independent. Moreover, (40) implies

$$m^{-\alpha/(\alpha p - 1)} S_{m^{\alpha p/(\alpha p - 1)}, m^{1/(\alpha p - 1)}}^* \leq m^{-\alpha/(\alpha p - 1)} S_{m^{\alpha p/(\alpha p - 1)}, m^{\alpha p/(\alpha p - 1)}}^* \xrightarrow{\text{a.c.}} 0$$

as $m \to \infty$, whence for all $\varepsilon > 0$

$$\infty > \sum_{m=1}^{\infty} \mathrm{P}\{S^*_{m^{\alpha p/(\alpha p - 1)}, m^{1/(\alpha p - 1)}} \geq m^{\alpha/(\alpha p - 1)}\varepsilon\}$$

$$= \sum_{m=1}^{\infty} \mathrm{P}\{S^*_{m^{1/(\alpha p - 1)}} \geq m^{\alpha/(\alpha p - 1)}\varepsilon\}$$

$$\geq \int_1^{\infty} \mathrm{P}\{S^*_{t^{1/(\alpha p - 1)}} \geq (t+1)^{\alpha/(\alpha p - 1)}\varepsilon\}dt$$

$$\geq (\alpha p - 1)\int_1^{\infty} x^{\alpha p - 2}\,\mathrm{P}\{S^*_x \geq \varepsilon' x^{\alpha}\}dx$$

for $\varepsilon' = 2^{\alpha/(\alpha p - 1)}\varepsilon$ and (39) follows. □

The converse to Theorem 8 appears in Section 6.4.

Corollary 5. *If* $\{X_n, n \geq 1\}$ *are* i.i.d. r.v.s *with* $\mathrm{E}\,X_1 = 0$, $\mathrm{E}|X_1|^p < \infty$ *for some* p *in* $[1, 2)$, *then*

$$\sum_{n=1}^{\infty} n^{p-2}\,\mathrm{P}\{|S_n| > n\varepsilon\} < \infty, \qquad \varepsilon > 0. \tag{41}$$

The convergence of series such as (39), (41) is enhanced when an explicit bound $C(\varepsilon, p)$ is obtained for the sum since uniform integrability of stopping or last times related to S_n may be deducible therefrom.

EXERCISES 5.2

1. Prove for i.i.d. r.v.s $\{X_n\}$ with $S_n = \sum_1^n X_i$ that $(S_n - C_n)/n \xrightarrow{\text{a.c.}} 0$ for some sequence of constants C_n iff $\mathrm{E}|X_1| < \infty$.
2. If $\{X_n\}$ are i.i.d. with $\mathrm{E}|X_1|^p = \infty$ for some $p \in (0, \infty)$, then $\mathrm{P}\{\overline{\lim}_n |S_n|/n^{1/p} = \infty\} = 1$.
3. Demonstrate for i.i.d. r.v.s $\{X_n\}$ that $\mathrm{E}\sup_n |X_n/n| < \infty$ iff $\mathrm{E}|X_1|\log^+|X_1| < \infty$.
4. If $S_n = \sum_1^n X_i$, where $\{X_n\}$ are i.i.d. $\mathscr{L}_p$ r.v.s for some $p \geq 1$, then $\mathrm{E}|S_n/n|^p \to |\mathrm{E}\,X|^p$. *Hint*: Recall Example 4.3.1.
5. If $\{X_n\}$ are i.i.d. r.v.s with $\mathrm{E}\,X_1 = 1$ and $\{a_n\}$ are bounded real numbers, then $(1/n)\sum_1^n a_j \to 1$ iff $(1/n)\sum_{j=1}^n a_j X_j \xrightarrow{\text{a.c.}} 1$.
6. Let $\{X_n\}$ be i.i.d. r.v.s with $(S_n/n) - C_n \xrightarrow{\mathrm{P}} 0$ for some sequence of constants $\{C_n\}$. Prove that

   i. $\int_{[\alpha n < |X_1| < \beta n]} |X_1| = o(1)$ whenever $0 < \alpha < \beta < \infty$,
   ii. $2\int_{[|X_1| \leq 2n]} X_1 - \int_{[|X_1 + X_2| \leq 2n]} (X_1 + X_2) = o(1)$.
7. Let $\{X, X_n, n \geq 1\}$ be i.i.d. $\mathscr{L}_p$ r.v.s for some p in $(0, 2)$ where $\mathrm{E}\,X = 0$ whenever $1 \leq p < 2$. If $\{a_n, n \geq 1\}$ are positive constants satisfying $a_n^p/\sum_{j=1}^n a_j^p = O(1/n)$, $\sum_{j=1}^n a_j^p \to \infty$, then

$$\frac{1}{\left(\sum_{j=1}^n a_j^p\right)^{1/p}} \sum_{j=1}^n a_j X_j \xrightarrow{\text{a.c.}} 0. \tag{*}$$

Conversely, if $a_n^p/\sum_{j=1}^n a_j^p \sim Cn$, then (*) ensures $X \in \mathscr{L}_p$ (and also $\mathrm{E}\, X = 0$ if $p = 1$). *Hint*: if $Y_n = a_n(\sum_{j=1}^n a_j^p)^{-1/p} X_n \cdot I_{[|X_n| \le n^{1/p}]}$, then $\mathrm{E}|Y_n|^\alpha \le Cn^{-\alpha/p}\mathrm{E}|X|^\alpha I_{[|X| \le n^{1/p}]}$ whence $\sum_{n=1}^\infty \mathrm{E}|Y_n|^\alpha < \infty$ for $\alpha > p$ and $\sum_{n=1}^\infty |\mathrm{E}\, Y_n| < \infty$ for $1 < p < 2$ as in Theorem 5.1.3. N.B. Exercise 7 with $p = 1$ and Theorem 5.2.3(ii) are related to summability.

8. For $r > 0$ and any r.v. X, prove that $X \in \mathscr{L}_r$ iff

$$\sum_{n=1}^\infty n^{r-1}(\log n)^r \mathrm{P}\{|X| \ge n \log n\} < \infty$$

Hint: Employ the techniques of Theorem 7.

9. (Klass–Teicher) If $\{X, X_n, n \ge 1\}$ are i.i.d. r.v.s and $\{b_n, n \ge 1\}$ are constants with (i) $b_n/n \uparrow$ or (ii) $b_n/n \downarrow$, $b_n/n^{1/2} \to \infty$, $\sum_1^n (b_j/j)^2 = O(b_n^2/n)$ and $b_n \uparrow$, then

$$\frac{1}{b_n}\sum_1^n X_i - \frac{n}{b_n}\mathrm{E}\, X I_{[|X| \le b_n]} \xrightarrow{\mathrm{P}} 0 \text{ iff } n\, \mathrm{P}\{|X| > b_n\} = o(1).$$

10. Prove that if $\{X_n, n \ge 1\}$ are independent r.v.s with $\mathrm{E}\, X_n = 0$, $\mathrm{E}\, X_n^2 = \sigma_n^2$, $s_n^2 = \sum_1^n \sigma_i^2 \to \infty$, then $s_n^{-1}(\log s_n^2)^{-\alpha} \sum_1^n X_i \xrightarrow{\text{a.c.}} 0$ for $\alpha > \frac{1}{2}$.

11. (Feller–Chung) Let $\mathrm{P}(x)$ be nonnegative and nonincreasing on $(0, \infty)$ and suppose that the positive, nondecreasing sequence $\{b_n, n \ge 1\}$ satisfies (*) $\underline{\lim}_{n\to\infty} b_{nr}/b_n > c > 1$ for some integer r or *a fortiori* either (**) $b_n/n^\beta \uparrow$ for some $\beta > 0$ or (***) $b_n^2 \sum_{j=n}^\infty b_j^{-2} = O(n)$. Then $\sum_{n=1}^\infty \mathrm{P}(xb_n)$ either diverges for all $x > 0$ or converges for all $x > 0$. (*Hint*: Any integer m satisfies $r^k \le m < r^{k+1}$, where $k = k(m) \to \infty$ as $m \to \infty$, whence $b_{nm} \ge c^k b_n$, all large n). Thus, if $\{X, X_n\}$ are i.i.d. and $\sum_{n=1}^\infty \mathrm{P}\{|X| > Cb_n\} = \infty$ for some $C > 0$ and $\{b_n\}$ as stipulated, $\overline{\lim}_{n\to\infty} |X_n|/b_n \overset{\text{a.c.}}{=\!=} \infty = \overline{\lim}_{n\to\infty} |S_n|/b_n$.

12. (Heyde–Rogozin) If $\{X, X_n, n \ge 1\}$ are i.i.d. with (*) $\underline{\lim}_{x\to\infty} x^2 \mathrm{P}\{|X| > x\}/\mathrm{E}\, X^2 I_{[|X| \le x]} > \alpha > 0$, then for every sequence $\{b_n\}$ satisfying $0 < b_n \uparrow \infty$ either $\overline{\lim}_{n\to\infty} |S_n|/b_n \overset{\text{a.c.}}{=\!=} \infty$ or $(1/b_n)(S_n - \sum_1^n \mathrm{E}\, X I_{[|X| \le b_j]}) \xrightarrow{\text{a.c.}} 0$. In particular, for symmetric i.i.d. r.v.s satisfying (*), $S_n/b_n \longrightarrow 0$ or $\overline{\lim}\, S_n/b_n \overset{\text{a.c.}}{=\!=} \infty$. *Hint*: $\mathrm{P}\{|X| > x\} \le (1 + \delta^2\alpha^{-1})\mathrm{P}\{|X| > \delta x\}$, $\delta > 1$, $x > x_0$, whence the series of Exercise 11 either converges for all $C > 0$ or diverges for all $C > 0$.

13. (Komlös–Revesz) Let $\{X_n, n \ge 1\}$ be independent r.v.s with means $\mathrm{E}\, X_n$ and positive variances σ_n^2 satisfying $\lim \mathrm{E}\, X_n = c$ and $\sum_{n=1}^\infty \sigma_n^{-2} = \infty$. Then

$$\frac{\sum_{j=1}^n X_j/\sigma_j^2}{\sum_{j=1}^n \sigma_j^{-2}} \xrightarrow{\text{a.c.}} c.$$

Hint: $\sum_{j=1}^n [(X_j - \mathrm{E}\, X_j)/\sigma_j^2 \sum_{i=1}^j \sigma_i^{-2})]$ converges a.c.

14. (Heyde) Let $\{X_n, n \ge 1\}$ be i.i.d. $\mathscr{L}_p$ r.v.s, $S_n = \sum_1^n X_i$, $\bar{S}_n = \max_{1 \le j \le n} S_j$. (i) If $1 \le p < 2$, then $n^{-1/p}(\bar{S}_n - n\mu) \xrightarrow{\text{a.c.}} 0$ or $n^{-1/p}\bar{S}_n \xrightarrow{\text{a.c.}} 0$ according as $\mu = \mathrm{E}\, X_1 \ge 0$ or $\mu < 0$. (ii) If $0 < p < 1$, then $n^{-1/p}\bar{S}_n \xrightarrow{\text{a.c.}} 0$.

15. (Derman–Robbins) $\{X_n\}$ i.i.d. with E X_1 nonexistent does not preclude $P\{S_n/n \to \infty\} = 1$. *Hint*: Let $\underline{\lim}_{x\to\infty} x^\alpha P\{X > x\} > 0$ and $E(X^-)^\beta < \infty$ for $1 > \beta > \alpha > 0$. The latter implies $n^{-1/\beta} \sum_1^n X_i^- \xrightarrow{\text{a.c.}} 0$, while the former entails $P\{\sum_1^n X_i^+ \le Cn^{1/\beta}\} \le P\{\max_{1\le i\le n} X_i^+ \le Cn^{1/\beta}\} \le \exp(-C'n^\gamma)$, $\gamma = 1 - \alpha\beta^{-1} > 0, C > 0$.

16. (Klass) Let $\{X_n\}$ be i.i.d. with E $X_1 = 0$ and $\sum_{n=1}^\infty P\{|X_n| > b_n\} < \infty$, where

$$b_n > 0, n^{-2}b_n^2 \downarrow, n^{-1}b_n^2 \uparrow \infty.$$

Then $E|S_n| = o(b_n)$. *Hint*: $X_j = Y_{n,j} + Z_{n,j}$, where $Y_{n,j} = X_j I_{[|X_j|\le b_n]}$. In particular, X_n i.i.d. with E $X_1 = 0$, $E|X_1|^p < \infty$, $p \in [1, 2)$, implies $E|S_n| = o(n^{1/p})$.

17. **Strong Laws for Arrays.** Let $\{X_{ni}, 1 \le i \le n\}$ be an array of r.v.'s that are identically distributed and *rowwise independent*, i.e., $\{X_{n1}, \ldots, X_{nn}\}$ are independent for $n \ge 2$. If $E|X_{11}|^q < \infty$, $0 < q < 2$, and E $X_{11} = 0$ whenever $1 \le q < 2$, then $n^{-2/q}\sum_{i=1}^n X_{ni} \xrightarrow{\text{a.s.}} 0$. (For an extension to $2 \le q < 4$, see Example 10.4.1.)

5.3 Stopping Times, Copies of Stopping Times, Wald's Equation

Let $(\Omega, \mathscr{F}, P)$ be a probability space and $\{\mathscr{F}_n, n \ge 1\}$ an increasing sequence of sub-σ-algebras of $\mathscr{F}$, that is, $\mathscr{F}_1 \subset \mathscr{F}_2 \subset \cdots \subset \mathscr{F}$. A measurable function $T = T(\omega)$ taking values $1, 2, \ldots, \infty$ is called a **stopping time** relative to $\{\mathscr{F}_n\}$ or simply an **$\{\mathscr{F}_n\}$-time** if $\{T = j\} \in \mathscr{F}_j$, $j = 1, 2 \ldots$. Clearly, If $T \equiv m$, then T is an $\mathscr{F}_n$-time. If T is an $\mathscr{F}_n$-time, then setting $\mathscr{F}_0 = \{\emptyset, \Omega\}$, $\mathscr{F}_\infty = \sigma(\bigcup_1^\infty \mathscr{F}_n)$,

$$\{T \ge n\} = \Omega - \bigcup_1^{n-1}\{T = j\} \in \mathscr{F}_{n-1}, \qquad 1 \le n \le \infty. \tag{1}$$

Moreover, since $\{T = \infty\} = \Omega - \{T < \infty\}$, a stopping time is completely determined by the sets $\{T = n\}$, $1 \le n < \infty$.

A stopping time T is said to be **finite** if $P\{T = \infty\} = 0$ and **defective** if $P\{T = \infty\} > 0$. A finite stopping time is also called a **stopping rule** or **stopping variable.** When $\mathscr{F}_n = \sigma(X_1, \ldots, X_n)$, $n \ge 1$, for some sequence of r.v.s $\{X_n\}$, an $\mathscr{F}_n$-time will generally be alluded to as an **$\{X_n\}$-time** or a stopping time relative to $\{X_n\}$. Stopping times and rules have already appeared incognito in Theorem 5.2.5, Lemmas 5.1.3, 5.1.4, Theorem 3.4.1, and Lemma 3.3.5.

The notion of a stopping time derives from gambling. Its definition is simply the mathematical formulation of the fact that, since an honest gambler cannot peer into the future, his decision to stop gambling at any time n must be based solely upon the outcomes $X_1, \ldots, X_n$ up to that time and not on subsequent outcomes X_j, $j > n$.

Let $\{X_n, n \ge 1\}$ constitute a sequence of r.v.s and $\{\mathscr{F}_n, n \ge 1\}$ an increasing sequence of sub-σ-algebras of $\mathscr{F}$ such that X_n is $\mathscr{F}_n$-measurable for each $n \ge 1$. Then $\{X_n, \mathscr{F}_n, n \ge 1\}$ will be called a **stochastic sequence**.

Clearly, for any sequence of r.v.s $\{X_n\}$, $\{X_n, \sigma(X_1, \ldots, X_n), n \geq 1\}$ is a stochastic sequence.

For any stochastic sequence $\{X_n, \mathscr{F}_n, n \geq 1\}$ and $\mathscr{F}_n$-time T, define

$$X_T = X_{T(\omega)}(\omega), \quad \text{where } X_\infty(\omega) = \overline{\lim_{n\to\infty}} X_n(\omega), \tag{2}$$

$$\mathscr{F}_\infty = \sigma\left(\bigcup_1^\infty \mathscr{F}_n\right), \qquad \mathscr{F}_T = \{A: A \in \mathscr{F}_\infty, A \cdot \{T = n\} \in \mathscr{F}_n, n \geq 1\}. \tag{3}$$

It is easy to verify that (i) $\mathscr{F}_T$ is a sub-σ-algebra of $\mathscr{F}_\infty$, (ii) X_T and T are $\mathscr{F}_T$-measurable, and (iii) $\mathscr{F}_T$ coincides with $\mathscr{F}_m$ when $T \equiv m$.

Since

$$\mathbf{P}\{|X_T| = \infty\} = \mathbf{P}\{T = \infty, |X_\infty| = \infty\},$$

X_T is a r.v. if either T is finite or $\overline{\lim}\, X_n$ is finite, a.c.

For $\{\mathscr{F}_n\}$-times T_1 and T_2 with $T_1 \leq T_2$,

$$\mathscr{F}_{T_1} \subset \mathscr{F}_{T_2} \tag{4}$$

since if $A \in \mathscr{F}_{T_1}$ then $A\{T_2 \leq n\} = A\{T_1 \leq n\} \cdot \{T_2 \leq n\} \in \mathscr{F}_n$, $1 \leq n < \infty$. If T is an $\{\mathscr{F}_n\}$-time, so is $T + m$ for every integer $m \geq 1$, and since $T \leq T + m$, necessarily $\mathscr{F}_T \subset \mathscr{F}_{T+1} \subset \cdots \subset \mathscr{F}_\infty$. Hence, $\mathscr{F}_\infty = \sigma(\bigcup_1^\infty \mathscr{F}_n) \subset \sigma(\bigcup_1^\infty \mathscr{F}_{T+n}) \subset \mathscr{F}_\infty$, that is, $\sigma(\bigcup_1^\infty \mathscr{F}_{T+n}) = \mathscr{F}_\infty$. Consequently, the sequence $\mathscr{G}_n = \mathscr{F}_{T+n}$, $n \geq 1$, may be utilized to define a new stopping time. Note, incidentally, that $\mathscr{F}_{T+n} = \mathscr{F}_{n+m}$ when $T \equiv m$.

Suppose that T_1 is an $\{\mathscr{F}_n, n \geq 1\}$-time and T_2 is a $\{\mathscr{G}_n, n \geq 1\}$-time, where $\mathscr{G}_n = \mathscr{F}_{T_1+n}$, $n \geq 1$. Then $T = T_1 + T_2$ is an $\{\mathscr{F}_n, n \geq 1\}$-time and $\mathscr{F}_T = \mathscr{G}_{T_2}$. Since for $1 \leq m < \infty$

$$\{T = m\} = \bigcup_{j=1}^m \{T_1 = j, T_2 = m - j\} \in \mathscr{F}_m,$$

T is an $\{\mathscr{F}_n, n \geq 1\}$-time. To prove $\mathscr{F}_T = \mathscr{G}_{T_2}$, let $A \in \mathscr{G}_{T_2}$. Then for $m = 1, 2, \ldots$ and $j = 1, \ldots, m - 1$,

$$A\{T_2 = m - j\} \in \mathscr{G}_{m-j} = \mathscr{F}_{T_1+m-j},$$

$$A\{T_1 = j, T_2 = m - j\} = A\{T_2 = m - j\}\{T_1 + m - j = m\} \in \mathscr{F}_m,$$

$$A\{T = m\} = \bigcup_{j=1}^{m-1} A\{T_1 = j, T_2 = m - j\} \in \mathscr{F}_m,$$

which implies that $A \in \mathscr{F}_T$. Conversely, let $A \in \mathscr{F}_T$. Then for $r = 1, 2, \ldots$, and $m = 0, 1, \ldots$,

$$A\{T = r + m\} \in \mathscr{F}_{r+m},$$

$$A\{T_2 = m\} \cdot \{T_1 + m = r + m\} = A\{T = r + m\}\{T_1 = r\} \in \mathscr{F}_{r+m},$$

$$A\{T_2 = m\} \in \mathscr{F}_{T_1+m} = \mathscr{G}_m,$$

which implies that $A \in \mathscr{G}_{T_2}$.

On the other hand, if T_1, T_2 are $\{\mathscr{F}_n, n \geq 1\}$-times with $T_1 < T_2$ and

$$\tau = \begin{cases} T_2 - T_1 & \text{if } T_1 < \infty \\ \infty & \text{if } T_1 = \infty, \end{cases} \tag{5}$$

then $T_2 = T_1 + \tau$ and τ is a $\{\mathscr{G}_n, n \geq 1\}$-time, where $\mathscr{G}_n = \mathscr{F}_{T_1+n}$. It suffices to note that when $1 \leq n < \infty$

$$\{\tau = n\}\{T_1 + n = j + n\} = \{T_1 = j\}\{T_2 = j + n\} \in \mathscr{F}_{j+n}, \qquad 1 \leq j < \infty,$$

implies $\{\tau = n\} \in \mathscr{F}_{T_1+n} = \mathscr{G}_n, n \geq 1$.

Lemma 1. *If T is an $\{X_n\}$-time for some sequence $\{X_n\}$ of* r.v.s, *there exists a sequence $\{C_n\}$ of disjoint Borel cylinder sets of $(R^\infty, \mathscr{B}^\infty)$ whose corresponding bases B_n are n-dimensional Borel sets, $n \geq 1$, such that*

$$\{\omega: T = n\} = \{\omega: (X_1, \ldots, X_n, \ldots) \in C_n\}, \qquad n = 1, 2 \ldots \tag{6}$$

Conversely, given any sequence $\{C_n, n \geq 1\}$ of disjoint Borel cylinder sets with n-dimensional Borel bases, an $\{X_n\}$-time T is defined by (6) *and $\{T = \infty\} = \Omega - \bigcup_1^\infty \{T = n\}$.*

PROOF. If T is an $\{X_n\}$-time, then $\{T = n\} \in \sigma(X_1, \ldots, X_n), n \geq 1$, whence by Theorem 1.4.4 there exists an n-dimensional Borel set B_n for which

$$\{T = n\} = \{(X_1, \ldots, X_n) \in B_n\}.$$

For each $n \geq 1$, let C'_n be the cylinder set in R^∞ with base B_n. Then

$$\{T = n\} = \{(X_1, \ldots, X_n, \ldots) \in C'_n\}, \qquad n \geq 1.$$

Moreover, $C_n = C'_n - \bigcup_{j=1}^{n-1} C'_j, n \geq 1$, are disjoint Borel cylinder sets with n-dimensional Borel bases. Since $\{T = m\} \cdot \{T = n\} = \varnothing$ for $m \neq n$, (6) follows.

Conversely, given a sequence of disjoint cylinders $C_n \in \mathscr{B}^\infty$ with n-dimensional Borel bases $B_n, n \geq 1$, if T is defined as stipulated, then

$$\{T = n\} = \{(X_1, \ldots, X_n) \in B_n\} \in \sigma(X_1, \ldots, X_n), \qquad n \geq 1,$$

so that T is an $\{X_n\}$-time. □

Lemma 2. *If $\{X_n, n \geq 1\}$ are* i.i.d. r.v.s *and T is a finite $\{\mathscr{F}_n\}$-time where $\mathscr{F}_n$ and $\sigma(X_j, j > n)$ are independent, $n \geq 1$, then $\mathscr{F}_T$ and $\sigma(X_{T+1}, X_{T+2}, \ldots)$ are independent and $\{X_{T+n}, n \geq 1\}$ are* i.i.d. *with the same distribution as X_1.*

PROOF. If $\lambda_1, \ldots, \lambda_n$ are real numbers and $A \in \mathscr{F}_T$,

$$P\left\{A \cdot \bigcap_{i=1}^n [X_{T+i} < \lambda_i]\right\} = \sum_{j=1}^\infty P\left\{A \cdot [T = j] \bigcap_{i=1}^n [X_{j+i} < \lambda_i]\right\}$$

$$= \sum_{j=1}^\infty P\{A \cdot [T = j]\} \cdot P\left\{\bigcap_{i=1}^n [X_{j+i} < \lambda_i]\right\}$$

since $A \cdot [T = j] \in \mathscr{F}_j$. Hence,

$$\mathrm{P}\left\{A \cdot \bigcap_{i=1}^{n} [X_{T+i} < \lambda_i]\right\} = \sum_{j=1}^{\infty} \mathrm{P}\{A \cdot [T = j]\} \prod_{i=1}^{n} \mathrm{P}\{X_{j+i} < \lambda_i\}$$

$$= \sum_{j=1}^{\infty} \mathrm{P}\{A \cdot [T = j]\} \prod_{i=1}^{n} \mathrm{P}\{X_i < \lambda_i\}$$

$$= \mathrm{P}\{A\} \prod_{i=1}^{n} \mathrm{P}\{X_i < \lambda_i\}. \tag{7}$$

Hence, taking $A = \Omega$,

$$\mathrm{P}\{X_{T+i} < \lambda_i, 1 \le i \le n\} = \prod_{i=1}^{n} \mathrm{P}\{X_i < \lambda_i\}$$

and, in particular, $\mathrm{P}\{X_{T+j} < \lambda_j\} = \mathrm{P}\{X_j < \lambda_j\} = \mathrm{P}\{X_1 < \lambda_j\}$, $1 \le j \le n$. Thus, since n is an arbitrary positive integer, $\{X_{T+n}, n \ge 1\}$ are i.i.d. with the same distribution as X_1. Consequently, from (7)

$$\mathrm{P}\left\{A \cdot \bigcap_{i=1}^{n} [X_{T+i} < \lambda_i]\right\} = \mathrm{P}\{A\} \prod_{i=1}^{n} \mathrm{P}\{X_{T+i} < \lambda_i\}, \qquad A \in \mathscr{F}_T,$$

and therefore, in view of the arbitrariness of $\lambda_1, \ldots, \lambda_n$ and n, the classes $\mathscr{F}_T$ and $\sigma(X_{T+1}, X_{T+2}, \ldots)$ are independent. □

Corollary 1. *$\sigma(T)$ and $\sigma(X_{T+1}, X_{T+2}, \ldots)$ are independent.*

PROOF. It suffices to recall that T is $\mathscr{F}_T$-measurable. □

Next, let $\{X_n, n \ge 1\}$ be i.i.d. r.v.s and T a finite $\{X_n\}$-time. Then, by Lemma 1 there exist disjoint cylinder sets $\{C_n, n \ge 1\}$ in $\mathscr{B}^\infty$ with n-dimensional Borel bases such that

$$\{T = n\} = \{(X_1, X_2, \ldots) \in C_n\}, \qquad 1 \le n < \infty.$$

Define $T^{(1)} = T_1 = T$ and $T^{(j+1)}, j \ge 1$, via $T_j = \sum_1^j T^{(i)}$, by

$$\{T^{(j+1)} = n\} = \{(X_{T_j+1}, X_{T_j+2}, \ldots) \in C_n\}, \qquad 1 \le n < \infty.$$

Then, as noted earlier for $j = 1$, $T^{(j+1)}$ is a finite $\{X_{T_j+n}, n \ge 1\}$-time, $j \ge 1$. The stopping variables $\{T^{(j+1)}, j \ge 1\}$ or $\{T^{(j)}, j \ge 1\}$ will be called **copies of T**. Moreover, as follows from earlier discussion, $T_m = \sum_{j=1}^m T^{(j)}$ is a (finite) $\{X_n\}$-time and $\mathscr{F}_{T_m} \subset \mathscr{F}_{T_{m+1}}$, $m \ge 1$.

In a sequence of Bernoulli trials with parameter $p = \frac{1}{2}$, i.e., $S_n = \sum_1^n X_i$, where $\{X_n\}$ are i.i.d. r.v.s with $\mathrm{P}\{X_i = \pm 1\} = \frac{1}{2}$, let $T = T_1 = T^{(1)} = \inf\{n \ge 1: S_n = 0\}$ and $T^{(j+1)} = \inf\{n \ge 1: S_{T_j+n} = 0\}$. Then $T_j = \sum_{i=1}^j T^{(i)}$, $j \ge 1$, are the return times to the origin and $T^{(j)}$ is the time between the $(j-1)$st and jth return. If, rather, $T = \inf\{n \ge 1: S_n = 1\}$, then T_j is the **first passage time** through the barrier at j, whereas $T^{(j)}$ is the amount of time required to pass from $j - 1$ to j. Either choice of T yields a finite $\{X_n\}$-time with infinite expectation (Exercises 3.4.1, 3.4.2).

Lemma 3. *Let $\{X_n, n \geq 1\}$ be* i.i.d. r.v.*s and T a finite $\{X_n\}$-time. If $T_0 = 0$, $T^{(1)} = T$, and $\{T^{(j)}, j > 1\}$ are copies of T, then, setting $T_m = \sum_{j=1}^{m} T^{(j)}$, the random vectors*

$$V_m = (T^{(m)}, X_{T_{m-1}+1}, X_{T_{m-1}+2}, \ldots, X_{T_m}), \qquad m \geq 1,$$

are i.i.d.

PROOF. As already noted, T_m is a finite $\{X_n\}$- time. Moreover, it is easy to see that V_m and hence also $(V_1, \ldots, V_m)$ is $\mathscr{F}_{T_m}$-measurable. By Lemma 2, $\sigma(X_{T_m+1}, X_{T_m+2}, \ldots)$ is independent of $\mathscr{F}_{T_m}$, $m \geq 1$, and, since $T^{(m+1)}$ and $(X_{T_m+1}, \ldots, X_{T_{m+1}})$ are $\sigma(X_{T_m+1}, X_{T_m+2}, \ldots)$-measurable, $\sigma(V_{m+1})$ and $\mathscr{F}_{T_m}$ are independent for $m \geq 1$. Thus V_{m+1} is independent of $(V_1, \ldots, V_m)$, $m \geq 1$, which is tantamount to independence of $\{V_m, m \geq 1\}$.

Furthermore, for all real λ_i, $1 \leq i \leq n$, and $m, n \geq 1$, if $\{C_n, n \geq 1\}$ are the Borel cylinder sets of (6) defining $T^{(1)}$,

$$\begin{aligned} q_m &= \mathrm{P}\{T^{(m)} = n, X_{T_{m-1}+1} < \lambda_1, X_{T_{m-1}+2} < \lambda_2, \ldots, X_{T_m} < \lambda_{T^{(m)}}\} \\ &= \mathrm{P}\{T^{(m)} = n, X_{T_{m-1}+1} < \lambda_1, \ldots, X_{T_{m-1}+n} < \lambda_n\} \\ &= \mathrm{P}\{(X_{T_{m-1}+1}, X_{T_{m-1}+2}, \ldots, X_{T_{m-1}+n}, \ldots) \in C_n, \\ &\qquad X_{T_{m-1}+1} < \lambda_1, \ldots, X_{T_{m-1}+n} < \lambda_n\}. \\ &= \mathrm{P}\{(X_1, X_2, \ldots) \in C_n, X_1 < \lambda_1, \ldots, X_n < \lambda_n\} = q_1 \end{aligned}$$

by Lemma 2, since T_m is a finite $\{X_n\}$-time. Thus, $\{V_m, m \geq 1\}$ are i.i.d. random vectors. □

Corollary 2. *If T is a finite $\{X_n\}$-time, where $\{X_n, n \geq 1\}$ are* i.i.d. *random variables, then the copies $\{T^{(n)}, n \geq 1\}$ of T are* i.i.d. *random variables.*

If $\{X_n, n \geq 1\}$ are r.v.s and T is an $\{X_n\}$-time, then the expectation of $|X_T|$ is

$$\begin{aligned} \mathrm{E}|X_T| &= \int_{\{T<\infty\}} |X_T| + \int_{\{T=\infty\}} |\overline{\lim}\, X_n| \\ &= \sum_{j=1}^{\infty} \int_{\{T=j\}} |X_j| + \int_{\{T=\infty\}} |\overline{\lim}\, X_n|. \end{aligned} \tag{8}$$

Analogously, if the expectation of X_T exists,

$$\mathrm{E}\, X_T = \mathrm{E}(X_T)^+ - \mathrm{E}(X_T)^- = \sum_{j=1}^{\infty} \int_{[T=j]} X_j + \int_{[T=\infty]} \overline{\lim}\, X_n. \tag{9}$$

As customary, X_T is called integrable if $|\mathrm{E}\, X_T| < \infty$. If a stopping time T is integrable, then necessarily $\mathrm{P}\{T \geq n\} = o(n^{-1})$ and *a fortiori* T is a finite stopping time or stopping variable.

Consider a sum of a random number of r.v.s $S_T = \sum_1^T X_i$, where T is an $\{X_n, n \geq 1\}$-time. If T is integrable and the $\{X_n\}$ are i.i.d. $\mathscr{L}_1$ r.v.s, the expectation of S_T has the natural form (10).

Theorem 1 (Wald's Equation). *Let* $\{X_n, n \geq 1\}$ *be* i.i.d. r.v.s *on* $(\Omega, \mathscr{F}, P)$, $S_n = \sum_1^n X_i$, $n \geq 1$, *and let* $\{\mathscr{F}_n, n \geq 1\}$ *constitute an increasing sequence of sub-σ-algebras of* $\mathscr{F}$ *with* (i) $\mathscr{F}_n$ *and* $\sigma(X_{n+1})$ *independent*, $n \geq 1$. *If* $E X_1$ *exists and* T *is an* $\{\mathscr{F}_n\}$*-time with* $E T < \infty$, *then*

$$E S_T = E X_1 \cdot E T. \tag{10}$$

PROOF. Suppose first that $E|X_1| < \infty$. Then by Theorem 4.3.3

$$E \sum_1^\infty |X_n| I_{[T \geq n]} = \sum_1^\infty E|X_n| I_{[T \geq n]} = \sum_1^\infty E|X_n| P\{T \geq n\} = E T \cdot E|X_1|,$$

whence by the dominated convergence theorem

$$E S_T = E \sum_1^\infty X_n \cdot I_{[T \geq n]} = \sum_1^\infty P\{T \geq n\} \cdot E X_n = E X_1 \cdot E T.$$

If, rather, $E X_1 = \infty$, then $E X_1^- < \infty$, $E X_1^+ = \infty$, whence by what has just been proved

$$E S_T^- \leq E \sum_1^\infty X_n^- I_{[T \geq n]} = E T \cdot E X_1^- < \infty.$$

Therefore, $E S_T$ exists and

$$\begin{aligned} E S_T &= E \sum_1^\infty (X_n^+ - X_n^-) I_{[T \geq n]} = E \sum_1^\infty X_n^+ I_{[T \geq n]} - E \sum_1^\infty X_n^- I_{[T \geq n]} \\ &= E T \cdot E X_1^+ - E T \cdot E X_1^- = E T \cdot E X_1. \end{aligned}$$

The proof for $E X_1 = -\infty$ is analogous. □

Corollary 3. *If* $\{X_n\}$ *are* i.i.d. r.v.s *for which* $E X_1$ *exists and* T *is an* $\{X_n\}$*-time with* $E T < \infty$, *then, setting* $S_n = \sum_1^n X_i$, (10) *obtains.*

The next result is a refinement of Theorem 1, with (i) due to Robbins–Samuel.

Theorem 2. *Let* $\{X_n\}$ *be* i.i.d. r.v.s, *let* $S_n = \sum_1^n X_i$, *and let* T *be a finite* $\{X_n\}$*-time for which* $E S_T$ *exists.* (i) *If* $E X_1$ *exists and either* $E X_1 \neq 0$ *or* $E T < \infty$, *then* (10) *holds.* (ii) *If* $P\{|X_1| > n\} = o(n^{-1})$ *and* $E T < \infty$, *then*

$$\frac{E S_T}{E T} = \lim_{n \to \infty} E X_1 I_{[|X_1| \leq n]}, \tag{11}$$

and when S_T *is integrable,*

$$\frac{S_n}{n} \xrightarrow{P} \frac{E S_T}{E T}. \tag{12}$$

PROOF. Let $T^{(j)}$, $j \geq 1$, be the copies of T and set $T_0 = 0$, $T_m = \sum_1^m T^{(j)}$. Then according to Corollary 2 and Lemma 3, $T^{(j)}$, $j \geq 1$, are i.i.d. and so are

$S^{(j)} = X_{T_{j-1}+1} + \cdots + X_{T_j}$ for $j \geq 1$. By Corollary 5.2.2

$$\frac{S_{T_m}}{m} = \frac{S^{(1)} + \cdots + S^{(m)}}{m} \xrightarrow{\text{a.c.}} \mathrm{E}\, S_T, \qquad \frac{T_m}{m} \xrightarrow{\text{a.c.}} \mathrm{E}\, T. \tag{13}$$

If $\mathrm{E}\, X_1$ exists, then by this same corollary and exercise 3.3.10

$$\frac{S_{T_m}}{T_m} \xrightarrow{\text{a.c.}} \mathrm{E}\, X_1,$$

which, in conjunction with (13), yields (10). On the other hand, if $\mathrm{P}\{|X_1| > n\} = o(n^{-1})$, then by Theorem 5.2.6

$$\frac{S_{T_m}}{T_m} - c_m \xrightarrow{\mathrm{P}} 0, \tag{14}$$

where $c_m = \mathrm{E}\, X_1 I_{[|X_1| \leq m]}$. Hence, any subsequence $\{n''\}$ of the positive integers has itself a subsequence, say $\{n'\}$, for which by (14) and Lemma 3.3.2

$$\frac{S_{T_{n'}}}{T_{n'}} - c_{n'} \xrightarrow{\text{a.c.}} 0. \tag{15}$$

Then, via (15) and (13), recalling that $\mathrm{E}\, T < \infty$,

$$c_{n'} = \left(c_{n'} - \frac{S_{T_{n'}}}{T_{n'}}\right) + \frac{S_{T_{n'}}}{n'} \cdot \frac{n'}{T_{n'}} \xrightarrow{\text{a.c.}} \frac{\mathrm{E}\, S_T}{\mathrm{E}\, T}.$$

Thus, every subsequence of $\{c_n\}$ has itself a subsequence whose limit is independent of the subsequences, yielding (11). Finally, (12) follows from (11) and Theorem 5.2.4. □

Corollary 4. *Let $\{X_n, n \geq 1\}$ be* i.i.d. r.v.s, *let $S_n = \sum_1^n X_i$ and let T be an integrable $\{X_n\}$-time. If $n\, \mathrm{P}[|X_1| > n] = o(1)$ and $\mathrm{E}\, X_1 I_{[|X_1| \leq n]}$ has no limit as $n \to \infty$, then $\mathrm{E}\, S_T$ does not exist.*

The next theorem is the second moment analogue of Wald's equation.

Theorem 3. *If $\{X_n\}$ are independent* r.v.s *with $\mathrm{E}\, X_n = 0$, $\mathrm{E}\, X_n^2 = \sigma^2 < \infty$, $S_n = \sum_1^n X_i$, $n \geq 1$ and T is an $\{\mathscr{F}_n\}$-time with $\mathrm{E}\, T < \infty$ where* (i) *$\mathscr{F}_n \supset \sigma(X_1 \cdots X_n)$ and* (ii) *$\mathscr{F}_n$ and $\sigma(X_{n+1})$ are independent, $n \geq 1$, then*

$$\mathrm{E}\, S_T^2 = \sigma^2\, \mathrm{E}\, T. \tag{16}$$

PROOF. If $T(n) = \min[T, n]$, then $T(n)$ is a finite stopping time and

$$\begin{aligned}\mathrm{E}\, S_{T(n)}^2 &= \mathrm{E}\left(\sum_1^n X_j I_{[T \geq j]}\right)^2 \\ &= \mathrm{E}\left(\sum_1^{n-1} X_j I_{[T \geq j]}\right)^2 + \mathrm{E}\, X_n^2 I_{[T \geq n]} + 2\, \mathrm{E}\, X_n I_{[T \geq n]} \sum_1^{n-1} X_j I_{[T \geq j]}.\end{aligned}$$

Since $I_{[T \geq n]} \sum_1^{n-1} X_j I_{[T \geq j]}$ is $\mathscr{F}_{n-1}$-measurable and $\sigma(X_n)$ is independent of

$\mathscr{F}_{n-1}$,

$$\mathrm{E}\, X_n I_{[T \geq n]} \sum_1^{n-1} X_j I_{[T \geq j]} = \mathrm{E}\, X_n \mathrm{E} \sum_1^{n-1} X_j I_{[T \geq j]} I_{[T \geq n]} = 0,$$

whence

$$\mathrm{E}\left(\sum_1^n X_j I_{[T \geq j]}\right)^2 - \mathrm{E}\left(\sum_1^{n-1} X_j I_{[T \geq j]}\right)^2 = \mathrm{E}\, X_n^2 I_{[T \geq n]} = \sigma^2 \mathrm{P}[T \geq n],$$

and summing,

$$\mathrm{E}\, S_{T(n)}^2 = \mathrm{E}\left(\sum_1^n X_j I_{[T \geq j]}\right)^2 = \sigma^2 \sum_{j=1}^n \mathrm{P}\{T \geq j\} = \sigma^2\, \mathrm{E}\, T(n).$$

Since $T(n) \uparrow T$,

$$\lim_{n \to \infty} \mathrm{E}\, S_{T(n)}^2 = \sigma^2\, \mathrm{E}\, T < \infty. \tag{17}$$

Moreover, in completely analogous fashion,

$$\mathrm{E}(S_{T(n)} - S_{T(m)})^2 = \mathrm{E}\left(\sum_{m+1}^n X_j I_{[T \geq j]}\right)^2 = \sigma^2[\mathrm{E}\, T(n) - \mathrm{E}\, T(m)]$$

$$= o(1) \quad \text{as } n > m \to \infty.$$

Thus, $S_{T(n)} \xrightarrow{\mathscr{L}_2} S$, and so by Corollary 4.2.5 and (17), $\mathrm{E}\, S^2 = \sigma^2\, \mathrm{E}\, T$. It remains to identify S with S_T, and this follows from the existence of a subsequence of $S_{T(n)}$ converging a.c. to S and $T(n) \xrightarrow{\text{a.c.}} T$. □

Corollary 5. *If* $\{X_n\}$ *are* i.i.d. r.v.s *with* $\mathrm{E}\, X_1 = 0$, $\mathrm{E}\, X_1^2 = \sigma^2 < \infty$ *and* T *is an* $\{X_n\}$*-time with* $\mathrm{E}\, T < \infty$, *then* (16) *holds.*

EXAMPLE 1. Let $\{X_n, n \geq 1\}$ be independent r.v.s with $\sup_{n \geq 1} \mathrm{E}|X_n|^r \leq M < \infty$, where either $r = 2$, $\mathrm{E}\, X_n = 0$, or $0 < r \leq 1$. If $\{a_n, n \geq 1\}$ are positive constants with $a_n \downarrow$, $A_n = \sum_{j=1}^n a_j$ and $S_n = \sum_1^n X_i$, then for any finite $\{X_n\}$-time T

$$\mathrm{E}\, a_T |S_T|^r \leq M\, \mathrm{E}\, A_T \tag{18}$$

and, moreover, for any α in (0, 1)

$$\mathrm{E}|S_T|^{r\alpha} \leq (M/\alpha)^\alpha\, \mathrm{E}\, T^\alpha. \tag{19}$$

PROOF. If T is a bounded stopping time, $S_0 = 0$, and $\delta_{r,2} = 2$ or 0 according as $r = 2$ or not, then via independence

$$\mathrm{E}\, a_T |S_T|^r = \sum_{j=1}^\infty \int_{[T=j]} a_j |S_j|^r$$

$$= \sum_{j=1}^\infty \sum_{n=1}^j \int_{[T=j]} (a_n |S_{n-1} + X_n|^r - a_{n-1} |S_{n-1}|^r)$$

$$\leq \sum_{n=1}^{\infty} \int_{[T \geq n]} a_n(|X_n|^r + \delta_{r,2} X_n S_{n-1}) \leq M \sum_{n=1}^{\infty} a_n \, \mathrm{P}\{T \geq n\}$$

$$= M \sum_{j=1}^{\infty} \sum_{n=1}^{j} a_n \, \mathrm{P}\{T = j\} = M \, \mathrm{E} \, A_T.$$

Hence, for any finite stopping time T, (18) holds for $T(n) = \min[T, n]$, yielding (18) as $n \to \infty$ by Fatou's lemma and monotone convergence. To prove (19), note that by Hölder's inequality

$$\mathrm{E}|S_T|^{r\alpha} = \mathrm{E}\frac{|S_T|^{r\alpha}}{T^{\alpha(1-\alpha)}} \cdot T^{\alpha(1-\alpha)} \leq \left(\mathrm{E}\frac{|S_T|^r}{T^{1-\alpha}}\right)^{\alpha} (\mathrm{E}\, T^{\alpha})^{1-\alpha},$$

and so, employing (18) with $a_n = 1/n^{1-\alpha}$, $A_n \leq n^{\alpha}/\alpha$, (19) follows. □

As an application of Example 1, if $\{X_n, n \geq 1\}$ are independent r.v.s that obey the Central Limit Theorem with $\mathrm{E}\, X_n = 0$, $\mathrm{E}\, X_n^2 = 1$, $S_n = \sum_1^n X_i$, $n \geq 1$ and

$$T_c = \inf\{n \geq 1 : |S_n| \geq c n^{1/2}\}, \quad c > 0,$$
$$U_m = \inf\{n \geq m : |S_n| \geq c_n\}, \quad \frac{c_n}{n^{1/2}} \to \infty, \tag{20}$$

then $\mathrm{E}\, T_c^{\alpha} = \infty$ for $c^2 > 1/\alpha$, $0 < \alpha < 1$, and $\mathrm{E}\, U_m^{\alpha} = \infty$ for $m > m_{\alpha}$, all $\alpha > 0$. The latter statement follows from the former which, in turn, results from (19) via

$$c^{2\alpha}\, \mathrm{E}\, T_c^{\alpha} \leq \mathrm{E}|S_{T_c}|^{2\alpha} \leq \left(\frac{1}{\alpha}\right)^{\alpha} \mathrm{E}\, T_c^{\alpha}.$$

The same conclusion holds if $\{X_n, n \geq 1\}$ are independent r.v.s with mean zero and variance one which obey the central limit theorem (Chapter 9), since T_c is a finite stopping time (Exercise 3.2.12).

Lemma 4. *Let $\{X_n, n \geq 1\}$ be i.i.d. random variables and $T = \inf\{n \geq 1 : X_n \in B\}$, where B is a linear Borel set such that $0 < \mathrm{P}\{X_1 \in B\} < 1$. If $T_0 = 0$, $T_n = \sum_{j=1}^n T^{(j)}$, $n \geq 1$, where $\{T^{(j)}, j \geq 1\}$ are copies of T, then setting*

$$Y_n = X_{T_n}, \quad Z_n = \sum_{T_{n-1} < j < T_n} X_j, \quad n \geq 1, \tag{21}$$

$\{Y_1, Z_1, Y_2, Z_2, \ldots\}$ is a sequence of independent variables.

Proof. If $n = 1$, Y_1 and Z_1 are independent according to exercise 5.3.12. For $n \geq 2$, via Lemma 3,

$$\mathrm{P}\left\{\bigcap_{i=1}^{n} [Y_i < \mu_i, Z_i < \lambda_i]\right\}$$

$$= \sum_{j=1}^{\infty} \mathrm{P}\left\{T^{(n)} = j, X_{T_{n-1}+j} < \mu_n, \sum_{T_{n-1}+1}^{T_{n-1}+j-1} X_i < \lambda_n, \bigcap_{i=1}^{n-1} [Y_i < \mu_i, Z_i < \lambda_i]\right\}$$

$$= \sum_{j=1}^{\infty} \mathrm{P}\left\{T^{(n)} = j, X_{T_{n-1}+j} < \mu_n, \sum_{T_{n-1}+1}^{T_{n-1}+j-1} X_i < \lambda_n\right\}$$

$$\cdot \mathrm{P}\left\{\bigcap_{i=1}^{n-1} [Y_i < \mu_i, Z_i < \lambda_i]\right\}$$

$$= \mathrm{P}\{Y_n < \mu_n, Z_n < \lambda_n\}\mathrm{P}\left\{\bigcap_{i=1}^{n-1} [Y_i < \mu_i, Z_i < \lambda_i]\right\}$$

$$= \mathrm{P}\{Y_n < \mu_n\}\mathrm{P}\{Z_n < \lambda_n\}\mathrm{P}\left\{\bigcap_{i=1}^{n-1} [Y_i < \mu_i, Z_i < \lambda_i]\right\}$$

since the joint distributions of (Y_n, Z_n) and (Y_1, Z_1) are identical. The conclusion then follows by induction. □

If $\{X, X_n, n \geq 1\}$ are i.i.d. random variables with $\mathrm{E}\, X = 0, \mathrm{E}\, X^2 < \infty$, then $\overline{\lim}_{n\to\infty} S_n/n^{1/2} = \infty$, a.c., as noted in exercise 3.2.12. This remains true even when $\mathrm{E}\, X^2 = \infty$, according to

Theorem 4 (Stone). *If* $S_n = \sum_{j=1}^{n} X_j, n \geq 1$, *where* $\{X_n, n \geq 1\}$ *are* i.i.d. *random variables with* $\mathrm{E}\, X_1 = 0, \mathrm{E}|X_1| > 0$, *then*

$$\overline{\lim_{n\to\infty}}\, S_n/n^{1/2} = \infty = -\underline{\lim_{n\to\infty}}\, S_n/n^{1/2}, \qquad \text{a.s.} \tag{22}$$

PROOF. It suffices to prove the first half of (22). To this end, choose finite constants a, b such that $a < 0 < b$ and $\mathrm{P}\{a < X_1 < 0\}\mathrm{P}\{0 < X_1 < b\} > 0$. Then $T = \inf\{n \geq 1: a < X_n < b\}$ is an $\{X_n\}$-time with finite expectation. Since, as noted earlier, the theorem holds when $\mathrm{E}\, X_1^2 < \infty$, it may be supposed that $\mathrm{E}\, X_1^2 = \infty$, whence $0 < \mathrm{P}\{a < X_1 < b\} < 1$. Set $T_0 = 0$, $T_n = \sum_{j=1}^{n} T^{(j)}$, $n \geq 1$, where $\{T^{(j)}, j \geq 1\}$ are copies of T and define

$$Y_n = X_{T_n}, \qquad Z_n = \sum_{T_{n-1}<j<T_n} X_j. \tag{23}$$

According to Lemma 3, $\{Y_j, j \geq 1\}$ and $Z_j, j \geq 1\}$ are each i.i.d. sequences and, moreover, via Lemma 4, $\{Y_j, j \geq 1\}$ is independent of $\{Z_j, j \geq 1\}$. By Wald's equation, $\mathrm{E}\, Y_1 + \mathrm{E}\, Z_1 = \mathrm{E} \sum_1^T X_j = 0$, whence

$$S_{T_n} = \sum_{j=1}^{n} (Y_j + Z_j) = \sum_{j=1}^{n} (Y_j - \mathrm{E}\, Y_j) + \sum_{j=1}^{n} (Z_j - \mathrm{E}\, Z_j). \tag{24}$$

By example 5.2.2, $\{Z_j - \mathrm{E}\, Z_j, j \geq 1\}$ is a recurrent sequence, and so $\mathrm{P}\{|\sum_{j=1}^{n} (Z_j - \mathrm{E}\, Z_j)| < M, \text{i.o.}\} = 1$ for some M in $(0, \infty)$. Define $\tau_m = \inf\{n \geq m: |\sum_1^n (Z_j - \mathrm{E}\, Z_j)| < M\}$, whence $\tau_m < \infty$, a.s. and $\tau_m \uparrow \infty$ as $m \to \infty$. Now, $\sigma^2 = \mathrm{E}(Y_1 - \mathrm{E}\, Y_1)^2 \in (0, \infty)$ whence

$$M + S_{T_{\tau_m}} \geq \sum_{j=1}^{\tau_m} (Y_j - \mathrm{E}\, Y_j) = \left[\frac{1}{\sigma \tau_m^{1/2}} \sum_{j=1}^{\tau_m} (Y_j - \mathrm{E}\, Y_j)\right] \sigma \tau_m^{1/2} \tag{25}$$

via (24), and since τ_m is independent of $\{Y_j, j \geq 1\}$,

$$\mathrm{P}\left\{\frac{1}{\sigma\tau_m^{1/2}} \sum_{j=1}^{\tau_m} (Y_j - \mathrm{E}\, Y_j) > x\right\}$$

$$= \sum_{k=m}^{\infty} \mathrm{P}\left\{\frac{1}{\sigma k^{1/2}} \sum_{j=1}^{k} (Y_j - \mathrm{E}\, Y_j) > x\right\} \mathrm{P}\{\tau_m = k\} \xrightarrow{m\to\infty} 1 - \Phi(x)$$

for all x via the Central Limit Theorem for i.i.d. random variables (Corollary 9.1.2).

Consequently, for any $x > 0$, as $n \to \infty$,

$$\mathrm{P}\left\{S_{T_{\tau_m}} > \frac{x}{2}\sigma\tau_m^{1/2}\right\} \geq 1 - \Phi(x) + o(1). \tag{26}$$

Now, $T_{\tau_m}/\tau_m = \dfrac{1}{\tau_m}\sum_{j=1}^{\tau_m} T^{(j)} \xrightarrow{\text{a.c.}} \mathrm{E}\, T < \infty$, implying via (26) that, for $x > 0$, as $m \to \infty$,

$$\mathrm{P}\left\{S_{T_{\tau_m}} > \frac{x\sigma}{3}(T_{\tau_m}/\mathrm{E}\, T)^{1/2}\right\} \geq 1 - \Phi(x) + o(1),$$

and so

$$\mathrm{P}\left\{S_{T_{\tau_m}}/T_{\tau_m}^{1/2} > \frac{x\sigma}{3}\mathrm{E}^{-1/2}T, \text{ i.o.}\right\}$$

$$= \lim_{m\to\infty} \mathrm{P}\left\{\bigcup_{j=m}^{\infty}\left[S_{T_{\tau_j}}/T_{\tau_j}^{1/2} > \frac{x\sigma}{3}\mathrm{E}^{-1/2}T\right]\right\}$$

$$\geq 1 - \Phi(x)$$

whence for all $x > 0$

$$Q_x = \mathrm{P}\left\{\overline{\lim_{n\to\infty}}\, S_n/n^{1/2} \geq \frac{x\sigma}{3}\mathrm{E}^{-1/2}T\right\} \geq 1 - \Phi(x).$$

Thus, $Q_x = 1$ by the Kolmogorov zero-one law and (22) follows as $x \to \infty$. □

Lemma 5. *Let* $S_0 = 0$, $S_n = \sum_{i=1}^n X_i$, $\bar{S}_n = \max_{0\leq j\leq n} S_j$, $n \geq 1$, *where* $\{X, X_n, n \geq 1\}$ *are* i.i.d. *random variables with* $\mathrm{E}\, X = 0$, $\mathrm{E}|X| > 0$. *Then, for* $x > 0$ *and any integer* $k \geq 1$,

$$\mathrm{P}\{\bar{S}_n \geq 2kx\} \leq n\mathrm{P}\{X > x\} + \mathrm{P}^k\{\bar{S}_n \geq x\}.$$

Proof. If $X_0 = 0$, $\bar{X}_n = \max_{0\leq j\leq n} X_j$, and $A_k = \{\bar{S}_n \geq 2kx, \bar{X}_n \leq x\}$, it suffices to prove that $\mathrm{P}\{A_k\} \leq \mathrm{P}^k\{\bar{S}_n \geq x\}$. To this end, define $T_0 = 0$, $T_i = \sum_{h=1}^i T^{(h)}$, $1 \leq i \leq k$, where

$$T^{(i)} = \inf\{j \geq 1: S_{T_{i-1}+j} - S_{T_{i-1}} \geq x\}, \qquad 1 \leq i \leq k.$$

By Corollary 5.4.2, T_1 is a finite stopping time and, by Corollary 5.3.2, $\{T^{(i)}, 1 \le i \le k\}$ are i.i.d. Clearly, $T_1 \le n$ on the set A_1 since otherwise $A_1 \cdot \{\bar{S}_n < x\} \ne \emptyset$, a contradiction. For $k \ge 2$, suppose, inductively, that $T_i \le n$ on A_i for $1 \le i \le m < k$. Then, on A_{m+1},

$$\bar{S}_{T_{m+1}-1} \le \sum_{i=1}^{m+1} \max_{1 \le j < T^{(i)}} (S_{T_{i-1}+j} - S_{T_{i-1}}) + \sum_{i=1}^{m} X_{T^{(i)}} < 2(m+1)x,$$

implying that $n > T_{m+1} - 1$, that is, $T_{m+1} \le n$ on A_{m+1}. Thus, for $m = k - 1$,

$$P\{A_k\} \le P\left\{\sum_{i=1}^{k} T^{(i)} \le n\right\} = P^k\{\bar{S}_n \ge x\}. \qquad \square$$

EXERCISES 5.3

1. Verify that if T is an $\{\mathscr{F}_n\}$-time, X_{T_m+j} is $\mathscr{F}_{T_{m+j}}$-measurable, $j \ge 1$, $m \ge 1$.
2. If T_1 and T_2 are $\mathscr{F}_n$-times, so are $T_1 T_2$, $\max(T_1, T_2)$, $\min(T_1, T_2)$, and kT_1, where k is a positive integer.
3. If T is an integrable $\{X_n\}$-time, where $\{X_n\}$ are independent r.v.s with $E X_n = 0$, $E|X_n| \le C < \infty$, $n \ge 1$, then Wald's equation holds.
4. If (X_i, Y_i), $i \ge 1$, are i.i.d. $\mathscr{L}_2$ random vectors with $E X_1 = E Y_1 = 0$, and $\mathscr{F}_n = \sigma(X_1, Y_1, \ldots, X_n, Y_n)$, $S_n = \sum_1^n X_i$, $U_n = \sum_1^n Y_i$, then for any integrable $\mathscr{F}_n$-time T, the identity $E S_T U_T = E T \cdot E X_1 Y_1$ holds.
5. Let $S_n = \sum_{i=1}^n X_i$ where $\{X_n, n \ge 1\}$ are i.i.d. r.v.s with $E X_1 = \mu > 0$ and $N = N_p$ an $\{X_n\}$-time (or a r.v. independent of $\{X_n, n \ge 1\}$) having the geometric distribution. Then $\lim_{p \to 0} P\{S_N / E S_N < x\} = 1 - e^{-x}$, $x > 0$.
6. Show that the condition $E T < \infty$ cannot be dropped in Corollary 3. *Hint*: Consider $P\{X_n = 1\} = P\{X_n = -1\} = \frac{1}{2}$ and $T = \inf\{n \ge 1 : S_n > 0\}$.
7. If $\{X_n, n \ge 1\}$ are independent random variables with $E X_n = 0$, $E X_n^2 = 1$ and T^* (resp. T_*) $= \inf\{n \ge 1 : |S_n| >$ (resp. $<$)$cn^{1/2}\}$, where $\inf \emptyset = \infty$, then $E T^* = \infty$, $c \ge 1$ and $E T_* = \infty$, $c \le 1$.
8. If $\{X_n, n \ge 1\}$ are independent r.v.s with $E X_n = 0$, $E X_n^2 = \sigma_n^2$, $n \ge 1$ and T is an $\mathscr{F}_n$-time for which $E \sum_1^T \sigma_j^2 < \infty$, where (i) $\mathscr{F}_n \supset \sigma(X_1, \ldots, X_n)$, (ii) $\mathscr{F}_n$ and $\sigma(X_{n+1})$ are independent, $n \ge 1$, then $E(\sum_1^T X_j)^2 = E \sum_1^T \sigma_j^2$.
9. (Yadrenko) If $\{X_n, n \ge 1\}$ are i.i.d. r.v.s uniformly distributed on $[0, 1]$ and $T = \inf\{n > 1 : S_n \ge 1\}$, where $S_n = \sum_1^n X_j$, prove that $E T = e = 2 E S_T$.
10. Utilize the method of Example 1 to give an alternative proof of Theorem 3.
11. Let $\{X_n, n \ge 1\}$ be i.i.d. with $P\{X_1 > M\} > 0$, all M in $(0, \infty)$. If $T_c = \inf\{n \ge 1 : X_n \ge c\}$, where $|c| < \infty$, prove that $E T_c^m < \infty$, $m \ge 1$, and $E X_{T_c} \le E X_{T_d}$ for $c < d$ and $E X_{T_c} = P^{-1}\{X \ge c\} \int_{[X \ge c]} X$.
12. If $\{X, X_n, n \ge 1\}$ are i.i.d., B is a linear Borel set with $P\{X \in B\} > 0$, and $T = \inf\{n \ge 1 : X_n \in B\}$, then X_T and $\sum_1^{T-1} X_i$ are independent r.v.s. Moreover, if $E|X| < \infty$, then $E X_T = E X I_{[X \in B]} \cdot E T$.

13. Show that $\limsup_{n\to\infty} n^{-1/2}|\sum_{i=1}^{n} X_i| = \infty$, a.c. for *any* sequence of i.i.d. random variables $\{X_n, n \geq 1\}$ with $E|X_1| > 0$.

14. Let $S_n = \sum_{i=1}^{n} X_i$, $n \geq 1$, where $\{X, X_n, n \geq 1\}$ are i.i.d. random variables with $E X = 0$, $E|X| > 0$. For $x > 0$, $\beta > 0$, and $k = 1, 2, \ldots$, show that

$$q_k \equiv P\left\{\max_{1\leq j\leq n} S_j/j^\beta \geq 2kx\right\} \leq P\left\{\max_{1\leq j\leq n} X_j/j^\beta \geq x\right\} + P^k\left\{\max_{1\leq j\leq n} S_j/j^\beta \geq x\right\},$$

and if $1 \leq r \leq 2$ and $\beta < 1/r$, there exists a universal constant $C_{r,\beta}$ such that

$$q_k \leq P\left\{\max_{1\leq j\leq n} X_j/j^\beta \geq x\right\} + [C_{r,\beta}(2x)^{-r}n^{1-r\beta}E|X|^r]^k.$$

Hint: This generalizes Lemma 5.3.5.

5.4 Chung–Fuchs Theorem, Elementary Renewal Theorem, Optimal Stopping

An instance when the generality of Theorem 5.3.1 (as opposed to the specificity of its corollary) is desirable is furnished by

Theorem 1. *Let* $\{X_n\}$ *be* i.i.d. r.v.s *with* $E|X_1| > 0$, $S_n = \sum_1^n X_i$, *and define*

$$T_+ = \inf\{n \geq 1: S_n \geq 0\}, \qquad T_c' = \inf\{n \geq 1: S_n > c > 0\}. \tag{1}$$

Then:

i. $P\{T_+ < \infty\} = 1$ iff $\overline{\lim}\, S_n \overset{\text{a.c.}}{=} \infty$, *in which case* $E\, S_{T_+} > 0$;
ii. $E\, T_+ < \infty$ iff $E\, T_c' < \infty$ *for all c in* $(0, \infty)$.

Proof. If under the hypothesis of (i), T_+ is a finite $\{X_n\}$-time, then S_{T_+} is a r.v. and $E\, S_{T_+} \geq E\, X_1^+ \geq 0$. Let $T^{(j)}, j \geq 1$, be copies of T_+ and set $T_0 = 0$, $T_n = \sum_1^n T^{(j)}, n \geq 1$. By Lemma 5.3.3, the r.v.s $S^{(m)} = X_{T_{m-1}+1} + \cdots + X_{T_m}$, $m \geq 1$, are i.i.d., whence by Corollary 5.2.2

$$\frac{S_{T_n}}{n} = \frac{S^{(1)} + \cdots + S^{(n)}}{n} \xrightarrow{\text{a.c.}} E\, S_{T_+} \geq E\, X_1^+.$$

Hence, $\overline{\lim}\, S_n \geq \overline{\lim}\, n\, E\, X_1^+ \geq 0$, a.c., and, moreover, $\overline{\lim}\, S_n = +\infty$, a.c., since $E\, X_1^+ > 0$ in view of the fact that the only permissible alternative, namely, $E\, X_1 < 0$, would imply (same corollary) that $S_n/n \xrightarrow{\text{a.c.}} E\, X_1 < 0$, contradicting $\overline{\lim}\, S_n \geq 0$, a.c. The converse is trivial.

Apropos of (ii), since via (i)

$$\frac{S^{(1)} + \cdots + S^{(n)}}{n} \xrightarrow{\text{a.c.}} E\, S_{T_+} > 0,$$

for any $c > 0$ there exists an integer k such that $P\{\sum_1^k S^{(j)} > c\} \ge \frac{1}{2}$. Setting $Z_n = \sum_{(n-1)k+1}^{nk} S^{(j)}$, the r.v.s $\{Z_n, n \ge 1\}$ are i.i.d. by Lemma 5.3.3 and clearly $Z_n \ge 0$, a.c., with $P\{Z_1 > c\} \ge \frac{1}{2}$. Define $\tau = \inf\{n \ge 1 : Z_n > c\}$. Then

$$P\{\tau \ge n\} = P\{Z_1 \le c, \ldots, Z_{n-1} \le c\} = P^{n-1}\{Z_1 \le c\},$$

whence

$$E\,\tau = \sum_{n=1}^{\infty} P\{\tau \ge n\} = \sum_{n=1}^{\infty} P^{n-1}\{Z_1 \le c\} = \frac{1}{P\{Z_1 > c\}} \le 2.$$

Moreover, since

$$c < \sum_1^{\tau} Z_i = \sum_1^{k\tau} S^{(j)} = \sum_{j=1}^{k\tau} \sum_{T_{j-1}+1}^{T_j} X_i = \sum_1^{T_{k\tau}} X_i = S_{T_{k\tau}},$$

necessarily $T'_c \le T_{k\tau}$, whence

$$E\,T'_c \le E\,T_{k\tau} = E \sum_1^{k\tau} T^{(j)} = E(k\tau) \cdot E\,T_+ < \infty$$

by Wald's equation since $k\tau$ is a $\mathscr{G}_n$-time, where $\mathscr{G}_n = \sigma(T^{(1)}, \ldots, T^{(n)}, S^{(1)}, \ldots, S^{(n)})$, i.e., $\{k\tau = kn\} \in \mathscr{G}_{nk}$ and $\sigma(T^{(n+1)})$ is independent of $\mathscr{G}_n$ (recall Lemma 5.3.3). Again, the converse is trivial since $T_+ \le T'_c$. □

The stopping rule T_+ and its partial analogue T'_-, defined in (2) below, are interconnected as in

Theorem 2. *Let* $\{X_n\}$ *be* i.i.d. r.v.s, $S_n = \sum_1^n X_i$, *and define*

$$T_+ = \inf\{n \ge 1 : S_n \ge 0\}, \qquad T_- = \inf\{n \ge 1 : S_n \le 0\},$$
$$T'_+ = \inf\{n \ge 1 : S_n > 0\}, \qquad T'_- = \inf\{n \ge 1 : S_n < 0\}. \tag{2}$$

Then, (i)

$$E\,T_+ = \frac{1}{P\{T'_- = \infty\}}, \qquad E\,T'_+ = \frac{1}{P\{T_- = \infty\}}, \tag{3}$$

$$P\{T_- = \infty\} = (1 - \xi)P\{T'_- = \infty\}, \tag{4}$$

where

$$\xi = \sum_1^{\infty} P\{S_1 > 0, \ldots, S_{n-1} > 0, S_n = 0\} = P\{S_{T_-} = 0, T_- < \infty\}. \tag{5}$$

Moreover, if $E|X_1| > 0$,

ii. $\xi < 1$,
iii. T_- *is defective* iff $S_n \xrightarrow{\text{a.c.}} \infty$.

PROOF. (i) For $k > 1$ and $1 \le n \le k$, define

$$A_1^k = \{S_1 \le S_2, \dots, S_1 \le S_k\},$$
$$A_n^k = \{S_1 > S_n, \dots, S_{n-1} > S_n \le S_{n+1}, \dots, S_n \le S_k\}.$$

Then, if $k \ge n \ge 1$,

$$\begin{aligned} P\{A_n^k\} &= P\{X_n < 0, \dots, X_2 + \cdots + X_n < 0\} \\ &\quad \cdot P\{X_{n+1} \ge 0, \dots, X_{n+1} + \cdots + X_k \ge 0\} \\ &= P\{T_+ \ge n\}P\{T'_- > k - n\}, \end{aligned}$$

and so when $k \ge 1$

$$1 = \sum_{n=1}^{k} P\{T_+ \ge n\}P\{T'_- > k - n\}, \tag{6}$$

yielding

$$1 \ge P\{T'_- = \infty\}E\, T_+, \tag{7}$$

whence $P\{T'_- = \infty\} > 0$ implies $E\, T_+ < \infty$. Conversely, if $E\, T_+ < \infty$, via (6)

$$1 \le \sum_{n=1}^{j} P\{T_+ \ge n\}P\{T'_- > k - j\} + \sum_{n=j}^{k} P\{T_+ \ge n\};$$

letting $k \to \infty$,

$$1 \le \sum_{n=1}^{j} P\{T_+ \ge n\}P\{T'_- = \infty\} + \sum_{n=j}^{\infty} P\{T_+ \ge n\},$$

and then letting $j \to \infty$,

$$1 \le P\{T'_- = \infty\}E\, T_+,$$

implying $P\{T'_- = \infty\} > 0$. Consequently, $P\{T'_- = \infty\}E\, T_+ = 1$ if either $P\{T'_- = \infty\} > 0$ or $E\, T_+ < \infty$, and so, without qualification,

$$E\, T_+ = \frac{1}{P\{T'_- = \infty\}}.$$

Similarly,

$$E\, T'_+ = \frac{1}{P\{T_- = \infty\}},$$

establishing (3). Next,

$$P\{T'_- = \infty\} - P\{T_- = \infty\} = P\{T'_- = \infty, T_- < \infty\}$$

$$= \sum_{n=1}^{\infty} P\{T_- = n, T'_- = \infty\}$$

$$= \sum_{n=1}^{\infty} P\{S_1 > 0, \ldots, S_{n-1} > 0, S_n = 0, T'_- = \infty\}$$

$$= \sum_{n=1}^{\infty} P\{S_1 > 0, \ldots, S_{n-1} > 0, S_n = 0, S_{n+1} \geq 0, S_{n+2} \geq 0, \ldots\}$$

$$= \sum_{n=1}^{\infty} P\{S_1 > 0, \ldots, S_{n-1} > 0, S_n = 0, X_{n+1} \geq 0, X_{n+1} + X_{n+2} \geq 0, \ldots\}$$

$$= \sum_{n=1}^{\infty} P\{S_1 > 0, \ldots, S_{n-1} > 0, S_n = 0\} P\{S_1 \geq 0, S_2 \geq 0, \ldots\}$$

$$= \xi\, P\{T'_- = \infty\},$$

yielding

$$(1 - \xi)P\{T'_- = \infty\} = P\{T_- = \infty\},$$

which completes the proof of (i).

(ii) Suppose that $E|X_1| > 0$. If $\xi = 1$, then $T_- < \infty$, a.c., and, replacing $\{X_n\}$ by $\{-X_n\}$ in Theorem 1(i), $E\, S_{T_-} < 0$. On the other hand, $\xi = 1$ entails $S_{T_-} = 0$, a.c., via (5), a flagrant contradiction.

(iii) If T_- and hence also T'_- is defective, $E\, T_+ < \infty$ by (3) and *a fortiori* $T_+ < \infty$, a.c., so that according to Theorem 1(i)

$$\overline{\lim}\, S_n \overset{\text{a.c.}}{=\!=} \infty. \tag{8}$$

Thus, $T = \inf\{n \geq 1 : S_n > K\} < \infty$, a.c. for all $K > 0$ and, via Lemma 5.3.2,

$$P\{\underline{\lim}_{n\to\infty} (S_{T+n} - S_T) \geq 0\} = P\{\underline{\lim}_{n\to\infty} S_n \geq 0\} = \varepsilon > 0.$$

Consequently, with probability at least ε,

$$\underline{\lim}_{n\to\infty} S_n = \underline{\lim}_{n\to\infty} S_{T+n} = S_T + \underline{\lim}_{n\to\infty} (S_{T+n} - S_T) > K,$$

implying that $P\{\underline{\lim}_{n\to\infty} S_n = \infty\} \geq \varepsilon > 0$ whence, by the Kolmogorov zero-one law, $\underline{\lim}_{n\to\infty} S_n = \infty$, a.c., that is, $S_n \xrightarrow{\text{a.c.}} \infty$.

Conversely, if $\underline{\lim}_{n\to\infty} S_n \overset{\text{a.c.}}{=\!=} \infty$, T cannot be finite since Theorem 1 (with $X \to -X$) then would entail $\underline{\lim}_{n\to\infty} S_n \overset{\text{a.c.}}{=\!=} -\infty$. □

A combination of Theorems 1 and 2 yields

Corollary 1. *If* $\{X_n\}$ *are* i.i.d. r.v.s, $S_n = \sum_1^n X_i$, *and*

$$T_+ = \inf\{n \geq 1 : S_n \geq 0\}, \qquad T_- = \inf\{n \geq 1 : S_n \leq 0\},$$

$$T'_c = \inf\{n \geq 1 : S_n > c > 0\}, \qquad T'_{-c} = \inf\{n \geq 1 : S_n < -c < 0\},$$

then either $S_n = 0, n \geq 1$, a.c., *or one of the following holds*:

i. T_- *is defective*, $\lim S_n \overset{\text{a.c.}}{=\!=} \infty$, $E\, T_+ < \infty$, $E\, T'_c < \infty$, $c > 0$;

ii. T_+ *is defective*, $\lim S_n \overset{\text{a.c.}}{=} -\infty$, $\mathrm{E}\,T_- < \infty$, $\mathrm{E}\,T'_{-c} < \infty$, $c > 0$;
iii. T_+ *and* T_- *are finite*, $\overline{\lim}\, S_n \overset{\text{a.c.}}{=} \infty$, $\underline{\lim}\, S_n \overset{\text{a.c.}}{=} -\infty$, $\mathrm{E}\,T_+ = \mathrm{E}\,T_- = \infty$.

PROOF. Supposing $\{S_n, n \geq 1\}$ nondegenerate, necessarily $\mathrm{E}|X_1| > 0$.

(i) If T_- is defective, T'_- is defective *a fortiori*, whence by Theorem 2 $S_n \xrightarrow{\text{a.c.}} \infty$ and $\mathrm{E}\,T_+ < \infty$. Moreover, by Theorem 1(ii), $\mathrm{E}\,T'_c < \infty$ for $c > 0$. Similarly for (ii).

If, as in (iii), both T_+ and T_- are finite, then Theorem 1(i) ensures that

$$\overline{\lim}\, S_n = \infty \quad \text{a.c.}, \qquad \underline{\lim}\, S_n = -\infty \quad \text{a.c.};$$

moreover, according to Theorem 2(ii) and (4), T'_- and (analogously) T'_+ are finite, whence by Theorem 2(i), $\mathrm{E}\,T_+ = \mathrm{E}\,T_- = \infty$. □

The following provides a criterion for category (iii) of Corollary 1.

Corollary 2 (Chung–Fuchs). *Let* $\{X_n, n \geq 1\}$ *be* i.i.d. r.v.*s with* $\mathrm{E}\,X_1 = 0$ *and* $\mathrm{E}|X_1| > 0$ *and let* $S_n = \sum_1^n X_j$. *Then*

$$\overline{\lim_{n\to\infty}}\, S_n = \infty \quad \text{a.c.}, \qquad \underline{\lim_{n\to\infty}}\, S_n = -\infty \quad \text{a.c.}$$

PROOF. By Corollary 1, it suffices to prove that $\mathrm{E}\,T_+ = \mathrm{E}\,T_- = \infty$. Suppose, on the contrary, that $\mathrm{E}\,T_+$ (say) $< \infty$. Then, by Wald's equation,

$$0 \leq \mathrm{E}(X_1^+) \leq \mathrm{E}\,S_{T_+} = \mathrm{E}\,T_+ \cdot \mathrm{E}\,X_1 = 0,$$

implying $\mathrm{E}\,X_1^+ = 0$. Since $\mathrm{E}\,X_1 = 0$, $\mathrm{E}\,X_1^- = 0$ and therefore $\mathrm{E}|X_1| = 0$, a contradiction. □

The next theorem, which extends Corollary 2, asserts that the same conclusion holds if $S_n/n \xrightarrow{P} 0$ and $\mathrm{E}|X_1| > 0$.

Theorem 3 (Chung–Ornstein). *If* $\{X_n\}$ *are* i.i.d. r.v.*s with* $\mathrm{E}|X_1| > 0$, $S_n = \sum_1^n X_i$, *and*

$$\frac{S_n}{n} \xrightarrow{P} 0 \tag{9}$$

then T_+ *and* T_- *are finite and*

$$\mathrm{P}\left\{\overline{\lim_{n}}\, S_n = \infty\right\} = 1 = \mathrm{P}\left\{\underline{\lim_{n}}\, S_n = -\infty\right\}. \tag{10}$$

PROOF. According to Corollary 1 it suffices to prove that T_+, T_- are finite and, by symmetry, merely that T_- is finite. Suppose contrariwise that T_- is defective, whence $\mathrm{E}\,T_+ < \infty$ by Theorem 2. Now the WLLN (Theorem 5.2.4) together with (9) implies

$$n\,\mathrm{P}\{|X_1| > n\} = o(1), \qquad \mathrm{E}\,X_1 I_{[|X_1| \leq n]} = o(1). \tag{11}$$

Since $\mathrm{E}\,S_{T_+}$ exists and $\mathrm{E}\,T_+ < \infty$ as noted above, Theorem 5.3.2(ii) ensures

$\mathrm{E}\, S_{T_+} = 0$. Thus, $\mathrm{E}\, X_1^+ \le \mathrm{E}\, S_{T_+} = 0$, entailing $\mathrm{E} X_1 I_{[|X_1| \le n]} \to -\mathrm{E}\, X_1^- < 0$, in contradiction of (11). Consequently T_- is finite. □

The same proof yields

Corollary 3. *Let* $\{X_n\}$ *be* i.i.d. r.v.s *with* $\mathrm{E}|X_1| > 0$, $S_n = \sum_1^n X_i$, *and* $n\, \mathrm{P}\{|X_1| > n\} \to 0$. (i) *If either* $\lim \mathrm{E}\, X_1 I_{[|X_1| \le n]}$ *does not exist or* $\lim \mathrm{E}\, X_1 I_{[|X_1| \le n]} \ge 0$, *then* $T_+ < \infty$, a.c. (ii) *If, moreover,* $\mathrm{E}\, X_1^+ = \mathrm{E}\, X_1^- = \infty$, *and if* $\lim \mathrm{E}\, X_1 \times I_{[|X_1| \le n]} = c$ *finite or* $\mathrm{E}\, X_1 I_{[|X_1| \le n]}$ *has no limit, then* $\mathrm{P}\{\overline{\lim}\, S_n = \infty\} = \mathrm{P}\{\underline{\lim}\, S_n = -\infty\} = 1$.

Corollary 1 implies that apart from degeneracy only three possible modes of behavior exist for sums S_n of i.i.d. r.v.s $\{X_n\}$. If, moreover, $\mathrm{E}\, X_1^+ = \mathrm{E}\, X_1^- = \infty$, the same trichotomy will be shown to exist for the averages S_n/n.

Lemma 1. (i) *If* $\{X_n, n \ge 1\}$ *are* i.i.d. r.v.s *with* $S_n = \sum_1^n X_i$, $n \ge 1$, *then* $S_n \xrightarrow{\text{a.c.}} \infty$ iff *there exists an* $\{X_n\}$*-time* T *for which* $\mathrm{E}\, T < \infty$ *and* $\mathrm{E}\, S_T > 0$. (ii) *Moreover, when* $\mathrm{E}\, X_1^+ = \infty$, *then* $(S_n/n) \xrightarrow{\text{a.c.}} \infty$ iff *there exists an* $\{X_n\}$*-time* T *with* $\mathrm{E}\, T < \infty$ *and* $\mathrm{E}\, S_T > -\infty$.

PROOF. Under (i), if $S_n \xrightarrow{\text{a.c.}} \infty$, then Corollary 1 ensures that $\mathrm{E}\, T_+ < \infty$, where $T_+ = \inf\{n \ge 1 : S_n \ge 0\}$, and clearly $\mathrm{E}\, S_{T_+} \ge \mathrm{E}\, X_1^+ > 0$. Conversely, if T is an $\{X_n\}$-time with $\mathrm{E}\, T < \infty$ and $\mathrm{E}\, S_T > 0$, let $T^{(j)}, j \ge 1$, be copies of T with $T_n = \sum_1^n T^{(j)}$, $T_0 = 0$. By Lemma 5.3.3, $V_n = (T^{(n)}, S^{(n)})$, $n \ge 1$, are i.i.d. random vectors, where $S^{(n)} = X_{T_{n-1}+1} + \cdots + X_{T_n}$. Define a $\{V_n\}$-time τ by

$$\tau = \inf\{n \ge 1 : S_{T_n} \ge 0\} = \inf\left\{n \ge 1 : \sum_1^n S^{(j)} \ge 0\right\}.$$

Since $\sum_1^n S^{(j)} \xrightarrow{\text{a.c.}} \infty$, Corollary 1 guarantees that $\mathrm{E}\, \tau < \infty$ and since $S_{T_\tau} \ge 0$, $T_+ \le T_\tau$. Hence, via Wald's equation

$$\mathrm{E}\, T_+ \le \mathrm{E}\, T_\tau = \mathrm{E} \sum_1^\tau T^{(j)} = \mathrm{E}\, \tau \cdot \mathrm{E}\, T^{(1)} < \infty$$

and, invoking Corollary 1 once more, $S_n \xrightarrow{\text{a.c.}} \infty$.

Apropos of (ii), let T be an $\{X_n\}$-time with $\mathrm{E}\, T < \infty$ and $\mathrm{E}\, S_T > -\infty$. For $K > 0$, define $X'_n = X_n - K$, $S'_n = \sum_1^n X'_j$, and

$$T' = T I_{[X'_1 \le 0]} + I_{[X'_1 > 0]}.$$

Then $T' \le T$, $\mathrm{E}(S'_{T'})^+ \ge \mathrm{E}(X'_1)^+ = \infty$, and

$$\mathrm{E}(S'_{T'})^- \le \mathrm{E}(S'_T)^- \le \mathrm{E}(S_T^- + KT) < \infty.$$

Consequently, by (i) $S_n - nK = S'_n \xrightarrow{\text{a.c.}} \infty$ for all $K > 0$, implying $S_n/n \xrightarrow{\text{a.c.}} \infty$. The remainder of (ii) follows trivially from (i). □

Theorem 4 (Kesten). *If* $\{X_n, n \geq 1\}$ *are* i.i.d. r.v.s *with* $\mathrm{E}\, X_1^+ = \mathrm{E}\, X_1^- = \infty$ *and* $S_n = \sum_1^n X_i$, $n \geq 1$, *then one of the following holds:*

i. $S_n/n \xrightarrow{\text{a.c.}} \infty$;

ii. $S_n/n \xrightarrow{\text{a.c.}} -\infty$;

iii. $\overline{\lim}(S_n/n) \overset{\text{a.c.}}{=\!=} \infty$, $\underline{\lim}(S_n/n) \overset{\text{a.c.}}{=\!=} -\infty$.

Proof. (i) If $S_n \xrightarrow{\text{a.c.}} \infty$, Corollary 1 ensures $\mathrm{E}\, T_+ < \infty$ and $\mathrm{E}\, S_{T_+} \geq \mathrm{E}\, X_1^+ = \infty$, where $T_+ = \inf\{n \geq 1 : S_n \geq 0\}$, so that $S_n/n \xrightarrow{\text{a.c.}} \infty$ by Lemma 1. Similarly, $S_n \xrightarrow{\text{a.c.}} -\infty$ guarantees (ii). Since the hypothesis precludes degeneracy, only the alternative $\overline{\lim}\, S_n = \infty$, a.c., and $\underline{\lim}\, S_n = -\infty$, a.c., remains, in which case by (ii) of Lemma 1 no $\{X_n\}$-time T with $\mathrm{E}\, T < \infty$ and $|\mathrm{E}\, S_T| \leq \infty$ exists. Hence, if $X'_n = X_n - K$ and $S'_n = \sum_1^n X'_1$, no $\{X_n\}$-time T' with $\mathrm{E}\, T' < \infty$ and $|\mathrm{E}\, S'_{T'}| \leq \infty$ exists for any finite constant K. Again invoking Lemma 1 and Corollary 1, $\overline{\lim}\, S'_n = \infty$, a.c., and $\underline{\lim}\, S'_n = -\infty$, a.c. In view of the arbitrariness of K, (iii) follows. □

Corollary 4. *If* $\{X_n\}$ *are* i.i.d. *random variables, then, for every* α *in* $(0, \frac{1}{2}]$, *one of the following holds:*

(i) $S_n/n^\alpha \xrightarrow{\text{a.c.}} \infty$.
(ii) $S_n/n^\alpha \xrightarrow{\text{a.c.}} -\infty$.
(iii) $\overline{\lim}\, S_n/n^\alpha \overset{\text{a.c.}}{=\!=} \underline{\lim}\, S_n/n^\alpha \overset{\text{a.c.}}{=\!=} -\infty$.

Proof. If $\mu = \mathrm{E}\, X$ exists, then $S_n/n \xrightarrow{\text{a.c.}} \mu \in [-\infty, \infty]$, and if $\mu \neq 0$, either (i) or (ii) holds. If, rather, $\mu = 0$, Theroem 5.3.4 asserts that (iii) obtains for $\alpha = 1/2$ (a fortiori, for $\alpha < \frac{1}{2}$). Finally, if $\mathrm{E}\, X^+ = \mathrm{E}\, X^- = \infty$, Corollary 4 follows from Theorem 4. □

Renewal theory is concerned with the so-called renewal function $\mathrm{E}\, N_c$, where

$$N_c = \max\left\{j : S_j = \sum_1^j X_i \leq c\right\}, \qquad c > 0,$$

the r.v.s $\{X_n, n \geq 1\}$ being i.i.d. with $\mu = \mathrm{E}\, X_1 \in (0, \infty]$. Although, N_c is not a stopping time, when $X_1 \geq 0$, a.c. $N_c + 1 = T'_c$. Thus, when $X_1 \geq 0$, a.c., (13) and (14) below hold with N_c replacing T'_c and the former is known as the **elementary renewal theorem**. A stronger result (due to Blackwell) asserts that $\mathrm{E}\, N_{c+\alpha} - \mathrm{E}\, N_c \to \alpha/\mu$ as $c \to \infty$ (modification being necessary when $\{X_n\}$ are lattice r.v.s).

Stopping times may be utilized to obtain the elementary renewal theorem. A first step in this direction is

Lemma 2 (Gundy-Siegmund). *Let* $\{X_n, n \geq 1\}$ *be independent, nonnegative* $\mathscr{L}_1$ r.v.s *and let* $\{T_m, m \geq 1\}$ *be a sequence of* $\{X_n\}$*-times satisfying* $\mathrm{E}\, T_m < \infty$, $m \geq 1$, *and* $\lim_{m\to\infty} \mathrm{E}\, T_m = \infty$. *If*

$$\sum_{j=1}^{n} \int_{[X_j > \varepsilon j]} X_j = o(n), \qquad \varepsilon > 0,$$

then $\mathrm{E}\, X_{T_n} = o(\mathrm{E}\, T_n)$ *as* $n \to \infty$.

PROOF. For $\varepsilon > 0$, choose $N \geq 1$ such that $\sum_{j=1}^{n} \int_{[X_j > \varepsilon j]} X_j < n\varepsilon$ for $n \geq N$. Then, if $n \geq N$,

$$\begin{aligned}
\mathrm{E}\, X_{T_n} &= \int_{[X_{T_n} \leq \varepsilon T_n]} X_{T_n} + \sum_{j=1}^{\infty} \int_{[T_n = j,\, X_j > \varepsilon j]} X_j \\
&\leq \varepsilon\, \mathrm{E}\, T_n + \sum_{j=1}^{N-1} \mathrm{E}\, X_j + \sum_{j=N}^{\infty} \mathrm{E}\, X_j I_{[X_j > \varepsilon j,\ T_n \geq j]} \\
&= \varepsilon\, \mathrm{E}\, T_n + O(1) + \sum_{j=N}^{\infty} \mathrm{P}\{T_n \geq j\} \mathrm{E}\, X_j I_{[X_j > \varepsilon j]} \\
&= \varepsilon\, \mathrm{E}\, T_n + O(1) + \sum_{k=N}^{\infty} \sum_{j=N}^{k} \mathrm{E}\, X_j I_{[X_j > \varepsilon j]}\, \mathrm{P}\{T_n = k\} \\
&\leq \varepsilon\, \mathrm{E}\, T_n + O(1) + \sum_{k=N}^{\infty} \varepsilon k\, \mathrm{P}\{T_n = k\} \leq 2\varepsilon\, \mathrm{E}\, T_n + O(1),
\end{aligned}$$

and so $\mathrm{E}\, X_{T_n} = o(\mathrm{E}\, T_n)$. □

Theorem 5 (Elementary Renewal Theorem). *If* $\{X_n, n \geq 1\}$ *are* i.i.d. *with* $\mu = \mathrm{E}\, X_1 \in (0, \infty]$, $S_n = \sum_1^n X_i$, *and*

$$T'_c = \inf\{n \geq 1 : S_n > c\}, \qquad c > 0, \tag{12}$$

then (i) $\mathrm{E}\, T'_c < \infty$ *and*

$$\lim_{c\to\infty} \frac{\mathrm{E}\, T'_c}{c} = \frac{1}{\mathrm{E}\, X_1}. \tag{13}$$

Moreover, (ii) *if* $\sigma_{X_1}^2 = \sigma^2 < \infty$, *then*

$$\lim_{c\to\infty} \frac{\sigma^2(T'_c)}{c} = \frac{\sigma^2}{\mu^3} \tag{14}$$

PROOF. By Corollary 5.2.2, $S_n/n \xrightarrow{\text{a.c.}} \mathrm{E}\, X_1 > 0$, whence $S_n \xrightarrow{\text{a.c.}} \infty$. Thus, Corollary 5.4.1 guarantees that the $\{X_n\}$-time T'_c has finite expectation.

For any m in $(0, \mu)$, choose the positive constant K sufficiently large to ensure $\mathrm{E}\, X_1 I_{[X_1 \le K]} > m$ and define

$$X'_n = X_n I_{[X_n \le K]}, \qquad S'_n = \sum_1^n X'_i, \qquad V = \inf\{n \ge 1 : S'_n > c\}.$$

Then $\{X'_n\}$ are i.i.d. and, as earlier, $\mathrm{E}\, V < \infty$. By Wald's equation

$$K + c \ge \mathrm{E}\, S'_V = \mathrm{E}\, V \cdot \mathrm{E}\, X'_1,$$

and so

$$\frac{\mathrm{E}\, T'_c}{K + c} \le \frac{\mathrm{E}\, V}{K + c} \le \frac{1}{\mathrm{E}\, X'_1} \le \frac{1}{m},$$

whence

$$\overline{\lim_{c \to \infty}} \frac{\mathrm{E}\, T'_c}{c} \le \frac{1}{m}$$

and $(m \to \mu)$

$$\overline{\lim_{c \to \infty}} \frac{\mathrm{E}\, T'_c}{c} \le \frac{1}{\mu}. \tag{15}$$

If $\mu = \infty$, (15) is tantamount to (13). If, rather, $\mu < \infty$, then by Wald's equation $\mu\, \mathrm{E}\, T'_c = \mathrm{E}\, S_{T'_c} > c$, implying

$$\underline{\lim_{c \to \infty}} \frac{\mathrm{E}\, T'_c}{c} \ge \frac{1}{\mu},$$

which, in conjunction with (15), yields (13).

Apropos of (ii), setting $T = T'_c$, by Theorem 5.3.3

$$\mathrm{E}(S_T - \mu T)^2 = \sigma^2\, \mathrm{E}\, T < \infty. \tag{16}$$

Since $T = T'_c \uparrow \infty$ as $c \uparrow \infty$ and $\int_{[X_1^2 > \varepsilon n]} X_1^2 = o(1)$, $\varepsilon > 0$, by Lemma 2 and part (i)

$$\mathrm{E}\, X_T^2 = o(\mathrm{E}\, T) = o(c) \quad \text{as } c \to \infty.$$

Hence,

$$\mathrm{E}(S_T - c)^2 \le \mathrm{E}(X_T^+)^2 \le \mathrm{E}\, X_T^2 = o(c) \quad \text{as } c \to \infty, \tag{17}$$

so that

$$\begin{gathered} \mathrm{E}\, S_T^2 < \infty, \qquad \mathrm{E}|X_T| = o(\sqrt{c}), \\ c \le \mu\, \mathrm{E}\, T = \mathrm{E}\, S_T \le c + \mathrm{E}|X_T| = c + o(\sqrt{c}), \end{gathered} \tag{18}$$

whence $\mathrm{E}\, T^2 < \infty$ via (16) and (18).

From (17) and (18)

$$\mathrm{E}(S_T - \mu\, \mathrm{E}\, T)^2 \le 2[\mathrm{E}(S_T - c)^2 + (c - \mu\, \mathrm{E}\, T)^2] = o(c), \tag{19}$$

and from (16), (19), and Schwarz's inequality

$$\mu^2 \,\mathrm{E}(T - \mathrm{E}\,T)^2 = \mathrm{E}(\mu T - S_T + S_T - \mu\,\mathrm{E}\,T)^2 = \sigma^2\,\mathrm{E}\,T + o(c).$$

Hence, by (13)

$$\sigma_T^2 = \mathrm{E}(T - \mathrm{E}\,T)^2 = \frac{\sigma^2}{\mu^2}\,\mathrm{E}\,T + o(c) = \frac{\sigma^2}{\mu^3}\,c + o(c),$$

which is (14). □

When $X_n, n \geq 1$ are i.i.d. with $\mathrm{E}\,X_1 = \mu > 0, \sigma^2 = \mathrm{E}\,X_1^2 \in (0, \infty), (\mu^3/c\sigma^2)^{1/2}$ $(T_c - (c/\mu))$ has a limiting normal distribution as $c \to \infty$ according to Theorem 9.4.2.

Example 1. If $\{X, X_n, n \geq 1\}$ are nonnegative i.i.d. r.v.s with $\mathrm{P}\{X > 0\} > 0$ and $T_x = \inf\{n \geq 1: S_n = \sum_1^n X_i > x\}$,

$$\tfrac{1}{2}\,E\,T_x \leq \frac{x}{\mathrm{E}\min(X, x)} \leq \mathrm{E}\,T_x, \qquad x > 0. \tag{20}$$

Remark. Since, setting $S_0 = 0$,

$$\mathrm{E}\,T_x = \sum_{n=1}^{\infty} \mathrm{P}\{T_x \geq n\} = \sum_{n=0}^{\infty} \mathrm{P}\{S_n \leq x\},$$

and since if F is the d.f. of X and

$$a(x) = \frac{x}{\int_0^x [1 - F(y)]dy}, \tag{21}$$

according to Corollary 6.2.1 and Example 6.2.2

$$\frac{x}{a(x)} = \int_0^x y\,dF(y) + x[1 - F(x)] = \mathrm{E}\min(X, x),$$

it follows that the inequalities of (20) may be transcribed as

$$\frac{1}{2}\sum_{n=0}^{\infty} \mathrm{P}\{S_n \leq x\} \leq a(x) \leq \sum_{n=0}^{\infty} \mathrm{P}\{S_n \leq x\}, \qquad x > 0. \tag{22}$$

Proof. Since $T_x \geq n$ iff $S_{n-1} \leq x$, the series expression for $\mathrm{E}\,T_x$ is immediate (see Exercise 4.1.4). Set

$$X_n' = \min(X_n, x), \qquad S_n' = \sum_1^n X_i',$$

and note that (omitting subscripts) $x \leq \min(S_T, x) \leq S_T'$, implying

$$x \leq \mathrm{E}\,S_T' \leq x + \mathrm{E}\,X_T' \leq 2x.$$

Moreover, by Wald's equation

$$\mathrm{E}\, S'_T = \mathrm{E}\, T \cdot \mathrm{E}\, X'_1 = \mathrm{E}\, T \cdot \mathrm{E} \min(X, x),$$

yielding (20). □

Consider the plight of a gambler who makes a series of plays with outcomes $X_n, n \geq 1$, where the $X_n, n \geq 1$, are i.i.d. r.v.s, and who has the option of stopping at any time $n \geq 1$ with the fortune $Y_n = \max_{1 \leq i \leq n} X_i - cn$, $c > 0$. Since the gambler is not clairvoyant his choice of a rule for cessation of play is a stopping time, i.e., his decision to stop at any specific time n must be based only on $X_1, \ldots, X_n$ (and not on X_j with $j > n$). Does there exist an **optimal stopping rule**, i.e., one which maximizes $\mathrm{E}\, Y_T$ over the class of all finite stopping times?

Theorem 6. *Let* $\{X_n, n \geq 1\}$ *be* i.i.d. r.v.s *with* $\mathrm{E}|X_1| < \infty$ *and*

$$Y_n = \max_{1 \leq i \leq n} X_i - cn, \qquad n \geq 1, \tag{23}$$

where c is a positive constant. If T is the finite $\{X_n\}$-time defined by

$$T = \inf\{n \geq 1 \colon X_n \geq \beta\}, \tag{24}$$

where β is the unique solution of

$$\mathrm{E}(X_1 - \beta)^+ = c, \tag{25}$$

then $\mathrm{E}\, T < \infty$ *and T is an optimal stopping rule, that is,*

$$\mathrm{E}\, Y_T = \beta \geq \mathrm{E}\, Y_{T'} \tag{26}$$

for every finite $\{X_n\}$-time T' for which $\mathrm{E}\, Y_{T'}$ *exists.*

Proof. Since $f(\beta) = \mathrm{E}(X_1 - \beta)^+$ is a decreasing continuous function of β with $\lim_{\beta \to \infty} f(\beta) = 0$, $\lim_{\beta \to -\infty} f(\beta) = \infty$, and is strictly decreasing when $f(\beta) > 0$, necessarily (25) has a unique solution β. Set $p = \mathrm{P}\{X_1 \geq \beta\}$ and $q = 1 - p$. By (25), $p > 0$. Then

$$\mathrm{P}\{T = n\} = pq^{n-1}, \qquad \mathrm{P}\{T < \infty\} = \sum pq^{n-1} = \frac{p}{1-q} = 1,$$

$$\mathrm{E}\, T = \sum_1^\infty npq^{n-1} = p \frac{d}{dq}\left(\sum_0^\infty q^n\right) = \frac{p}{(1-q)^2} = \frac{1}{p},$$

and, since $X_j \geq \beta$ on $\{T = j\}$,

$$\mathrm{E}\, X_T = \sum_1^\infty \mathrm{E}\, X_j I_{[T=j]}$$

$$= \sum_1^\infty \mathrm{E}\, X_j I_{[T \geq j]} I_{[X_j \geq \beta]}$$

$$= \sum_1^\infty \mathrm{P}\{T \geq j\} \mathrm{E}\, X_1 I_{[X_1 \geq \beta]}$$

$$= \sum_1^\infty [\mathrm{E}(X_1 - \beta)^+ + \beta p] \mathrm{P}\{T \geq j\}$$

$$= (c + \beta p)\mathrm{E}\, T = \frac{c}{p} + \beta.$$

Moreover,

$$\mathrm{E}\, Y_T^- = \mathrm{E}(X_T - cT)^- \leq \mathrm{E}\, X_T^- + c\,\mathrm{E}\, T \leq \beta^- + c\,\mathrm{E}\, T < \infty,$$

whence $\mathrm{E}\, Y_T$ exists and

$$\mathrm{E}\, Y_T = \mathrm{E} \max_{1 \leq j \leq T} X_j - c\,\mathrm{E}\, T = \mathrm{E}\, X_T - \frac{c}{p} = \beta,$$

yielding the first part of (26). Moreover, if T' is any finite $\{X_n\}$-time for which $\mathrm{E}\, Y_{T'}$ exists, it may be supposed without loss of generality that $\mathrm{E}\, Y_{T'} > -\infty$. Since for any $b > \beta$

$$Y_n = \max_{1 \leq i \leq n} X_i - cn \leq b + \sum_1^n [(X_i - b)^+ - c]$$

and $\mathrm{E}[(X_1 - b)^+ - c] < 0$ by (25), Theorem 5.3.2 ensures that

$$\mathrm{E}\, Y_{T'} \leq b + \mathrm{E} \sum_1^{T'} [(X_i - b)^+ - c] < b.$$

Hence, $\mathrm{E}\, Y_{T'} \leq \beta$, which is the second portion of (26). □

The stopping rule T_c in (28) is of especial interest since the finiteness of its moments depends on the value of the parameter c. According to Exercise 5.3.7 (also the application of Corollary 5.3.5) $\mathrm{E}\, T_c = \infty$ for $c \geq 1$ whereas Example 2 demonstrates under the Lindeberg condition (27) (see Chapter 9 for a more general form) that $\mathrm{E}\, T_c < \infty$ for $0 < c < 1$. For $\{X_n, n \geq 1\}$ as in Example 2 there exist constants $c_k \downarrow 0$ (c_k is the smallest positive root of the Hermite polynomial of degree $2k$) such that $\mathrm{E}\, T_c^k(m) < \infty$ for $c < c_k$, all $m \geq 1$ while $\mathrm{E}\, T_c^k(m) = \infty$ for all large m when $c > c_k$ [B. Brown].

EXAMPLE 2. Let $S_n = \sum_{j=1}^n X_j$ where $\{X_j, j \geq 1\}$ are independent r.v.s with $\mathrm{E}\, X_n = 0$, $\mathrm{E}\, X_n^2 = 1$, $n \geq 1$ obeying

$$\sum_{j=1}^n \int_{[X_j^2 > \varepsilon j]} X_j^2 = o(n) \tag{27}$$

and define

$$T_c = T_c(1),\ T_c(m) = \inf\{n \geq m: |S_n| > cn^{1/2}\},\ c > 0,\ m = 1, 2, \ldots. \quad (28)$$

Then for all $m \geq 1$, $\mathrm{E}\,T_c(m) < \infty$ for all c in $(0, 1)$.

PROOF. The argument for $m > 1$ requires only minor modifications from that for $m = 1$ which will therefore be supposed. Let $c \in (0, 1)$ and assume that $\mathrm{E}\,T_c = \infty$. If $V = \min(T_c, n)$, clearly $\mathrm{E}\,V < \infty$ and $\mathrm{E}\,V \to \infty$ as $n \to \infty$. Thus, $\mathrm{E}\,X_V^2 = o(\mathrm{E}\,V)$ by Lemma 2 and via Theorem 5.3.3

$$\begin{aligned} \mathrm{E}\,V = \mathrm{E}\,S_V^2 &= \mathrm{E}\,S_{V-1}^2 + 2\,\mathrm{E}\,S_{V-1}X_V + \mathrm{E}\,X_V^2 \\ &\leq c^2\,\mathrm{E}\,V + 2c[\mathrm{E}\,V \cdot \mathrm{E}\,X_V^2]^{1/2} + \mathrm{E}\,X_V^2 \\ &= c^2\,\mathrm{E}\,V + o(\mathrm{E}\,V). \end{aligned}$$

Hence, $(1 - c^2)\,\mathrm{E}\,V \leq o(\mathrm{E}\,V)$ yielding a contradiction as $n \to \infty$. Thus, $\mathrm{E}\,T_c < \infty$ for $0 < c < 1$. □

EXERCISES 5.4

1. Verify the second equality of (3).
2. Prove that the stopping rule T of (24) remains optimal when $Y_n = X_n - cn$, $c > 0$.
3. Prove Corollary 3.
4. Prove that if $\{X_n\}$ are i.i.d. with $\mathrm{E}\,X_1 = \mu \in (0, \infty]$ and $N_c = \sup\{n \geq 1: S_n \leq c\}$, then $N_c/c \xrightarrow{\text{a.c.}} 1/\mu$.
5. Let $\{X_n\}$ be i.i.d. r.v.s with $\mathrm{E}\,X_1 = \mu > 0$ and
$$T = T_c' = \inf\{n \geq 1: S_n > c\} \quad \text{for } c > 0.$$
Prove that $T/c \xrightarrow{c \to \infty} 1/\mu$, a.c. If, moreover, $\mathrm{E}\,X_1^2 < \infty$, then
$$\frac{S_T - \mu\,\mathrm{E}\,T}{\sqrt{T}} \xrightarrow{\text{P}} 0 \quad \text{as } c \to \infty.$$
6. Prove that if $\{X_n\}$ are i.i.d. with $\mathrm{E}\,X_1 = \mu \in (0, \infty]$ and $T_c = \inf\{n \geq 1: S_n > cn^\alpha\}$, $c > 0$, $0 < \alpha < 1$, then
$$\frac{T_c^{1-\alpha}}{c} \xrightarrow{\text{a.c.}} \frac{1}{\mu} \quad \text{as} \quad c \to \infty \quad \text{and } \mathrm{E}\,T_c \sim \left(\frac{c}{\mu}\right)^{1/(1-\alpha)}.$$
7. (Chow–Robbins) Let $\{Y_n, n \geq 1\}$ be positive r.v.s with $\lim Y_n = 1$, a.c., and $\{a_n, n \geq 1\}$ positive constants with $a_n \to \infty$, $a_n/a_{n-1} \to 1$. For $c > 0$ define $N = N_c = \inf\{n \geq 1: Y_n \leq a_n/c\}$. Prove that $\mathrm{P}\{N < \infty, \lim_c a_N/c = 1\} = 1$ and, if $\mathrm{E}\sup_{n \geq 1} Y_n < \infty$, then $\lim_{c \to \infty} \mathrm{E}\,a_N/c = 1$.
8. If $S_n = \sum_1^n X_i$, where $\{X_n, n \geq 1\}$ are nondegenerate i.i.d. r.v.s, $T = \inf\{j \geq 1: S_j < -a < 0 \text{ or } S_j > b > 0\}$ is of interest in sequential analysis (Wald). Prove that T is a stopping variable with finite moments of all orders. *Hint*: $\mathrm{P}\{T > rn\} \leq \mathrm{P}\{|S_{jr} - S_{r(j-1)}| < a + b, 1 \leq j \leq n\}$ for all integers $r > 0$.

9. (Alternative proof that $T'_+ < \infty$, a.c., implies $\mathrm{E}\, T_- = \infty$.) Let $\{T^{(j)}, j \geq 1\}$ be copies of T'_+ and set $T_n = \sum_1^n T^{(j)}$. Then $Z = \sum_1^\infty I_{[T_n < \infty]}$ = number of times S_n exceeds S_i, $i < n$ (take $S_0 = 0$) and $\mathrm{P}\{T_- \geq n + 1\} = \mathrm{P}\{S_n > S_i, 0 \leq i < n\} = \mathrm{P}\{\bigcup_1^n [T_j = n]\}$, whence $\mathrm{E}\, T_- - 1 = \mathrm{E}\, Z = \infty$.

10. If $\{X_n, n \geq 1\}$ are i.i.d. with $S_n = \sum_1^n X_i$, then (i) $\overline{\lim}\, S_n = \infty$, a.c., iff there is a finite $\{X_n\}$-time T with $\mathrm{E}\, S_T > 0$. (ii) Moreover, when $\mathrm{E}\, X_1^+ = \infty$ $\overline{\lim}\, S_n/n = \infty$, a.c., iff there exists a finite $\{X_n\}$-time T with $\mathrm{E}\, S_T > -\infty$.

11. If $\{X_n, n \geq 1\}$ are i.i.d. r.v.s with $\mathrm{E}\, X_1 = \mu > -\infty$ and $S_n = \sum_1^n X_i$, $T_c = \inf\{n \geq 1: S_n > cn\}$, then $\mathrm{E}\, T_c < \infty$ for $c < \mu$ and $\mathrm{E}\, T_c = \infty$ for $c \geq \mu$.

12. Let $\{S_n = \sum_1^n X_i, n \geq 1, S_0 = 0\}$ be a random walk on the line. A real number c is said to be **recurrent** if $\mathrm{P}\{|S_n - c| < \varepsilon, \text{i.o.}\} = 1$ for all $\varepsilon > 0$ and is called **possible** if for every $\varepsilon > 0$ there is an integer n such that $\mathrm{P}\{|S_n - c| < \varepsilon\} > 0$. Prove that if c is possible and b is recurrent, then $b - c$ is recurrent. Since every recurrent value is clearly possible, the set Q of recurrent values is an additive group. Show that Q is closed. Thus, if Q is nonempty, $Q = (-\infty, \infty)$ or $Q = \{nc: c \neq 0, n = 0, \pm 1, \pm 2, \ldots\}$ or $Q = \{0\}$ (the latter only if X is degenerate at 0).

References

J. H. Abbott and Y. S. Chow, "Some necessary conditions for a.s. convergence of sums of independent r.v.s.," *Bull. Institute Math. Academia Sinica* **1** (1973), 1–7.

L. E. Baum and M. Katz, "Convergence rates in the law of large numbers," *Trans. Amer. Math. Soc.* **120** (1965), 108–123.

B. Brown, "Moments of a stopping rule related to the central limit theorem," *Ann. Math. Stat.* **40** (1969), 1236–1249.

D. L. Burkholder, "Independent sequences with the Stein property," *Ann. Math. Stat.* **39** (1968), 1282–1288.

Y. S. Chow, "Local convergence of martingales and the law of large numbers," *Ann. Math. Stat.* **36** (1965), 552–558.

Y. S. Chow, "Delayed sums and Borel summability of independent, identically distributed random variables," *Bull. Inst. Math., Academia Sinica* **1** (1973), 207–220.

Y. S. Chow and H. Robbins, "On the asymptotic theory of fixed-width sequential confidence intervals for the mean." *Ann. Math. Stat.* **36** (1965), 457–462.

Y. S. Chow and H. Teicher, "Almost certain summability of i.i.d. random variables," *Ann. Math. Stat.* **42** (1971), 401–404.

Y. S. Chow, H. Robbins, and D. Siegmund, *Great Expectations: The Theory of Optimal Stopping*, Houghton Mifflin, Boston, 1972.

Y. S. Chow, H. Robbins, and H. Teicher, "Moments of randomly stopped sums," *Ann. Math. Stat.* **36** (1965), 789–799.

K. L. Chung, "Note on some strong laws of large numbers," *Amer. Jour. Math.* **69** (1947), 189–192.

K. L. Chung, *A Course in Probability Theory*, Harcourt Brace, New York, 1968; 2nd ed., Academic Press, New York, 1974.

K. L. Chung and W. H. J. Fuchs, "On the distribution of values of sums of random variables," *Mem. Amer. Math. Soc.* **6** (1951).

K. L. Chung and D. Ornstein, "On the recurrence of sums of random variables," *Bull. Amer. Math. Soc.* **68** (1962), 30–32.

C. Derman and H. Robbins, "The SLLN when the first moment does not exist," *Proc. Nat. Acad. Sci. U.S.A.* **41** (1955), 586–587.

J. L. Doob, *Stochastic Processes*, Wiley, New York, 1953.

K. B. Erickson, "The SLLN when the mean is undefined," *Trans. Amer. Math. Soc.* **185** (1973), 371–381.

W. Feller, "Über das Gesetz der grossen Zahlen," *Acta Univ. Szeged., Sect. Sci. Math.* **8** (1937), 191–201.

W. Feller, "A limit theorem for random variables with infinite moments," *Amer. Jour. Math.* **68** (1946), 257–262.

W. Feller, *An Introduction to Probability Theory and its applications*, Vol. 2, Wiley, New York, 1966.

R. Gundy and D. Siegmund, "On a stopping rule and the central limit theorem," *Ann. Math. Stat.* **38** (1967), 1915–1917.

C. C. Heyde, "Some renewal theorems with applications to a first passage problem," *Ann. Math. Stat.* **37** (1966), 699–710.

T. Kawata, *Fourier Analysis in Probability Theory*, Academic Press, New York, 1972.

H. Kesten, "The limit points of a random walk," *Ann. Math. Stat.* **41** (1970), 1173–1205.

A. Khintchine and A. Kolmogorov, "Über Konvergenz von Reichen, derem Glieder durch den Zufall bestimmt werden," *Rec. Math.* (*Mat. Sbornik*) **32** (1924), 668–677.

A. Kolmogorov, "Über die Summen durch den Zufall bestimmer unabbängiger Grössen," *Math. Ann.* **99** (1928), 309–319; **102** (1930), 484–488.

M. J. Klass, "Properties of optimal extended-valued stopping rules," *Ann. Prob.* **1** (1973), 719–757.

M. Klass and H. Teicher, "Iterated logarithm laws for random variables barely with or without finite mean," *Ann. Prob.* **5** (1977), 861–874.

K. Knopp, *Theory and Application of Infinite Series*, Stechert-Hafner, New York, 1928.

P. Lévy, *Theorie de l'addition des variables aléatoires*, Gauthier-Villars, Paris, 1937; 2nd ed., 1954.

M. Loève, "On almost sure convergence," *Proc. Second Berkeley Symp. Math. Stat. Prob.*, pp. 279–303, Univ. of California Press, 1951.

M. Loève, *Probability Theory*, 3rd ed., Van Nostrand, Princeton, 1963; 4th ed., Springer-Verlag, Berlin and New York, 1977–1978.

J. Marcinkiewicz and A. Zygmund, "Sur les fonctions indépendantes," *Fund. Math.* **29** (1937), 60–90.

P. Revesz, *The Laws of Large Numbers*, Academic Press, New York, 1968.

H. Robbins and E. Samuel, "An extension of a lemma of Wald," *J. Appl. Prob.* **3** (1966), 272–273.

F. Spitzer, "A combinatorial lemma and its applications to probability theory," *Trans. Amer. Math. Soc.* **82** (1956), 323–339.

C. J. Stone, "The growth of a recurrent random walk," *Ann. Math. Stat.* **37** (1966), 1040–1041.

H. Teicher, "Almost certain convergence in double arrays," *Z. Wahr. verw. Gebiete* **69** (1985), 331–345.

A. Wald, "On cumulative sums of random variables," *Ann. Math. Stat.* **15** (1944), 283–296.

6 Measure Extensions, Lebesgue–Stieltjes Measure, Kolmogorov Consistency Theorem

6.1 Measure Extensions, Lebesgue–Stieltjes Measure

A salient underpinning of probability theory is the one-to-one correspondence between distribution functions on R^n and probability measures on the Borel subsets of R^n. Verification of this correspondence involves the notion of measure extension.

Recall that a measure μ on a class $\mathscr{A}$ of subsets of a space Ω is σ-finite if $\Omega = \bigcup_{n=1}^{\infty} \Omega_n$ with $\Omega_n \in \mathscr{A}$, $\mu\{\Omega_n\} < \infty$, $n \geq 1$. Moreover, if μ and v are set functions on classes $\mathscr{G}$ and $\mathscr{H}$ respectively with $\mathscr{G} \subset \mathscr{H}$ and $\mu\{A\} = v\{A\}$ for each $A \in \mathscr{G}$, then v is dubbed an extension of μ to $\mathscr{H}$ while μ is called the restriction of v to $\mathscr{G}$, the latter being denoted by $\mu = v|_{\mathscr{G}}$.

Theorem 1. *If μ is a measure on a semi-algebra $\mathscr{S}$ of subsets of Ω, there exists a measure extension v of μ to $\sigma(\mathscr{S})$. Moreover, if μ is a probability or σ-finite measure, then v is likewise and the extension is unique.*

PROOF. For each subset $A \subset \Omega$, define

$$v\{A\} = \inf\left\{\sum_{1}^{\infty} \mu\{S_n\} : \bigcup_{1}^{\infty} S_n \supset A,\ S_n \in \mathscr{S},\ n \geq 1\right\}, \tag{1}$$

$$\mathscr{M} = \{D \subset \Omega : v\{D \cdot A\} + v\{D^c \cdot A\} = v\{A\} \text{ for all } A \subset \Omega\}. \tag{2}$$

Clearly, for $A \subset B \subset \Omega$, $0 = v\{\varnothing\} \leq v\{A\} \leq v\{B\}$; moreover, $\Omega \in \mathscr{M}$, and $D^c \in \mathscr{M}$ whenever $D \in \mathscr{M}$.

(i) $\mu = v|_{\mathscr{S}}$.

If $S \in \mathscr{S}$, then by (1), $\nu\{S\} \le \mu\{S\} + \mu\{\varnothing\} + \mu\{\varnothing\} + \cdots = \mu\{S\}$, while if $\bigcup_1^\infty S_n \supset S$, where $S_n \in \mathscr{S}$, $n \ge 1$, then, since μ is a measure on $\mathscr{S}$, by Corollary 1.5.1 $\mu\{S\} = \mu\{\bigcup_1^\infty S_n S\} \le \sum_1^\infty \mu\{S_n\}$, and so via (1), $\mu\{S\} \le \nu\{S\}$.

(ii) ν is subadditive, that is, for $A_0 \subset \bigcup_1^\infty A_n \subset \Omega$

$$\nu\{A_0\} \le \sum_1^\infty \nu\{A_n\}. \tag{3}$$

In proving (3) it may be supposed that $\nu\{A_n\} < \infty$, $n \ge 1$. For any $\varepsilon > 0$ and $n \ge 1$, choose $S_{n,m} \in \mathscr{S}$, $m \ge 1$, with

$$A_n \subset \bigcup_1^\infty S_{n,m}, \qquad \sum_1^\infty \mu\{S_{n,m}\} \le \nu\{A_n\} + \frac{\varepsilon}{2^n}.$$

Since $\bigcup_{n,m} S_{n,m} \supset \bigcup_n A_n \supset A_0$, by (1)

$$\nu\{A_0\} \le \sum_{n,m} \mu\{S_{n,m}\} \le \sum_1^\infty \nu\{A_n\} + \varepsilon,$$

and, letting $\varepsilon \to 0$, (3) obtains.

(iii) $\mathscr{S} \subset \mathscr{M}$.

Let $S \in \mathscr{S}$ and $A \subset \Omega$. By (ii) and the definition of $\mathscr{M}$ it suffices to verify that

$$\nu\{A\} \ge \nu\{S \cdot A\} + \nu\{S^c \cdot A\}, \tag{4}$$

and in so doing it may clearly be supposed that $\nu\{A\} < \infty$.

For any $\varepsilon > 0$, choose $S_n \in \mathscr{S}$, $n \ge 1$, such that $\bigcup_1^\infty S_n \supset A$ and

$$\nu\{A\} + \varepsilon \ge \sum_1^\infty \mu\{S_n\}. \tag{5}$$

Since $\mathscr{S}$ is a semi-algebra, for each $n \ge 1$ there exists a finite partition of $S^c S_n$ in $\mathscr{S}$, say $(S_{n,m}, m = 1, \ldots, m_n)$. Then $(S \cdot S_n, S_{n,m}, m = 1, 2, \ldots, m_n)$ is a partition of S_n in $\mathscr{S}$, and by additivity of μ on $\mathscr{S}$

$$\mu\{S_n\} = \mu\{S \cdot S_n\} + \sum_{m=1}^{m_n} \mu\{S_{n,m}\}, \; n \ge 1. \tag{6}$$

But $\bigcup_{n=1}^\infty \bigcup_{m=1}^{m_n} S_{m,n} = \bigcup_{n=1}^\infty S^c \cdot S_n \supset A \cdot S^c$ and $\bigcup_1^\infty SS_n \supset SA$, whence by (1)

$$\sum_{n=1}^\infty \sum_{m=1}^{m_n} \mu\{S_{n,m}\} \ge \nu\{A \cdot S^c\}, \qquad \sum_{n=1}^\infty \mu\{S \cdot S_n\} \ge \nu\{S \cdot A\}. \tag{7}$$

Consequently, from (5), (6) and (7)

$$\nu\{A\} + \varepsilon \ge \sum_1^\infty \mu\{S \cdot S_n\} + \sum_{n=1}^\infty \sum_{m=1}^{m_n} \mu\{S_{n,m}\} \ge \nu\{S \cdot A\} + \nu\{S^c \cdot A\},$$

which is tantamount to (4) as $\varepsilon \to 0$.

(iv) $\mathscr{M}$ is an algebra and for every $D \in \mathscr{M}$ and $A \subset \Omega$ and finite partition $\{D_n, n = 1, 2, \ldots, m\}$ of D in $\mathscr{M}$,

$$\nu\{A \cdot D\} = \sum_1^m \nu\{A \cdot D_n\}. \tag{8}$$

Now for $A \subset \Omega$ and $D_i \in \mathscr{M}$, $i = 1, 2$,

$$\begin{aligned} \nu\{A\} &= \nu\{A \cdot D_1\} + \nu\{A \cdot D_1^c\} \\ &= \nu\{A \cdot D_1 D_2\} + \nu\{AD_1 D_2^c\} + \nu\{AD_1^c D_2\} \\ &\quad + \nu\{AD_1^c D_2^c\}. \end{aligned} \tag{9}$$

Replacing A by $A(D_1 \cup D_2)$ in (9),

$$\nu\{A(D_1 \cup D_2)\} = \nu\{A \cdot D_1 D_2\} + \nu\{AD_1 D_2^c\} + \nu\{AD_1^c D_2\}, \tag{10}$$

whence from (9) and (10)

$$\nu\{A\} = \nu\{A(D_1 \cup D_2)\} + \nu\{AD_1^c D_2^c\},$$

and so, via the definition, $D_1 \cup D_2 \in \mathscr{M}$. Since, as noted at the outset, $\mathscr{M}$ is closed under complements, $\mathscr{M}$ is an algebra, and, moreover, if $D_1 D_2 = \varnothing$, (10) yields

$$\nu\{A(D_1 \cup D_2)\} = \nu\{AD_1\} + \nu\{AD_2\},$$

which is precisely (8) when $m = 2$. The general statement of (8) follows by induction.

(v) $\mathscr{M}$ is a σ-algebra, and for every $D \in \mathscr{M}$ and $A \subset \Omega$ and σ-partition $\{D_n, n \geq 1\}$ of D in $\mathscr{M}$,

$$\nu\{A \cdot D\} = \sum_1^\infty \nu\{A \cdot D_n\}. \tag{11}$$

Let $\{D_n, n \geq 1\}$ be the stated σ-partition of D and set $E_n = \bigcup_1^n D_i$. By (iv), $E_n \in \mathscr{M}$ for every positive integer n, whence for any $A \subset \Omega$,

$$\begin{aligned} \nu\{A\} &= \nu\{A\,E_n\} + \nu\{A\,E_n^c\} \geq \nu\{A \cdot E_n\} + \nu\{AD^c\} \\ &= \sum_1^n \nu\{A \cdot D_i\} + \nu\{A \cdot D^c\}. \end{aligned}$$

Hence, by (3)

$$\nu\{A\} \geq \sum_1^\infty \nu\{A \cdot D_i\} + \nu\{A \cdot D^c\} \geq \nu\{A \cdot D\} + \nu\{A \cdot D^c\}, \tag{12}$$

and so equality holds throughout (12), yielding (11) upon replacement of A by AD. Moreover, if $\{D_j, j \geq 1\}$ is any sequence of disjoint sets of $\mathscr{M}$ and $D = \bigcup_1^\infty D_j$, (12) remains intact, whence it is clear that (see Exercise 1.3.2) $\mathscr{M}$ is a σ-algebra. Clearly, if μ is finite or σ-finite, ν inherits the characteristic.

It follows directly via (i), (iii), and (v) that (vi) $\mathscr{M} \supset \sigma(\mathscr{S})$, ν is a measure on $\mathscr{M}$, and $\mu = \nu|_{\mathscr{S}}$.

Finally, to prove uniqueness, let ν^* be any measure extension of μ to $\sigma(\mathscr{S})$ and define

$$\mathscr{E} = \{E: E \in \sigma(\mathscr{S}) \text{ and } \nu(E) = \nu^*(E)\}.$$

If $\mu\{\Omega\} < \infty$, then $\nu\{\Omega\} = \nu^*\{\Omega\} < \infty$ and it is easily verified that $\mathscr{E}$ is a λ-class containing the π-class $\mathscr{S}$, whence $\nu = \nu^*$ on $\sigma(\mathscr{S})$. If μ is merely σ-finite on $\mathscr{S}$, there exist sets $\Omega_n \in \mathscr{S}$ with $\bigcup_1^\infty \Omega_n = \Omega$ and $\mu\{\Omega_n\} < \infty$, $n \geq 1$. Then, as just seen, $\nu = \nu^*$ on $\Omega_n \cap \sigma(\mathscr{S})$, $n \geq 1$, and so $\nu = \nu^*$ on $\sigma(\mathscr{S})$. □

The set function ν defined by (1) on the σ-algebra of all subsets of Ω is called the **outer measure induced by** μ, while the σ-algebra $\mathscr{M}$ described in (2) is termed the **σ-algebra of ν-measurable sets.**

Any measure μ on a σ-algebra $\mathscr{A}$ is called **complete** if $B \subset A \in \mathscr{A}$ and $\mu\{A\} = 0$ imply $B \in \mathscr{A}$ (and necessarily $\mu\{B\} = 0$). A **complete measure space** is a measure space whose measure is complete.

The outer measure ν stipulated in (1) defines a complete measure extension of μ to $\mathscr{M}$, the σ-algebra of all ν-measurable sets (Exercise 2).

As an important application of Theorem 1, the special cases of Lebesgue–Stieltjes and Lebesgue measure on the line will be considered. Let $R = [-\infty, \infty]$ and $\mathscr{S}$ denote the semi-algebra of all intervals of the form $[a, b)$, $-\infty \leq a \leq b \leq \infty$ and R, $\{\infty\}$ in addition. A class of measures known as Lebesgue–Stieltjes measures will be generated on $\mathscr{S}$ via monotone functions.

For any mapping f of $R = [-\infty, \infty]$ into R, set

$$f(t+) = \lim_{s \to t+} f(s) = \lim_{t < s \to t} f(s), \qquad t \in [-\infty, \infty), \qquad f(\infty+) = f(\infty),$$

$$f(t-) = \lim_{s \to t-} f(s) = \lim_{t > s \to t} f(s), \qquad t \in (-\infty, \infty], \qquad f(-\infty-) = f(-\infty),$$

when these limits exist. If $f(t) = f(t-)$, then f is said to be continuous from the left or **left continuous at** $t \in R$. A function which is left continuous at all points of some set $T \subset R$ is called left continuous on T and when $T = R$, simply **left continuous**. Similarly, $f(t) = f(t+)$ defines right continuity at t and the analogous terms are employed.

If f is a function with $f(t-)$ existing for each $t \in R$, then $g(t) = f(t-)$ is left continuous (Exercise 3). In particular, if f is a monotone function on R, $f(t-)$ exists and $g(t) = f(t-)$ is left continuous. Since the set of discontinuities of a monotone function is countable and two left-continuous functions are identical if they coincide except for a countable set, every monotone function m on R defines a unique left-continuous function $F = F_m$ via $F(t) = m(t-)$.

Lemma 1. *Let F be a nondecreasing, left continuous function on $R = [-\infty, \infty]$ with $F(-\infty) = F(-\infty+)$ and $|F(t)| < \infty$, $|t| < \infty$. If*

$$\mu\{[a, b)\} = F(b) - F(a), \qquad -\infty \leq a \leq b \leq \infty,$$
$$\mu(\{\infty\}) = 0 = \mu\{\varnothing\}, \qquad \mu\{R\} = F(\infty) - F(-\infty), \tag{13}$$

then μ is a measure on the semi-algebra $\mathscr{S}$.

PROOF. Clearly, μ is nonnegative and additive on $\mathscr{S}$. To verify σ-additivity, consider $S \in \mathscr{S}$ with $S \neq \{\infty\}$ or $\varnothing$ and let $\{S_n, n \geq 1\}$ be a σ-partition of S in $\mathscr{S}$. By Corollary 1.5.1

$$\mu\{S\} \geq \sum_1^\infty \mu\{S_n\}. \tag{14}$$

In proving the reverse inequality, since $\mu\{R\} = \mu\{[-\infty, \infty)\}$, it may be supposed that $S = [a, b)$, $-\infty \leq a < b \leq \infty$. Moreover, in view of the hypothesized equality $F(-\infty) = F(-\infty+)$ and left continuity, it suffices to establish that

$$\mu\{[c, d)\} \leq \sum_1^\infty \mu\{S_n \cdot [c, d)\} \quad \text{whenever } a \leq c < d < b \leq \infty, c \neq -\infty. \tag{15}$$

Since $\infty \notin [a, b)$, necessarily $S_n \neq \{\infty\}$, $n \geq 1$ whence $S_n \cdot [c, d) = [a_n, b_n)$, where $c \leq a_n \leq b_a \leq d, n \geq 1$.

For any $\varepsilon > 0$, set $J_n = (a_n - \delta_n, b_n)$ where $\delta_n > 0$ satisfies $F(a_n) - F(a_n - \delta_n) < \varepsilon/2^n$, $n \geq 1$ via left continuity. Then $[c, d] = \bigcup_1^\infty [a_n, b_n) \subset \bigcup_1^\infty J_n$ whence by the Heine–Borel theorem $[c, d) \subset \bigcup_1^m J_{n_k} \subset \bigcup_1^m [a_{n_k} - \delta_{n_k}, b_{n_k})$ for some finite set of integers $n_1, \ldots, n_m$. By Corollary 1.5.1

$$\begin{aligned}\mu\{[c, d)\} &\leq \sum_1^m (F(b_{n_k}) - F(a_{n_k} - \delta_{n_k})) \leq \sum_1^m (F(b_{n_k}) - F(a_{n_k}) + \varepsilon 2^{-n_k}) \\ &\leq \sum_1^m (F(b_{n_k}) - F(a_{n_k})) + \varepsilon \leq \sum_1^\infty \mu\{S_k\} + \varepsilon.\end{aligned}$$

Thus, (15) obtains as $\varepsilon \to 0$, and μ is a measure on $\mathscr{S}$. □

Theorem 2. *Any nondecreasing, finite function m on $(-\infty, \infty)$ determines a complete measure ν_m on the σ-algebra $\mathscr{M}_m$ of all ν_m-measurable subsets of $R = [-\infty, \infty]$ with*

$$\nu_m\{[a, b)\} = m(b-) - m(a-), \qquad -\infty < a \leq b < \infty,$$
$$\nu_m(\{\infty\}) = \nu_m(\{-\infty\}) = 0. \tag{16}$$

Moreover, $\mathscr{M}_m \supset$ the σ-algebra $\mathscr{B}$ of all Borel sets of R and ν_m is unique on $\mathscr{B}$.

PROOF. Set $F(t) = m(t-)$ for $-\infty < t \leq \infty$ and $F(-\infty) = m(-\infty+)$. Then F is defined and left continuous on R with $F(-\infty+) = F(-\infty)$. By Lemma 1, the set function μ defined on $\mathscr{S}$ by (13) is a measure thereon, whence Theorem 1 guarantees the existence of a measure extension $\tilde{\nu}_m$ of μ to $\mathscr{M}_m \supset \mathscr{B} = \sigma(\mathscr{S})$, where $\tilde{\nu}_m$ and $\mathscr{M}_m$ are defined by (1) and (2) respectively. According to Exercise 2, $\tilde{\nu}_m$ is complete on $\mathscr{M}_m$.

Hence, $\nu_m\{A\} = \tilde{\nu}_m\{A - \{-\infty\}\}$, $A \in \mathscr{M}_m$ is a complete measure on $\mathscr{M}_m$ satisfying (16).

To verify uniqueness on $\mathscr{B}$, let ν^* be any other measure on $\mathscr{B}$ satisfying

(16). Then

$$\nu^*\{\{\infty\}\} = \nu^*\{\{-\infty\}\} = \nu_m\{\{\infty\}\} = \nu_m\{\{-\infty\}\}$$

$$\nu^*\{[a, b)\} = m(b-) - m(a-) = \nu_m\{[a, b)\}, \ -\infty < a \le b < \infty,$$

and for $-\infty < b < \infty$

$$\nu^*\{[-\infty, b)\} = \nu^*\{(-\infty, b)\} = \lim_{a \to -\infty} \nu^*\{[a, b)\} = \lim \nu_m\{[a, b)\}$$
$$= \nu_m\{[-\infty, b)\};$$

similarly, for $-\infty < a < \infty$

$$\nu^*\{[a, \infty)\} = \nu_m\{[a, \infty)\}$$

Thus, taking $a = b$,

$$\nu^*\{R\} = \nu^*\{[-\infty, \infty)\} = \nu_m\{[-\infty, \infty)\} = \nu_m\{R\}.$$

Hence, ν_m and ν^* coincide on $\mathscr{S}$ whence by Theorem 1 these σ-finite measures also coincide on $\mathscr{B} = \sigma(\mathscr{S})$. □

For any finite, nondecreasing function m on $(-\infty, \infty)$ the corresponding measure ν_m is called the **Lebesgue–Stieltjes measure determined by m**. Similarly, the complete measure space $(R, \mathscr{M}_m, \nu_m)$ is referred to as the **Lebesgue–Stieltjes measure space determined by m**.

In the important special case $m(t) = t$ for $t \in (-\infty, \infty)$, the corresponding measure $\nu = \nu_m$ is the renowned **Lebesgue measure**, generalizing the notion of length; the sets of $\mathscr{M} = \mathscr{M}_m$ are called **Lebesgue-measurable sets** and $(R, \mathscr{M}, \nu)$ is the **Lebesgue measure space** (of R).

EXERCISES 6.1

1. Let μ be a measure on a σ-algebra $\mathscr{A}$ and define $\bar{\mathscr{A}} = \{A \,\Delta\, N : A \in \mathscr{A}, N \subset B \in \mathscr{A}, \mu\{B\} = 0\}$. Then $\bar{\mathscr{A}}$ is a σ-algebra, $\bar{\mathscr{A}} \supset \mathscr{A}$, and $\bar{\mu}$ is a complete measure on $\bar{\mathscr{A}}$, where $\bar{\mu}\{A \,\Delta\, N\} = \mu\{A\}$ for $A \,\Delta\, N \in \bar{\mathscr{A}}$.

2. Prove that the extension ν (as defined in Theorem 1) of a measure μ on $\mathscr{S}$ to the σ-algebra $\mathscr{M}$ of all ν-measurable sets is complete.

3. Let f map $R = [-\infty, \infty]$ into R with $f(t-)$ existing for every $t \in R$. Then $g(t) = f(t-)$, $t \in R$, is left continuous, i.e., $g(t) = g(t-)$, $t \in R$.

4. If ν_m is the Lebesgue–Stieltjes measure determined by a finite nondecreasing function m on $(-\infty, \infty)$ and $F(t) = m(t-)$, $-\infty \le t \le \infty$, prove that for $-\infty < a < b < \infty$

$$\nu\{[a, b)\} = F(b-) - F(a-), \qquad \nu\{(a, b]\} = F(b+) - F(a+),$$
$$\nu\{[a, b]\} = F(b+) - F(a-), \qquad \nu\{(a, b)\} = F(b) - F(a+),$$
$$\nu\{\{\infty\}\} = \nu\{\{-\infty\}\} = 0, \qquad \nu\{\{a\}\} = F(a+) - F(a-).$$

5. If $m(t)$ is as in Exercise 4 and $G(t) = m(t+)$, $-\infty \le t \le \infty$, then the nondecreasing, right continuous function G determines a measure $\tilde{\mu}$ on the semi-algebra $\tilde{\mathscr{S}}$ of all

finite or infinite intervals of the form $(a, b]$, $a < b$, and also $\varnothing$, $R = [-\infty, \infty]$, $\{-\infty\}$ via $\tilde{\mu}\{(a, b]\} = G(b) - G(a)$, $\tilde{\mu}\{\{-\infty\}\} = \mu\{\varnothing\} = 0$, $\tilde{\mu}\{R\} = G(\infty) - G(-\infty)$.

6. There is a 1–1 correspondence between d.f.s and probability measures on the Borel sets of the line.
7. If $(R, \mathscr{M}, \nu)$ is the Lebesgue measure space (of R), $\Omega = [0, 1]$, $\mathscr{A} = \Omega \cdot \mathscr{M}$, and $P = \nu|_{\mathscr{A}}$, then $(\Omega, \mathscr{A}, P)$ is a probability space.
8. If X is a r.v. with d.f. F, then $P\{X^+ = 0\} = F(0+)$, $P\{X^+ < x\} = F(x)$, $x > 0$, and $P\{X^- = 0\} = 1 - F(0)$, $P\{X^- < x\} = 1 - F(-x+)$, $x > 0$. Find the d.f. of $|X|$.
9. Give an example to show that the uniqueness assertion of Theorem 1 is not true without the restriction of σ-finiteness on $\mathscr{S}$. *Hint*: Take $\Omega = \{r: r \text{ rational}, 0 \le r < 1\}$, $\mathscr{S} = \{\Omega \cdot [a, b): 0 \le a \le b \le 1\}$, $\mu(\varnothing) = 0$ and $\mu(A) = \infty$ if $\varnothing \ne A \in \mathscr{S}$, $\nu(A) =$ number of elements in A for $A \in \sigma(\mathscr{S})$, $\nu^*(A) = 2\nu(A)$ for $A \in \sigma(\mathscr{S})$.
10. If $(R, \mathscr{M}, \nu)$ is the Lebesgue measure space of the real line and E is a Lebesgue measurable set, so is $E + x = \{y + x: y \in E\}$ and, moreover, $\nu\{E + x\} = \nu\{E\}$ for every $x \in (-\infty, \infty)$.
11. For any real x, y consider the equivalence relation $x \sim y$ if $x - y = r =$ rational. Let the subset E of $[0, 1)$ contain exactly one point of each equivalence class. Then E is a non-Lebesgue-measurable set. *Hint*: (i) if $x \in (0, 1)$, then $x \in E + r$ for some r in $(-1, 1)$. (ii) $(E + r) \cap (E + s) = \varnothing$ for distinct rationals r, s. Thus, if E is Lebesgue measurable so is $S = \bigcup_{r \in F}(E + r)$, where F is the set of rationals in $(-1, 1)$ via Exercise 10. Then $S \subset (-1, 2)$, whence $3 \ge \nu\{S\} = \sum_1^\infty \nu\{E\}$, implying $\nu\{S\} = \sum \nu\{E\} = 0$. However, by (i), $[0, 1) \subset \bigcup(E + r) = S$, so that $\nu\{S\} \ge 1$, a contradiction.

6.2 Integration in a Measure Space

Let (S, Σ, μ) constitute a measure space, that is, S is an arbitrary nonempty set, Σ is a σ-algebra of subsets of S, and μ is a measure on Σ. In the event that μ is σ-finite, (S, Σ, μ) will be termed a σ-finite measure space. Any property that obtains except on a set of μ-measure zero will be said to hold **almost everywhere** (abbreviated a.e.).

Let X be a Σ-measurable function, that is, a mapping from S to $R = [-\infty, \infty]$ for which $X^{-1}(\mathscr{B}) \subset \Sigma$, where $\mathscr{B}$ is the class of Borel subsets of R. Then, paralleling the approach in Chapter 4, the integral

$$E\,X = \int_S X\,d\mu$$

may be defined as follows:

(i) If $X \ge 0$, a.e., and $\mu\{X = \infty\} > 0$, then $E\,X = \infty$, while if $X \ge 0$, a.e., and $\mu\{X = \infty\} = 0$,

$$E\,X = \lim_{n \to \infty} \sum_{i=1}^{\infty} \frac{i}{2^n}\,\mu\left\{\frac{i}{2^n} < X \le \frac{i+1}{2^n}\right\}. \tag{1}$$

(ii) In general, if either $\mathrm{E}\,X^+ < \infty$ or $\mathrm{E}\,X^- < \infty$,

$$\mathrm{E}\,X = \mathrm{E}\,X^+ - \mathrm{E}\,X^-, \tag{2}$$

in which case E X is said to exist, denoted by $|\mathrm{E}\,X| \leq \infty$. If $|\mathrm{E}\,X| < \infty$, that is, E X exists and is finite, X is called an **integrable function**, or, simply, **integrable**.

If $|\mathrm{E}\,X| \leq \infty$, the indefinite integral of X is defined by

$$\int_A X\,d\mu = \mathrm{E}\,XI_A, \qquad A \in \Sigma.$$

As in Chapter 4, it is easy to verify that the limit in (1) always exists and that:

$$0 \leq \mathrm{E}\,X \leq \infty \quad \text{if } X \geq 0, \text{a.e.,} \qquad \mathrm{E}\,1 = \mu(S);$$

$$|X| < \infty, \text{a.e.,} \quad \text{if } \mathrm{E}|X| < \infty; \qquad X = 0, \text{a.e.,} \quad \text{iff } \mathrm{E}|X| = 0;$$

$$\mathrm{E}\,X = \mathrm{E}\cdot Y \quad \text{if } X = Y, \text{a.e., and } |\mathrm{E}\,X| \leq \infty.$$

From (1) and (2) it is readily seen that if (S, Σ_i, μ_i), $i = 1, 2$, are measure spaces with $\mu_1 = \mu_2|_{\Sigma_1}$ and X is a Σ_1-measurable function, then

$$\int_S X\,d\mu_2 = \int_S X\,d\mu_1 \tag{3}$$

in the sense that if one of the integrals exists, so does the other and the two are equal.

Associated with any measure space (S, Σ, μ) are the spaces $\mathscr{L}_p = \mathscr{L}_p(S, \Sigma, \mu)$, $p > 0$, of all measurable functions X for which $\mathrm{E}|X|^p < \infty$. For any measurable function X, and especially when $X \in \mathscr{L}_p$, the $\mathscr{L}_p$-norm of X is defined by

$$\|X\|_p = \mathrm{E}^{1/p}|X|^p. \tag{4}$$

Let $\{X_n, n \geq 1\}$ be a sequence of measurable functions on the measure space (S, Σ, μ). If $\mu\{\underline{\lim}\, X_n \neq \overline{\lim}_n\, X\} = 0$, set $X = \underline{\lim}\, X_n$ whence $\lim X_n = X$, a.e., denoted by $X_n \xrightarrow{\text{a.e.}} X$. If X is finite a.e., X_n will be said to **converge** a.e., denoted by $X_n \xrightarrow{\text{a.e.}} X$, finite. Alternatively, if X_n is finite a.e., $n \geq 1$, then X_n **converges in measure** to X, written $X_n \xrightarrow{\mu} X$, if $\lim \mu\{|X_n - X| > \varepsilon\} = 0$ for all $\varepsilon > 0$. These are obvious analogues of a.c. convergence and convergence in probability on a probability space, but the correspondence is not without limitation (see Exercise 6.2.4). In case $\|X_n - X\|_p = o(1)$, the earlier notation $X_n \xrightarrow{\mathscr{L}_p} X$ will be employed.

In dealing with the basic properties of the integral in Chapter 4, the proof of Theorem 4.1.1(i) utilized the fact that for any nonnegative integrable r.v. X on a probability space $(\Omega, \mathscr{F}, \mathrm{P})$

$$\mathrm{E}\,X = \lim_{n\to\infty} \sum_{i=0}^{\infty} \frac{i+1}{2^n}\,\mathrm{P}\left\{\frac{i}{2^n} < X \leq \frac{i+1}{2^n}\right\}. \tag{5}$$

The counterpart of (5) in the case of a nonnegative, integrable function X on a measure space (S, Σ, μ) is

$$\mathrm{E}\,X = \lim_{n\to\infty} \sum_{i=1}^{\infty} \frac{i+1}{2^n}\,\mu\left\{\frac{i}{2^n} < X \le \frac{i+1}{2^n}\right\}. \tag{6}$$

To prove (6), note that, setting $s_n = \sum_{i=1}^{\infty} (i/2^n)\mu\{i/2^n < X \le (i+1)/2^n\}$,

$$s_{2n} - s_n \ge \sum_{i=1}^{2^n-1} \frac{i}{2^{2n}}\,\mu\left\{\frac{i}{2^{2n}} < X \le \frac{i+1}{2^{2n}}\right\} \ge 2^{-2n}\mu\{2^{-2n} < X \le 2^{-n}\} = o(1)$$

since $\mathrm{E}\,X = \lim_{n\to\infty} s_n$. Moreover, $2^{-2n}\mu\{X > 2^{-n}\} \le 2^{-n}\mathrm{E}\,X = o(1)$, so that

$$2^{-2n}\mu\{X > 2^{-2n}\} = o(1),$$

whence

$$2^{-(2n-1)}\mu\{X > 2^{-(2n-1)}\} \le 2\cdot 2^{-2n}\mu\{X > 2^{-2n}\} = o(1).$$

Consequently, the difference of the right and left sides of (6), viz., $\lim_{n\to\infty} 2^{-n}\mu\{X > 2^{-n}\}$, is zero.

It may be noted via (6) or (1) that if X is a nonnegative, integrable function on (S, Σ, μ), then μ is σ-finite on $S\{X > 0\}$.

Theorem 1. *Let X, Y, X_n, Y_n be measurable functions on the measure space (S, Σ, μ).*

i. *If $X = \sum_1^\infty x_n I_{A_n} \ge 0$, where $\{A_n, n \ge 1\}$ are disjoint measurable sets, then*

$$\mathrm{E}\,X = \sum_1^\infty x_n \mu\{A_n\} \quad \text{where } 0\cdot\infty = \infty\cdot 0 = 0. \tag{7}$$

ii. (a) *If $X \ge 0$, a.e., there exist elementary functions X_n with $0 \le X_n \uparrow X$, a.e., and $\mathrm{E}\,X_n \uparrow \mathrm{E}\,X$.* (b) *If, moreover, $X < \infty$, a.e., then $X - X_n \le 2^{-n}$, $n \ge 1$, a.e., is attainable.*

iii. *If $X \ge Y \ge 0$, a.e., then $\mathrm{E}\,X \ge \mathrm{E}\,Y$ and for $0 < a < \infty$*

$$\mathrm{E}\,X \ge a\mu\{X \ge a\}, \tag{8}$$

iv. *If $0 \le X_n \uparrow X$, a.e., then $\mathrm{E}\,X_n \uparrow \mathrm{E}\,X$ (monotone convergence).*

v. *If $\mathrm{E}\,X$, $\mathrm{E}\,Y$, and $\mathrm{E}\,X + \mathrm{E}\,Y$ are defined, then so is $\mathrm{E}(X+Y)$ and*

$$\mathrm{E}(X+Y) = \mathrm{E}\,X + \mathrm{E}\,Y.$$

vi. *If $X \ge Y$, a.e., $|\mathrm{E}\,X| \le \infty$, $|\mathrm{E}\,Y| \le \infty$, then $\mathrm{E}\,X \ge \mathrm{E}\,Y$.*

vii. *If $\mathrm{E}\,X$, $\mathrm{E}\,Y$, $a\,\mathrm{E}\,X + b\,\mathrm{E}\,Y$ are defined for finite a, b, then*

$$\mathrm{E}(aX + bY) = a\,\mathrm{E}\,X + b\,\mathrm{E}\,Y. \tag{9}$$

viii. *X is integrable iff $|X|$ is integrable and if $X \ge 0$, $Y \ge 0$, a.e., then $\mathrm{E}(X+Y)^p \le 2^p(\mathrm{E}\,X^p + \mathrm{E}\,Y^p)$, $p > 0$.*

PROOF. The argument follows that of Theorem 4.1.1 with two modifications. Firstly, in proving (i) write $I_{ni} = (i/2^n, (i+1)/2^n]$, $\mu_{ni} = \mu\{X' \in I_{ni}\}$, and replace (13) of Theorem 4.1.1 by

$$\sum_{i=1}^{\infty} \frac{i+1}{2^n} \mu_{ni} \geq \sum_{i=1}^{\infty} \sum_{j: x'_j \in I_{ni}} x'_j \mu\{A_j\} = \sum_{j: x'_j > 2^{-n}} x'_j \mu\{A_j\} \xrightarrow{n \to \infty} \sum_{j=1}^{\infty} x'_j \mu\{A_j\}.$$

Consequently, $\mathrm{E}\, X = \sum_{j=1}^{\infty} x'_j \mu\{A_j\}$ via (6).

Secondly, in the proof of (iv), i.e., monotone convergence, replace (15), (16) of Section 4.1 by

$$\lim_{m \to \infty} \mathrm{E}\, X_m \geq \sum_{i=1}^{k} \frac{i}{2^n} \varliminf_{m \to \infty} \mu\{X_m \in I_{ni}\} \geq \sum_{i=1}^{k} \frac{i}{2^n} \mu\{X \in I_{ni}\} > a,$$

utilizing Exercise 1.5.2. □

The next theorem incorporates analogues of Lemma 4.2.1, Corollaries 4.2.2 and 4.2.3, and Theorem 4.3.2.

Theorem 2. *Let $\{X_n, n \geq 1\}$ be a sequence of measurable functions on the measure space (S, Σ, μ).*

i. *If $X_n \geq 0$, a.e., $n \geq 1$, then*

$$\mathrm{E} \varliminf_{n \to \infty} X_n \leq \varliminf_{n \to \infty} \mathrm{E}\, X_n \qquad (Fatou). \tag{10}$$

ii. *If $|X_n| \leq Y$, a.e., where $\mathrm{E}\, Y < \infty$ and either $X_n \xrightarrow{\text{a.e.}} X$ or $X_n \xrightarrow{\mu} X$, then*

$$\mathrm{E}|X_n - X| \to 0 \qquad (\textit{Lebesgue dominated convergence theorem}). \tag{11}$$

iii. *If $|X|^p \in \mathscr{L}_1$, $|Y|^{p'} \in \mathscr{L}_1$, where $p > 1$, $p' > 1$, $(1/p) + (1/p') = 1$, then $XY \in \mathscr{L}_1$ and*

$$\mathrm{E}|XY| \leq \|X\|_p \|Y\|_{p'} \qquad (\textit{Hölder inequality}). \tag{12}$$

iv. *If $|\mathrm{E}\, X| \leq \infty$, the indefinite integral $\nu\{A\} = \int_A X\, d\mu$ is a σ-additive set function on Σ.*

PROOF. The proofs of Hölder's inequality and (iv) are identical with those of Chapter 4, while the argument in Fatou's lemma is precisely that used for X'_n in Theorem 4.2.2(ii). Apropos of (ii), suppose first that $X_n \xrightarrow{\text{a.e.}} X$. The hypothesis ensures that $X_n \in \mathscr{L}_1$, $X \in \mathscr{L}_1$, whence by (10)

$$\mathrm{E}(Y \pm X) = \mathrm{E} \varliminf_{n \to \infty} (Y \pm X_n) \leq \varliminf_{n \to \infty} \mathrm{E}(Y \pm X_n),$$

implying $\mathrm{E}\, X \leq \varliminf_{n \to \infty} \mathrm{E}\, X_n \leq \varlimsup_{n \to \infty} \mathrm{E}\, X_n \leq \mathrm{E}\, X$, that is, $|X_n| \leq Y \in \mathscr{L}_1$ and $X_n \xrightarrow{\text{a.e.}} X$ imply

$$\mathrm{E}\, X = \lim_{n \to \infty} \mathrm{E}\, X_n. \tag{13}$$

Since $0 \leq |X_n - X| \leq 2Y$, (13) ensures that $\mathrm{E}|X_n - X| \to 0$.

On the other hand, if $X_n \xrightarrow{\mu} X$, then (Exercise 3.3.13) every subsequence of X_n, say X'_n, satisfies $\sup_{m>n} \mu\{|X'_m - X'_n| > \varepsilon\} = o(1)$, all $\varepsilon > 0$, and hence has a further subsequence, say X_{n_k}, with $X_{n_k} \xrightarrow{\text{a.e.}} Y_0$, finite. By the portion already proved, $\mathrm{E}|X_{n_k} - Y_0| = o(1)$, whence $X_{n_k} \xrightarrow{\mu} Y_0$ via (8). Thus $X = Y_0$, a.e., and $\mathrm{E}|X_{n_k} - X| = o(1)$. Consequently, every subsequence of $\mathrm{E}|X_n - X|$ has a further subsequence converging to zero and so $\mathrm{E}|X_n - X| = o(1)$. □

Although most properties of a probability space carry over to a measure space, Jensen's inequality and the usefulness of uniform integrability are conspicuous exceptions (see Exercises 12, 13, and 4).

Let $(R, \mathscr{M}_m, v_m^*)$ be the Lebesgue–Stieltjes measure space determined by a finite, nondecreasing function m on $(-\infty, \infty)$ and let X be a Borel function on R with $|\mathrm{E}\, X| \le \infty$. Then

$$\int_A X\, dm \quad \text{or} \quad \int_A X(t)dm(t), \qquad A \in \mathscr{M}_m,$$

is just an alternative notation for the indefinite Lebesgue–Stieltjes integral $\int_A X\, dv_m^*$, $A \in \mathscr{M}_m$, that is, for the indefinite integral on the Lebesgue–Stieltjes measure space determined by m. According to (3), if $v_m = v_m^*|_{\mathscr{B}}$, then

$$\int_A X\, dm = \int_A X\, dv_m^* = \int_A X\, dv_m, \qquad A \in \mathscr{B},$$

since X is a Borel function. Consequently, when dealing with Borel sets B, the Lebesgue–Stieltjes integral $\int_B X\, dm$ may be envisaged as being defined either in the Lebesgue–Stieltjes measure space $(\mathrm{R}, \mathscr{M}_m, v_m^*)$ or in the Borel measure space $(R, \mathscr{B}, v_m)$. Since $v_m\{\{\infty\}\} = v_m\{\{-\infty\}\} = 0$, the integral $\int_{-\infty}^{\infty} X\, dm$ may be unambiguously defined as

$$\int_{[-\infty,\infty]} X\, dm = \int_{(-\infty,\infty)} X\, dm = \int_{[-\infty,\infty)} X\, dm = \int_{(-\infty,\infty]} X\, dm.$$

However, if $-\infty < a < b < \infty$, $\int_a^b X\, dm$ is not, in general, well defined since

$$\int_{[a,b]} X\, dm = [m(a+) - m(a-)]X(a) + [m(b+) - m(b-)]X(b) + \int_{(a,b)} X\, dm$$

via additivity of the indefinite integral and Exercise 6.1.4. On the other hand, if a and b are **continuity points** of m, i.e., $m(b+) = m(b-)$ and $m(a+) = m(a-)$, then $\int_a^b X\, dm$ may be interpreted as the common value

$$\int_{[a,b]} X\, dm = \int_{(a,b)} X\, dm = \int_{(a,b]} X\, dm = \int_{[a,b)} X\, dm.$$

Thus, only when a and b are continuity points (including $\pm\infty$) of m will the notation $\int_a^b X\, dm$ be utilized for a Lebesgue–Stieltjes integral.

As in the case of Riemann integrals, if $a < b$,

$$\int_b^a X\,dm = -\int_a^b X\,dm$$

by fiat (under the prior proviso).

In the important special case $m(t) = t$, $-\infty < t < \infty$, the Lebesgue–Stieltjes integral $\int_A X\,dm$ is denoted by $\int_A X(t)dt$ and is called the **Lebesgue integral**.

Let X be a measurable transformation from a measure space (S, Σ, μ) to a measurable space $(T, \mathscr{A})$. Then, as in Section 1.6, the measure μ induces a measure ν on $\mathscr{A}$ via

$$\nu\{A\} = \mu\{X^{-1}(A)\}, \qquad A \in \mathscr{A}. \tag{14}$$

In fact, $\nu\{\varnothing\} = 0$, $v\{T\} = \mu\{S\}$, and ν is σ-additive since if $\{A_n, n \geq 1\}$ are disjoint sets of $\mathscr{A}$, then $X^{-1}(A_n)$, $n \geq 1$, are disjoint sets of Σ and $X^{-1}(\bigcup_1^\infty A_n) = \bigcup_1^\infty X^{-1}(A_n)$. The measure ν, induced by μ and the measurable transformation X, will be denoted by μX^{-1}. The next result might well be called the change of variable theorem since it justifies a technique immortalized by integral calculus.

Theorem 3. *Let X be a measurable transformation from a measure space (S, Σ, μ) to a measurable space $(T, \mathscr{A})$ and let $\nu = \mu X^{-1}$ be the induced measure on $\mathscr{A}$. If g is a real $\mathscr{A}$-measurable function on T, then*

$$\int_T g\,d\nu = \int_S g(X)d\mu \tag{15}$$

in the sense that if either integral exists, so does the other and the two are equal.

Proof. Note that $g(X)$ is a real Σ-measurable function. Since $g = g^+ - g^-$ and $g^\pm$ are measurable functions on $(T, \mathscr{A})$, it suffices to prove that every nonnegative $\mathscr{A}$-measurable function is in $\mathscr{G}$, where

$$\mathscr{G} = \{g: g \text{ is a nonnegative } \mathscr{A}\text{-measurable function for which (15) holds}\}.$$

Now by monotone convergence and linearity, $\mathscr{G}$ is a monotone system (see Section 1.4), and if $g = I_A$ for any $A \in \mathscr{A}$, then

$$\int_T g\,d\nu = \int_A d\nu = \mu\{X^{-1}(A)\} = \int_S I_A(X)d\mu.$$

Thus, $\mathscr{G}$ contains all indicator functions of sets of the σ-algebra $\mathscr{A}$, whence by Theorem 1.4.3, $\mathscr{G}$ contains all nonnegative $\mathscr{A}$-measurable functions. □

As an application of Theorem 3, let $(\Omega, \mathscr{F}, \mathrm{P})$ be a probability space and X a r.v. thereon. Then X is a measurable transformation from $(\Omega, \mathscr{F}, \mathrm{P})$ to $(R, \mathscr{B})$ (as usual, $R = [-\infty, \infty]$) and in conjunction with P induces a probability measure $\nu_X = \mathrm{P}\,X^{-1}$ on $\mathscr{B}$ via

$$\nu_X\{B\} = \mathrm{P}\{X^{-1}(B)\}, \qquad B \in \mathscr{B}. \tag{16}$$

In particular, for $-\infty < a < b < \infty$

$$\nu_X\{[a, b)\} = P\{a \le X < b\} = F_X(b) - F_X(a),$$

where F_X is the d.f. of X. Since F_X is left continuous, Theorem 6.1.2 ensures that the measure ν_X coincides with the restriction to $\mathscr{B}$ of the Lebesgue–Stieltjes measure determined by F_X. Actually, if ν_X is completed to $\overline{\mathscr{B}}$ as in Exercise 6.1.1, then ν_X on $\overline{\mathscr{B}}$ is precisely the Lebesgue–Stieltjes measure on $\mathscr{M}_F$ determined by F_X (for a proof, see Halmos (1950, p. 56)).

Corollary 1. *If X is a* r.v. *on a probability space* $(\Omega, \mathscr{F}, P)$ *with* d.f. F_X *and g is a finite Borel function,*

$$E\, g(X) = \int_{-\infty}^{\infty} g(t)dF_X(t) = \int_{-\infty}^{\infty} t\, dF_{g(X)}(t) \tag{17}$$

in the sense that if either of the integrals exists, so does the other and the two are equal.

PROOF. Let ν_X be defined on $\mathscr{B}$ by (16). Then, as noted earlier, ν_X is the restriction to $\mathscr{B}$ of the Lebesgue–Stieltjes measure determined by F_X. By Theorem 3

$$E\, g(X) = \int_\Omega g(X)\, dP = \int_R g\, d\nu_X = \int_{-\infty}^{\infty} g(t)dF_X(t) \tag{18}$$

in the sense delineated therein. Replacing $g(t)$ by t and X by $g(X)$ in (18) yields

$$E\, g(X) = \int_{-\infty}^{\infty} t\, dF_{g(X)}(t), \tag{19}$$

again in the aforementioned sense. Thus (17) and hence the corollary is proved. □

It may be noted that E X has so far been calculated only when X is an elementary function. Corollary 1 says explicitly (and Theorem 3 implicitly) that E X may be replaced by a Lebesgue–Stieltjes integral, but this does not resolve the problem of evaluation. Theorem 4 asserts that under modest conditions the Lebesgue–Stieltjes and Riemann–Stieltjes integrals coincide, and fortunately the latter is susceptible of calculation via (21)(x), (vii), etc. (see below), as illustrated by Lemma 1.

A finite nondecreasing function m on $(-\infty, \infty)$ elicits, in addition to the Lebesgue–Stieltjes integral $\int_A X\, dm$ (defined for Borel functions X and Borel sets A whenever $\int_R X\, dm$ exists), also the Riemann–Stieltjes integral

$$\int_a^b f(t)dm(t), \qquad -\infty < a < b < \infty.$$

(This will never be denoted by $\int_D f(t)dm(t)$, where D is a closed, open, or half-open interval with end points a and b, such notation being reserved for Lebesgue–Stieltjes integrals).

If $f(t)$ and $m(t)$ are bounded functions on $[a, b]$ the **Riemann–Stieltjes integral** of f with respect to m from a to b is defined by

$$\int_a^b f(t)dm(t) = \lim \sum_{k=1}^{n} f(\xi_k)[m(t_k) - m(t_{k-1})] \tag{20}$$

provided the limit, which is taken as $\max_{1 \le k \le n} (t_k - t_{k-1}) \to 0$, exists independently of the choice of ξ_k in $[t_{k-1}, t_k]$, where

$$a = t_0 < t_1 < \cdots < t_n = b, \qquad t_{k-1} \le \xi_k \le t_k, \qquad 1 \le k \le n.$$

When f is continuous and m is finite and nondecreasing on $[a, b]$, both $\int_a^b f(t)dm(t)$ and $\int_a^b m(t)df(t)$ exist (see Widder (1961, Chapter 5)). In the special case $m(t) = t$, the integral defined by (20) coalesces to the **Riemann integral** of f from a to b and is denoted $\int_a^b f(t)dt$.

Let f, f_1, f_2 be continuous functions and m, m_1, m_2 finite, nondecreasing functions on the finite interval $[a, b]$. Then (Widder (1961, Chapter 5)), if c denotes a finite constant,

i. $\displaystyle\int_a^b cf(t)dm(t) = c\int_a^b f(t)dm(t), \int_a^b f(t)d(m(t) + c) = \int_a^b f(t)dm(t),$

ii. $\displaystyle\int_a^b dm(t) = m(b) - m(a),$

$\displaystyle\int_a^b (f_1(t) + f_2(t))dm(t) = \int_a^b f_1(t)dm(t) + \int_a^b f_2(t)dm(t),$

iii. $\displaystyle\int_a^b f(t)d(m_1(t) + m_2(t)) = \int_a^b f(t)dm_1(t) + \int_a^b f(t)dm_2(t),$

iv. $\displaystyle\int_a^b f(t)dm(t) = \int_a^c f(t)dm(t) + \int_c^b f(t)dm(t), a < c < b,$

v. $\displaystyle\int_a^b f_1(t)dm(t) \le \int_a^b f_2(t)dm(t) \text{ if } f_1(t) \le f_2(t) \text{ on } [a, b],$ (21)

vi. $\displaystyle\left|\int_a^b f(t)dm(t)\right| \le \int_a^b |f(t)|dm(t) \le [m(b) - m(a)] \max_{a \le t \le b} |f(t)|,$

vii. $\displaystyle\int_a^b f(t)dm(t) + \int_a^b m(t)df(t) = m(b)f(b) - m(a)f(a),$

viii. $\displaystyle\int_a^b m(t)df(t) = m(a)\int_a^\xi df(t) + m(b)\int_\xi^b df(t) \text{ for some } \xi \text{ in } [a, b],$

ix. $\displaystyle\int_a^b f(t)m(t)dt = m(a)\int_a^\xi f(t)dt + m(b)\int_\xi^b f(t)dt \text{ for some } \xi \text{ in } [a, b],$

x. $\displaystyle\int_a^b f(t)dm(t) = \int_a^b f(t)m'(t)dt,$

provided m has a continuous derivative m' on $[a, b]$.

The Riemann–Stieltjes integral $\int_a^b f(t)dm(t)$ has been defined for finite closed intervals $[a, b]$ and exists if f is continuous and m is finite and nondecreasing on $[a, b]$. Moreover, if f is continuous and m is finite and nondecreasing on $[a, \infty)$ for some finite constant a, the definition may be extended via

$$\int_a^\infty f(t)dm(t) = \lim_{b\to\infty} \int_a^b f(t)dm(t) \tag{22}$$

provided the limit on the right exists. Analogously, for any finite constant b, if $f(t)$ is continuous and $m(t)$ is finite and nondecreasing on $(-\infty, b]$ (resp. on $(-\infty, \infty)$) define

$$\begin{aligned} \int_{-\infty}^b f(t)dm(t) &= \lim_{a\to-\infty} \int_a^b f(t)dm(t), \\ \int_{-\infty}^\infty f(t)dm(t) &= \lim_{\substack{b\to\infty \\ a\to-\infty}} \int_a^b f(t)dm(t) \end{aligned} \tag{23}$$

provided the limits on the right exist (independently of the manner in which $b \to \infty$, $a \to -\infty$). If $a = -\infty$ or $b = \infty$ or both, $\int_a^b f(t)dm(t)$ is frequently alluded to as an improper Riemann–Stieltjes integral.

The relationship between the Riemann–Stieltjes integral $\int_a^b f(t)dm(t)$ and the Lebesgue–Stieltjes integral $\int_{[a,b)} f(t)dm(t)$ is embodied in

Theorem 4. *If $f(t)$ is a continuous function and $m(t)$ is a finite nondecreasing function on $(-\infty, \infty)$, then if $m(a) = m(a-)$ and $m(b) = m(b-)$,*

$$\int_{[a,b)} f(t)dm(t) = \int_a^b f(t)dm(t), \qquad -\infty < a < b < \infty; \tag{24}$$

moreover, if $|\int_{(-\infty,\infty)} f(t)dm(t)| \le \infty$, then the Riemann–Stieltjes integral $\int_{-\infty}^\infty f(t)dm(t)$ exists and

$$\begin{aligned} \int_{(-\infty,\infty)} f(t)dm(t) &= \int_{-\infty}^\infty f(t)dm(t), \\ \int_{(-\infty,b)} f(t)dm(t) &= \int_{-\infty}^b f(t)dm(t) \quad \text{if } m(b) = m(b-), \\ \int_{[a,\infty)} f(t)dm(t) &= \int_a^\infty f(t)dm(t) \quad \text{if } m(a-) = m(a). \end{aligned} \tag{25}$$

Proof. Choose $t_j^{(n)}$ such that $m(t_j^{(n)}) = m(t_j^{(n)}-)$, $1 \le j \le n$,

$$a = t_0^{(n)} < t_1^{(n)} < \cdots < t_n^{(n)} = b, \qquad \max_{1\le k\le n} (t_k^{(n)} - t_{k-1}^{(n)}) < \frac{2(b-a)}{n}.$$

On $[a, b)$ define

$$f_n(t) = \sum_{k=1}^{n} f(t_{k-1}^{(n)}) I_{[t_{k-1}^{(n)}, t_k^{(n)})}(t),$$

and note that $f_n \to f$ on $[a, b)$ by the continuity of f.

Since the Lebesgue–Stieltjes measure $\nu = \nu_m$ determined by m is finite on the finite interval $[a, b)$ and $\max_{a \le t \le b} |f(t)| < \infty$, by the Lebesgue dominated convergence theorem

$$\begin{aligned}
\int_{[a,b)} f(t) dm(t) &= \int_{[a,b)} f \, d\nu = \lim_n \int_{[a,b)} f_n \, d\nu \\
&= \lim_n \mathrm{E} \sum_{k=1}^{n} f(t_{k-1}^{(n)}) I_{[t_{k-1}^{(n)}, t_k^{(n)})} \\
&= \lim_n \sum_{k=1}^{n} f(t_{k-1}^{(n)}) [m(t_k^{(n)}) - m(t_{k-1}^{(n)})] = \int_a^b f(t) dm(t)
\end{aligned}$$

since the Riemann–Stieltjes integral exists.

To prove (25), note that via the portion already proved

$$\begin{aligned}
\int_{[a,\infty)} f^{\pm}(t) dm(t) &= \lim_{\substack{b \to \infty \\ m(b) = m(b-)}} \int_{[a,b)} f^{\pm}(t) dm(t) = \lim_{\substack{b \to \infty \\ m(b) = m(b-)}} \int_a^b f^{\pm}(t) dm(t) \\
&= \int_a^{\infty} f^{\pm}(t) dm(t),
\end{aligned}$$

and the comparable statement with $[a, \infty)$ replaced by $(-\infty, b)$ follows analogously. The remaining portion of (25) obtains by letting $a \to -\infty$ subject to $m(a) = m(a-)$. □

The Lebesgue–Stieltjes and Riemann–Stieltjes integrals can be extended to nonincreasing functions m in an obvious fashion by defining

$$\begin{aligned}
\int_B f(t) dm(t) &= -\int_B f(t) d(-m(t)), \quad B \in \mathscr{B}, \\
\int_a^b f(t) dm(t) &= -\int_a^b f(t) d(-m(t)), \quad -\infty \le a < b \le \infty,
\end{aligned} \tag{26}$$

whenever the integrals on the right are defined. In the special case where $m = 1 - F$ with F a d.f., the latter may also be expressed as

$$\int_a^b f(t) d[1 - F(t)] = -\int_a^b f(t) dF(t). \tag{27}$$

It may be noted via the definition that whenever $\int_a^b f(t) dm(t)$ exists for $-\infty < a < b < \infty$, so does $\int_{-b}^{-a} f(-t) dm(-t)$ and the two are equal.

The absolute moment $\mathrm{E}|X|^r$, $r > 0$ always exists for any r.v. X, and its finiteness has been shown equivalent to convergence of the series $\sum_1^\infty P\{|X| \geq n^{1/r}\}$ in Corollary 4.1.2. It may be evaluated as a Riemann integral involving the tails of the distribution of X as follows from Corollary 2 of

Lemma 1. *If X is a* r.v. *with* d.f. *F and $m(t)$ is a continuous nondecreasing function on $(-\infty, \infty)$ with $m(0) = 0$, then*

i. $E|m(X)| = \int_{-\infty}^0 F(t)dm(t) + \int_0^\infty [1 - F(t)]dm(t)$,
ii. if $|E\, m(X)| < \infty$, $E\, m(X) = \int_0^\infty [1 - F(t)]\, dm(t) - \int_{-\infty}^0 F(t)\, dm(t)$.
iii. $E\, m(|X|) = \int_0^\infty [1 - F(t) + F(-t)]dm(t)$.

PROOF. By Theorem 4, the discussion following, and integration by parts ((21)(vii)), if $C > 0$ and $G = 1 - F$,

$$\begin{aligned} E|m(X)| &= \int_{-\infty}^{\infty} |m(t)|dF(t) = \int_0^\infty m(t)dF(t) - \int_{-\infty}^0 m(t)dF(t) \\ &= \lim_{C\to\infty}\left[-\int_0^C m(t)d[1 - F(t)] - \int_{-C}^0 m(t)dF(t)\right] \\ &= \lim_{C\to\infty}\left[m(-C)F(-C) + \int_{-C}^0 F(t)dm(t) - m(C)G(C)\right. \\ &\quad \left. + \int_0^C G(t)dm(t)\right] \leq \int_{-\infty}^0 F(t)dm(t) + \int_0^\infty G(t)dm(t). \quad (28) \end{aligned}$$

Thus, in proving (i) it may be supposed that $E|m(X)| < \infty$, and so, since

$$\begin{aligned} 0 \leq m(C)G(C) - m(-C)F(-C) &\leq \int_C^\infty m(t)dF(t) - \int_{-\infty}^{-C} m(t)dF(t) \\ &= \int_{[|t|\geq C]} |m(t)|dF(t) = o(1) \end{aligned}$$

as $C \to \infty$, (i) follows from (28).

Part (ii) follows by applying (i) separately to m^+ and m^-. Likewise (i) implies (iii) in view of $1 - F_{|X|}(t) = P\{|X| \geq t\} = G(t) + F(-t+)$ and Exercise 6.2.6.

Corollary 2. *For any* r.v. *X with* d.f. *F,*

$$E|X|^r = r\int_0^\infty t^{r-1}[1 - F(t) + F(-t)]dt, \qquad r > 0, \quad (29)$$

and if $E\, X^{2k+1} < \infty$ *for some* $k = 0, 1, \ldots,$ *then*

$$E\, X^{2k+1} = (2k + 1)\int_0^\infty t^{2k}[1 - F(t) - F(-t)]dt. \quad (30)$$

Proof. The first statement is an immediate consequence of (iii) and (x) of (21), while the second follows similarly from (ii). □

The following combines Example 2.1.1 and Corollary 4.2.4:

Example 1. Let $\{X_n, n \geq 0\}$ be nonnegative functions with $X_n \in \mathscr{L}_1(S, \Sigma, \mu)$ for $n \geq 0$ and $\mathrm{E}\, X_n \to \mathrm{E}\, X_0$. If either $X_n \xrightarrow{\mu} X_0$ or $X_n \xrightarrow{\text{a.e.}} X_0$, then $\mathrm{E}|X_n - X_0| \to 0$.

Proof. If $Y_n = (X_0 - X_n)^+$, then $Y_n \leq X_0$, $n \geq 1$. Since either $Y_n \xrightarrow{\mu} 0$ or $Y_n \xrightarrow{\text{a.e.}} 0$, by Lebesgue's dominated convergence theorem $\mathrm{E}(X_0 - X_n)^+ = \mathrm{E}\, Y_n \to 0$. Then $\mathrm{E}(X_0 - X_n) \to 0$ implies $\mathrm{E}(X_0 - X_n)^- \to 0$, and so $\mathrm{E}|X_0 - X_n| \to 0$. □

Two alternative types of truncation of r.v.s have already been utilized in Theorem 4.2.2 and Theorem 5.1.3, 5.2.4, etc., and it is of interest to compare the corresponding expectations.

Example 2. If X is a nonnegative r.v. with d.f. F and

$$Y = \min(X, s), \qquad Z = XI_{[X<s]}, \tag{31}$$

then for any choice of r and s in $(0, \infty)$

$$\mathrm{E}\, Y^r = r \int_0^s t^{r-1}[1 - F(t)]dt, \tag{32}$$

$$\mathrm{E}\, Z^r = \mathrm{E}\, Y^r - s^r[1 - F(s)] = \int_0^s t^r\, dF(t). \tag{33}$$

Proof. If G is the d.f. of Y, then

$$G(x) = F(x)I_{[x<s]} + I_{[x \geq s]},$$

whence by Corollary 2

$$\mathrm{E}\, Y^r = r \int_0^\infty t^{r-1}[1 - G(t)]dt = r \int_0^s t^{r-1}[1 - F(t)]dt.$$

Since $Z = YI_{[X<s]}$, via (21)(vii)

$$\begin{aligned} \mathrm{E}\, Z^r &= \mathrm{E}\, Y^r I_{[X<s]} = \mathrm{E}\, Y^r - \mathrm{E}\, Y^r I_{[X \geq s]} \\ &= r \int_0^s t^{r-1}[1 - F(t)]dt - s^r[1 - F(s)] = \int_0^s t^r\, dF(t). \end{aligned}$$ □

Exercises 6.2

1. If μ_1, μ_2 are measures on $(\Omega, \mathscr{F})$ and $\int X\, d(\mu_1 + \mu_2)$ exists so does $\int X\, d\mu_i$, $i = 1, 2$, and $\int X\, d(\mu_1 + \mu_2) = \int X\, d\mu_1 + \int X\, d\mu_2$.
2. The integral of a nonnegative measurable function over a set of measure zero has the value 0; also, if $\int_A g\, d\mu = 0$ for every measurable set A, then $g = 0$, a.e.

3. If S is the set of positive integers, Σ is the class of all subsets of S, and $\mu(A)$ = number of integers in $A \in \Sigma$, then (S, Σ, μ) is a nonfinite measure space and convergence in measure is equivalent to uniform convergence everywhere.
4. Demonstrate in a nonfinite measure space (S, Σ, μ) that $X_n \xrightarrow{\text{a.e.}} X$ does not necessarily imply $X_n \xrightarrow{\mu} X$. *Hint*: Utilize Exercise 3. If $X_n \xrightarrow{\mu} X$, does $X_n Y \xrightarrow{\mu} XY$?
5. If $X_n \in \mathscr{L}_p(S, \Sigma, \mu)$ and $X_n \xrightarrow{\mathscr{L}_p} X$ for some $p > 0$, then $X \in \mathscr{L}_p$ and $\mathrm{E}|X_n|^p \to \mathrm{E}|X|^p$.
6. If f is a finite, nondecreasing function and m is a continuous nondecreasing function on $[a, b]$, where $-\infty < a < b < \infty$, then

$$\int_a^b f(t)dm(t) = \int_a^b f(t+)dm(t) = \int_a^b f(t-)dm(t).$$

7. Prove **Minkowski's inequality**: If $X_i \in \mathscr{L}_p$, $i = 1, 2$, and $p \geq 1$, $\|X_1 + X_2\|_p \leq \|X_1\|_p + \|X_2\|_p$.
8. If $F_{|X|}$ is the d.f. of $|X|$, verify that for every $c > 0$

$$\int_c^\infty t\, dF_{|X|}(t) = c\,\mathrm{P}\{|X| \geq c\} + \int_c^\infty \mathrm{P}\{|X| \geq t\}dt.$$

   (i) Show that a sequence of r.v.'s $\{X_n, n \geq 1\}$ is u.i. iff $\sup_{n\geq 1} \int_c^\infty \mathrm{P}\{|X_n| \geq t\}\, dt \to 0$ as $c \to \infty$.
   A r.v. X is said to be **stochastically larger** than a r.v. Y if $\mathrm{P}\{X \geq x\} \geq \mathrm{P}\{Y \geq x\}$ for all x. (ii) If the r.v.s X_n, $n \geq 1$, are u.i. and $|X_n|$ is stochastically larger than $|Y_n|$, $n \geq 1$, then $\{Y_n, n \geq 1\}$ is u.i.
9. If $\{X_n, n \geq 1\}$ are $\mathscr{L}_1$ r.v.s with a common distribution, then $\mathrm{E}\max_{1\leq i\leq n}|X_i| = o(n)$. *Hint*: Use Exercise 8 to establish u.i.
10. Show that the analogue of (21)(iv) for Lebesgue–Stieltjes integrals is not true in general. Construct an example for which the Riemann–Stieltjes integral over a finite interval $[a, b]$ fails to exist.
11. Establish that $g(t) = (\sin t)/t$ is Riemann but not Lebesgue integrable over $(-\infty, \infty)$ and find a function $g(t)$ which is Lebesgue integrable but not Riemann integrable.
12. Let $S = \{1, 2\}$, $\Sigma = \{\{1\}, \{2\}, \varnothing, S\}$ and μ = counting measure. If $X(s) \equiv 1$, Jensen's inequality fails for the convex function X^2.
13. Let $S = \{1, 2, \ldots\}$, $\Sigma = \{A : A \subset S\}$, μ = counting measure on Σ. If $X_n(s) = n^{-1}I_{[1\leq s\leq n]}$, then $X_n \xrightarrow[\text{u.c.}]{\mu} 0$ and $\mathrm{E}\,X_n \equiv 1$ despite the fact that for $\mu\{A\} < 1$, $\mathrm{E}\,X_n I_A \equiv 0$. Thus, Theorem 4.2.3(i) may fail in a σ-finite measure space.
14. If f, m are finite and nondecreasing on $(-\infty, \infty)$ with f continuous, prove that $\int_{[a,b]} f(t)dm(t) + \int_{[a,b]} m(t)df(t) = f(b)m(b+) - f(a)m(a-)$.
15. Let $p_n \in (0, 1)$, $q_n = 1 - p_n$, where $np_n \to \lambda \in (0, \infty)$. Let μ be counting measure on the class of all subsets of $\Omega = \{1, 2, \ldots\}$. If $X(j) = \lambda^j e^{-\lambda}/j!$ and $X_n(j) = \binom{n}{j}p_n^j q_n^{n-j}$, prove that $X_n \xrightarrow{\mu} X$ and $X_n \xrightarrow{\mathscr{L}_p} X$, $p \geq 1$. *Hint:* Apply Example 1 or Example 2.1.1.
16. (Erickson) In Example 5.4.1 the function $a(x) = x/\int_0^x [1 - F(y)]dy$ was encountered, where F is the d.f. of a nonnegative r.v. X. (i) Show that $a(x)$ is nondecreasing. (ii) Prove that $\mathrm{E}\,X = \infty$ implies $\mathrm{E}\,a(X) = \infty$. *Hint*: $\mathrm{E}\,a(X) < \infty$ entails $a(x) = o((1 - F(x))^{-1})$ and hence $\mathrm{E}\,X/\int_0^X y\,dF(y) < \infty$, contradicting the Abel–Dini theorem.

17. Random variables $\{|X_n|^p, n \geq 1\}$ are u.i. iff $\sup_{n\geq 1} \int_K^\infty t^{p-1} P\{|X_n| > t\} dt \to 0$ as $K \to \infty$.

18. (Young's inequality) Let φ be a continuous, strictly increasing function on $[0, \infty)$ with $\varphi(0) = 0$. If ψ is the inverse function (which therefore satisfies the same conditions) then for any $a \geq 0, b \geq 0$

$$ab \leq \int_0^a \varphi(x)dx + \int_0^b \psi(x)dx$$

and equality holds iff $b = \varphi(a)$.

6.3 Product Measure, Fubini's Theorem, n-Dimensional Lebesgue–Stieltjes Measure

Let $(\Omega_i, \mathscr{F}_i, \mu_i)$, $i = 1, 2$, denote two measure spaces. Ignoring for a moment the measures, the spaces engender (Section 1.3) a product measurable space $(\Omega, \mathscr{F})$, where $\Omega = \times_{i=1}^2 \Omega_i = \Omega_1 \times \Omega_2$ and $\mathscr{F} = \times_{i=1}^2 \mathscr{F}_i = \mathscr{F}_1 \times \mathscr{F}_2 = \sigma(\{A_1 \times A_2 : A_i \in \mathscr{F}_i, i = 1, 2\})$.

For any set $A \in \mathscr{F}$, the sections

$$A^{(1)}(\omega_1) = \{\omega_2 : (\omega_1, \omega_2) \in A\}, \qquad A^{(2)}(\omega_2) = \{\omega_1 : (\omega_1, \omega_2) \in A\}$$

are $\mathscr{F}_2$- and $\mathscr{F}_1$-measurable respectively according to Theorem 1.4.2. Thus, $\mu_2\{A^{(1)}(\omega_1)\}$ and $\mu_1\{A^{(2)}(\omega_2)\}$ are well-defined real functions, the first on Ω_1 and the second on Ω_2. For notational simplicity these will be denoted by $\mu_2\{A^{(1)}\}$ and $\mu_1\{A^{(2)}\}$ respectively.

Now if μ_1 and μ_2 are finite measures,

$$\mathscr{A} = \{A \in \mathscr{F} : \mu_{3-i}\{A^{(i)}\} \text{ is } \mathscr{F}_i\text{-measurable}, i = 1, 2\}$$

is a λ-class containing all rectangles with measurable sides, whence $\mathscr{A} \supset \mathscr{F}$, that is, $\mu_{3-i}\{A^{(i)}\}$ is $\mathscr{F}_i$-measurable $i = 1, 2$; this carries over to the case where μ_i is σ-finite, $i = 1, 2$, since if $\Omega_i = \bigcup_{j=1}^\infty B_{i,j}$, where B_{ij} are disjoint sets of $\mathscr{F}_i$, $\mu_i\{B_{ij}\} < \infty, j \geq 1, i = 1, 2$, then $\Omega_1 \times \Omega_2 = \bigcup_{j_1, j_2} (B_{1j_1} \times B_{2j_2})$ and every measurable set $A = \bigcup_{j_1, j_2} A(B_{1j_1} \times B_{2j_2})$ with $\mu_i\{A^{(3-i)}\} = \sum_{j_1, j_2} \mu_i\{[A(B_{1j_1} \times B_{2j_2})]^{(3-i)}\}$ being $\mathscr{F}_{3-i}$-measurable, $i = 1, 2$. Thus, when $(\Omega_i, \mathscr{F}_i, \mu_i)$ are σ-finite measure spaces, $i = 1, 2$, and $A \in \mathscr{F} = \mathscr{F}_1 \times \mathscr{F}_2$, $\mu_{3-i}\{A^{(i)}\}$ is $\mathscr{F}_i$-measurable, $i = 1, 2$.

Theorem 1. *If $(\Omega_i, \mathscr{F}_i, \mu_i)$, $i = 1, 2$, are σ-finite measure spaces and $\Omega = \Omega_1 \times \Omega_2$, $\mathscr{F} = \mathscr{F}_1 \times \mathscr{F}_2$, then there exists a σ-finite measure space $(\Omega, \mathscr{F}, \mu)$ for which*

$$\mu\{B_1 \times B_2\} = \mu_1\{B_1\} \cdot \mu_2\{B_2\} \quad \text{if } B_i \in \mathscr{F}_i, \qquad i = 1, 2; \tag{1}$$

moreover, the measure μ is uniquely determined by (1) and, furthermore, for every $A \in \mathscr{F}$, $\mu_{3-i}\{A^{(i)}\}$ is $\mathscr{F}_i$-measurable, $i = 1, 2$, and

$$\mu\{A\} = \int_{\Omega_2} \mu_1\{A^{(2)}\} d\mu_2 = \int_{\Omega_1} \mu_2\{A^{(1)}\} d\mu_1. \tag{2}$$

PROOF. For $A \in \mathscr{F}$, define μ by the first equality in (2). Then $\mu \geq 0$ and $\mu\{\varnothing\} = 0$. Since $(\bigcup_1^\infty A_n)^{(2)} = \bigcup_1^\infty A_n^{(2)}$ for every sequence $\{A_n\}$ of $\mathscr{F}$ sets, and disjointness of $\{A_n\}$ entails that of $\{A_n^{(2)}\}$, the set function μ so defined is a measure on $\mathscr{F}$, whence $(\Omega, \mathscr{F}, \mu)$ is a measure space. Moreover, if $A = B_1 \times B_2$, where $B_i \in \mathscr{F}_i$, $i = 1, 2$, then

$$\mu\{A\} = \int_{\Omega_2} \mu_1\{A^{(2)}\} d\mu_2 = \int_{B_2} \mu_1\{A^{(2)}\} d\mu_2 = \int_{B_2} \mu_1\{B_1\} d\mu_2 = \prod_{i=1}^{2} \mu_i\{B_i\},$$

so that (1) holds. Now (Exercise 1.5.3),

$$\mathscr{G} = \{B_1 \times B_2 : B_i \in \mathscr{F}_i, i = 1, 2\}$$

is a semi-algebra and, furthermore, since μ_1 and μ_2 are σ-finite, $\tilde{\mu} = \mu|_{\mathscr{G}}$ is a σ-finite measure on $\mathscr{G}$.

By Theorem 6.1.1 the extension μ of $\tilde{\mu}$ is σ-finite on $\mathscr{F} = \sigma(\mathscr{G})$ and uniquely determined by $\tilde{\mu}$. Finally, if

$$\mu^*\{A\} = \int_{\Omega_1} \mu_2\{A^{(1)}\} d\mu_1, \qquad A \in \mathscr{F},$$

then, by analogy with the preceding, μ^* is a measure on $\mathscr{F}$ satisfying (1), whence by uniqueness $\mu = \mu^*$, i.e., (2) holds. □

The measure μ defined in (2) is called the **product measure** of the σ-finite measures μ_1 and μ_2 and is denoted by $\boldsymbol{\mu_1 \times \mu_2}$. The σ-finite measure space $(\Omega_1 \times \Omega_2, \mathscr{F}_1 \times \mathscr{F}_2, \mu_1 \times \mu_2)$ is referred to as the **product measure space** (of the σ-finite measure spaces $(\Omega_i, \mathscr{F}_i, \mu_i)$, $i = 1, 2$).

In situations where more than one measure is floating around, almost everywhere statements must be qualified. Thus, **a.e.** $[\mu_1]$ abbreviates the statement "except for a set of μ_1-measure zero."

Corollary 1. *If $(\Omega, \mathscr{F}, \mu)$ is the product measure space of the σ-finite measure spaces $(\Omega_i, \mathscr{F}_i, \mu_i)$, $i = 1, 2$, and $\mu\{A\} = 0$ for some $A \in \mathscr{F}$, then $\mu_1\{A^{(2)}(\omega_2)\} = 0$, a.e. $[\mu_2]$, and $\mu_2\{A^{(1)}(\omega_1)\} = 0$, a.e. $[\mu_1]$.*

If $X = X(\omega_1, \omega_2)$ is a function on the product measurable space $(\Omega_1 \times \Omega_2, \mathscr{F}_1 \times \mathscr{F}_2)$, then the functions $X_{\omega_1}^{(1)}(\omega_2)$ defined for each $\omega_1 \in \Omega_1$ by

$$X_{\omega_1}^{(1)}(\omega_2) = X(\omega_1, \omega_2)$$

are called **sections of $\boldsymbol{X}$ at $\boldsymbol{\omega_1}$**. Analogously, the functions $X_{\omega_2}^{(2)}(\omega_1)$ are **sections of $\boldsymbol{X}$ at $\boldsymbol{\omega_2}$**. It follows directly from Theorem 1.4.2(ii) that every section at ω_i of an $\mathscr{F}_1 \times \mathscr{F}_2$-measurable function X is an $\mathscr{F}_{3-i}$-measurable function, $i = 1, 2$. Define for $i = 1, 2$

$$A_{3-i} = \left\{\omega_{3-i} : \int_{\Omega_i} X^+(\omega_1, \omega_2) d\mu_i = \int_{\Omega_i} X^-(\omega_1, \omega_2) d\mu_i = \infty\right\}. \qquad (3)$$

Theorem 2 (Fubini). *If $(\Omega, \mathscr{F}, \mu)$ is the product measure space of the σ-finite measure spaces $(\Omega_i, \mathscr{F}_i, \mu_i)$, $i = 1, 2$, and X is an $\mathscr{F}$-measurable function whose integral exists, then*

$$\int_{\Omega_i} X(\omega_1, \omega_2)d\mu_i \equiv \left[\int_{\Omega_i} X^+(\omega_1, \omega_2)d\mu_i - \int_{\Omega_i} X^-(\omega_1, \omega_2)d\mu_i\right] I_{A^c_{3-i}}$$

is $\mathscr{F}_{3-i}$-measurable, $i = 1, 2$, and

$$\int_{\Omega} X\, d\mu = \int_{\Omega_1} d\mu_1 \int_{\Omega_2} X(\omega_1, \omega_2)d\mu_2 = \int_{\Omega_2} d\mu_2 \int_{\Omega_1} X(\omega_1, \omega_2)d\mu_1; \quad (4)$$

moreover, in the case where X is integrable so are almost all sections of X at ω_i for $i = 1, 2$.

PROOF. The general case follows from consideration of X^+, X^-, and so $X \geq 0$ may be so supposed; moreover, μ_1 and μ_2 may be assumed finite. Let $\mathscr{H} = \{X: X \geq 0, X \text{ is } \mathscr{F}\text{-measurable}, \int_{\Omega_i} X(\omega_1, \omega_2)d\mu_i \text{ is } \mathscr{F}_{3-i}\text{-measurable}, i = 1, 2, \text{and (4) holds})\}$. Then $\mathscr{H}$ contains all indicator functions of sets of $\mathscr{F}$ by Theorem 1. Since $\mathscr{H}$ is a monotone system, $\mathscr{H}$ contains all nonnegative $\mathscr{F}$-measurable functions by Theorem 1.4.3. Moreover, if $X \geq 0$ is integrable, it is clear from (4) that so are $\int_{\Omega_2} X(\omega_1, \omega_2)d\mu_2$ and $\int_{\Omega_1} X(\omega_1, \omega_2)d\mu_1$ for almost all ω_1 and ω_2 respectively. □

If $(\Omega_i, \mathscr{F}_i, \mu_i)$, $1 \leq i \leq n$, are σ-finite measure spaces and $(\Omega, \mathscr{F}) = \times_{i=1}^n (\Omega_i, \mathscr{F}_i)$ is the product measurable space, it can be proved inductively that there is one and only one measure μ on $\times_{i=1}^n \mathscr{F}_i$ such that for every measurable rectangle $B_1 \times \cdots \times B_n$

$$\mu\{B_1 \times \cdots \times B_n\} = \prod_{i=1}^n \mu_i\{B_i\}.$$

Then μ is called the product measure and denoted by $\mu_1 \times \cdots \times \mu_n$ or $\times_{i=1}^n \mu_i$. Associativity of ordinary multiplication together with Theorem 6.1.1 guarantees that $\times_{i=1}^n \mu_i = (\times_{i=1}^m \mu_i) \times (\times_{m+1}^n \mu_i)$ for any integer m in $[1, n)$. The measure space $(\Omega, \mathscr{F}, \mu) = (\times_1^n \Omega_i, \times_{i=1}^n \mathscr{F}_i, \times_{i=1}^n \mu_i)$ is alluded to as the (σ-finite) **product measure space** of $(\Omega_i, \mathscr{F}_i, \mu_i)$, $1 \leq i \leq n$, and is denoted by $\times_{i=1}^n (\Omega_i, \mathscr{F}_i, \mu_i)$. The extension of Fubini's theorem is immediate, and so the integral of any nonnegative $\mathscr{F}$-measurable or μ-integrable function on the product measure space $\times_{i=1}^n (\Omega_i, \mathscr{F}_i, \mu_i)$ may be evaluated by iterated integrals **in any order** (i.e., via an n-fold iteration of integrals analogous to (4), where $n = 2$).

In the important special case where $\Omega_i = R = [-\infty, \infty]$, $\mathscr{F}_i = \mathscr{B} =$ the class of all (linear) Borel measurable sets and $\mu_i = \mu =$ Lebesgue measure, $1 \leq i \leq n$, the corresponding product measure space $(R^n, \mathscr{B}^n, \mu^n)$ is the ***n*-dimensional Borel measure space**.

In Section 1.6 it was observed that the d.f. F_X of a r.v. X on a probability space $(\Omega, \mathscr{F}, P)$ is a d.f. in the sense that it is a nondecreasing, left-continuous function on R with

$$F(-\infty) = F(-\infty+) = 0, \qquad F(\infty) = F(\infty-) = 1.$$

The converse was mentioned there but its proof need be deferred no longer. If G is an arbitrary d.f. on the line, there always exists a r.v. X on some probability space with $F_X = G$. It suffices to take $\Omega = R$, $\mathscr{F} = \mathscr{B} =$ linear Borel sets, $X(\omega) = \omega$ (the **coordinate r.v.**) and to define $P\{[a, b)\} = G(b) - G(a)$, $-\infty \le a \le b \le \infty$. By Theorem 6.1.2 this ensures a unique probability measure P on $\mathscr{B}$, and, clearly,

$$F_X(x) = P\{X(\omega) < x\} = P\{\omega : \omega < x\} = G(x).$$

This statement and indeed the notions of Lebesgue–Stieltjes measure, Lebesgue–Stieltjes integral, etc., extend readily to R^n. An ***n*-dimensional distribution function** or **d.f. on R^n** is a function F satisfying

$$\lim_{x_j \to -\infty} F(x_1, \ldots, x_n) = F(x_1, \ldots, x_{j-1}, -\infty, x_{j+1}, \ldots, x_n) = 0, \qquad 1 \le j \le n \tag{5, i}$$

$$\lim_{x_j > y_j \to x_j} F(x_1, \ldots, x_{j-1}, y_j, x_{j+1}, \ldots, x_n) = F(x_1, \ldots, x_j, \ldots, x_n), \qquad 1 \le j \le n, \tag{5, ii}$$

$$\begin{aligned} \Delta_n^{a,b} F \equiv F(b_1, \ldots, b_n) &- \sum_{j=1}^{n} F(b_1, \ldots, b_{j-1}, a_j, b_{j+1}, \ldots, b_n) \\ &+ \sum_{1 \le j < k \le n} F(b_1, \ldots, b_{j-1}, a_j, b_{j+1}, \ldots, b_{k-1}, a_k, b_{k+1}, \ldots, b_n) - \cdots \\ &+ (-1)^n F(a_1, \ldots, a_n) \ge 0, \end{aligned} \tag{5, iii}$$

$$\lim_{\substack{x_j \to \infty \\ 1 \le j \le n}} F(x_1, \ldots, x_n) = F(\infty, \ldots, \infty) = 1, \tag{5, iv}$$

whenever $-\infty \le a_i \le b_i \le \infty$, $1 \le i \le n$.

Let $\mathscr{S}$ be the class of all finite or infinite intervals $[a, b)$ together with $\varnothing$, R, $\{\infty\}$ and define

$$\mathscr{S}^n = \{S : S = S_1 \times S_2 \times \cdots \times S_n, S_i \in \mathscr{S}, 1 \le i \le n\}$$

and

$$P\{S\} = \begin{cases} \Delta_n^{a,b} F & \text{when } S = \bigtimes_{i=1}^{n} S_i \text{ with either } S_i = [a_i, b_i) \text{ or} \\ & S_i = [a_i, b_i], \quad b_i = -a_i = \infty \\ 0 & \text{if } S = \bigtimes_{i=1}^{n} S_i \text{ with some } S_i = \varnothing \text{ or } \{\infty\}. \end{cases} \tag{6}$$

where F satisfies (5, i)–(5, iii) and is finite on $(-\infty, \infty)^n$. As in Lemma 6.1.1, P is a measure on the semi-algebra $\mathscr{S}^n$, and so by Theorem 6.1.1, P can be extended to $\mathscr{M}^n$, the class of all P-measurable sets, the extension being unique on the class $\mathscr{B}^n$ of n-dimensional Borel sets. The measure P and the measure space $(R^n, \mathscr{M}^n, \mathrm{P})$ are called respectively the ***n*-dimensional Lebesgue–Stieltjes measure** and the ***n*-dimensional Lebesgue-Stieltjes measure space determined by *F***. The same terms respectively are used for the restriction of P to $\mathscr{B}^n$ and $(R^n, \mathscr{B}^n, \mathrm{P})$. If F is a d.f., that is, satisfies (5, iv) also, then $(R^n, \mathscr{B}^n, P)$ is a probability space.

If $X_1, \ldots, X_n$ are r.v.s on a probability space $(\Omega, \mathscr{F}, \mathrm{P})$, their joint d.f.

$$F_{X_1, \ldots, X_n}(x_1, \ldots, x_n) = \mathrm{P}\{X_1 < x_1, \ldots, X_n < x_n\}$$

is readily verified to be a d.f. on R^n in the sense of (5). Conversely, given any n-dimensional d.f. F as in (5, i)–(5, iv), there always exist n r.v.s $X_1, \ldots, X_n$ on some probability space $(\Omega, \mathscr{F}, \mathrm{P})$ whose joint d.f. is the preordained F. It suffices to choose $\Omega = R^n$, $\mathscr{F} = \mathscr{B}^n$, $X_i(\omega) = i$th coordinate of ω, $1 \le i \le n$, and to define P on $\mathscr{S}^n$ via (6). According to the prior discussion, P is uniquely determined on the Lebesgue–Stieltjes measure space $(R^n, \mathscr{B}^n, \mathrm{P})$. Moreover,

$$\begin{aligned}\mathrm{P}\{X_1 < x_1, \ldots, X_n < x_n\} &= \mathrm{P}\{\omega : \omega_i < x_i, 1 \le i \le n\} \\ &= \lim_{a \to -\infty} \Delta_n^{a, x} F = F(x_1, \ldots, x_n).\end{aligned}$$

Thus, there is a one-to-one correspondence between distribution functions on R^n and probability measures on $(R^n, \mathscr{B}^n)$.

If $Y = Y(\omega_1, \ldots, \omega_n)$ is an integrable function on the Lebesgue–Stieltjes measure space $(R^n, \mathscr{M}^n, \mathrm{P})$, the integral of $Y \cdot I_A$ for any $A \in \mathscr{B}^n$ is denoted by $\int YI_A \, d\mathrm{P}$ or by

$$\int \cdots \int_A Y(\omega_1, \ldots, \omega_n) dF(\omega_1, \ldots, \omega_n) \tag{7}$$

and abbreviated (when the context is clear) by $\int_A Y \, dF$.

The analogue of the first part of Corollary 6.2.1 is

Theorem 3. *If $X = (X_1, \ldots, X_m)$ is a random vector with d.f. F_X and g is a Borel function on R^m,*

$$\mathrm{E}\, g(X) = \int_{R^m} g(t_1, \ldots, t_m) dF_X(t_1, \ldots, t_m)$$

in the sense that if either of the integrals exists, so does the other and the two are equal.

It should be noted that the iterated integrals (4) of Fubini's theorem may also be written as in

$$\int_{\Omega_1} \int_{\Omega_2} X(\omega_1, \omega_2) d\mu_2(\omega_2) d\mu_1(\omega_1), \qquad \int_{\Omega_1} \int_{\Omega_2} X \, d\mu_2 \, d\mu_1. \tag{8}$$

Definition. Convolution of d.f.s is a binary operation denoted by $*$, which associates with every pair F_1, F_2 of d.f.s, a d.f. F, denoted by $F_1 * F_2$, and defined by

$$F(x) = (F_1 * F_2)(x) = \int_{-\infty}^{\infty} F_1(x - y)dF_2(y). \tag{9}$$

The dominated convergence theorem permits limits to be taken inside the integral and thus the convolution $F_1 * F_2$ of any two d.f.s is again a d.f.

The primary interest of convolution for probability theory stems from

Theorem 4. *If X_1 and X_2 are independent* r.v.s *on some probability space* $(\Omega, \mathscr{F}, \mathrm{P})$ *with corresponding* d.f.s F_1 *and* F_2, *then their sum* $X_1 + X_2$ *has the* d.f. $F_1 * F_2$.

PROOF. Since, via the independence of X_1 and X_2, the mapping $\omega \to (X_1(\omega), X_2(\omega))$ takes P-measure on $\mathscr{F}$ sets into the product $(F_1 \times F_2)$-measure on the Borel sets of the plane, by Theorem 3 and Fubini's theorem

$$\begin{aligned}\mathrm{P}\{X_1 + X_2 < x\} &= \int_{\Omega} I_{[X_1+X_2<x]}(\omega)\, d\mathrm{P}(\omega) = \int_{[z+y<x]} d(F_1 \times F_2)(z, y) \\ &= \int_{-\infty}^{\infty}\int_{-\infty}^{\infty} I_{[z+y<x]}(z, y)dF_1(z)dF_2(y) \\ &= \int_{-\infty}^{\infty} F_1(x - y)dF_2(y),\end{aligned}$$

and so the d.f. of $X_1 + X_2$ is $F_1 * F_2$. □

The convolution formula (9) may be paralleled by a convolution of the measures $F_1\{\cdot\}, F_2\{\cdot\}$ determined by the d.f.s F_1, F_2. To this end, let the translation of the Borel set B by the amount y be denoted by $B - y = \{x - y : x \in B\}$ and define

$$F\{B\} = \int_{-\infty}^{\infty} F_1\{B - y\}dF_2(y). \tag{10}$$

It is a simple matter to check that F is a probability measure on the Borel sets of the line. Moreover, when $B = (-\infty, x)$, (10) reduces to (9). Then by Theorem 6.1.1 the measure $F\{\cdot\}$ of (10) is that determined by the d.f. $F_1 * F_2$ of (9), thereby justifying the common name F. Furthermore, for any Borel function g integrable relative to $F_1 * F_2$,

$$\int_{-\infty}^{\infty} g(z)d(F_1 * F_2)(z) = \int_{-\infty}^{\infty}\int_{-\infty}^{\infty} g(x + y)dF_1(x)dF_2(y). \tag{11}$$

When g is the indicator function I_B of some Borel set B, then $g(x + y) = I_{B-y}$ for every real y, whence (11) coalesces to (10). Hence, by Theorem 1.4.3, (11) holds for any $(F_1 * F_2)$-integrable function.

Example 1. Let $(\Gamma, \mathscr{F})$ and $(\Lambda, \mathscr{G})$ be measurable spaces with ν a measure on $\mathscr{G}$ and $\mu_\lambda\{A\}$ a function on $\mathscr{F} \times \Lambda$ such that $\mu_\lambda\{\cdot\}$ is a measure on $\mathscr{F}$ for almost all $\lambda \in \Lambda$ and moreover $\mu_\lambda\{A\}$ is $\mathscr{G}$-measurable for every $A \in \mathscr{F}$. Define

$$\mu\{A\} = \int_\Lambda \mu_\lambda\{A\} d\nu(\lambda), \quad A \in \mathscr{F}$$

and for any $\mathscr{F}$-measurable function f

$$\int_\Gamma f(\gamma) d\mu_\lambda(\gamma) = \left[\int_\Gamma f^+(\gamma) d\mu_\lambda(\gamma) - \int_\Gamma f^-(\gamma) d\mu_\lambda(\gamma)\right] I_{[\int_\Gamma f^+ d\mu_\lambda = \infty = \int_\Gamma f^- d\mu_\lambda]^c}$$

Then μ is a measure on $\mathscr{F}$, $\int_\Gamma f(\gamma) d\mu_\lambda(\gamma)$ is $\mathscr{G}$-measurable and

$$\int_\Gamma f(\gamma) d\mu(\gamma) = \int_\Lambda \left(\int_\Gamma f(\gamma) d\mu_\lambda(\gamma)\right) d\nu(\lambda) \tag{12}$$

if either integral exists.

Proof. Let $\Lambda' = \{\lambda: \mu_\lambda$ is a measure on $\mathscr{F}\}$. Then $\mu\{\Lambda - \Lambda'\} = 0$ whence $\mu\{A\} = \int_{\Lambda'} \mu_\lambda\{A\} d\nu(\lambda)$. Via the monotone convergence theorem, μ is a measure on $\mathscr{F}$. As usual, in verifying (12), it suffices to consider non-negative f. Let $\mathscr{H} = \{f \geq 0$: f is $\mathscr{F}$-measurable, $\int_\Gamma f(\gamma) d\mu_\lambda(\gamma)$ is $\mathscr{G}$-measurable and (12) holds$\}$. Then $I_A \in \mathscr{H}$ for all $A \in \mathscr{F}$ by definition and moreover by Theorem 6.2.1 $\mathscr{H}$ is a monotone system. Thus, by Theorem 1.4.3, $\mathscr{H}$ contains all non-negative $\mathscr{F}$-measurable functions. □

Exercises 6.3

1. Let $(\Omega_i, \mathscr{F}_i, P_i)$ be a probability space, where $\Omega_i = (-\infty, \infty)$ and $\mathscr{F}_i = \{B: B$ is a Borel subset of $(-\infty, \infty)\}$, $i = 1, 2$. If $(\Omega, \mathscr{F}, P) = (\Omega_1 \times \Omega_2, \mathscr{F}_1 \times \mathscr{F}_2, P_1 \times P_2)$ and if $X_i(\omega) = \omega_i$, $i = 1, 2$, for $\omega = (\omega_1, \omega_2) \in \Omega$, then X_1 and X_2 are independent r.v.s on $(\Omega, \mathscr{F}, P)$.

2. If $\Omega_1 = \Omega_2 = [0, 1]$, $\mathscr{F}_1 = \mathscr{F}_2 = [0, 1] \cdot \mathscr{B}$, $\mu_1 =$ Lebesgue measure, and $\mu_2\{A\} =$ number of points in A, then $(\Omega_i, \mathscr{F}_i, \mu_i)$, $i = 1, 2$, are measure spaces. The set $A = \{(\omega_1, \omega_2): \omega_1 = \omega_2\} \in \mathscr{F}_1 \times \mathscr{F}_2$, but

$$\int_{\Omega_2} \mu_1(A^{(2)}) d\mu_2 = 0 = 1 - \int_{\Omega_1} \mu_2(A^{(1)}) d\mu_1,$$

whence Theorem 1 as well as the uniqueness part of Theorem 6.1.1 is invalid if σ-finiteness is omitted.

3. For any (not necessarily σ-finite) measure spaces $(\Omega_i, \mathscr{F}_i, \mu_i)$, $i = 1, 2$, and any integrable function X on the product measure space $(\Omega, \mathscr{F}, \mu)$,

$$\int_\Omega X \, d\mu = \int_{\Omega_1} d\mu_1 \int_{\Omega_2} X \, d\mu_2 = \int_{\Omega_2} d\mu_2 \int_{\Omega_1} X \, d\mu_1.$$

Hint: Consider $X \geq 0$ and utilize the remark just prior to Theorem 6.2.1.

4. For $j \geq 1$, define $a_{i,2j-1} = (2^j - 1)^{i-1}/2^{ji}$ and $a_{i,2j} = -a_{i,2j+1}$, $i \geq 1$. The row sums of the double series $\sum_j a_{ij}$ are simply the elements of the first column. What are the column sums? Taking μ_1 and μ_2 as counting measure, does this example contradict Fubini's theorem?

5. Show that $g(\omega_1, \omega_2) = e^{-\omega_1\omega_2} - 2e^{-2\omega_1\omega_2}$ is Lebesgue integrable (i) over $\Omega_1 = [1, \infty)$ for each ω_2 and (ii) over $\Omega_2 = (0, 1]$ for each ω_1, but that Fubini's theorem fails. Why?

6. An alternative construction of a r.v. X on a probability space with a preassigned d.f. F is to take $\Omega = [0, 1]$, $\mathscr{F}$ = Borel subsets of $[0, 1]$, P = Lebesgue measure on $[0, 1]$, and $X(\omega) = F^{-1}(\omega)$, where $F^{-1}(\omega) = \sup\{x: F(x) < \omega\}$.

7. Prove that the random vectors $X = (X_1, \ldots, X_m)$ and $Y = (Y_1, \ldots, Y_n)$ on $(\Omega, \mathscr{F}, \mathrm{P})$ are independent of one another iff their joint d.f. $F_{X,Y} = F_X \cdot F_Y$ and conclude that X and Y independent entails $(R^{m+n}, \mathscr{B}^{m+n}, \nu_{X,Y}) = (R^m, \mathscr{B}^m, \nu_X) \times (R^n, \mathscr{B}^n, \nu_Y)$, where ν_X, ν_Y, and $\nu_{X,Y}$ are the Lebesgue–Stieltjes measures determined by F_X, F_Y, $F_{X,Y}$ respectively.

8. Random variables X_1, X_2 have a **bivariate Poisson distribution** if

$$\mathrm{P}\{X_1 = j, X_2 = k\} = e^{-(a_1+a_2+a_{12})} \sum_{i=0}^{\min(j,k)} \frac{a_{12}^i a_1^{j-i} a_2^{k-i}}{i!(j-i)!(k-i)!}$$

for any pair of nonnegative integers (j, k), where a_1, a_2, a_{12} are nonnegative parameters. Define a probability space and r.v.s X_1, X_2 on it whose joint distribution is bivariate Poisson and show that X_j is a Poisson r.v. with mean $a_j + a_{12}$. Prove that the correlation $\rho(X_1, X_2) \geq 0$ and that X_1 and X_2 are independent iff $a_{12} = 0$.

9. Random variables $X_1, \ldots, X_k$ have a **multinomial distribution** if $\mathrm{P}\{X_i = x_i, 1 \leq i \leq k\} = n! \prod_{i=1}^k (p_i^{x_i}/(x_i!))$ for any nonnegative integers x_i, $1 \leq i \leq k$, with $\sum_1^k x_i = n$ and zero otherwise. Here, n is a positive integer and $\sum_1^k p_i = 1$, $p_i > 0$, $1 \leq i \leq k$. Prove that if $\{A_i, 1 \leq i \leq k\}$ is a partition of Ω in $\mathscr{F}$ with $p_i = \mathrm{P}\{A_i\}$ and X_i = number of occurrences of A_i in n independent trials, $1 \leq i \leq k$, then $X_1, \ldots, X_k$ have a multinomial distribution. Show that X_i is a binomial r.v. with parameters p_i, n and that $\rho(X_i, X_j) < 0$.

10. Random variables $\{X_i, 1 \leq i \leq n\}$ are called **interchangeable** if their joint d.f. is a symmetric function, i.e., invariant under permutations, and r.v.s $\{X_n, n \geq 1\}$ are interchangeable if every finite subset is interchangeable. Prove that if $\{X_n, n \geq 1\}$ are $\mathscr{L}_2$ interchangeable r.v.s, then $\rho(X_1, X_2) \geq 0$. *Hint*: $\sigma^2(\sum_1^n X_j) \geq 0$.

6.4 Infinite-Dimensional Product Measure Space, Kolmogorov Consistency Theorem

In determining a product measure on an infinite-dimensional product measurable space (Section 1.3) it is well to keep in mind the distinguished role played by the number 1 in infinite (and even finite) products.

Let $(\Omega_i, \mathscr{F}_i, P_i)$, $i \geq 1$, be a sequence of probability spaces and define

$$\Omega = \bigtimes_{i=1}^{\infty} \Omega_i, \qquad \Omega'_n = \bigtimes_{i=1}^{n} \Omega_i, \qquad \Omega''_n = \bigtimes_{n+1}^{\infty} \Omega_i,$$

$$\mathscr{F} = \bigtimes_{i=1}^{\infty} \mathscr{F}_i, \qquad \mathscr{F}'_n = \bigtimes_{1}^{n} \mathscr{F}_i, \qquad \mathscr{F}''_n = \bigtimes_{n+1}^{\infty} \mathscr{F}_i,$$

$$\mathscr{G}_n = \{A: A = B_n \times \Omega''_n \text{ where } B_n \in \mathscr{F}'_n\}, \qquad \mathscr{G} = \bigcup_{1}^{\infty} \mathscr{G}_n.$$

Then for each $n \geq 1$, $(\Omega''_n, \mathscr{F}''_n)$ is a measurable space, $\{\mathscr{G}_n\}$ is an increasing sequence of sub-σ-algebras of $\mathscr{F}$, and $\mathscr{G}$ is the algebra of cylinders of $\mathscr{F}$. For $A = B_n \times \Omega''_n \in \mathscr{G}_n$, define

$$P\{A\} = (P_1 \times P_2 \times \cdots \times P_n)\{B_n\}.$$

Note that P is well defined on $\mathscr{G}$, since if $A = B_n \times \Omega''_n = B_m \times \Omega''_m \in \mathscr{G}_n \cdot \mathscr{G}_m$ where, say $m > n$, then necessarily $B_m = B_n \times \Omega_{n+1} \times \cdots \times \Omega_m$, whence

$$\begin{aligned}(P_1 \times \cdots \times P_m)\{B_m\} &= (P_1 \times \cdots \times P_n)\{B_n\} \cdot \prod_{i=n+1}^{m} P_i\{\Omega_i\} \\ &= (P_1 \times \cdots \times P_n)\{B_n\}.\end{aligned}$$

Clearly, P is σ-additive on $\mathscr{G}_n$, $n \geq 1$, and additive on the algebra $\mathscr{G}$, and, moreover, $P\{\Omega\} = 1$.

Now $\Omega = \Omega'_n \times \Omega''_n$, $\mathscr{F} = \mathscr{F}'_n \times \mathscr{F}''_n$, $n \geq 1$, that is, if $\omega \in \Omega$, $\omega = (\omega'_n, \omega''_n)$, where $\omega'_n \in \Omega'_n$, $\omega''_n \in \Omega''_n$ for $n \geq 1$. For any $A \subset \Omega$, set

$$A(\omega'_n) = \{\omega''_n: (\omega'_n, \omega''_n) \in A\}, \qquad A(\omega''_n) = \{\omega'_n: (\omega'_n, \omega''_n) \in A\}.$$

By Theorem 1.4.2, if $A \in \mathscr{F}$, then $A(\omega'_n)$ is $\mathscr{F}''_n$-measurable and if $A \in \mathscr{G}$, say $A = B_m \times \Omega''_m \in \mathscr{G}_m$, where $m > n$, then

$$\begin{aligned}A(\omega'_n) &= \{\omega''_n: (\omega'_n, \omega''_n) \in B_m \times \Omega''_m\} \\ &= \{(\omega_{n+1}, \ldots, \omega_m): (\omega_1, \ldots, \omega_m) \in B_m\} \times \Omega''_m \qquad (1) \\ &= B_m(\omega'_n) \times \Omega''_m\end{aligned}$$

is an $(m - n)$-dimensional cylinder in $(\Omega''_n, \mathscr{F}''_n)$ so that if $P^{(n)}$ is the analogue of P for the space $(\Omega''_n, \mathscr{F}''_n)$

$$P^{(n)}\{A(\omega'_n)\} = (P_{n+1} \times \cdots \times P_m)\{B_m(\omega'_n)\}.$$

Hence, via Theorem 6.3.1 and Fubini's theorem,

$$\begin{aligned}P\{A\} &= (P_1 \times \cdots \times P_m)\{B_m\} \\ &= \int_{\Omega'_n} (P_{n+1} \times \cdots \times P_m)\{B_m(\omega'_n)\} d(P_1 \times \cdots \times P_n) \\ &= \int_{\Omega_n} \cdots \int_{\Omega_1} P^{(n)}\{A(\omega'_n)\} dP_1(\omega_1) \cdots dP_n(\omega_n). \qquad (2)\end{aligned}$$

Theorem 1. *If* $(\Omega_i, \mathscr{F}_i, P_i)$, $i \geq 1$, *is a sequence of probability spaces, there exists a unique probability measure* P *on the product σ-algebra* $\times_{i=1}^{\infty} \mathscr{F}_i$ *such that for every cylinder* $A = B_n \times \Omega_{n+1} \times \Omega_{n+2} \times \cdots$ *with* $B_n \in \times_1^n \mathscr{F}_i$,

$$P\{A\} = (P_1 \times \cdots \times P_n)\{B_n\}. \tag{3}$$

PROOF. Clearly, P, as defined by (3), is nonnegative and additive on the algebra $\mathscr{G}$ with $P\{\Omega\} = 1$. In view of Theorem 6.1.1 it is sufficient to verify σ-additivity of P on $\mathscr{G}$ and thus by Theorem 1.5.2(iv) to prove that $A_n \in \mathscr{G}$, $n \geq 1$, with $A_n \downarrow$ and $\inf_{n \geq 1} P\{A_n\} = \varepsilon > 0$ imply $\bigcap_{n=1}^{\infty} A_n \neq \varnothing$. To this end, set $D_n = \{\omega_1 : P^{(1)}\{A_n(\omega_1)\} > \varepsilon/2\}$, noting that the prime in (2) is superfluous since $\Omega_1' = \Omega_1$. Since

$$\begin{aligned}
\varepsilon \leq P\{A_n\} &= \int_{\Omega_1} P^{(1)}\{A_n(\omega_1)\}\, dP_1 \\
&= \int_{D_n} P^{(1)}\{A_n(\omega_1)\}\, dP_1 + \int_{D_n^c} P^{(1)}\{A_n(\omega_1)\}\, dP_1 \\
&\leq P_1\{D_n\} + \frac{\varepsilon}{2},
\end{aligned}$$

necessarily $P_1\{D_n\} \geq \varepsilon/2$, $n \geq 1$. Now $\{D_n\}$ is a decreasing sequence of $\mathscr{F}_1$ sets, whence, since P_1 is a probability measure on $\mathscr{F}_1$, Theorem 1.5.2(iv) ensures the existence of a point $\omega_1^* \in \Omega_1$ with $\omega_1^* \in \bigcap_1^{\infty} D_n$. Thus, $A_n(\omega_1^*)$, $n \geq 1$, is a decreasing sequence of $\mathscr{G}^{(1)}$ sets with $P^{(1)}\{A_n(\omega_1^*)\} \geq \varepsilon/2$, $n \geq 1$ and the argument just offered with Ω, $\{A_n\}$, ε applies to Ω_1'', $\{A_n(\omega_1^*)\}$, $\varepsilon/2$, and yields a point $\omega_2^* \in \Omega_2$ with $P^{(2)}\{A_n(\omega_1^*, \omega_2^*)\} \geq \varepsilon/4$, $n \geq 1$. Continuing in this fashion a sequence $(\omega_1^*, \omega_2^*, \ldots)$ is obtained with $\omega_m^* \in \Omega_m$ and

$$P^{(m)}\{A_n(\omega_1^*, \ldots, \omega_m^*)\} \geq \frac{\varepsilon}{2^m}, \qquad m = 1, 2, \ldots; n = 1, 2, \ldots. \tag{4}$$

To prove that $\omega^* = (\omega_1^*, \omega_2^*, \ldots) \in \bigcap_1^{\infty} A_n$, note that since for each $n \geq 1$, $A_n = B_m \times \Omega_m''$ for some $m \geq 1$, necessarily

$$A_n(\omega_1^*, \ldots, \omega_m^*) = \begin{cases} \Omega_m'' & \text{if } (\omega_1^*, \ldots, \omega_m^*) \in B_m \\ \varnothing & \text{if } (\omega_1^*, \ldots, \omega_m^*) \notin B_m. \end{cases}$$

But (4) ensures that $A_n(\omega_1^*, \ldots, \omega_m^*) \neq \varnothing$. Thus, $(\omega^*, \ldots, \omega_m^*) \in B_m$, whence $(\omega_1^*, \ldots, \omega_m^*, \omega_{m+1}^*, \ldots) \in B_m \times \Omega_m'' = A_n$, $n \geq 1$. □

Theorem 1 establishes that a sequence of probability spaces $(\Omega_i, \mathscr{F}_i, P_i)$ $i \geq 1$, engenders an **infinite-dimensional product measure space** $(\times_{i=1}^{\infty} \Omega_i, \times_{i=1}^{\infty} \mathscr{F}_i, \times_{i=1}^{\infty} P_i)$ such that for every $n = 1, 2, \ldots$, denoting $\times_{i=1}^{\infty} P_i$ by P,

$$P\left\{A_1 \times \cdots \times A_n \times \left(\mathop{\times}_{n+1}^{\infty} \Omega_i\right)\right\} = \prod_{i=1}^{n} P_i\{A_i\}, \qquad A_i \in \mathscr{F}_i, 1 \leq i \leq n.$$

The proof is based on the fact that if $\mathscr{G}_n = \{A : A = B_n \times \Omega_n''\}$ is the class of cylinders with n-dimensional bases $B_n \in \times_{i=1}^{n} \mathscr{F}_i$ and

$$\hat{P}_n\{A\} = (P_1 \times \cdots \times P_n)\{B_n\},$$

then $\hat{P}_n$ is a probability measure on the σ-algebra $\mathscr{G}_n$ with $\hat{P}_n = \hat{P}_{n+1}|_{\mathscr{G}_n}$ and, moreover, if $P\{A\} = \lim_n \hat{P}_n\{A\}$ for $A \in \mathscr{G} = \bigcup_1^\infty \mathscr{G}_n$, then P is σ-additive on the algebra $\mathscr{G}$, whence there is a unique extension to $\sigma(\mathscr{G}) = \times_1^\infty \mathscr{F}_i$.

The following question then poses itself: If $(\Omega, \mathscr{F}_i, P_i)$, $i \geq 1$, is a sequence of probability spaces with $\mathscr{F}_i \subset \mathscr{F}_{i+1}$ and $P_i = P_{i+1}|_{\mathscr{F}_i}$, is $P\{A\} = \lim_n P_n\{A\}$, $A \in \mathscr{G} = \bigcup_1^\infty \mathscr{F}_i$, necessarily σ-additive on the algebra $\mathscr{G}$? The answer is, in general, negative; see Example 3. However, if $\Omega = R^\infty$ and $\mathscr{F}_n$ is the class of cylinders of $(R^\infty, \mathscr{B}^\infty)$ with n-dimensional bases, the answer becomes affirmative.

Theorem 2. *Let $(R^n, \mathscr{B}^n, P_n)$, $n \geq 1$, be a sequence of probability spaces with*

$$P_{n+1}\{A_n \times R\} = P_n\{A_n\}, \qquad A_n \in \mathscr{B}^n, n \geq 1, \tag{5}$$

and let $\tilde{\mathscr{B}}^n$ be the σ-algebra of cylinders in $(R^\infty, \mathscr{B}^\infty)$ with n-dimensional Borel bases. If $\mathscr{G} = \bigcup_1^\infty \tilde{\mathscr{B}}^n$ and for each $A = A_n \times \times_{n+1}^\infty R$ with $A_n \in \mathscr{B}^n$,

$$P\{A\} = P_n\{A_n\}, \tag{6}$$

then P is σ-additive on the algebra $\mathscr{G}$ and determines a unique probability measure extension $\tilde{P}$ to $\mathscr{B}^\infty$.

PROOF. In view of Theorem 6.1.1, it suffices to prove that P is well defined and a probability measure on $\mathscr{G}$. The former is an immediate consequence of (5) and, clearly, P is nonnegative and additive on $\mathscr{G}$ with $P\{R^\infty\} = 1$.

Let $\mathscr{D}_n$ and $\tilde{\mathscr{D}}_n$ denote the classes of all sets of the form $J_1 \times \cdots \times J_n$ and $J_1 \times \cdots \times J_n \times R \times R \times \cdots$ respectively, where J_i signifies an interval of R, $1 \leq i \leq n$, i.e., $J_i = [a_i, b_i]$, $[a_i, b_i)$, (a_i, b_i), or $(a_i, b_i]$ for $-\infty \leq a_i \leq b_i \leq \infty$. Then the classes $\mathscr{H}_n$ and $\tilde{\mathscr{H}}_n$ of all finite unions of sets of $\mathscr{D}_n$ and $\tilde{\mathscr{D}}_n$ respectively are algebras. If $\tilde{\mathscr{H}} = \bigcup_1^\infty \tilde{\mathscr{H}}_n$, then $\tilde{\mathscr{H}}$ is a subalgebra of $\mathscr{G}$, whence P is defined and additive on $\tilde{\mathscr{H}}$.

To check σ-additivity on $\tilde{\mathscr{H}}$, let $\{\tilde{A}_n, n \geq 1\}$ be a decreasing sequence of sets of $\tilde{\mathscr{H}}$ with $\inf_n P\{\tilde{A}_n\} = \varepsilon > 0$, whence $\tilde{A}_n = A_{m_n} \times R \times R \times \cdots$ for some $A_{m_n} \in \mathscr{H}_{m_n}$, and $m_{n+1} > m_n$, $n \geq 1$. Since P_n is a probability measure on $\mathscr{H}_n$, every interval $J_1 \times \cdots \times J_n$ of $\mathscr{D}_n$ contains a closed subinterval whose P_n-measure is arbitrarily close to that of $J_1 \times \cdots \times J_n$. Thus, there is a closed set B_n (which is a finite union of closed intervals) with

$$P_{m_n}\{A_{m_n} - B_n\} < \frac{\varepsilon}{2^{n+1}}, \qquad B_n \subset A_{m_n}. \tag{7}$$

Let $\tilde{B}_n = B_n \times R \times R \cdots$, whence (7) holds with $\tilde{A}_n$, $\tilde{B}_n$, P replacing A_{m_n}, B_n, P_{m_n} respectively. Consequently, if $\tilde{C}_n = \tilde{B}_1 \cdot \tilde{B}_2 \cdots \tilde{B}_n$,

$$P\{\tilde{A}_n - \tilde{C}_n\} = P\left\{\bigcup_{j=1}^n (\tilde{A}_n - \tilde{B}_j)\right\} \leq P\left\{\bigcup_1^n (\tilde{A}_j - \tilde{B}_j)\right\} < \frac{\varepsilon}{2},$$

whence

$$P\{\tilde{C}_n\} > P\{\tilde{A}_n\} - \frac{\varepsilon}{2} \ge \frac{\varepsilon}{2}$$

and $\tilde{C}_n \neq \varnothing$. Let $\omega^{(n)} = (\omega_1^{(n)}, \omega_2^{(n)}, \ldots) \in \tilde{C}_n$. Since $\tilde{C}_n \downarrow$, necessarily $\omega^{(n+p)} \in \tilde{C}_n \subset \tilde{B}_n$, implying $(\omega_1^{(n+p)}, \ldots, \omega_{m_n}^{(n+p)}) \in B_n$, $p = 0, 1, \ldots$. Choose a subsequence $\{n_{1k}\}$ of the positive integers for which $\omega_1^{(n_{1k})} \to$ a limit $\omega_1^{(0)}$ (finite or infinite) as $k \to \infty$. Likewise, there is a subsequence $\{n_{2k}\}$ of $\{n_{1k}\}$ with $\omega_2^{(n_{2k})} \to \omega_2^{(0)}$ as $k \to \infty$, etc. Then

$$\omega^{(n_{kk})} = (\omega_1^{(n_{kk})}, \omega_2^{(n_{kk})}, \ldots) \to (\omega_1^{(0)}, \omega_2^{(0)}, \ldots) \equiv \omega^{(0)}$$

and so as $k \to \infty$

$$(\omega_1^{(n_{kk})}, \ldots, \omega_{m_n}^{(n_{kk})}) \to (\omega_1^{(0)}, \ldots, \omega_{m_n}^{(0)}) \in B_n, \qquad n \ge 1.$$

Therefore $\omega^{(0)} \in \tilde{B}_n \subset \tilde{A}_n$, $n \ge 1$, so that $\bigcap_1^\infty \tilde{A}_n \neq \varnothing$. Consequently, P is σ-additive on the algebra $\tilde{\mathscr{H}}$ with $P\{R^\infty\} = 1$, whence by Theorem 6.1.1, P has a unique extension $\tilde{P}$ to $\mathscr{B}^\infty = \sigma(\tilde{\mathscr{H}})$. Clearly, $\tilde{P}$ is σ-additive on $\mathscr{G}$, and since $\tilde{P} = P$ on $\tilde{\mathscr{H}}_n$, $\tilde{P} = P$ on $\sigma(\tilde{\mathscr{H}}_n) = \tilde{\mathscr{B}}_n$, $n \ge 1$, whence $\tilde{P} = P$ on $\mathscr{G} = \bigcup_1^\infty \tilde{\mathscr{B}}_n$. □

Is it always possible to define a sequence of r.v.s $\{X_n, n \ge 1\}$ on some probability space $(\Omega, \mathscr{F}, P)$ such that the joint d.f.s of all finite subsets of r.v.s $X_1, \ldots, X_n$ coincide with d.f.s $F_{1,\ldots,n}$ assigned *a priori*? The answer is yes provided the assignment is not internally contradictory.

A family $\{F_{1,\ldots,n}(x_1, \ldots, x_n)\}$ of n-dimensional d.f.s defined for all $n \ge 1$ will be called **consistent** if for all $n \ge 1$

$$F_{1,\ldots,n}(x_1, \ldots, x_n) = \lim_{x_{n+1} \to +\infty} F_{1,\ldots,n+1}(x_1, \ldots, x_{n+1}). \tag{8}$$

Theorem 3 (Kolmogorov Consistency Theorem). *If $\{F_{1,\ldots,n}, n \ge 1\}$ is a consistent family of* d.f.s, *then there exists a probability measure* P *on* $(R^\infty, \mathscr{B}^\infty)$ *such that the* d.f.s *of the coordinate* r.v.s $X_1, \ldots, X_n$ *on* $(R^\infty, \mathscr{B}^\infty, P)$ *coincide with the preassigned* d.f.s $F_{1,\ldots,n}$, *that is, such that if*

$$X_k(\omega) = \omega_k, \quad k = 1, 2, \ldots, \quad \textit{for } \omega = (\omega_1, \omega_2, \ldots) \in R^\infty \tag{9}$$

then for all $n \ge 1$,

$$P\{X_1 < x_1, \ldots, X_n < x_n\} = F_{1,\ldots,n}(x_1, \ldots, x_n). \tag{10}$$

PROOF. If A_n is an n-dimensional set of $\mathscr{B}^n$, define

$$P_n\{A_n\} = \int \cdots \int_{A_n} dF_{1,\ldots,n}(t_1, \ldots, t_n), \tag{11}$$

whence $(R^n, \mathscr{B}^n, P_n)$ is a probability space, $n \ge 1$. Employing the notation of (6) of Section 3, for all pairs of real numbers $a_i < b_i$, $1 \le i \le n$,

$$\mathrm{P}_{n+1}\left\{\bigtimes_{i=1}^{n}[a_i, b_i) \times R\right\} = \lim_{\substack{a_{n+1}\to -\infty \\ b_{n+1}\to\infty}} \mathrm{P}_{n+1}\left\{\bigtimes_{i=1}^{n+1}[a_i, b_i)\right\}$$

$$= \lim_{\substack{a_{n+1}\to -\infty \\ b_{n+1}\to\infty}} \Delta_{n+1}^{a,b} F_{1,\dots,n+1} = \Delta_n^{a,b} F_{1,\dots,n} = \mathrm{P}_n\left\{\bigtimes_{i=1}^{n}[a_i, b_i)\right\}$$

in view of (8). Hence, $\mathrm{P}_{n+1}\{A_n \times R\} = \mathrm{P}_n\{A_n\}$ for all $A_n \in \mathscr{B}^n$, $n \geq 1$, and so by Theorem 2 there is a probability measure P on $(R^\infty, \mathscr{B}^\infty)$ such that for all $A_n \in \mathscr{B}^n$

$$\mathrm{P}\{A_n \times R \times R \times \cdots\} = \mathrm{P}_n\{A_n\}, \qquad n \geq 1.$$

The coordinate functions defined by (9) are r.v.s and by (9) and (11)

$$\begin{aligned}\mathrm{P}\{X_1 < x_1, \dots, X_n < x_n\} &= \mathrm{P}\{\omega\colon \omega_1 < x_1, \dots, \omega_n < x_n\} \\ &= F_{1,\dots,n}(x_1, \dots, x_n).\end{aligned}$$

□

Corollary 1. *If* $\{X_n, n \geq 1\}$ *is a sequence of* r.v.s *on some probability space* $(\Omega, \mathscr{F}, \mathrm{P})$, *there exists a sequence of coordinate* r.v.s $\{X'_n, n \geq 1\}$ *on the probability space* $(R^\infty, \mathscr{B}^\infty, \mathrm{P}')$ *such that the joint* d.f.s *of* $X_1, \dots, X_n$ *and* $X'_1, \dots, X'_n$ *are identical for all* $n \geq 1$.

Theorem 4. *If* $\{X_n, n \geq 1\}$ *are* r.v.s *on some probability space* $(\Omega, \mathscr{F}, \mathrm{P})$, *there exist two sequences of coordinate* r.v.s $\{X''_n, n \geq 1\}$ *and* $\{X'_n, n \geq 1\}$ *on* $(R^\infty, \mathscr{B}^\infty)$ *each having the same joint distributions as* $\{X_n, n \geq 1\}$, i.e.,

$$F_{X''_1,\dots,X''_n} = F_{X'_1,\dots,X'_n} = F_{X_1,\dots,X_n}, \qquad n \geq 1, \tag{12}$$

such that the stochastic processes

$$\{X''_n, n \geq 1\} \quad \textit{and} \quad \{X'_n, n \geq 1\} \tag{13}$$

are independent of one another.

PROOF. For $\omega = (\omega_1, \omega_2, \dots) \in R^\infty$, define $Y_j(\omega) = \omega_j$, $j \geq 1$, and $\hat{F}_n$, $\hat{\mathrm{P}}$ by

$$\begin{aligned}\hat{F}_{2n} = \hat{\mathrm{P}}\{Y_1 < y_1, \dots, Y_{2n} < y_{2n}\} &= F_{X_1,X_2,\dots,X_n}(y_1, y_3, \dots, y_{2n-1}) \\ &\quad \cdot F_{X_1,X_2,\dots,X_n}(y_2, y_4, \dots, y_{2n}),\end{aligned}$$

$$\begin{aligned}\hat{F}_{2n-1} &= \hat{\mathrm{P}}\{Y_1 < y_1, \dots, Y_{2n-1} < y_{2n-1}\} \\ &= F_{X_1,X_2,\dots,X_n}(y_1, y_3, \dots, y_{2n-1}) \\ &\quad \cdot F_{X_1,X_2,\dots,X_{n-1}}(y_2, y_4, \dots, y_{2n-2})\end{aligned}$$

for $n \geq 1$. Then $\{\hat{F}_n, n \geq 1\}$ is a consistent sequence of d.f.s and hence determines a probability measure $\hat{\mathrm{P}}$ on $(R^\infty, \mathscr{B}^\infty)$ which clearly satisfies (12) and (13) if $X''_n = Y_{2n-1}$, $X'_n = Y_{2n}$, $n \geq 1$. □

The r.v.s X^*_n, $n \geq 1$, defined by

$$X^*_n = X''_n - X'_n = Y_{2n-1} - Y_{2n}, \qquad n \geq 1,$$

are called the **symmetrized** X_n. Given any sequence of r.v.s $\{X_n, n \geq 1\}$, in subsequent allusion to the symmetrized X_n, namely X_n^*, the distinction between X_n and X_n'' (which are probabilistically indistinguishable) will be glossed over and X_n^* will be written

$$X_n^* = X_n - X_n', \qquad n \geq 1,$$

where $\{X_n', n \geq 1\}$ is independent of $\{X_n, n \geq 1\}$ and possesses the same joint distributions.

Since the $\{X_n^*, n \geq 1\}$ are symmetric r.v.s, it is frequently easier (especially for independent $\{X_n, n \geq 1\}$) to prove a result for $\{X_n^*\}$ rather than attempt a direct argument with $\{X_n\}$ (see Chapter 10).

The following examples complement the three series theorem, the second furnishing a typical exploitation of symmetrization. For any r.v.s $\{X_n, n \geq 1\}$ and positive constant c, define

$$s_n^2(c) = \sum_{j=1}^{n} \sigma^2(X_j I_{[|X_j| \leq c]}) = \sum_{j=1}^{n} (\mathrm{E}\, X_j^2 I_{[|X_j| \leq c]} - \mathrm{E}^2\, X_j I_{[|X_j| \leq c]}). \tag{14}$$

EXAMPLE 1. Let $S_n = \sum_1^n X_i, n \geq 1$, where $\{X_n, n \geq 1\}$ are independent, symmetric r.v.s and let $s_n(c)$ be as in (14).

If for some $c > 0$, $s_n^2(c) \to \infty$, then

$$\varlimsup_{n\to\infty} \frac{S_n}{s_n(c)} \overset{\text{a.c.}}{=\!=} \infty \overset{\text{a.c.}}{=\!=} -\varliminf_{n\to\infty} \frac{S_n}{s_n(c)}. \tag{15}$$

If $\sum_{n=1}^{\infty} \mathrm{P}\{|X_n| > c\} = \infty$, all $c > 0$, then

$$\varlimsup_{n\to\infty} S_n \overset{\text{a.c.}}{=\!=} \infty \overset{\text{a.c.}}{=\!=} -\varliminf_{n\to\infty} S_n. \tag{16}$$

If $s_n^2(c) = O(1)$, all $c > 0$ and $\sum_{n=1}^{\infty} \mathrm{P}\{|X_n| > c\} < \infty$, some $c > 0$ then

$$S_n \text{ converges a.c.} \tag{17}$$

PROOF. Define

$$X_n^c = X_n I_{[|X_n| \leq c]}. \tag{18}$$

Then symmetry guarantees that $\mathrm{E}\, X_n^c = 0$, $n \geq 1$, and the hypothesis of (15) ensures that the uniformly bounded r.v.s $\{X_n^c, n \geq 1\}$ obey a central limit theorem (Theorem 9.1.1 or Exercise 9.1.1). Thus if Φ denotes the normal d.f., for all $b > 0$

$$\lim_{n\to\infty} \mathrm{P}\left\{\bigcup_{j=n}^{\infty}\left[\sum_{i=1}^{j} X_i^c > b s_j(c)\right]\right\} \geq \lim_{n\to\infty} \mathrm{P}\left\{\sum_{i=1}^{n} X_i^c > b s_n(c)\right\} = 1 - \Phi(b) > 0,$$

implying by the Kolmogorov zero–one law that for all $b > 0$

$$\mathrm{P}\left\{\varlimsup_{n\to\infty} \frac{1}{s_n(c)} \sum_{i=1}^{n} X_i^c > \frac{b}{2}\right\} \geq \mathrm{P}\left\{\varlimsup_{n\to\infty} \frac{1}{s_n(c)} \sum_{i=1}^{n} X_i^c \geq b\right\} = 1.$$

Since $s_n(c) \to \infty$, Lemma 4.2.6 ensures

$$P\left\{\overline{\lim_{n\to\infty}} \frac{S_n}{s_n(c)} \geq \frac{b}{2}\right\} = 1, \tag{19}$$

and so, in view of the arbitrariness of b, (19) obtains with $b = \infty$. The analogous statement for the lower limit follows by symmetry.

Apropos of (16), via $|X_n| \leq |S_n| + |S_{n-1}|$ and the Borel–Cantelli theorem

$$2\,\overline{\lim_{n\to\infty}}\,|S_n| \geq \overline{\lim_{n\to\infty}}\,|X_n| = \infty, \text{ a.c.,}$$

and by symmetry $\overline{\lim}_{n\to\infty} S_n = \infty = -\underline{\lim}\, S_n$, a.c. The final statement, (17), follows immediately from the three series theorem (Theorem 5.1.2). □

Corollary 2. *For independent, symmetric* r.v.s $\{X_n\}$ *either* $\sum_1^n X_i$ *converges* a.c. *or* $\overline{\lim}_{n\to\infty} \sum_1^n X_i = \infty = -\underline{\lim}_{n\to\infty} \sum_1^n X_i$, a.c.

EXAMPLE 2. Let $\{X_n, n \geq 1\}$ be independent r.v.s and $S_n = \sum_1^n X_i$, $n \geq 1$. If $\sum_{n=1}^{\infty} P\{|X_n| > c\} = \infty$, all $c > 0$, then

$$\overline{\lim_{n\to\infty}}\,|S_n| = \infty, \quad \text{a.c.} \tag{20}$$

If for some $c > 0$, $\sum_{n=1}^{\infty} P\{|X_n| > c\} < \infty$, and $s_n^2(c) \to \infty$, then

$$\overline{\lim_{n\to\infty}} \frac{|S_n|}{s_n(c)} \overset{\text{a.c.}}{=} \infty. \tag{21}$$

Suppose that for some $c > 0$,

$$\sum_{n=1}^{\infty} P\{|X_n| > c\} < \infty \quad \text{and} \quad s_n^2(c) = O(1). \tag{22}$$

If $\overline{\lim}_{n\to\infty} \sum_{j=1}^n E\,X_j^c = \infty$ (resp. $\underline{\lim}_{n\to\infty} \sum_{j=1}^n E\,X_j^c = -\infty$), then $\overline{\lim}_{n\to\infty} S_n = \infty$ (resp. $\underline{\lim}_{n\to\infty} S_n = -\infty$), a.c. If $-\infty < \underline{\lim}_{n\to\infty} \sum_1^n E\,X_j^c \leq \overline{\lim}_{n\to\infty} \sum_1^n E\,X_j^c < \infty$, then $\overline{\lim}_{n\to\infty} |S_n| < \infty$, a.c.

PROOF. Since the proof of (16) did not utilize symmetry, (20) is immediate. To establish (21), let $\{X_n'', n \geq 1\}$ and $\{X_n', n \geq 1\}$ be i.i.d. stochastic processes on on some probability space whose joint distributions (of $X_1', \ldots, X_n'$ or $X_1'', \ldots, X_n''$) are identical with those of $X_1, \ldots, X_n$. The double prime will be deleted in what follows. Since, in the notation of (18),

$$2\sigma_{X_n^c}^2 = \sigma_{X_n^c - X_n'^c}^2 = E[X_n^c - X_n'^c]^2 = \int_{[|X_n| \leq c,\, |X_n'| \leq c]} (X_n - X_n')^2$$

$$+ 2\int_{[|X_n| \leq c,\, |X_n'| > c]} X_n^2$$

$$= \int_{[|X_n - X_n'| \le 2c]} (X_n - X_n')^2 - \int_{[|X_n - X_n'| \le 2c, |X_n'| > c]} (X_n - X_n')^2$$

$$+ 2 \int_{[|X_n| \le c, |X_n'| > c]} X_n^2 - \int_{[|X_n - X_n'| \le 2c, |X_n'| \le c, |X_n| > c]} (X_n - X_n')^2$$

necessarily

$$2\sigma^2_{X_n^c} \le \sigma^2_{(X_n - X_n')^{2c}} + 2c^2 \, \mathrm{P}\{|X_n| > c\},$$

$$\sigma^2_{(X_n - X_n')^{2c}} \le 2\sigma^2_{X_n^c} + 8c^2 \, \mathrm{P}\{|X_n| > c\},$$

and so $\sum_{j=1}^n \sigma^2_{(X_j - X_j')^{2c}} \sim 2s_n^2(c)$. It follows from (15) that $\overline{\lim}_{n\to\infty} |S_n - S_n'|/s_n(c) = \infty$, a.c., and this, in turn, ensures the conclusion of (21).

Finally, the hypothesis of (22) guarantees that $\sum_{j=1}^n (X_j - \mathrm{E}\, X_j^c)$ converges a.c. The final conclusion of (22) follows from the identity

$$\sum_{j=1}^n X_j = \sum_{j=1}^n (X_j - \mathrm{E}\, X_j^c) + \sum_{j=1}^n \mathrm{E}\, X_j^c,$$

while the prior ones follow upon replacing n in this identity by suitable subsequences $\{n_i\}$. □

Since Levy's inequality acquired a very simple form for symmetric r.v.s. (Corollary 3.35), symmetrization is especially useful in proving the following converse to Theorem 5.2.8.

Theorem 5. *Let* $\alpha p \ge 1$ *and* $S_n = \sum_{i=1}^n X_i$, $n \ge 1$, *where* $\{X, X_n, n \ge 1\}$ *are non-degenerate* i.i.d. *random variables. If, for some* $\varepsilon > 0$,

$$\sum_{n=1}^\infty n^{\alpha p - 2} \mathrm{P}\{|S_n| \ge \varepsilon n^\alpha\} < \infty, \tag{23}$$

then $\alpha > 0$ *and* $\mathrm{E}|X|^p < \infty$. *Moreover,* $\mathrm{E}\, X = 0$ *if either* $\alpha < 1$ *or* $\alpha = 1$ *and* (23) *holds for all* $\varepsilon > 0$.

Proof. By Theorem 4, there exists a sequence $\{X', X_n', n \ge 1\}$ such that $\{X, X_n, n \ge 1\}$ and $\{X', X_n', n \ge 1\}$ are i.i.d. Set $Y = X - X'$, $Y_n = X_n - X_n'$, and $T_n = \sum_{j=1}^n Y_j$, $n \ge 1$. Then, for some $\varepsilon > 0$,

$$\sum_{n=1}^\infty n^{\alpha p - 2} \mathrm{P}\{|T_n| \ge 2\varepsilon n^\alpha\} < \infty, \tag{24}$$

so that, as $m \to \infty$,

$$\sum_{n=m}^{2m} n^{\alpha p - 2} \mathrm{P}\{|T_n| \ge 2\varepsilon(1 + 2^\alpha) m^\alpha\} \le \sum_{n=m}^{2m} n^{\alpha p - 2} \mathrm{P}\{|T_n| \ge 2\varepsilon n^\alpha\} = o(1). \tag{25}$$

Now, if $T_n^* = \max_{1 \le j \le n} |T_j|$ and $Y_n^* = \max_{1 \le j \le n} |Y_j|$, for any $\delta > 0$, via Lévy's inequality,

$$2\mathrm{P}\{|T_n| \ge \delta\} \ge \mathrm{P}\{T_n^* \ge \delta\} \ge \mathrm{P}\{Y_n^* \ge 2\delta\} = 1 - \mathrm{P}^n\{|Y| < 2\delta\}$$

$$= \mathrm{P}\{|Y| \geq 2\delta\} \sum_{j=0}^{n-1} \mathrm{P}^j\{|Y| < 2\delta\}$$

$$\geq \mathrm{P}\{|Y| \geq 2\delta\} \sum_{j=0}^{n-1} [1 - 2\mathrm{P}\{|T_j| \geq \delta\}]. \tag{26}$$

Consequently, choosing $\delta = \eta n^\alpha$, where $\eta = 2\varepsilon$, (24) implies that

$$\sum_{n=1}^{\infty} n^{\alpha p-2}\mathrm{P}\{|Y| \geq 2\eta n^\alpha\} \sum_{j=0}^{n-1} [1 - 2\mathrm{P}\{|T_j| \geq \eta n^\alpha\}] < \infty, \tag{27}$$

while (25) and the initial inequality of (26) ensure that, as $m \to \infty$,

$$m^{\alpha p-1}\mathrm{P}\{T_m^* \geq \eta(1 + 2^\alpha)m^\alpha\} \leq \sum_{n=m}^{2m} n^{\alpha p-2}\mathrm{P}\{T_n^* \geq \eta(1 + 2^\alpha)m^\alpha\} = o(1).$$

Since $\alpha p \geq 1$, as $m \to \infty$,

$$\mathrm{P}\{T_m^* \geq \eta' m^\alpha\} = o(1), \tag{28}$$

where $\eta' = \eta(1 + 2^\alpha) = 2\varepsilon(1 + 2^\alpha)$. In view of $\mathrm{P}\{T_m^* > 0\} \geq \mathrm{P}\{T_1^* > 0\} = \mathrm{P}\{|Y| \neq 0\} > 0$, necessarily $\alpha \geq 0$, so that $\alpha p \geq 1$ guarantees $\alpha > 0$.

Clearly, (23) holds when ε is replaced by $\varepsilon(1 + 2^\alpha)$, whence (27) likewise holds with η replaced by η'. Thus, since $\mathrm{P}\{|T_j| \geq \eta' n^\alpha\} \leq \mathrm{P}\{T_n^* \geq \eta' n^\alpha\}$ for $1 \leq j \leq n$, (28) ensures that the second sum in (27) with η replaced by η' is $n + o(n)$, implying that

$$\sum_{n=1}^{\infty} n^{\alpha p-1}\mathrm{P}\{|Y| \geq 2\eta' n^\alpha\} < \infty,$$

and hence also that

$$\infty > \int_1^\infty t^{\alpha p-1}\mathrm{P}\{|Y| \geq 2\eta' t^\alpha\}\,dt = C \int_{2\eta'}^\infty u^{p-1}\mathrm{P}\{|Y| \geq u\}\,du$$

for some positive constant C, yielding $\mathrm{E}|Y|^p < \infty$. Hence, via independence of X and X',

$$\infty > \mathrm{E}|X - X'|^p \geq \mathrm{E}[(X - K)^+]^p\mathrm{P}\{X' < K\},$$

and choosing K large enough so that $\mathrm{P}\{X' < K\} > 0$, necessarily $\mathrm{E}(X^+)^p < \infty$. Similarly, $\mathrm{E}(X^-)^p < \infty$ and so $\mathrm{E}|X|^p < \infty$.

When $\alpha \leq 1$, clearly $p \geq 1$, so that $\mathrm{E}|X| < \infty$. By Theorem 5.2.8, for all $\varepsilon' > 0$,

$$\sum_{n=1}^{\infty} n^{\alpha p-2}\mathrm{P}\{|S_n - n\,\mathrm{E}\,X| \geq \varepsilon' n^\alpha\} < \infty,$$

which, in conjunction with (23), yields

$$\sum_{n=1}^{\infty} n^{\alpha p-2}\mathrm{P}\{|\mathrm{E}\,X| \geq 2\varepsilon n^{\alpha-1}\} < \infty.$$

Thus, $\mathrm{E}\,X = 0$ if either $\alpha < 1$ or $\alpha = 1$ and (23) holds for all $\varepsilon > 0$. □

Actually, $\alpha > 1/2$ in Theorem 5, since the alternative, $\alpha \le 1/2$, requires $p \ge 2$, whence the Central Limit Theorem for i.i.d. random variables (Corollary 9.1.2) implies that

$$P\{|T_n| \ge 2\varepsilon n^{\alpha}\} \ge P\{|T_n| \ge 2\varepsilon n^{1/2}\} \to 2[1 - \Phi(2\varepsilon)] > 0,$$

which is incompatible with (24).

EXAMPLE 3. Let $\Omega = \{1, 2, \ldots\}$, $\mathscr{F}_n = \sigma(\{1\}, \{2\}, \ldots, \{n\})\uparrow$, $P_n\{A\} = 0$ or 1 according as A is a finite set or not. Since no two infinite sets of $\mathscr{F}_n$ are disjoint, P_n is a probability measure on $\mathscr{F}_n$ and moreover, $P_{n+1}|\mathscr{F}_n = P_n$. However, $P\{A\} = \lim P_n\{A\}$ is not σ-additive since

$$P\{\{m\}\} = \lim_{n\to\infty} P_n\{\{m\}\} = 0, \quad m = 1, 2, \ldots.$$

whereas

$$P\{\Omega\} = \lim_{n\to\infty} P_n\{\Omega\} = 1 \ne \sum_{n=1}^{\infty} P\{\{n\}\}$$

EXERCISES 6.4

1. Verify that it is possible to define a sequence of independent r.v.s $\{X_n\}$ with specified d.f.s F_n on a probability space.

2. Let $(\Omega, \mathscr{B}, \nu)$ be the probability space consisting of Lebesgue measure on the Borel subsets of $[0, 1]$. Each ω in $[0, 1]$ may be written in binary expansion as $\omega = X_1X_2\cdots = \sum_1^\infty 2^{-n}X_n$, where $X_n = X_n(\omega) = 0$ or 1 and this expansion is unique except for a set of ω of the form $m/2^n$ which is countable and hence has probability (Lebesgue measure) zero. For definiteness regarding such ω, only the expansion consisting of infinitely many ones will be used. Show that for all $n \ge 1$, $\{\omega: X_n(\omega) = 1\} \in \mathscr{B}$ and that $\{X_n, n \ge 1\}$ is a sequence of independent r.v.s. Describe the r.v.s Y_n where $\omega = \sum_{n=1}^\infty r^{-n}Y_n$ and r is an integer >2.

3. In Exercise 2, define $Z_n = Z_n(\omega) = 1$ or -1 according as the integer i for which $(i-1)/2^n \le \omega < i/2^n$ is odd or even. Show that any two but not all three of Z_1, Z_2, $Z_1 \cdot Z_2$ are independent. What is the relation between Z_n and X_n? The Z_n, $n \ge 1$, are known as the **Rademacher** functions.

4. Let $\Omega_0 = \{0, 1\}$, $\mathscr{F}_0 =$ class of subsets of Ω_0, $\mu_0(\{1\}) = \frac{1}{2} = \mu_0(\{0\})$, and define $\mu_i \equiv \mu_0$, $i \ge 1$, and $(\Omega, \mathscr{F}, \mu) = (\Omega_0^\infty, \mathscr{F}_0^\infty, \times_1^\infty \mu_i)$. Prove the following. (i) Each point of Ω is an $\mathscr{F}$-set of μ-measure zero. (ii) The set D of all points of Ω with only finitely many coordinates equal to 1 has μ-measure 0. (iii) Define $\Omega' = \Omega - D$, $\mathscr{F}' = \mathscr{F}\Omega'$, $\mu'\{A \cdot \Omega'\} = \mu\{A\}$, $A \in \mathscr{F}$. (iv) For each $\omega \in \Omega'$, $z(\omega) = \sum_{i=1}^\infty \omega_i 2^{-i}$ is a 1–1 map of Ω' onto $Z = (0, 1)$. (v) If $C = \{z: 0 \le a \le z < b \le 1\}$, $A = \{\omega: z(\omega) \in C\}$, then A is measurable and $\mu'\{A\} = b - a$. *Hint*: It suffices to take binary rational a, b. (vi) For any Borel subset B of Z, the set $A = \{\omega: z(\omega) \in B\}$ is measurable and $\mu'\{A\} =$ Lebesgue measure of B.

5. Let $T = [a, b]$, $-\infty \le a < b \le \infty$, be an index set, $\Omega = R^T = \{\omega: \omega = \omega(t), t \in T\} =$ space of real functions on T. If B_n is a Borel set of R^n, then $A = \{\omega(t): (\omega(t_1), \ldots, \omega(t_n)) \in B_n\}$ is a Borel cylinder set for any choice of distinct $t_1, \ldots, t_n$ in T. The class

$\mathscr{A}^T$ of Borel cylinder sets is an algebra. Define $\mathscr{B}^T = \sigma(\mathscr{A}^T)$ and let $T = [a, b]$, $-\infty \le a < b \le \infty$. Do the sets $\{\omega(t): \omega(t)$ is bounded on $T\}$, $\{\omega(t): \omega(t)$ is continuous on $T\}$ belong to $\mathscr{B}^T$? If $A^* = \{\omega(t): \omega(t_i) \in B_i, i = 1, 2, \ldots\}$, and $\mathscr{A}^*$ is the class of all such set A^* (as t_i and B_i vary), set $\mathscr{B}^* = \sigma(\mathscr{A}^*)$. Is $\mathscr{B}^T = \mathscr{B}^*$?

6. If $S_n = \sum_1^n X_i$ where $\{X, X_n, n \ge 1\}$ are i.i.d. then $\sum_1^\infty n^{-1} \mathrm{P}\{|S_n| > \varepsilon n\} < \infty$ all $\varepsilon > 0$ iff $\mathrm{E}\, X = 0$. *Hint*: Sufficiency is contained in Theorem 5.2.7. For necessity, define $S_n^* = \sum_1^n X_j^*$ where $\{X_n^*, n \ge 1\}$ is the symmetrized sequence. The hypothesis ensures convergence of $\sum_1^\infty n^{-1} \mathrm{P}\{|S_n^*| > n\varepsilon\}$ and hence also $\sum_1^\infty n^{-1} \mathrm{P}\{\max_{1 \le j \le n} |X_j^*| > n\varepsilon\}$, $\varepsilon > 0$.

6.5 Absolute Continuity of Measures, Distribution Functions; Radon–Nikodym Theorem

Let $(\Omega, \mathscr{F}, \mu)$ be a measure space and T an arbitrary nonempty set. The **essential supremum** g of a family $\{g_t, t \in T\}$ of measurable functions from Ω into $R = [-\infty, \infty]$, denoted by $\mathrm{esup}_{t \in T}\, g_t$, is defined by the properties:

i. g is measurable,
ii. $g \ge g_t$, a.e., for each $t \in T$,
iii. For any h satisfying (i) and (ii), $h \ge g$, a.e.

Clearly, if such a g exists, it is unique in the sense that two such essential suprema of the same family are equal a.e.

Lemma 1. *Let $(\Omega, \mathscr{F}, \mu)$ be a σ-finite measure space and $\{g_t, t \in T\}$ a nonempty family of real, measurable functions. Then there exists a countable subset $T_0 \subset T$ such that*

$$\sup_{t \in T_0} g_t = \operatorname*{esup}_{t \in T} g_t.$$

PROOF. Since μ is σ-finite, it suffices to prove the theorem when μ is finite; moreover, by considering $\tan^{-1} g_t$ if necessary, it may be supposed that $|g_t| \le C < \infty$ for all $t \in T$. Let $\mathscr{I}$ signify the class of all countable subsets $I \subset T$ and set

$$\alpha = \sup_{I \in \mathscr{I}} \mathrm{E}\left(\sup_{t \in I} g_t\right),$$

whence α is finite. Choose $I_n \in \mathscr{I}$, $n \ge 1$ for which $\alpha = \sup_{n \ge 1} \mathrm{E}(\sup_{t \in I_n} g_t)$ and let $T_0 = \bigcup_1^\infty I_n$. Then T_0 is a countable subset of T and clearly $\alpha = \mathrm{E}[\sup_{t \in T_0} g_t]$. The measurable function $g = \sup_{t \in T_0} g_t$ satisfies (ii) since otherwise for some $t \in T$ necessarily $\alpha < \mathrm{E} \max(g, g_t) \le \alpha$. Obviously, (iii) holds, and so $g = \mathrm{esup}_{t \in T}\, g_t$. □

Definition. If $(\Omega, \mathscr{F}, \mu_i)$, $i = 1, 2$, are two measure spaces and $\mu_1\{A\} = 0$ whenever $\mu_2\{A\} = 0$, then μ_1 is said to be **absolutely continuous with respect to μ_2** or simply **μ_2-continuous**. If, rather $\mu_1\{N^c\} = 0$ for some set $N \in \mathscr{F}$ with $\mu_2\{N\} = 0$, then μ_1 is called **μ_2-singular** (or the pair μ_1, μ_2 is dubbed singular).

If g is a nonnegative integrable function on $(\Omega, \mathscr{F}, \mu)$, the indefinite integral $v_g\{A\} = \int_A g\, d\mu$ is absolutely continuous relative to μ. This is tantamount to saying that the integral of a nonnegative function g over a set A of measure zero has the value zero (Exercise 6.2.2). The Radon–Nikodym theorem asserts under modest assumptions that the indefinite integral v_g is the prototype of a measure absolutely continuous with respect to μ. The crucial step in proving this is

Lemma 2. *Let $(\Omega, \mathscr{F}, \mu)$ be a σ-finite measure space and v a σ-finite measure on $\mathscr{F}$, and let $\mathscr{H}$ denote the family of all measurable functions $h \geq 0$ satisfying $\int_A h\, d\mu \leq v\{A\}$, $A \in \mathscr{F}$. Then*

$$v\{A\} = \psi\{A\} + \int_A g\, d\mu, \qquad A \in \mathscr{F} \tag{1}$$

where ψ is a μ-singular measure and

$$g = \operatorname*{esup}_{h \in \mathscr{H}} h. \tag{2}$$

PROOF. Since μ and v are σ-finite, it suffices by a standard argument to consider the case where both are finite and, furthermore, the trivial case $\mu \equiv 0$ may be eliminated. According to Lemma 1, there exists a sequence $h_n \in \mathscr{H}$, $n \geq 1$, for which $g = \operatorname{esup}_{h \in \mathscr{H}} h = \sup_{n \geq 1} h_n$. Now if $h_1, h_2 \in \mathscr{H}$, then $h = \max(h_1, h_2) \in \mathscr{H}$ since

$$\int_A h\, d\mu = \int_{A[h_1 \geq h_2]} h_1\, d\mu + \int_{A[h_1 < h_2]} h_2\, d\mu \leq v\{A\}, \qquad A \in \mathscr{F},$$

and so it may be supposed that $h_n \leq h_{n+1}$, $n \geq 1$. Then $g = \lim_n h_n$, whence by the monotone convergence theorem

$$\int_A g\, d\mu \leq v\{A\}, \qquad A \in \mathscr{F}.$$

Consequently, ψ as defined by (1) is a measure.

Next, for $n \geq 1$ and $A \in \mathscr{F}$ with $\mu\{A\} > 0$,

$$\mathscr{D}_n(A) = \left\{B \in \mathscr{F} : B \subset A, \psi\{B\} < \frac{1}{n}\mu\{B\}\right\}$$

is nonempty; otherwise, the choice $h_0 = (1/n)I_A$ would guarantee for all $B \in \mathscr{F}$ that

$$\int_B h_0\, d\mu = \frac{1}{n}\mu\{AB\} \leq \psi\{AB\} \leq \psi\{B\} = v\{B\} - \int_B g\, d\mu,$$

implying $h_0 + g \in \mathscr{X}$ and thus violating $g = \operatorname{esup}_{h \in \mathscr{H}} h$.

Choose $B_{1,n} \in \mathscr{D}_n(\Omega)$ with

$$\mu\{B_{1,n}\} \geq \tfrac{1}{2} \sup\{\mu\{B\}: B \in \mathscr{D}_n(\Omega)\}$$

$= \alpha_{1,n}$ (say). If $\mu\{B_{1,n}^c\} = 0$, stop; otherwise, choose $B_{2,n} \in \mathscr{D}_n(B_{1,n}^c)$ with

$$\mu\{B_{2,n}\} \geq \tfrac{1}{2} \sup\{\mu\{B\}: B \in \mathscr{D}_n(B_{1,n}^c)\}$$

$= \alpha_{2,n}$ (say). If $\mu\{B_{1,n}^c \cdot B_{2,n}^c\} = 0$, stop; otherwise, choose $B_{3,n} \in \mathscr{D}_n(B_{1,n}^c \cdot B_{2,n}^c)$ with

$$\mu\{B_{3,n}\} \geq \tfrac{1}{2} \sup\{\mu\{B\}: B \in \mathscr{D}_n(B_{1,n}^c \cdot B_{2,n}^c)\} = \alpha_{3,n}$$

and so on. If the process terminates at some finite step k_n, set $B_{j,n} = \varnothing$, $j > k_n$.

Since $\mathscr{D}_n(A_1) \subset \mathscr{D}_n(A_2)$ for $A_1 \subset A_2$, necessarily $B_{j,n} \in \mathscr{D}_n(\Omega)$ for $j \geq 1$ if $B_{j,n} \neq \varnothing$, and since $\mathscr{D}_n(A)$ is closed under countable disjoint unions, $M_n = \bigcup_{j=1}^\infty B_{j,n} \in \mathscr{D}_n(\Omega)$, $n \geq 1$. Now, if $\mu\{M_n^c\} > 0$, for some $n \geq 1$, there exists some $D \in \mathscr{D}_n(M_n^c)$, whence $\mu\{D\} > 0$. Moreover, $\alpha_{m,n} \to 0$ as $m \to \infty$ via disjointness of $\{B_{j,n}, j \geq 1\}$ and finiteness of μ. However, for all m

$$2\alpha_{m,n} = \sup\left\{\mu\{B\}: B \in \mathscr{D}_n\left(\bigcap_1^{m-1} B_{j,n}^c\right)\right\}$$

$$\geq \sup\{\mu\{B\}: B \in \mathscr{D}_n(M_n^c)\} \geq \mu\{D\} > 0,$$

a contradiction. Thus, $\mu\{M_n^c\} = 0$, $n \geq 1$, and $\psi\{M_n\} < (1/n)\mu\{M_n\} = (1/n)\mu\{\Omega\}$. Consequently,

$$\psi\left\{\bigcap_{n=1}^\infty M_n\right\} \leq \lim_{n\to\infty} \psi\{M_n\} = 0$$

$$\mu\left\{\left(\bigcap_{n=1}^\infty M_n\right)^c\right\} = \mu\left\{\bigcup_{n=1}^\infty M_n^c\right\} \leq \sum_{n=1}^\infty \mu\{M_n^c\} = 0. \qquad \square$$

Corollary 1 (Lebesgue Decomposition Theorem). *If μ, ν are σ-finite measures on a measurable space $(\Omega, \mathscr{F})$, there exist two uniquely determined measures λ_1, λ_2 such that $\nu = \lambda_1 + \lambda_2$, where λ_2 is μ-continuous and λ_1 is μ-singular.*

PROOF. It suffices to verify uniqueness. Let $\lambda_1 + \lambda_2 = \lambda_1' + \lambda_2'$, where λ_1, λ_1' are μ-singular and λ_2, λ_2' are μ-continuous. If $\lambda_1 \neq \lambda_1'$, there exists $A \in \mathscr{F}$ with $\mu\{A\} = 0$ and $\lambda_1\{A\} \neq \lambda_1'\{A\}$. But then $\lambda_2\{A\} \neq \lambda_2'\{A\}$, violating absolute continuity. □

The Lebesgue decomposition may be used to prove

Theorem 1 (Radon–Nikodym). *If ν_1, ν_2, μ are σ-finite measures on a measurable space $(\Omega, \mathscr{F})$ with ν_i being μ-continuous, $i = 1, 2$, and if $\nu = \nu_1 - \nu_2$ is well defined on $\mathscr{F}$ (i.e., $\nu_1\{\Omega\}$ and $\nu_2\{\Omega\}$ are not both ∞) then there exists an $\mathscr{F}$-*

measurable function g, finite a.e. $[\mu]$, *such that*

$$\nu\{A\} = \int_A g \, d\mu, \qquad A \in \mathscr{F}, \tag{3}$$

and g is unique to within sets of μ-measure zero.

PROOF. Let g_i and ψ_i be defined as in Lemma 2, $i = 1, 2$. Then both ν_i and $\int_A g_i \, d\mu$ are μ-continuous and hence also ψ_i, $i = 1, 2$. Since according to Lemma 2, ψ_i is μ-singular, $i = 1, 2$, necessarily (Exercise 1) $\psi_i \equiv 0$, whence

$$\nu_i\{A\} = \int_A g_i \, d\mu, \qquad A \in \mathscr{F}, \ i = 1, 2.$$

Moreover, $g = g_1 - g_2$ is $\mathscr{F}$-measurable and so

$$\nu\{A\} = \nu_1\{A\} - \nu_2\{A\} = \int_A g \, d\mu, \qquad A \in \mathscr{F},$$

which is (3). In proving uniqueness, it may be assumed that μ is finite. If g^* is any other $\mathscr{F}$-measurable function satisfying (3), then for any $C > 0$

$$A = \{C > g^* > g > -C\} \in \mathscr{F},$$

whence

$$\int_A g^* \, d\mu = \nu\{A\} = \int_A g \, d\mu,$$

necessitating $\mu\{A\} = 0$, all $C > 0$ and hence $\mu\{g^* > g\} = 0$. Analogously, $\mu\{g^* < g\} = 0$ and so $g^* = g$, a.e. Finally, when ν is finite, g is μ-integrable and hence finite a.e. $[\mu]$, whence the latter also obtains when ν is σ-finite. □

Corollary 2. $|\nu\{A\}| < \infty$ *for all* $A \in \mathscr{F}$ iff *g is μ-integrable and ν is a measure* iff $g \geq 0$ a.e. $[\mu]$.

A function g defined by (3) is called the **Radon–Nikodym derivative of ν with respect to μ** and is denoted, in suggestive fashion, by $d\nu/d\mu$. Thus, if ν is a (well-defined) difference of two μ-continuous, σ-finite measures, (3) may be restated as

$$\nu\{A\} = \int_A \frac{d\nu}{d\mu} \, d\mu, \qquad A \in \mathscr{F}. \tag{3'}$$

Theorem 2. *Let μ be a σ-finite measure and ν a μ-continuous, σ-finite measure on the measurable space $(\Omega, \mathscr{F})$. If X is an $\mathscr{F}$-measurable function whose integral $\int_\Omega X \, d\nu$ exists, then for every $A \in \mathscr{F}$,*

$$\int_A X \, d\nu = \int_A X \frac{d\nu}{d\mu} \, d\mu \tag{4}$$

PROOF. It may be supposed that μ is finite and (via $X = X^+ - X^-$) that $X \geq 0$. Let $\mathscr{H}$ be the class of nonnegative $\mathscr{F}$-measurable functions for which (4) obtains. Then $\mathscr{H}$ is a monotone system which contains the indicator functions of all sets in $\mathscr{F}$. By Theorem 1.4.3, $\mathscr{H}$ contains all nonnegative $\mathscr{F}$-measurable functions. □

Corollary 3. *If ν, μ, λ are σ-finite measures on a measurable space $(\Omega, \mathscr{F})$ with ν being μ-continuous and μ being λ-continuous, then ν is λ-continuous and*

$$\frac{d\nu}{d\lambda} = \frac{d\nu}{d\mu} \cdot \frac{d\mu}{d\lambda}, \qquad \text{a.e. } [\lambda].$$

PROOF. Clearly, ν is λ-continuous and $d\nu/d\mu$ is $\mathscr{F}$-measurable with $\int_\Omega (d\nu/d\mu)d\mu$ extant. Thus, by Theorem 2, for all $A \in \mathscr{F}$

$$\int_A \frac{d\nu}{d\mu}\frac{d\mu}{d\lambda}\, d\lambda = \int_A \frac{d\nu}{d\mu}\, d\mu = \nu\{A\} = \int_A \frac{d\nu}{d\lambda}\, d\lambda,$$

which is tantamount to the conclusion. □

If F is a d.f. on R for which there exists a Borel function f on $(-\infty, \infty)$ with

$$F(x) = \int_{(-\infty, x)} f(t)dt = \int_{-\infty}^{x} f(t)dt, \qquad -\infty < x < \infty \tag{5}$$

where dt signifies Lebesgue measure on $(R, \mathscr{B})$, then, as noted in Section 1.6, F is said to be absolutely continuous and f is called the density function of F. In particular, if F is the d.f. attached to some r.v. X on a probability space, then f is also called the density of X. Clearly, when the f of (5) exists, it is unique to within sets of Lebesgue measure zero.

Corollary 4. *If X is a* r.v. *on a probability space $(\Omega, \mathscr{F}, \mathrm{P})$ with* d.f. *F and density f, and g is a Borel function on $R = [-\infty, \infty]$ such that $\mathrm{E}\, g(X)$ exists, then for any linear Borel set B*

$$\int_{[X \in B]} g(X)\, d\mathrm{P} = \int_B g(t) f(t) dt \tag{6}$$

and, in particular,

$$\mathrm{P}\{X \in B\} = \int_B f(t)dt. \tag{7}$$

PROOF. Let μ denote the restriction to $\mathscr{B}$ of the Lebesgue–Stieltjes measure determined by F. From (5), for $-\infty < a < b < \infty$

$$\int_{[a,b)} f(t)dt = F(b) - F(a) = F(b-) - F(a-),$$

and so by the uniqueness of the restriction of Lebesgue–Stieltjes measure to $\mathscr{B}$

$$\mu\{B\} = \int_B f(t)dt, \qquad B \in \mathscr{B}.$$

Thus, μ is absolutely continuous with respect to Lebesgue measure, whence by Theorem 2

$$\int_B g(t)f(t)dt = \int_B g(t)d\mu(t) = \int_{-\infty}^{\infty} g(t)I_B(t)d\mu(t)$$

$$= \mathrm{E}\, I_B(X)g(X) = \int_{[X \in B]} g(X)\, d\mathrm{P},$$

recalling Corollary 6.2.1. □

Analogously, if (i) F is an n-dimensional d.f. or (ii) F is the joint d.f. of r.v.s $X_1, \ldots, X_n$ on some probability space and there exists a Borel function f on R^n with

$$F(x_1, \ldots, x_n) = \int_{-\infty}^{x_n} \cdots \int_{-\infty}^{x_1} f(t_1, \ldots, t_n)dt_1 \cdots dt_n, \tag{8}$$

where $dt_1 \cdots dt_n$ signifies Lebesgue measure on $(R^n, \mathscr{B}^n)$, then F is declared absolutely continuous and f is called the density function of F in case (i) and the joint density of the r.v.s $X_1, \ldots, X_n$ in case (ii). Again, f is unique to within sets of n-dimensional Lebesgue measure zero. Under (ii), if g is any Borel function on R^n for which $\mathrm{E}\, g(X_1, \ldots, X_n)$ exists, then for any $B \in \mathscr{B}^n$

$$\int_{[(X_1, \ldots, X_n) \in B]} g(X_1, \ldots, X_n)\, d\mathrm{P} = \int_B g(t_1, \ldots, t_n)f(t_1, \ldots, t_n)dt_1 \cdots dt_n \tag{9}$$

and, in particular,

$$\mathrm{P}\{(X_1, \ldots, X_n) \in B\} = \int_B f(t_1, \ldots, t_n)dt_1 \cdots dt_n. \tag{10}$$

Exercises 6.5

1. If ψ and μ are measures such that ψ is both μ-continuous and μ-singular, then $\psi \equiv 0$.
2. Two measures μ, ν are called **equivalent**, denoted $\mu \equiv \nu$, if each is absolutely continuous with respect to the other. Verify that this is an equivalence relation. If $(\Omega, \mathscr{F}, \mu)$ is a probability space, X_i is a nonnegative $\mathscr{L}_1$ random variable and μ_i is the indefinite integral of X_i, $i = 1, 2$, then, if $\mu\{[X_1 = 0] \,\Delta\, [X_2 = 0]\} = 0$, the two indefinite integrals are equivalent measures.
3. If $(\Omega, \mathscr{F}, \mu_i)$ is a measure space, $i = 1, 2$, then μ_1 is absolutely continuous relative to $\mu_1 + \mu_2$.
4. If $F(x; a, b) = (b - a)^{-1}(x - a)$, $a \le x \le b$, the corresponding measure $F\{\cdot\}$ is absolutely continuous relative to Lebesgue measure with Radon–Nikodym derivative $f(x) = (b - a)^{-1}I_{[a \le x \le b]}$ and F is called the **uniform** distribution on

$[a, b]$. When $a = 0, b = 1$, $F\{\cdot\}$ coincides with Lebesgue measure on $[0, 1]$ and a r.v. with d.f. F is said to be **uniformly distributed**. Show that if $X_1, \ldots, X_n$ are i.i.d. uniformly distributed r.v.s, the measure determined by the d.f. of $X = (X_1, \ldots, X_n)$ is the n-dimensional Lebesgue measure on the hypercube $0 \le x_i \le 1, 1 \le i \le n$, of R^n.

5. A completely additive set function ν on a measurable space $(\Omega, \mathscr{F})$ which assumes at most one of the values $+\infty$ and $-\infty$ and satisfies $\nu\{\varnothing\} = 0$ is sometimes called a **signed measure.** If ν is a signed measure on a measurable space $(\Omega, \mathscr{F})$ and $\nu^+\{A\} = \sup\{\nu\{B\}: A \supset B \in \mathscr{F}\}$, then ν^+ is a measure satisfying $\nu^+\{A\} \ge \nu\{A\}$, $A \in \mathscr{F}$. Likewise, $\nu^-\{A\} = -\inf\{\nu\{B\}: A \supset B \in \mathscr{F}\}$ is a measure with $\nu^-\{A\} \ge -\nu\{A\}$; the measures ν^+ and ν^- are called the upper and lower variations respectively and the representation $\nu = \nu^+ - \nu^-$ is the **Jordan decomposition** of ν. If ν is σ-finite, so are ν^+ and ν^-.

6. If $\nu = \nu^+ - \nu^-$ is the Jordan decomposition (Exercise 5) of the signed measure ν, then $\bar{\nu} = \nu^+ + \nu^-$ is a measure called the **total variation** of ν. Clearly, $|\nu\{A\}| \le \bar{\nu}\{A\}$, $A \in \mathscr{F}$.

7. If ν is a signed measure on a measurable space $(\Omega, \mathscr{F})$ with total variation $\bar{\nu}$ (see Exercise 6) and X is integrable relative to $\bar{\nu}$, one may define $\int X\,d\nu = \int X\,d\nu^+ - \int X\,d\nu^-$. Prove that if ν is finite, $\bar{\nu}\{A\} = \sup\{|\int_A X\,d\nu|: X$ is measurable and $|X| \le 1\}$.

8. Let V denote a linear functional on $\mathscr{L}_p(\Omega, \mathscr{F}, \mu)$, i.e., range $V = (-\infty, \infty)$ and $V(af + bg) = aV(f) + bV(g)$ for all $f, g \in \mathscr{L}_p$. It is continuous if $V(f_n) \to V(f)$ whenever $\|f_n - f\|_p \to 0$, and V is bounded if $|V(f)| \le C_p\|f\|_p$ for all $f \in \mathscr{L}_p$, where C_p is a finite constant. Prove that a continuous linear functional is bounded. *Hint*: Otherwise, there exist $f_n \in \mathscr{L}_p$ with $|V(f_n)| > n\|f_n\|_p$, whence if $g_n = f_n/(n\|f_n\|_p)$, $\|g_n\|_p = o(1)$ but $|V(g_n)| > 1$.

9. If $(\Omega, \mathscr{F}, \mu)$ is a σ-finite measure space and V is a continuous linear functional on $\mathscr{L}_p(\Omega, \mathscr{F}, \mu)$, $p > 1$, there exists $g \in \mathscr{L}_q$, where $(1/p) + (1/q) = 1$, such that $V(f) = \int f \cdot g\,d\mu$ for all $f \in \mathscr{L}_p$. This is known as the (Riesz) representation theorem. *Hint*: Let μ be finite. Since $I_A \in \mathscr{L}_p(\Omega, \mathscr{F}, \mu)$ for all $A \in \mathscr{F}$, a set function on $\mathscr{F}$ is defined by $\nu\{A\} = V(I_A)$. It is finitely additive and, moreover, σ-additive by continuity of V and the fact that $V(0) = 0$. Further, ν is finite since V is bounded (Exercise 8); ν is absolutely continuous with respect to μ. By the Radon–Nikodym theorem, there is an $\mathscr{F}$-measurable g, finite a.e., with $V(I_A) = \nu\{A\} = \int_A g\,d\mu$.

10. Set functions $\{\nu_n, n \ge 1\}$ on $(\Omega, \mathscr{F}, \mu)$ are **uniformly absolutely** continuous relative to μ if for all $\varepsilon > 0$, $\mu\{A\} < \delta_\varepsilon$ implies $|\nu_n\{A\}| < \varepsilon$ for all $n \ge 1$. The sequence $\{\nu_n, n \ge 1\}$ is **equicontinuous from above at** $\varnothing$ if for all $\varepsilon > 0$ and $A_m \downarrow \varnothing$, $|\nu_n\{A_m\}| < \varepsilon$ for all $n \ge 1$ wherever $m \ge m_\varepsilon$. Prove that if measures $\{\nu_n, n \ge 1\}$ are equicontinuous from above at $\varnothing$ and also absolutely continuous relative to μ, then $\{\nu_n, n \ge 1\}$ are uniformly absolutely continuous relative to μ.

11. If $f_n \in \mathscr{L}_p(\Omega, \mathscr{F}, \mu)$, $n \ge 1$, then $\|f_n - f_m\|_p = o(1)$ as $n, m \to \infty$ iff (i) $f_n - f_m \xrightarrow{\mu} 0$ as, $n, m \to \infty$ and (ii) $\int_A |f_n|^p\,d\mu$, $n \ge 1$, are equicontinuous from above at $\varnothing$.

12. Random variables $X_1, \ldots, X_n$ have a (nonsingular) joint **normal** distribution if their d.f. is absolutely continuous with density defined by $f(x_1, \ldots, x_n) = (2\pi)^{-n/2}$

$\sqrt{|A|}\exp\{-\frac{1}{2}\sum_{i,j=1}^{n} a_{ij}(x_i - \theta_i)(x_j - \theta_j)\}$, where $A = \{a_{ij}\}$ is a positive definite matrix of order n and $|A|$ signifies the determinant of A. Here, $\theta = (\theta_1, \ldots, \theta_n)$ is a real vector. Verify that this yields a bona fide probability measure and that $\mathrm{E}\, X_i = \theta_i$, $\rho(X_i, X_j) = \sigma_{ij}/\sigma_i\sigma_j$, where $\{\sigma_{ij}\}$ is the inverse matrix of A.

References

J. L. Doob, *Stochastic Processes*, Wiley, New York, 1953.

P. R. Halmos, *Measure Theory*, Van Nostrand, Princeton, 1950; Springer-Verlag, Berlin and New York, 1974.

G. H. Hardy, J. E. Littlewood, and G. Polya, *Inequalities*, Cambridge Univ. Press, London, 1934.

A. N. Kolmogorov, Foundations of Probability (Nathan Morrison, translator), Chelsea, New York, 1950.

M. Loève, *Probability Theory*, 3rd ed., Van Nostrand, Princeton, 1963; 4th ed., Springer-Verlag, Berlin and New York, 1977–1978.

E. J. McShane, *Integration*, Princeton Univ. Press, Princeton, 1944.

M. E. Monroe, *Introduction to Measure and Integration*, Addison–Wesley, Cambridge, Mass., 1953.

H. Robbins, "Mixture of Distributions," *Ann. Math. Statist.* **19** (1948), 360–369.

S. Saks, *Theory of the Integral* (L. C. Young, translator), Stechert–Hafner, New York, 1937.

J. L. Snell, "Applications of martingale system theorems," *Trans. Amer. Math. Soc.* **73** (1952), 293–312.

D. V. Widder, *Advanced Calculus*, 2nd ed., Prentice–Hall, Englewood Cliffs, New Jersey, 1961.

7 Conditional Expectation, Conditional Independence, Introduction to Martingales

7.1 Conditional Expectations

From a theoretical vantage point, conditioning is a useful means of exploiting auxiliary information. From a practical vantage point, conditional probabilities reflect the change in unconditional probabilities due to additional knowledge.

The latter is represented by a sub-σ-algebra $\mathscr{G}$ of the basic σ-algebra $\mathscr{F}$ of events occurring in the underlying probability space $(\Omega, \mathscr{F}, \mathrm{P})$. Associated with any measurable function X on Ω whose integral is defined, i.e., $|\mathrm{E}\, X| \leq \infty$, is a function Y on Ω with $|\mathrm{E}\, Y| \leq \infty$ satisfying

i. Y is $\mathscr{G}$-measurable,
ii. $\int_A Y\, d\mathrm{P} = \int_A X\, d\mathrm{P}$, all $A \in \mathscr{G}$.

Such a function Y is called the conditional expectation of X given $\mathscr{G}$ and is denoted by $\mathrm{E}\{X \mid \mathscr{G}\}$. In view of (i) and (ii), any $\mathscr{G}$-measurable function Z which differs from $Y = \mathrm{E}\{X \mid \mathscr{G}\}$ on a set of probability zero also qualifies as $\mathrm{E}\{X \mid \mathscr{G}\}$. In other words, the conditional expectation $\mathrm{E}\{X \mid \mathscr{G}\}$ is only defined to within an equivalence, i.e., is any representative of a class of functions whose elements differ from one another only on sets of probability measure zero—an unpleasant feature.

To establish the existence of $\mathrm{E}\{X \mid \mathscr{G}\}$ for any $\mathscr{F}$-measurable function X with $|\mathrm{E}\, X| \leq \infty$, define the set functions $\lambda, \lambda^+, \lambda^-$ on $\mathscr{F}$ by

$$\lambda = \lambda^+ - \lambda^-, \qquad \lambda^{\pm}\{A\} = \int_A X^{\pm}\, d\mathrm{P}, \qquad A \in \mathscr{F}. \tag{1}$$

The measures $\lambda^{\pm}$ are P-continuous on $\mathscr{F}$ and so, if their restrictions $\lambda_{\mathscr{G}}^{\pm} \equiv \lambda^{\pm}|_{\mathscr{G}}$ are σ-finite, the Radon–Nikodym theorem (Theorem 6.5.1) ensures the existence of $Y = d\lambda_{\mathscr{G}}/d\mathrm{P}_{\mathscr{G}}$ satisfying

$$\int_A Y\, d\mathrm{P} = \lambda\{A\} = \int_A X\, d\mathrm{P}, \qquad A \in \mathscr{G}.$$

Thus, it suffices when $\lambda_{\mathscr{G}}^{\pm}$ are σ-finite to equate

$$\mathrm{E}\{X \mid \mathscr{G}\} = \frac{d\lambda_{\mathscr{G}}}{d\mathrm{P}_{\mathscr{G}}} \tag{2}$$

and to recall that the Radon–Nikodym derivative of (2) is unique to within sets of measure zero.

The second of the following lemmas shows that a similar procedure may be employed even when σ-finiteness is lacking.

Lemma 1. *If ν is a* P*-continuous measure on $\mathscr{F}$, there exists a set $\mathrm{E} \in \mathscr{F}$ such that ν is σ-finite on $\mathscr{F} \cap \mathrm{E}$ and for each $A \in \mathscr{F} \cap \mathrm{E}^c$*

$$\nu(A) = 0 = \mathrm{P}\{A\} \quad \text{or} \quad \nu\{A\} = \infty > \mathrm{P}\{A\} > 0. \tag{3}$$

PROOF. Set

$$\mathscr{D} = \{D: D \in \mathscr{F}, \nu \text{ is } \sigma\text{-finite on } \mathscr{F} \cap D\}$$

and then choose $D_n \in \mathscr{D}$, $n \geq 1$, with $\sup_{n \geq 1} \mathrm{P}\{D_n\} = \sup_{D \in \mathscr{D}} \mathrm{P}\{D\} = \alpha$ (say). Clearly, $\mathrm{E} = \bigcup_1^\infty D_n \in \mathscr{D}$, whence $\mathrm{P}\{\mathrm{E}\} = \alpha$. Moreover, for $D \in \mathscr{F} \cap \mathrm{E}^c$ either $\nu\{D\} < \infty$, implying $D \cup \mathrm{E} \in \mathscr{D}$ and hence $\alpha \geq \mathrm{P}\{D \cup \mathrm{E}\} = \mathrm{P}\{D\} + \alpha$, that is, $\mathrm{P}\{D\} = 0 = \nu\{D\}$, or alternatively $\nu\{D\} = \infty$, whence $\mathrm{P}\{D\} > 0$ by the P-continuity of ν. □

Lemma 2. *If ν is a* P*-continuous measure on $\mathscr{F}$, there exists an $\mathscr{F}$-measurable function $d\nu/d\mathrm{P} \geq 0$, a.c., with*

$$\nu\{A\} = \int_A \frac{d\nu}{d\mathrm{P}}\, d\mathrm{P}, \qquad A \in \mathscr{F}. \tag{4}$$

Moreover, if ν is σ-finite, then $d\nu/d\mathrm{P}$ is a random variable.

PROOF. Choose E as in Lemma 1 and set $\nu' = \nu|_{\mathscr{F} \cap \mathrm{E}}$ and $\mathrm{P}' = \mathrm{P}|_{\mathscr{F} \cap \mathrm{E}}$. Then ν' is σ-finite and P′-continuous, whence by the Radon–Nikodym theorem (Theorem 6.5.1) $d\nu'/d\mathrm{P}'$ exists on E. Define

$$\frac{d\nu}{d\mathrm{P}} = \begin{cases} \dfrac{d\nu'}{d\mathrm{P}'} & \text{on E} \\ \infty & \text{on } \mathrm{E}^c. \end{cases}$$

Then $d\nu/dP$ is $\mathscr{F}$-measurable and (4) holds for $A \in \mathscr{F} \cap E$. If, rather, $A \in \mathscr{F} \cap E^c$, then (3) ensures that (4) still obtains. Finally, if $d\nu/dP$ were infinite on a set B of positive P-measure, then for every measurable subset $A \subset B$

$$\nu\{A\} = \int_A \frac{d\nu}{dP}\, dP = \infty \quad \text{or} \quad 0$$

and ν would not be σ-finite. In other words, if ν is σ-finite, $d\nu/dP$ is a r.v. □

Theorem 1. *If X is an $\mathscr{F}$-measurable function with $|E\,X| \le \infty$ and $\mathscr{G}$ is a sub-σ-algebra of $\mathscr{F}$, then* (i) *there exists a $\mathscr{G}$-measurable function* $E\{X \mid \mathscr{G}\}$ *unique to within sets of measure zero, called the* **conditional expectation of X relative to $\mathscr{G}$**, *for which*

$$\int_A E\{X \mid \mathscr{G}\}\, dP = \int_A X\, dP \quad \textit{for all } A \in \mathscr{G}. \tag{5}$$

(ii) *If X is integrable and Z is an integrable, $\mathscr{G}$-measurable random variable such that for some π-class $\mathscr{D}$ with $\sigma(\mathscr{D}) = \mathscr{G}$*

$$E\,Z = E\,X, \qquad \int_A Z\, dP = \int_A X\, dP, \qquad A \in \mathscr{D}, \tag{6}$$

then $Z = E\{X \mid \mathscr{G}\}$, a.c.

PROOF. (i) For $A \in \mathscr{G}$, define the measures $\nu^{\pm}\{A\} = \int_A X^{\pm}\, dP$. By Lemma 2, $d\nu^{\pm}/dP_{\mathscr{G}}$ exists, where $P_{\mathscr{G}} = P|_{\mathscr{G}}$. Since $|E\,X| \le \infty$, at least one of the pair $E\,X^+$, $E\,X^-$ is finite, and so at least one of $d\nu^{\pm}/dP_{\mathscr{G}}$ is integrable. Thus, if

$$Y = \begin{cases} \dfrac{d\nu^+}{dP_{\mathscr{G}}} - \dfrac{d\nu^-}{dP_{\mathscr{G}}} & \text{on } \min\left(\dfrac{d\nu^+}{dP_{\mathscr{G}}}, \dfrac{d\nu^-}{dP_{\mathscr{G}}}\right) < \infty \\ 0, & \text{otherwise,} \end{cases} \tag{7}$$

the function Y is $\mathscr{G}$-measurable, $|E\,Y| \le \infty$, and for $A \in \mathscr{G}$

$$\int_A Y\, dP = \int_A \frac{d\nu^+}{dP_{\mathscr{G}}}\, dP_{\mathscr{G}} - \int_A \frac{d\nu^-}{dP_{\mathscr{G}}}\, dP_{\mathscr{G}}$$

$$= \int_A X^+\, dP - \int_A X^-\, dP = \int_A X\, dP,$$

so that (5) holds with $E\{X \mid \mathscr{G}\} = Y$. As for uniqueness, let Y_1 and Y_2 be $\mathscr{G}$-measurable and satisfy (5). Then $P[Y_1 > r > Y_2] = 0$ for every finite rational number r. Hence, $P\{Y_1 > Y_2\} = 0$. Similarly $P\{Y_2 > Y_1\} = 0$ and so $Y_1 = Y_2$, a.c.

Apropos of (ii), if

$$\mathscr{A} = \left\{A\colon A \in \mathscr{F}, \int_A Z\, dP = \int_A X\, dP\right\},$$

then $\Omega \in \mathscr{A} \supset \mathscr{D}$ by (6). Since X and Z are integrable, $\mathscr{A}$ is a λ-class, whence by Theorem 1.3.2 $\mathscr{A} \supset \sigma(\mathscr{D}) = \mathscr{G}$ and so $Z = \mathrm{E}\{X \mid \mathscr{G}\}$, a.c. □

An immediate consequence of Theorem 1 is

Corollary 1. *Let $\mathscr{G}_1, \mathscr{G}_2$ be σ-algebras of events and let X, Y be $\mathscr{L}_1$ random variables. If $\sigma(X)$ and $\mathscr{G}_1$ are independent, then $\mathrm{E}\{X \mid \mathscr{G}_1\} = \mathrm{E}\, X$ a.c. and if*

$$\int_{A_1 A_2} X \, d\mathrm{P} = \int_{A_1 A_2} Y \, d\mathrm{P}, \qquad A_i \in \mathscr{G}_i,\ i = 1, 2,$$

then $\mathrm{E}\{X \mid \sigma(\mathscr{G}_1 \cup \mathscr{G}_2)\} = \mathrm{E}\{Y \mid \sigma(\mathscr{G}_1 \cup \mathscr{G}_2)\}$, a.c.

A concrete construction of a conditional expectation appears in

Corollary 2. *Let the random vectors $X = (X_1, \ldots, X_m)$ and $Y = (Y_1, \ldots, Y_n)$ on $(\Omega, \mathscr{F}, \mathrm{P})$ be independent of one another and let f be a Borel function on $R^m \times R^n$ with $|\mathrm{E} f(X, Y)| \le \infty$. If, for $x \in R^m$,*

$$g(x) = \begin{cases} \mathrm{E} f(x, Y) & \text{if } |\mathrm{E} f(x, Y)| \le \infty \\ 0, & \text{otherwise,} \end{cases} \tag{8}$$

then g is a Borel function on R^m with

$$g(X) = \mathrm{E}\{f(X, Y) \mid \sigma(X)\}, \qquad \text{a.c.} \tag{9}$$

Proof. Let F_X, F_Y, and $F_{X,Y}$ be the joint distribution functions of the random vectors, X, Y and (X, Y) respectively, and denote the corresponding Lebesgue–Stieltjes measures by ν_X, ν_Y, and $\nu_{X,Y}$. Since f is a Borel function on R^{m+n}, Theorem 1.4.2 ensures that $f(x, y)$ is a Borel function on R^n for each fixed $x \in R^m$, and so by Theorem 6.3.3 and Fubini's theorem

$$g_{\pm}(x) \equiv \mathrm{E}\, f^{\pm}(x, Y) = \int_{R^n} f^{\pm}(x, y) d\nu_Y(y)$$

are Borel functions on R^m. Thus

$$D = \{x \in R^m : g_+(x) = \infty = g_-(x)\}$$

is a Borel set and $g(x) = [g_+(x) - g_-(x)] I_{D^c}(x)$ is a Borel function whence by Theorem 1.4.4, $g(X)$ is $\sigma(X)$-measurable.

By independence $(R^{m+n}, \mathscr{B}^{m+n}, \nu_{X,Y}) = (R^m, \mathscr{B}^m, \nu_X) \times (R^n, \mathscr{B}^n, \nu_Y)$. If $A \in \sigma(X)$, Theorem 1.4.4 guarantees the existence of $B \in \mathscr{B}^m$ such that $A = \{X \in B\}$, and once more via Theorem 6.3.3 and Fubini's theorem

$$\begin{aligned}\int_A g_\pm(X)\,dP &= E\, I_A g_\pm(X) = \int_B g_\pm(x) d\nu_X(x)\\ &= \int_B \int_{R^n} f^\pm(x, y) d\nu_Y(y) d\nu_X(x)\\ &= \int_{B \times R^n} f^\pm \, d\nu_{X,Y}\\ &= \int_A f^\pm(X, Y)\,dP.\end{aligned}$$

Since $|E\, f(X, Y)| \le \infty$, either $E\, g_+(X)$ or $E\, g_-(X)$ is finite whence $g_+(X) < \infty$, a.c. or $g_-(X) < \infty$, a.c. Hence, $I_D(X) = 0$, a.c. Thus, for $A \in \sigma(X)$,

$$\begin{aligned}\int_A g(X)\,dP &= \int_A [g_+(X) - g_-(X)] I_{D^c}\,dP = \int_A g_+(X)\,dP - \int_A g_-(X)\,dP\\ &= \int_A f^+(X, Y)\,dP - \int_A f^-(X, Y)\,dP = \int_A f(X, Y)\,dP,\end{aligned}$$

and so (9) holds by Theorem 1(i). □

As will be seen in Section 2, the conditional probability of an event A given $\mathscr{G}$, denoted $P\{A \mid \mathscr{G}\}$, may be defined as $E\{I_A \mid \mathscr{G}\}$. A typical exploitation of conditional expectation appears in

EXAMPLE 1 (Kesten). If $\{S_n = \sum_1^n X_i, n \ge 1\}$, where $\{X, X_n \ge 1\}$ are non-negative i.i.d. r.v.s with $E\, X = \infty$, then $\overline{\lim}_{n\to\infty} X_n/S_{n-1} = \infty$, a.c.

PROOF. Set $A_i = \{S_{i-1} \le \varepsilon X_i\}$, $\varepsilon > 0$, where $S_0 = 0$, and note that $A_i A_j \subset A_i\{\sum_{h=i+1}^{j-1} X_h \le \varepsilon X_j\}$ for $i < j$, implying for $k \ge 1$ via independence and identical distributions

$$P\left\{A_i \bigcup_{j=i+k}^{\infty} A_j\right\} \le P\{A_i\} P\left\{\bigcup_{j=k}^{\infty} A_j\right\},$$

so that (12) of Lemma 4.2.4 obtains with $B_j = A_j$. Moreover, if F is the d.f. of X and $a(0) = 0$,

$$a(x) = \frac{x}{\int_0^x [1 - F(y)]dy}, \qquad x > 0,$$

then, clearly, $a(x)/x \downarrow$ and (Exercise 6.2.16) $a(x) \uparrow$ and $E\, a(X) = \infty$. It follows that $E\, a(\varepsilon X) = \infty$, all $\varepsilon > 0$. However, via Corollary 2 and Examples 5.4.1, 6.2.2

$$\begin{aligned}\sum_{n=1}^{\infty} P\{A_n\} &= E \sum_{n=1}^{\infty} P\{S_{n-1} < \varepsilon X_n \mid X_n\} = \int_0^\infty \sum_{n=1}^{\infty} P\{S_{n-1} \le \varepsilon x\} dF(x)\\ &\ge \int_0^\infty a(\varepsilon x) dF(x) = E\, a(\varepsilon X),\end{aligned}$$

and so Lemma 4.2.4 ensures $P\{A_n, \text{ i.o.}\} = 1$, all $\varepsilon > 0$, which is tantamount to the conclusion of Example 1.

It follows immediately from this example that for any i.i.d. r.v.s $\{X, X_n, n \geq 1\}$ with $E|X| = \infty$,

$$\overline{\lim_{n\to\infty}} \frac{|X_n|}{|\sum_1^{n-1} X_i|} = \infty, \quad \text{a.c.} \qquad \square$$

Let X be an $\mathscr{F}$-measurable function with $|E X| \leq \infty$ and $\{Y_\lambda, \lambda \in \Lambda\}$, $\{\mathscr{G}_\lambda, \lambda \in \Lambda\}$, nonempty families of random variables and σ-algebras of events respectively. It is customary to define

$$\begin{aligned} E\{X \mid Y_\lambda, \lambda \in \Lambda\} &= E\{X \mid \sigma(Y_\lambda, \lambda \in \Lambda)\}, \\ E\{X \mid \mathscr{G}_\lambda, \lambda \in \Lambda\} &= E\{X \mid \sigma(\mathscr{G}_\lambda, \lambda \in \Lambda)\}, \end{aligned} \tag{10}$$

and, in particular,

$$\begin{aligned} E\{X \mid Y_1, \ldots, Y_n\} &= E\{X \mid \sigma(Y_1, \ldots, Y_n)\}, \\ E\{X \mid Y\} &= E\{X \mid \sigma(Y)\}. \end{aligned} \tag{11}$$

Since, by definition $E\{X \mid Y_1, \ldots, Y_n\}$ is $\sigma(Y_1, \ldots, Y_n)$-measurable, Theorem 1.4.4 guarantees that for some Borel function g on R^n

$$E\{X \mid Y_1, \ldots, Y_n\} = g(Y_1, \ldots, Y_n).$$

Conversely, if g is a Borel function on R^n such that $|E\, g(Y_1, \ldots, Y_n)| \leq \infty$ and for every $A \in \sigma(Y_1, .., Y_n)$

$$\int_A g(Y_1, \ldots, Y_n)\, dP = \int_A X\, dP,$$

then $g(Y_1, \ldots, Y_n) = E\{X \mid Y_1, \ldots, Y_n\}$, a.c.

In particular, if $Y = I_A$ for some $A \in \mathscr{F}$, then $\sigma(Y) = \{\emptyset, A, A^c, \Omega\}$ and every version of $E\{X \mid Y\}$ must be constant on each of the sets A, A^c, necessitating

$$E\{X \mid Y\}(\omega) = \begin{cases} \dfrac{1}{P\{A\}} \displaystyle\int_A X\, dP & \text{if } \omega \in A \\[2ex] \dfrac{1}{P\{A^c\}} \displaystyle\int_{A^c} X\, dP & \text{if } \omega \in A^c, \end{cases} \tag{12}$$

where either of the constants on the right can be construed as any number in $[-\infty, \infty]$ when the corresponding set A or A^c has probability zero.

More generally, if $\{A_n, n \geq 1\}$ is a σ-partition of Ω in $\mathscr{F}$ with $P\{A_n\} > 0$, $n \geq 1$, and $\mathscr{G} = \sigma(A_n, n \geq 1)$, then for any measurable function X with $|E X| \leq \infty$

$$E\{X \mid \mathscr{G}\} = \sum_1^\infty \left(\frac{1}{P\{A_n\}} \int_{A_n} X\, dP \right) I_{A_n}, \quad \text{a.c.} \tag{13}$$

Moreover, this remains valid even if $P\{A_m\} = 0$ for certain indices m, the quantity in the parenthesis being then interpreted as an arbitrary constant.

Some simple consequences of the definition of conditional expectation are

$$E\{1 \mid \mathscr{G}\} = 1, \text{a.c.}, \quad E\{X \mid \mathscr{G}\} \geq 0, \text{a.c.} \quad \text{if } X \geq 0, \text{a.c.}, \tag{14, i}$$

$$E\{cX \mid \mathscr{G}\} = c\,E\{X \mid \mathscr{G}\}, \text{a.c.}, \quad \text{if} \mid E\,X \mid \leq \infty \text{ and } c \text{ is a finite constant}, \tag{14, ii}$$

$$E\{X + Y \mid \mathscr{G}\} = E\{X \mid \mathscr{G}\} + E\{Y \mid \mathscr{G}\}, \text{a.c.}, \quad \text{if } E(X^- + Y^-) < \infty \text{ or } E(X^+ + Y^+) < \infty. \tag{14, iii}$$

These properties assert roughly that if $TX = E\{X \mid \mathscr{G}\}$, then T is linear, order preserving (monotone), and $T1 = 1$.

$$E\{X \mid \mathscr{G}\} = X, \text{a.c. if } X \text{ is } \mathscr{G}\text{-measurable}. \tag{14, iv}$$

If $\mathscr{G}_1, \mathscr{G}_2$ are σ-algebras with $\mathscr{G}_1 \subset \mathscr{G}_2 \subset \mathscr{F}$ and $|E\,X| \leq \infty$, then

$$E\{E\{X \mid \mathscr{G}_2\} \mid \mathscr{G}_1\} = E\{X \mid \mathscr{G}_1\} = E\{E\{X \mid \mathscr{G}_1\} \mid \mathscr{G}_2\}, \text{a.c.} \tag{14, v}$$

Proof. Since $E\{E\{X \mid \mathscr{G}_2\} \mid \mathscr{G}_1\}$ is $\mathscr{G}_1$-measurable and for $A \in \mathscr{G}_1$

$$\int_A E\{E\{X \mid \mathscr{G}_2\} \mid \mathscr{G}_1\} dP = \int_A E\{X \mid \mathscr{G}_2\} dP = \int_A X\, dP,$$

the first equality of (14, v) follows. Since $E\{X \mid \mathscr{G}_1\}$ is $\mathscr{G}_i$-measurable for $i = 1$ and hence $i = 2$, the second equality of (14, v) follows from (14, iv). □

Theorem 2. *Let* $\{X_n, n \geq 1\}$ *and* Y *be random variables with* $E|Y| < \infty$ *and* $\mathscr{G}$ *a* σ*-algebra of events.*

i. (*Monotone convergence theorem for conditional expectations*). *If* $Y \leq X_n \uparrow X$, a.c., *then* $E\{X_n \mid \mathscr{G}\} \uparrow E\{X \mid \mathscr{G}\}$, a.c.
ii. (*Fatou lemma for conditional expectations*). *If* $Y \leq X_n$, $n \geq 1$, a.c., *then* $E\{\underline{\lim}\, X_n \mid \mathscr{G}\} \leq \underline{\lim}\, E\{X_n \mid \mathscr{G}\}$, a.c.
iii. (*Lebesgue dominated convergence theorem for conditional expectations*). *If* $X_n \xrightarrow{\text{a.c.}} X$ *and* $|X_n| \leq |Y|$, $n \geq 1$, a.c., *then* $E\{X_n \mid \mathscr{G}\} \xrightarrow{\text{a.c.}} E\{X \mid \mathscr{G}\}$.

Proof. (i) By the monotone property of conditional expectations, $E\{Y \mid \mathscr{G}\} \leq E\{X_n \mid \mathscr{G}\} \uparrow$ some function Z, a.c. For $A \in \mathscr{G}$, by ordinary monotone convergence

$$\int_A Z\, dP = \lim_n \int_A E\{X_n \mid \mathscr{G}\} dP = \lim_n \int_A X_n\, dP = \int_A X\, dP,$$

and, since Z is $\mathscr{G}$-measurable, $Z = E\{X \mid \mathscr{G}\}$, a.c.

Apropos of (ii), set $Y_n = \inf_{m \ge n} X_m$. Then $Y \le Y_n \uparrow \underline{\lim}_{m\to\infty} X_m$, whence $\underline{\lim}_n \mathrm{E}\{X_n|\mathscr{G}\} \ge \underline{\lim}_n \mathrm{E}\{Y_n|\mathscr{G}\} = \mathrm{E}\{\underline{\lim}_n X_n|\mathscr{G}\}$ by (i). Finally, (iii) is a consequence of (ii) via

$$\pm\mathrm{E}\{X|\mathscr{G}\} = \mathrm{E}\{\pm X|\mathscr{G}\} \le \underline{\lim}\, \mathrm{E}\{\pm X_n|\mathscr{G}\} = \underline{\lim}(\pm \mathrm{E}\{X_n|\mathscr{G}\}), \qquad \text{a.c.} \quad \square$$

An extremely useful fact about conditional expectations is furnished by

Theorem 3. *Let X be a random variable with $|\mathrm{E}\,X| \le \infty$ and $\mathscr{G}$ a σ-algebra of events. If Y is a finite-valued $\mathscr{G}$-measurable random variable such that $|\mathrm{E}\,XY| \le \infty$, then*

$$\mathrm{E}\{XY|\mathscr{G}\} = Y\,\mathrm{E}\{X|\mathscr{G}\}, \qquad \text{a.c.} \tag{15}$$

Proof. By separate consideration of $X^{\pm}$ and $Y^{\pm}$ it may be supposed that $X \ge 0$ and $Y \ge 0$. Moreover, by the monotone convergence theorem for conditional expectations it may even be assumed that X and Y are bounded r.v.s. Set

$$\nu\{A\} = \int_A XY\,d\mathrm{P}, \qquad \mu\{A\} = \int_A X\,d\mathrm{P}, \qquad A \in \mathscr{F}.$$

Then both μ and ν are finite, P-continuous measures on $\mathscr{F}$ and, denoting as usual the restrictions of P, ν, μ to $\mathscr{G}$ by $\mathrm{P}_{\mathscr{G}}$, $\nu_{\mathscr{G}}$, $\mu_{\mathscr{G}}$,

$$\frac{d\nu_{\mathscr{G}}}{d\mathrm{P}_{\mathscr{G}}} = \mathrm{E}\{XY|\mathscr{G}\}, \qquad \frac{d\mu_{\mathscr{G}}}{d\mathrm{P}_{\mathscr{G}}} = \mathrm{E}\{X|\mathscr{G}\}, \qquad \frac{d\mu}{d\mathrm{P}} = X, \qquad \text{a.c.}$$

For $A \in \mathscr{G}$, by Theorem 6.5.2

$$\int_A Y\,d\mu_{\mathscr{G}} = \int_A Y \frac{d\mu_{\mathscr{G}}}{d\mathrm{P}_{\mathscr{G}}}\,d\mathrm{P}_{\mathscr{G}} = \int_A Y\,\mathrm{E}\{X|\mathscr{G}\}\,d\mathrm{P}_{\mathscr{G}},$$

and for $A \in \mathscr{F}$

$$\int_A Y\,d\mu = \int_A Y \frac{d\mu}{d\mathrm{P}}\,d\mathrm{P} = \int_A YX\,d\mathrm{P}.$$

Consequently,

$$\int_A Y\,\mathrm{E}\{X|\mathscr{G}\}\,d\mathrm{P} = \int_A XY\,d\mathrm{P}, \qquad A \in \mathscr{G},$$

and since $Y\,\mathrm{E}\{X|\mathscr{G}\}$ is $\mathscr{G}$-measurable, (15) follows. $\square$

Theorem 4 (Jensen's Inequality for Conditional Expectations). *Let Y be $\mathscr{F}$-measurable with $|\mathrm{E}\,Y| \le \infty$ and g any finite convex function on $(-\infty, \infty)$ with $|\mathrm{E}\,g(Y)| \le \infty$. If for some σ-algebra $\mathscr{G}$ of events,* (i) $X = \mathrm{E}\{Y|\mathscr{G}\}$, a.c. *or* (ii) *$X$ is $\mathscr{F}$-measurable with $X \le \mathrm{E}\{Y|\mathscr{G}\}$,* a.c. *and $g\uparrow$, then*

$$g(X) \le \mathrm{E}\{g(Y)|\mathscr{G}\}, \quad \text{a.c.} \tag{16}$$

PROOF. Since (ii) follows directly from (i), it suffices to prove the latter. To this end, define

$$g^*(t) = \lim_{s \to t^-} \frac{g(s) - g(t)}{s - t},$$

whence g^* is a finite, nondecreasing function on $(-\infty, \infty)$. Now the secant line of a convex function is always above the one-sided tangent line, that is,

$$g(t) \geq g(s) + (t - s)g^*(s), \qquad -\infty < s, t < \infty, \tag{17}$$

whence if $A = \{|X| \leq M, g^*(X) \geq 0\}$, $0 < M < \infty$ *and* $B = \{|X| < \infty, g^*(X) \geq 0\}$ both $g(X)$ and $g^*(X)$ are bounded on $A \in \mathscr{G}$, so by (17)

$$I_A g(Y) \geq I_A g(X) + I_A(Y - X)g^*(X), \quad \text{a.c.}$$

Since $|\mathrm{E}\, I_A(Y - X)g^*(X)| \leq \infty$, by Theorem 3,

$$I_A \mathrm{E}\{g(Y)|\mathscr{G}\} = \mathrm{E}\{I_A g(Y)|\mathscr{G}\} \geq I_A g(X), \quad \text{a.c.}$$

As $M \to \infty$, $I_A \xrightarrow{\text{a.c.}} I_B$, so (16) holds on B. Similarly, for $\{g^*(X) \leq 0, |X| < \infty\}$.

Consider next $D = \{X = \infty\}$. If $g^*(s) > 0$, some s in $(-\infty, \infty)$, then (17) ensures

$$I_D \mathrm{E}\{g(Y)|\mathscr{G}\} \geq g(s)I_D + I_D \mathrm{E}\{Y - s|\mathscr{G}\}g^*(s) = \infty = g(X)\cdot I_D$$

If $g^*(s) \leq 0$, all s in $(-\infty, \infty)$, then $g \downarrow$ whence $g(\infty) \leq g(X)$, a.c. and (16) holds on D since

$$I_D \mathrm{E}\{g(Y)|\mathscr{G}\} \geq I_D g(\infty) = I_D g(X).$$

Let $D' = \{X = -\infty\}$. If $g^*(s) > 0$, all s in $(-\infty, \infty)$, then $g \uparrow$ whence

$$I_{D'} \mathrm{E}\{g(Y)|\mathscr{G}\} \geq I_{D'} g(-\infty) = g(X)I_{D'}$$

whereas if $g^*(s) < 0$ for some s in $(-\infty, \infty)$, then via (17) $g(Y) \geq g(s) + (Y - s)g^*(s)$ implying

$$\begin{aligned}\mathrm{E}\{g(Y)|\mathscr{G}\} &\geq g(s) + [\mathrm{E}\{Y|\mathscr{G}\} - s]g^*(s)\\ &= g(s) + [X - s]g^*(s) = \infty, \quad \text{a.c. on } D'\end{aligned}$$

and (16) holds on D'. □

Corollary 3. *For X and $\mathscr{G}$ as in the theorem, with probability one*

$$|\mathrm{E}\{X|\mathscr{G}\}| \leq \mathrm{E}\{|X||\mathscr{G}\}, \qquad \mathrm{E}^r\{|X||\mathscr{G}\} \leq \mathrm{E}\{|X|^r|\mathscr{G}\}, \qquad r \geq 1,$$

$$\mathrm{E}\{\max(a, X)\,|\,\mathscr{G}\} \geq \max\{a, \mathrm{E}\{X\,|\,\mathscr{G}\}\}, \qquad -\infty < a < \infty.$$

Theorem 5 (Extended Fatou Lemma For Conditional Expectations). *Let $\{X_n, n \geq 1\}$ be random variables with $|\mathrm{E}\, X_n| \leq \infty$, $n \geq 1$, and $\mathscr{G}$ a σ-algebra of events. If $|\mathrm{E}\, \underline{\lim}_n X_n| \leq \infty$ and*

$$\sup_{n \geq 1} \mathrm{E}\{X_n^- I_{[X_n^- > k]}\,|\,\mathscr{G}\} \xrightarrow{\text{a.c.}} 0 \quad \text{as } k \to \infty,$$

then $\mathrm{E}\{\underline{\lim}_n X_n\,|\,\mathscr{G}\} \leq \underline{\lim}_n \mathrm{E}\{X_n\,|\,\mathscr{G}\}$, a.c.

Proof. If $Y_k = \sup_{n\ge 1} \mathrm{E}\{X_n^- I_{[X_n^- > k]} \mid \mathscr{G}\}$, $k > 0$, then $Y_k \xrightarrow{\text{a.c.}} 0$ by hypothesis. Since with probability one

$$\mathrm{E}\{X_n \mid \mathscr{G}\} = \mathrm{E}\{X_n(I_{[X_n^- \le k]} + I_{[X_n^- > k]}) \mid \mathscr{G}\} \ge \mathrm{E}\{X_n I_{[X_n^- \le k]} \mid \mathscr{G}\} - Y_k,$$

it follows via Theorem 2(ii) that for all $k > 0$

$$\begin{aligned} \varliminf_n \mathrm{E}\{X_n \mid \mathscr{G}\} &\ge \varliminf_n \mathrm{E}\{X_n I_{[X_n^- \le k]} \mid \mathscr{G}\} - Y_k \\ &\ge \mathrm{E}\{\varliminf_n X_n I_{[X_n^- \le k]} \mid \mathscr{G}\} - Y_k \\ &\ge \mathrm{E}\{\varliminf X_n \mid \mathscr{G}\} - Y_k, \qquad \text{a.c.,} \end{aligned}$$

which yields the theorem upon letting $k \to \infty$. □

Corollary 4. *Let $\mathscr{G}$ be a σ-algebra of events, $\{X_n, n \ge 1\}$ random variables with $|\mathrm{E}\, X_n| \le \infty$, $n \ge 1$, and $\lim_{k\to\infty} \mathrm{E}\{|X_n| I_{[|X_n| > k]} \mid \mathscr{G}\} = 0$, uniformly in n with probability one. If $X_n \xrightarrow{\text{a.c.}} X$, where $|\mathrm{E}\, X| \le \infty$, then $\mathrm{E}\{X_n \mid \mathscr{G}\} \xrightarrow{\text{a.c.}} \mathrm{E}\{X \mid \mathscr{G}\}$.*

Proof. Applying the theorem to X_n and $-X_n$,

$$\mathrm{E}\{X \mid \mathscr{G}\} \le \varliminf_n \mathrm{E}\{X_n \mid \mathscr{G}\} \le \varlimsup \mathrm{E}\{X_n \mid \mathscr{G}\} \le \mathrm{E}\{X \mid \mathscr{G}\}, \qquad \text{a.c.} \quad □$$

Corollary 5. *Let $\mathscr{G}$ be a σ-algebra of events and $\{X_n, n \ge 1\}$ random variables with $|\mathrm{E}\, X_n| \le \infty$, $n \ge 1$. If $X_n \xrightarrow{\text{a.c.}} X_0$, where $|\mathrm{E}\, X_0| \le \infty$ and for some $r > 1$*

$$\sup_{n\ge 1} \mathrm{E}\{|X_n|^r \mid \mathscr{G}\} < \infty, \qquad \text{a.c.,}$$

then $\mathrm{E}\{X_n \mid \mathscr{G}\} \xrightarrow{\text{a.c.}} \mathrm{E}\{X_0 \mid \mathscr{G}\}$.

Proof. Let $X_n' = X_n I_A$, $n \ge 0$, where $A = \{\sup_{n\ge 1} \mathrm{E}\{|X_n|^r \mid \mathscr{G}\} < C < \infty\}$. Since $\mathrm{E}|X_n'|^r = \mathrm{E}[\mathrm{E}\{|X_n'|^r \mid \mathscr{G}\}] \le C$, Fatou's lemma ensures that $\mathrm{E}|X_0'|^r \le C$. Moreover, for $K > 0$,

$$\mathrm{E}\{|X_n'| I_{[|X_n'| > K]} \mid \mathscr{G}\} \le K^{1-r} \mathrm{E}\{|X_n'|^r \mid \mathscr{G}\} \le CK^{1-r}$$

uniformly in n with probability one, whence $\mathrm{E}\{X_n' \mid \mathscr{G}\} \xrightarrow{\text{a.c.}} \mathrm{E}\{X_0' \mid \mathscr{G}\}$ by Corollary 4 and the conclusion follows as $C \to \infty$. □

Example 2. Let $S_n = \sum_{j=1}^n X_j$, where $\{X_n, n \ge 1\}$ are independent r.v.s and let $\{\alpha_n, n \ge 1\}$ be constants such that $\mathrm{P}\{S_n < \alpha_n\} > 0$, $n \ge 1$. Then

$$\mathrm{P}\left\{\bigcup_{n=1}^{\infty} [S_n \ge \alpha_n]\right\} = 1 \tag{18}$$

entails

$$\mathrm{P}\{S_n \ge \alpha_n, \text{ i.o.}\} = 1. \tag{19}$$

Proof. Set $A_N = \bigcup_{n=N}^{\infty} [S_n \ge \alpha_n]$, $N \ge 1$, and suppose that $\mathrm{P}\{A_N\} = 1$ for some $N \geq 1$. Then $1 = \text{E}\,\text{P}\{A_N | S_N\}$, implying $\text{P}\{A_N | S_N\} = 1$, a.c., and so on the set $\{S_N < \alpha_N\}$

$$\text{P}\{A_{N+1} \mid S_N\} = \text{P}\{A_N \mid S_N\} = 1, \qquad \text{a.c.} \tag{20}$$

Next, if $h(x) = \text{P}\{\bigcup_{n=N+1}^{\infty} [S_n - S_N \geq \alpha_n - x]\}$, then $h\uparrow$ and by Corollary 7.1.2

$$h(S_N) = \text{P}\{A_{N+1} | S_N\}, \qquad \text{a.c.} \tag{21}$$

According to (20), $h(S_N) = 1$, a.c., on $\{S_N < \alpha_N\}$. Thus

$$\int_{(-\infty, \alpha_N)} [1 - h(x)]\, d\text{P}\{S_N < x\} = \text{P}\{S_N < \alpha_N\} - \int_{[S_N < \alpha_N]} h(S_N)\, d\text{P} = 0,$$

and so $h(x) = 1$ for every $x < \alpha_n$ that is a point of increase of $\text{P}\{S_N < x\}$. Since $\text{P}\{S_N < \alpha_N\} > 0$, there must be at least one such x. Then monotonicity of h guarantees $h(x) = 1$ for all $x \geq \alpha_N$ and (21) ensures

$$\text{P}\{A_{N+1} \mid S_N\} = 1, \qquad \text{a.c.}$$

on $[S_N \geq \alpha_N]$. Consequently, recalling (20), $\text{P}\{A_{N+1} \mid S_N\} = 1$, a.c., and so $\text{P}\{A_{N+1}\} = 1$. Since $\text{P}\{A_1\} = 1$ by hypothesis, it follows inductively that $\text{P}\{A_N\} = 1$ for all $N \geq 1$, which, in turn, yields (19). □

Corollary 6. *If* $S_n = \sum_1^n X_i$, *where* $\{X_n, n \geq 1\}$ *are independent* r.v.s *with* $\text{P}\{X_n < 0\} > 0, n \geq 1$, *then for nonnegative, constants* $\{\alpha_n, n \geq 1\}$, (18) *implies* (19).

PROOF. $\text{P}\{S_n < \alpha_n\} \geq \text{P}\{S_n < 0\} \geq \text{P}\{X_1 < 0\} \cdots \text{P}\{X_n < 0\} > 0$.

Remark. Clearly, both equality signs can be deleted from (18) and (19) provided $\text{P}\{S_n < \alpha_n\} > 0, n \geq 1$, is modified to $P\{S_n \leq \alpha_n\} > 0, n \geq 1$.

EXERCISES 7.1

1. If X is a r.v. with $|\text{E}\,X| \leq \infty$ and $\mathscr{G}$ a σ-algebra of events with $\sigma(X)$ and $\mathscr{G}$ independent classes, then $\text{E}\{X \mid \mathscr{G}\} = \text{E}\,X$, a.c. In particular, if $\{X_n\}$ are independent r.v.s with $|\text{E}\,X_n| \leq \infty, n \geq 1$, then $\text{E}\{X_n \mid X_1, \ldots, X_{n-1}\} = \text{E}\{X_n \mid X_{n+1}, X_{n+2}, \ldots\} = \text{E}\,X_n$, a.c.

2. Let $\{A_n, n \geq 1\}$ be σ-partition of Ω in $\mathscr{F}$. Verify that if $\mathscr{G} = \sigma(A_n, n \geq 1)$ and $|\text{E}\,X| \leq \infty$, then
$$\text{E}\{X | \mathscr{G}\} = \sum_1^{\infty} I_{A_n}\left(\frac{\int_{A_n} X\, d\text{P}}{\text{P}\{A_n\}}\right).$$
where the parenthetical quantity is an arbitrary real number if $\text{P}\{A_n\} = 0$.

3. Show that Corollary 5 remains true if $\text{E}|X| < \infty$ and $C = \infty$.

4. Let $\mathscr{D}$ be a semi-algebra of events and X, Y r.v.s with $|E X| < \infty$, $|E Y| < \infty$. If $\int_D X \, dP \le \int_D Y \, dP$, $D \in \mathscr{D}$, then $E\{X \mid \sigma(\mathscr{D})\} \le E\{Y \mid \sigma(\mathscr{D})\}$, a.c.

5. Let (X_1, X_2) be jointly normally distributed (Exercise 6.5.12) with p.d.f.

$$[2\pi\sigma_1\sigma_2(1-\rho^2)^{1/2}]^{-1} \exp\left\{-\frac{1}{2(1-\rho^2)}\left[\frac{x_1^2}{\sigma_1^2} - \frac{2\rho x_1 x_2}{\sigma_1\sigma_2} + \frac{x_2^2}{\sigma_2^2}\right]\right\}.$$

Find $E\{X_1 \mid X_2\}$ and the conditional variance.

6. Prove that
   i. if X is an $\mathscr{L}_2$ r.v. and Y is a r.v. such that $E\{X \mid Y\} = Y$, a.c. and $E\{Y \mid X\} = X$, a.c., then $X = Y$, a.c.
   ii. Let $\mathscr{G}_1, \mathscr{G}_2$ be σ-algebras of events and X an $\mathscr{L}_1$ r.v. If $X_1 = E\{X \mid \mathscr{G}_1\}$, $X_2 = E(X_1 \mid \mathscr{G}_2)$, and $X = X_2$, a.c., then $X_1 = X_2$, a.c.
   iii. If X, Y are $\mathscr{L}_1$ r.v.s with $E\{X \mid Y\} = Y$, a.c., and $E\{Y \mid X\} = X$, a.c., then $X = Y$, a.c.

7. Prove that the set E of lemma 1 and the function $d\nu/dP$ are unique to within an equivalence.

8. Let X be an $\mathscr{L}_2$ r.v. and $\mathscr{G}$ a σ-algebra of events. Prove that (i) $\sigma^2(E\{X \mid \mathscr{G}\}) \le \sigma^2(X)$, (ii) if for any α in $(-\infty, \infty)$, $Y = \min(X, \alpha)$,

$$E\{[X - E\{X \mid \mathscr{G}\}]^2 \mid \mathscr{G}\} \ge E\{[Y - E\{Y \mid \mathscr{G}\}]^2 \mid \mathscr{G}\}, \qquad \text{a.c.}$$

(iii) $E(X - E\{X \mid \mathscr{G}\})^2 \le E(X - Y)^2$ for any $\mathscr{G}$-measurable r.v. Y.

9. Let $\mathscr{G}$ be an σ-algebra of events and $\{X_n, n \ge 1\}$ r.v.s. If for some $p \ge 1$, $X_n \xrightarrow{\mathscr{L}_p} X$, then $E\{X_n \mid \mathscr{G}\} \xrightarrow{\mathscr{L}_p} E\{X \mid \mathscr{G}\}$.

10. Let $\mathscr{G}$ be a σ-algebra of events and X a nonnegative r.v. Prove that $E\{X \mid \mathscr{G}\} = \operatorname{esup}\{h: h$ is $\mathscr{G}$-measurable, $h \ge 0$, a.c., and $\int_A h \, dP \le \int_A X \, dP$, all $A \in \mathscr{G}\}$.

11. Show via an example that in Theorem 2(iii) $X_n \xrightarrow{P} X$ cannot replace $X_n \xrightarrow{\text{a.c.}} X$.

12. If $\{X_n, n \ge 1\}$ are $\mathscr{L}_1$ interchangeable r.v.s (Exercise 6.3.10) and $\mathscr{F}_n = \sigma(\sum_{i=1}^j X_i, j \ge n)$, prove that $E\{X_i \mid \mathscr{F}_n\} = (1/n)\sum_1^n X_j$, a.c., $1 \le i \le n$. More generally, if $\{X_n, n \ge 1\}$ are interchangeable r.v.s, φ is a symmetric Borel function on R^m with $E|\varphi(X_1, \ldots, X_m)| < \infty$, and $\mathscr{F}_n = \sigma(U_{m,j}, j \ge n)$, where

$$\binom{n}{m} U_{m,n} = \sum_{1 \le i_1 < \cdots < i_m \le n} \varphi(X_{i_1}, \ldots, X_{i_m}), \qquad n \ge m,$$

then for $1 \le i_1 < \cdots < i_m \le n+1$

$$E\{\varphi(X_{i_1}, \ldots, X_{i_m}) \mid \mathscr{F}_{n+1}\} = E\{\varphi(X_1, \ldots, X_m) \mid \mathscr{F}_{n+1}\}, \qquad \text{a.c.}$$

13. (Chow–Robbins) Let $S_n = \sum_1^n X_i$, $n \ge 1$, where $\{X, X_n, n \ge 1\}$ are i.i.d. with $E|X| = \infty$. Then for any positive constants $\{b_n\}$, either $P\{\underline{\lim}_{n\to\infty} b_n^{-1}|S_n| = 0\} = 1$ or (*) $P\{\overline{\lim}_{n\to\infty} b_n^{-1}|S_n| = \infty\} = 1$. *Hint*: If (*) fails, by the zero-one law $\overline{\lim}\, b_n^{-1}|S_n| < \infty$, a.c., whence $P\{\overline{\lim}\, b_n^{-1}|S_{n-1}| < \infty\} = P\{\overline{\lim}\, b_n^{-1}|\sum_{i=2}^n X_i| < \infty\} = 1$, entailing $P\{\overline{\lim}\, b_n^{-1}|X_{n+1}| < \infty\} \ge P\{\overline{\lim}\, b_n^{-1}(|S_n| + |S_{n-1}|) < \infty\} = 1$. Now apply $\underline{\lim}\, b_n^{-1}|S_n| \le \underline{\lim}\,|S_n|/(1 + |X_{n+1}|) \cdot \overline{\lim}(1 + |X_{n+1}|)/b_n$ and Example 1.

7.2 Conditional Probabilities, Conditional Probability Measures

Let $(\Omega, \mathscr{F}, \mathrm{P})$ be a fixed probability space and $\mathscr{G}$ a sub-σ-algebra of $\mathscr{F}$. For $A, B \in \mathscr{F}$, define

$$\mathrm{P}\{A \mid \mathscr{G}\} = \mathrm{E}\{I_A \mid \mathscr{G}\}, \qquad \mathrm{P}\{A \mid B\} = \mathrm{E}\{I_A \mid I_B\}. \tag{1}$$

The former, $\mathrm{P}\{A \mid \mathscr{G}\}$, is called the **conditional probability of the event A given $\mathscr{G}$** and according to the prior section is a $\mathscr{G}$-measurable function on Ω satisfying

$$\int_G \mathrm{P}\{A|\mathscr{G}\}\, d\mathrm{P} = \int_G I_A\, d\mathrm{P} = \mathrm{P}\{A \cdot G\}, \qquad G \in \mathscr{G}. \tag{2}$$

The latter, $\mathrm{P}\{A \mid B\}$, is called the **conditional probability of the event A given the event B** and according to 7.1(12), if $\mathrm{P}\{B\} > 0$,

$$\mathrm{P}\{A \mid B\} = \frac{1}{\mathrm{P}\{B\}} \int_B I_A\, d\mathrm{P} = \frac{\mathrm{P}\{AB\}}{\mathrm{P}\{B\}}, \qquad \omega \in B. \tag{3}$$

The properties 7.1(14) of conditional expectation and monotone convergence yield:

$$0 \le \mathrm{P}\{A \mid \mathscr{G}\} \le 1, \text{ a.c.}; \tag{4, i}$$

$$\mathrm{P}\{A \mid \mathscr{G}\} = 0, \text{ a.c., iff } \mathrm{P}\{A\} = 0, \qquad \mathrm{P}\{A \mid \mathscr{G}\} = 1, \text{ a.c., iff } \mathrm{P}\{A\} = 1; \tag{4, ii}$$

If $\{A_n, n \ge 1\}$ are disjoint sets of $\mathscr{F}$, then

$$\mathrm{P}\left\{\bigcup_1^\infty A_n \,\middle|\, \mathscr{G}\right\} = \sum_{n=1}^\infty \mathrm{P}\{A_n \mid \mathscr{G}\}, \quad \text{a.c.}; \tag{4, iii}$$

If $A_n \in \mathscr{F}$, $n \ge 1$, and $\lim_{n\to\infty} A_n = A$, then

$$\lim_{n\to\infty} \mathrm{P}\{A_n \mid \mathscr{G}\} = \mathrm{P}\{A \mid \mathscr{G}\}, \quad \text{a.c.} \tag{4, iv}$$

Property (iii) asserts that for each sequence $\{A_n\}$ of disjoint events

$$\mathrm{P}\left\{\bigcup_1^\infty A_n \,\middle|\, \mathscr{G}\right\}(\omega) = \sum_{n=1}^\infty \mathrm{P}\{A_n \mid \mathscr{G}\}(\omega)$$

except for ω in a null set which may well depend on the particular sequence $\{A_n\}$. It does **not** stipulate that there exists a **fixed** null set N such that

$$\mathrm{P}\left\{\bigcup_1^\infty A_n \,\middle|\, \mathscr{G}\right\}(\omega) = \sum_{n=1}^\infty \mathrm{P}\{A_n \mid \mathscr{G}\}(\omega), \qquad \omega \in N^c$$

for every sequence $\{A_n\}$ of disjoint events. In fact, the later is false (Halmos, 1950, p. 210).

Definition. Let $\mathscr{F}_1$, $\mathscr{G}$ be σ-algebras of events. A **conditional probability measure** or **regular conditional probability on** $\mathscr{F}_1$ **given** $\mathscr{G}$ is a function $P(A, \omega)$ defined for $A \in \mathscr{F}_1$ and $\omega \in \Omega$ such that

i. for each $\omega \in \Omega$, $P(A, \omega)$ is a probability measure on $\mathscr{F}_1$,
ii. for each $A \in \mathscr{F}_1$, $P(A, \omega)$ is a $\mathscr{G}$-measurable function on Ω coinciding with the conditional probability of A given $\mathscr{G}$, i.e., $P(A, \omega) = P\{A \mid \mathscr{G}\}(\omega)$, a.c.

One advantage of regular conditional probabilities is that conditional expectations may then be envisaged as ordinary expectations relative to the conditional probability measure, as stated in

Theorem 1. *If for some pair* $\mathscr{F}_1, \mathscr{G}$ *of σ-algebras of events,* $P_\omega\{A\} \equiv P(A, \omega)$ *is a regular conditional probability on* $\mathscr{F}_1$ *given* $\mathscr{G}$ *and* X *is an* $\mathscr{F}_1$*-measurable function with* $|E\,X| \le \infty$, *then*

$$E\{X \mid \mathscr{G}\}(\omega) = \int_\Omega X\, dP_\omega, \qquad \text{a.c.} \tag{5}$$

PROOF. By separate consideration of X^+ and X^-, it may be supposed that $X \ge 0$. Let

$$\mathscr{H} = \{X : X \ge 0, X \text{ is } \mathscr{F}_1\text{-measurable, and (5) holds}\}.$$

By definition of conditional probability measure, $I_A \in \mathscr{H}$ for $A \in \mathscr{F}_1$. Since, as is readily verified, $\mathscr{H}$ is a monotone system, Theorem 1.4.3 ensures that $\mathscr{H}$ contains all nonnegative $\mathscr{F}_1$-measurable functions. □

In general, a regular conditional probability measure does not exist (Doob, 1953, p. 624). A simple case where it does is

EXAMPLE 1. Let $\mathscr{F}_1 = \mathscr{F}$, $\mathscr{G} = \sigma(A_n, n \ge 1)$, where $\{A_n, n \ge 1\}$ is a σ-partition of Ω in $\mathscr{F}$. Then, if

$$P(A, \omega) = \begin{cases} P\{AA_n\}/P\{A_n\}, & \omega \in A_n,\ P\{A_n\} > 0 \\ P\{A\}, & \omega \in A_n,\ P\{A_n\} = 0, \end{cases}$$

$P(A, \omega) = P\{A \mid \mathscr{G}\}$ a.c. (as in Exercise 7.1.2), and so $P(A, \omega)$ is a regular conditional probability relative to $\mathscr{F}$ given $\mathscr{G}$.

A more interesting and important illustration is that of

EXAMPLE 2. Let X_1, X_2 be coordinate r.v.s on a probability space $(R^2, \mathscr{B}^2, P)$ with an absolutely continuous d.f. $F(x_1, x_2)$, i.e., for some Borel function f

$$F(x_1, x_2) = \int_{-\infty}^{x_2} \int_{-\infty}^{x_1} f(s, t)\, ds\, dt, \qquad (x_1, x_2) \in R^2. \tag{6}$$

Let $\mathscr{F}_1 = \mathscr{B}^2 = \sigma(X_1, X_2)$, $\mathscr{G} = \sigma(X_2) = R \times \mathscr{B} = \{R \times B: B \in \mathscr{B}\}$, and for $(x_1, x_2) \in R^2$ define

$$f_1(x_1) = \int_{-\infty}^{\infty} f(x_1, t)dt, \qquad f_2(x_2) = \int_{-\infty}^{\infty} f(s, x_2)ds, \tag{7}$$

$$f_1(x_1 \mid x_2) = \begin{cases} f(x_1, x_2)/f_2(x_2) & \text{if } f_2(x_2) > 0 \\ f_1(x_1) & \text{if } f_2(x_2) = 0. \end{cases} \tag{8}$$

By Fubini's theorem, $f_i(x_i)$ is a Borel function on R^1 for $i = 1, 2$, and so $f_1(x_1 \mid x_2)$ is a Borel function on R^2. For $B \in \mathscr{B}^2$ and $x = (x_1, x_2) \in R^2$ define

$$\mathrm{P}(B, x) = \int_{[s:\,(s, x_2) \in B]} f_1(s \mid x_2)ds. \tag{9}$$

Then for each $x \in R^2$, $\mathrm{P}(B, x)$ is a probability measure on $\mathscr{B}^2$, and for each $B \in \mathscr{B}^2$, $\mathrm{P}(B, x)$ is a Borel function in x_2 and hence $\sigma(X_2)$-measurable. Moreover, for $B \in \mathscr{B}^2$ and $A_2 = R \times B_2 \in \sigma(X_2)$

$$\begin{aligned}
\int_{A_2} \mathrm{P}(B, x)\, d\mathrm{P} &= \int_{B_2} \int_{-\infty}^{\infty} \mathrm{P}(B, (s, t)) f(s, t) ds\, dt \\
&= \int_{B_2} \int_{-\infty}^{\infty} \left[\int_{[u:\,(u, t) \in B]} f_1(u \mid t) du \right] f(s, t) ds\, dt \\
&= \int_{B_2} \int_{[u:\,(u, t) \in B]} f_1(u \mid t) f_2(t) du\, dt \\
&= \int_{B_2} \int_{[u:\,(u, t) \in B]} f(u, t) du\, dt = \int_{B_2} \int_{-\infty}^{\infty} I_B f(u, t) du\, dt \\
&= \mathrm{P}\{B \cdot A_2\},
\end{aligned}$$

and so according to (2), $\mathrm{P}(B, x) = \mathrm{P}\{B \mid X_2\}(x)$, a.c., for each $B \in \mathscr{B}^2$. Consequently, $\mathrm{P}(B, x)$ is a regular conditional probability measure on $\mathscr{B}^2$ given $\sigma(X_2)$. Hence, by Theorem 1 for any Borel function h on R^2 with

$$|\mathrm{E}\, h(X_1, X_2)| \le \infty$$

$$\mathrm{E}\{h(X_1, X_2) \mid X_2\} = \int_{-\infty}^{\infty} h(s, X_2) f_1(s \mid X_2) ds, \quad \text{a.c.} \tag{10}$$

The Borel function $f_1(x_1 \mid x_2)$ is called the **conditional density** of X_1 given $X_2 = x_2$, while $\int_{-\infty}^{\infty} h(s, x_2) f_1(s \mid x_2) ds$ is termed the conditional expectation of $h(X_1, X_2)$ given that $X_2 = x_2$ and is denoted by

$$\mathrm{E}\{h(X_1, X_2) \mid X_2 = x_2\}.$$

Analogously, $f_2(x_2 \mid x_1)$ is the conditional density of X_2 given $X_1 = x_1$.

Moreover,

$$F_1(x_1 \mid x_2) = \int_{[u < x_1]} f_1(u \mid x_2)du, \qquad F_2(x_2 \mid x_1) = \int_{[u < x_2]} f_2(u \mid x_1)du$$

are the **conditional d.f.s** of X_1 given $X_2 = x_2$ and of X_2 given $X_1 = x_1$ respectively, and from (10)

$$\mathrm{P}\{X_1 < x_1 \mid X_2\} = F_1(x_1 \mid X_2), \qquad \mathrm{P}\{X_2 < x_2 \mid X_1\} = F_2(x_2 \mid X_1).$$

As noted in Chapter 1, any random vector $X = (X_1, \ldots, X_n)$ on $(\Omega, \mathscr{F}, \mathrm{P})$ induces a probability measure P_X on the class $\mathscr{B}^n$ of n-dimensional Borel sets via $\mathrm{P}_X\{B\} = \mathrm{P}\{X^{-1}(B)\}$, $B \in \mathscr{B}^n$. For any σ-algebra $\mathscr{G} \subset \mathscr{F}$ if $\mathrm{P}_X\{B \mid \mathscr{G}\}$ is defined by $\mathrm{P}_X\{B \mid \mathscr{G}\} = \mathrm{P}\{X^{-1}(B) \mid \mathscr{G}\}$, $B \in \mathscr{B}^n$, then $\mathrm{P}_X\{B \mid \mathscr{G}\}$ is a $\mathscr{G}$-measurable r.v. for each $B \in \mathscr{B}^n$. Moreover, for any sequence of disjoint sets $B_m \in \mathscr{B}^n$, $m \geq 1$, there exists a null set $N \in \mathscr{G}$ depending in general on $\{B_m, m \geq 1\}$ such that

$$\mathrm{P}_X\left\{\bigcup_1^\infty B_m \,\middle|\, \mathscr{G}\right\}(\omega) = \sum_{m=1}^\infty \mathrm{P}_X\{B_m \mid \mathscr{G}\}(\omega), \qquad \omega \in N^c. \tag{11}$$

Can a version of $\mathrm{P}_X\{B \mid \mathscr{G}\}$ be chosen for each $B \in \mathscr{B}^n$ so that on the complement of a single null set $N \in \mathscr{G}$, (11) holds for all disjoint sequences $\{B_m, m \geq 1\} \subset \mathscr{B}^n$? An affirmative answer to this case has been given by Doob in Theorem 2 (below)

Definition. Let $X = (X_1, \ldots, X_n)$ or $(X_1, X_2, \ldots)$ be a stochastic process on $(\Omega, \mathscr{F}, \mathrm{P})$ and $\mathscr{G}$ a σ-algebra of events. A function $\mathrm{P}_X(B, \omega)$ on $\mathscr{B}^n \times \Omega$, $1 \leq n \leq \infty$, is termed a **regular conditional distribution for** X **given** $\mathscr{G}$ provided that for each $\omega \in \Omega$, $\mathrm{P}_X(B, \omega)$ is a probability measure on $\mathscr{B}^n$ and that for $B \in \mathscr{B}^n$, $\mathrm{P}_X(B, \omega)$ is $\mathscr{G}$-measurable with

$$\mathrm{P}_X(B, \omega) = \mathrm{P}\{X^{-1}(B) \mid \mathscr{G}\}, \qquad \text{a.c.} \tag{12}$$

Theorem 2 (Doob). *If $X = (X_1, \ldots, X_n)$ is a random vector on a probability space $(\Omega, \mathscr{F}, \mathrm{P})$ and $\mathscr{G}$ is a σ-algebra of events, there exists a regular conditional distribution for X given $\mathscr{G}$.*

PROOF. For any rational numbers λ_i, $1 \leq i \leq n$, define

$$F_n^\omega(\lambda_1, \ldots, \lambda_n) = \mathrm{P}\left\{\bigcap_{i=1}^n [X_i < \lambda_i] \,\middle|\, \mathscr{G}\right\}(\omega). \tag{13}$$

By the properties enumerated in (4) there is a null set $N \in \mathscr{G}$ such that for $\omega \in N^c$ and all rational numbers λ_i, λ_i', r_{im}

$$F_n^\omega(\lambda_1, \ldots, \lambda_n) \geq F_n^\omega(\lambda_1', \ldots, \lambda_n') \quad \text{if } \lambda_i > \lambda_i', \ 1 \leq i \leq n, \tag{i}$$

$$F_n^\omega(\lambda_1, \ldots, \lambda_n) = \lim_{\substack{r_{im} \uparrow \lambda_i \\ 1 \leq i \leq n}} F_n^\omega(r_{1m}, \ldots, r_{nm}), \tag{ii}$$

$$\lim_{\lambda_i \to -\infty} F_n^\omega(\lambda_1, \ldots, \lambda_n) = 0, \qquad 1 \le i \le n, \quad \lim_{\substack{\lambda_i \to \infty \\ 1 \le i \le n}} F_n^\omega(\lambda_1, \ldots, \lambda_n) = 1, \tag{iii}$$

$$\Delta_n^{\lambda, \lambda'} F_n^\omega \ge 0 \quad \text{if } \lambda \le \lambda', \tag{iv}$$

where the notation in (iv) is that of (5) of Section 6.3; also, $\lambda \le \lambda'$ signifies that the inequality holds for each coordinate. Define for any real numbers μ_i and $\omega \in N^c$

$$F_n^\omega(\mu_1, \ldots, \mu_n) = \lim_{\substack{\lambda_i \to \mu_i - \\ i \le i \le n}} F_n^\omega(\lambda_1, \ldots, \lambda_n), \qquad \lambda_i \text{ rational}, 1 \le i \le n, \tag{14}$$

while for $\omega \in N$, set

$$F_n^\omega(\mu_1, \ldots, \mu_n) = \mathrm{P}\left\{\bigcap_{i=1}^n [X_i < \mu_i]\right\}. \tag{15}$$

Then for each $\omega \in \Omega$, $F_n^\omega(\mu_1, \ldots, \mu_n)$ is an n-dimensional d.f. and hence determines a Lebesgue–Stieltjes measure μ_ω on $\mathscr{B}^n$ with $\mu_\omega\{R^n\} \equiv 1$. For $B \in \mathscr{B}^n$ and $\omega \in \Omega$ define

$$\mathrm{P}_X(B, \omega) = \mu_\omega\{B\}. \tag{16}$$

If

$$\mathscr{H} = \{B: B \in \mathscr{B}^n, \mathrm{P}_X(B, \omega) \text{ is } \mathscr{G}\text{-measurable}, \mathrm{P}_X(B, \omega) = \mathrm{P}\{X^{-1}(B) | \mathscr{G}\}, \text{a.c.}\},$$

$$\mathscr{D} = \left\{B: B = \mathop{\times}_{i=1}^n [-\infty, \lambda_i), \text{ rational } \lambda_i \in (-\infty, \infty), 1 \le i \le n\right\}$$

then $\mathscr{H}$ is a λ-class, $\mathscr{D}$ is a π-class, and by (13) through (16) $\mathscr{H} \supset \mathscr{D}$. Hence, by Theorem 1.3.2, $\mathscr{H} \supset \sigma(\mathscr{D}) = \mathscr{B}^n$, that is, $\mathrm{P}_X(B, \omega)$ is a regular conditional distribution for X given $\mathscr{G}$. □

Corollary 1. *If $X = (X_1, X_2, \ldots)$ is a (countable) stochastic process on $(\Omega, \mathscr{F}, \mathrm{P})$ and $\mathscr{G}$ is a σ-algebra of events, there exists a regular conditional distribution for X given $\mathscr{G}$.*

Proof. For all $n \ge 1$ and rational λ_i, $1 \le i \le n$, choose $F_n^\omega(\lambda_1, \ldots, \lambda_n)$ as in (13). Select the null set $N \in \mathscr{G}$ such that in addition to properties (i)–(iv), for $\omega \in N^c$

$$\lim_{\lambda_{n+1} \to \infty} F_{n+1}^\omega(\lambda_1, \ldots, \lambda_{n+1}) = F_n^\omega(\lambda_1, \ldots, \lambda_n), \qquad n \ge 1. \tag{v}$$

Define $F_n^\omega(\mu_1, \ldots, \mu_n)$ as in (14), (15). Then for each $\omega \in \Omega$, $\{F_n^\omega, n \ge 1\}$ is a consistent family of d.f.s and hence by Theorem 6.4.3 determines a measure μ_ω on $(R^\infty, \mathscr{B}^\infty)$. Define $\mathrm{P}_X(B, \omega) = \mu_\omega\{B\}$, $B \in \mathscr{B}^\infty$, and

$$\mathscr{H} = \{B: B \in \mathscr{B}^\infty, \mathrm{P}_X(B, \omega) \text{ is } \mathscr{G}\text{-measurable},$$
$$\mathrm{P}_X(B, \omega) = \mathrm{P}\{X^{-1}(B) | \mathscr{G}\}, \text{a.c.}\},$$

$$\mathscr{D} = \bigcup_{n=1}^\infty \left\{B: B \in \mathscr{B}^\infty, B = \mathop{\times}_{i=1}^n [-\infty, \lambda_i) \times R \times R \times \cdots \text{rational } \lambda_i \in (-\infty, \infty)\right\},$$

whence $\mathscr{H}$ is a λ-class, $\mathscr{D}$ is a π-class, and $\mathscr{H} \supset \mathscr{D}$, so that by Theorem 1.3.2 $\mathscr{H} \supset \sigma(\mathscr{D}) = \mathscr{B}^\infty$. □

Corollary 2. *Let $X = (X_1, \ldots, X_n)$ or $(X_1, X_2, \ldots)$ be a stochastic process on $(\Omega, \mathscr{F}, \mathrm{P})$ and $\mathscr{G}$ a σ-algebra of events. If $\mathrm{P}_X^\omega\{B\} = \mathrm{P}_X(B, \omega)$ is a regular conditional distribution for X given $\mathscr{G}$ and h is a Borel function on R^n, $1 \le n \le \infty$, with $|\mathrm{E}\, h(X)| \le \infty$, then*

$$\mathrm{E}\{h(X) \mid \mathscr{G}\}(\omega) = \int_{R^n} h(x)\, d\mathrm{P}_X^\omega(x), \qquad \text{a.c.},$$

where $\mathrm{P}_X^\omega(x) = \mathrm{P}_X^\omega\{\times_{i=1}^n (-\infty, x_i)\}$ for $x = (x_1, x_2, \ldots, x_n)$.

PROOF. As in Theorem 1. □

The question naturally poses itself whether a regular conditional distribution for X given $\mathscr{G}$ engenders a corresponding regular conditional probability on $\sigma(X)$ given $\mathscr{G}$. The answer is positive under the proviso of

Theorem 3. *Let $X = (X_1, \ldots, X_n)$ or $(X_1, X_2, \ldots)$ be a stochastic process on $(\Omega, \mathscr{F}, \mathrm{P})$ and $\mathscr{G}$ a σ-algebra of events. If the range of X, i.e., $\{X(\omega); \omega \in \Omega\}$ is a Borel set, then there exists a regular conditional probability on $\sigma(X)$ given $\mathscr{G}$.*

PROOF. Let $\mathrm{P}_X(B, \omega)$ be a regular conditional distribution for X given $\mathscr{G}$ and set $S = \{X(\omega): \omega \in \Omega\}$. Then

$$\mathrm{P}_X(S, \omega) = \mathrm{P}\{X^{-1}(S) \mid \mathscr{G}\}(\omega) = \mathrm{P}\{\Omega \mid \mathscr{G}\}(\omega) = 1$$

except for $\omega \in N \in \mathscr{G}$ with $\mathrm{P}\{N\} = 0$. If for $A \in \sigma(X)$ there are two sets B_1, $B_2 \in \mathscr{B}^n$ with $A = X^{-1}(B_1) = X^{-1}(B_2)$, then $B_1 \,\Delta\, B_2 \subset S^c$, implying

$$\mathrm{P}_X(B_1, \omega) = \mathrm{P}_X(B_1 B_2, \omega) = \mathrm{P}_X(B_2, \omega), \qquad \omega \in N^c.$$

Hence, for $A \in \sigma(X)$ it is permissible to define

$$\mathrm{P}(A, \omega) = \mathrm{P}_X(B, \omega), \qquad A = X^{-1}(B), \qquad \omega \in N^c,$$

and for any fixed $\omega_0 \in N^c$ set

$$\mathrm{P}(A, \omega) = \mathrm{P}_X(B, \omega_0), \qquad \omega \in N.$$

Then for each $\omega \in \Omega$, $\mathrm{P}(A, \omega)$ is a probability measure on $\sigma(X)$, and for each $A \in \sigma(X)$, $\mathrm{P}(A, \omega)$ is $\mathscr{G}$-measurable with

$$\mathrm{P}(A, \omega) = \mathrm{P}_X(B, \omega) = \mathrm{P}\{A \mid \mathscr{G}\}(\omega), \qquad \text{a.c.} \qquad □$$

The preceding notions will be utilized to prove the conditional Hölder inequality.

Theorem 4. *If X, Y are random variables on $(\Omega, \mathscr{F}, \mathrm{P})$, $\mathscr{G}$ is a σ-algebra of events and $1 < p < \infty$, $(1/p) + (1/q) = 1$, then*

$$\mathrm{E}\{|XY| \mid \mathscr{G}\} \le \mathrm{E}^{1/p}\{|X|^p \mid \mathscr{G}\} \cdot \mathrm{E}^{1/q}\{|Y|^q \mid \mathscr{G}\}, \qquad \text{a.c.} \tag{17}$$

PROOF. For $B \in \mathscr{B}^2$ and $\omega \in \Omega$, let $\hat{P}_\omega\{B\} = P(B, \omega)$ be a regular conditional distribution for (X, Y) given $\mathscr{G}$. By Corollary 2

$$E\{|XY| \,|\, \mathscr{G}\}(\omega) = \int_{R^2} |x_1 x_2| \, d\hat{P}_\omega, \qquad \text{a.c.,}$$

$$E^{1/p}\{|X|^p \,|\, \mathscr{G}\}(\omega) = \left(\int_{R^2} |x_1|^p \, d\hat{P}_\omega\right)^{1/p}, \qquad \text{a.c.,}$$

$$E^{1/q}\{|Y|^q \,|\, \mathscr{G}\}(\omega) = \left(\int_{R^2} |x_2|^q \, d\hat{P}_\omega\right)^{1/q}, \qquad \text{a.c.}$$

Since for each $\omega \in \Omega$, $\hat{P}_\omega$ is a probability measure on $\mathscr{B}^2$, (17) follows from the ordinary Hölder inequality. □

EXERCISES 7.2

1. If the events $\{B_j\}$ of $(\Omega, \mathscr{F}, P)$ constitute a finite or σ-partition of Ω in $\mathscr{F}$ and have positive probabilities, prove **Bayes' theorem**, i.e., for any event **A**

$$P\{B_i \,|\, A\} = \frac{P\{A \,|\, B_i\}P\{B_i\}}{\sum_j P\{A \,|\, B_j\}P\{B_j\}}.$$

2. If in Example 2, $F_i(x_i) = \lim_{x_{3-i} \to \infty} F(x_1, x_2)$, $i = 1, 2$, prove that the d.f. F_1 is a mixture of the family $F(x_1 \,|\, x_2)$ in the sense of Exercise 8.2.3.

3. If $X = (X_1, \ldots, X_n)$ is a random vector whose d.f. F is absolutely continuous relative to Lebesgue measure with density f find the conditional density of $(X_1, \ldots, X_j)$ given $X_{j+1} = x_{j+1}, \ldots, X_n = x_n$, where $1 \le j < n$ and verify the existence of a regular conditional probability measure on $\mathscr{B}^n$ given $\sigma(X_{j+1}, \ldots, X_n)$.

4. If r.v.s X and Y have a joint normal distribution (Exercise 6.5.12), show that the conditional distributions of X given Y and Y given X are both normal. Is the converse true?

5. If $X_1, \ldots, X_k$ have a multinomial distribution (Exercise 6.3.9) find the conditional distribution of X_1 given X_2, the conditional mean $E\{X_1 \,|\, X_2\}$, and the conditional variance $\sigma^2(X_1 | X_2)$.

6. (D. G. Kendall) If $A_1, \ldots, A_n$ are interchangeable events on $(\Omega, \mathscr{F}, P)$ and $S_n =$ number of events $A_1, \ldots, A_n$ that occur, prove for $1 \le i_1 < \cdots < i_j \le n$ and $1 \le j \le N \le n$ that

$$P\{A_{i_1} \cdots A_{i_j} \,|\, S_n = N\} = \binom{n-j}{N-j} \Big/ \binom{n}{N}.$$

7. Let $(\Omega, \mathscr{F}, P)$ be a probability space and $\mathscr{G}$ a σ-algebra of events. If $A \in \mathscr{F}$ and $B = \{\omega : P\{A \,|\, \mathscr{G}\} > 0\}$, verify that (i) $B \in \mathscr{G}$, (ii) $P\{A - B\} = 0$, and, moreover, if B' satisfies (i) and (ii) then $P\{B - B'\} = 0$ (i.e., B is a $\mathscr{G}$-measurable cover of A).

8. Prove the conditional Minkowski inequality (Exercise 6.2.7) and also the conditional Markov inequality $E\{|Y| \,|\, \mathscr{G}\} \ge cP\{|Y \ge c| \mathscr{G}\}$ for any $c > 0$, $Y \in \mathscr{L}_1$, and sub-σ-algebra $\mathscr{G}$.

7.3 Conditional Independence, Interchangeable Random Variables

The notion of conditional independence parallels that of unconditional or ordinary independence. As usual, all events and random variables pertain to the basic probability space $(\Omega, \mathscr{F}, P)$.

Definition. Let $\mathscr{G}$ be a σ-algebra of events and $\{\mathscr{G}_n, n \geq 1\}$ a sequence of classes of events. Two such classes $\mathscr{G}_1$ **and** $\mathscr{G}_2$ are said to be **conditionally independent given** $\mathscr{G}$ if for all $A_i \in \mathscr{G}_i$, $i = 1, 2$,

$$P\{A_1 A_2 \mid \mathscr{G}\} = P\{A_1 \mid \mathscr{G}\} \cdot P\{A_2 \mid \mathscr{G}\}, \quad \text{a.c.} \tag{1}$$

More generally, the sequence $\{\mathscr{G}_n, n \geq 1\}$ is declared **conditionally independent given** $\mathscr{G}$ if for all choices of $A_m \in \mathscr{G}_{k_m}$, where $k_i \neq k_j$ for $i \neq j$, $m = 1, 2, \ldots, n$, and $n = 2, 3, \ldots$,

$$P\{A_1 A_2 \cdots A_n \mid \mathscr{G}\} = \prod_{i=1}^{n} P\{A_i \mid \mathscr{G}\}, \quad \text{a.c.} \tag{2}$$

Finally, a sequence $\{X_n, n \geq 1\}$ of random variables is called **conditionally independent given** $\mathscr{G}$ if the sequence of classes $\mathscr{G}_n = \sigma(X_n)$, $n \geq 1$, is conditionally independent given $\mathscr{G}$.

If $\mathscr{G} = \{\varnothing, \Omega\}$, then conditional independence given $\mathscr{G}$ coalesces to ordinary (unconditional) independence, while if $\mathscr{G} = \mathscr{F}$, then every sequence of classes of events is conditionally independent given $\mathscr{G}$.

Independent r.v.s $\{X_n\}$ may lose their independence under conditioning; for example, if $\{X_n\}$ are Bernoulli trials with parameter p and $S_n = \sum_1^n X_i$, then $P\{X_i = 1 \mid S_2\} > 0$, $i = 1, 2$ for $S_2 = 0$ or 2, whereas $P\{X_1 = 1, X_2 = 1 \mid S_2\} = 0$ when $S_2 = 0$.

On the other hand, dependent r.v.s may gain independence under conditioning, i.e., become conditionally independent. Thus, if $\{X_n\}$ are independent integer valued r.v.s (see Exercise 7.3.3) and $S_n = \sum_1^n X_i$, clearly the r.v.s $\{S_n, n \geq 1\}$ are dependent. However, given that the event $\{S_2 = k\}$ (of positive probability) occurs,

$$\begin{aligned} P\{S_1 = i, S_3 = j \mid S_2\} &= \frac{P\{S_1 = i, S_2 = k, S_3 = j\}}{P\{S_2 = k\}} \\ &= \frac{P\{S_1 = i\}P\{X_2 = k - i\}P\{X_3 = j - k\}}{P\{S_2 = k\}} \\ &= P\{S_1 = i \mid S_2\} \frac{P\{X_3 = j - k, S_2 = k\}}{P\{S_2 = k\}} \\ &= P\{S_1 = i \mid S_2\}P\{S_3 = j \mid S_2\}. \end{aligned}$$

If the subscript n of S_n is interpreted as "time" and the r.v.s S_1, S_2, S_3 are envisaged as chance occurrences of the past, present and future times respectively, then the prior relationship may be picturesquely stated as, "The past and future are conditionally independent given the present." In fact, designating the r.v. S_n as "the present," the r.v.s $S_1, \ldots, S_{n-1}$ as "the past," and the r.v.s $S_{n+1}, \ldots, S_{n+m}$ as "the future," it may be verified for any $n \geq 2$, $m > 0$, that the past and the future are conditionally independent given the present. This property holds not only for sums S_n of independent r.v.s but more generally when the r.v.s $\{S_n, n \geq 1\}$ constitute a Markov chain (Exercise 7.3.1).

Theorem 1. *If $\mathscr{G}_i$, $i = 1, 2, 3$, are σ-algebras of events, then conditional independence of $\mathscr{G}_1$ and $\mathscr{G}_2$ given $\mathscr{G}_3$ is* **equivalent** *to any* **one** *of the following*:

i. For all $A_1 \in \mathscr{G}_1$,

$$P\{A_1 \mid \sigma(\mathscr{G}_2 \cup \mathscr{G}_3)\} = P\{A_1 \mid \mathscr{G}_3\}, \quad \text{a.c.} \tag{3}$$

ii. For every $D \in \mathscr{D}$ where $\mathscr{D}$ is a π-class with $\mathscr{G}_1 = \sigma(\mathscr{D})$,

$$P\{D \mid \sigma(\mathscr{G}_2 \cup \mathscr{G}_3)\} = P\{D \mid \mathscr{G}_3\}, \quad \text{a.c.} \tag{4}$$

iii. For every $D_i \in \mathscr{D}_i$ where $\mathscr{D}_i$ is a π-class with $\mathscr{G}_i = \sigma(\mathscr{D}_i)$, $i = 1, 2$,

$$P\{D_1 D_2 \mid \mathscr{G}_3\} = P\{D_1 \mid \mathscr{G}_3\} \cdot P\{D_2 \mid \mathscr{G}_3\}, \quad \text{a.c.} \tag{5}$$

iv. For every $\sigma(\mathscr{G}_1 \cup \mathscr{G}_3)$-measurable function X with $|E\,X| \leq \infty$,

$$E\{X \mid \sigma(\mathscr{G}_2 \cup \mathscr{G}_3)\} = E\{X \mid \mathscr{G}_3\}, \quad \text{a.c.} \tag{6}$$

PROOF. (i) (1) ⇒ (3): Let $A_i \in \mathscr{G}_i$, $i = 1, 2, 3$. By the definition of conditional expectation, (1), and Theorem 7.1.3,

$$\int_{A_2 A_3} I_{A_1} = \int_{A_3} I_{A_1 A_2} = \int_{A_3} P\{A_1 A_2 \mid \mathscr{G}_3\} = \int_{A_3} P\{A_1 \mid \mathscr{G}_3\} \cdot P\{A_2 \mid \mathscr{G}_3\}$$

$$= \int_{A_3} I_{A_2} P\{A_1 \mid \mathscr{G}_3\} = \int_{A_2 A_3} P\{A_1 \mid \mathscr{G}_3\},$$

whence (3) obtains by Theorem 7.1.1(ii). Conversely, by 7.1, (14)(v), Theorem 7.1.3, and (3),

$$\begin{aligned} P\{A_1 A_2 \mid \mathscr{G}_3\} &= E\{P\{A_1 A_2 \mid \sigma(\mathscr{G}_2 \cup \mathscr{G}_3)\} \mid \mathscr{G}_3\} \\ &= E\{I_{A_2} P\{A_1 \mid \sigma(\mathscr{G}_2 \cup \mathscr{G}_3)\} \mid \mathscr{G}_3\} \\ &= E\{I_{A_2} P\{A_1 \mid \mathscr{G}_3\} \mid \mathscr{G}_3\} = P\{A_1 \mid \mathscr{G}_3\} P\{A_2 \mid \mathscr{G}_3\}, \quad \text{a.c.} \end{aligned}$$

Apropos of (ii), since (3) ⇒ (4) trivially, it suffices to prove the reverse implication. If

$$\mathscr{A} = \{A : A \in \mathscr{F}, P\{A \mid \sigma(\mathscr{G}_2 \cup \mathscr{G}_3)\} = P\{A \mid \mathscr{G}_3\}, \text{a.c.}\},$$

then $\mathscr{A}$ is a λ-class $\supset \mathscr{D}$ via (4), and so by Theorem 1.3.2, $\mathscr{A} \supset \sigma(\mathscr{D}) = \mathscr{G}_1$, which is tantamount to (3).

Concerning (iii), it suffices to prove (5) $\Rightarrow$ (1), the converse being trivial. Since

$$\mathscr{A} = \{A_2 \in \mathscr{F} : \mathrm{P}\{D_1 A_2 \mid \mathscr{G}_3\} = \mathrm{P}\{D_1 \mid \mathscr{G}_3\} \cdot \mathrm{P}\{A_2 \mid \mathscr{G}_3\}, \text{ a.c., all } D_1 \in \mathscr{D}_1\},$$

$$\mathscr{A}^* = \{A_1 \in \mathscr{F} : \mathrm{P}\{A_1 A_2 | \mathscr{G}_3\} = \mathrm{P}\{A_1 | \mathscr{G}_3\} \cdot \mathrm{P}\{A_2 | \mathscr{G}_3\}, \text{ a.c., all } A_2 \in \mathscr{G}_2\}$$

are both λ-classes and $\mathscr{A} \supset \mathscr{D}_2$ by (5), Theorem 1.3.2 ensures that $\mathscr{A} \supset \sigma(\mathscr{D}_2) = \mathscr{G}_2$. Then a reapplication of this theorem yields $\mathscr{A}^* \supset \sigma(\mathscr{D}_1) = \mathscr{G}_1$ and (1) obtains.

Finally, to verify (iv) note that, taking $X = I_{A_1}$, (6) $\Rightarrow$ (3), and hence it suffices to prove (3) $\Rightarrow$ (6). For $A_i \in \mathscr{G}_i$, $i = 1, 3$ via (3),

$$\begin{aligned} \mathrm{P}\{A_1 A_3 | \sigma(\mathscr{G}_2 \cup \mathscr{G}_3)\} &= I_{A_3} \mathrm{P}\{A_1 | \sigma(\mathscr{G}_2 \cup \mathscr{G}_3)\} = I_{A_3} \mathrm{P}\{A_1 | \mathscr{G}_3\} \\ &= \mathrm{P}\{A_1 A_3 | \mathscr{G}_3\}, \quad \text{a.c.} \end{aligned}$$

If $\mathscr{A}$ denotes the class of non-negative $\mathscr{F}$-measurable functions for which (6) holds, then $\mathscr{A}$ is a λ-system which, as just shown, contains the indicator functions of all sets of a π-class generating $\sigma(\mathscr{G}_1 \cup \mathscr{G}_3)$. Hence, by Theorem 1.4.3, $\mathscr{A}$ contains all non-negative $\sigma(\mathscr{G}_1 \cup \mathscr{G}_3)$-measurable functions and therefore (6) holds. □

Corollary 1. *Random variables $X_1, \ldots, X_n$ on $(\Omega, \mathscr{F}, \mathrm{P})$ are conditionally independent given a σ-algebra $\mathscr{G}$ of events iff for all $(x_1, \ldots, x_n) \in R^n$*

$$\mathrm{P}\left\{\bigcap_{i=1}^{n} [X_i < x_i] \,\middle|\, \mathscr{G}\right\} = \prod_{i=1}^{n} \mathrm{P}\{X_i < x_i \mid \mathscr{G}\}, \qquad \text{a.c.}$$

PROOF. This follows easily from (iii). □

A direct application of (6) yields

Corollary 2. *If random variables X_1 and X_2 are conditionally independent given a σ-algebra $\mathscr{G}$ of events and $|\mathrm{E}\, X_1| \le \infty$, then*

$$\mathrm{E}\{X_1 \mid \sigma(\mathscr{G} \cup \sigma(X_2))\} = \mathrm{E}\{X_1 \mid \mathscr{G}\}, \qquad \text{a.c.}$$

Corollary 3. *Let $\mathscr{G}_i$ be a σ-algebra of events, $i = 1, 2, 3$. If $\sigma(\mathscr{G}_1 \cup \mathscr{G}_3)$ is independent of $\mathscr{G}_2$, then $\mathscr{G}_1$ and $\mathscr{G}_2$ are conditionally independent given $\mathscr{G}_3$ (as well as unconditionally independent).*

PROOF. Since (3) with the subscripts 1 and 2 interchanged holds via Exercise 7.1.1, this follows directly from Theorem 1. □

Recall that r.v.s $\{X_n, n \ge 1\}$ are called **interchangeable** if the joint distribution of every finite subset of k of these r.v.s depends only upon k and not the particular subset, $k \ge 1$.

Definition. A mapping $\pi = (\pi_1, \pi_2, \ldots)$ from the set N of all positive integers onto itself is called a **finite permutation** if π is one-to-one and $\pi_n = n$ for all but a finite number of integers. Let Q denote the set of all finite permutations π and let $\mathscr{B}^\infty$ be the class of Borel subsets of $R^\infty = R \times R \times \cdots$ and $X = (X_1, X_2, \ldots)$ a sequence of r.v.s on $(\Omega, \mathscr{F}, \mathrm{P})$. Define $\pi X = (X_{\pi_1}, X_{\pi_2}, \ldots)$ for $\pi = (\pi_1, \pi_2, \ldots)$. Then

$$\mathscr{S} = \{X^{-1}(B): B \in \mathscr{B}^\infty, \mathrm{P}\{X^{-1}(B) \,\Delta\, (\pi X)^{-1}(B)\} = 0, \text{all } \pi \in Q\}, \tag{7}$$

is called the **σ-algebra of permutable events** (of X).

Theorem 2. *Random variables $X_n, n = 1, 2, \ldots$, on $(\Omega, \mathscr{F}, \mathrm{P})$ are interchangeable* iff *they are conditionally independent and identically distributed given some σ-algebra $\mathscr{G}$ of events. Moreover, $\mathscr{G}$ can be taken to be either the σ-algebra $\mathscr{S}$ of permutable events or the tail σ-algebra $\mathscr{T}$, and*

$$\mathrm{P}\{X_1 < x_1 \mid \mathscr{S}\} = \mathrm{P}\{X_1 < x_1 \mid \mathscr{T}\}, \qquad \text{a.c.} \tag{8}$$

PROOF. Sufficiency is immediate since if

$$\mathrm{P}\{X_1 < x_1, \ldots, X_m < x_m \mid \mathscr{G}\} = \prod_{j=1}^m \mathrm{P}\{X_1 < x_j \mid \mathscr{G}\}, \qquad \text{a.c., } m \geq 1,$$

it follows upon taking expectations that $\{X_n, n \geq 1\}$ are interchangeable.

Apropos of necessity, for $n \geq 1$ and any real x, define

$$\xi_n(x) = \frac{1}{n} \sum_{i=1}^n I_{[X_i < x]}. \tag{9}$$

Then

$$\mathrm{E}[\xi_n(x) - \xi_m(x)]^2 = \frac{|m - n|}{m \cdot n} (\mathrm{P}\{X_1 < x\} - \mathrm{P}\{X_1 < x, X_2 < x\}) \to 0$$

as $m, n \to \infty$, whence $\xi_n(x) \xrightarrow{\mathrm{P}}$ some $\mathscr{T}$-measurable r.v. $\xi(x)$, implying (Exercise 3.3.1(iv))

$$\prod_{j=1}^m \xi_n(x_j) \xrightarrow{\mathrm{P}} \prod_{j=1}^m \xi(x_j), \qquad m \geq 1. \tag{10}$$

Let $\alpha_1, \ldots, \alpha_m$ denote positive integers and set

$$M = \{\alpha: \alpha = (\alpha_1, \ldots, \alpha_m), \text{ where } 1 \leq \alpha_i \leq n \text{ for } 1 \leq i \leq m\},$$

$$N = \{\alpha: \alpha \in M, \alpha_i \neq \alpha_j, i \neq j, 1 \leq i, j \leq m\},$$

$$I(\alpha) = I_{[X_{\alpha_1} < x_1, \ldots, X_{\alpha_m} < x_m]}.$$

Whenever $n > m$, taking cognizance of (9),

$$\prod_{j=1}^m \xi_n(x_j) = \frac{1}{n^m} \prod_{j=1}^m \sum_{i=1}^n I_{[X_i < x_j]} = \frac{1}{n^m} \sum_{\alpha \in M} I(\alpha)$$

$$= \frac{1}{n^m} \left(\sum_{\alpha \in M - N} + \sum_{\alpha \in N} \right) I(\alpha). \tag{11}$$

Since as $n \to \infty$

$$\frac{1}{n^m} \sum_{\alpha \in M-N} I(\alpha) \le \frac{1}{n^m} \sum_{\alpha \in M-N} 1 = \frac{n^m - n(n-1)\cdots(n-m+1)}{n^m} \to 0$$

for each element $A = X^{-1}(B) \in \mathscr{S}$, (11) entails

$$\int_A \prod_{j=1}^m \xi_n(x_j)\, dP = n^{-m} \sum_{\alpha \in N} \int_{X^{-1}(B)} I(\alpha)\, dP + o(1). \tag{12}$$

Now for any $\alpha \in N$, if $\pi \in Q$ is chosen so that $\pi_i = \alpha_i$, $1 \le i \le m$, then by definition of $\mathscr{S}$ and interchangeability

$$\int_{X^{-1}(B)} I(\alpha)\, dP = \int_{(\pi X)^{-1}(B)} I_{[X_{\alpha_1} < x_1, \ldots, X_{\alpha_m} < x_m]}\, dP$$

$$= \int_{X^{-1}(B)} I_{[X_1 < x_1, \ldots, X_m < x_m]}\, dP.$$

Thus, from (12)

$$\int_A \prod_{j=1}^m \xi_n(x_j)\, dP = \frac{n(n-1)\cdots(n-m+1)}{n^m}$$

$$\times \int_{X^{-1}(B)} I_{[X_1 < x_1, \ldots, X_m < x_m]}\, dP + o(1)$$

$$\to \int_A I_{[X_1 < x_1, \ldots, X_m < x_m]}\, dP. \tag{13}$$

On the other hand, the left side of (13) tends to $\int_A \prod_{j=1}^m \xi(x_j)\, dP$ by (10) and dominated convergence. Hence for any $A \in \mathscr{S}$, all real $x_1, \ldots, x_m$, and positive integers m,

$$\int_A \prod_{j=1}^m \xi(x_j)\, dP = \int_A I_{[X_1 < x_1, \ldots, X_m < x_m]}\, dP,$$

and therefore

$$E\left\{\prod_{j=1}^m \xi(x_j) \,\middle|\, \mathscr{S}\right\} = P\{X_1 < x_1, \ldots, X_m < x_m \mid \mathscr{S}\}, \qquad \text{a.c.} \tag{14}$$

Since $\mathscr{S} \supset \mathscr{T}$ and $\xi(x)$ is $\mathscr{T}$-measurable for every x, (14) entails

$$P\{X_1 < x_1, \ldots, X_m < x_m \mid \mathscr{S}\} = \prod_1^m \xi(x_j), \qquad \text{a.c.} \tag{15}$$

$$P\{X_1 < x_1, \ldots, X_m < x_m \mid \mathscr{T}\} = E\left\{\prod_1^m \xi(x_j) \,\middle|\, \mathscr{T}\right\} = \prod_1^m \xi(x_j), \qquad \text{a.c.} \tag{16}$$

In particular, for $m = 1$, $\xi(x_1) = P\{X_1 < x_1 \mid \mathscr{S}\} = P\{X_1 < x_1 \mid \mathscr{T}\}$, a.c., and by (15), (16)

$$\begin{aligned}
P\{X_1 < x_1, \ldots, X_m < x_m \mid \mathscr{S}\} &= \prod_1^m P\{X_1 < x_i \mid \mathscr{S}\} = \prod_1^m P\{X_1 < x_i \mid \mathscr{T}\} \\
&= P\{X_1 < x_1, \ldots, X_m < x_m \mid \mathscr{T}\}, \qquad \text{a.c.}
\end{aligned} \tag{17}$$

Finally, for any $j = 2, \ldots, m$, letting $x_i \to \infty$ for $i \neq j$, $1 \leq i \leq m$, in (17),

$$\begin{aligned}
P\{X_j < x_j \mid \mathscr{S}\} &= P\{X_1 < x_j \mid \mathscr{S}\} = P\{X_1 < x_j \mid \mathscr{T}\} \\
&= P\{X_j < x_j \mid \mathscr{T}\}, \qquad \text{a.c.}
\end{aligned} \tag{18}$$

Together, (17) and (18) assert that $\{X_n, n \geq 1\}$ are conditionally i.i.d. given either $\mathscr{S}$ or $\mathscr{T}$ and that (8) holds. □

Corollary 4 (de Finetti). *If* $\{X_n, n \geq 1\}$ *are interchangeable* r.v.s, *there exists a* σ*-algebra* $\mathscr{G}$ *of events such that for all* $m \geq 1$

$$P\{X_1 < x_1, \ldots, X_m < x_m\} = \int \prod_{j=1}^m P\{X_1 < x_j \mid \mathscr{G}\}\, dP.$$

Corollary 5. *If* r.v.s $\{X_n, n \geq 1\}$ *are conditionally independent given some* σ*-algebra* $\mathscr{G}$ *of events, there exists a regular conditional distribution, say* P^ω, *for* $X = (X_1, X_2, \ldots)$ *given* $\mathscr{G}$ *such that for each* $\omega \in \Omega$ *the coordinate* r.v.s $\{\xi_n, n \geq 1\}$ *of the probability space* $(R^\infty, \mathscr{B}^\infty, P^\omega)$ *are independent. Moreover, if* $X_j \in \mathscr{L}_1$, $1 \leq j \leq n$, *and* $E[X_1 X_2 \cdots X_n]$ *exists,*

$$E\{X_1 X_2 \cdots X_n \mid \mathscr{G}\} = \prod_{i=1}^n E\{X_i \mid \mathscr{G}\}, \qquad \text{a.c., } n \geq 2. \tag{19}$$

PROOF. For any rational λ, define $F_i^\omega(\lambda) = P\{X_i < \lambda \mid \mathscr{G}\}(\omega)$, $i \geq 1$. There exists a null set $N \in \mathscr{G}$ such that for $\omega \in N^c$, all positive integers n, and all rational $\lambda, \lambda', r, r_1, \ldots, r_n$:

$$P\left\{\bigcap_{i=1}^n [X_i < r_i] \,\middle|\, \mathscr{G}\right\}(\omega) = \prod_{i=1}^n F_i^\omega(r_i);$$

$$F_i^\omega(\lambda) \geq F_i^\omega(\lambda'), \quad \lambda > \lambda'; \qquad F_i^\omega(\lambda) = \lim_{r \uparrow \lambda} F_i^\omega(r), \quad 1 \leq i \leq n;$$

$$\lim_{\lambda \to -\infty} F_i^\omega(\lambda) = 0, \qquad \lim_{\lambda \to \infty} F_i^\omega(\lambda) = 1, \qquad 1 \leq i \leq n.$$

Define for $\omega \in N^c$, rational $r_1, \ldots, r_n$, and all $n \geq 2$,

$$F_{n,\omega}(r_1, \ldots, r_n) = \prod_{i=1}^n F_i^\omega(r_i).$$

Then properties (i)–(v) of Theorem 7.2.2 and Corollary 7.2.1 hold, so that defining $F_{n,\omega} = \prod_{j=1}^n F_{X_j}$ for $\omega \in N$ and continuing as in the proof of Corollary 7.2.1, existence is ensured of a regular conditional distribution P^ω for X given $\mathscr{G}$.

Let $\{\xi_n, n \geq 1\}$ be coordinate r.v.s on $(R^\infty, \mathscr{B}^\infty, P^\omega)$. Since for all $n \geq 2$ and real $x_1, \ldots, x_n$

$$P^\omega\{\xi_1 < x_1, \ldots, \xi_n < x_n\} = \prod_{i=1}^n F_i^\omega(x_i -) = \prod_{i=1}^n P^\omega\{\xi_i < x_i\}$$

for $\omega \in N^c$, while

$$P^\omega\{\xi_1 < x_1, \ldots, \xi_n < x_n\} = F_{n,\omega}(x_1, \ldots, x_n) = \prod_{i=1}^n P\{X_i < x_i\}$$

$$= \prod_{i=1}^n P^\omega\{\xi_i < x_i\}$$

for $\omega \in N$, it follows that $\{\xi_n, n \geq 1\}$ are independent r.v.s on $(R^\infty, \mathscr{B}^\infty, P^\omega)$ for each $\omega \in \Omega$.

Apropos of (19), if P_X^ω is a regular conditional distribution for $X = (X_1, X_2, \ldots)$ given $\mathscr{G}$, then since P_X is a product measure, say, $P_1 \times P_2 \times \cdots$ on $\mathscr{B}^\infty$, via Corollary 7.2.2 and Fubini's theorem

$$E\{X_1 X_2 \cdots X_n | \mathscr{G}\} = \int_{R^\infty} x_1 x_2 \cdots x_n \, dP_X^\omega.$$

$$= \prod_{i=1}^n \int_R x_i \, dP_i(x_i) = \prod_{i=1}^n E\{X_i | \mathscr{G}\}, \qquad \text{a.c.} \qquad \square$$

EXAMPLE 1. Let $\{X_n, n \geq 1\}$ be interchangeable r.v.s on $(\Omega, \mathscr{F}, P)$ and $\mathscr{G} = \mathscr{T}$ or $\mathscr{S}$ (as in Theorem 2). (i) If $X_1 \in \mathscr{L}_1$, then $1/n \sum_{i=1}^n X_i \xrightarrow{\text{a.c.}} E\{X_1 | \mathscr{G}\}$ (ii) If $n \cdot P\{|X_1| > n\} = o(1)$, then $1/n \sum_{i=1}^n X_i - E\{X_1 | \mathscr{G}\} \xrightarrow{P} 0$.

PROOF. Let P^ω be a regular conditional distribution for $X = (X_1, X_2, \ldots)$ given $\mathscr{G}$ such that the coordinate r.v.s $\{\xi_n, n \geq 1\}$ of the probability space $(R^\infty, \mathscr{B}^\infty, P^\infty\}$ are i.i.d. for each $\omega \in \Omega$. If $S_n = \sum_1^n X_i$ and $E^\omega \xi_1 = \int_{R^\infty} \xi_1 \, dP^\omega$, then by Corollary 7.2.2 $E\{X_1 | \mathscr{G}\} = E^\omega \xi_1$, a.c. By Corollaries 7.2.2 and 7.3.5 for any $\varepsilon > 0$

$$\lim_{m \to \infty} P\left\{\bigcup_{n=m}^\infty \left[\left|\frac{1}{n} S_n - E\{X_1 | \mathscr{S}\}\right| > \varepsilon\right]\right\} = \lim_{m \to \infty} P\left\{\bigcup_{n=m}^\infty \left[\left|\frac{1}{n} S_n(\omega) - E^\omega \xi_1\right| > \varepsilon\right]\right\}$$

$$= \lim_{m \to \infty} E\, P^\omega\left\{\bigcup_{n=m}^\infty \left[\left|\frac{1}{n} \sum_{i=1}^n \xi_i - E^\omega \xi_1\right| > \varepsilon\right]\right\} = 0. \qquad \square$$

Apropos of (ii), set $S_n' = \sum_{j=1}^n X_j'$, $T_n' = \sum_{j=1}^n \xi_j$, where $X_j' = X_j I_{[|X_j| \leq n]}$, $\xi_j' = \xi_j I_{[|\xi_j| \leq n]}$, $1 \leq j \leq n$. Then $P\{S_n \neq S_n'\} \leq \sum_{j=1}^n P\{|X_j| > n\} = o(1)$ and

$$E(S_n' - n\, E\{X_1' | \mathscr{G}\})^2 = EE^\omega(T_n' - n\, E^\omega \xi_1')^2 = EE^\omega\left[\sum_{j=1}^n (\xi_j' - E^\omega \xi_1')\right]^2$$

$$= n\, EE^\omega(\xi_1' - E^\omega \xi_1')^2 \leq n\, E\, \xi_1'^2 = o(n^2)$$

according to (17) of Theorem 5.2.4. Thus, $S_n'/n - E\{X_1 | \mathscr{G}\} \xrightarrow{P} 0$, and so $S_n/n - E\{X_1 | \mathscr{G}\} \xrightarrow{P} 0$. $\square$

In view of Corollaries 4 and 5 it seems plausible that many results valid for sequences of i.i.d. random variables carry over to sequences of interchangeable r.v.s (see Chapter 9).

Theorem 3. *Let $\{X_n, n \geq 1\}$ be conditionally independent given a σ-algebra $\mathscr{G}$ of events and let $\mathscr{T} = \bigcap_{n=1}^{\infty} \sigma(X_j, j \geq n)$ denote the tail σ-algebra. Then for any $T \in \mathscr{T}$ there exists some $G \in \mathscr{G}$ with* $\mathrm{P}\{G \,\Delta\, T\} = 0$.

PROOF. Via Corollary 1, $\sigma(X_1, \ldots, X_n)$ and $\sigma(X_j, j > n)$ are conditionally independent given $\mathscr{G}$ for any $n \geq 1$. Then $\bigcup_{n=1}^{\infty} \sigma(X_1, \ldots, X_n)$ and $\mathscr{T}$ are conditionally independent given $\mathscr{G}$, whence by Theorem 1(iii), $\sigma(X_1, X_2, \ldots)$ and $\mathscr{T}$ are conditionally independent given $\mathscr{G}$. Hence, for $T \in \mathscr{T}$, $\mathrm{P}\{T \mid \mathscr{G}\} = \mathrm{P}^2\{T \mid \mathscr{G}\}$, a.c., implying $\mathrm{P}\{\mathrm{P}\{T \mid \mathscr{G}\} = 0 \text{ or } 1\} = 1$. Thus,

$$\mathrm{P}\{T \mid \mathscr{G}\} = I_G, \qquad \text{a.c.,}$$

for some $G \in \mathscr{G}$, implying

$$\mathrm{P}\{T \cdot G\} = \int_G \mathrm{P}\{T \mid \mathscr{G}\}\, d\mathrm{P} = \mathrm{P}\{G\}$$

or $\mathrm{P}\{G \cdot T^c\} = 0$. Analogously, $\mathrm{P}\{T \cdot G^c\} = 0$, and so $\mathrm{P}\{G \,\Delta\, T\} = 0$. □

Theorem 2 states that interchangeable r.v.s $\{X_n, n = 1, 2, \ldots\}$ are conditionally i.i.d. given either the tail σ-algebra $\mathscr{F}$ or the σ-algebra $\mathscr{S}$ of permutable events. What is the relationship between $\mathscr{S}$ and $\mathscr{T}$? More generally, if $\{X_n, n \geq 1\}$ are conditionally i.i.d. given $\mathscr{G}$ and also given $\mathscr{H}$, what is the connection between $\mathscr{G}$ and $\mathscr{H}$?

Definition. If $(\Omega, \mathscr{F}, \mathrm{P})$ is a probability space and $\mathscr{G}$ is a sub-σ-algebra of $\mathscr{F}$, the **completion** $\mathscr{G}^*$ of $\mathscr{G}$ is a sub-σ-algebra of $\mathscr{F}$ (Exercise 1.5.9) defined by

$$\mathscr{G}^* = \{G \,\Delta\, N : G \in \mathscr{G}, N \in \mathscr{F} \text{ with } \mathrm{P}\{N\} = 0\}.$$

This definitition of completion differs slightly from the customary one which replaces $\mathscr{G}^*$ by $\mathscr{G}_* = \{G \,\Delta\, M : G \in \mathscr{G},\ M \subset N \in \mathscr{F},\ \mathrm{P}\{N\} = 0\}$. However, $\mathscr{G}_*$ may not be a sub-σ-algebra of $\mathscr{F}$, whereas $\mathscr{G}^* \subset \mathscr{F}$.

Lemma 1. *Let $\mathscr{G}$ be a σ-algebra of events and $\mathscr{G}^*$ its completion. If X^* is any $\mathscr{G}^*$-measurable function, there exists a $\mathscr{G}$-measurable function X such that* $X = X^*$, a.c.

PROOF. Let

$$\mathscr{H} = \{X^* : X^* \geq 0,\ X^* \text{ is } \mathscr{G}^*\text{-measurable, and } X^* = X, \text{ a.c., for some } \mathscr{G}\text{-measurable function } X\}.$$

Then $\mathscr{H}$ is a monotone system and $I_G \in \mathscr{H}$ for all $G \in \mathscr{G}^*$. Hence, $\mathscr{H}$ contains all nonnegative $\mathscr{G}^*$-measurable functions. The general case now follows from $X^* = X^{*+} - X^{*-}$. □

Lemma 2. *Let $\mathscr{G}^*$ be the completion of a σ-algebra $\mathscr{G}$ of events.* (i) *If* $|\mathrm{E}\,X| \le \infty$, *then* $\mathrm{E}\{X \mid \mathscr{G}^*\} = \mathrm{E}\{X \mid \mathscr{G}\}$, a.c. (ii) r.v.s $\{X_n,\ n \ge 1\}$ *are conditionally independent (resp. conditionally* i.i.d.) *given* $\mathscr{G}$ iff $\{X_n, n \ge 1\}$ *are conditionally independent (resp. conditionally* i.i.d.) *given* $\mathscr{G}^*$.

PROOF. (i) Since by Lemma 1, $\mathrm{E}\{X \mid \mathscr{G}^*\} = Y$, a.c., where Y is $\mathscr{G}$-measurable, $\mathrm{E}\{X \mid \mathscr{G}^*\} = \mathrm{E}\{X \mid \mathscr{G}\}$, a.c., whence (ii) follows immediately from (i). □

Theorem 4. *Let* $\{X_n,\ n \ge 1\}$ *be a sequence of* r.v.s, $\mathscr{G}$ *a sub-σ-algebra of* $\sigma(X_n, n \ge 1)$, *and* $\mathscr{T}$ *the tail σ-algebra. If* $\mathscr{G}^*$ *and* $\mathscr{T}^*$ *are the completions of* $\mathscr{G}$ *and* $\mathscr{T}$ *respectively and* $\{X_n, n \ge 1\}$ *are conditionally* i.i.d. *given* $\mathscr{G}$, *then* $\mathscr{G}^* = \mathscr{T}^*$.

PROOF. According to Theorem 3, $\mathscr{G}^* \supset \mathscr{T}$ and hence $\mathscr{G}^* \supset \mathscr{T}^*$. By Theorem 2, $\{X_n, n \ge 1\}$ are interchangeable and the proof of this theorem shows that

$$\xi_n(x) = \frac{1}{n} \sum_{j=1}^{n} I_{[X_j < x]} \xrightarrow{\mathscr{L}_2} \mathrm{P}\{X_1 < x \mid \mathscr{T}\},$$

whence (Exercise 7.1.9)

$$\mathrm{E}\left\{\frac{1}{n} \sum_{j=1}^{n} I_{[X_j<x]} \middle| \mathscr{G}^*\right\} \xrightarrow{\mathrm{P}} \mathrm{E}\{\mathrm{P}\{X_1 < x|\mathscr{T}\}|\mathscr{G}^*\} = \mathrm{P}\{X_1 < x|\mathscr{T}\}.$$

By Lemma 2, $\{X_n\}$ are conditionally i.i.d. given $\mathscr{G}^*$, implying

$$\mathrm{E}\left\{\frac{1}{n} \sum_{j=1}^{n} I_{[X_j < x]} \middle| \mathscr{G}^*\right\} = \mathrm{P}\{X_1 < x \mid \mathscr{G}^*\} = \mathrm{P}\{X_1 < x \mid \mathscr{T}\}, \qquad \text{a.c.}$$

Consequently, for $A \in \mathscr{G}$ and $(x_1, \ldots, x_m) \in R^m$, employing Lemma 2 again,

$$\mathrm{P}\left\{A \cdot \bigcap_1^m [X_i < x_i] \middle| \mathscr{T}\right\} = \mathrm{E}\left[I_A \mathrm{P}\left\{\bigcap_1^m [X_i < x_i] \middle| \mathscr{G}^*\right\} \middle| \mathscr{T}\right]$$

$$= \mathrm{E}\left[I_A \prod_{i=1}^{m} \mathrm{P}\{X_i < x_i \mid \mathscr{G}^*\} \middle| \mathscr{T}\right] = \mathrm{E}\left[I_A \prod_{i=1}^{m} \mathrm{P}\{X_i < x_i \mid \mathscr{T}\} \middle| \mathscr{T}\right]$$

$$= \mathrm{P}\{A \mid \mathscr{T}\} \cdot \mathrm{P}\left\{\bigcap_1^m [X_i < x_i] \middle| \mathscr{T}\right\}, \qquad \text{a.c.}$$

Hence, by Theorem 1, $\mathscr{G}$ and $\sigma(X_1, \ldots, X_m)$ are conditionally independent given $\mathscr{T}$ for any positive integer m. Therefore, $\mathscr{G}$ and $\sigma(X_1, X_2, \ldots)$ are conditionally independent given $\mathscr{T}$. Since $\mathscr{G} \subset \sigma(X_1, X_2, \ldots)$, $\mathrm{P}\{G \mid \mathscr{T}\} = 0$ or 1, a.c., for $G \in \mathscr{G}$. For any $G \in \mathscr{G}$ if $T = \{\omega\colon \mathrm{P}\{G \mid \mathscr{T}\} = 1\}$, then $T \in \mathscr{T}$ and

$$\mathrm{P}\{TG\} = \mathrm{E}\,\mathrm{P}\{TG \mid \mathscr{T}\} = \mathrm{E}\,I_T\,\mathrm{P}\{G \mid \mathscr{T}\} = \mathrm{P}\{T\},$$
$$\mathrm{P}\{G\} = \mathrm{E}\,\mathrm{P}\{G \mid \mathscr{T}\} = \mathrm{E}\,I_T\,\mathrm{P}\{G \mid \mathscr{T}\} = \mathrm{P}\{TG\}.$$

Therefore, $\mathrm{P}\{T \,\Delta\, G\} = 0$, whence $G \in \mathscr{T}^*$. Hence, $\mathscr{G} \subset \mathscr{T}^*$, implying $\mathscr{G}^* \subset \mathscr{T}^*$. Consequently, $\mathscr{G}^* = \mathscr{T}^*$. □

It follows directly from Theorem 4 that if $\{X_n, n \geq 1\}$ are conditionally i.i.d. given $\mathscr{G}$ and also given $\mathscr{H}$, then $\mathscr{G}^* = \mathscr{H}^*$.

Definition. A σ-algebra $\mathscr{G}$ of events is called **degenerate** if $P\{G\} = 0$ or 1, for all $G \in \mathscr{G}$.

Corollary 6. *If $\{X_n, n \geq 1\}$ are conditionally* i.i.d. *given a σ-algebra $\mathscr{G} \subset \sigma(X_n, n \geq 1)$ and the tail σ-algebra is degenerate, then $\mathscr{G}$ is degenerate.*

Corollary 7. *If $\{X_n, n \geq 1\}$ are interchangeable* r.v.s *with a degenerate tail σ-algebra, then the σ-algebra of permutable events is degenerate.*

In particular, Corollary 7 in conjunction with the Kolmogorov zero–one law yields

Corollary 8 (Hewitt–Savage Zero–One Law). *If $\{X_n, n \geq 1\}$ are* i.i.d. r.v.s, *then the σ-algebra of permutable events is degenerate.*

EXERCISES 7.3

1. Let S denote the set of positive or nonnegative or all integers and let $\{p_n, n \in S\}$ be a probability density function. Define $P = \{p_{ij}\}$ to be a **stochastic matrix**, that is, $p_{ij} \geq 0$ for all $i, j \in S$, and $\sum_{j \in S} p_{ij} = 1$ for all $i \in S$. (i) Verify that $P\{\bigcap_{j=0}^n [X_j = i_j]\} = p_{i_0} \prod_{j=0}^{n-1} p_{i_j, i_{j+1}}$ satisfies the consistency requirement of Theorem 6.4.3. The r.v.s $\{X_n, n \geq 0\}$ are called a temporally homogeneous **Markov chain** with state space S. (ii) Check that $p_{ij} = P\{X_{n+1} = j \mid X_n = i\}$, $n \geq 0$ for all $i, j \in S$, and verify for $n \geq 1$, $m > 0$, that "the past" $X_0, \ldots, X_{n-1}$ and "the future" $X_{n+1}, \ldots, X_{n+m}$ are conditionally independent given "the present" X_n. (iii) Verify that the product of two stochastic matrices is a stochastic matrix and interpret the equation $P^{m+n} = P^m \cdot P^n$ probabilistically.
2. Verify that the class $\mathscr{S}$ of permutable events as defined in (7) is a σ-algebra and prove Corollary 1.
3. If $S_n = \sum_1^n X_i$ where $\{X_n, n \geq 1\}$ are independent r.v.s then S_1 and S_3 are conditionally independent given S_2. *Hint*: Via Corollary 7.1.2 if F_3 is the d.f. of X_3
$$P\{S_1 < y_1, S_3 < y_3 | S_1, S_2\} = I_{[S_1 < y_1]} \cdot F_3(y_3 - S_2)$$
4. If $\{X_n, n \geq 1\}$ are interchangeable r.v.s, then for all $r > 0$
$$E|X_1 X_2 \cdots X_n|^r = \int E^n\{|X_1|^r \mid \mathscr{S}\}\, dP, \qquad n = 1, 2, \ldots,$$
where $\mathscr{S}$ is the σ-algebra of permutable events.
5. There exist interchangeable r.v.s $\{X_n, n \geq 1\}$ not i.i.d. with $E X_1^2 = \infty$ but
$$E X_1 X_2 \cdots X_k$$
finite for all $k \geq 1$ (in particular, with zero covariances).

6. If a r.v. X is independent of the i.i.d. r.v.s $\{X_n, n \geq 1\}$, show that $Y_n = X_n + X$, $n \geq 1$, are interchangeable r.v.s. Express that joint d.f. of $(Y_1, \ldots, Y_n)$ in terms of the d.f.s of X and X_1.

7. For any $n \geq 2$, find n interchangeable r.v.s $\{X_1, \ldots, X_n\}$ which cannot be embedded in a collection of $n + 1$ interchangeable r.v.s $\{X_1, \ldots, X_n, X_{n+1}\}$. *Hint*: Recall Exercise 6.3.10.

8. If $\{X_n, n \geq 1\}$ are $\mathscr{L}_1$ interchangeable r.v.s with $\rho = \mathrm{E}(X_1 - \mathrm{E}\, X_1)(X_2 - \mathrm{E}\, X_1)$, then $\mathrm{E}\, X_1 X_2 = 0$ iff $\mathrm{E}\, X_1 = 0$ and $\rho = 0$.

9. If $\{X_n, n \geq 1\}$ are $\mathscr{L}_2$ interchangeable r.v.s and $\{Y_n, n \geq 1\}$ is defined by $X_n = Y_n + \mathrm{E}\{X_1 | \mathscr{S}\}$, then $\{Y_n, n \geq 1\}$ constitute uncorrelated interchangeable r.v.s.

10. Prove that if $\{X_n\}$ are $\mathscr{L}_1$ interchangeable r.v.s then

$$\mathrm{E}\left|\frac{1}{n}\sum_1^n X_j\right| \leq \mathrm{E}\left|\frac{1}{n-1}\sum_1^{n-1} X_j\right|, \qquad n \geq 2.$$

11. Prove that if the r.v.'s $\{X_n, n \geq 1\}$ of Corollary 5 are conditionally i.i.d. given $\mathscr{G}$, then the coordinate r.v.'s $\{\xi_n, n \geq 1\}$ are i.i.d.

7.4 Introduction to Martingales

The indebtedness of probability theory to gambling is seldom as visible as in the concept and development of martingales. The underlying notion is that of a fair game in which regardless of the whims of chance in assigning outcomes to the past and present, one's (conditional) total expected future fortune is precisely one's current aggregate. Analogously, submartingales and supermartingales correspond to favorable and unfavorable games, respectively. Although martingales were first discussed by Lévy, the realization of their potential, the fundamental development of the subject and indeed most of the theorems presented are due to Doob.

Let $(\Omega, \mathscr{F}, \mathrm{P})$ be a probability space and N some subset of the integers and $\pm\infty$, that is, $N \subset \{-\infty, \ldots, -2, -1, 0, 1, 2, \ldots, \infty\}$. A sequence $\mathscr{F}_n$ of sub-σ-algebras of $\mathscr{F}$ with indices in N will be called a stochastic basis if it is increasing, i.e., $\mathscr{F}_m \subset \mathscr{F}_n$ for $m < n$. If $\{\mathscr{F}_n, n \in N\}$ is a stochastic basis and S_n is an $\mathscr{F}_n$-measurable function for each $n \in N$, then $\{S_n, \mathscr{F}_n, n \in N\}$ is called a **stochastic sequence**. An $\mathscr{L}_p$ **stochastic sequence**, $p > 0$, is one for which $\mathrm{E}|S_n|^p < \infty$, $n \in N$, and correspondingly an $\mathscr{L}_p$ **bounded** stochastic sequence is one satisfying $\sup_{n \in N} \mathrm{E}|S_n|^p < \infty$.

Definition. A **submartingale** is a stochastic sequence $\{S_n, \mathscr{F}_n, n \in N\}$ with $|\mathrm{E}\, S_n| \leq \infty$, $n \in N$, such that for all $m, n \in N$ with $m < n$

$$\mathrm{E}\{S_n | \mathscr{F}_m\} \geq S_m, \qquad \text{a.c.} \tag{1}$$

If $\{-S_n, \mathscr{F}_n, n \in N\}$ is a submartingale, then $\{S_n, \mathscr{F}_n, n \in N\}$ is called a **supermartingale**. Moreover, if $\{S_n, \mathscr{F}_n, n \in N\}$ is both a submartingale and a supermartingale, it is termed a **martingale**.

Definition. If $\{\mathscr{F}_n, n \in N\}$ is a stochastic basis, a **stopping time** or $\mathscr{F}_n$-time T is a function from Ω to $N \cup \{\infty\}$ such that $\{T = n\} \in \mathscr{F}_n$, $n \in N$, and

$$P\{[T \in N] \cup [T = \infty]\} = 1.$$

When $N = \{1, 2, \ldots, \infty\}$, the preceding coincides with the definition of Section 5.3 and T is a **finite stopping time** or **stopping variable** if

$$P\{T = \infty\} = 0.$$

If $\{S_n, \mathscr{F}_n, n \in N\}$ is a submartingale and there exists a measurable function R with $|E\, R| \leq \infty$ (resp. an $\bigcap_{n \in N} \mathscr{F}_n$-measurable function L with $|E\, L| \leq \infty$) such that for each $n \in N$

$$E\{R \mid \mathscr{F}_n\} \geq S_n \quad \left(\text{resp. } E\left\{S_n \,\middle|\, \bigcap_{n \in N} \mathscr{F}_n\right\} \geq L\right), \qquad \text{a.c.,} \tag{2}$$

then $\{S_n, \mathscr{F}_n, N\}$ is said to be **closed on the right** (resp. left). A **closed submartingale** $\{S_n, \mathscr{F}_n, N\}$ is one which is closed both on the right and left. A submartingale $\{S_n, \mathscr{F}_n, N\}$ is declared to have a **last** (resp., first) **element** if N has a maximum (resp. minimum). Obviously, a submartingale with a last (resp. first) element is closed on the right (resp. left). A martingale $\{S_n, \mathscr{F}_n, N\}$ is said to be **closed on the right** by R (resp. on left by L) if (2) holds with equality.

The submartingale $\{S_n, \mathscr{F}_n, n \geq 1\}$ will be closed iff it is closed on the right since it is automatically closed on the left by S_1. An analogous statement holds for $\{S_n, \mathscr{F}_n, n \leq -1\}$.

The important special case of a submartingale or martingale $\{S_n, \mathscr{F}_n, N\}$ with $\mathscr{F}_n = \sigma(S_m, m \leq n, m \in N)$, $n \in N$, will be denoted by $\{S_n, N\}$ or $\{S_n, n \in N\}$.

Simple properties of conditional expectations readily entail the following:

i. The stochastic sequence $\{S_n, \mathscr{F}_n, -\infty < n < \infty\}$ is a submartingale iff for every finite integer n, $|E\, S_n| \leq \infty$ and $E\{S_{n+1} \mid \mathscr{F}_n\} \geq S_n$, a.c.
ii. If $\{S_n, \mathscr{F}_n, N\}$ is a submartingale and $\{S_n, \mathscr{G}_n, n \in N\}$ is a stochastic sequence with $\mathscr{G}_n \subset \mathscr{F}_n$, $n \in N$, then $\{S_n, \mathscr{G}_n, N\}$ is a submartingale.
iii. If both $\{S_n, \mathscr{F}_n, N\}$ and $\{S'_n, \mathscr{F}_n, N\}$ are submartingales and $E\, S_n + E\, S'_n$ exists for each $n \in N$, then $\{S_n + S'_n, \mathscr{F}_n, N\}$ is a submartingale.
iv. An $\mathscr{L}_1$ stochastic sequence $\{S_n = \sum_1^n X_i, \mathscr{F}_n, -\infty < n < \infty\}$ is a submartingale (resp. martingale) iff for $-\infty < n < \infty$

$$E\{X_{n+1} \mid \mathscr{F}_n\} \geq 0 \quad (\text{resp. } E\{X_{n+1} \mid \mathscr{F}_n\} = 0), \qquad \text{a.c.;}$$

the r.v.s $\{X_n\}$ are called **martingale differences** in the latter case.
v. If $\{S_n = \sum_{i=1}^n X_i, \mathscr{F}_n, n \geq 1\}$ is an $\mathscr{L}_2$ martingale, then

$$E\, S_n^2 = \sum_{j=1}^n E\, X_j^2, \qquad n \geq 1.$$

The condition of (iv) corroborates the view of S_n as an aggregate of outcomes X_n of favorable or fair games.

If, as in Example 3, $\{S_n, \mathscr{F}_n, n \geq m\}$ is a stochastic sequence such that $\{S_n^*, \mathscr{F}_n^*, n \leq -m\}$ is a martingale, where $S_n^* = S_{-n}, \mathscr{F}_n^* = \mathscr{F}_{-n}, n \leq -m$, then $\{S_n, \mathscr{F}_n, n \geq m\}$ is sometimes alluded to as a downward (or reverse) martingale, in contradistinction to the more standard upward martingale $\{S_n, \mathscr{F}_n, n \geq m\}$ of Example 1.

The following examples attest to the pervasiveness of martingales.

Example 1. Let $\{X_n, n \geq 1\}$ be independent $\mathscr{L}_1$ random variables with $S_n = \sum_{i=1}^n X_i$. Then $\{S_n, n \geq 1\}$ is a submartingale if $\mathrm{E}\, X_n \geq 0, n > 1$, and a martingale if $\mathrm{E}\, X_n = 0, n > 1$.

Example 2. Let $\{X_n, n \geq 1\}$ be independent $\mathscr{L}_1$ random variables with $\mathrm{E}\, X_n = 0, n \geq 1$. For any integer $k \geq 1$, if

$$S_{k,n} = \sum_{1 \leq i_1 < \cdots < i_k \leq n} X_{i_1} X_{i_2} \cdots X_{i_k}, \qquad n \geq k,$$

then $\{S_{k,n}, n \geq k\}$ is an $\mathscr{L}_1$ martingale. The important special case $k = 1$ is subsumed in Example 1. More generally, if $\{X_n, n \geq 1\}$ are $\mathscr{L}_k$ random variables with $\mathrm{E}\{X_{n+1} | \mathscr{F}_n\} = 0, n \geq 1$, where $\{X_n, \mathscr{F}_n, n \geq 1\}$ is a stochastic sequence, then $\{S_{k,n}, \mathscr{F}_n, n \geq k\}$ is an $\mathscr{L}_1$ martingale with $\mathrm{E}\, S_{k,k} = 0$.

Example 3. Let $\{X_n, n \geq 1\}$ be interchangeable random variables and $\varphi(\cdot)$, a symmetric Borel function on R^m with $\mathrm{E}|\varphi(X_1, \ldots, X_m)| < \infty$. A sequence of so-called U-statistics is defined for any integer $m \geq 1$ by

$$U_{m,n} = \binom{n}{m}^{-1} \sum_{1 \leq i_1 < \cdots < i_m \leq n} \varphi(X_{i_1}, \ldots, X_{i_m}), \qquad n \geq m.$$

If $\mathscr{F}_n = \sigma(U_{m,j}, j \geq n)$, then for $1 \leq i_1 < \cdots < i_m \leq n + 1$ and $B \in \mathscr{F}_{n+1}$

$$\int_B \varphi(X_{i_1}, \ldots, X_{i_m})\, d\mathrm{P} = \int_B \varphi(X_1, \ldots, X_m)\, d\mathrm{P}$$

via symmetry and interchangeability (Exercise 7.1.12), implying

$$\mathrm{E}\{\varphi(X_{i_1}, \ldots, X_{i_m}) | \mathscr{F}_{n+1}\} = \mathrm{E}\{\varphi(X_1, \ldots, X_m) | \mathscr{F}_{n+1}\}, \qquad \text{a.c.},$$

for $1 \leq i_1 < \cdots < i_m \leq n+1$, and hence

$$\mathrm{E}\{U_{m,n} | \mathscr{F}_{n+1}\} = \mathrm{E}\{U_{m,n+1} | \mathscr{F}_{n+1}\} = U_{m,n+1}, \qquad \text{a.c.}$$

Consequently, if $U_n^* = U_{m,-n}$ and $\mathscr{F}_n^* = \mathscr{F}_{-n}$, $n \leq -m$, then $\{U_n^*, \mathscr{F}_n^*, n \leq -m\}$ is a martingale closed on the right. The important special cases $m = 1$, $\varphi(x) = x$ and $m = 2$, $\varphi(x_1, x_2) = (x_1 - x_2)^2/2$ yield the arithmetic mean $U_{1,n} = (1/n) \sum_{i=1}^n X_i = \bar{X}_n$ (say) and the sample variance $U_{2,n} = (n-1)^{-1} \sum_{i=1}^n (X_i - \bar{X}_n)^2$ respectively.

EXAMPLE 4. Let $\{X_n, n \geq 1\}$ be $\mathscr{L}_1$ r.v.s which are conditionally independent given $\mathscr{G}$ with $E\{X_n \mid \mathscr{G}\} \geq 0$ a.c. for $n \geq 1$. If $S_n = \sum_1^n X_i$ and $\mathscr{F}_n = \sigma(\mathscr{G} \cup \sigma(X_1, \ldots, X_n))$, then, recalling Theorem 7.3.1(iv), $\{S_n, \mathscr{F}_n, n \geq 1\}$ is a submartingale. In particular, if $\{X_n, n \geq 1\}$ are $\mathscr{L}_1$ interchangeable r.v.s with $E\,X_1X_2 = 0$, then $\{S_n, \mathscr{F}_n, n \geq 1\}$ is a martingale since, invoking Theorem 7.3.2 and Corollary 7.3.5, $0 = E\,X_1X_n = E[E\{X_1X_n \mid \mathscr{G}\}] = E[E^2\{X_n \mid \mathscr{G}\}]$, $n \geq 1$.

EXAMPLE 5. Let $\{S_n = \sum_1^n X_j, \mathscr{F}_n, n \geq 1\}$ be a submartingale with $E\,X_n^+ < \infty$, $n \geq 1$, and let T be an $\mathscr{F}_n$-time. If $T_n = \min(T, n)$ and $U_n = S_{T_n} = \sum_1^n X_j I_{[T \geq j]}$, then $\{U_n, \mathscr{F}_n, n \geq 1\}$ is a submartingale since U_n is $\mathscr{F}_n$-measurable, $E\,U_n^+ \leq \sum_1^n E\,X_j^+ < \infty$, and

$$\begin{aligned} E\{U_n \mid \mathscr{F}_{n-1}\} &= U_{n-1} + E\{X_n I_{[T \geq n]} \mid \mathscr{F}_{n-1}\} \\ &= U_{n-1} + I_{[T \geq n]} E\{X_n \mid \mathscr{F}_{n-1}\} \geq U_{n-1}, \text{ a.c., } n > 1. \end{aligned}$$

EXAMPLE 6. Let $(\Omega, \mathscr{F}, P)$ be the probability space consisting of Lebesgue measure on the Lebesgue-measurable sets of $\Omega = [0, 1]$ and for each $n = 1, 2, \ldots$ let the points $0 = \omega_{n,-1} < \omega_{n,0} < \omega_{n,1} < \cdots < \omega_{n,2^n} = 1$ engender a partition, say Q_n, of Ω into disjoint intervals such that Q_{n+1} is a subpartition of Q_n. Choose $\mathscr{F}_n =$ the σ-algebra generated by the intervals $\Lambda_{n,0} = [0, \omega_{n,0}]$, $\Lambda_{n,j} = (\omega_{n,j-1}, \omega_{n,j}]$, $1 \leq j \leq 2^n$, and for any finite, real-valued function g on Ω define

$$U_n(\omega) = \frac{g(\omega_{n,j}) - g(\omega_{n,j-1})}{\omega_{n,j} - \omega_{n,j-1}}, \qquad \omega \in \Lambda_{n,j},\ 0 \leq j \leq 2^n.$$

Since, as is readily verified, for every $n \geq 1$

$$\int_\Lambda U_n \, dP = \int_\Lambda U_{n+1} \, dP, \qquad \Lambda \in \mathscr{F}_n, \tag{3}$$

$\{U_n, \mathscr{F}_n, n \geq 1\}$ is a martingale. Note that for $N = \{1, 2, \ldots\}$, $E|U_n| < \infty$, $n \geq 1$, and (3) constitute an alternative definition of an $\mathscr{L}_1$ martingale $\{U_n, \mathscr{F}_n, n \geq 1\}$.

EXAMPLE 7. Let $\{X_n, n \geq 1\}$ be any sequence of $\mathscr{L}_1$ r.v.s and define $X_n' = X_n - E\{X_n \mid X_1, \ldots, X_{n-1}\}$, $n \geq 2$, and $X_1' = X_1$ or $X_1 - E\,X_1$. If $S_n' = \sum_1^n X_j'$, then $\{S_n', n \geq 1\}$ is a martingale. In particular, if $\{S_n = \sum_1^n X_j, n \geq 1\}$ is an $\mathscr{L}_1$ submartingale and

$$S_n' = S_n - \sum_{j=2}^n E\{X_j \mid X_1 \ldots, X_{j-1}\}, \quad n \geq 2, \quad S_1' = S_1 = X_1, \tag{4}$$

then $\{S_n', n \geq 1\}$ is a martingale and $0 \leq \sum_{j=2}^n E\{X_j \mid X_1, \ldots, X_{j-1}\} \uparrow$ a.c.

in n. Since $|S'_n| \le |S_n| + \sum_{j=2}^n \mathrm{E}\{X_j \mid X_1 \cdots X_{j-1}\}$,

$$\mathrm{E}|S'_n| \le \mathrm{E}|S_n| + \sum_{j=2}^{n} \mathrm{E}\, X_j = \mathrm{E}|S_n| + \mathrm{E}\, S_n - \mathrm{E}\, X_1 \le 3 \sup_{n \ge 1} \mathrm{E}|S_n|.$$

Hence, every $\mathscr{L}_1$ submartingale $\{S_n, n \ge 1\}$ can be expressed as $S_n = S'_n + S''_n$, $n \ge 1$, where $\{S'_n, n \ge 1\}$ is an $\mathscr{L}_1$ martingale and $0 \le S''_n \uparrow$ a.c. Moreover, if $\{S_n, n \ge 1\}$ is $\mathscr{L}_1$ bounded so are $S'_n, n \ge 1$, and $S''_n, n \ge 1$.

The first question to be explored is convergence of submartingales.

Theorem 1. *If $\{S_n, \mathscr{F}_n, n \ge 1\}$ is a submartingale such that for every $\{\mathscr{F}_n\}$-time T*

$$\int_{[T<\infty]} S_T \, d\mathrm{P} \ne \infty,$$

then $\lim_{n\to\infty} S_n$ *exists* a.c.

Proof. If $\mu_n\{A\} = \int_A S_n \, d\mathrm{P}$, $A \in \mathscr{F}_n$, then by the submartingale property, $\mu_{n+1}\{A\} \ge \mu_n\{A\}$ for $A \in \mathscr{F}_n$. Suppose that for some pair of real numbers α, β

$$V = \left\{ \overline{\lim_{n\to\infty}}\, S_n > \alpha > \beta > \underline{\lim_{n\to\infty}}\, S_n \right\}$$

has positive probability, say $\mathrm{P}\{V\} > \delta > 0$. It may and will be supposed, replacing S_n by $(\alpha - \beta)^{-1}(S_n - \beta)$, that $\alpha = 1$, $\beta = 0$.

Set $m_0 = 1$, $B_0 = \Omega$, $V_0 = B_0 V$ and $C_n^0 = B_0\{S_n > 1, S_j \le 1$, for $m_0 \le j < n\}$. Define $A_1 = \bigcup_{m_0}^{n_1} C_n^0$, where n_1 is large enough to ensure $\mathrm{P}\{V_0 - A_1\} < \delta/4$.

Next, define $D_n^0 = A_1\{S_n < 0, S_j \ge 0, n_1 \le j < n\}$, $B_1 = \bigcup_{n_1}^{m_1} D_n^0$, where m_1 is large enough to guarantee $\mathrm{P}\{V_0 - B_1\} < \delta/4$ and note that

$$\begin{aligned}
\mu_{m_1}\{A_1 - B_1\} &\ge \mu_{m_1}\{A_1 - B_1\} + \sum_{j=n_1}^{m_1} \mu_j\{D_j^0\} \\
&= \mu_{m_1}\left\{A_1 - \bigcup_{n_1}^{m_1} D_j^0\right\} + \sum_{j=n_1}^{m_1} \mu_j\{D_j^0\} \\
&= \mu_{m_1}\left\{A_1 - \bigcup_{n_1}^{m_1-1} D_j^0\right\} + \sum_{j=n_1}^{m_1-1} \mu_j\{D_j^0\} \\
&\ge \mu_{m_1-1}\left\{A_1 - \bigcup_{n_1}^{m_1-1} D_j^0\right\} + \sum_{j=n_1}^{m_1-1} \mu_j\{D_j^0\} \ge \cdots \ge \\
&\ge \mu_{n_1}\{A_1\} = \sum_{n=m_0}^{n_1} \mu_{n_1}\{C_n^0\} \\
&\ge \mathrm{P}\{A_1\} \ge \mathrm{P}\{V_0\} - \mathrm{P}\{V_0 - A_1\} > 3\delta/4.
\end{aligned}$$

Furthermore, if $V_1 = B_1 \cdot V_0 = B_1 B_0 V$, then $P\{V_1\} = P\{V_0\} - P\{V_0 - B_1\} > 3\delta/4$.

If $C_n^1 = B_1\{S_n > 1, S_j \le 1, m_1 \le j < n\}$, $A_2 = \bigcup_{m_1}^{n_2} C_n^1$, then n_2 may be chosen so that $P\{V_1 - A_2\} < \delta/2^3$. Set $D_n^1 = A_2\{S_n < 0, S_j \ge 0, n_2 \le j < n\}$, and

$$B_2 = \bigcup_{n_2}^{m_2} D_n^1 = A_2 \bigcup_{n_2}^{m_2} \{S_n < 0\} = \bigcup_1^{n_1} \{S_i > 1\} \cdot \bigcup_{n_1}^{m_1} \{S_j < 0\}$$

$$\cdot \bigcup_{m_1}^{n_2} \{S_k > 1\} \cdot \bigcup_{n_2}^{m_2} \{S_n < 0\},$$

whence, analogously, $P\{V_1 - B_2\} < \delta/2^3$ if m_2 is large and, moreover,

$$\mu_{m_2}\{A_2 - B_2\} \ge P\{A_2\} \ge P\{V_1\} - P\{V_1 - A_2\} \ge \frac{5\delta}{8}.$$

Proceeding inductively, for $k = 1, 2, \ldots$ there exist integers $m_{k+2} > m_{k+1} > n_{k+1} > n_k$ and sets $A_k \in \mathscr{F}_{n_k}$, $B_k \in \mathscr{F}_{m_k}$ with $A_k \supset B_k \supset A_{k+1}$ such that

$$\mu_{m_k}\{A_k - B_k\} > \frac{(2^k + 1)\delta}{2^{k+1}} > \frac{\delta}{2}.$$

Now the disjoint sets $C_k = A_k - B_k \in \mathscr{F}_{m_k}$, $k \ge 1$, and so setting $\{T = m_k\} = C_k\{S_{m_k} \ge 0\}$, $k \ge 1$, and $\{T = \infty\} = (\bigcup_1^\infty C_k[S_{m_k} \ge 0])^c$, it follows since $A_k - B_k \subset \{S_{m_k} \ge 0\}$ that

$$\int_{[T<\infty]} S_T^- = 0, \qquad \int_{[T<\infty]} S_T^+ = \sum_1^\infty \int_{C_j[S_{m_j} \ge 0]} S_{m_j}^+ \ge \sum_1^\infty \mu_{m_j}\{C_j\} = \infty,$$

contradicting the hypothesis. Consequently, $P\{V\} = 0$ for any choice of $\alpha > \beta$, implying the conclusion of the theorem. □

Lemma 1. *If $\{S_n, \mathscr{F}_n, n \in N\}$ is a martingale and φ is any real convex function with $|E\varphi(S_n)| \le \infty$, then $\{\varphi(S_n), \mathscr{F}_n, n \in N\}$ is a submartingale. If $\{S_n, \mathscr{F}_n, n \in N\}$ is merely a submartingale but φ is, in addition, nondecreasing, then $\{\varphi(S_n), \mathscr{F}_n, n \in N\}$ is likewise a submartingale.*

PROOF. By the conditional Jensen inequality, Theorem 7.1.4, for $m < n$ and $m, n \in N$,

$$E\{\varphi(S_n) \mid \mathscr{F}_m\} \ge \varphi(E\{S_n \mid \mathscr{F}_m\}) = \varphi(S_m), \quad \text{a.c.}$$

yielding the first statement. The submartingale hypothesis combined with φ nondecreasing converts the prior equality to $\ge$, thereby preserving the conclusion. □

In particular, if $\{S_n, n \ge 1\}$ is a martingale, then $\{|S_n|^p, n \ge 1\}$ is a submartingale for any $p \ge 1$, and if $\{S_n, n \ge 1\}$ is a submartingale, so is

$$\{\max(S_n, K), n \ge 1\} \quad \text{for } K \in (-\infty, \infty).$$

Lemma 2. *If* $\{S_n, \mathcal{F}_n, n \geq 1\}$ *is a submartingale with* $\sup_{n\geq 1} \mathrm{E}\, S_n^+ = M < \infty$, *then for any stopping time* T

$$\int_{[T<\infty]} S_T^+ \, d\mathrm{P} \leq M, \qquad \int_{[T<\infty]} |S_T| \, d\mathrm{P} \leq 2M - \mathrm{E}\, S_1 \leq 3 \sup_{n\geq 1} \mathrm{E}|S_n|. \quad (5)$$

PROOF. Set $T' = \min(T, n)$, $n \geq 1$. Since $\{S_n^+, \mathcal{F}_n, n \geq 1\}$ is a submartingale by Lemma 1,

$$\mathrm{E}\, S_{T'}^+ = \sum_{j=1}^{n} \int_{[T=j]} S_j^+ + \int_{[T>n]} S_n^+ \leq \sum_{j=1}^{n} \int_{[T=j]} S_n^+ + \int_{[T>n]} S_n^+ = \mathrm{E}\, S_n^+,$$

so that $\mathrm{E}\, S_{T'}^+ \leq M$, whence the first part of (5) follows from Fatou's lemma. Next,

$$\begin{aligned}\mathrm{E}\, S_1 &= \int_{[T=1]} S_1 + \int_{[T>1]} S_1 \leq \int_{[T=1]} S_1 + \int_{[T>1]} S_2 \\ &\leq \int_{[T=1]} S_1 + \int_{[T=2]} S_2 + \int_{[T>2]} S_2 \leq \cdots \\ &\leq \int_{[1\leq T\leq n]} S_T + \int_{[T>n]} S_n = \mathrm{E}\, S_{T'},\end{aligned}$$

implying

$$\mathrm{E}|S_{T'}| = 2\,\mathrm{E}\, S_{T'}^+ - \mathrm{E}\, S_{T'} \leq 2M - \mathrm{E}\, S_1,$$

and the remainder of (5) follows once more by Fatou. □

The special case $X_n \equiv X_0$ of the next corollary, due to Doob (1940), may be considered a milestone in the development of martingales. When X_0 is bounded, this had been obtained by Lévy (1937).

Corollary 1. *Let* $\{\mathcal{F}_n, n \geq 1\}$ *be a stochastic basis and* $\{X_n, n \geq 0\}$ *a sequence of* $\mathcal{L}_1$ r.v.s *with* $X_n \xrightarrow{\text{a.c.}} X_0$ *and* $\mathrm{E} \sup_{n\geq 1} |X_n| < \infty$. *Then, if*

$$\mathcal{F}_\infty = \sigma\left(\bigcup_{n=1}^{\infty} \mathcal{F}_n\right),$$

$$\mathrm{E}\{X_n \mid \mathcal{F}_n\} \xrightarrow{\text{a.c.}} \mathrm{E}\{X_0 \mid \mathcal{F}_\infty\}.$$

PROOF. (i) It will first be demonstrated for any integrable r.v. X_0 that

$$\mathrm{E}\{X_0 \mid \mathcal{F}_n\} \xrightarrow{\text{a.c.}} \mathrm{E}\{X_0 \mid \mathcal{F}_\infty\}.$$

By considering X_0^+ and X_0^- separately, it may and will be supposed that $X_0 \geq 0$. Then, setting $S_n = \mathrm{E}\{X_0 \mid \mathcal{F}_n\}$, $\{S_n, \mathcal{F}_n, n \geq 1\}$ is a nonnegative martingale with $\sup_{n\geq 1} \mathrm{E}|S_n| = \mathrm{E}\, S_1 = \mathrm{E}\, X_0 < \infty$. Hence, by Theorem 1 and Lemma 2, $\lim_{n\to\infty} S_n = S_\infty$ exists, a.c., and, moreover, S_∞ is an integrable r.v. by Fatou's lemma. Since

$$P\{S_n > C\} \le C^{-1}\,E\,S_n = C^{-1}\,E\,X_0 \to 0 \text{ as } C \to \infty,$$

$$\int_{[S_n > C]} S_n = \int_{[S_n > C]} E\{X_0 \mid \mathscr{F}_n\} = \int_{[S_n > C]} X_0 \to 0$$

as $C \to \infty$, and so $\{S_n, n \ge 1\}$ is u.i., whence $S_n \xrightarrow{\mathscr{L}_1} S_\infty$ by Corollary 4.2.4. Thus, for all $n \ge 1$ and $A \in \mathscr{F}_n$, if $m > n$,

$$\int_A X_0 = \int_A S_n = \int_A S_m \xrightarrow{m \to \infty} \int_A S_\infty,$$

implying $\int_A X_0 = \int_A S_\infty$ for all $A \in \bigcup_1^\infty \mathscr{F}_n$, so that $E\{X_0 \mid \mathscr{F}_\infty\} = S_\infty = \lim S_n$, a.c.

Next, set $Y_m = \sup_{n \ge m} |X_n - X_0|$. For every integer m and all $n \ge m$

$$\begin{aligned} D_n &\equiv |E\{X_n \mid \mathscr{F}_n\} - E\{X_0 \mid \mathscr{F}_\infty\}| \\ &\le |E\{(X_n - X_0) \mid \mathscr{F}_n\}| + |E\{X_0 \mid \mathscr{F}_n\} - E\{X_0 \mid \mathscr{F}_\infty\}| \\ &\le E\{Y_m \mid \mathscr{F}_n\} + |E\{X_0 \mid \mathscr{F}_n\} - E\{X_0 \mid \mathscr{F}_\infty\}|, \end{aligned}$$

whence for every integer m

$$\overline{\lim_{n \to \infty}}\, D_n \le E\{Y_m \mid \mathscr{F}_\infty\}$$

by the part (i) already proved. Since $Y_m \xrightarrow{\text{a.c.}} 0$ and $|Y_m| \le 2 \sup |X_n| \in \mathscr{L}_1$, it follows that $E\{Y_m \mid \mathscr{F}_\infty\} \xrightarrow{\text{a.c.}} 0$ as $m \to \infty$ by Theorem 7.1.2, whence $D_n \xrightarrow{\text{a.c.}} 0$. □

Theorem 2. *Let $\{S_n, \mathscr{F}_n, n \ge 1\}$ be a submartingale and $\mathscr{F}_\infty = \sigma(\bigcup_1^\infty \mathscr{F}_n)$.*

i. *If $\sup_{n \ge 1} E\,S_n^+ < \infty$, then $S_\infty = \lim S_n$ exists* a.c. *with S_∞ finite* a.c. *on $\{S_1 > -\infty\}$. Moreover, if $\sup_{n \ge 1} E|S_n| < \infty$, then $E|S_\infty| < \infty$.*
ii. *If $\{S_n^+, n \ge 1\}$ are* u.i., *then $S_\infty = \lim S_n$ exists* a.c. *and $\{S_n, \mathscr{F}_n, 1 \le n \le \infty\}$ is a submartingale.*
iii. *If $\{S_n, \mathscr{F}_n, n \ge 1\}$ is a submartingale closed by some $\mathscr{F}$-measurable function S with $E\,S^+ < \infty$, then $\{S_n^+, n \ge 1\}$ are* u.i., *so that*

$$S_\infty = \lim_{n \to \infty} S_n$$

exists a.c. *and, moreover, $\{S_n, \mathscr{F}_n, 1 \le n \le \infty\}$ is a submartingale closed by S.*
iv. *The* r.v.s *$\{S_n, n \ge 1\}$ are* u.i. iff *$\{S_n, \mathscr{F}_n, 1 \le n \le \infty\}$ is an $\mathscr{L}_1$ submartingale with $\lim_{n \to \infty} E\,S_n = E\,S_\infty$* iff *$S_n \xrightarrow{\mathscr{L}_1} S_\infty$, where $S_\infty = \lim S_n$.*

PROOF. (i) By Lemma 2 and Theorem 1, $S_n \xrightarrow{\text{a.c.}} S_\infty$. Moreover, if $\sup_{n \ge 1} E|S_n| < \infty$, Fatou's lemma guarantees $E|S_\infty| < \infty$. Next, for any $k > 0$ set $S_n' = S_n I_{[S_1 > -k]}$. Then $\{S_n', \mathscr{F}_n, n \ge 1\}$ is a submartingale with $\sup_{n \ge 1} E\,S_n'^+ < \infty$ and $E\,S_1' \ge -k$. Lemma 2 with $T \equiv n$ ensures $\sup_{n \ge 1} E|S_n'| < \infty$, whence S_∞ is finite a.c. on $\{S_1 > -k\}$. Letting $k \to \infty$, the remaining portion of (i) obtains.

Apropos of (ii), the hypothesis implies $S_n \xrightarrow{\text{a.c.}} S_\infty$ via (i). Moreover, for $A \in \mathscr{F}_m$ and $n \geq m$, applying Theorem 4.2.2(ii) to $-S_n I_A$,

$$\int_A S_m \leq \int_A S_n \leq \overline{\lim_n} \int_A S_n \leq \int_A S_\infty,$$

whence $E\{S_\infty \mid \mathscr{F}_m\} \geq S_m$, a.c., for $m \geq 1$.

In case (iii), the hypothesis and Lemma 1 ensure that $\{S_n^+, \mathscr{F}_n, n \geq 1\}$ is a submartingale closed by S^+, whence for $n \geq 1$ and $k > 0$

$$\int_{[S_n > k]} S_n^+ \leq \int_{[S_n > k]} S^+.$$

Since $P\{S_n > k\} \leq k^{-1} E\, S_n^+ \leq k^{-1} E\, S^+ \to 0$ uniformly in n as $k \to \infty$, it follows that

$$\lim_{k \to \infty} \int_{[S_n > k]} S_n^+ \leq \lim_{k \to \infty} \int_{[S_n > k]} S^+ = 0,$$

uniformly in n as $k \to \infty$. Now each S_n^+ is integrable and so $\{S_n^+, n \geq 1\}$ are u.i. By (ii) $\{S_n, \mathscr{F}_n, 1 \leq n \leq \infty\}$ is a submartingale. To verify that it is closed by S, define $S_n^{(k)} = \max(S_n, -k)$, $1 \leq n \leq \infty$, and $S^{(k)} = \max(S, -k)$, where $k > 0$. Then $\{S_n^{(k)}, n \geq 1\}$ are u.i., $S_n^{(k)} \xrightarrow{\text{a.c.}} S_\infty^{(k)}$, and by Lemma 1 $\{S_n^{(k)}, \mathscr{F}_n, n \geq 1\}$ is a submartingale closed by $S^{(k)}$. Hence, for $A \in \mathscr{F}_n$, $n \geq 1$,

$$\int_A S^{(k)} \geq \int_A S_n^{(k)} \xrightarrow{n \to \infty} \int_A S_\infty^{(k)} \geq \int_A S_\infty.$$

Since $E\, S^+ < \infty$ and $-S^{(1)} \leq -S^{(k)} \uparrow -S$,

$$\int_A S = \lim_{k \to \infty} \int_A S^{(k)} \geq \int_A S_\infty,$$

implying $E\{S \mid \mathscr{F}_\infty\} \geq S_\infty$, a.c.

In part (iv), if $\{S_n, n \geq 1\}$ are u.i., (i) and (ii) ensure that $S_n \xrightarrow{\text{a.c.}} S_\infty$ and $\{S_n, \mathscr{F}_n, 1 \leq n \leq \infty\}$ is an $\mathscr{L}_1$ submartingale. Then, by u.i., $\lim_{n \to \infty} E\, S_n = E\, S_\infty$.

Conversely, if $\{S_n, \mathscr{F}_n, 1 \leq n \leq \infty\}$ is an $\mathscr{L}_1$ submartingale with $\lim E\, S_n = E\, S_\infty$, then $\{S_n^+, n \geq 1\}$ are u.i. by (iii). Hence, $E\, S_n^+ \to E\, S_\infty^+$, and so $E\, S_n^- \to E\, S_\infty^-$. Since $S_n^- \xrightarrow{\text{a.c.}} S_\infty^-$, Corollary 4.2.4 ensures that $\{S_n^-, n \geq 1\}$ are u.i.

Finally, if $\{S_n, n \geq 1\}$ are u.i., $S_n \xrightarrow{\mathscr{L}_1} S_\infty$ by Theorem 4.2.3, and this same theorem also yields the converse. □

Theorem 3. *Let $\{S_n, \mathscr{F}_n, n \geq 1\}$ be a martingale and $\mathscr{F}_\infty = \sigma(\bigcup_{n=1}^\infty \mathscr{F}_n)$*

i. *If $\sup E|S_n| < \infty$, then $S_n \xrightarrow{\text{a.c.}} S_\infty \in \mathscr{L}_1$.*
ii. *If $\{S_n, n \geq 1\}$ is u.i., then $S_n \xrightarrow{\text{a.c.}} S_\infty \in \mathscr{L}_1$ and $\{S_n, \mathscr{F}_n, 1 \leq n \leq \infty\}$ is a martingale.*

iii. *If the martingale* $\{S_n, \mathscr{F}_n, n \geq 1\}$ *is closed by some* r.v. $S \in \mathscr{L}_1$, *then* $\{S_n, n \geq 1\}$ *is* u.i., *so that* $S_n \xrightarrow{\text{a.c.}} S_\infty \in \mathscr{L}_1$ *and, moreover,* $\mathrm{E}\{S \mid \mathscr{F}_n\} = S_n$, a.c., $1 \leq n \leq \infty$.

iv. *The* r.v.s $\{S_n, n \geq 1\}$ *are* u.i. iff $\{S_n, \mathscr{F}_n, 1 \leq n \leq \infty\}$ *is an* $\mathscr{L}_1$ *martingale with* $\lim \mathrm{E}\, S_n = \mathrm{E}\, S_\infty$ iff $S_n \xrightarrow{\mathscr{L}_1} S_\infty$, *where* $S_\infty = \lim S_n$.

PROOF. Parts (i), (ii), (iii), (iv) follow directly from their counterparts in Theorem 2 (applied in the latter three cases to both S_n and $-S_n$). □

Corollary 2. *If* $\{S_n, \mathscr{F}_n, n \geq 1\}$ *is a positive (or negative)* $\mathscr{L}_1$ *martingale,* $S_n \xrightarrow{\text{a.c.}} S_\infty \in \mathscr{L}_1$.

The next theorem illustrates how global martingale convergence in conjunction with stopping times yields local martingale convergence.

Theorem 4. *If*

$$\left\{S_n = \sum_{i=1}^{n} X_i, \mathscr{F}_n, n \geq 1\right\}$$

is a submartingale with $\mathrm{E}\sup_{n\geq 1} X_n^+ < \infty$, *then* S_n *converges* a.c. *on the set* $\{X_1 > -\infty, \sup_{n\geq 1} S_n < \infty\}$.

PROOF. For any $c > 0$, define $T = T_c = \inf\{n \geq 1 : S_n > c\}$. Then T is a stopping time and $\{T_c = \infty\} = \{\sup_{n\geq 1} S_n \leq c\} \to \{\sup_{n\geq 1} S_n < \infty\}$ as $c \to \infty$. As seen in Example 5, $\{U_n = \sum_{j=1}^{n} X_j I_{[T\geq j]}, \mathscr{F}_n, n \geq 1\}$ is a submartingale and

$$\begin{aligned}\mathrm{E}\, U_n^+ &= \mathrm{E}\left(\sum_{j=1}^{n} X_j I_{[T\geq j]}\right)^+ \leq \mathrm{E}\left(\sum_{j=1}^{n} X_j I_{[T>j]}\right)^+ + \mathrm{E}\left(\sum_{1}^{n} X_j I_{[T=j]}\right)^+ \\ &\leq C + \mathrm{E}\sup X_n^+ < \infty.\end{aligned}$$

Hence, by Theorem 2, U_n converges a.c. on $\{X_1 > -\infty\}$. Therefore, $S_n = \sum_{j=1}^{n} X_j$ converges a.c. on $\{X_1 > -\infty, T_c = \infty\}$. Consequently, letting $c \to \infty$, S_n converges a.c. on $\{X_1 > -\infty, \sup_{n\geq 1} S_n < \infty\}$. □

An issue of considerable importance and unquestionable utility in probability theory is the effect on expected values of randomly stopping a stochastic sequence.

Theorem 5. *Let* $\{S_n, \mathscr{F}_n, n \geq 1\}$ *be a submartingale.*

i. *If* T *is a finite* $\{\mathscr{F}_n\}$*-time with*

$$|\mathrm{E}\, S_T| \leq \infty, \qquad \varliminf_{n\to\infty} \int_{[T>n]} S_n^+ = 0, \tag{6}$$

then for $n \geq 1$

$$\mathrm{E}\{S_T \mid \mathscr{F}_n\} \geq S_n, \text{ a.c. on } \{T \geq n\} \quad \text{and} \quad \mathrm{E}\, S_T \geq \mathrm{E}\, S_1. \tag{7}$$

ii. *If* $\{T_n, n \geq 1\}$ *is a sequence of finite* $\{\mathscr{F}_n\}$*-times with* $T_1 \leq T_2 \leq \cdots$ *satisfying*

$$|\mathrm{E}\, S_{T_m}| \leq \infty, \qquad \varliminf_{n \to \infty} \int_{[T_m > n]} S_n^+ = 0, \qquad m \geq 1, \tag{8}$$

and $\mathscr{F}_{T_n} = \{B \subset \Omega: B[T_n = j] \in \mathscr{F}_j, j \geq 1\}$, *then* $\{S_{T_n}, \mathscr{F}_{T_n}, n \geq 1\}$ *is a submartingale.*

PROOF. To prove the first part of (7) it suffices to verify for $n \geq 1$ and $A \in \mathscr{F}_n$ that

$$\int_{A[T \geq n]} S_T \geq \int_{A[T \geq n]} S_n. \tag{9}$$

Now

$$\begin{aligned}\int_{A[T \geq n]} S_n &= \int_{A[T = n]} S_n + \int_{A[T > n]} S_n \\ &\leq \int_{A[T = n]} S_n + \int_{A[T > n]} \mathrm{E}\{S_{n+1} \mid \mathscr{F}_n\} \\ &= \int_{A[T = n]} S_T + \int_{A[T \geq n+1]} S_{n+1}.\end{aligned}$$

The last term on the right is the first term on the left with n replaced by $n + 1$, so, repeating the argument $m - n - 1$ times,

$$\int_{A[T \geq n]} S_n \leq \int_{A[n \leq T < m]} S_T + \int_{A[T \geq m]} S_m = \int_{A[n \leq T \leq m]} S_T + \int_{A[T > m]} S_m. \tag{10}$$

Noting that $\int_{A[T > m]} S_m \leq \int_{[T > m]} S_m^+$, the desired conclusion (9) follows via (6) since $|\mathrm{E}\, S_T| \leq \infty$. The remaining portion of (7) follows from (9) with $n = 1$, $A = \Omega = [T \geq 1]$.

Apropos of (ii), since S_{T_n} is $\mathscr{F}_{T_n}$-measurable, it suffices to prove that

$$\mathrm{E}\{S_{T_{n+1}} \mid \mathscr{F}_{T_n}\} \geq S_{T_n}, \qquad \text{a.c.}, \qquad n \geq 1. \tag{11}$$

Let $B \in \mathscr{F}_{T_n}$ and $B_m = B[T_n = m]$. Then $B_m \in \mathscr{F}_m$ and $T_{n+1} \geq m$ a.c. on B_m. By (i)

$$\int_{B_m} S_{T_n} = \int_{B_m} S_m \leq \int_{B_m} \mathrm{E}\{S_{T_{n+1}} \mid \mathscr{F}_m\} = \int_{B_m} S_{T_{n+1}},$$

whence, summing on m,

$$\int_B S_{T_n} \leq \int_B S_{T_{n+1}},$$

implying (11). □

Corollary 3. *If $\{S_n, \mathscr{F}_n, n \in N\}$ is a submartingale with N having a finite last element and T is a finite $\mathscr{F}_n$-time with $|\mathrm{E}\, S_T| \le \infty$, then* (7) *holds for $n \in N$.*

PROOF. If m is the last element of N, the last term of (10) vanishes. □

Corollary 4. *If $\{S_n, \mathscr{F}_n, n \ge 1\}$ is a martingale and T is a finite $\{\mathscr{F}_n\}$-time with*

$$|\mathrm{E}\, S_T| \le \infty, \qquad \lim_{n\to\infty} \int_{[T>n]} |S_n| = 0, \tag{12}$$

then for $n \ge 1$

$$\mathrm{E}\{S_T \mid \mathscr{F}_n\} = S_n, \text{ a.c. on } \{T \ge n\} \quad \text{and} \quad \mathrm{E}\, S_T = \mathrm{E}\, S_1. \tag{13}$$

Moreover, under the hypothesis of (ii) *of Theorem* 5 *with S_n^+ replaced by $|S_n|$ in* (8), *$\{S_{T_n}, \mathscr{F}_{T_n}, n \ge 1\}$ is a martingale; in particular, for any $\mathscr{F}_n$-time T, $\{S_{\min[T,n]}, n \ge 1\}$ is a martingale.*

In the age-old problem of a gambler's ruin (Section 3.4) i.e., symmetric random walk on the line with absorbing barriers at integers $b > 0$ and $-a < 0$, a particle starting at the origin moves at each stage one unit to the right or left with equal probabilities, the successive moves X_i, $i = 1, 2, \ldots$, being independent. Then $\{S_n = \sum_1^n X_i, n \ge 1\}$ is a martingale and $T = \inf\{n \ge 1: S_n = b \text{ or } -a\}$ is a stopping variable satisfying (12) and $T \ge 1$. According to Corollary 4, $\{S_1, S_T\}$ is a martingale, so that

$$b\, \mathrm{P}\{S_T = b\} - a\, \mathrm{P}\{S_T = -a\} = \mathrm{E}\, S_T = \mathrm{E}\, S_1 = 0.$$

Since the sum of these two probabilities is unity,

$$\mathrm{P}\{S_T = -a\} = \frac{b}{a+b}.$$

As another illustration of the preceding, consider the following generalization of the so-called ballot problem, in which r votes for the incumbent and s votes for his rival are cast successively in random order with $s > r$. The probability that the winner was ahead at every stage of the voting is $(s - r)/(s + r)$ and can be obtained from

EXAMPLE 8. Let $\{X_j, 1 \le j \le n\}$ be nonnegative integer-valued $\mathscr{L}_1$ interchangeable random variables and set $S_j = \sum_1^j X_i$, $1 \le j \le n$. Then

$$\mathrm{P}\{S_j < j, 1 \le j \le n \mid S_n\} = \left(1 - \frac{S_n}{n}\right)^+.$$

PROOF. Since the above is trivially true when $S_n \ge n$, suppose $S_n < n$. If $Y_{-j} = S_j/j$, $\mathscr{F}_{-j} = \sigma(S_j, \ldots, S_n)$, $1 \le j \le n$, then, as noted in Example 3, $\{Y_j, \mathscr{F}_j, -n \le j \le -1\}$ is a martingale. Furthermore, if

$$T = \inf\{j: -n \le j \le -1, Y_j \ge 1\}$$

and $T = -1$ if no such j exists, then T is a stopping rule or finite $\{\mathscr{F}_j\}$-time and, moreover, a bounded r.v. Since by definition $Y_T = 1$ on $\bigcup_1^n \{S_j \geq j\}$ and zero elsewhere, Corollary 3 implies that on the set $\{S_n < n\}$

$$\mathrm{P}\left\{\bigcup_1^n [S_j \geq j] \,\middle|\, S_n\right\} = \mathrm{E}\{Y_T \mid \mathscr{F}_{-n}\} = Y_{-n} = \frac{S_n}{n},$$

which is tantamount to the proposition. □

Setting $X_j = 2$ or 0 according as the jth vote goes to the loser or his rival, note that if r_j (resp. s_j) of the first j votes are cast for the loser (resp. rival), then given that $S_n = 2r$ the event $S_j < j$, $1 \leq j \leq n = r + s$, is tantamount to $r_j < s_j$, $1 \leq j \leq n$.

Theorem 6. *Let $\{S_n = \sum_1^n X_j, \mathscr{F}_n, n \geq 1\}$ be a submartingale with* $\mathrm{E}\,X_n^+ < \infty$, *$n \geq 1$ and let T be a finite $\{\mathscr{F}_n\}$-time and $\mathscr{F}_0 = (\varnothing, \Omega)$. If* (i)

$$\mathrm{E}\sum_{n=1}^{T} \mathrm{E}\{X_n^+ \mid \mathscr{F}_{n-1}\} < \infty \tag{14}$$

or (ii) *$\{S_n^+, n \geq 1\}$ are uniformly integrable, then for $n \geq 1$*

$$\mathrm{E}\{S_T \mid \mathscr{F}_n\} \geq S_n, \text{ a.c. } on\ \{T \geq n\}, \quad and \quad \mathrm{E}\,S_T \geq \mathrm{E}\,S_1. \tag{15}$$

PROOF. Under (i)

$$\mathrm{E}\sum_1^T X_n^+ = \mathrm{E}\sum_1^\infty X_n^+ I_{[T\geq n]} = \sum_1^\infty \mathrm{E}[I_{[T\geq n]}\,\mathrm{E}\{X_n^+ \mid \mathscr{F}_{n-1}\}]$$

$$= \mathrm{E}\sum_1^\infty I_{[T\geq n]}\,\mathrm{E}\{X_n^+ \mid \mathscr{F}_{n-1}\} = \mathrm{E}\sum_1^T \mathrm{E}\{X_n^+ \mid \mathscr{F}_{n-1}\} < \infty.$$

Hence, $\mathrm{E}\,S_T^+ \leq \mathrm{E}\sum_1^T X_n^+ < \infty$ and, moreover,

$$\int_{[T>n]} S_n^+ \leq \int_{[T>n]} \sum_{j=1}^n X_j^+ \leq \int_{[T>n]} \sum_{j=1}^T X_j^+ = o(1)$$

as $n \to \infty$. Thus, (6) and consequently the conclusion (7) of Theorem 5 hold.

Under (ii), $\sup_{n\geq 1} \mathrm{E}\,S_n^+ < \infty$ whence $\mathrm{E}\,S_T^+ < \infty$ by Lemma 2. Since $\mathrm{P}\{T > n\} = o(1)$, the remainder of (6) and consequently (15) follow from u.i. □

Corollary 5. *If $\{S_n = \sum_1^n X_i, \mathscr{F}_n, n \geq 1\}$ is an $\mathscr{L}_1$ martingale and T is a finite $\{\mathscr{F}_n\}$-time satisfying*

$$\mathrm{E}\sum_{n=1}^{T} \mathrm{E}\{|X_n| \mid \mathscr{F}_{n-1}\} < \infty \tag{16}$$

(in particular, if T is a bounded r.v.) *or if $\{S_n, \mathscr{F}_n, n \geq 1\}$ is* u.i., *then for any*

$n \geq 1$

$$\mathrm{E}\{S_T \mid \mathscr{F}_n\} = S_n, \text{ a.c. } on \ \{T \geq n\}, \quad and \quad \mathrm{E}\, S_T = \mathrm{E}\, S_1. \tag{17}$$

Corollary 6. *If* $\{S_n = \sum_1^n X_j, \mathscr{F}_n, n \geq 1\}$ *is an* $\mathscr{L}_1$ *martingale and* T *is a finite* $\{\mathscr{F}_n\}$*-time satisfying* (12) *or* (16), *then for any* $r \geq 1$

$$\mathrm{E}\{|S_T|^r \mid \mathscr{F}_n\} \geq |S_n|^r, \text{ a.c. } on \ \{T \geq n\}, \quad and \quad \mathrm{E}|S_T|^r \geq \mathrm{E}|S_1|^r. \tag{18}$$

PROOF. By Lemma 1, $\{|S_n|, \mathscr{F}_n, n \geq 1\}$ is a submartingale, and so, by Theorem 6

$$(|S_n| I_{[T \geq n]})^r \leq \mathrm{E}^r\{|S_T| I_{[T \geq n]} \mid \mathscr{F}_n\} \leq \mathrm{E}\{|S_T|^r I_{[T \geq n]} \mid \mathscr{F}_n\},$$

recalling the conditional Jensen inequality. □

Next, martingale generalizations of Theorem 5.3.1 (Wald's equation) and Theorem 5.3.3 will be obtained. Let $\{S_n = \sum_{j=1}^n X_j, \mathscr{F}_n, n \geq 1\}$ be an $\mathscr{L}_2$ martingale. As noted in Example 2, $\{S_n^2 - \sum_1^n X_j^2 = 2\sum_{1 \leq i < j \leq n} X_i X_j, \mathscr{F}_n, n \geq 2\}$ is an $\mathscr{L}_1$ martingale with expectation zero, whence for any stopping variable T, setting $T(n) = \min(T, n)$,

$$\mathrm{E}\, S_{T(n)}^2 = \mathrm{E} \sum_1^{T(n)} X_j^2 \tag{19}$$

by Corollary 4. Since $T(n) \uparrow T$, it therefore follows from Fatou's lemma that

$$\mathrm{E}\, S_T^2 \leq \mathrm{E} \sum_1^T X_j^2. \tag{20}$$

Thus, if $\mathrm{E}\, S_T^2 = \infty$, equality holds in (20). In the contrary case $\mathrm{E}\, S_T^2 < \infty$, in view of (20), equality will hold therein recalling (19) if,

$$\mathrm{E}\, S_T^2 \geq \mathrm{E}\, S_{T(n)}^2 \tag{21}$$

Hence, it suffices to verify (21) to establish equality in (20).

Lemma 3. *If* $\{S_n, \mathscr{F}_n, n \geq 1\}$ *is an* $\mathscr{L}_2$ *martingale and* T *is any finite* $\{\mathscr{F}_n\}$*-time, then* (20) *obtains. Moreover, if*

$$\varliminf_{n\to\infty} \int_{[T > n]} S_n^2 < \infty \tag{22}$$

or

$$\varliminf_{n\to\infty} \int_{[T > n]} |S_n| = 0 \tag{23}$$

holds, then $\mathrm{E}\, S_T^2 = \mathrm{E} \sum_1^T X_j^2$.

PROOF. In view of the prior discussion, it suffices to show that (22) ⇒ (23)

$\Rightarrow$ (21). Suppose that $\underline{\lim} \int_{[T>n]} |S_n| = c > 0$. For any k in $(0, \infty)$

$$\underline{\lim}_{n\to\infty} \int_{[T>n]} S_n^2 \geq k \, \underline{\lim}_{n\to\infty} \int_{[T>n, |S_n|>k]} |S_n| = k \cdot c \to \infty$$

as $k \to \infty$, and so (22) $\Rightarrow$ (23). Next, supposing, as is permissible, that $\mathrm{E}\, S_T^2 < \infty$, necessarily $\mathrm{E}|S_T| < \infty$, and so via Theorem 5, (23) implies that on $\{T \geq n\}$

$$\mathrm{E}\{|S_T| \mid \mathscr{F}_n\} \geq |S_n|,$$

whence on $\{T \geq n\}$

$$\mathrm{E}\{S_T^2 \mid \mathscr{F}_n\} \geq S_n^2 = S_{T(n)}^2.$$

Since $\mathrm{E}\{S_T^2 \mid \mathscr{F}_n\} = \mathrm{E}\{S_{T(n)}^2 \mid \mathscr{F}_n\} = S_{T(n)}^2$ on $\{T < n\}$, (21) follows. □

Theorem 7. *If $\{S_n = \sum_1^n X_j, \mathscr{F}_n, n \geq 1\}$ is an $\mathscr{L}_2$ martingale and T is any finite $\{\mathscr{F}_n\}$-time, then $\mathrm{E}\, S_T^2 \leq \mathrm{E} \sum_1^T X_j^2$. Moreover, if any one of the four conditions*

$$\underline{\lim}_{n\to\infty} \int_{[T>n]} |S_n| = 0, \qquad \underline{\lim}_{n\to\infty} \int_{[T>n]}^{\bullet} S_n^2 < \infty,$$
$$\mathrm{E} \sum_{n=1}^{T} X_n^2 < \infty, \qquad \mathrm{E} \sum_{n=1}^{T} |X_n| < \infty, \tag{24}$$

holds, then, setting $\mathscr{F}_0 = \{\varnothing, \Omega\}$,

$$\mathrm{E}\, S_T^2 = \mathrm{E} \sum_{j=1}^{T} X_j^2 = \mathrm{E} \sum_{j=1}^{T} \mathrm{E}\{X_j^2 \mid \mathscr{F}_{j-1}\}. \tag{25}$$

Either of the last two conditions of (24) implies $\mathrm{E}\, S_T = \mathrm{E}\, X_1$. If $\{S_n = \sum_1^n X_i, \mathscr{F}_n, n \geq 1\}$ is merely an $\mathscr{L}_1$ martingale, the last condition of (24) entails $\mathrm{E}\, S_T = \mathrm{E}\, X_1$.

Proof. First, $\mathrm{E} \sum_1^T X_j^2 < \infty \Rightarrow$ (22) since, recalling (19),

$$\int_{[T>n]} S_n^2 \leq \mathrm{E}\, S_{T\wedge n}^2 = \mathrm{E} \sum_1^{T\wedge n} X_j^2 \leq \mathrm{E} \sum_1^{T} X_j^2. \tag{26}$$

and so in view of Lemma 3 and Corollary 4 it suffices to note that the final condition of (24) ensures that

$$\mathrm{E}|S_T| \leq \mathrm{E} \sum_1^T |X_n| < \infty$$

and

$$\mathrm{E}|S_n|I_{[T>n]} \leq \mathrm{E}\sum_1^n |X_j|I_{[T>n]} \leq \mathrm{E}\sum_1^T |X_j|I_{[T>n]} = o(1). \qquad \square$$

Corollary 7. *If* $\{X_n, n \geq 1\}$ *are independent* r.v.s *with* $\mathrm{E}\,X_n = 0$, $n \geq 1$, *and* T *is a finite* $\{X_n\}$*-time, then, setting* $S_n = \sum_{i=1}^n X_i$,

$$\mathrm{E}\sum_1^T \mathrm{E}|X_j| < \infty \tag{27}$$

implies $\mathrm{E}\,S_T = 0$. *If* $\sigma_n^2 = \mathrm{E}\,X_n^2 < \infty$, $n \geq 1$, *then either* (27) *or*

$$\mathrm{E}\sum_1^T \sigma_j^2 < \infty \tag{28}$$

implies

$$\mathrm{E}\,S_T^2 = \mathrm{E}\sum_1^T X_j^2 = \mathrm{E}\sum_1^T \sigma_j^2. \tag{29}$$

PROOF. This follows directly from Theorem 7. $\square$

A useful device in obtaining Doob's maximal inequality (33) and a martingale generalization (36) of Kolmogorov's inequality is

Lemma 4. *Let* $\{S_n = \sum_{j=1}^n X_j, \mathscr{F}_n, n \geq 1\}$ *be an* $\mathscr{L}_1$ *stochastic sequence and* $\{v_n, \mathscr{F}_{n-1}, n \geq 1\}$ *a stochastic sequence with* $v_n \in \mathscr{L}_\infty$, $n \geq 1$. *Then for any bounded* $\{\mathscr{F}_n\}$*-time* T,

$$\mathrm{E}\,v_T S_T = \mathrm{E}\sum_1^T [v_j\,\mathrm{E}\{X_j \mid \mathscr{F}_{j-1}\} + (v_j - v_{j-1})S_{j-1}], \qquad S_0 = 0,\ \mathscr{F}_0 = \mathscr{F}_1, \tag{30}$$

and, moreover, if $(v_{j+1} - v_j)S_j \leq 0$, a.c., $j \geq 0$,

$$\mathrm{E}\,v_T S_T \leq \mathrm{E}\sum_1^T v_j\,\mathrm{E}\{X_j \mid \mathscr{F}_{j-1}\}. \tag{31}$$

PROOF. If $U_n = v_n S_n - \sum_1^n [v_j\,\mathrm{E}\{X_j \mid \mathscr{F}_{j-1}\} + (v_j - v_{j-1})S_{j-1}]$, then $\{U_n, \mathscr{F}_n, n \geq 1\}$ is a martingale and (30) follows from Corollary 4 or 5. Then (31) follows directly from (30). $\square$

Corollary 8 (Dubins–Freedman). *If* $\{S_n = \sum_1^n X_j, \mathscr{F}_n, n \geq 1\}$ *is an* $\mathscr{L}_2$ *martingale with* $\mathrm{E}\,X_1 = 0$ *and* $Y_n = \mathrm{E}\{X_n^2 \mid \mathscr{F}_{n-1}\}$, $n \geq 1$, *where* $\mathscr{F}_0 = \{\varnothing, \Omega\}$, *then for any stopping time* T *and real numbers* a, b *with* $b > 0$

$$\int_{[T<\infty]} \left(\frac{a + S_T}{b + Y_1 + \cdots + Y_T}\right)^2 \leq \frac{a^2 + Y_1}{(b + Y_1)^2} + \frac{1}{b + Y_1}. \tag{32}$$

PROOF. If $v_n^{-1} = (b + Y_1 + \cdots + Y_n)^2$, then $\{v_n, \mathscr{F}_{n-1}, n \geq 1\}$ is a stochastic sequence with $v_{n+1} \leq v_n$. Since $\{(a + S_n)^2, \mathscr{F}_n, n \geq 1\}$ is a nonnegative submartingale, by (31) (with $S_0 = -a$)

$$\int_{[T\le n]}\left(\frac{a+S_T}{b+Y_1+\cdots+Y_T}\right)^2 \le \sum_{j=1}^n \mathrm{E}\, v_j\, \mathrm{E}\{(a+S_j)^2-(a+S_{j-1})^2 \mid \mathscr{F}_{j-1}\}$$
$$= v_1(a^2+Y_1)+\mathrm{E}\sum_{j=2}^n v_j Y_j.$$

Since

$$v_j Y_j = \frac{Y_j}{(b+Y_1+\cdots+Y_j)^2} \le \frac{1}{b+Y_1+\cdots+Y_{j-1}} - \frac{1}{b+Y_1+\cdots+Y_j},$$

the conclusion (32) follows as $n\to\infty$. □

Theorem 8. *If* $\{S_n=\sum_1^n X_j, \mathscr{F}_n, n\ge 1\}$ *is a nonnegative* $\mathscr{L}_1$ *submartingale and* $\{v_n, \mathscr{F}_{n-1}, n\ge 1\}$ *a stochastic sequence of* $\mathscr{L}_\infty$ r.v.s *with* $v_n\ge v_{n+1}\ge 0$, a.c., *then for any* $\lambda>0$, (i)

$$\lambda\, \mathrm{P}\left\{\max_{1\le j\le n} v_j S_j \ge \lambda\right\} + \int_{[\max v_j S_j<\lambda]} v_n S_n \le \sum_{j=1}^n \mathrm{E}\, v_j X_j \tag{33}$$

and (ii) (Doob Inequalities)

$$\mathrm{P}\left\{\max_{1\le j\le n} S_j \ge \lambda\right\} \le \frac{1}{\lambda}\int_{[\max S_j\ge\lambda]} S_n, \tag{34}$$

whence

$$\|S_n\|_p \le \left\| \max_{1<j\le n} S_j \right\|_p \le \frac{p}{p-1}\|S_n\|_p, \qquad p>1,$$
$$\left\| \max_{1\le j\le n} S_j \right\|_p < \frac{e}{e-1}(1+\|S_n \log^+ S_n\|_p), \qquad p=1. \tag{35}$$

(iii) (Hájek–Rényi Inequality) *If* $\{U_n=\sum_{j=1}^n u_j, \mathscr{F}_n, n\ge 1\}$ *is an* $\mathscr{L}_2$ *martingale and* $\{b_n, n\ge 1\}$ *is a positive, nondecreasing real sequence, then for any* $\lambda>0$

$$\mathrm{P}\left\{\max_{1\le j\le n}\left|\frac{U_j}{b_j}\right|\ge\lambda\right\} \le \frac{1}{\lambda^2}\sum_{j=1}^n \frac{\mathrm{E}\, u_j^2}{b_j^2}. \tag{36}$$

PROOF. Let $T_i=\inf\{n\ge 1: v_n S_n \ge \lambda^i\}$ and $T_i'=\min(T_i, n)$, $n\ge 1$, $i=1, 2$. From (31) of Lemma 4

$$\lambda^i\, \mathrm{P}\left\{\max_{1\le j\le n} v_j S_j \ge \lambda^i\right\} + \int_{\left[\max_{1\le j\le n} v_j S_j<\lambda^i\right]} v_n S_n$$
$$\le \int_{[T_i\le n]} v_{T_i} S_{T_i} + \int_{[T_i>n]} v_n S_n = \mathrm{E}\, v_{T_i'} S_{T_i'}$$
$$\le \mathrm{E}\sum_1^{T_i'} v_j\, \mathrm{E}\{X_j \mid \mathscr{F}_{j-1}\} \le \sum_{j=1}^n \mathrm{E}\, v_j X_j, \tag{37}$$

yielding (33) for $i = 1$. To obtain (36), take $i = 2$, $v_j = b_j^{-2}$ in (37), noting that $S_n = U_n^2$, $n \geq 1$, is a nonnegative $\mathscr{L}_1$ submartingale. Since (34) is an immediate consequence of (33), it remains to prove (35). Setting $S_n^* = \max_{1 \leq j \leq n} S_j$, if $p > 1$ it follows from Corollary 6.2.2 and (34) that

$$\mathrm{E}\, S_n^{*p} = p \int_0^\infty \lambda^{p-1}\, \mathrm{P}\{S_n^* \geq \lambda\} d\lambda \leq p \int_0^\infty \lambda^{p-2} \int_{[S_n^* \geq \lambda]} S_n \, d\mathrm{P}\, d\lambda$$

$$= p\, \mathrm{E}\, S_n \int_0^{S_n^*} \lambda^{p-2}\, d\lambda = \frac{p}{p-1} \mathrm{E}\, S_n (S_n^*)^{(p-1)}.$$

Hence, if $(p-1)q = p$, by Hölder

$$\mathrm{E}\, S_n^{*p} \leq \frac{p}{p-1} \|S_n\|_p \|S_n^{*(p-1)}\|_q = \frac{p}{p-1} \|S_n\|_p\, \mathrm{E}^{1/q} S_n^{*p},$$

yielding the first part of (35). If, rather, $p = 1$, again via (34)

$$\mathrm{E}\, S_n^* - 1 \leq \mathrm{E}(S_n^* - 1)^+ = \int_0^\infty \mathrm{P}\{S_n^* - 1 \geq \lambda\} d\lambda$$

$$\leq \int_0^\infty \frac{1}{\lambda + 1} \int_{[S_n^* \geq \lambda + 1]} S_n \, d\mathrm{P}\, d\lambda$$

$$= \mathrm{E}\, S_n \int_0^{(S_n^* - 1)^+} \frac{d\lambda}{\lambda + 1} = \mathrm{E}\, S_n \log^+ S_n^*.$$

Since for constants $a \geq 0$, $b > 0$ necessarily $a \log b \leq a \log^+ a + be^{-1}$,

$$\mathrm{E}\, S_n^* - 1 \leq \mathrm{E}\, S_n \log^+ S_n + e^{-1}\, \mathrm{E}\, S_n^*,$$

from which the second portion of (35) is immediate. □

Example 9. If $\{S_n, \mathscr{F}_n, n \geq 1\}$ is a submartingale and h is any nonnegative, increasing, convex function, then for any positive t and real x

$$\mathrm{P}\left\{\max_{1 \leq j \leq n} S_j \geq x\right\} \leq \frac{\mathrm{E}\, h(tS_n)}{h(tx)} \tag{38}$$

and, in particular,

$$\mathrm{P}\left\{\max_{1 \leq j \leq n} S_j \geq x\right\} \leq e^{-tx}\, \mathrm{E}\, e^{tS_n}, \qquad t > 0. \tag{39}$$

Proof. Since $\{h(tS_j), \mathscr{F}_j, 1 \leq j \leq n\}$ is a nonnegative submartingale via Lemma 1, (34) ensures that

$$\mathrm{P}\left\{\max_{1 \leq j \leq n} S_j \geq x\right\} \leq \mathrm{P}\left\{h\left(\max_{1 \leq j \leq n} tS_j\right) \geq h(tx)\right\} = \mathrm{P}\left\{\max_{1 \leq j \leq n} h(tS_j) \geq h(tx)\right\}$$

$$\leq \frac{\mathrm{E}\, h(tS_n)}{h(tx)}.$$

□

The next example generalizes Example 5.2.1.

EXAMPLE 10. If $S_n = \sum_{j=1}^{n} X_j$, where $\{X_n, n \geq 1\}$ are i.i.d. $\mathscr{L}_p$ r.v.'s for some p in $(0, 2)$ with $\mathrm{E}\, X_1 = 0$ whenever $1 \leq p < 2$, then

$$\sum_{j=2}^{\infty} j^{-2/p} X_j S_{j-1} \text{ converges a.c.} \tag{40}$$

PROOF. If $Y_n = n^{-1/p} X_n I_{[|X_n| \leq n^{1/p}]}$, $n \geq 1$, and $T_n = n^{-1/p} S_{n-1} I_{[|S_{n-1}| \leq n^{1/p}]}$, $n \geq 2$, then $\{T_n(Y_n - \mathrm{E}\, Y_n), n \geq 1\}$ is a martingale difference sequence wtih

$$\sum_{n=2}^{\infty} \mathrm{E}\, T_n^2(Y_n - \mathrm{E}\, Y_n)^2 \leq \sum_{n=2}^{\infty} \mathrm{E}(Y_n - \mathrm{E}\, Y_n)^2 \leq \sum_{n=2}^{\infty} \mathrm{E}\, Y_n^2 < \infty$$

by (13) of Theorem 5.1.3. Then $\sum_{n=2}^{\infty} T_n(Y_n - \mathrm{E}\, Y_n)$ converges a.c. via Exercise 7.4.1. Moreover, $\sum_{n=2}^{\infty} \mathrm{E}\, |T_n \mathrm{E}\, Y_n| \leq \sum_{n=2}^{\infty} |\mathrm{E}\, Y_n| < \infty$, and $\{Y_n\}$ and $\{n^{-1/p} X_n\}$ are equivalent sequences according to Theorem 5.1.3. Hence, $\sum_{n=2}^{\infty} T_n \mathrm{E}\, Y_n$ converges a.c., and so $\sum_{n=2}^{\infty} T_n Y_n$ converges a.c.

By the Marcinkiewicz–Zygmund strong law of large numbers, $\mathrm{P}\{|S_{n-1}| > n^{1/p}, \text{i.o.}\} = 0$ whence $\{T_n Y_n, n \geq 1\}$ and $\{S_{n-1}/n^{1/p} \cdot X_n/n^{1/p}, n \geq 1\}$ are equivalent sequences and (40) follows. □

EXERCISES 7.4

1. If $\{S_n, n \geq 1\}$ is a martingale satisfying (i) $\sum_{j=1}^{\infty} \mathrm{E}\, X_j^2 < \infty$ or (ii) $\sum_1^{\infty} \mathrm{E}(|X_j| I_{[|X_j| > 1]} + X_j^2 I_{[|X_j| \leq 1]}) < \infty$ or (iii) $\sum_{j=1}^{\infty} \mathrm{E}\, |X_j|^p < \infty$ for some p in $[1, 2]$, then S_n converges a.c. *Hint*: For (ii) consider $X_j' = X_j I_{[|X_j| \leq 1]} - \mathrm{E}\{X_j I_{[|X_j| \leq 1]} | X_1, \ldots, X_{j-1}\}$ and $X_j'' = X_j - X_j'$.

2. In statistical problems, likelihood ratios $U_n = g_n(X_1, \ldots, X_n)/f_n(X_1, \ldots, X_n)$ are encountered, where f_n, g_n are densities, each being a candidate for the actual density of r.v.s $X_1, \ldots, X_n$. If $\{X_n, n \geq 1\}$ are coordinate r.v.s on $(R^\infty, \mathscr{B}^\infty, \mathrm{P})$ and g_n vanishes whenever f_n does, show that $\{U_n, n \geq 1\}$ is a martingale when f_n is the true density.

3. There exist martingales $\{S_n, \mathscr{F}_n, n \geq 1\}$ and stopping variables T for which (13) fails. Let $\{X_n, n \geq 1\}$ be i.i.d. with $\mathrm{E}\, X_1 = 0$ and set $T = \inf\{n \geq 1 : S_n = \sum_1^n X_i > 0\}$. Then for $n = 1$, $\mathrm{E}\{S_T - S_n | \mathscr{F}_n\} > 0$ on $[T > 1]$.

4. If $\{S_n, \mathscr{F}_n, n \geq 1\}$ is a martingale or positive submartingale, then for any stopping time T, $\mathrm{E}\, |S_T| \leq \lim_{n \to \infty} \mathrm{E}\, |S_n|$. In particular, if $\{S_n, \mathscr{F}_n, n \geq 1\}$ is $\mathscr{L}_1$ bounded, $\mathrm{E}\, |S_T| < \infty$ for every stopping time T.

5. If $\mathrm{E}\, |S_T| < \infty$, it is spurious generality to have $\underline{\lim}_{n \to \infty}$ rather than $\lim_{n \to \infty}$ in (6) or (12). *Hint*: If $V_n = S_n^+$ in the first case and $|S_n|$ in the second, then, as in the proof of Theorem 5,
$$\mathrm{Z} \int_{A[T \geq n]} V_n \leq \int_{A[n \leq T \leq m]} V_T + \int_{A[T > m]} V_m.$$

6. (i) Find a positive $\mathscr{L}_1$ martingale which is not u.i. (ii) If Y_n, $n \geq 1$, are r.v.s on $(\Omega, \mathscr{F}, \mathrm{P})$ and $A \in \sigma(Y_1, Y_2, \ldots)$, then $\mathrm{P}\{A | Y_1, \ldots, Y_n\} \xrightarrow{\text{a.c.}} I_A$. *Hint*: Apply Corollary 1. (iii) If $\{\mathscr{F}_n, n \geq 1\}$ is a stochastic basis and $\{X_n, n \geq 1\}$ are r.v.'s with

$\mathrm{E}\sup_{n \geq 1} |X_n| < \infty$, then

$$\mathrm{E}\{\underline{\lim}_{n\to\infty} X_n | \mathscr{F}_\infty\} \leq \underline{\lim}_{n\to\infty} \mathrm{E}\{X_n | \mathscr{F}_\infty\}.$$

7. If $\{A_n, n \geq 1\}$ is a sequence of events and $A_n \in \mathscr{F}_n \uparrow$, prove the following version of the Borel–Cantelli theorem:

$$\mathrm{P}\left\{A_n, \text{ i.o. } \Delta \left[\sum_{n=1}^{\infty} \mathrm{P}\{A_{n+1} | \mathscr{F}_n\} = \infty\right]\right\} = 0.$$

8. If $\{S_n = \sum_{i=1}^n X_i, n \geq 1\}$ is an $\mathscr{L}_p$ martingale for some p in $[1, 2]$ and $0 < b_n \uparrow \infty$ then $\lim S_n/b_n = 0$, a.c. on the set A where $\sum_{n=2}^\infty b_n^{-p}\mathrm{E}\{|X_n|^p | \mathscr{F}_{n-1}\} < \infty$. In particular, $S_n/b_n \xrightarrow{\text{a.c.}} 0$ if $\mathrm{P}\{A\} = 1$. *Hint*: Set $T_n = T \wedge n$ where $T = \inf\{n \geq 1: \sum_{j=1}^{n+1} b_j^{-p}\mathrm{E}\{|X_j|^p | \mathscr{F}_{j-1}\} > K > 0$ and apply Exercise 1 to $\sum_1^{T_n} X_j/b_j$.

9. Let Y be an $\mathscr{L}_1$ r.v. and $\{\mathscr{G}_n, -\infty < n < \infty\}$ a stochastic basis. If $\mathscr{G}_\infty = \sigma(\bigcup_{-\infty}^\infty \mathscr{G}_n)$, $\mathscr{G}_{-\infty} = \bigcap_{-\infty}^\infty \mathscr{G}_n$, and $U_n = \mathrm{E}\{Y | \mathscr{G}_n\}$, $-\infty \leq n \leq \infty$, then $\{U_n, \mathscr{G}_n, -\infty \leq n \leq \infty\}$ is a martingale.

10. Let $\{S_n = \sum_1^n X_j, \mathscr{F}_n, n \geq 1\}$ be a martingale with bounded conditional variance, i.e., $\mathrm{E}\{S_n - \mathrm{E}\{S_n | \mathscr{F}_{n-1}\}]^2 | \mathscr{F}_{n-1}\} = \mathrm{E}\{X_n^2 | \mathscr{F}_{n-1}\} \leq \sigma_n^2$, $n \geq 1$, where σ_n^2 is a finite constant. If $U_{k,n}$ is as defined in Example 2, show that $\mathrm{E}(U_{k,n} - U_{k,n-1})^2 \leq \sigma_n^2 \mathrm{E}\, U_{k-1,n-1}^2$, and hence that $\mathrm{E}\, U_{2,n}^2 \leq \sum_{j=1}^{n-1} \sigma_{j+1}^2 \sum_{i=1}^j \sigma_i^2$. If $\sigma_n^2 = \sigma^2$, $\mathrm{E}\, U_{k,n}^2 \leq \binom{n}{k}\sigma^{2k}$ and $\mathrm{E}(U_{k,n} - U_{k,n-1})^2 \leq n^{k-1}\sigma^{2k}$, $k \geq 1$.

11. Show, if $\{S_n, \mathscr{F}_n, n \geq 1\}$ is a nonnegative martingale with $\mathrm{E}\, S_1 = 1$, that $\mathrm{P}\{S_n \geq \lambda$, some $n \geq 1\} \leq 1/\lambda$.

12. Prove that an $\mathscr{L}_1$ stochastic sequence $\{\sum_{i=1}^n X_i, \sigma(X_1, \ldots, X_n), n \geq 1\}$ is a martingale iff $\mathrm{E}\, X_{n+1}\varphi(X_1, \ldots, X_n) = 0$ for all $n \geq 1$ and all bounded Borel functions φ, whereas the $\mathscr{L}_1$ r.v.s $\{X_n, n \geq 1\}$ are independent iff for all $n \geq 1$

$$\mathrm{E}\,\psi(X_{n+1})\varphi(X_1, \ldots, X_n) = 0$$

for all bounded Borel functions φ, ψ with $\mathrm{E}\,\psi(X_{n+1}) = 0$.

13. (Harris) A branching process is a sequence $\{Z_n, n \geq 0\}$ of nonnegative integer-valued r.v.s with $Z_0 = 1$ and such that the conditional distribution of Z_{n+1} given $(Z_0, \ldots, Z_n)$ is that of a sum of Z_n i.i.d. r.v.s each with the distribution of Z_1. If $\mathrm{E}\, Z_1 = m \in (0, \infty)$, verify that $\{W_n = Z_n/m^n, n \geq 1\}$ is a convergent martingale.

14. (Breiman) Let A, B be linear Borel sets and $\{X_n, n \geq 1\}$ r.v.s. If

$$\mathrm{P}\left\{\bigcup_{j=n}^\infty [X_j \in B] \,\middle|\, X_1, \ldots, X_n\right\} \geq \delta I_{[X_n \in A]}, \delta > 0,$$

prove that $\mathrm{P}\{X_n \in A, \text{ i.o.}\} \leq \mathrm{P}\{X_n \in B, \text{ i.o.}\}$. *Hint*: If $F_N = \bigcup_{j=N}^\infty [X_j \in B]$, $\lim_{n\to\infty} \mathrm{P}\{F_N | X_1 \cdots X_n\} = I_{F_N}$.

15. (Doob) If $\{X_n n \geq 1\}$ are independent r.v.s with $\mathrm{E}\, X_n = 0$, $n \geq 1$, and $S_n = \sum_1^n X_i$, $S_n^* = \max_{1 \leq i \leq n} |S_i|$, then $\mathrm{E}\, S_n^* \leq 8\, \mathrm{E}|S_n|$, thereby improving (33) for $p = 1$. *Hint*: Via Ottaviani's inequality (Exercise 3.3.16)

$$\int_0^\infty \mathrm{P}\{S_n^* > 2y\}dy \leq 2\,\mathrm{E}|S_n| + \int_{2\mathrm{E}|S_n|}^\infty 2\,\mathrm{P}\{|S_n| > y\}dy.$$

7.5. U-Statistics

Let h be a measurable symmetric function on R^k, $k \geq 1$ (i.e., invariant under the permutations of its arguments for $k \geq 2$) and $\{X_n, n \geq 1\}$ a sequence of i.i.d. random variables. Then (Example 7.4.3) a sequence of U-statistics $U_{k,n}(h)$ and their corresponding sums $S_{k,n}(h)$ are defined by

$$\binom{n}{k} U_{k,n}(h) = S_{k,n}(h) = \sum_{1 \leq i_1 < \cdots < i_k \leq n} h(X_{i_1}, \ldots, X_{i_k}), \qquad n \geq k \geq 1. \quad (1)$$

For $h \in \mathscr{L}_1$, that is, $\mathrm{E}|h| = \mathrm{E}|h(X_1, \ldots, X_k)| < \infty$, the U-statistic $U_{k,n}(h)$ and its "kernel" h are said to be **degenerate** of order $i - 1$ where $2 \leq i \leq k$ if $\mathrm{E}\{h(X_1, \ldots, X_k)|X_1, \ldots, X_j\} = 0$, a.s. for $j = i - 1$ but not for $j = i$; otherwise (i.e., if $\mathrm{E}\{h(X_1, \ldots, X_k)|X_1\}$ is not a.s. zero), they are non-degenerate. In the particular case $i = k$, they are called **completely degenerate**. It is sometimes convenient to express non-degeneracy as "degeneracy of order zero."

Let I denote the identity operator, that is, $If = f$, and define $\hat{f}_j$ and operators Q_j, $1 \leq j \leq k$ by

$$\hat{f}_j = Q_j f = \mathrm{E}\{f(X_1, \ldots, X_k)|X_\alpha, 1 \leq \alpha \leq k, \alpha \neq j\}$$

for any function f on R^k with $\mathrm{E}|f| < \infty$.

Lemma 1. *$Q_i^2 f = Q_i f$, $1 \leq i \leq k$ and for $1 \leq i \neq j \leq k$*

$$Q_i Q_j f = Q_j Q_i f = \mathrm{E}\{f(X_1, \ldots, X_k)|X_\alpha, \alpha \neq i, j\}.$$

Proof. Since $Q_i^2 f = Q_i[Q_i f]$, the initial statement is immediate via (14, iv) of Section 7.1. Next, set $\mathscr{G}_1 = \sigma(X_i)$, $\mathscr{G}_2 = \sigma(X_j)$, and $\mathscr{G}_3 = \sigma(X_\alpha, \alpha \neq i, j)$. Then $\hat{f}_j$ is $\sigma(\mathscr{G}_1 \cup \mathscr{G}_3)$-measurable and, moreover, $\mathscr{G}_2$ and $\sigma(\mathscr{G}_1 \cup \mathscr{G}_3)$ are independent. By Corollary 7.3.3, $\mathscr{G}_1$ and $\mathscr{G}_2$ are conditionally independent given $\mathscr{G}_3$ whence, by Theorem 7.3.1 (iv), $\mathrm{E}\{\hat{f}_j|\sigma(\mathscr{G}_2 \cup \mathscr{G}_3)\} = \mathrm{E}\{\hat{f}_j|\mathscr{G}_3\}$, a.c. Hence,

$$\begin{aligned} Q_i Q_j f &= \mathrm{E}\{\hat{f}_j|\sigma(\mathscr{G}_2 \cup \mathscr{G}_3)\} = \mathrm{E}\{\hat{f}_j|\mathscr{G}_3\} \\ &= \mathrm{E}\{\mathrm{E}\{f|X_\alpha, \alpha \neq j\}|X_\beta, \beta \neq i, j\} = \mathrm{E}\{f|X_\beta, \beta \neq i, j\} = Q_j Q_i f. \quad \square \end{aligned}$$

Setting $f_j^* = \mathrm{E}\{f|X_1, \ldots, X_j\}$, Lemma 1 ensures that

$$f_j^* = \mathrm{E}\{f|X_\alpha, \alpha \neq j + 1, \ldots, k\} = Q_{j+1} \ldots Q_k f.$$

Define for $1 \leq j \leq k$,

$$f_j = f_j(X_1, \ldots, X_j) = (I - Q_1) \cdots (I - Q_j) Q_{j+1} \cdots Q_k f = \prod_{i=1}^{j} (I - Q_i) f_j^* \quad (2)$$

whence, via Exercise 7.1.1,

$$f_1(X_1) = (I - Q_1)\mathrm{E}\{f|X_1\} = \mathrm{E}\{f(X_1, \ldots, X_k)|X_1\} - \mathrm{E}f$$

and, for $r \geq 1$,

$$f_1(X_r) = \mathrm{E}\{f(X_r, X_{r+1}, \ldots, X_{r+k-1})|X_r\} - \mathrm{E}f$$

are i.i.d. random variables.

Lemma 2. *The functions $f_j = f_j(X_1, \ldots, X_j)$ defined in (2) are completely degenerate for $2 \leq j \leq k$ and*

$$\begin{aligned} f(X_1, \ldots, X_k) = \mathrm{E}f &+ \sum_{i=1}^{k} f_1(X_i) \\ &+ \sum_{1 \leq i_1 < i_2 \leq k} f_2(X_{i_1} X_{i_2}) + \cdots \\ &+ \sum_{1 \leq i_1 < \cdots < i_{k-1} \leq k} f_{k-1}(X_{i_1}, \ldots, X_{i_{k-1}}) \\ &+ f_k(X_1, \ldots, X_k). \end{aligned}$$

PROOF. Since $Q_j(I - Q_j)f_j^* = 0$ and the operators commute,

$$\mathrm{E}\{f_j|X_1, \ldots, X_{j-1}\} = Q_j \cdots Q_k f_j = Q_j \cdots Q_k \prod_{i=1}^{j} (I - Q_i)f_j^* = 0, \qquad \text{a.c.}$$

whence $f_j(X_1, \ldots, X_j)$ is completely degenerate, $2 \leq j \leq k$. The final statement follows from the representation

$$f = \prod_{i=1}^{k} [(I - Q_i) + Q_i]f,$$

noting via Lemma 1 that

$$Q_1 \cdots Q_k f = \mathrm{E}\{f|X_\alpha, \alpha \neq 1, \ldots, k\} = \mathrm{E}f. \qquad \square$$

Lemma 3. *Let $f \in R^k$, $k \geq 2$, with $|\mathrm{E}f| \leq \infty$ and let $1 \leq i_1 < \cdots < i_k \leq n$ and $1 \leq \beta_1 < \cdots < \beta_m \leq n$ be two sets of integers.* (i) *If $\{i_1, \ldots, i_k\} \cap \{\beta_1, \ldots, \beta_m\} = \{\delta_1, \ldots, \delta_j\}$, where $1 \leq j \leq k \wedge m$, then*

$$\mathrm{E}\{f(X_{i_1}, \ldots, X_{i_k})|X_{\beta_1}, \ldots, X_{\beta_m}\} = \mathrm{E}\{f(X_{i_1}, \ldots, X_{i_k})|X_{\delta_1}, \ldots, X_{\delta_j}\}.$$

(ii) *If $i_k \notin \{\beta_1, \ldots, \beta_m\}$, then*

$$\begin{aligned} &\mathrm{E}\{f(X_{i_1}, \ldots, X_{i_k})|X_{\beta_1}, \ldots, X_{\beta_m}, X_{i_1}, \ldots, X_{i_{k-1}}\} \\ &\quad = \mathrm{E}\{f(X_{i_1}, \ldots, X_{i_k})|X_{i_1}, \ldots, X_{i_{k-1}}\}. \end{aligned}$$

PROOF. Let $\{\alpha_{j+1}, \ldots, \alpha_k\} = \{i_1, \ldots, i_k\} - \{\delta_1, \ldots, \delta_j\}$ and $\{\gamma_{j+1}, \ldots, \gamma_m\} = \{\beta_1, \ldots, \beta_m\} - \{\delta_1, \ldots, \delta_j\}$. Set $\mathscr{G}_1 = \sigma(X_{\alpha_{j+1}}, \ldots, X_{\alpha_k})$, $\mathscr{G}_2 = \sigma(X_{\gamma_{j+1}}, \ldots, X_{\gamma_m})$, and $\mathscr{G}_3 = \sigma(X_{\delta_1}, \ldots, X_{\delta_j})$. Then $\sigma(\mathscr{G}_1 \cup \mathscr{G}_3)$ is independent of $\mathscr{G}_2$ whence $\mathscr{G}_1$ and $\mathscr{G}_2$ are conditionally independent given $\mathscr{G}_3$ by Corollary 7.3.3. Since $f(X_{i_1}, \ldots, X_{i_k})$ is $\sigma(\mathscr{G}_1 \cup \mathscr{G}_3)$-measurable, (i) follows from Theorem 7.3.1 (iv).

Apropos of (ii), if A is the set of distinct integers among $\beta_1, \ldots, \beta_m, i_1, \ldots, i_{k-1}$, then $\{i_1, \ldots, i_{k-1}\} = A \cap \{i_1, \ldots, i_k\}$, so that (ii) follows from (i). $\square$

Corollary 1. *If h is a completely degenerate kernel on R^k, $k \geq 2$, and $\mathscr{F}_n = \sigma(X_1, \ldots, X_n)$ then, for any constants $\{a_j, j \geq 1\}$,*

$$\left\{\sum_{j=k}^{n} a_j \sum_{1 \leq i_1 < \cdots < i_{k-1} < j} h(X_{i_1}, \ldots, X_{i_{k-1}}, X_j), \mathscr{F}_n, n \geq k\right\}$$

is a martingale and, in particular ($a_j = 1, j \geq 1$), so is $\{S_{k,n}(h), \mathscr{F}_n, n \geq k\}$.

An important aspect of non-degenerate U-statistics is their decomposition into an average of i.i.d. random variables plus a finite linear combination of completely degenerate U-statistics. This is an immediate consequence of

Theorem 1 (Hoeffding's decomposition). *If $\{X_n, n \geq 1\}$ are* i.i.d. *random variables and h is a symmetric function on R^k, $k \geq 2$, with* $\mathrm{E}\, h(X_1, \ldots, X_k) = 0$, *then* (i)

$$U_{k,n}(h) = \sum_{j=1}^{k} \binom{k}{j} U_{j,n}(h_j), \tag{3}$$

where h_j is as in (2). *Moreover, if h is degenerate of order $i - 1$, where $2 \leq i \leq k$, the first $i - 1$ terms of the sum in* (3) *vanish. Furthermore,* (ii) *if $S_{j,n}(h_j) = \binom{n}{j} U_{j,n}(h_j)$, $1 \leq j \leq k$, then $\{S_{j,n}(h_j), \mathscr{F}_n = \sigma(X_1, \ldots, X_n), n \geq j\}$ is a martingale, $1 \leq j \leq k$, and, if* $\mathrm{E}\, h^2 < \infty$, *then* $\mathrm{E}\, S_{l,n}(h_l) S_{m,n}(h_m) = 0$ *for $m \neq l$.*

PROOF. Replacing $f, X_1, \ldots, X_k$ by $h, X_{t_1}, \ldots, X_{t_k}$ in Lemma 2 and summing,

$$\binom{n}{k} U_{k,n}(h) = S_{k,n}(h_k) + \sum_{j=1}^{k-1} \left[\sum_{1 \leq t_1 < \cdots < t_k \leq n} \; \sum_{1 \leq i_1 < \cdots < i_j \leq k} h(X_{t_{i_1}}, \ldots, X_{t_{i_j}}) \right].$$

Each of the $\binom{n}{j}$ terms $h_j(X_{i_1}, \ldots, X_{i_j})$ appears the same number of times in the bracketed double sum, and all together there are $\binom{n}{k}\binom{k}{j}$ terms. Thus, $h_j(X_{i_1}, \ldots, X_{i_j})$ is repeated $\binom{n}{k}\binom{k}{j}/\binom{n}{j}$ times. Hence,

$$\binom{n}{k} U_{k,n}(h) = S_{k,n}(h_k) + \sum_{j=1}^{k-1} \frac{\binom{n}{k}\binom{k}{j}}{\binom{n}{j}} S_{j,n}(h_j) = \binom{n}{k} \sum_{j=1}^{k} \binom{k}{j} U_{j,n}(h_j),$$

which is tantamount to (3). According to Lemma 2, h_j is completely degenerate for $j \geq 2$, whereas $U_{1,n}(h_1) = 1/n \sum_{r=1}^{n} \mathrm{E}\{h \mid X_r\}$ is an average of i.i.d. random variables. Moreover, if h is degenerate of order $i - 1$, where $2 \leq i \leq k$, then $h_j^* = 0 = h_j$ for $1 \leq j \leq i - 1$ via (2), whence the first $i - 1$ terms of (3) vanish.

Furthermore, since h_j is completely degenerate for $j \geq 2$, Corollary 1 ensures that $\{S_{j,n}(h_j), \mathscr{F}_n, n \geq k\}$ is a martingale when $j \geq 2$, and clearly the same conclusion holds for $j = 1$.

Finally, if $l \neq m$, it follows from Lemma 4(i) that

$$\mathrm{E}\, h_l(X_{i_1}, \ldots, X_{i_l}) h_m(X_{\beta_1}, \ldots, X_{\beta_m}) = 0,$$

implying that $\mathrm{E}\, S_{l,n}(h_l) S_{m,n}(h_m) = 0$ for $l \neq m$. □

Lemma 4. *Let $1 \le i_1 < \cdots < i_k \le n$ and $1 \le \beta_1 < \cdots < \beta_m \le n$ be two sets of integers with $\{i_1, \ldots, i_k\} \cap \{\beta_1, \ldots, \beta_m\} = \{i_1, \ldots, i_j\}$, where $1 \le j < k$. (i) If f is a completely degenerate function on R^k, $k \ge 2$, and g is a function on R^m with $|\mathrm{E}\, g| \le \infty$, $|\mathrm{E}\, fg| \le \infty$, then*

$$\mathrm{E}\, f(X_{i_1}, \ldots, X_{i_k}) g(X_{\beta_1}, \ldots, X_{\beta_m}) = 0.$$

(ii) *If $m = k$ and h is a symmetric function on R^k, $k \ge 2$, with $\mathrm{E}\, h^2 < \infty$, then*

$$\mathrm{E}\, h(X_{i_1}, \ldots, X_{i_k}) h(X_{\beta_1}, \ldots, X_{\beta_k}) = \mathrm{E}[\mathrm{E}\{h(X_1, \ldots, X_k) | X_1, \ldots, X_j\}]^2,$$

and the right side is an increasing function of j.

PROOF. Employing Lemma 3(i),

$$\begin{aligned} &\mathrm{E}\{f(X_{i_1}, \ldots, X_{i_k}) g(X_{\beta_1}, \ldots, X_{\beta_m}) | X_{i_1}, \ldots, X_{i_j}\} \\ &\quad = \mathrm{E}\{f(X_{i_1}, \ldots, X_{i_k}) \mathrm{E}\{g(X_{\beta_1}, \ldots, X_{\beta_m}) | X_{i_1}, \ldots, X_{i_k}\} | X_{i_1}, \ldots, X_{i_j}\} \\ &\quad = \mathrm{E}\{f(X_{i_1}, \ldots, X_{i_k}) \mathrm{E}\{g(X_{\beta_1}, \ldots, X_{\beta_m}) | X_{i_1}, \ldots, X_{i_j}\} | X_{i_1}, \ldots, X_{i_j}\} \\ &\quad = \mathrm{E}\{g(X_{\beta_1}, \ldots, X_{\beta_m}) | X_{i_1}, \ldots, X_{i_j}\} \mathrm{E}\{f(X_{i_1}, \ldots, X_{i_k}\} | X_{i_1}, \ldots, X_{i_j}\}, \end{aligned} \tag{4}$$

and the last term equals zero a.s. when f is completely degenerate, so that (i) is an immediate consequence of (4).

Apropos of (ii), taking $m = k$, $g = f = h$ in (4),

$$\begin{aligned} &\mathrm{E}\{h(X_{i_1}, \ldots, X_{i_k}) h(X_{\beta_1}, \ldots, X_{\beta_k}) | X_{i_1}, \ldots, X_{i_j}\} \\ &\quad = \mathrm{E}\{h(X_{i_1}, \ldots, X_{i_k}) | X_{i_1}, \ldots, X_{i_j}\} \mathrm{E}\{h(X_{\beta_1}, \ldots, X_{\beta_k}) | X_{i_1}, \ldots, X_{i_j}\} \\ &\quad = \mathrm{E}^2\{h(X_1, \ldots, X_k) | X_1, \ldots, X_j\} \end{aligned}$$

since $\{X_n, n \ge 1\}$ are i.i.d. whence (ii) follows upon taking expectations. Moreover, if $Y_j = \mathrm{E}\{h(X_1, \ldots, X_k) | X_1, \ldots, X_j\}$, $1 \le j \le k$ and $\mathscr{F}_j = \sigma(X_1, \ldots, X_j)$, then $\{Y_j^r, \mathscr{F}_j, 1 \le j \le k\}$ is a martingale for $r = 1$ and hence a submartingale for $r = 2$. Thus, $\mathrm{E}\, Y_j^2$, $1 \le j \le k$ is an increasing sequence. □

Corollary 2. *Let $\mathrm{E}\, h^2 < \infty$. If the U-statistic $U_{k,n}(h)$ is degenerate of order $i - 1$, where $2 \le i \le k$ or non-degenerate ($i = 1$) with $\mathrm{E}\, h = 0$, then its variance is*

$$\sigma^2_{U_{k,n}(h)} = \binom{n}{k}^{-1} \sum_{j=i}^{k} \binom{k}{j} \binom{n-k}{k-j} \mathrm{E}[\mathrm{E}^2\{h(X_1, \ldots, X_k) | X_1, \ldots, X_j\}],$$

and so

$$\sigma^2_{U_{n,k}(h)} = \frac{i! \binom{k}{i}^2 e^i}{n^i} \mathrm{E}[\mathrm{E}^2\{h(X_1, \ldots, X_k) | X_1, \ldots, X_i\}] + O\left(\frac{1}{n^{i+1}}\right).$$

PROOF. Set $q_j = \mathrm{E}[\mathrm{E}^2\{h(X_1, \ldots, X_k)|X_1, \ldots, X_j\}]$. Since the number of pairs $(i_1, \ldots, i_k), (\beta_1, \ldots, \beta_k)$ with exactly j integers in common in $\binom{n}{k}\binom{k}{j}\binom{n-k}{k-j}$, Lemma 4 yields

$$\sigma^2_{U_{k,n}(h)} = \binom{n}{k}^{-2} \sum_{1 \le i_1 < \cdots < i_k \le n} \; \sum_{1 \le \beta_1 < \cdots < \beta_k \le n} \mathrm{E}\, h(X_{i_1}, \ldots, X_{i_k})h(X_{\beta_1}, \ldots, X_{\beta_k})$$
$$= \binom{n}{k}^{-1} \sum_{j=1}^{k} \binom{k}{j}\binom{n-k}{k-j} q_j,$$

and, if h is degenerate of order $i - 1$, that first $i - 1$ terms of the sum vanish.

For large n, the dominant term of the sum is

$$\binom{k}{i}\binom{n-k}{k-i}\binom{n}{k}^{-1} q_i = i!\binom{k}{i}^2 q_i \frac{[(n-k)!]^2}{n!(n-2k+i)!}$$
$$= i!\binom{k}{i}^2 q_i \left(\frac{e}{n}\right)^i \left(1 + O\left(\frac{1}{n}\right)\right)$$

via Stirling's formula (Lemma 2.3.1). □

Since (Example 7.4.3) $\{U_{k,n}(h), \sigma(U_{k,j}, j \ge n), n \ge k\}$ is a reverse martingale for $h \in \mathscr{L}_1$, it follows (see the discussion after Theorem 11.1.1) under this proviso that $U_{k,n}(h) \xrightarrow{\text{a.s.}} \mathrm{E}\, h$ or, equivalently, $[1/\binom{n}{k}]S_{k,n}(h) \xrightarrow{\text{a.s.}} \mathrm{E}\, h$. The next theorem provides an analogue of the Marcinkiewicz–Zygmund strong law of large numbers for U-statistics.

Define

$$p_j = \frac{jp}{k - (k-j)p}, \qquad 1 \le j \le k. \tag{5}$$

Then $p_k = p$ and for p in $(1, 2k/(2k - j))$, the sequence $\{p_j, 1 \le j \le k\}$ is decreasing, and $p_j \in (1, 2)$.

Theorem 2. *Let $\{X_n, n \ge 1\}$ be* i.i.d. *random variables on some probability space $(\Omega, \mathscr{F}, P)$ and h a symmetric function on R^k, $k \ge 2$. If* (i) *h is degenerate of order $i - 1$ and $1 < p < 2k/(2k - i)$, where $2 \le i \le k$, then*

$$n^{k(p-1)/p}|U_{k,n}(h) - \mathrm{E}\, h| \xrightarrow{\text{a.s.}} 0, \quad \textit{that is, } n^{-k/p}S_{k,n}(h - \mathrm{E}\, h) \xrightarrow{\text{a.s.}} 0 \tag{6}$$

provided

$$\mathrm{E}\{h(X_1, \ldots, X_k)|X_1, \ldots, X_j\} \in \mathscr{L}_{p_j} \qquad i \le j \le k. \tag{7}$$

(ii) *Alternatively,* (6) *holds for non-degenerate h if $1 < p < k/(k - 1)$ and* (7) *is satisfied for $i = 1$.* (iii) *Moreover, if $0 < p < 1$ and $h \in \mathscr{L}_p$, then* $n^{-k/p}S_{k,n}(h) \xrightarrow{\text{a.s.}} 0$.

PROOF. Suppose initially that $i = k$ and $h \in \mathscr{L}_p$ whence, according to (i), $1 < p < 2$. Hence, $\alpha = k/p \in (k/2, k)$. Set

$$g_j(X_1, \ldots, X_k) = h(X_1, \ldots, X_k)I_{[|h| \le j^\alpha]} - \mathrm{E}\{h(X_1, \ldots, X_k)I_{[|h| \le j^\alpha]} | X_1, \ldots, X_{k-1}\}$$

and define

$$\tilde{S}_{k,n}(h) = \sum_{j=k}^{n} j^{-\alpha} \sum_{1 \le i_1 < \cdots < i_{k-1} < j} h(X_{i_1}, \ldots, X_{i_{k-1}}, X_j),$$

$$T_{k,n}(h) = \sum_{j=k}^{n} j^{-\alpha} \sum_{1 \le i_1 < \cdots < i_{k-1} < j} g_j(X_{i_1}, \ldots, X_{i_{k-1}}, X_j).$$

If $\mathscr{F}_n = \sigma(X_1, \ldots, X_n)$, Corollary 1 asserts that $\{\tilde{S}_{k,n}(h), \mathscr{F}_n, n \ge k\}$ is a martingale and the same is true of $\{T_{k,n}(h), \mathscr{F}_n, n \ge k\}$ since, via Lemma 3(i),

$$\mathrm{E}\{g_n(X_{i_1}, \ldots, X_{i_{k-1}}, X_n) | \mathscr{F}_{n-1}\} = \mathrm{E}\{g_n(X_{i_1}, \ldots, X_{i_{k-1}}, X_n) | X_{i_1}, \ldots, X_{i_{k-1}}\} \overset{\text{a.s.}}{=} 0.$$

Hence, employing Lemma 4(i) and setting $A_n = \{(n-1) < |h|^{1/\alpha} \le n\}, n \ge 1$,

$$\begin{aligned}
\mathrm{E}[T_{k,n}(h)]^2 &= \sum_{j=k}^{n} j^{-2\alpha}\binom{j-1}{k-1}\mathrm{E}\, g_j^2 \le \sum_{j=k}^{\infty} j^{k-1-2\alpha}\mathrm{E}\, h^2 I_{[|h| \le j^\alpha]} \\
&= \sum_{j=k}^{\infty} j^{k-1-2\alpha} \sum_{n=1}^{j} \int_{A_n} h^2\, dP \le \sum_{n=1}^{\infty} \left(\frac{1}{n} + \frac{1}{2\alpha - k}\right) n^{k-2\alpha} \int_{A_n} h^2\, dP \\
&\le \left(\frac{2\alpha - k + 1}{2\alpha - k}\right) \sum_{n=1}^{\infty} \int_{A_n} |h|^{k/\alpha}\, dP = \left(\frac{2\alpha - k + 1}{2\alpha - k}\right) E|h|^p < \infty.
\end{aligned}$$

Moreover, since h is completely degenerate,

$$\begin{aligned}
\sup_{n \ge k} \mathrm{E}|\tilde{S}_{k,n}(h) - T_{k,n}(h)| &\le 2\sum_{j=k}^{\infty} j^{-\alpha}\binom{j-1}{k-1}\mathrm{E}|h| I_{[|h| > j^\alpha]} \\
&\le 2\mathrm{E}\left(|h| \sum_{j \le |h|^{1/\alpha}} j^{k-1-\alpha}\right) \le C E|h|^{k/\alpha} < \infty.
\end{aligned}$$

Thus, in view of

$$\sup_{n \ge k} \mathrm{E}|\tilde{S}_{k,n}(h)| \le \sup_{n \ge k} \mathrm{E}|\tilde{S}_{k,n}(h) - T_{k,n}(h)| + \sup_{n \ge k} \mathrm{E}|T_{k,n}(h)|,$$

the martingale $\{\tilde{S}_{k,n}(h), \mathscr{F}_n, n \ge k\}$ is $\mathscr{L}_1$-bounded and hence convergent so that Kronecker's lemma yields

$$n^{-k/p} S_{k,n-1}(h) = n^{-\alpha} \sum_{j=k}^{n} \sum_{1 \le i_1 < \cdots < i_{k-1} < j} h(X_{i_1}, \ldots, X_{i_{k-1}}, X_j) \xrightarrow{\text{a.s.}} 0. \tag{8}$$

Therefore, (8) holds whenever h is a completely degenerate kernel belonging to $\mathscr{L}_p$, $1 < p < 2$ and k is any integer ≥ 2. Clearly, (8) is tantamount to (6) since $\mathrm{E}\,h = 0$.

Suppose next that $2 \leq i < k$. Then, via Hoeffding's decomposition (Theorem 1) and since again $\mathrm{E}\,h = 0$,

$$\begin{aligned} n^{-k/p}|S_{k,n}(h)| &= n^{-k/p}\binom{n}{k}\left|\sum_{j=i}^{k}\binom{k}{j}S_{j,n}(h_j)\Big/\binom{n}{j}\right| \\ &\leq n^{-k/p}\sum_{j=i}^{k}\frac{1}{(k-j)!}\frac{(n-j)!}{(n-k)!}|S_{j,n}(h_j)| \\ &\leq \sum_{j=i}^{k}\frac{1}{(k-j)!}n^{k-j-k/p}|S_{j,n}(h_j)| \\ &= \sum_{j=i}^{k}\frac{1}{(k-j)!}|S_{j,n}(h_j)|/n^{j/p_j}, \end{aligned} \tag{9}$$

where p_j is as in (5) and h_j is completely degenerate for $j \geq i \geq 2$. In view of the cited properties of $\{p_j, 1 \leq j \leq k\}$, the hypothesis $1 < p < 2k/(2k - i)$ implies that $p_i \in (1, 2)$ whence $p_j \in (1, 2)$ for $j \geq i$. Hence, if $h_j \in \mathscr{L}_{p_j}$ for $j \geq i$, (8) in conjunction with (9) ensures that $n^{-k/p}S_{k,n}(h) \xrightarrow{\text{a.s.}} 0$.

Now, $\mathrm{E}\{h(X_1, \ldots, X_k)|X_1, \ldots, X_j\}$, $1 \leq j \leq k$ is a martingale whence $|\mathrm{E}\{h(X_1, \ldots, X_k)|X_1, \ldots, X_j\}|^p$, $1 \leq j \leq k$ is a submartingale for any $p \geq 1$, implying that

$$\mathrm{E}|\mathrm{E}\{h|X_1, \ldots, X_{j-1}\}|^p \leq \mathrm{E}|\mathrm{E}\{h|X_1, \ldots, X_j\}|^p, \qquad p \geq 1.$$

Next, either employ Exercise 7.5.4 or note via (2) that

$$h_j = (I - Q_1)\cdots(I - Q_j)\mathrm{E}\{h|X_1, \ldots, X_j\},$$

where I is the identity operator and $Q_r f = \mathrm{E}\{f|X_\alpha, \alpha \neq r\}$. It follows that $\mathrm{E}\{h|X_1, \ldots, X_j\} \in \mathscr{L}_{p_j}$ entails $h_j \in \mathscr{L}_{p_j}$, thereby proving (i).

If, rather, h is non-degenerate and $1 < p < k/(k - 1)$, suppose without loss of generality that $\mathrm{E}\,h = 0$. Then, exactly as in case (i), all terms of the sum in (9) for which $j \geq 2$ converge a.s. to zero, and when $j = 1$,

$$\begin{aligned} &n^{-[k-(k-1)p]/p}S_{1,n}(h_1) \\ &\quad = n^{-[k-(k-1)p]/p}\sum_{r=1}^{n}\mathrm{E}\{h(X_r, X_{r+1}, \ldots, X_{r+k-1})|X_r\} \xrightarrow{\text{a.s.}} 0 \end{aligned}$$

by the classical Marcinkiewicz–Zygmund strong law (Theorem 5.2.2) so that (6) holds in case (ii).

Finally, when $0 < p < 1$, it suffices to prove that $n^{-k/p}S_{k,n}(|h|) \xrightarrow{\text{a.s.}} 0$ whence h may be supposed non-negative. Set $h_j = hI_{[h \leq j^\alpha]}$, where $\alpha = k/p$ and

$$D_{k-1,j}(h) = \sum_{1 \leq i_1 < \cdots < i_{k-1} < j} h(X_{i_1}, \ldots, X_{i_{k-1}}, X_j).$$

Now,

$$\sum_{j=k}^{\infty} P\{D_{k-1,j}(h) \neq D_{k-1,j}(h_j)\}$$

$$\leq \sum_{j=k}^{\infty} P\{h(X_{i_1}, \ldots, X_{i_{k-1}}, X_j) > j^{\alpha}$$

$$\text{for some choice of } 1 \leq i_1 < \cdots < i_{k-1} < j\}$$

$$\leq \sum_{j=k}^{\infty} \binom{j-1}{k-1} P\{h(X_1, \ldots, X_k) > j^{\alpha}\} \leq \sum_{j=k}^{\infty} j^{k-1} \sum_{n=j+1}^{\infty} P\{A_n\}$$

$$\leq \sum_{n=1}^{\infty} (n-1)^k \mathrm{E}\, I_{A_n} \leq \mathrm{E} \sum_{n=1}^{\infty} h^{k/\alpha} I_{A_n} = \mathrm{E}\, h^p < \infty.$$

Hence, $P\{D_{k-1,n}(h) \neq D_{k-1,n}(h_n), \text{ i.o.}\} = 0$. Moreover, since $\alpha > k$,

$$\mathrm{E} \sum_{j=k}^{n} j^{-\alpha} D_{k-1,j}(h_j) = \sum_{j=k}^{n} j^{-\alpha} \binom{j-1}{k-1} \mathrm{E}\, h_j \leq \sum_{j=k}^{\infty} j^{k-1-\alpha} \sum_{n=1}^{j} \mathrm{E}\, h I_{A_n}$$

$$\leq \frac{(\alpha - k + 1)}{\alpha - k} \sum_{n=1}^{\infty} \frac{1}{n^{\alpha - k}} \mathrm{E}\, h I_{A_n} \leq \left(\frac{\alpha - k + 1}{\alpha - k}\right) \mathrm{E}\, h^{k/\alpha} < \infty.$$

Thus, the series $\sum_{j=k}^{\infty} j^{-\alpha} D_{k-1,j}(h_j)$ and hence also $\sum_{j=k}^{\infty} j^{-\alpha} D_{k-1,j}(h)$ converges a.s. whence $n^{-k/p} S_{k,n}(h) \xrightarrow{\text{a.s.}} 0$. □

Corollary 3. *Let $h \in \mathscr{L}_p$ with h degenerate of order $i - 1$ and $\mathrm{E}\{h(X_1, \ldots, X_k) | X_1, \ldots, X_{k-1}\} \in \mathscr{L}_{p_i}$ where p_i is as in (5). If* (i) $1 < p < 2k/(2k - i)$ *when $i \geq 2$ or* (ii) $1 < p < k/(k-1)$ *when $i = 1$ and $\mathrm{E}\, h = 0$, then $n^{-k/p} S_{k,n}(h) \xrightarrow{\text{a.s.}} 0$.*

Corollary 4. *If $\{U_{k,n}(h), n \geq k \geq 2\}$ is a sequence of completely degenerate U-statistics with $\mathrm{E}|h|^{2k/(2k-1)} < \infty$, then, as $n \to \infty$,*

$$n^{1/2} U_{k,n}(h) \xrightarrow{\text{a.s.}} 0. \tag{10}$$

PROOF. For completely degenerate $h \in \mathscr{L}_p$, $1 < p < 2$, Theorem 2 ensures that $n^{k(p-1)/p} U_{k,n}(h) \xrightarrow{\text{a.s.}} 0$. Since $k(p-1)/p = 1/2$ for $p = 2k/(2k-1)$, (10) follows. □

Lemma 5. *Let $\{U_{k,n}(h), n \geq k \geq 2\}$ be a sequence of non-degenerate U-statistics with $\mathrm{E}\, h = 0$. If*

$$\mathrm{E}\{h(X_1, \ldots, X_k) | X_1, \ldots, X_j\} \in \mathscr{L}_{2j/(2j-1)}, \qquad 2 \leq j \leq k \tag{11}$$

(a fortiori, if $\mathrm{E}|h|^{4/3} < \infty$), then

$$n^{1/2} U_{k,n}(h) = \frac{k}{n^{1/2}} \sum_{r=1}^{n} \mathrm{E}\{h(X_r, X_{r+1}, \ldots, X_{r+k-1}) | X_r\} + o(1), \qquad \text{a.s.} \tag{12}$$

PROOF. Via Hoeffding's decomposition,

$$n^{1/2}U_{k,n}(h) = \frac{k}{n^{1/2}} \sum_{r=1}^{n} \mathrm{E}\{h(X_r, \ldots, X_{r+k-1})|X_r\} + n^{1/2} \sum_{j=2}^{k} \binom{k}{j} U_{j,n}(h_j).$$

As in the proof of Theorem 2, (11) implies that $h_j \in \mathscr{L}_{2j/(2j-1)}$, $2 \le j \le k$ whence (12) follows from Corollary 4. Clearly, $\mathrm{E}|h|^{4/3} < \infty$ ensures (11). □

A Central Limit Theorem and Law of the Iterated Logarithm for non-degenerate U-statistics will be proved in Sections 9.1 and 10.2.

EXERCISE 7.5

1. If $\tilde{S}_{k,n} = \sum_{1 \le i_1 < \cdots < i_k \le n} X_{i_1} \cdots X_{i_k}$, $n \ge k \ge 1$ and $\tilde{S}_{0,n} = 1$, $\tilde{S}_{1,n} = \sum_{i=1}^{n} X_i$, and $S_n^{(j)} = \sum_{i=1}^{n} X_i^j$, $j \ge 1$, prove that (i)

$$(k+1)\tilde{S}_{k+1,n} = \sum_{j=0}^{k} (-1)^j \tilde{S}_{k-j,n} S_n^{(j+1)}, \qquad k \ge 1,$$

and (ii) if $\{X, X_n, n \ge 1\}$ are i.i.d. with $\mathrm{E}\, X = 0$, then $\tilde{S}_{k,n}/\binom{n}{k}$ is a completely degenerate U-statistic for $k \ge 2$.

2. Let $\tilde{S}_{k,n}$, $n \ge k$ and $S_n^{(j)}$, $j \ge 1$ be as in Exercise 1, where $\{X, X_n, n \ge 1\}$ are i.i.d., and let $\{b_n, n \ge 1\}$ be constants such that $0 < b_n/n^{1/2} \uparrow \infty$. If (i) $(1/b_n)\sum_1^n X_i \xrightarrow{\text{a.s.}} 0$ and (ii) $\sum_{n=1}^{\infty} \mathrm{P}\{|X| > b_n\} < \infty$, prove that $\tilde{S}_{k,n}/b_n^k \xrightarrow{\text{a.s.}} 0$, $k \ge 2$. In particular, if $X \in \mathscr{L}_p$, $0 < p < 2$ and $\mathrm{E}\, X = 0$ whenever $p \ge 1$, then $\tilde{S}_{k,n}/n^{k/p} \xrightarrow{\text{a.s.}} 0$. *Hint:* Employ the Newton identities

$$k\tilde{S}_{k,n} = (-1)^k \sum \prod_{i=1}^{k} (S_n^{(j)})^{m_j}/j^{m_j} m_j!,$$

where the sum is over all non-negative integers m_j, $1 \le j \le k$ satisfying $\sum_{j=1}^{k} j m_j = k$.

3. Find a non-degenerate kernel h such that $\mathrm{E}|h|^p = \infty$, some $p > 1$, but $\mathrm{E}\{h(X_1, \ldots, X_k)|X_1\} \in \mathscr{L}_q$, all $q > 0$.

4. Show that if $\mathrm{E}\, h(X_1, \ldots, X_k) = 0$, for $2 \le j \le k$

$$\begin{aligned}\mathrm{E}\{h(X_1, \ldots, X_k)|X_1, \ldots, X_j\} &= h_j(X_1, \ldots, X_j) + \sum_{i=1}^{j} h_1(X_i) \\ &\quad + \sum_{1 \le i_1 < i_2 \le j} h_2(X_{i_1}, X_{i_2}) + \cdots \\ &\quad + \sum_{1 \le i_1 < \cdots < i_{j-1} \le j} h_{j-1}(X_{i_1}, \ldots, X_{i_{j-1}}).\end{aligned}$$

5. (Kemperman) For any countable partition of $(-\infty, \infty)$ into Borel sets B_j, $j = 1, 2 \ldots$, define the non-negative, symmetric function h by $h(X_1, \ldots, X_k) = \sum_{j=1}^{\infty} u_j \prod_{i=1}^{k} I_{[X_i \in B_j]}$, $k \ge 2$ and let $\{X, X_n, n \ge 1\}$ be i.i.d. random variables with $\mathrm{P}\{X \in B_j\} = \pi_j > 0$, $j \ge 1$, where $\sum_1^{\infty} \pi_j = 1$. Set $\pi_j = c/j^{\alpha}$, $u_j = j^{\beta}$ with $\alpha > 1$, $p > 1$. Prove that if $k\alpha - \beta > 1 \ge k\alpha - p\beta$ and $\alpha[1 + (k-1)p] > 1 + p\beta$, then $h \in \mathscr{L}_1$,

$h \notin \mathscr{L}_p$, $\mathrm{E}\{h(X_1, \ldots, X_k)|X_1\} \in \mathscr{L}_p$. These inequalities hold, for example, if $\beta = k - 1$ and $1 < \alpha < [1 + (k-1)p]/k$.

6. If $1 < p < (2k/2k - i)$, there exist degenerate kernels of order $i = 1$, $2 \le i \le k$ such that $n^{k(p-1)/p} U_{k,n} \xrightarrow{\text{a.s.}} 0$ despite $\mathrm{E}|\mathrm{E}\{h(X_1, \ldots, X_k)|X_1, \ldots, X_i\}|^{p_i} = \infty$, where p_i is as in (5). *Hint:* Let $\{X_n, n \ge 1\}$ be symmetric i.i.d. random variables with $\mathrm{P}\{|X| > t\} = ct^{-p}(\log t)^{-1}$, $t \ge 2$, where $1 < p < 2$. Then

$$g_i = \sum_{1 \le \alpha_1 < \cdots < \alpha_{k-i} \le k} \left(\prod_{r=1}^{k-i} |x_{\alpha_r}| \right) \prod_{s \ne \alpha_1, \ldots, \alpha_{k-i}} x_s$$

is degenerate of order $i - 1$, $2 \le i \le k$. Hint: Set $X'_l = X_l I_{[|X_l| \le v \cdot 2^{v/p_i}]}$, $l \ge 1$, $\alpha > 0$.

References

Y. V. Borovskich and V. S. Korolyuk, *Theory of U-Statistics*, Kluwer Academic Publishers, Boston, 1994.

L. Breiman, *Probability*, Addison-Wesley, Reading, Mass., 1968.

H. Bühlman, "Austauschbare stochastiche Variabeln und ihre Grenzwartsatze," *Univ. of California Publications in Statistics*, **3** (1960), 1–36.

Y. S. Chow, "A martingale inequality and the law of large numbers," *Proc. Amer. Math. Soc.* **11** (1960), 107–111.

Y. S. Chow, H. Robbins, and D. Siegmund, *Great Expectations: The Theory of Optimal Stopping*, Houghton Mifflin, Boston, 1972.

Y. S. Chow, H. Robbins, and H. Teicher, "Moments of randomly stopped sums," *Ann. Math. Stat.* **36** (1965), 789–799.

K. L. Chung, *A course in Probability Theory*, Harcourt Brace, New York, 1968; 2nd ed., Academic Press, New York, 1974.

J. L. Doob, "Regular properties of certain families of chance variables," *Trans. Amer. Math. Soc.* **47** (1940), 455–486.

J. L. Doob, *Stochastic Processes*, Wiley, New York, 1953.

L. E. Dubins and D. A. Freedman, "A sharper form of the Borel–Cantelli lemma and the strong law," *Ann. Math. Stat.* **36** (1965), 800–807.

E. B. Dynkin and A. Mandelbaum, "Symmetric statistics, Poisson point processes and multiple Wiener integrals," *Ann. Statist.* **11** (1983), 739–745.

B. de Finetti, La prévision; ses lois logiques, ses sources subjectives," *Annales de l'Institut Henri Poincaré* **7** (1937), 1–68.

E. Giné and J. Zinn, "Marcinkiewicz-type laws of large numbers and convergence of moments for *U*-statistics," *Probability in Banach Spaces* **8** (1992), 273–291, Birkhauser, Boston.

J. Hájek and A. Rényi, "Generalization of an inequality of Kolmogorov," *Acta Math. Acad. Sci. Hung.* **6** (1955), 281–283.

P. R. Halmos, *Measure Theory*, Van Nostrand, Princeton, N. J., 1950; Springer-Verlag, Berlin and New York, 1974.

E. Hewitt and L. J. Savage, "Symmetric measures on Cartesian products," *Trans. Amer. Math. Soc.* **80** (1955), 470–501.

W. Hoeffding, "The strong low of large numbers for *U*-statistics," Institute of Statistics Mimeo Series 302 (1961), University of North Carolina, Chapel Hill, NC.

D. G. Kendall, "On finite and infinite sequences of exchangeable events," *Studia Scient. Math. Hung.* **2** (1967), 319–327.

M. J. Klass, "Properties of optimal extended-valued stopping rules," *Ann. Prob.* **1** (1973), 719–757.

K. Krickeberg, *Probability Theory*, Addison–Wesley, Reading, Mass., 1965.

P. Lévy, *Théorie de l'addition des variables aléatoires*, Gauthier–Villars, Paris, 1937; 2nd ed., 1954.

M. Loève, *Probability Theory*, 3rd ed., Van Nostrand, Princeton, 1963; 4th ed., Springer-Verlag, Berlin and New York, 1977–1978.

P. K. Sen, "On $\mathscr{L}_p$ convergence of *U*-statistics," *Ann. Inst. Statist. Math.* **26** (1974), 55–60.

R. J. Serfling, *Approximation Theorems of Mathematical Statistics*, Wiley, New York.

H. Teicher, "On the Marcinkiewicz-Zygmund strong law for *U*-statistics," *J. Theoret. Probability* **10** (1997),

8 Distribution Functions and Characteristic Functions

8.1 Convergence of Distribution Functions, Uniform Integrability, Helly–Bray Theorem

Distribution functions are mathematical artifacts with properties that are independent of any probabilistic setting. Notwithstanding, most of the theorems of interest are geared to d.f.s of r.v.s and the majority of proofs are simpler and more intuitive when couched in terms of r.v.s having, or probability measures determined by, the given d.f.s. Since r.v.s possessing preassigned d.f.s can always be defined on some probability space, the language of r.v.s and probability will be utilized in many of the proofs without further ado.

Recall that a d.f. on the line is a nondecreasing, left-continuous function on $R = [-\infty, \infty]$ with $F(\infty) = \lim_{x\to\infty} F(x) = 1$, $F(-\infty) = \lim_{x\to-\infty} F(x) = 0$. A discrete d.f. was defined in Section 1.6 as roughly tantamount to a "step function" with a finite or denumerable number of jumps. As such, it determines and is determined by a probability density function (p.d.f.), say f, and a nonempty countable subset S of $(-\infty, \infty)$ with f positive on S and vanishing on S^c. Absolutely continuous d.f.s were encountered in Section 6.5. A d.f. F is absolutely continuous iff $F(x) = \int_{-\infty}^{x} f(t)dt$, $-\infty < x < \infty$, for some Borel function $f \geq 0$, a.e., with $\int_{-\infty}^{\infty} f(t)dt = 1$. A d.f. F is termed **singular** if it is continuous and its corresponding probability measure is singular with respect to Lebesgue measure (Exercise 5).

The first proposition states that any d.f. on R is a convex linear combination of these three types.

Also, a d.f. is **degenerate** or **improper** if it has only a single point of increase (Exercise 1.6.4) and otherwise **nondegenerate** or **proper**.

Proposition 1. *If F is an arbitrary* d.f. *on $R = [-\infty, \infty]$, then $F = \alpha_1 F_1 + \alpha_2 F_2 + \alpha_3 F_3$ where $\sum_{i=1}^{3} \alpha_i = 1, \alpha_i \geq 0$, $i = 1, 2, 3$, and F_1, F_2, F_3 are discrete, absolutely continuous and singular* d.f.s *respectively.*

PROOF. If F is discrete, $F = F_1, \alpha_1 = 1$, while if F is continuous, F coincides with F^* in what follows. Set $S_1 = \{x: F(x+) - F(x) > 0\}$ so that if F is neither discrete nor continuous, $\alpha_1 = \mathrm{P}\{S_1\} \in (0, 1)$ where P is the measure induced by F. Hence, if P_1 is the probability measure determined by

$$\mathrm{P}_1\{\{x\}\} = \frac{1}{\alpha_1} \mathrm{P}\{\{x\}\}, \qquad x \in S_1,$$

$$\mathrm{P}_1\{B\} = 0, \qquad B \in S_1^c \cdot \mathscr{B},$$

then the d.f. corresponding to P_1, say F_1, is discrete. Moreover, $\mathrm{P}^* = [1/(1 - \alpha_1)](\mathrm{P} - \alpha_1 \mathrm{P}_1)$ is a probability measure vanishing on all one-point sets, whence its corresponding d.f. $F^* = [1/(1 - \alpha_1)](F - \alpha_1 F_1)$ is continuous. If P^* is absolutely continuous (resp. singular) relative to Lebesgue measure, its d.f. F^* may be taken as F_2 (resp. F_3) and $\alpha_3 = 0$ (resp. $\alpha_2 = 0$). Otherwise, by Corollary 6.5.1, $F^* = \beta F_2 + (1 - \beta)F_3, 0 < \beta < 1$, where F_2 is absolutely continuous and, moreover, F_3 is singular. Thus, $F - \alpha_1 F_1 = (1 - \alpha_1)F^* = \beta(1 - \alpha_1)F_2 + (1 - \beta)(1 - \alpha_1)F_3$ is the asserted decomposition. □

The **support** (Exercise 1.6.4) or **spectrum** of an arbitrary d.f. F is the closed set S defined by

$$S = \{x: F(x + \varepsilon) - F(x - \varepsilon) > 0, \text{all } \varepsilon > 0\}.$$

and the elements of S are called **points of increase**.

An instance of convergence of a sequence of d.f.s to a d.f. occurred in Corollary 2.3.1, but the situation there was too specialized to furnish clues to the general problem.

For any real function G, let **C(*G*)** denote the **set of continuity points** of G, that is, $C(G) = \{x: -\infty < x < \infty, G(x-) = G(x+) = G(x)\}$. Note that if G is monotone, $C(G)$ is the complement of a countable set and *a fortiori* dense in $(-\infty, \infty)$.

Definition. A sequence of nondecreasing functions G_n on $(-\infty, \infty)$ is said to **converge weakly** to a nondecreasing function G on $(-\infty, \infty)$, denoted by $G_n \xrightarrow{w} G$, if $\lim_{n\to\infty} G_n(x) = G(x)$ for all $x \in C(G)$. If, in addition, $G_n(\infty) \to G(\infty)$ and $G_n(-\infty) \to G(-\infty)$ where, as usual, $G(\pm\infty) = \lim_{x\to\pm\infty} G(x)$, then $\{G_n\}$ is said to **converge completely** to G, denoted by $G_n \xrightarrow{c} G$.

In the special case of d.f.s F_n, complete convergence of $\{F_n\}$ guarantees that the "limit function" F, if left continuous (as may and will be supposed via Lemma 8.2.1 even when merely $F_n \xrightarrow{w} F$), is a d.f.

If $\{X_n, n \geq 1\}$ is a sequence of r.v.s on some probability space $(\Omega, \mathscr{F}, P)$ with d.f.s F_{X_n} that converge **completely** to F, the r.v.s $\{X_n\}$ are said to **converge in distribution or law**, symbolized by $X_n \xrightarrow{d} X_F$. Here X_F is, in general, a **fictitious** r.v. with d.f. F. It is not asserted that any such "r.v." exists on $(\Omega, \mathscr{F}, P)$, but, of course, one can always define a r.v. X with d.f. F on another probability space; rather, $X_n \xrightarrow{d} X_F$ is simply a convenient alternative notation for $F_{X_n} \xrightarrow{c} F$. Clearly, convergence in distribution is a property of the d.f.s of the r.v.s in question and not of the r.v.s themselves.

However, if $X_n \xrightarrow{P} X$ (*a fortiori*, if $X_n \xrightarrow{\text{a.c.}} X$ or $X_n \xrightarrow{\mathscr{L}_p} X$), then the following Corollary 1 asserts that a bona fide r.v. X_F on $(\Omega, \mathscr{F}, P)$ does exist and coincides with X. Such a case may be denoted simply by $X_n \xrightarrow{d} X$, that is, $F_{X_n} \xrightarrow{c} F_X$.

Theorem 1 (Slutsky). *If $\{X_n, n \geq 1\}$ and $\{Y_n, n \geq 1\}$ are* r.v.s *on some probability space with $X_n - Y_n \xrightarrow{P} 0$ and $Y_n \xrightarrow{d} X_F$, then $X_n \xrightarrow{d} X_F$.*

PROOF. Let $x, x \pm \varepsilon \in C(F)$, where $\varepsilon > 0$ and $x \in (-\infty, \infty)$. Then

$$\begin{aligned} P\{X_n < x\} &= P\{X_n < x, |X_n - Y_n| < \varepsilon\} + P\{X_n < x, |X_n - Y_n| \geq \varepsilon\} \\ &\leq F_{Y_n}(x + \varepsilon) + P\{|X_n - Y_n| \geq \varepsilon\} \end{aligned}$$

and, analogously,

$$F_{Y_n}(x - \varepsilon) \leq P\{X_n < x\} + P\{|X_n - Y_n| \geq \varepsilon\}.$$

Thus,

$$F(x - \varepsilon) \leq \varliminf_n F_{X_n}(x) \leq \varlimsup_n F_{X_n}(x) \leq F(x + \varepsilon),$$

and letting $\varepsilon \to 0$ subject to $x \pm \varepsilon$ in $C(F)$, the conclusion follows. □

Corollary 1. *If $\{X, X_n, n \geq 1\}$ are* r.v.s *on some probability space with $X_n \xrightarrow{P} X$, then $X_n \xrightarrow{d} X$.*

Corollary 2. *If $\{X_n\}, \{Y_n\}, \{Z_n\}$ are sequences of* r.v.s *on* $(\Omega, \mathscr{F}, P)$ *with $X_n \xrightarrow{d} X_F$, $Y_n \xrightarrow{P} a$, $Z_n \xrightarrow{P} b$, where a, b are finite constants, then*

$$X_n Y_n + Z_n \xrightarrow{d} aX_F + b.$$

Note. Here, $aX_F + b$ is a fictitious r.v. whose distribution coincides with that of $aX + b$ when X is a bona fide r.v. with d.f. F.

PROOF. By the theorem it suffices to prove $X_n Y_n + b \xrightarrow{d} aX_F + b$ or equivalently that $X_n Y_n \xrightarrow{d} aX_F$. Since it is obvious that $aX_n \xrightarrow{d} aX_F$, applying the theorem once more it suffices to prove that $X_n(Y_n - a) \xrightarrow{P} 0$ or, renotating, that $X_n U_n \xrightarrow{P} 0$ if $X_n \xrightarrow{d} X_F$, $U_n \xrightarrow{P} 0$. To this end, for any $\delta > 0$, choose $\pm h \in C(F)$

such that $F(h) - F(-h) \geq 1 - \delta$. Then, for all sufficiently large n and any $\varepsilon > 0$,

$$\begin{aligned} P\{|U_n X_n| > \varepsilon\} &\leq P\{|U_n X_n| > \varepsilon, 0 < |X_n| \leq h\} + 2\delta \\ &\leq P\{|U_n| > \varepsilon/h\} + 2\delta \xrightarrow{n \to \infty} 2\delta \end{aligned}$$

and the result follows as $\delta \to 0$. □

Corollary 3. *If* $\{a, b, a_n, b_n, n \geq 1\}$ *are finite constants with* $a_n \to a$, $b_n \to b$ *and the* r.v.s $X_n \xrightarrow{d} X_F$, *then* $a_n X_n + b_n \xrightarrow{d} aX_F + b$.

If $\{X_n\}$ is a sequence of r.v.s and b_n is a sequence of positive constants such that $X_n/b_n \xrightarrow{P} 0$, it is natural, paralleling the classical notation, to designate this by $X_n = o_p(b_n)$. Analogously, $X_n = O_p(b_n)$ will signify that X_n/b_n is bounded in probability, i.e., for every $\varepsilon > 0$, there are constants C_ε, N_ε such that

$$P\{|X_n| \geq C_\varepsilon b_n\} \leq \varepsilon$$

for $n > N_\varepsilon$. In this notation, Theorem 1 says that if $X_n \xrightarrow{d} X_F$, then $X_n + o_p(1) \xrightarrow{d} X_F$. A calculus paralleling that of o and O exists for o_p and O_p. For example, the Taylor expansion

$$f(x) = \sum_{j=0}^{k} \frac{(x - c)^j}{j!} f^{(j)}(c) + o(|x - c|^k),$$

valid as $x \to c$ under the hypothesis below (Cramér, 1946, p. 290) leads directly to

Corollary 4. *If* $f(x)$ *has* k *derivatives at* $x = c$ *and the* r.v.s X_n *satisfy* $X_n = c + o_p(b_n)$, *where* $b_n = 1$ *or* $b_n = o(1)$, *then*

$$f(X_n) = \sum_{j=0}^{k} \frac{(X_n - c)^j}{j!} f^{(j)}(c) + o_p(b_n^k).$$

It will be advantageous to prove the ensuing for nondecreasing functions G_n on $(-\infty, \infty)$. In the special case where the G_n are d.f.s, the condition which conjoined with weak convergence yields complete convergence is (iii)(γ) in

Lemma 1. *Let* $\{G_n, n \geq 0\}$ *be finite, nondecreasing functions on* $(-\infty, \infty)$ *with* $G_n \xrightarrow{w} G_0$. *Set* $\Delta G_n = G_n(\infty) - G_n(-\infty)$, $n \geq 0$, *where* $G(\infty) = G(\infty -)$ *and* $G(-\infty) = G(-\infty +)$. *Then*

i. $\overline{\lim}_{n\to\infty} G_n(-\infty) \leq G_0(-\infty) \leq G_0(\infty) \leq \underline{\lim}_{n\to\infty} G_n(\infty)$,
ii. $\Delta G_0 \leq \underline{\lim}_{n\to\infty} \Delta G_n$.

Moreover, if $\Delta G_n(a) = G_n(a) - G_n(-a)$ *for* $n \geq 0$, $0 < a < \infty$, *and if* $\Delta G_n < \infty$ *for* $n \geq 1$, *then*

iii. (α) $\lim G_n(\pm\infty) = G_0(\pm\infty)$, *finite* iff ($\beta$) $\lim_{n\to\infty} \Delta G_n = \Delta G_0 < \infty$ iff (γ) $\sup_{n\geq 1}[\Delta G_n - \Delta G_n(a)] = o(1)$ *as* $a \to \infty$.

PROOF. Since $G_n(-\infty) \le G_n(x) \le G_n(\infty)$, taking $x \in C(G_0)$ and letting $n \to \infty$,

$$\overline{\lim_{n\to\infty}}\, G_n(-\infty) \le G_0(x) \le \underline{\lim_{n\to\infty}}\, G_n(\infty),$$

yielding (i) as $x \to \pm\infty$. Then (ii) follows immediately from (i). That (α) implies (β) is trivial. For the reverse implication, let $\Delta G_n \to \Delta G_0 < \infty$. Then $G_0(\pm\infty)$ are finite and by (i)

$$\begin{aligned}\overline{\lim_{n\to\infty}}\, G_n(\infty) &= \overline{\lim}[\Delta G_n + G_n(-\infty)] \\ &= \Delta G_0 + \overline{\lim}\, G_n(-\infty) \le G_0(\infty) \le \underline{\lim_{n\to\infty}}\, G_n(\infty),\end{aligned}$$

whence $\lim_{n\to\infty} G_n(\infty) = G_0(\infty)$, finite, and so $\lim G_n(-\infty) = G_0(-\infty)$, finite.

Under (γ), for any $\varepsilon > 0$ choose $\bar{a} = \bar{a}(\varepsilon) > 0$ such that $\Delta G_n - \Delta G_n(a) < \varepsilon$, $n \ge 1$, for $a \ge \bar{a}$. Then if $\pm a \in C(G_0)$,

$$\overline{\lim_{n}}\, \Delta G_n \le \Delta G_0(a) + \varepsilon < \infty,$$

ensuring $\Delta G_0 < \infty$ by (ii), and since ε is arbitrary, $\overline{\lim}_n \Delta G_n \le \Delta G_0$. In conjunction with (ii), this yields (β). Conversely, under (β), for any $\varepsilon > 0$ choose the integer n_1 such that $n \ge n_1$ entails $\Delta G_n - \Delta G_0 < \varepsilon$ and select $\bar{a} > 0$ with $\pm\bar{a} \in C(G_0)$ such that $\Delta G_0 - \Delta G_0(\bar{a}) < \varepsilon$. Then for $n \ge$ some integer n_2, $\Delta G_0(\bar{a}) - \Delta G_n(\bar{a}) < \varepsilon$, implying for $n \ge n_0 = \max(n_1, n_2)$ that

$$\Delta G_n - \Delta G_n(\bar{a}) < 3\varepsilon.$$

Choose a_j such that $\Delta G_j - \Delta G_j(a_j) < 3\varepsilon$, $1 \le j < n_0$, whence for $a \ge a' = \max(\bar{a}, a_1, \ldots, a_{n_0-1})$

$$\sup_{n\ge 1}[\Delta G_n - \Delta G_n(a)] < 3\varepsilon,$$

which is tantamount to (γ). □

Lemma 2 (Helly–Bray). *If $\{F_n, n \ge 1\}$ is a sequence of* d.f.s *with $F_n \xrightarrow{w} F$ and $a \in C(F)$, $b \in C(F)$, then for every real, continuous function g on $[a, b]$*

$$\lim_{n\to\infty} \int_a^b g\, dF_n = \int_a^b g\, dF. \tag{1}$$

PROOF. As the notation indicates, the integrals in (1) are Riemann–Stieltjes, although they may also be interpreted (Theorem 6.2.4) as Lebesgue–Stieltjes integrals over $[a, b)$. For $\varepsilon > 0$, choose by uniform continuity $\delta > 0$ so that $|g(x) - g(y)| < \varepsilon$ for $|x - y| < \delta$, $x, y \in [a, b]$. Select $x_i \in C(F)$, $1 < i \le k$, such that $a = x_1 < x_2 < \cdots < x_{k+1} = b$ and $\max_{1\le i\le k}(x_{i+1} - x_i) < \delta$.

Then

$$H_n \equiv \int_a^b g\,dF_n - \int_a^b g\,dF$$

$$= \sum_{i=1}^{k}\left\{\left[\int_{x_i}^{x_{i+1}} g(x)dF_n(x) - \int_{x_i}^{x_{i+1}} g(x_i)dF_n(x)\right]\right.$$

$$+ \left[\int_{x_i}^{x_{i+1}} g(x_i)dF_n(x) - \int_{x_i}^{x_{i+1}} g(x_i)dF(x)\right]$$

$$\left. + \left[\int_{x_i}^{x_{i+1}} g(x_i)dF(x) - \int_{x_i}^{x_{i+1}} g(x)dF(x)\right]\right\}$$

$$= \sum_{i=1}^{k}\left\{\int_{x_i}^{x_{i+1}} [g(x) - g(x_i)]dF_n(x) + \int_{x_i}^{x_{i+1}} [g(x_i) - g(x)]dF(x)\right.$$

$$\left. + g(x_i)[F_n(x_{i+1}) - F_n(x_i) - F(x_{i+1}) + F(x_i)]\right\}.$$

Hence,

$$|H_n| \le \varepsilon + \varepsilon + \sum_{i=1}^{k} |g(x_i)||F_n(x_{i+1}) - F_n(x_i) - F(x_{i+1}) + F(x_i)| \to 2\varepsilon$$

as $n \to \infty$. Since ε is arbitrary, (1) follows. □

Lemma 3. (i) *If $\{G_n, n \ge 0\}$ are finite, nondecreasing functions on $(-\infty, \infty)$ with $\lim_{n\to\infty} G_n(x) = G_0(x)$ for $x \in$ some dense subset D of $(-\infty, \infty)$, then $G_n \xrightarrow{w} G_0$.*

(ii) *Let $\{F_n, n \ge 1\}$ be d.f.s with $F_n \xrightarrow{w} F_0$ and g a nonnegative continuous function on $(-\infty, \infty)$. For $n \ge 0$, $a \in C(F_0)$, and $x \in [-\infty, \infty]$, define*

$$G_n(x) = \int_a^x g\,dF_n.$$

Then G_n is finite and nondecreasing on $(-\infty, \infty)$, $n \ge 0$, and

(α) $G_n \xrightarrow{w} G_0$,

(β) $\underline{\lim} \int_a^\infty g\,dF_n \ge \int_a^\infty g\,dF_0$, $\quad \underline{\lim} \int_{-\infty}^a g\,dF_n \ge \int_{-\infty}^a g\,dF_0$.

PROOF. (i) If $x \in C(G_0)$ and $\varepsilon > 0$, choose $\delta > 0$ such that $|G_0(y) - G_0(x)| < \varepsilon$ for $|x - y| < \delta$. Select $x_i \in D$, $i = 1, 2$, with $x - \delta < x_1 < x < x_2 < x + \delta$. Then

$$G_0(x) - \varepsilon < G_0(x_1) \leftarrow G_n(x_1) \le G_n(x) \le G_n(x_2) \to G_0(x_2) < G_0(x) + \varepsilon,$$

whence $\lim G_n(x) = G_0(x)$ for $x \in C(G_0)$. Apropos of (ii), note that by (the Helly–Bray) Lemma 2 and part (i) of the current lemma $G_n \xrightarrow{w} G_0$. Then (β) follows directly from Lemma 1(i). □

Definition. If $\{F_n, n \geq 1\}$ is a sequence of d.f.s on R, and g is a real, continuous function on $(-\infty, \infty)$, then g is called **uniformly integrable** (u.i) **relative to** $\{F_n\}$ if

$$\sup_{n\geq 1} \int_{[|y|\geq a]} |g(y)|\,dF_n(y) = o(1) \quad \text{as } a \to \infty. \tag{2}$$

Furthermore, $\{F_n, n \geq 1\}$ is said to be **tight** if the function 1 is u.i. relative to $\{F_n\}$.

Clearly, (i) if f and g are u.i. relative to $\{F_n\}$, so are f^+ and $af + bg$ for any finite real numbers a, b. (ii) if f, g are continuous, $|f| \leq |g|$, and g is u.i. relative to $\{F_n\}$, so is f.

Thus, in the case of d.f.s, Lemma 1(iii) may be rephrased as follows:

If the d.f.s $F_n \xrightarrow{w} F$, *then* $F_n \xrightarrow{c} F$ iff $\{F_n\}$ *is tight* iff *every bounded continuous function* g *is* u.i. *relative to* $\{F_n\}$.

Theorem 2. *If* $\{F_n, n \geq 1\}$ *is a sequence of* d.f.s *on* R *with* $F_n \xrightarrow{w} F$ *and* g *is a nonnegative, continuous function on* $(-\infty, \infty)$ *for which* $\int_{-\infty}^{\infty} g\,dF_n < \infty$, $n \geq 1$, *then*

$$\lim_{n\to\infty} \int_{-\infty}^{\infty} g\,dF_n = \int_{-\infty}^{\infty} g\,dF < \infty \tag{3}$$

iff g *is* u.i. *relative to* $\{F_n\}$.

Proof. For any $a \in C(F)$ and $x \in [-\infty, \infty]$ define $G_n(x) = \int_a^x g\,dF_n$, $G(x) = \int_a^x g\,dF$. By Lemma 3, $G_n \xrightarrow{w} G$. If g is u.i. relative to $\{F_n\}$, then (iii)(γ) of Lemma 1 holds, whence by (iii)(α) thereof

$$G_n(\pm\infty) \to G(\pm\infty), \qquad \text{finite,}$$

which is virtually (3).

Conversely, if (3) holds, so does (iii)(β) of Lemma 1 for G, G_n as defined, whence by (iii)(γ), g is u.i. relative to $\{F_n\}$. □

Corollary 5. *If the* d.f.s $F_n \xrightarrow{w} F$ *and* g *is a continuous function on* $(-\infty, \infty)$ *which is* u.i. *relative to* $\{F_n\}$, *then* (3) *holds and* $\int_{-\infty}^{\infty} |g|\,dF < \infty$.

Corollary 6 (Helly–Bray Theorem). (i) *If the* d.f.s $F_n \xrightarrow{c} F$ *and* g *is a bounded, continuous function on* $(-\infty, \infty)$, *then*

$$\lim_{n\to\infty} \int_{-\infty}^{\infty} g\,dF_n = \int_{-\infty}^{\infty} g\,dF. \tag{4}$$

(ii) *If the* d.f.s $F_n \xrightarrow{w} F$ *and* g *is continuous on* $(-\infty, \infty)$ *with* $\lim_{y\to\pm\infty} g(y) = 0$, *then* Eq. (4) *holds.*

PROOF. Since $|g| \le M < \infty$ and 1 is u.i. relative to $\{F_n\}$ by (iii)(γ) of Lemma 1, necessarily g is u.i. relative to $\{F_n\}$, and the conclusion follows from Corollary 5.

In case (ii), for any $\varepsilon > 0$ and sufficiently large a, $|g(y)| < \varepsilon$ for $|y| \ge a$ and so g is u.i. relative to $\{F_n\}$. □

Corollary 7. *If* d.f.s $F_n \xrightarrow{w} F$ *and* $\{\int |x|^s \, dF_n(x), n \ge 1\}$ *is a bounded sequence for some* $s > 0$, *then*

i. $F_n \xrightarrow{c} F$,
ii. $\int |x|^r \, dF_n(x) \to \int |x|^r \, dF(x)$, $0 \le r < s$, *and*
iii. $\int x^k \, dF_n(x) \to \int x^k \, dF(x)$, $k = 1, 2, \ldots [s]$, $k \ne s$.

PROOF. This follows from Corollary 5 since for $0 \le r < s$ and some C in $(0, \infty)$

$$\int_{[|x| \ge a]} |x|^r \, dF_n(x) \le \frac{1}{a^{s-r}} \int |x|^s \, dF_n(x) < \frac{C}{a^{s-r}}, \qquad n \ge 1. \qquad \square$$

The Helly–Bray theorem (Corollary 6(i)) is extremely useful and clearly complements the earlier Helly–Bray lemma.

The notion of a function u.i. relative to d.f.s $\{F_n\}$ is redolent of that of uniform integrability of r.v.s $\{X_n\}$ encountered in Chapter 4. The connection between these is elucidated in

Proposition 2. *Let g be a continuous function on $(-\infty, \infty)$ and let $\{X_n\}$ be* r.v.s *on a probability space* $(\Omega, \mathcal{F}, P)$ *with* d.f.s $\{F_n\}$. *If g is* u.i. *relative to* $\{F_n\}$, *then the* r.v.s $\{g(X_n)\}$ *are* u.i. *Conversely, if the* r.v.s $\{g(X_n)\}$ *are* u.i. *and either* (i) $|g(t)| \to \infty$ *as* $|t| \to \infty$ *or* (ii) $\{F_n\}$ *is tight, then g is* u.i. *relative to* $\{F_n\}$.

PROOF. Throughout, in addition to any other requirements choose $a > 0$ so that $\pm a \in \bigcap_1^\infty C(F_n)$. If g is u.i. relative to $\{F_n\}$ and $\varepsilon > 0$, select a so that for $n \ge 1$ it also satisfies the first inequality of

$$\varepsilon > \int_{[|y| \ge a]} |g(y)| \, dF_n = \int_{[|X_n| \ge a]} |g(X_n)| \, dP \ge \int_{[|g(X_n)| > b]} |g(X_n)| \, dP, \qquad (5)$$

where $b = \max\{|g(y)| : |y| \le a\}$ and the equality holds via Theorem 6.2.4 and Corollary 6.2.1. Thus $\{g(X_n)\}$ are u.i.

Conversely, in case (i), as $a \to \infty$ there exists $K = K_a \to \infty$ such that

$$\int_{[|y| \ge a]} |g(y)| \, dF_n = \int_{[|X_n| \ge a]} |g(X_n)| \, dP \le \int_{[|g(X_n)| \ge K]} |g(X_n)| \, dP,$$

whence u.i. of $\{g(X_n)\}$ implies that of g relative to $\{F_n\}$. Under (ii), for any $\varepsilon > 0$ choose $b > 0$ such that

$$\sup_{n \ge 1} \int_{[|g(X_n)| > b]} |g(X_n)| \, dP < \varepsilon$$

and then select $a > 0$ so that $\sup_{n\geq 1} \mathrm{P}\{|X_n| \geq a\} < \varepsilon/b$. Then, for $n \geq 1$

$$\int_{[|y|\geq a]} |g(y)|\,dF_n(y) = \int_{[|X_n|\geq a]} |g(X_n)|\,d\mathrm{P} \leq \varepsilon + b\,\mathrm{P}\{|X_n| \geq a\} < 2\varepsilon,$$

whence g is u.i. relative to $\{F_n\}$. □

Proposition 2 in conjunction with Theorem 2 yields the following improvement of Theorem 4.2.3(i):

Corollary 8. *For some $p > 0$, let $\{X_n, n \geq 1\}$* be $\mathscr{L}_p$ r.v.s *on $(\Omega, \mathscr{F}, \mathrm{P})$ with $X_n \xrightarrow{\mathrm{d}} X_F$. Then $\mathrm{E}|X_n|^p \to \mathrm{E}|X_F|^p$, finite* iff *$\{|X_n|^p, n \geq 1\}$ is* u.i.

If $\mathscr{F}^*$ denotes the class of all d.f.s on R, many distances may be defined on $\mathscr{F}^*$. One prominent choice is $d^*[F, G] = \sup_{x\in R}|F(x) - G(x)|$ (see Exercise 2). The Lévy distance $d[F, G]$ corresponds to the maximum distance between F and G measured along lines of slope -1 (in contemplating this, draw vertical lines connecting F and also G at any discontinuities) multiplied by the factor $1/\sqrt{2}$. Formally,

$$d[F_n, F] = \inf\{h > 0\colon F(x - h) - h \leq F_n(x) \leq F(x + h) + h, \text{ all } x\}. \quad (6)$$

Theorem 3. *Let $\{F, F_n, n \geq 1\}$ be* d.f.s. *Then* (i) $F_n \xrightarrow{\mathrm{c}} F$ iff (ii) $\int g\,dF_n \to \int g\,dF$ *for every bounded, continuous function g* iff (iii) $d[F_n, F] \to 0$ iff (iv) $\overline{\lim}\, F_n\{C\} \leq F\{C\}$, $\underline{\lim}\, F_n\{V\} \geq F\{V\}$ for all closed sets C and open sets V, where $F_n\{\cdot\}$, $F\{\cdot\}$ are the probability measures determined by F_n, F respectively.

PROOF. That (i) implies (ii) is the Helly–Bray theorem, Corollary 6. To show that (i) implies (iii), for any $\varepsilon > 0$ choose $a, b \in C(F)$ such that $\varepsilon/2$ exceeds both $F(a)$ and $1 - F(b)$ and then select $a_j \in C(F)$, $0 \leq j \leq m$, with $a_0 = a < a_1 < \cdots < a_m = b$ and $|a_j - a_{j-1}| < \varepsilon$, $1 \leq j \leq m$. Determine N_j, $0 \leq j \leq m$, so that $n \geq N_j$ entails $|F_n(a_j) - F(a_j)| < \varepsilon/2$ and set $N = \max_{0\leq j\leq m} N_j$. Let $n > N$. If $x \leq a_0$,

$$F_n(x) \leq F_n(a_0) < F(a_0) + \frac{\varepsilon}{2} \leq \varepsilon + F(x + \varepsilon),$$

$$F_n(x) \geq 0 > F(x) - \frac{\varepsilon}{2} \geq F(x - \varepsilon) - \varepsilon,$$

and, analogously, $F(x - \varepsilon) - \varepsilon \leq F_n(x) \leq F(x + \varepsilon) + \varepsilon$ for $x \geq a_m$. Moreover, if $a_{j-1} \leq x \leq a_j$ for some j, $1 \leq j \leq m$,

$$F_n(x) \leq F_n(a_j) < F(a_j) + \frac{\varepsilon}{2} \leq F(x + \varepsilon) + \varepsilon,$$

$$F_n(x) \geq F_n(a_{j-1}) > F(a_{j-1}) - \frac{\varepsilon}{2} \geq F(x - \varepsilon) - \varepsilon.$$

Combining these, $d[F_n, F] < \varepsilon$ and (iii) follows.

To verify that (iii) implies (i), for any $x_0 \in C(F)$ and $\varepsilon > 0$ choose $\delta > 0$ such that $|x - x_0| \le \delta$ entails $|F(x) - F(x_0)| < \varepsilon$. Set $h = \min(\varepsilon, \delta)$ and select N so that $d[F_n, F] < h$ when $n \ge N$. Then, for $n \ge N$ from (6),

$$F_n(x_0) \le F(x_0 + h) + h \le F(x_0 + \delta) + \varepsilon \le F(x_0) + 2\varepsilon,$$
$$F_n(x_0) \ge F(x_0 - h) - h \ge F(x_0 - \delta) - \varepsilon \ge F(x_0) - 2\varepsilon,$$

and (i) follows.

To obtain (i) from (ii), retain the choice of x_0, δ, ε; define

$$h(x) = \begin{cases} 1, & x \le x_0 - \delta \\ \dfrac{x_0 - x}{\delta}, & x_0 - \delta \le x \le x_0 \\ 0, & x > x_0, \end{cases}$$

and set $h_1(x) = h(x)$, $h_2(x) = h(x - \delta)$. For any d.f. G

$$G(x_0 - \delta) = \int_{-\infty}^{x_0 - \delta} h_1\, dG \le \int_{-\infty}^{\infty} h_1\, dG \le \int_{-\infty}^{x_0} dG \le G(x_0)$$

$$G(x_0) = \int_{-\infty}^{x_0} h_2\, dG \le \int_{-\infty}^{\infty} h_2\, dG \le \int_{-\infty}^{x_0 + \delta} dG \le G(x_0 + \delta),$$

and so, taking $G = F$ and then $G = F_n$,

$$\int h_1\, dF - \int h_1\, dF_n \ge F(x_0 - \delta) - F_n(x_0) \ge F(x_0) - F_n(x_0) - \varepsilon,$$

$$\int h_2\, dF - \int h_2\, dF_n \le F(x_0 + \delta) - F_n(x_0) \le F(x_0) - F_n(x_0) + \varepsilon,$$

whence via (ii), for all sufficiently large n

$$|F_n(x_0) - F(x_0)| \le \varepsilon + \sum_{i=1}^{2} \left| \int h_i\, dF - \int h_i\, dF_n \right| < 3\varepsilon.$$

It remains to establish the equivalence of (i) and (iv). Under the latter, for any $a, x \in C(F)$ with $a < x$

$$\underline{\lim}_{n\to\infty} F_n(x) \ge \underline{\lim}[F_n(x) - F_n(a)] \ge \underline{\lim}\, F_n\{(a, x)\} \ge F\{(a, x)\}$$

$$= F(x) - F(a) \xrightarrow{a \to -\infty} F(x),$$

$$1 - \overline{\lim}_{n\to\infty} F_n(a) = \underline{\lim}[1 - F_n(a)] \ge \underline{\lim}[F_n(x) - F_n(a)]$$

$$\ge F(x) - F(a) \xrightarrow{x \to \infty} 1 - F(a).$$

Hence,

$$F(x) \ge \overline{\lim}_{n\to\infty} F_n(x) \ge \underline{\lim}_{n\to\infty} F_n(x) \ge F(x) \quad \text{for } x \in C(F).$$

Finally, to confirm that (i) entails (iv), it suffices by considering complements to verify the assertion about open sets, and these may be supposed subsets of $(-\infty, \infty)$. For any $-\infty < a < b < \infty$, choose $\varepsilon > 0$ so that $a + \varepsilon$, $b - \varepsilon$ are in $C(F)$. Then

$$\begin{aligned}\underline{\lim}_{n\to\infty} F_n\{(a, b)\} &\geq \underline{\lim}[F_n(b - \varepsilon) - F_n(a + \varepsilon)] \\ &= F(b - \varepsilon) - F(a + \varepsilon) = F\{(a + \varepsilon, b - \varepsilon)\}.\end{aligned}$$

As $\varepsilon \downarrow 0$, $(a + \varepsilon, b - \varepsilon) \uparrow (a, b)$, and so $\underline{\lim}\, F_n\{(a, b)\} \geq F\{(a, b)\}$. Since every open set of $(-\infty, \infty)$ is a countable disjoint union of finite open intervals, the second statement of (iv) follows. □

If X_F is a fictitious r.v. with d.f. F and g is a finite Borel function on $(-\infty, \infty)$, it is natural to signify by $g(X_F)$ a fictitious r.v. with d.f. $G(x) = F\{g^{-1}(-\infty, x)\}$, where, as earlier, $F\{\cdot\}$ represents the probability measure determined by the d.f. $F(\cdot)$.

Corollary 9. *If $\{X_n, n \geq 1\}$ is a sequence of* r.v.s *on* $(\Omega, \mathscr{F}, \mathrm{P})$ *with $X_n \xrightarrow{d} X_F$, and $F\{D\} = 0$, where D is the discontinuity set of the Borel function g, then $g(X_n) \xrightarrow{d} g(X_F)$.*

PROOF. Let F_n, G_n denote the d.f.s of X_n, $g(X_n)$ respectively. By (iv), for any closed set C of $(-\infty, \infty)$, if $\bar{A}$ = closure of A,

$$\begin{aligned}\overline{\lim}_{n\to\infty} G_n\{C\} &= \overline{\lim}\, F_n\{g^{-1}(C)\} \leq \overline{\lim}\, F_n\{\overline{g^{-1}(C)}\} \leq F\{\overline{g^{-1}(C)}\} \\ &\leq F\{g^{-1}(C) \cup D\} = F\{g^{-1}(C)\} = G\{C\},\end{aligned}$$

and so the conclusion follows from Theorem 3. □

Corollary 10. *If $\{X_n, n \geq 1\}$ is a sequence of* r.v.s *on* $(\Omega, \mathscr{F}, \mathrm{P})$ *with $X_n \xrightarrow{d} X_F$, then $g(X_n) \xrightarrow{d} g(X_F)$ for any continuous function g.*

EXERCISES 8.1

1. Let F be the d.f. of a r.v. Y, where $\mathrm{P}\{Y = 1\} = \mathrm{P}\{Y = 0\} = \frac{1}{2}$ and define $X_n \equiv Y$, $X = 1 - Y$. Verify that $X_n \xrightarrow{d} X$ but $X_n \overset{P}{\nrightarrow} X$. Also prove that if $X_n \xrightarrow{d} X_F$, where F is degenerate at c (i.e., $F\{\{c\}\} = 1$), then $X_n \xrightarrow{P} c$.

2. If F_n is the d.f. of X_n, $n \geq 0$, where $\mathrm{P}\{X_n = -(1/n)\} = 1 - \mathrm{P}\{X_n = 0\} = \frac{1}{2}$, $n \geq 1$, and $\mathrm{P}\{X_0 = 0\} = 1$, verify that $F_n \xrightarrow{c} F_0$, $\lim F_n(0) \neq F_0(0)$,
$$d^*\{F_n, F_0] = \sup_x |F_n(x) - F_0(x)| \nrightarrow 0.$$

3. If X and Y are r.v.s with d.f.s F and G and $\mathrm{P}\{|X - Y| \geq \varepsilon\} < \varepsilon$, then the Levy distance $d[F, G] \leq \varepsilon$.

4. If d.f.s $F_n \xrightarrow{c} F_0$ and m_n is the unique median of F_n, $n \geq 0$, prove that $m_n \to m_0$. Can an analogous statement be made if the medians are not unique?

5. Define $X = 2\sum_{j=1}^{\infty} X_j/3^j$, where the $\{X_j\}$ are i.i.d. r.v.s on some probability space with $P\{X_1 = 1\} = P\{X_1 = 0\} = \frac{1}{2}$. Then $0 \le X \le 1$, and if $X_1 = 1$, then $X \ge \frac{2}{3}$, while if $X_1 = 0$, $X \le 2\sum_{j=2}^{\infty} 3^{-j} = \frac{1}{3}$. Verify that the d.f. F of X satisfies $F(x) = 2^{-k}F(3^k x)$, $0 < x < 3^{-k}$, $k \ge 1$, and that F is a singular d.f.

6. If $\{S_n, n \ge 1\}$ is a sequence of binomial r.v.s with p.d.f. $b(k; n, p)$, find the density and d.f. F such that $F_n \xrightarrow{c} F$, where F_n is the d.f. of $(npq)^{-1}(S_n - np)^2$.

7. If $g(x) = x^\alpha$, $\alpha > 0$, and $P\{X_n = a_n\} = 1/n = 1 - P\{X_n = 0\}$, is g u.i. relative to $F_n = F_{X_n}$ if $a_n = a \cdot n^{1/\alpha}$, $a > 0$; if $a_n \equiv a$?

8. (Chernoff) Suppose that $f_{nj}(a_n) = O(b_{nj})$, $1 \le j \le m$, and $f_{nj}(a_n) = o(b_{nj})$, $m < j \le k$, imply $g_n(a_n) = o(b_n)$ for some constants a_n and $b_n > 0$, $b_{nj} > 0$, and Borel functions g_n, f_{nj}. If $\{X_n\}$ are r.v.s with $f_{nj}(X_n) = O_p(b_{nj})$ or $o_p(b_{nj})$ according as $1 \le j \le m$ or $m < j \le k$, then $g_n(X_n) = o_p(b_n)$. If, rather, $g_n(a_n) = O(b_n)$, then $g_n(X_n) = O_p(b_n)$.

9. If $\{F_n, n \ge 1\}$ is a sequence of d.f.s with $F_m(x) - F_n(x) \to 0$ as $m, n \to \infty$ for all $x \in (-\infty, \infty)$, does $F_n \xrightarrow{c}$ some F?

10. If r.v.s $X_n \xrightarrow{P} X$ and $|g|^p$ is u.i. relative to F_{X_n}, then $g(X_n) \xrightarrow{\mathscr{L}_p} g(X)$.

11. If $\{F_n(x) = \int_{-\infty}^{x} f_n(y)dy, n \ge 0\}$ are absolutely continuous d.f.s with $f_n(x) \to f_0(x)$, a.e., then $F_n\{B\} \to F_0\{B\}$ uniformly in all Borel sets B. *Hint*: $\int_{-\infty}^{\infty} |f_n(x) - f_0(x)|dx \to 0$.

12. Find d.f.s $F_n \xrightarrow{c} F$ and a Borel set B such that $F\{B\} = 0$, $F_n\{B\} \equiv 1$. *Hint*: If $Y_n, n \ge 1$ is a sequence of binomial r.v.s with parameters n and $p = \frac{1}{2}$, let F_n be the d.f. of $[Y_{n^2} - (n^2/2)]/(n/2)$.

13. If d.f.s $F_n \xrightarrow{c} F$, where F is continuous, then F_n converges to F uniformly on $(-\infty, \infty)$.

14. If for $n \ge 1$ and $|u| < u_0 \in (0, \infty)$, the m.g.f.s $\varphi_n(u) = \int_{-\infty}^{\infty} e^{ux}\, dF_n(x) < g(u) < \infty$ and $F_n \xrightarrow{c} F$, then $\varphi_n(u) \to \int_{-\infty}^{\infty} e^{ux}\, dF(x)$, finite for $|u| < u_0$.

15. Let $S_n = \sum_1^n X_i$ where $\{X_n, n \ge 1\}$ are i.i.d. r.v.s with $E\, X_1 = \mu > 0$. If $N = N_p$ is an $\{X_n\}$ – time (or a r.v. independent of $\{X_n, n \ge 1\}$) having the geometric distribution then $\lim_{p\to 0} P\{S_N/E\, S_N < x\} = 1 - e^{-x}$, $x > 0$. *Hint*: Recall Exercise 3.1.15.

16. Let $H_k(\cdot)$, $k \ge 1$, denote the Hermite polynomial of degree k, which satisfies $H_{k+1}(x) = xH_k(x) - kH_{k-1}(x)$, $k \ge 1$, with $H_0(x) = 1$, $H_1(x) = x$. If $\tilde{S}_{k,n} = \sum_{1 \le i_1 < \cdots < i_k \le n} X_{i_1} \cdots X_{i_k}$, $n \ge k \ge 1$, where $\{X, X_n, n \ge 1\}$ are i.i.d. with $E\, X = 0$, $E\, X^2 = 1$, prove that $k!\,\tilde{S}_{k,n}/n^{k/2} \xrightarrow{d} H_k(Z)$, where Z is a standard normal random variable. *Hint:* Use induction, Exercise 7.5.1, and Theorem 8.1.1.

8.2 Weak Compactness, Fréchet–Shohat, Glivenko–Cantelli Theorems

The Fréchet–Shohat and Glivenko–Cantelli theorems are of special interest and use in statistics. It is advantageous that neither in the proof or application of the former to r.v.s X_n is any supposition, such as the X_n being independent, interchangeable, or constituting a martingale, needed. A first step in the direction of proof is the notion of (sequential) weak compactness. Recall that $y \to x-$ means that $y < x$ and $y \to x$, and analogously for $y \to x+$.

Lemma 1. *If G is a bounded nondecreasing function on D, a dense subset of $(-\infty, \infty)$, then*

$$F(x) = \lim_{\substack{y \in D \\ y \to x-}} G(y)$$

is a left-continuous, nondecreasing function on $(-\infty, \infty)$ with $C(F) \supset C(G)$ and $F(x) = G(x)$ for $x \in C(G)$.

PROOF. Let $F(x) = a$. For any $\varepsilon > 0$ there exists $x' \in D$ with $x' < x$ and $G(x') > a - \varepsilon$. Hence, $F(y) > a - \varepsilon$ for $y \in D \cap (x', x)$, implying $F(x-) \geq a - \varepsilon$, and thus $F(x-) \geq a$. Since F inherits the monotonicity of G, necessarily $F(x-) \leq a$, whence $F(x-) = a = F(x)$. Moreover, if $y_n \in D$, $y_n \uparrow x$, $x_n \in D$, $x_n \downarrow x \in C(G)$, it follows that

$$G(x) \leftarrow G(y_n) \leq F(y_{n+1}) \leq F(x) \leq F(x_n) \leq G(x_n) \to G(x), \qquad \square$$

yielding the final statement of the lemma.

Lemma 2. (i) *Every sequence of* d.f.s *is weakly compact, that is, contains a subsequence converging weakly to a left-continuous function.* (ii) *A sequence of* d.f.s $\{F_n\}$ *is completely compact,* i.e., *every subsequence contains a further subsequence converging completely,* iff $\{F_n, n \geq 1\}$ *is tight.*

(iii) *A sequence of* d.f.s $\{F_n\}$ *converges completely (resp. weakly) to F* iff *every subsequence of $\{F_n\}$ has itself a subsequence converging completely (resp. weakly) to the same function F.*

PROOF. Let $D = \{r_j\}$ be a countable dense set, say the rationals. Since $0 \leq F_n(r_1) \leq 1$, there exists a convergent subsequence $\{F_{n_{1j}}(r_1), j \geq 1\}$. Then $0 \leq F_{n_{1j}}(r_2) \leq 1$ and there exists a convergent subsequence $\{F_{n_{2j}}(r_2)\}$ of $\{F_{n_{1j}}(r_2)\}$. Continuing in this fashion, the diagonal sequence $\{F_{n_{jj}}, j \geq 1\}$ converges to a bounded nondecreasing function G on D, whence by Lemma 1

$$F(x) = \lim_{\substack{y \to x- \\ y \in D}} G(y)$$

is left continuous on $(-\infty, \infty)$. If now $x \in C(F)$, choose $x_m \in D$, $x'_m \in D$ with $x_m \downarrow x$ strictly and $x'_m \to x-$. Then,

$$\begin{aligned} F(x) \xleftarrow{m \to \infty} G(x'_m) \xleftarrow{j \to \infty} F_{n_{jj}}(x'_m) \leq F_{n_{jj}}(x) \leq F_{n_{jj}}(x_m) &\xrightarrow{j \to \infty} G(x_m) \\ &\leq F(x_{m-1}) \xrightarrow{m \to \infty} F(x), \end{aligned}$$

and so $F_{n_{jj}}(x) \to F(x)$ for $x \in C(F)$, establishing (i).

Sufficiency in (ii) is an immediate consequence of (i) and Lemma 8.1.1(iii). For necessity, suppose 1 is not u.i, relative to $\{F_n\}$. Then for some $\varepsilon > 0$ there exists a sequence $a_n \to \infty$ and a subsequence $\{n_j\}$ of $\{n\}$ such that $\Delta F_{n_j} - \Delta F_{n_j}(a_j) > \varepsilon$ for $j \geq 1$. Hence, by Lemma 8.1.1(iii) no subsequence of F_{n_j} can converge completely.

Apropos of (iii), let $x \in C(F)$ and suppose that $F_n(x)$ does not tend toward $F(x)$ as $n \to \infty$. Then there exists a subsequence $F_{n'}$ for which $\lim F_{n'}(x)$ exists but differs from $F(x)$, thereby precluding $F_{n'}$ from having a subsequence converging completely (or weakly) to F. The remaining portion of (iii) is trivial. □

Lemma 8.1.3 ensures the following

Corollary 1. *If $\{F_n, n \geq 0\}$ are* d.f.s *with $\lim_{n\to\infty} F_n(x) = F_0(x)$ for x in a dense subset of $(-\infty, \infty)$, then $F_n \xrightarrow{c} F_0$.*

Theorem 1 (Fréchet–Shohat). *If $\{F_n\}$ is a sequence of* d.f.s *whose moments $\alpha_{n,k} = \int_{-\infty}^{\infty} x^k \, dF_n(x) \xrightarrow{n\to\infty} \alpha_k$ finite, $k = 1, 2, \ldots$, where $\{\alpha_k\}$ are the moments of a uniquely determined* d.f. *F, then $F_n \xrightarrow{c} F$.*

Proof. By (i) of Lemma 2, any subsequence of $\{F_n\}$ has a further subsequence, say $\{F_{n_i}\}$, converging weakly to a left-continuous function F^*, whence by hypothesis and Corollary 8.1.7, $F_{n_i} \xrightarrow{c} F^*$ and

$$\alpha_k = \lim_{i\to\infty} \alpha_{n_i,k} = \int_{-\infty}^{\infty} x^k \, dF^*(x), \qquad k = 1, 2, \ldots.$$

Since, by hypothesis, F is uniquely determined, $F^* = F$ and so (iii) of Lemma 2 ensures that $F_n \xrightarrow{c} F$. □

This raises the question of when a d.f. is uniquely determined by its moments, if indeed they exist. A partial answer appears in Proposition 8.4.6.

The next lemma is of interest in its own right and instrumental in proving the Glivenko–Cantelli theorem.

Lemma 3. *If* d.f.s *$F_n \xrightarrow{c} F$ and $F_n(x\pm) \to F(x\pm)$ at all discontinuity points x of F, then F_n converges uniformly to F in $(-\infty, \infty)$. In particular, if the* d.f.s *F_n converge to a continuous* d.f. *F, the convergence is uniform throughout $(-\infty, \infty)$.*

Proof. By hypothesis, $F_n(x) \to F(x)$ for all x. For any positive integer k, let x_{jk} be the smallest x for which $j/k \leq F(x+)$, $1 \leq j < k$, and set $x_{0k} = -\infty$, $x_{kk} = \infty$. Then $F(x_{jk}) \leq j/k$, $0 \leq j < k$ and for $x_{jk} < x < x_{j+1,k}$, $0 \leq F(x_{j+1,k}) - F(x_{jk}+) \leq 1/k$ so that

$$\begin{aligned} F_n(x_{jk}+) - F(x_{jk}+) - \frac{1}{k} &\leq F_n(x_{jk}+) - F(x_{j+1,k}) \leq F_n(x) - F(x) \\ &\leq F_n(x_{j+1,k}) - F(x_{jk}+) \\ &\leq F_n(x_{j+1,k}) - F(x_{j+1,k}) + \frac{1}{k}. \end{aligned}$$

Hence,

$$\begin{aligned} \sup_{-\infty<x<\infty} |F_n(x) - F(x)| \leq{}& \max_{1\leq j<k} |F_n(x_{jk}) - F(x_{jk})| \\ &+ \max_{0\leq j<k} |F_n(x_{jk}+) - F(x_{jk}+)| + \frac{1}{k}, \end{aligned}$$

and since as $n \to \infty$ the right side $\to 1/k$, which is arbitrarily small for large k, the left side $\to 0$ as $n \to \infty$. □

If $X_1, X_2, \ldots,$ are i.i.d. r.v.s, the **sample** or **empirical d.f.** F_n^ω based on $X_1, \ldots, X_n$ is defined by $F_n^\omega(x) = (1/n)\sum_{j=1}^n I_{[X_j < x]}(\omega)$. Note that for all n and x, $F_n^\omega(x)$ is a r.v. and for every n and almost all ω, $F_n^\omega(x)$ is a d.f. Therefore, for almost all ω, $\{F_n^\omega(x), n \geq 1\}$ is a sequence of d.f.s.

Theorem 2 (Glivenko–Cantelli). *If* $\{X_n, n \geq 1\}$ *are* i.i.d. r.v.s *on a probability space* $(\Omega, \mathscr{F}, \mathrm{P})$ *with* d.f. F *and* F_n^ω *is the empirical* d.f. *based on* $X_1, \ldots, X_n$, *then* $\sup_{-\infty < x < \infty} |F_n^\omega(x) - F(x)| \xrightarrow{\text{a.c.}} 0$.

PROOF. For every x in $(-\infty, \infty)$, $Y_j = I_{[X_j < x]}$ and $Z_j = I_{[X_j \leq x]}$, $j \geq 1$, constitute Bernoulli trials with success probabilities $p = F(x)$ and $F(x+)$ respectively. By the strong law of large numbers for Bernoulli r.v.s, $F_n^\omega(x) = (1/n)\sum_1^n Y_j \xrightarrow{\text{a.c.}} F(x)$ and $F_n^\omega(x+) = (1/n)\sum_1^n Z_j \xrightarrow{\text{a.c.}} F(x+)$. Thus, if $C = \{c_j, j \geq 1\}$ = set of rational numbers enlarged by any irrational discontinuity points of F and

$$A_1 = \{\omega: F_n^\omega(c_j \pm) \to F(c_j \pm), j \geq 1\},$$
$$A_2 = \{\omega: F_n^\omega(x) \text{ is a d.f.}, n \geq 1\},$$

it follows that $\mathrm{P}\{A_1\} = 1$. Moreover, as noted above, $\mathrm{P}\{A_2\} = 1$ and so $\mathrm{P}\{A_1 A_2\} = 1$. Since for $\omega \in A_1 A_2$, $\{F_n^\omega(x), n \geq 1\}$ is a sequence of d.f.s with $F_n^\omega(c_j \pm) \to F(c_j \pm)$, $j \geq 1$, Corollary 1 ensures that $F_n^\omega \xrightarrow{c} F$ for $\omega \in A_1 A_2$, whence by Lemma 3, $F_n^\omega(x)$ converges uniformly to $F(x)$ for $\omega \in A_1 A_2$. □

In problems of interest, d.f.s are likely to be attached to r.v.s X_n on a probability space $(\Omega, \mathscr{F}, \mathrm{P})$ and, as in the case of a.c. convergence or convergence in probability, it is generally necessary to normalize the r.v.s X_n, say to $Y_n = (X_n - b_n)/a_n$, $(a_n > 0)$, to achieve convergence in distribution. This converts the d.f. F_n of X_n to

$$F_{Y_n}(x) = \mathrm{P}\{X_n < a_n x + b_n\} = F_n(a_n x + b_n)$$

and raises some questions in the uniqueness realm.

Definition. Two d.f.s F and G are said to be of the **same type** if for some positive a and real b

$$G(x) = F(ax + b). \tag{1}$$

Theorem 3. *If* $\{F_n\}$ *are* d.f.s *such that for some positive constants* a_n *and real constants* b_n

$$F_n(x) \xrightarrow{c} F(x), \qquad F_n(a_n x + b_n) \xrightarrow{c} G(x),$$

where F and G are nondegenerate d.f.s, *then F and G are of the same type, that is,* (1) *holds, and* $a_n \to a$, $b_n \to b$.

PROOF. Note at the outset that if H is any nondegenerate d.f. with $H(a_1 x + b_1) = H(a_2 x + b_2)$ for all x, where a_j and b_j are real, $j = 1, 2$, and $a_1 + a_2 \neq 0$, then $a_1 = a_2$, $b_1 = b_2$. For if $a_1 x + b_1 \neq a_2 x + b_2$, then

$$x' = \tfrac{1}{2}[(a_1 + a_2)x + b_1 + b_2]$$

cannot be a point of increase of $H(x)$. Thus, if x' is a point of increase, $a_1 x + b_1 = a_2 x + b_2$ and, since there are two distinct points of increase, necessarily $a_1 = a_2$, $b_1 = b_2$.

The hypothesis (and proof) is easily couched in terms of r.v.s, namely, $X_n \xrightarrow{d} X_F$ (nondegenerate) and $Y_n = a_n^{-1}(X_n - b_n) \xrightarrow{d} Y_G$ (nondegenerate), where $a_n > 0$. Let a be a finite or infinite limit point of $\{a_n\}$ and $\{n'\}$ a subsequence of the positive integers $\{n\}$ with $a_{n'} \to a$; let b be a finite or infinite limit point of $\{b_{n'}\}$ and $\{n''\}$ a subsequence of $\{n'\}$ with $b_{n''} \to b$.

If $a = \infty$, $X_{n'}/a_{n'} \xrightarrow{P} 0$, whence by Theorem 8.1.1 $-b_{n''}/a_{n''} = Y_{n''} - (X_{n''}/a_{n''}) \xrightarrow{d} Y_G$, which is impossible since Y_G is nondegenerate. Likewise, $a = 0$ is precluded since this would entail $b_{n''} = X_{n''} - a_{n''} Y_{n''} \xrightarrow{d} X_F$. Thus, $0 < a < \infty$ and, since $(a_{n''} - a)Y_{n''} \xrightarrow{P} 0$,

$$aY_{n''} + b_{n''} = X_{n''} - (a_{n''} - a)Y_{n''} \xrightarrow{d} X_F,$$

which ensures b finite in view of $aY_{n''} \xrightarrow{d} aY_G$. Thus, by Corollary 8.1.3

$$Y_G \xleftarrow{d} Y_{n''} = a_{n''}^{-1}(X_{n''} - b_{n''}) \xrightarrow{d} a^{-1}(X_F - b)$$

or $G(x) = F(ax + b)$. By the remark at the outset, no other subsequence of $\{b_{n'}\}$ can have an alternative limit point b', that is, $\lim b_{n'} = b$. Analogously, if $\{n^*\}$ is a subsequence of $\{n\}$ with $a_{n^*} \to a^*$, $b_{n^*} \to b^*$, the prior reasoning requires $F(ax + b) = F(a^*x + b^*)$, which, by the initial comment, entails $a = a^*$, $b = b^*$. Thus $a_n \to a$, $b_n \to b$. □

Corollary 2. *If $F_n(a_n x + b_n) \xrightarrow{c} F(x)$, $F_n(\alpha_n x + \beta_n) \xrightarrow{c} G(x)$, where F, G are nondegenerate, $a_n, \alpha_n > 0$, then F and G are related by (1) and $\alpha_n/a_n \to a$, $(\beta_n - b_n)/a_n \to b$. In particular, if $G = F$, $\alpha_n \sim a_n$, $\beta_n - b_n = o(a_n)$.*

As seen in Section 6.3, the class of distribution functions on R is closed under the convolution operation $*$. It is likewise closed under the more general operation of mixture (Exercise 3).

EXERCISES 8.2

1. Give an example showing that Theorem 3 is invalid without the nondegeneracy hypothesis.
2. Verify that the class $\mathscr{F}^*$ of all d.f.s on R is a complete metric space under the Levy distance.
3. Let Λ be a Borel subset of $\tilde{R}^m$ where $\tilde{R} = (-\infty, \infty)$, $m \geq 1$, and for every $\lambda \in \Lambda$ let $F(x; \lambda)$ be a d.f. on R such that $F(x; \lambda)$ is a Borel function on R^{m+1}. If G is any d.f. whose support $\subset \Lambda$, show that $H(x) = \int_\Lambda F(x; \lambda) dG(\lambda)$ is a d.f. on R. It is called a

G-mixture of the family $\mathscr{F} = \{F(x; \lambda), \lambda \in \Lambda\}$ or simply a **mixture**. Convolution is the special case $F(x; \lambda) = F(x - \lambda)$, $m = 1$. If E_H and E_{F_λ} denote integration relative to H and $F(x; \lambda)$ respectively, show for any Borel function φ with $\mathrm{E}_H|\varphi| < \infty$ that $\mathrm{E}_H[\varphi] = \int \mathrm{E}_{F_\lambda}[\varphi]dG(\lambda)$.

4. Let Q be the planar region bounded by the quadrilateral with vertices $(0, 0)$, $(1, 1)$, $(0, \frac{1}{2})$, $(\frac{1}{2}, 1)$ and the triangle with vertices $(\frac{1}{2}, 0)$, $(1, 0)$, $(1, \frac{1}{2})$. If X, Y are jointly uniformly distributed on Q, that is, P is Lebesgue measure on the Borel subsets of Q, prove that $F_{X+Y} = F_X * F_Y$ despite the nonindependence of X and Y.

5. The convolution of two discrete distributions F_j, $j = 1, 2$, is discrete, and if the support $S(F_j)$ contains n_j points, $j = 1, 2$, $S(F_1 * F_2)$ contains at most $n_1 \cdot n_2$ and at least $n_1 + n_2 - 1$ points.

6. The convolution of two d.f.s is absolutely continuous if at least one of the components is absolutely continuous. The converse is false (Exercise 8.4.6).

7. If $\{X_n, n \geq 1\}$ are i.i.d. with a uniform distribution on $[0, 1]$, that is, $F(x) = x$ for $0 \leq x \leq 1$, show that $n(\max_{1 \leq i \leq n} X_i - 1) \xrightarrow{d} X_G$.

8. The Lévy **concentration function** of a d.f. F is defined by

$$Q(a) = \sup_{x \in R}[F((x + a)+) - F(x)], \quad a \geq 0.$$

Demonstrate that if $F = F_1 * F_2$, $Q_1(a/2) \cdot Q_2(a/2) \leq Q(a) \leq \min_{i=1,2} Q_i(a)$, and deduce that $F_1 * F_2$ is continuous iff F_1 or F_2 is continuous.

9. Let F_i be a discrete d.f. with maximum jump q_i, $i \geq 1$, and suppose that $G_n = F_1 * F_2 * \cdots * F_n \xrightarrow{c} G$. Prove that if G is continuous, $\prod_{i=1}^{n} q_i = o(1)$.

10. If f_1, f_2 are densities, their convolution is defined by

$$f = f_1 * f_2 = \int f_1(x - y) f_2(y) dy.$$

Verify that if F_j is an absolutely continuous d.f. with density f_j, $j = 1, 2$, then $F = F_1 * F_2$ is absolutely continuous with density $f = f_1 * f_2$.

11. (Chow–Robbins) Prove that if F is a d.f. with $F(0) = 0$, then $G(x) = \prod_1^\infty F(x + n)$ is a d.f. iff $\int x\, dF(x) < \infty$. More generally, G has a finite kth moment iff F has a finite $(k + 1)$st moment.

12. If X and Y are independent r.v.s such that $X + Y$ and X have identical d.f.s, then $Y = 0$, a.c. *Hint*: Employ the Lévy concentration function of Exercise 8.

8.3 Characteristic Functions, Inversion Formula, Lévy Continuity Theorem

Any r.v. X on a probability space $(\Omega, \mathscr{F}, \mathrm{P})$ has an associated characteristic function (abbreviated c.f.) $\varphi_X(t)$ defined for all real t by

$$\varphi(t) = \varphi_X(t) = \mathrm{E}\, e^{itX} = \mathrm{E} \cos tX + i\, \mathrm{E} \sin tX. \tag{1}$$

On the other hand, any d.f. F on R has a corresponding Fourier–Stieltjes transform $\varphi_F(t)$ defined for all real t by

$$\varphi(t) = \varphi_F(t) = \int_{-\infty}^{\infty} e^{itx}\, dF(x) = \int_{-\infty}^{\infty} \cos tx\, dF(x) + i \int_{-\infty}^{\infty} \sin tx\, dF(x), \quad (1')$$

where the integrals are of the improper Riemann–Stieltjes variety.

When $F = F_X$ is the d.f. of a r.v. X on some probability space $(\Omega, \mathscr{F}, \mathrm{P})$, then according to Theorem 6.2.4 the two integrals coincide. Thus, the probability integral of (1) may be envisaged as the Riemann–Stieltjes (or Lebesgue–Stieltjes) integral of (1′) with $F = F_X$ and vice versa in view of oft-repeated comments about construction of a r.v. X on $(\Omega, \mathscr{F}, \mathrm{P})$ with a preassigned d.f. F. Thus, the term c.f. may also be used in conjunction with d.f.

Clearly, $\varphi(0) = 1$, $|\varphi(t)| \le 1$, $\varphi(t) = \overline{\varphi(-t)}$, where $\bar{\varphi}$ signifies the complex conjugate of φ, and φ is uniformly continuous on $(-\infty, \infty)$ since by the dominated convergence theorem

$$|\varphi(t+h) - \varphi(t)| = |\mathrm{E}\, e^{itX}(e^{ihX} - 1)| \le \mathrm{E}|e^{ihX} - 1| \xrightarrow{h \to 0} 0$$

independently of t. Moreover, for any constants a, b

$$\varphi_{aX+b}(t) = e^{ibt}\varphi_X(at). \quad (2)$$

The importance of c.f.s in probability theory stems on the one hand from the one-to-one correspondence between c.f.s and d.f.s, stated as a corollary to the Lévy inversion formula below, and on the other from the ease of operation with c.f.s, due largely to the parallelism between convolution of d.f.s and multiplication of (the corresponding) c.f.s as formulated in Theorem 2 (below). In delving into probablistic problems one may, therefore, operate interchangeably with d.f.s or their c.f.s.

Theorem 1 (Lévy Inversion Formula). *If X is a* r.v. *with* c.f. φ, *then for* $-\infty < a < b < \infty$

$$\lim_{C \to \infty} \frac{1}{2\pi} \int_{-C}^{C} \frac{e^{-ita} - e^{-itb}}{it} \varphi(t)dt$$

$$= \mathrm{P}\{a < X < b\} + \frac{\mathrm{P}\{X = a\} + \mathrm{P}\{X = b\}}{2}. \quad (3)$$

If $a, b \in C(F)$, where $F = F_X$, the right side of (3) reduces to $F(b) - F(a)$. The integrand of (3) is defined at $t = 0$ by continuity and is not, in general, absolutely integrable over $(-\infty, \infty)$.

PROOF. Set

$$I(C) = \frac{1}{2\pi} \int_{-C}^{C} \frac{e^{-ita} - e^{-itb}}{it} \varphi(t)dt = \frac{1}{2\pi} \int_{-C}^{C} \frac{e^{-ita} - e^{-itb}}{it} \mathrm{E}\, e^{itX}\, dt.$$

Since $[(e^{-ita} - e^{-itb})/it]e^{itX}$ is bounded for all ω and t, by Fubini's theorem

$$I(C) = \frac{1}{2\pi} \mathrm{E} \int_{-C}^{C} \frac{e^{it(X-a)} - e^{it(X-b)}}{it} dt$$

$$= \frac{1}{\pi} \mathrm{E} \int_{0}^{C} \frac{\sin t(X-a) - \sin t(X-b)}{t} dt$$

$$= \frac{1}{\pi} \mathrm{E}\left[\int_0^{C(X-a)} \frac{\sin t}{t} dt - \int_0^{C(X-b)} \frac{\sin t}{t} dt\right] = \mathrm{E}\, J_C(X),$$

where

$$J_C(u) = \frac{1}{\pi} \int_{C(u-b)}^{C(u-a)} \frac{\sin t}{t} dt \xrightarrow{C \to \infty} \begin{cases} 1, & a < u < b \\ \frac{1}{2}, & u = a, b \\ 0, & u < a, u > b. \end{cases}$$

Since $|J_C(u)| \le 2$ for all $-\infty < u, C < \infty$, by dominated convergence

$$\lim_{C \to \infty} I(C) = \lim_{C \to \infty} \mathrm{E}\, J_C(X) = \mathrm{E} \lim_{C \to \infty} J_C(X)$$

$$= \mathrm{E}[\tfrac{1}{2} I_{[X = a \text{ or } b]} + I_{[a < X < b]}] = \frac{\mathrm{P}\{X = a\} + \mathrm{P}\{X = b\}}{2} + \mathrm{P}\{a < X < b\}.$$

□

Corollary 1. *There is a one-to-one correspondence between* d.f.s *on R and their* c.f.s.

PROOF. The very definition (1′) shows that identical d.f.s have identical c.f.s, while the converse follows from

$$F(b) = \lim_{a \to -\infty} \lim_{C \to \infty} \frac{1}{2\pi} \int_{-C}^{C} \frac{e^{-ita} - e^{-itb}}{it} \varphi(t) dt, \qquad b \in C(F).$$ □

Corollary 2. *If $\int_{-\infty}^{\infty} |\varphi(t)| dt < \infty$, then for $-\infty < a < b < \infty$*

$$\frac{1}{2\pi} \int_{-\infty}^{\infty} \frac{e^{-ita} - e^{-itb}}{it} \varphi(t) dt = \mathrm{P}\{a < X < b\} + \frac{\mathrm{P}\{X = a\} + \mathrm{P}\{X = b\}}{2}$$

$$= F(b) - F(a), \tag{3′}$$

where the corresponding d.f. *F is absolutely continuous with a bounded, continuous density*

$$f(x) = F'(x) = \frac{1}{2\pi} \int_{-\infty}^{\infty} e^{-itx} \varphi(t) dt. \tag{4}$$

PROOF. In view of the hypothesis, (3) transcribes as the first equality in (3′), whence, letting $b \to a+$, Lebesgue's dominated convergence theorem ensures $\mathrm{P}\{X = a\} = 0$. Hence, F is continuous, yielding the second equality. Since for b approaching a the integrand of (3′) divided by $b - a$ tends to $e^{-ita}\varphi(t)$, (4) follows by dominated convergence. Boundedness of f is apparent from

(4), and continuity of f follows once more by dominated convergence. To verify absolute continuity, note that via Fubini's theorem for $-\infty < a < b < \infty$

$$\begin{aligned}\int_a^b f(x)dx &= \frac{1}{2\pi}\int_a^b dx \int_{-\infty}^{\infty} e^{-itx}\varphi(t)dt \\ &= \frac{1}{2\pi}\int_{-\infty}^{\infty}\left(\int_a^b e^{-itx}\,dx\right)\varphi(t)dt \\ &= \frac{1}{2\pi}\int_{-\infty}^{\infty}\frac{e^{-ita}-e^{-itb}}{it}\varphi(t)dt = F(b) - F(a)\end{aligned}$$

by (3′). Since f is continuous, $f(x) \geq 0$ on $(-\infty, \infty)$ and, letting $b \to \infty$, $a \to -\infty$, $f \in \mathscr{L}_1(-\infty, \infty)$ and $\|f\| = 1$. □

Theorem 2. *If F_i is a* d.f. *with corresponding* c.f. *φ_i, $i = 0, 1, 2$, then $F_0 = F_1 * F_2$* iff *$\varphi_0 = \varphi_1 \cdot \varphi_2$.*

PROOF. Suppose $F_0 = F_1 * F_2$. Since e^{itz} is bounded and continuous, by (11) of Section 6.3

$$\begin{aligned}\varphi_0(t) &= \int e^{itz}\,d(F_1 * F_2)(z) = \int_{-\infty}^{\infty}\int_{-\infty}^{\infty} e^{it(x+y)}\,dF_1(x)dF_2(y) \\ &= \varphi_1(t)\cdot\varphi_2(t).\end{aligned}$$

Conversely, if $\varphi_0 = \varphi_1 \cdot \varphi_2$ and $F = F_1 * F_2$ has c.f. φ, then by the portion of the theorem just proved $\varphi = \varphi_1 \cdot \varphi_2 = \varphi_0$, whence by Corollary 1, $F_0 = F = F_1 * F_2$. □

Corollary 3. *If X_1, X_2 are independent* r.v.s *on some probability space with* c.f.s *φ_{X_1}, φ_{X_2}, then the* c.f. *of their sum is*

$$\varphi_{X_1+X_2} = \varphi_{X_1}\cdot\varphi_{X_2}. \tag{5}$$

PROOF. This is an immediate consequence of Theorem 2 and Theorem 6.3.4. □

Corollary 4. *Convolution is associative and commutative and so the class $\mathscr{F}^*$ of all* d.f.s *on R is an Abelian semigroup.*

What is the analogue vis-à-vis c.f.s of complete convergence of d.f.s? This and more is answered by

Theorem 3 (Levy continuity theorem). *Let $\{F_n\}$ be a sequence of* d.f.s *on R with corresponding* c.f.s *$\{\varphi_n\}$. If $F_n \xrightarrow{c} F$, then $\lim_{n\to\infty}\varphi_n(t) = \varphi_F(t)$ uniformly in $|t| \leq T$ for all $T > 0$. Conversely, if $\varphi_n(t)$ converges to a limit $g(t)$ on $(-\infty, \infty)$ which is continuous at $t = 0$, then g is the* c.f. *of some* d.f. *F and $F_n \xrightarrow{c} F$.*

PROOF. For arbitrary $\varepsilon > 0$, choose $\pm M \in C(F) \cdot \bigcap_1^\infty C(F_n)$ for which $F(-M) + 1 - F(M) < \varepsilon$ and then select N_1 such that $n \ge N_1$ entails

$$F_n(-M) < F(-M) + \varepsilon < 2\varepsilon, \qquad 1 - F_n(M) < 1 - F(M) + \varepsilon < 2\varepsilon.$$

Then

$$|\varphi_n(t) - \varphi(t)| = \left| \int_{-\infty}^{\infty} e^{itx}\, dF_n(x) - \int_{-\infty}^{\infty} e^{itx}\, dF(x) \right| \le |I_1| + |I_2| + |I_3|, \tag{6}$$

where

$$|I_1| = \left| \int_{-\infty}^{-M} e^{itx}\, dF_n(x) - \int_{-\infty}^{-M} e^{itx}\, dF(x) \right| \le F_n(-M) + F(-M) < 3\varepsilon,$$

$$|I_2| = \left| \int_{M}^{\infty} e^{itx}\, dF_n(x) - \int_{M}^{\infty} e^{itx}\, dF(x) \right| \le 1 - F_n(M) + 1 - F(M) < 3\varepsilon.$$

for $n \ge N_1$, and

$$I_3 = \int_{-M}^{M} e^{itx}\, dF_n(x) - \int_{-M}^{M} e^{itx}\, dF(x)$$

$$= [F_n(x) - F(x)]e^{itx} \Big|_{-M}^{M} - it \int_{-M}^{M} e^{itx}[F_n(x) - F(x)]\, dx,$$

whence for fixed but arbitrary $T > 0$ and $|t| \le T$

$$|I_3| \le |F_n(M) - F(M)| + |F_n(-M) - F(-M)| + T \int_{-M}^{M} |F_n(x) - F(x)|\, dx. \tag{7}$$

Since $\lim F_n(x) = F(x)$ except on a set of Lebesgue measure zero, the Lebesgue dominated convergence theorem ensures that for $n \ge N_2$ the last term in (7) is less than ε. Hence, for $n \ge N = \max(N_1, N_2)$ and $|t| \le T$, $|I_3| < 7\varepsilon$, and so from (6), $\varphi_n(t) \to \varphi(t)$ uniformly in $|t| \le T$.

Conversely, suppose that $\varphi_n(t) \to g(t)$ where g is continuous at $t = 0$. By Lemma 8.2.2 there is a monotone, left-continuous function F and a subsequence $\{F_{n_i}\}$ of $\{F_n\}$ such that $F_{n_i} \xrightarrow{w} F$. Then for any $\delta > 0$

$$\frac{1}{2\delta} \int_{-\delta}^{\delta} \varphi_{n_i}(t)dt = \frac{1}{2\delta} \int_0^{\delta} dt \int_{-\infty}^{\infty} (e^{itx} + e^{-itx})dF_{n_i}(x)$$

$$= \frac{1}{\delta} \int_0^{\delta} dt \int_{-\infty}^{\infty} \cos tx\, dF_{n_i}(x) = \int_{-\infty}^{\infty} \frac{\sin \delta x}{\delta x}\, dF_{n_i}(x). \tag{8}$$

By dominated convergence and (ii) of Corollary 8.1.6, it follows as $i \to \infty$ that

$$\frac{1}{2\delta} \int_{-\delta}^{\delta} g(t)dt = \int_{-\infty}^{\infty} \frac{\sin \delta x}{\delta x}\, dF(x),$$

and so, as $\delta \to 0$, by continuity of g at the origin and, once more, dominated convergence,

$$F(\infty) - F(-\infty) = g(0) = \lim \varphi_n(0) = 1.$$

Thus, $F_{n_i} \xrightarrow{c} F$, whence by the portion of the theorem already proved $\varphi_{n_i} \to \varphi_F$, implying $g = \varphi_F$. Analogously, every subsequence of $\{F_n\}$ has itself a subsequence converging completely to a d.f. F^* and $\varphi_{F^*} = g = \varphi_F$, entailing $F^* = F$. Consequently, Lemma 8.2.2(iii) ensures $F_n \xrightarrow{c} F$. □

Corollary 5. *If* $\{F, F_n, n \geq 1\}$ *are* d.f.s *with corresponding* c.f.s $\{\varphi, \varphi_n, n \geq 1\}$, *then* $F_n \xrightarrow{c} F$ iff $\lim \varphi_n = \varphi$.

The few but powerful theorems concerning c.f.s already established permit the extension of Theorem 3.3.1 via the ensuing

Lemma 1 (Doob). *If X is a* r.v. *with* c.f. φ, *then for any positive C and δ*

$$P\{|X| \geq C\} \leq \frac{(1 + 2\pi/C\delta)^2}{\delta} \int_0^\delta [1 - \mathscr{R}\{\varphi(t)\}]dt \tag{9}$$

PROOF. It suffices to prove

$$\int_0^\delta (1 - \cos Cu)du \geq \delta^3 \left(\delta + \frac{2\pi}{C}\right)^{-2} \tag{10}$$

since (9) then follows from

$$\begin{aligned}\int_0^\delta [1 - \mathscr{R}\{\varphi(t)\}]dt &= \int_0^\delta E(1 - \cos tX)dt \\ &= E\int_0^\delta (1 - \cos tX)dt \geq \int_{[|X| \geq C]} dP \int_0^\delta (1 - \cos tX)dt \\ &\geq \int_{[|X| \geq C]} \delta^3 \left(\delta + \frac{2\pi}{|X|}\right)^{-2} dP \\ &\geq \delta^3 \left(\delta + \frac{2\pi}{C}\right)^{-2} P\{|X| \geq C\}.\end{aligned}$$

To this end, choose the minimal $\delta_1 \geq \delta$ such that $C\delta_1 = 2k\pi$ for some positive integer k. Then $\delta_1 < \delta + 2\pi/C$ and

$$\int_0^{C\delta} (1 - \cos u)du = \sum_{j=1}^{k} \int_{2(j-1)\pi\delta/\delta_1}^{2j\pi\delta/\delta_1} (1 - \cos u)du.$$

Now for $0 < b < 2\pi$ and any real a

$$\int_a^{a+b} \cos u\, du \leq \int_{-b/2}^{b/2} \cos u\, du = 2\int_0^{b/2} \cos u\, du,$$

whence, since $y - \sin y \geq y^3/\pi^2$ for $0 \leq y \leq \pi$,

$$\int_0^{C\delta}(1-\cos u)du \geq 2k\int_0^{\pi\delta/\delta_1}(1-\cos u)du = 2k\left[\left(\frac{\pi\delta}{\delta_1}\right)-\sin\left(\frac{\pi\delta}{\delta_1}\right)\right]$$

$$\geq 2k\pi\left(\frac{\delta}{\delta_1}\right)^3 = \frac{C\delta^3}{\delta_1^2} \geq \frac{C\delta^3}{(\delta+2\pi/C)^2}$$

and (10) follows. □

Corollary 6. *If $\{X_n\}$ is a sequence of* r.v.s *on $(\Omega, \mathscr{F}, \mathrm{P})$ and for some $T > 0$ the* c.f. *of $X_m - X_n$, say $\varphi_{m,n}(t)$, converges to* 1 *uniformly in $|t| \leq T$ as $m > n \to \infty$, then $X_n \xrightarrow{\mathrm{P}}$ some* r.v. *X on $(\Omega, \mathscr{F}, \mathrm{P})$.*

Proof. For every $C > 0$, from (9)

$$\sup_{m>n} \mathrm{P}\{|X_m - X_n| > C\} \leq \sup_{m>n} T^{-3}\left(T + \frac{2\pi}{C}\right)^2 \int_0^T |1 - \varphi_{m,n}(t)|dt = o(1),$$

and so, by Lemma 3.3.2, $X_n \xrightarrow{\mathrm{P}} X$. □

Theorem 4. *If $\{X_n\}$ are independent* r.v.s *on a probability space $(\Omega, \mathscr{F}, \mathrm{P})$, $S_n = \sum_{i=1}^n X_i$, $n \geq 1$ and $S_n \xrightarrow{\mathrm{d}} S_F$, then there exists a* r.v. *S on $(\Omega, \mathscr{F}, \mathrm{P})$ such that $S_n \xrightarrow{\mathrm{a.c.}} S$.*

Proof. For $n \geq 1$, let φ_n, ψ_n, and φ be the c.f.s of S_n, X_n and F respectively. By hypothesis and Theorem 3, $\varphi_n(t) \to \varphi(t)$ uniformly in every bounded interval. Hence, for any ε in $(0, \frac{1}{4})$ there exists $T > 0$ and an integer $N > 0$ such that for $|t| \leq T$ and $n \geq N$

$$|\varphi(t)| > \tfrac{1}{2}, \qquad |\varphi_n(t) - \varphi(t)| < \varepsilon, \qquad |\varphi_{n+k}(t) - \varphi_n(t)| < 2\varepsilon, \qquad k \geq 1, \quad (11)$$

recalling that $\varphi(0) = 1$ and φ is continuous. By Corollary 3, $\varphi_n(t) = \prod_{j=1}^n \psi_j(t)$, and so for $|t| \leq T$ and $n \geq N$, (11) implies

$$\left|1 - \prod_{n+1}^{n+m} \psi_j(t)\right| = \frac{1}{|\varphi_n(t)|}|\varphi_n(t) - \varphi_{n+m}(t)| < 4\varepsilon, \qquad m \geq 1.$$

Therefore, $\lim_{n\to\infty} \varphi_{S_k - S_n}(t) = 1$ uniformly for $|t| \leq T$ and $k > n$. By Corollary 6, there exists a r.v. S on $(\Omega, \mathscr{F}, \mathrm{P})$ with $S_n \xrightarrow{\mathrm{P}} S$ and, consequently, by Lévy's theorem (Theorem 3.3.1) $S_n \xrightarrow{\mathrm{a.c.}} S$. □

Exercises 8.3

1. If φ is a c.f. so are $|\varphi(t)|^2$ and $\mathscr{R}\{\varphi(t)\}$ where $\mathscr{R}\{z\}$ denotes the real part of z. Find the c.f. of the singular d.f. of Exercise 8.1.5.

2. If F_n, G_n are d.f.s with $F_n \xrightarrow{\mathrm{c}} F$, $G_n \xrightarrow{\mathrm{c}} G$, then $F_n * G_n \xrightarrow{\mathrm{c}} F * G$.

3. If H is a G-mixture of $\mathscr{F} = \{F(x; \lambda), \lambda \in \Lambda\}$ as in Exercise 8.2.3 and $\varphi(t, \lambda)$ is the c.f. corresponding to $F(x; \lambda)$, show that $\varphi_H(t) = \int_\Lambda \varphi(t; \lambda)dG(\lambda)$. Thus, if $\{\varphi_n(t), n \geq 1\}$ is a sequence of c.f.s and $\sum_1^\infty c_n = 1$, $c_n \geq 0$, $\varphi(t) = \sum_1^\infty c_i\varphi_i$ is a c.f. Verify that $\exp\{\lambda[\varphi(t) - 1]\}$, $\lambda > 0$, is a c.f. if φ is.

4. Prove that if φ is a c.f., $1 - \mathscr{R}\{\varphi(2t)\} \leq 4[1 - \mathscr{R}\{\varphi(t)\}]$.

5. Prove that a real-valued c.f. φ satisfies $\varphi(t+h) \le \varphi(h)$ for $t > 0$ and sufficiently small $h > 0$.

6. Verify that the c.f. corresponding to a uniform distribution on $(-\alpha, \alpha)$ is $\varphi_\alpha(t) = (\sin \alpha t)/\alpha t$, $\alpha > 0$, and that $\lim \varphi_n(t)$ exists but $F_n \not\to$ to a d.f.

7. Find the c.f. φ_X if $P\{X = \alpha\} = P\{X = -\alpha\} = \frac{1}{2}$. Show by iteration of an elementary trigonometric identity that

$$\frac{\sin t}{t} = \frac{\sin t/2^n}{t/2^n} \prod_{j=1}^{n} \cos \frac{t}{2^j} \to \prod_{j=1}^{\infty} \cos \frac{t}{2^j}$$

and utilize this to state a result on convergence in law of sums of independent r.v.s.

8. If φ is the c.f. of the d.f. F, prove that

$$\lim_{C\to\infty} \frac{1}{2C} \int_{-C}^{C} e^{-itx} \varphi(t)\,dt = F(x+) - F(x),$$

and if X is an integer valued r.v., show that

$$\frac{1}{2\pi} \int_{-\pi}^{\pi} e^{-itj} \varphi_X(t)\,dt = P\{X = j\}.$$

Give an analogous formula for $S_n = \sum_{i=1}^n X_i$, where $\{X_i\}$ are i.i.d. integer-valued r.v.s.

9. Prove that $\lim_{C\to\infty} (1/2C) \int_{-C}^{C} |\varphi(t)|^2\,dt = \sum_{x \in R} [F(x+) - F(x)]^2$. *Hint*: if X_1, X_2 are independent with common d.f. F, apply Exercise 8 at $x = 0$ to $X_1 - X_2$.

10. Prove that

$$F(x+) + F(x-) = 1 - \frac{1}{\pi} ⨍ \frac{e^{-itx}\varphi(t)}{it}\,dt$$

where

$$⨍ = \lim_{\substack{\varepsilon\to 0\\ C\to\infty}} \left(\int_{-C}^{-\varepsilon} + \int_{\varepsilon}^{C} \right),$$

and hence that

$$F(x) = \frac{1}{2} - \frac{1}{2\pi} ⨍ \frac{e^{-itx}\varphi(t)}{it}\,dt$$

for F continuous.

11. Utilize c.f.s to give an alternative proof of Corollary 8.1.10.

12. Prove for any d.f. with c.f. φ that for all real x and $h > 0$

$$\frac{1}{2h} \int_{x}^{x+2h} F(y)\,dy - \frac{1}{2h} \int_{x-2h}^{x} F(y)\,dy = \frac{1}{\pi} \int_{-\infty}^{\infty} \left(\frac{\sin u}{u} \right)^2 e^{-iux/h} \varphi(u/h)\,du.$$

Hint: Apply the Levy inversion formula to $F * G_h$, where G_h is the uniform distribution on $[-h, h]$.

13. Utilize Exercise 12 to prove the converse portion of Theorem 3.

14. Let $H_i(x) = \int_\Lambda F(x; \lambda) dG_i(\lambda)$ be a G_i-mixture (Exercise 8.2.3) of the additively closed family $\mathscr{F} = \{F(x; \lambda), \lambda \in \Lambda \subset R^m\}$ (Exercise 8.4.5). Verify that the convolution $H_1 * H_2$ is a $(G_1 * G_2)$-mixture of $\mathscr{F}$. *Hint*: Utilize Exercise 3.

15. If, in Exercise 14, $m = 1$ and $\Lambda = [0, \infty)$ or $[0, 1, 2, \ldots]$ or {nonnegative rationals}, then the mapping of $\mathscr{G} = \{G: G\{\Lambda\} = 1\}$ onto $\mathscr{H} = \{H: H(x) = \int_\Lambda F(x; \lambda) dG(\lambda), G \in \mathscr{G}\}$ is one-to-one. Then $\mathscr{H}$ is said to be **identifiable**. *Hint*: $\psi(z; G) = \int_\Lambda z^\lambda \, dG(\lambda)$ is analytic in $0 < |z| < 1$ and $\varphi(t; \lambda) = \varphi^\lambda(t)$, $\lambda \in \Lambda$ where $\varphi(t) = \varphi(t; 1)$.

8.4 The Nature of Characteristic Functions, Analytic Characteristic Functions, Cramér–Lévy Theorem

In view of the crucial importance of c.f.s in probability theory, it is desirable to amass some information concerning their nature. A first step in this direction is a tally of the more popular d.f.s and the corresponding c.f.s.

Distribution	Density	Support	c.f.
Degenerate		$\{\alpha\}$	$e^{i\alpha t}$
Symmetric Bernoulli	$\frac{1}{2}$	$\{1, -1\}$	$\cos t$
Binomial	$\binom{n}{x} p^x (1-p)^{n-x}$	$x \in \{0, 1, \ldots, n\}$	$(pe^{it} + 1 - p)^n, 0 < p < 1$
Poisson	$\lambda^x \dfrac{e^{-\lambda}}{x!}$	$x \in \{0, 1, \ldots\}$	$\exp\{\lambda(e^{it} - 1)\}, \lambda > 0$
Normal	$\dfrac{1}{\sigma\sqrt{2\pi}} \exp\left\{\dfrac{-(x-\theta)^2}{2\sigma^2}\right\}$	$(-\infty, \infty)$	$e^{i\theta t - \sigma^2 t^2/2}, \sigma > 0$
Symmetric Uniform	$\dfrac{1}{2\alpha}$	$[-\alpha, \alpha]$	$\dfrac{\sin \alpha t}{\alpha t}, \alpha > 0$
Triangular	$\dfrac{1}{\alpha}\left(1 - \dfrac{\lvert x\rvert}{\alpha}\right)$	$[-\alpha, \alpha]$	$\dfrac{2(1 - \cos \alpha t)}{\alpha^2 t^2}, \alpha > 0$
Inverse Triangular	$\dfrac{1 - \cos \alpha x}{\pi \alpha x^2}$	$(-\infty, \infty)$	$\left[1 - \dfrac{\lvert t\rvert}{\alpha}\right]^+, \alpha > 0$
Gamma	$\dfrac{\alpha^\lambda}{\Gamma(\lambda)} x^{\lambda - 1} e^{-\alpha x}$	$(0, \infty)$	$\left(1 - \dfrac{it}{\alpha}\right)^{-\lambda}, \lambda > 0, \alpha > 0$
Cauchy	$\dfrac{\alpha}{\pi(\alpha^2 + x^2)}$	$(-\infty, \infty)$	$e^{-\alpha\lvert t\rvert}, \alpha > 0$

Characteristic functions and moments are intimately related , as will be seen via the preliminary

Lemma 1. *For $t \in (-\infty, \infty)$ and nonnegative integers n*

$$e^{it} - \sum_{j=0}^{n} \frac{(it)^j}{j!} = \frac{(it)^{n+1}}{n!} \int_0^1 e^{itu}(1-u)^n \, du$$

$$= i^{n+1} \int_0^t dt_{n+1} \int_0^{t_{n+1}} dt_n \cdots \int_0^{t_2} e^{it_1} \, dt_1, \tag{1}$$

and for any δ in $[0, 1]$

$$\left| e^{it} - \sum_{j=0}^{n} \frac{(it)^j}{j!} \right| \le \frac{2^{1-\delta} |t|^{n+\delta}}{(1+\delta)(2+\delta)\cdots(n+\delta)}, \tag{2}$$

where the denominator of the right side of (2) is unity for $n = 0$.

PROOF. Since for $n \ge 1$

$$\frac{(it)^{n+1}}{n!} \int_0^1 e^{itu}(1-u)^n \, du = \frac{-(it)^n}{n!} + \frac{(it)^n}{(n-1)!} \int_0^1 e^{itu}(1-u)^{n-1} \, du,$$

the first part of (1) follows by summing while the second is obtained inductively.

To prove (2), let I_n denote the left side of (1). Since $|e^{it} - 1| = 2|\sin t/2| \le 2^{1-\delta}|t|^\delta$ for $0 \le \delta \le 1$, let $n \ge 1$, whence from (1)

$$|I_n| \le \int_0^{|t|} \int_0^{t_{n+1}} \cdots \int_0^{t_3} |e^{it_2} - 1| \, dt_2 \cdots dt_{n+1}$$

$$\le 2^{1-\delta} \int_0^{|t|} \int_0^{t_{n+1}} \cdots \int_0^{t_3} t_2^\delta \, dt_2 \cdots dt_{n+1} = \frac{2^{1-\delta}|t|^{n+\delta}}{(1+\delta)\cdots(n+\delta)}. \qquad \square$$

Corollary 1. $|e^z - 1| \le (e^{|z|} - 1)$ *for any complex z.*

PROOF. (2) with $\delta = 1$, $n = 0$ yields the bound $2(e^{|z|} - 1)$. However, directly

$$|e^z - 1| = \left| \sum_1^\infty z^j/j! \right| \le \sum_1^\infty |z|^j/j! = e^{|z|} - 1. \qquad \square$$

Theorem 1. *If X is a* r.v. *with* c.f. *φ and $\mathrm{E}|X|^{n+\delta} < \infty$ for some nonnegative integer n and some δ in $[0, 1]$, then φ has continuous derivatives $\varphi^{(k)}$ of orders $k \le n$ and*

$$\varphi^{(k)}(t) = i^k \,\mathrm{E}\, X^k e^{itX}, \qquad \varphi^{(k)}(0) = i^k \,\mathrm{E}\, X^k, \qquad 1 \le k \le n, \tag{3}$$

$$\varphi(t) = \sum_{j=0}^{n} \frac{(it)^j \,\mathrm{E}\, X^j}{j!} + O(|t|^{n+\delta}), \qquad O(|t|^{n+\delta}) \le \frac{2^{1-\delta}\,\mathrm{E}|X|^{n+\delta}|t|^{n+\delta}}{(1+\delta)\cdots(n+\delta)}, \tag{4}$$

$$\varphi(t) = \sum_{j=0}^{n} \frac{(it)^j \,\mathrm{E}\, X^j}{j!} + o(|t|^n) \quad \text{as } t \to 0. \tag{5}$$

Conversely, if $\varphi^{(2k)}(0)$ exists and is finite for some $k = 1, 2, \ldots,$ then

$$\mathrm{E}\, X^{2k} < \infty.$$

PROOF. (4) follows easily from (2). To prove (3), note that via (1)

$$e^{itX} - \sum_{j=0}^{k-1} \frac{(itX)^j}{j!} = (iX)^k \int_0^t \int_0^{t_k} \cdots \int_0^{t_2} e^{it_1 X}\, dt_1 \cdots dt_k, \tag{6}$$

so that

$$\psi_k(t) \equiv \varphi(t) - \sum_{j=0}^{k-1} \frac{(it)^j\, \mathrm{E}\, X^j}{j!} = i^k\, \mathrm{E}\, X^k \int_0^t \int_0^{t_k} \cdots \int_0^{t_2} e^{it_1 X}\, dt_1 \cdots dt_k$$

and

$$\frac{\psi_k(t+h) - \psi_k(t)}{h} = \mathrm{E}(iX)^k \frac{1}{h} \int_t^{t+h} \int_0^{t_k} \cdots \int_0^{t_2} e^{it_1 X}\, dt_1 \cdots dt_k.$$

Since

$$\left| \frac{1}{h} \int_t^{t+h} \int_0^{t_k} \cdots \int_0^{t_2} e^{it_1 X}\, dt_1 \cdots dt_k \right| \le (|t| + |h|)^{k-1} = O(1) \quad \text{as } h \to 0,$$

by the dominated convergence theorem, for $1 \le k \le n$

$$\psi_k'(t) = i^k\, \mathrm{E}\, X^k \int_0^t \int_0^{t_{k-1}} \cdots \int_0^{t_2} e^{it_1 X}\, dt_1 \cdots dt_{k-1} \tag{7}$$

and, in particular,

$$\varphi'(t) = \psi_1'(t) = i\, \mathrm{E}\, X e^{itX}.$$

Repeating the previous argument, it follows via (7) that for $2 \le k \le n$

$$\psi_k^{(2)}(t) = \frac{d^2 \psi_k(t)}{dt^2} = i^k\, \mathrm{E}\, X^k \int_0^t \int_0^{t_{k-2}} \cdots \int_0^{t_2} e^{it_1 X}\, dt_1 \cdots dt_{k-2},$$

and, continuing in this fashion, for $k \le n$

$$\psi_k^{(k-1)}(t) = i^k\, \mathrm{E}\, X^k \int_0^t e^{it_1 X}\, dt_1,$$

$$\varphi^{(k)}(t) = \psi_k^{(k)}(t) = i^k\, \mathrm{E}\, X^k e^{itX},$$

which is (3). Now by (3),

$$\varphi^{(n)}(t+h) - \varphi^{(n)}(t) = i^n\, \mathrm{E}\, X^n e^{itX}(e^{ihX} - 1),$$

and since $|e^{ihX} - 1| \le 2$, once more by dominated convergence

$$\lim_{h \to 0} \varphi^{(n)}(t+h) = \varphi^{(n)}(t)$$

and φ has a continuous nth derivative, whence (5) follows from Taylor's theorem.

To prove the converse, define for any finite function g for $-\infty < x, h < \infty$

$$\Delta_h^{(1)} g(x) = g(x+h) - g(x-h),$$
$$\Delta_h^{(n)} g(x) = \Delta_h^{(n-1)} g(x+h) - \Delta_h^{(n-1)} g(x-h), \qquad n = 2, 3, \ldots.$$

Then $\Delta_h^{(n)}[ag_1(x) + bg_2(x)] = a\Delta_h^{(n)} g_1(x) + b\Delta_h^{(n)} g_2(x)$ for $-\infty < a, b < \infty$ and

$$(2h)^{-2n}\Delta_h^{(2n)}\varphi(t) = \mathrm{E}\,(2h)^{-2n}\Delta_h^{(2n)} e^{itX} = \mathrm{E}\, e^{itX}\left(\frac{e^{ihX} - e^{-ihX}}{2h}\right)^{2n}.$$

If $g(x)$ is a polynomial of degree n, $\Delta_h g(x)$ is a polynomial of degree at most $n-1$. Hence $\Delta_h^{(m)} g(x) = 0$ if g is a polynomial of degree $< m$ and, moreover, $\Delta_h^{(2m)} x^{2m} = (2h)^{2m}(2m)!$ Since $\varphi^{(2n)}(0)$ is finite, as $t \to 0$ (Hardy, 1952, p. 290)

$$\varphi(t) = \varphi(0) + t\varphi'(0) + \cdots + \frac{t^{2n}}{(2n)!}\,[\varphi^{(2n)}(0) + o(1)],$$

so that $\varphi^{(2n)}(0) = \lim_{h\to 0}(2h)^{-2n}\Delta_h^{(2n)}\varphi(t)|_{t=0}$, whence

$$|\varphi^{(2n)}(0)| = \left|\lim_{h\to 0} \mathrm{E}\left(\frac{e^{ihX} - e^{-ihX}}{2h}\right)^{2n}\right| = \lim_{h\to 0} \mathrm{E}\left(\frac{\sin hX}{h}\right)^{2n} \geq \mathrm{E}\, X^{2n}$$

by Fatou's lemma, completing the proof. □

In Section 3.3, a r.v. X was defined to be symmetric if X and $-X$ have the same distribution. The property in question belongs to the d.f. and is tantamount to saying that a d.f. is symmetric if $F(x) = 1 - F(-x+)$, $x \in R$.

Proposition 1. *A* c.f. *$\varphi(t)$ is real valued (for real t) iff its* d.f. *F is symmetric.*

PROOF. The proof is most easily couched in terms of r.v.s. If F is symmetric, that is, X and $-X$ are identically distributed,

$$\varphi(t) = \varphi_X(t) = \varphi_{-X}(t) = \varphi_X(-t) = \overline{\varphi(t)},$$

so that φ is real. Conversely, if φ is real,

$$\varphi_X(t) = \overline{\varphi_X(t)} = \varphi_X(-t) = \varphi_{-X}(t),$$

and so by the one-to-one correspondence between c.f.s and d.f.s, X and $-X$ are identically distributed, that is, F is symmetric. □

A function $\varphi(t)$ on the real line is said to be **periodic** with a period t_0 if $\varphi(t + t_0) = \varphi(t)$ for all real t.

Proposition 2. *If φ is the* c.f. *of the* d.f. *F and $|\varphi(t_0)| = 1$ for some $t_0 \neq 0$, then F is discrete with support S consisting of a subset of points in arithmetic progression and conversely. In this case, $S \subset \{(2\pi j + \theta_0)/t_0, j = 0, \pm 1, \ldots\}$ for some real θ_0 and $e^{-it\theta_0/t_0}\varphi(t)$ is a periodic function with period t_0.*

PROOF. Since $|\varphi(t_0)| = 1$, $\varphi(t_0) = e^{i\theta_0}$ for some real θ_0, whence $1 = e^{-i\theta_0}\varphi(t_0) = \int e^{i(t_0 x - \theta_0)}\,dF(x) = \int \cos(t_0 x - \theta_0)dF(x)$. In r.v. parlance,

$$\mathrm{E}[1 - \cos(t_0 X - \theta_0)] = 0,$$

necessitating

$$\mathrm{P}\{t_0 X - \theta_0 = 2j\pi, j = 0, \pm 1, \ldots\} = P\left\{X = \frac{2j\pi + \theta_0}{t_0}, j = 0, \pm 1, \ldots\right\} = 1.$$

Conversely, the latter entails, setting $p_j = \mathrm{P}\{X = (2j\pi + \theta_0)/t_0\} \geq 0$,

$$\varphi(t) = \sum_j p_j e^{it(2j\pi + \theta_0)/t_0} = e^{i\theta_0 t/t_0} \sum_j p_j e^{2j\pi it/t_0}, \tag{8}$$

whence $|\varphi(t_0)| = 1$ and $e^{-it\theta_0/t_0}\varphi(t)$ is periodic with a period t_0. □

Corollary 2. *A* c.f. *$\varphi(t)$ satisfies either* (i) $|\varphi(t)| < 1$ *for all* $t \neq 0$, (ii) $|\varphi(t)| \equiv 1$, *or* (iii) $|\varphi(t)| = 1$ *for countably many isolated values of* t.

PROOF. It suffices to prove that if $|\varphi(t_n)| = 1$ for $t_n \to s$ (finite), $0 \neq t_n \neq s$, then $|\varphi(t)| \equiv 1$. Now, via (8) with t_n replacing t_0 and then t_m replacing t_0, for $k, m, n = 1, 2, \ldots$

$$|\varphi(k(t_m - t_n))| = |\varphi(kt_m)| = 1.$$

Since $t_m - t_n \to 0$, every interval (a, b) contains a point c with $|\varphi(c)| = 1$, whence by continuity $|\varphi(t)| \equiv 1$. □

Corollary 3. *If a* c.f. *φ satisfies $|\varphi(t_0)| = |\varphi(\alpha t_0)| = 1$ for some $t_0 \neq 0$ and irrational α, then φ is degenerate, that is, $\varphi(t) = e^{i\theta t}$ for some real θ.*

PROOF. If φ is nondegenerate, by Proposition 2, $\mathrm{P}\{X = (2\pi j_i + \theta_0)/t_0\} > 0$ for some θ_0 and integers $j_1 \neq j_2$. Then for some real θ_1 and integers $k_1 \neq k_2$

$$\frac{2\pi j_i + \theta_0}{t_0} = \frac{2\pi k_i + \theta_1}{\alpha t_0}, \qquad i = 1, 2,$$

$$2\pi\left(j_1 - \frac{k_1}{\alpha}\right) = \frac{\theta_1}{\alpha} - \theta_0 = 2\pi\left(j_2 - \frac{k_2}{\alpha}\right),$$

whence $k_1 - k_2 = \alpha(j_1 - j_2)$, a contradiction. □

Corollary 2 in combination with Theorem 8.3.3. elicits some interesting conclusions about convergence in distribution of normalized sums of i.i.d. r.v.s.

Theorem 2. *If $\{X_n\}$ are* i.i.d. *with nondegenerate* d.f. *F, $S_n = \sum_1^n X_i$, and for some positive constants a_n and real b_n and nondegenerate* d.f. *G*

$$\frac{S_n - b_n}{a_n} \xrightarrow{\mathrm{d}} S_G,$$

then

$$a_n \to \infty, \qquad \frac{a_n}{a_{n-1}} \to 1 \quad \textit{and} \quad b_n - b_{n-1} = o(a_n).$$

PROOF. If $a_n \not\to \infty$, there is a subsequence $\{n'\}$ of the positive integers $\{n\}$ (which, for notational simplicity, will be replaced by $\{n\}$) such that $a_n \to a$, finite. By Corollary 8.1.3, $S_n - b_n = a_n(S_n - b_n)/a_n \xrightarrow{d} aS_G$. Thus, if ψ and φ are c.f.s respectively of X_1 and aS_G, by Theorem 8.3.3 $e^{-ib_nt}\psi^n(t) \to \varphi(t)$. Clearly, this entails $\varphi(t) = 0$ for all t such that $|\psi(t)| < 1$. By Corollary 2, $\varphi(t) = 0$ for all except perhaps countably many isolated values of t, whence $\varphi(t) \equiv 0$ by continuity, a contradiction. Thus, $a_n \to \infty$, whence $X_1/a_n \xrightarrow{P} 0$ and likewise $X_n/a_n \xrightarrow{P} 0$. Theorem 8.1.1 then ensures that $(S_{n-1} - b_n)/a_n \xrightarrow{d} S_G$, which in conjunction with $(S_{n-1} - b_{n-1})/a_{n-1} \xrightarrow{d} S_G$ necessitates

$$\frac{a_n}{a_{n-1}} \to 1, \qquad \frac{b_{n-1} - b_n}{a_n} \to 0$$

by Corollary 8.2.2. □

The next proposition reveals *inter alia* that $\varphi_\alpha(t) = e^{-|t|^\alpha}$ is a c.f. for $0 < \alpha \le 1$ but fails to detect that φ_α is likewise a c.f. for $1 < \alpha \le 2$, a fact which will be established in Theorem 12.3.2.

Proposition 3 (Polya). *A nonnegative, even function ψ convex and decreasing on $(0, \infty)$ with $\psi(0+) = \psi(0) = 1$ is a* c.f.

PROOF. Apart from the trivial case $\psi(t) \equiv 1$, if $\psi(t) \equiv c$ for all $t > t_0$, then $\psi = \lim_{n\to\infty} \psi_n$, where ψ_n is strictly decreasing (to c) for $t > t_0$. Thus, by Theorem 8.3.3 it suffices to prove the proposition for ψ strictly decreasing. For $m = n2^n$ and $n = 1, 2, \ldots$, set $t_j = j2^{-n}$, $\psi_j = \psi(t_j)$, $j = 0, 1, \ldots, m$. If $(t, y_j(t))$ is a point on the line determined by the points (t_{j-1}, ψ_{j-1}) and (t_j, ψ_j), then for $1 \le j \le m$

$$y_j(t) = \frac{\psi_j - \psi_{j-1}}{t_j - t_{j-1}}\, t + \frac{\psi_{j-1}t_j - \psi_j t_{j-1}}{t_j - t_{j-1}}$$

and $y_j(t_j) = y_{j+1}(t_j)$. Define c_m, λ_m by

$$c_m = y_m(0), \qquad \frac{c_m}{\lambda_m} = \frac{\psi_{m-1} - \psi_m}{t_m - t_{m-1}} = -y_m'(t),$$

whence, by monotonicity, $1 \ge y_m(0) = c_m > 0$ and $\lambda_m \ge t_m$. Note that $y_m(t) = c_m(1 - (t/\lambda_m))^+$ for $t \le t_m \le \lambda_m$. Define c_j, λ_j inductively for $j = m - 1, \ldots, 2, 1$ via

$$\sum_{i=j}^m c_i = y_j(0), \qquad \sum_{i=j}^m \frac{c_i}{\lambda_i} = \frac{\psi_{j-1} - \psi_j}{t_j - t_{j-1}} = -y_j'(t).$$

Then

$$y_j(t) = -t \sum_{i=j}^{m} \frac{c_i}{\lambda_i} + \sum_{i=j}^{m} c_i,$$

and so $y_j(t) = y_{j+1}(t)$ entails $t = \lambda_j = t_j$ for $j < m$. Moreover,

$$\psi_{j+1} + \psi_{j-1} \geq 2\psi_j = \left(\frac{t_{j+1} + t_{j-1}}{t_j}\right)\psi_j$$

by convexity, implying $y_j(0) \geq y_{j+1}(0)$. Thus, $c_j = y_j(0) - y_{j+1}(0) \geq 0$ and $\sum_{j=1}^{m} c_j = y_1(0) = 1$. Furthermore, $\psi_m(t) \equiv \sum_{j=1}^{m} c_j(1 - (|t|/\lambda_j))^+$ coincides with $y_j(|t|)$ for $t_{j-1} \leq |t| \leq t_j$, $1 \leq j \leq m$, and $\psi_m(t)$ is a c.f., $m \geq 1$, via Exercise 8.3.3 and the table of c.f.s. By hypothesis $\psi(0) = \psi(0+)$, whence $\psi(t) = \lim_{n\to\infty} \psi_n(t)$, and so ψ is a c.f. by Theorem 8.3.3. □

Corollary 4. *Two different* c.f.s *may coincide in any fixed finite interval* $(-T, T)$, $T > 0$.

PROOF. If suffices to change the c.f. $e^{-|t|}$ in $(T, T + 1)$ and $(-T - 1, -T)$ by replacing its arcs therein by line segments preserving continuity. According to Proposition 3, the modified function is still a c.f. □

A necessary and sufficient condition for a continuous complex function φ on $(-\infty, \infty)$ with $\varphi(0) = 1$ to be a c.f. exists (Exercise 16) but is too unwieldy to constitute a working criterion.

One of the most beautiful and striking theorems of probability theory, conjectured by Lévy and proved by Cramér, asserts that if a normal d.f. is the convolution of two nondegenerate d.f.s, each must be normal.

To place this in perspective, an algebraic viewpoint is helpful. If d.f.s F, F_1, F_2 are in the relation $F = F_1 * F_2$, it is natural to call F_1 and F_2 factors or divisors of F. The corresponding c.f.s φ_1 and φ_2 are literal divisors of φ. If J_α denotes the d.f. degenerate at α, then J_α is a factor of every d.f. F since $\varphi_F(t) = e^{i\alpha t}[e^{-i\alpha t}\varphi_F(t)]$, but these are trivial and unwanted factors. Clearly, all divisors of discrete d.f.s are discrete, but the analogous statement for continuous d.f.s is false, as revealed by the trigonometric identity

$$\frac{\sin 2t}{2t} = \frac{\sin t}{t} \cos t.$$

Definition. A family $\mathscr{F}'$ of d.f.s is **factor closed** if for all $F \in \mathscr{F}'$ the convolution relation $F = G_1 * G_2$ with G_i a nondegenerate d.f. implies that $G_i \in \mathscr{F}'$, $i = 1, 2$.

The theorem of Cramér–Lévy may be rephrased by saying that the family of normal d.f.s is factor closed. The only known method of proof is via c.f.s and requires a discussion of analytic c.f.s.

A c.f. φ is called analytic or more specifically r-analytic, $r > 0$ (resp. entire), if φ can be represented by a convergent power series in $-r < t < r$ (resp. in $-\infty < t < \infty$), i.e.,

$$\varphi(t) = \sum_{j=0}^{\infty} \frac{a_j t^j}{j!}, \qquad -r < t < r.$$

Proposition 4. *If F is a* d.f. *with corresponding* c.f. φ, *the following conditions are equivalent*:

i. φ *is r-analytic*;
ii. $\varphi(t) = \sum_{j=0}^{\infty} \alpha_j (it)^j/j!$, $-r < t < r$, where $\alpha_j = \int x^j \, dF(x)$ (*finite*);
iii. $\sum_{j=0}^{\infty} \alpha_{2j} t^{2j}/(2j)! < \infty$, $0 < t < r$;
iv. $\int e^{t|x|} \, dF(x) < \infty$, $0 < t < r$.

PROOF. Suppose as in (i) that for some $r > 0$ and complex numbers $\{a_n\}$ $\varphi(t) = \sum_0^{\infty} a_n t^n/n!$, $-r < t < r$. Then, $\varphi^{(k)}(t) = \sum_{n=k}^{\infty} a_n t^{n-k}/(n-k)!$, $k = 1, 2, \ldots$, whence by Theorem 8.4.1, $a_k = \varphi^{(k)}(0) = i^k \int_{-\infty}^{\infty} x^k \, dF(x) = i^k \alpha_k$ (say), yielding (ii). Since (ii) $\Rightarrow$ (iii) is trivial, set $\beta_j = \int |x|^j \, dF(x)$ and suppose $\sum_{j=0}^{\infty} \beta_{2j} t^{2j}/(2j)! < \infty$, which is (iii). Via Jensen's inequality,

$$\beta_{2j-1}^{1/(2j-1)} \le \beta_{2j}^{1/2j},$$

whence

$$\sum_{j=1}^{\infty} \frac{\beta_{2j-1} t^{2j-1}}{(2j-1)!} \le \sum_{j=1}^{\infty} \frac{(1 + \beta_{2j}) t^{2j-1}}{(2j-1)!} \le e^t + \frac{d}{dt} \sum_{1}^{\infty} \frac{\beta_{2j} t^{2j}}{(2j)!} < \infty$$

for $0 < t < r$, implying in this range that

$$\int e^{t|x|} \, dF(x) = \int_{-\infty}^{\infty} \sum_{j=0}^{\infty} \frac{|x|^j t^j}{j!} \, dF(x) = \sum_{j=0}^{\infty} \frac{\beta_j t^j}{j!} < \infty, \tag{9}$$

so that (iii) $\Rightarrow$ (iv). Finally, if (iv) obtains, then, employing the language of random variables, (9) implies that

$$\mathrm{E} \sum_{j=0}^{\infty} \frac{|(itX)^j|}{j!} = \mathrm{E}\, e^{|tX|} < \infty \quad \text{for } |t| < r,$$

whence by dominated convergence for $|t| < r$

$$\varphi(t) = \mathrm{E}\, e^{itX} = \sum_{j=0}^{\infty} \mathrm{E} \frac{(itX)^j}{j!} = \sum_{j=0}^{\infty} \frac{\alpha_j (it)^j}{j!}. \qquad \square$$

Corollary 5. *A* c.f. φ *with* d.f. F *is entire* iff $\int_{-\infty}^{\infty} e^{r|x|} \, dF(x) < \infty$ *for all* $r > 0$.

For any complex number $z = x + iy$ define

$$\varphi(z) = \int_{-\infty}^{\infty} e^{izu} \, dF(u) \tag{10}$$

provided $\int_{-\infty}^{\infty} e^{-yu} \, dF(u) < \infty$.

The next proposition asserts that a c.f. φ is r-analytic iff $\varphi(z)$ is defined and analytic in the strip $|\mathscr{I}(z)| = |y| < r$ of the complex plane. Consequently, the statement that φ is r-analytic may also be interpreted as asserting that $\varphi(z)$ as defined by (10) is analytic in $|\mathscr{I}(z)| < r$.

Proposition 5. *A* c.f. φ *with* d.f. F *is r-analytic* iff $\varphi(z)$ *(in* (10)) *is an analytic function in* $|\mathscr{I}(z)| < r$, *and in this case for* $k = 0, 1, 2, \ldots,$

$$\varphi^{(k)}(z) = \sum_{j=k}^{\infty} \frac{i^j \alpha_j z^{j-k}}{(j-k)!}, \qquad |z| < r, \tag{11}$$

where α_j *is the* jth *moment of* F *and* $\varphi^{(0)} = \varphi$.

Proof. If φ is r-analytic and $C(F)$ is the continuity set of F, for $a\, b \in C(F)$ by uniform convergence of the exponential series

$$g(z) = \int_a^b e^{izx}\, dF(x) = \sum_{j=0}^{\infty} \frac{(iz)^j}{j!} \int_a^b x^j\, dF(x),$$

and since the latter series converges uniformly in $|z| \le M$ for all $0 < M < \infty$, $g(z)$ is entire and for $k = 0, 1, \ldots$

$$g^{(k)}(z) = \sum_{j=k}^{\infty} \frac{i^j z^{j-k}}{(j-k)!} \int_a^b x^j\, dF(x) = \int_a^b (ix)^k e^{izx}\, dF(x).$$

Choose $\pm a_n \in C(F)$ with $0 < a_n \uparrow \infty$, $n \ge 1$, and define

$$g_1(z) = \int_{-a_1}^{a_1} e^{izx}\, dF(x), \qquad g_n(z) = \int_{[a_{n-1} \le |x| < a_n]} e^{izx}\, dF(x), \qquad n > 1.$$

Then by the preceding, $g_n(z)$ is entire, $n \ge 1$, and for $|\mathscr{I}(z)| \le s < r$

$$\sum_{n=m}^{\infty} |g_n(z)| \le \int_{[|u| \ge a_{m-1}]} e^{|u \mathscr{I}(z)|}\, dF(u) \le \int_{[|u| \ge a_{m-1}]} e^{s|u|}\, dF(u) < \infty$$

by Proposition 4. Thus, $\sum_1^\infty g_n(z)$ converges uniformly in $|\mathscr{I}(z)| \le s < r$, whence (Titchmarsh, 1932, p. 95) $\varphi(z) = \int_{-\infty}^{\infty} e^{izu}\, dF(u) = \sum_1^\infty g_n(z)$ is analytic in $|\mathscr{I}(z)| < r$. Moreover, for $|\mathscr{I}(z)| < r$ and $k = 0, 1, \ldots,$ setting $a_0 = 0$,

$$\varphi^{(k)}(z) = \sum_{n=1}^{\infty} g_n^{(k)}(z) = \sum_{n=1}^{\infty} \sum_{j=k}^{\infty} \frac{i^k (iz)^{j-k}}{(j-k)!} \int_{[a_{n-1} \le |x| < a_n]} x^j\, dF(x)$$

$$= \sum_{n=1}^{\infty} \int_{[a_{n-1} \le |x| < a_n]} (ix)^k e^{izx}\, dF(x) = \mathrm{E}\,(iX)^k e^{izX}$$

and so, via (and in the notation of) Proposition 4, for $|z| \le s < r$

$$\sum_{n=1}^{\infty} \sum_{j=k}^{\infty} \left| \frac{i^k (iz)^{j-k}}{(j-k)!} \int_{[a_{n-1} \le |u| < a_n]} u^j \, dF(u) \right|$$

$$\le \sum_{j=k}^{\infty} \frac{s^{j-k}}{(j-k)!} \beta_j = \frac{d^k}{ds^k} \sum_{j=0}^{\infty} \frac{\beta_j s^j}{j!} < \infty,$$

whence

$$\varphi^{(k)}(z) = \sum_{j=k}^{\infty} \frac{i^k (iz)^{j-k}}{(j-k)!} \alpha_j, \qquad |z| < r.$$

Conversely, if $\varphi(z)$ is defined in $|\mathscr{I}(z)| < r$, then (iv) of Proposition 4 holds whence φ is r-analytic. □

Proposition 6. (i) *If F is a* d.f. *whose* c.f. *φ is r-analytic for some $r > 0$, then F is uniquely determined by its moment sequence $\{\alpha_j = \int x^j \, dF(x), j \ge 1\}$.*

(ii) *A* c.f. *φ with moment sequence $\{\alpha_j, j \ge 1\}$ is r-analytic for some $r > 0$* iff

$$\overline{\lim_{n\to\infty}} \frac{(\alpha_{2n})^{1/2n}}{2n} < \infty. \tag{12}$$

PROOF. (i) Denote F, φ by F_1, φ_1 and suppose for $j \ge 1$ that $\alpha_j = \int x^j \, dF_k(x)$, $k = 1, 2$, where F_2 is a d.f. with c.f. φ_2. By hypothesis and Proposition 4, φ_2 is r-analytic, whence by Proposition 5, $\varphi_k(z)$ is analytic in $|\mathscr{I}(z)| < r$, $k = 1, 2$, and

$$\varphi_1(z) = \sum_{j=0}^{\infty} \frac{\alpha_j (iz)^j}{j!} = \varphi_2(z), \qquad |z| < r.$$

Therefore (Titchmarsh, 1932, p. 88), $\varphi_1(z) \equiv \varphi_2(z)$ in $|\mathscr{I}(z)| < r$, whence $\varphi_1(t) = \varphi_2(t)$ for $-\infty < t < \infty$. Then $F_1 = F_2$ by Corollary 8.3.1.

(ii) By Stirling's formula (Section 2.3)

$$\frac{1}{n} \log n! = \frac{1}{n}\left[\left(n + \frac{1}{2}\right)\log n - n + C + o(1)\right] = \log n + O(1)$$

whence, if $b_n = (|\alpha_n|/n!)^{-1/n}$, necessarily $\log b_n = -\log(|\alpha_n|^{1/n}/n) + O(1)$. Since the radius of convergence of a power series $\sum c_n z^n$ is $\underline{\lim} |c_n|^{-1/n}$ (Titchmarsh, 1932, p. 213), (ii) follows from Proposition 4. Conversely, (12) and $|\alpha_{2n-1}|^{1/(2n-1)} \le \beta_{2n-1}^{1/(2n-1)} \le \alpha_{2n}^{1/2n}$ imply (11) for $k = 0$ and some $r > 0$, so that *a fortiori* (ii) of Proposition (4) holds □

On the other hand a d.f. may be uniquely determined by its moment sequence without having an analytic c.f. To verify this, it is necessary to invoke the celebrated Carleman criterion (Shohat and Tamarkin, 1943, p. 19) which asserts that a d.f. F is uniquely determined by its moment sequence $\{\alpha_j, j \ge 1\}$ if

$$\sum_{j=1}^{\infty} \frac{1}{(\alpha_{2j})^{1/2j}} = \infty. \tag{13}$$

EXAMPLE 1. Let X be a random variable with $P\{X = n\} = Ce^{-n/\log n}$, $n \geq 3$, where C is such that $P\{X \geq 3\} = 1$. Then for $x \geq 3$ and $m \geq 4$, set

$$h(x) = x^m e^{-x/\log x}.$$

Now

$$\frac{h'(x)}{h(x)} = \frac{m}{x} - \frac{1}{\log x} + \frac{1}{\log^2 x} = 0$$

iff $x = x_m$, where

$$m = \frac{x_m}{\log x_m}\left(1 - \frac{1}{\log x_m}\right) \leq \frac{x_m}{\log x_m}.$$

As $m \to \infty$, $x_m \to \infty$, and therefore $m \log x_m / x_m \to 1$, whence $(m/2)\log m \leq x_m \leq 2m \log m$ for $m \geq m_0$. Define

$$M_m = \max_{x \geq 3} x^m e^{-x/\log x} = x_m^m e^{-x_m/\log x_m}.$$

Now for $m \geq m_0$

$$\begin{aligned}\alpha_{2m} = C\sum_{n=3}^{\infty} n^{2m} e^{-n/\log n} &\leq CM_{2m+2}\sum_{n=3}^{\infty}\frac{1}{n^2} \\ &\leq CM_{2m+2} \leq C[(4m+4)\log(2m+2)]^{2m+2}e^{-2m-2},\end{aligned}$$

and therefore for all large m

$$(\alpha_{2m})^{1/2m} \leq 4C^{1/2m}[(2m+2)\log(2m+2)]^{1+(1/m)} \leq 8(2m+2)\log(2m+2).$$

Hence, $\sum_1^{\infty} (\alpha_{2j})^{-1/2j} = \infty$, and moreover $\overline{\lim}_{m\to\infty} (\alpha_{2m})^{1/2m}/2m = \infty$ since

$$(\alpha_{2m})^{1/2m} > (CM_{2m})^{1/2m} > e^{-1}\left(\frac{m}{2}\log m\right) \text{ for all large } m.$$

Thus φ is not r-analytic for any $r > 0$ by Proposition 6, whereas the Carleman criterion ensures that the distribution in question is uniquely determined by its moments. □

Proposition 7. *If $\varphi, \varphi_1, \varphi_2$ are* c.f.s *with $\varphi(t) = \varphi_1(t)\varphi_2(t)$ for all real t and φ is r_0-analytic, then φ_i is r_0-analytic, $i = 1, 2$, and $\varphi(z) = \varphi_1(z) \cdot \varphi_2(z)$ for $|z| < r_0$.*

PROOF. If F, F_1, F_2 are the corresponding d.f.s, by Proposition 4

$$\int_{-\infty}^{\infty} e^{r|x|}\, dF(x) < \infty$$

for $0 < r < r_0$, whence by Section 6.3, (11),

$$\int_{-\infty}^{\infty} e^{ru}\, dF(u) = \int_{-\infty}^{\infty} e^{rx}\, dF_1(x) \int_{-\infty}^{\infty} e^{ry}\, dF_2(y), \qquad |r| < r_0.$$

Since neither integral on the right vanishes, $\int_{-\infty}^{\infty} e^{rx}\, dF_i(x) < \infty$, $i = 1, 2$, for $|r| < r_0$, so that for $i = 1, 2$

$$\int_0^{\infty} e^{rx}\, dF_i(x) < \infty, \qquad 0 < r < r_0,$$

$$\int_{-\infty}^{0} e^{rx}\, dF_i(x) < \infty, \qquad -r_0 < r < 0,$$

whence $\int_{-\infty}^{\infty} e^{r|x|}\, dF_i(x) < \infty$ for $0 < r < r_0$ and $i = 1, 2$. Again invoking Proposition 4, φ_i is r_0-analytic, $i = 1, 2$. Finally, since $\varphi(z)$ and $\varphi_1(z) \cdot \varphi_2(z)$ are both analytic in $|z| < r_0$ and coincide on the real axis, they must coincide in $|z| < r_0$. □

Corollary 6. *If $\varphi, \varphi_1, \varphi_2$ are* c.f.s *with $\varphi(t) = \varphi_1(t) \cdot \varphi_2(t)$ for all real t and φ is an entire* c.f., *so is φ_i, $i = 1, 2$, and $\varphi(z) = \varphi_1(z) \cdot \varphi_2(z)$ for all complex z.*

Theorem 3 (Cramér–Lévy). *The family of normal distributions is factor closed.*

PROOF. It suffices to prove that if φ_0 is the c.f. of the standard normal d.f. and $e^{-t^2/2} = \varphi_0(t) = \varphi_1(t) \cdot \varphi_2(t)$, then the c.f. φ_j is normal (or degenerate), $j = 1, 2$. Now $\varphi_0(z)$ is an entire function, and so by Corollary 6, $\varphi_j(z)$ is likewise an entire function $j = 1, 2$. Thus, the corresponding d.f. F_j has a finite mean, and so replacing φ_j by $\exp[(-1)^j it\alpha]\varphi_j$, it may be supposed that F_j has mean zero, $j = 1, 2$. In the parlance of random variables, $X_0 = X_1 + X_2$ with X_1, X_2 independent and $\mathrm{E}\, X_j = 0$, $j = 0, 1, 2$. Then $\mathrm{E}\{X_0 | X_j\} = X_j$, $j = 1, 2$, whence, employing the conditional Jensen inequality, for $j = 1, 2$ and all complex z

$$\begin{aligned}
|\varphi_j(z)| &= |\mathrm{E}\, e^{izX_j}| \le \mathrm{E}\, e^{|zX_j|} = \mathrm{E}\, e^{|z\mathrm{E}\{X_0|X_j\}|} \\
&\le \mathrm{E}\, e^{|z|\mathrm{E}\{|X_0|\,|X_j\}} \le \mathrm{E}\, e^{|zX_0|} \\
&= \frac{2}{\sqrt{2\pi}} \int_0^{\infty} e^{|zx| - x^2/2}\, dx = \frac{2e^{|z|^2/2}}{\sqrt{2\pi}} \int_0^{\infty} e^{-(1/2)(x-|z|)^2}\, dx \\
&\le 2e^{|z|^2/2}. \qquad (14)
\end{aligned}$$

Again invoking Corollary 6, $e^{-z^2/2} = \varphi_0(z) = \varphi_1(z) \cdot \varphi_2(z)$ for all complex z. Thus, $\varphi_j(z)$ is a nonvanishing entire function, whence by the ensuing Lemma 2, $\varphi_j(z) = \exp\{g_j(z)\}$ with $g_j(z)$ an entire function $j = 1, 2$. Hence, recalling (14), for all complex z

$$\mathscr{R}\{g_j(z)\} = \log|\varphi_j(z)| \le 1 + \tfrac{1}{2}|z|^2.$$

If $g_j(z) = \sum_0^{\infty} a_{nj} z^n$, by Lemma 3 which follows

$$|a_{nj} r^n| \le 4(\tfrac{1}{2}r^2 + 1), \qquad 0 < r < \infty, n \ge 3,$$

implying $a_{nj} = 0$, $n \ge 3$, $j = 1, 2$. Thus, $g_j(z) = a_{0j} + a_{1j}z + a_{2j}z^2$. Since $\varphi_j(0) = 1$ and $\mathrm{E}\, X_j = 0$, necessarily $a_{0j} = 0 = a_{1j}$ for $j = 1, 2$. Finally,

$\mathrm{E}\,X_j^2 = -\varphi_j''(0) = -2a_{2j}$ implies $\varphi_j(t) = e^{-\sigma_j^2 t^2/2}$, where $\sigma_j^2 = \mathrm{E}\,X_j^2 \geq 0$ and $\sigma_1^2 + \sigma_2^2 = 1$. □

Lemma 2 (Titchmarsh, 1932, p. 247). *If $\varphi(z)$ is an entire, nonvanishing function, then $\varphi(z) = \exp\{g(z)\}$, where $g(z)$ is entire.*

PROOF. If $h(z) = \varphi'(z)/\varphi(z)$, then h is entire, whence $g(z) = \int_0^z h(w)dw$ is entire. Since

$$\frac{d}{dz}[\varphi(z)e^{-g(z)}] = e^{-g(z)}\left[\varphi'(z) - \varphi(z)\cdot\frac{\varphi'(z)}{\varphi(z)}\right] = 0$$

necessarily, $\varphi(z) = Ce^{g(z)} = e^{g(z)}$ upon noting that $g(0) = 0$. □

Lemma 3 (Titchmarsh, 1932, p. 176). *If $\varphi(z) = \sum_0^\infty a_n z^n$ is analytic in $|z| < r_0$ and $A(r) = \max_{|z|=r} \mathscr{R}\{\varphi(z)\}$, $0 < r < r_0$, then $|a_n|r^n \leq 4A^+(r) - 2\mathscr{R}\{\varphi(0)\}$, $n > 0$.*

PROOF. Set $z = re^{i\theta}$, $0 < r < r_0$, and $a_n = \alpha_n + i\beta_n$, $n \geq 0$, whence $\varphi(z) = \sum_0^\infty a_n z^n = u(r, \theta) + iv(r, \theta)$ say. Since

$$u(r, \theta) = \sum_{n=0}^{\infty} (\alpha_n \cos n\theta - \beta_n \sin n\theta)r^n$$

converges uniformly in θ, for $n > 0$,

$$\frac{1}{\pi}\int_0^{2\pi} u(r, \theta)\cos n\theta\, d\theta = \alpha_n r^n, \qquad \frac{1}{\pi}\int_0^{2\pi} u(r, \theta)\sin n\theta\, d\theta = -\beta_n r^n$$

and, recalling that $a_n = \alpha_n + i\beta_n$,

$$|a_n r^n| = \left|\frac{1}{\pi}\int_0^{2\pi} u(r, \theta)e^{-in\theta}\, d\theta\right| \leq \frac{1}{\pi}\int_0^{2\pi} |u(r, \theta)|\, d\theta, \qquad n > 0.$$

Now $\mathscr{R}\{\varphi(0)\} = \alpha_0 = (1/2\pi)\int_0^{2\pi} u(r, \theta)d\theta$, implying for $r > 0$ that $|a_n r^n| + 2\alpha_0 \leq 1/\pi \int_0^{2\pi} [u(r, \theta) + |u(r, \theta)|]\, d\theta \leq 4A^+(r)$, which is Lemma 3. □

A theorem of Raikov–Ottaviani asserts that the Poisson family is factor closed (to within translations), but the proof closely parallels that of Theorem 3 (see Exercise 12).

Theorem 4. *The family of binomial distributions is factor closed (to within translations).*

PROOF. Suppose that $\varphi_1(t)\cdot\varphi_2(t) = \varphi(t) = (pe^{it} + q)^n$, $0 < p < 1$, $q = 1 - p$, n = positive integer. Clearly, the "translations" $e^{i\alpha t}\varphi_1(t)$ and $e^{-i\alpha t}\varphi_2(t)$ are also factors, but the theorem asserts that for some choice of α these are both binomial. Since $1 = |\varphi(2\pi)| = \prod_{j=1}^2 |\varphi_j(2\pi)| \leq 1$, Proposition 2 guarantees that F_j is discrete with support $S_j \subset \{c_j' + k: k \text{ an integer}, c_j' \text{ real}\}$, $j = 1, 2$. Clearly (Exercise 8.2.5), S_j is bounded since the support of the binomial distribution is. Thus, without loss of generality, $\varphi_j(t) = e^{ic_j t} \sum_{k=0}^{n_j} p_{jk} e^{ikt}$ with c_j real, $0 \le p_{jk} \le 1$, $p_{j0} > 0$. Since φ is entire, setting $z = ir$, r real, in Corollary 6,

$$(pe^{-r} + q)^n = e^{-r(c_1+c_2)} \prod_{j=1}^{2} \sum_{k=0}^{n_j} p_{jk} e^{-rk},$$

whence $c_1 + c_2 = 0$ via $r \to \pm\infty$. Consequently,

$$(pw + q)^n = \sum_{k=0}^{n_1} p_{1k} w^k \cdot \sum_{k=0}^{n_2} p_{2k} w^k,$$

implying $n_1 + n_2 = n$ and $\sum_{k=0}^{n_j} p_{jk} w^k = (pw + q)^{n_j}$, $j = 1, 2$, which is tantamount to the conclusion of the theorem. □

EXERCISES 8.4

1. (i) Since the triangular distribution (see c.f. table) is the convolution of two uniforms, the c.f. of the former follows readily. Utilize Corollary 8.3.2 to obtain the c.f. of the inverse triangular distribution.
(ii) Use contour integration to obtain the normal c.f.

2. Prove that if a c.f. $\varphi(t) = 1 + w(t) + o(t^2)$, where $w(t) = -w(-t)$, then $\varphi(t) \equiv 1$. *Hint*: Apply Lemma 8.3.1.

3. If φ is a c.f., which of $e^{i \sin t}$, $(1 - c)/(1 - c\varphi(t))$, $0 < c < 1$, $|\varphi(t)|^2$, $\mathscr{R}\{\varphi(t)\}$, $1/(1 + t^4)$, $(1 - |t|^\alpha)I_{[|t|<1]}$, $1 > \alpha > 0$, are c.f.s?

4. Find the d.f. whose c.f. is $C \sum_{j=2}^{\infty} [\cos jt/(j^2 \log j)]$ and utilize this c.f. to show that the converse of Theorem 1 is false for odd derivatives (e.g., φ').

5. A family of d.f.s $\mathscr{F}_0 = \{F(x; \lambda), \lambda \in \Lambda\}$ with Λ an Abelian semigroup under addition, is called reproductive or additively closed if $F(x; \lambda_1) * F(x; \lambda_2) = F(x; \lambda_1 + \lambda_2)$ for all $\lambda_j \in \Lambda$, $j = 1, 2$. If $\Lambda = \{0, 1, 2, \ldots\}$, the corresponding family of c.f.s $\varphi(t; \lambda) = [\varphi(t)]^\lambda$ for some c.f. φ. Show that if $\varphi(t)$ and $1/\varphi(t)$ are both c.f.s, then φ is degenerate. Hence, if $\Lambda \subset (-\infty, \infty)$, usually $\Lambda \subset [0, \infty)$.

6. Prove that the convolution of 2 singular d.f.s may be absolutely continuous. *Hint*:
$$\frac{\sin t}{t} = \prod_1^\infty \cos\frac{t}{2^{2j-1}} \cdot \prod_1^\infty \cos\frac{t}{2^{2j}}$$

7. Show that
$$\varphi_n(t) = \frac{e^{it} - \sum_{j=0}^{n-1}[(it)^j/j!]}{(it)^n/n!}$$
is a c.f., $n \ge 1$.

8. If $\{X_n, n \ge 1\}$ are independent r.v.s uniformly distributed on $[-\alpha_n, \alpha_n]$, $n \ge 1$ and $s_n^2 = \sum_1^n \alpha_1^2 \to \infty$, find the limit d.f. of $(1/s_n)\sum_1^n X_i$ as $n \to \infty$.

9. Prove that if F is absolutely continuous with c.f. φ, then $\lim_{|t|\to\infty} \varphi(t) = 0$. This is the **Riemann–Lebesgue lemma**. *Hint*: Suppose initially that density vanishes outside some finite interval and in general approximate. Show that if F has an absolutely continuous component, then $\overline{\lim}_{|t|\to\infty} |\varphi(t)| < 1$.

10. If $F_j, j = 1, 2, 3$, are d.f.s with $F_1 * F_3 = F_2 * F_3$, it does not follow that $F_1 = F_2$. *Hint*: If φ is the c.f. of a r.v. X with $P\{X = 0\} = \frac{1}{2}$, $P\{X = \pm k\pi\} = 2/(k^2\pi^2)\}$, $k = 1, 3, 5, \ldots$, then φ is a periodic c.f. coinciding with $\varphi_1(t) = (1 - |t|)I_{[|t| \le 1]}$ in $[-1, 1]$.

11. There exist uncountably many absolutely continuous d.f.s with support $(0, \infty)$ and identical moments v_k, $k = 1, 2, \ldots$. *Hint*: For $0 < \alpha < \frac{1}{2}$, $r > 0$, and a nonnegative integer k, set $c = (k + 1)/\alpha$, $b = r + is$ in the well-known formula $b^{-c}\Gamma(c) = \int_0^\infty y^{c-1}e^{-by}\,dy$. Convert this to a similar example with support $(-\infty, \infty)$.

12. Prove the Raikov–Ottaviani theorem that the family of Poisson d.f.s is factor closed. *Hint*: It may be supposed, without loss of generality, that the support S_j (of any factor F_j) $\subset \{0, 1, 2, \ldots\}$. Rather than c.f.s, utilize the corresponding probability generating functions $\psi(w) = e^{\lambda(w-1)}$, $\psi_j(w) = E\, w^{X_j} = \sum_{i=0}^{\infty} P\{X_j = i\}w^i$, $j = 1, 2$.

13. If c.f.s φ_1, φ_2 satisfy $\prod_{j=1}^2 [\varphi_j(t)]^{\alpha_j} = (pe^{it} + q)^n$, where $\alpha_j > 0$, $j = 1, 2$, then $\varphi_j(t) = e^{itc_j}(pe^{it} + q)^{n_j}$ with n_1, n_2 positive integers such that $\alpha_1 n_1 + \alpha_2 n_2 = n$; also $c_1\alpha_1 + c_2\alpha_2 = 0$.

14. If $1 - \varphi(t) = O(|t|^\alpha)$ as $t \to 0$ for some α in $(0, 2]$, then $P\{|X| > c\} = O(c^{-\alpha})$ as $c \to \infty$. *Hint*: Integrate $\int_{[|x|>c]} (1 - \cos tx)dF(x) \le kt^\alpha$ over $(0, c^{-1})$ or use Lemma 8.3.1.

15. Prove that

$$\frac{2}{\pi}\int_{-\infty}^{\infty} \frac{1 - \mathscr{R}\{\varphi(t)\}}{t^2}\,dt = \int_{-\infty}^{\infty} |x|\,dF(x).$$

16. Show that a c.f. is **positive definite** i.e., $\sum_{j,k=1}^n \varphi(t_j - t_k)\rho_j\bar{\rho}_k \ge 0$ for any $n \ge 1$, real $t_1, \ldots, t_n$ and complex $\rho_1, \ldots, \rho_n$. Conversely, a continuous, positive definite function φ on $(-\infty, \infty)$ with $\varphi(0) = 1$ is a c.f. This is **Bochner's theorem**.

8.5 Remarks on k-Dimensional Distribution Functions and Characteristic Functions

Multidimensional d.f.s (Section 6.3) lack the simplicity of their one-dimensional counterparts and it is frequently easier to deal with the corresponding measure or c.f. The c.f. of a random vector $X = (X_1, \ldots, X_k)$ on a probability space $(\Omega, \mathscr{F}, P)$ is defined by

$$\varphi(t) = \varphi(t_1, \ldots, t_k) = E\exp\left(i\sum_1^k t_j X_j\right) \tag{1}$$

and is also expressible as a Fourier–Stieltjes transform

$$\varphi(t_1, \ldots, t_k) = \int \exp\left(i\sum_1^k t_j x_j\right) dF(x_1, \ldots, x_k),$$

where F is the d.f. of X.

It is easy to verify that a joint d.f. $F(x_1, \ldots, x_k)$ is continuous at a point $(x_1, \ldots, x_k)$ if its **marginal d.f.**

$$F_j(y) = \lim_{\substack{y_i \to \infty \\ i \neq j}} F(y_i, \ldots, y_{j-1}, y, y_{j+1}, \ldots, y_k)$$

is continuous at $y = x_j$ for each $j = 1, 2, \ldots, k$. However, in contradistinction to the case $k = 1$, F may be discontinuous even though the corresponding measure $F\{\cdot\} = P\,X^{-1}$ assigns probability zero to every point of R^k. For example F may allocate probability one to some hyperplane or hypersurface of R^m, $1 \leq m < k$, without assigning positive probability to any point therein (see Exercise 1).

For any bounded increasing function F on R^k, define

$$C(F) = \{x = (x_1, \ldots, x_k) : F_j(x_j+) = F_j(x_j-), 1 \leq j \leq k\},$$

where F_j is the "marginal function" of the prior paragraph. As noted above, $C(F)$ need not coincide with the set of continuity points of F when $k > 1$.

Definition. A sequence $\{F_n, n \geq 1\}$ of d.f.s on R^k converges weakly to a function F on R^k, denoted by $F_n \xrightarrow{w} F$, if $\lim F_n(x) = F(x)$ for all $x \in C(F)$. If, moreover, F is a d.f., then F_n is said to converge completely to F, denoted by $F_n \xrightarrow{c} F$.

A straightforward generalization of Theorem 8.3.1 yields

Theorem 1. *If $X = (X_1, \ldots, X_k)$ is a random vector with* c.f. *φ and* d.f. *F, then*

$$\lim_{c \to \infty} \frac{1}{(2\pi)^k} \int_{-c}^{c} \cdots \int_{-c}^{c} \prod_{j=1}^{k} \frac{e^{-it_j a_j} - e^{-it_j b_j}}{it_j} \varphi(t_1, \ldots, t_k) dt_1 \cdots dt_k$$

$$= E \prod_{j=1}^{k} [\tfrac{1}{2} I_{[X_j = a_j \text{ or } b_j]} + I_{[a_j < X_j < b_j]}]$$

and the right side reduces to $P\{a_j < X_j < b_j, 1 \leq j \leq k\}$ *when $a = (a_1, \ldots, a_k)$ and $b = (b_1, \ldots, b_k)$ are in $C(F)$.*

The reformulation and proof of Theorem 8.3.2 for R^k are immediate. The statement of the Lévy continuity theorem (Theorem 8.3.3) carries over verbatim to R^k and the proof generalizes readily.

It is worthy of note that independence may be characterized in terms of c.f.s as well as d.f.s.

Theorem 2. *Random variables $X_1, \ldots, X_k$ on some probability space are independent iff their joint* c.f. *is the product of their marginal* c.f.s, *i.e.,*

$$\varphi_{X_1, \ldots, X_k}(t_1, \ldots, t_k) = \prod_{j=1}^{k} \varphi_{X_j}(t_j). \tag{2}$$

PROOF. If $X_1, \ldots, X_k$ are independent, by Theorem 4.3.3

$$\varphi_{X_1,\ldots,X_k}(t_1,\ldots,t_k) = \mathrm{E}\prod_{j=1}^{k} e^{it_jX_j} = \prod_{j=1}^{k}\mathrm{E}\, e^{it_jX_j} = \prod_{j=1}^{k}\varphi_{X_j}(t_j).$$

Conversely, since the c.f. of the product measure $F_1 \times F_2 \times \cdots \times F_k$ is

$$\int \exp\left(i\sum_1^k t_jx_j\right) d(F_1 \times \cdots \times F_k)$$
$$= \int \cdots \int \exp\left(i\sum_1^k t_jx_j\right) dF_1(x_1)\cdots dF_k(x_k)$$
$$= \prod_{j=1}^{k}\int e^{it_jx_j}\, dF_j(x_j) = \prod_{j=1}^{k}\varphi_{X_j}(t_j)$$

via Fubini's theorem, if (2) holds, the c.f.s of $F_{X_1,\ldots,X_k}$ and the product measure $F_1 \times \cdots \times F_k$ are identical. But then by the uniqueness theorem for d.f.s and c.f.s on R^k so are the d.f.s, that is,

$$F_{X_1,\ldots,X_k} = \prod_{j=1}^{k} F_{X_j},$$

which ensures independence. □

The question may be posed as to whether the collection of one-dimensional d.f.s of $cX' = \sum_{j=1}^k c_jX_j$ for all choices of the constant vector $c = (c_1, \ldots, c_k)$ determines a unique d.f. for $X = (X_1, \ldots, X_k)$. If the assignment of distributions to cX' is compatible with the existence of a joint distribution, the latter is necessarily unique since, denoting the joint c.f. of X by $\varphi_X(t)$, for any scalar u in $(-\infty, \infty)$

$$\varphi_X(ut) = \mathrm{E}\, e^{i(ut)X'} = \mathrm{E}\, e^{iu(tX')} = \varphi_{tX'}(u),$$

and so, setting $u = 1$, the family of univariate c.f.s on the right determines the multivariate c.f. on the left.

EXERCISES 8.5

1. Let $F(x_1, x_2)$ be the d.f. corresponding to a uniform density over the interval (0, 1) of the x_2 axis of R^2. Verify that (i) F is discontinuous at all points $(0, x_2)$ with $x_2 > 0$, (ii) the marginal d.f. $F_1(x_1)$ is discontinuous iff $x_1 = 0$, (iii) $F_2(x_2)$ is continuous. Note that $F\{\cdot\}$ assigns 0 probability to all points of R^2. Construct an F for which $C(F) \neq$ discontinuity set of F.
2. If φ is a c.f. on R^1 with d.f. F, what is the d.f. corresponding to $\psi(t_1, \ldots, t_k) = \varphi(\sum_1^k t_i)$?
3. Prove the d.f. $F(x)$ is continuous at $x = (x_1, \ldots, x_2)$ if $x \in C(F)$. Construct a discontinuous density f on R^2 with continuous marginal densities f_1 and f_2.

4. If F_i, $1 \le i \le k$, are d.f.s on R^1, show that for any α in $[-1, 1]$, $F(x_1, \ldots, x_k) = [1 + \alpha \prod_1^k (1 - F_i(x_i))] \prod_{i=1}^k F_i(x_i)$ is a d.f. with the given marginal d.f.s.

5. Prove that $X = (X_1, \ldots, X_k)$ has a multivariate normal distribution with mean vector $\theta = (\theta_1, \ldots, \theta_k)$ and covariance matrix $\Sigma = \{\sigma_{ij}\}$ iff every linear combination $cX' = \sum_1^k c_j X_j$ has a normal distribution on R^1 with mean $c\mu'$ and variance $c\Sigma c'$.

6. If $X_n = (X_{n1}, \ldots, X_{nk})$, $n \ge 1$, is a sequence of random vectors for which every linear combination $cX_n' \xrightarrow{d} N_{c\mu', c\Sigma c'}$, where $N_{\mu, \alpha}$ is a fictitious normal r.v. with mean μ and variance α, prove that the d.f. of X_n converges to the normal d.f. on R^k with mean vector μ and covariance matrix Σ.

7. Prove the Cramér–Lévy theorem (Theorem 8.4.3) in R^k. *Hint*: Use the result for $k = 1$.

8. Generalize Theorem 8.4.4 to the multinomial distribution.

9. Prove Theorem 8.5.1 and deduce the one-to-one correspondence between d.f.s and c.f.s on R^k.

10. Prove the k-dimensional analogue of the continuity theorem (Theorem 8.3.3).

11. Verify that if $F_n \xrightarrow{c} F$, the marginal d.f.s $F_{n,j} \xrightarrow{c} F_j$, $1 \le j \le k$.

12. Let the random vectors $X_n \xrightarrow{d} X_0$, where $X_0 = (X_{01}, \ldots, X_{0k})$ is a possibly fictitious random vector with d.f. F. If $\{Y_n, n \ge 0\}$ are k-dimensional random vectors whose ith component is $g_i(X_{n,1}, \ldots, X_{n,k})$, $1 \le i \le k$, $n \ge 0$, where $\{g_i, 1 \le i \le k\}$ are continuous functions on R^k, then $Y_n \xrightarrow{d} Y_0$.

References

H. E. Bray, "Elementary properties of the Stieltjes integral," *Ann. Math.* **20** (1919), 177–186.

F. P. Cantelli, "Una teoria astratta del calcola delle probabilitá," *Ist. Ital. Attuari* **3** (1932).

H. Chernoff. "Large sample theory: Parametric case, *Ann. Math. Stat.* **27** (1956), 1–22.

Y. S. Chow and H. Robbins, "On optimal stopping rules," *Z. Wahr.* **2** (1963), 33–49.

K. L. Chung, *A Course in Probability Theory*, Harcourt Brace, New York, 1968; 2nd ed., Academic Press, New York, 1974.

H. Cramér, "Über eine Eigenschaft der normalen Verteilungsfunktion," *Math. Z.* **41** (1936), 405–414.

H. Cramér, *Mathematical Methods of Statistics*, Princeton Univ. Press, Princeton, 1946.

H. Cramér, *Random Variables and Probability Distributions*, Cambridge Tracts Math. No. 36, Cambridge Univ. Press, London, 1937; 3rd ed., 1970.

J. L. Doob, *Stochastic Processes*, Wiley, New York, 1953.

W. Feller, *An Introduction to Probability Theory and Its Applications*, Vol. 2, Wiley, New York, 1966.

M. Fréchet and J. Shohat, "A proof of the generalized second limit theorem in the theory of probability," *Trans. Amer. Math. Soc.* **33** (1931).

J. Glivenko, *Stieltjes Integral*, 1936 [in Russian].

B. V. Gnedenko and A. N. Kolmogorov, *Limit Distributions for Sums of Independent Random Variables* (K. L. Chung, translator), Addison–Wesley, Reading, Mass., 1954.

G. H. Hardy, *A course of Pure Mathematics*, 10th ed., Cambridge Univ. Press, New York, 1952.

E. Helly, "Uber lineare Funktionaloperationen," *Sitz. Nat. Kais. Akad. Wiss.* **121**, IIa (1921), 265–277.

P. Lévy, *Calcul des probabilités*, Gauthier–Villars, Paris, 1925.

P. Lévy, *Théorie de l'addition des variables aléatoires*, Gauthier–Villars, Paris, 1937; 2nd ed., 1954.

M. Loève, *Probability Theory*, 3rd ed., Van Nostrand, Princeton, 1963; 4th ed., Springer-Verlag, Berlin and New York, 1977–1978.

E. Lukacs, *Characteristic Functions*, 2nd ed., Hoffner, New York, 1970.

G. Polya, "Remarks on characteristic functions," *Proc. 1st Berkeley Symp. Stat. and Prob., 1949*, 115–123.

D. A. Raikov, "On the decomposition of Gauss and Poisson laws," *Izv. Akad. Nauk, USSR* (*Ser. Mat.*) **2** (1938a), 91–124 [in Russian].

D. A. Raikov, "Un théorème de la théorie des fonctions caracteristiques analytiques," *Izvest. Fak. Mat. Mek. Univ. Tomsk NII* **2** (1938b), 8–11.

H. Robbins, "Mixture of distributions," *Ann. Math. Stat.* **19** (1948), 360–369.

S. Saks, *Theory of the Integral* (L. C. Young, translator), Stechert–Hafner, New York, 1937.

N. A. Sapogov, "The stability problem for a theorem of Cramer," *Izv. Akad. Nauk, USSR* (*ser. Mat.*) **15** (1951), 205–218. [See also selected translations, *Math. Stat. and Prob.* **1**, 41–53, Amer. Math. Soc.]

H. Scheffé, "A useful convergence theorem for probability distributions," *Ann. Math. Stat.* **18** (1947), 434–438.

J. A. Shohat and J. D. Tamarkin, "The problem of moments," Math. Survey No. 1, Amer. Math. Soc., New York, 1943.

E. Slutsky, "Uber stochastiche Asymptoten und Grenzwerte," Metron **5** (1925), 1–90.

H. Teicher, "On the factorization of distributions," Ph.D. Thesis, 1950. [See also *Ann. Math. Stat.* **25** (1954), 769–774.]

H. Teicher, "Sur les puissances de fonctions caracteristiques," *Comptes Rendus* **246** (1958), 694–696.

H. Teicher, "On the mixture of distributions," *Ann. Math. Stat.* **31** (1960), 55–73.

H. Teicher, "Identifiability of mixtures," *Ann. Math. Stat.* **32** (1961), 244–248.

E. C. Titchmarsh, *The Theory of Functions*, Oxford Univ. Press, 1932; 2nd ed., 1939.

9 Central Limit Theorems

Central limit theorems have played a paramount role in probability theory starting—in the case of independent random variables—with the DeMoivre–Laplace version and culminating with that of Lindeberg–Feller. The term "central" refers to the pervasive, although nonunique, role of the normal distribution as a limit of d.f.s of normalized sums of (classically independent) random variables. Central limit theorems also govern various classes of dependent random variables and the cases of martingales and interchangeable random variables will be considered.

9.1 Independent Components

Consider at the outset a sequence $\{X_n, n \geq 1\}$ of independent random variables with finite variances $\{\sigma_n^2, n \geq 1\}$. No generality is lost and much convenience is gained by supposing (as will be done) that $\mathrm{E}\, X_n = 0, n \geq 1$. Set

$$S_n = \sum_{i=1}^{n} X_i, \qquad s_n^2 = \mathrm{E}\, S_n^2 = \sum_{i=1}^{n} \sigma_i^2, \qquad n \geq 1. \tag{1}$$

The problem, of analytic rather than measure-theoretic character, is to determine when S_n, suitably normalized (to S_n/s_n), converges in law to the standard normal distribution. The solution is linked to the following

Definition. Random variables $\{X_n, n \geq 1\}$ with $\mathrm{E}\, X_n = 0$, $\mathrm{E}\, X_n^2 = \sigma_n^2 < \infty$, and d.f.s F_n are said to obey the **Lindeberg condition** if $s_n^2 = \sum_1^n \sigma_i^2 > 0$ for some n and

$$\sum_{j=1}^{n} \int_{[|x| > \varepsilon s_j]} x^2 \, dF_j(x) = o(s_n^2), \quad \text{all } \varepsilon > 0. \tag{2}$$

Condition (2) requires that $s_n \to \infty$ and is equivalent to the classical form of the Lindeberg condition, viz.,

$$\sum_{j=1}^{n} \int_{[|x| > \varepsilon s_n]} x^2 \, dF_j(x) = o(s_n^2) \quad \text{for all } \varepsilon > 0. \tag{2'}$$

Monotonicity of s_n yields (2) $\Rightarrow$ (2'), while the reverse implication follows by noting that for all $\varepsilon > 0$ and arbitrarily small $\delta > 0$

$$s_n^{-2} \sum_{j=1}^{n} \int_{[|x| > \varepsilon s_j]} x^2 \, dF_j(x) = s_n^{-2} \left[\sum_{j: s_j \le \delta s_n} + \sum_{j: s_j > \delta s_n} \right]$$

$$\le \delta^2 + s_n^{-2} \sum_{j=1}^{n} \int_{[|x| > \varepsilon \delta s_n]} x^2 \, dF_j \to \delta^2.$$

Despite their equivalence, (2) is structurally simpler than (2'). Moreover, (2) or (2') entails

$$\max_{1 \le j \le n} \frac{\sigma_j^2}{s_n^2} = o(1), \tag{3}$$

since for arbitrary $\varepsilon > 0$

$$\max_j \sigma_j^2 s_n^{-2} \le \max_j s_n^{-2} \left[\varepsilon^2 s_n^2 + \int_{[|x| > \varepsilon s_n]} x^2 \, dF_j \right] = \varepsilon^2 + o(1).$$

If $\{X_n\}$ are independent with $\mathrm{E}\, X_n = 0$, $\mathrm{E}\, X_n^2 = \sigma_n^2$, either $s_n^2 = \sum_1^n \sigma_i^2 \to \infty$ or $s_n^2 \uparrow s^2 < \infty$. The latter contingency is devoid of interest in the current context since, if N_{μ, σ^2} denotes a fictitious normal random variable with mean μ, variance σ^2, and both $S_n/s_n \xrightarrow{d} N_{0,1}$ and $s_n \uparrow s$, then $S_n \xrightarrow{d} N_{0, s^2}$ by Corollary 8.1.3. In terms of the characteristic function of X_j, say $\varphi_j(t)$, this entails

$$\varphi_1(t) \cdot \prod_{j=2}^{\infty} \varphi_j(t) = \lim_{n \to \infty} \prod_{j=1}^{n} \varphi_j(t) = \exp\left\{ -\frac{s^2 t^2}{2} \right\}.$$

By the Cramér–Lévy theorem (Theorem 8.4.3) both of the c.f.s $\varphi_1(t)$ and $\prod_{j=2}^{\infty} \varphi_j(t)$ must be normal. Isolating $\varphi_2(t)$ analogously, it follows that X_2 and eventually all X_n are normally distributed.

Theorem 1 (Lindeberg). *If $\{X_n, n \ge 1\}$ are independent random variables with zero means, variances $\{\sigma_n^2\}$, and distribution functions $\{F_n\}$ satisfying* (2), *then the distribution functions of the normalized sums S_n/s_n tend to the standard normal. Conversely* (Feller), *convergence of these distribution functions to the standard normal and*

$$\frac{\sigma_n}{s_n} \to 0, \qquad s_n \to \infty \tag{4}$$

imply (2).

PROOF. Let t, ε be fixed but arbitrary real numbers, the latter being positive. Set

$$Y_j(t) = e^{itX_j} - 1 - itX_j + \frac{t^2 X_j^2}{2}, \quad a_j(t) = e^{-\sigma_j^2 t^2/2} - 1 + \frac{\sigma_j^2 t^2}{2}$$

and note that $|Y_j(t)| \le \min[t^2 X_j^2, (1/6)|tX_j|^3]$ by Lemma 8.4.1 with $\delta = 1$ and $n = 1, 2$. Consequently, recalling that $\mathrm{E}\, X_j = 0$,

$$|\mathrm{E}\, e^{itX_j/s_n} - e^{-\sigma_j^2 t^2/2s_n^2}| = \left| \mathrm{E}\, Y_j\left(\frac{t}{s_n}\right) - a_j\left(\frac{t}{s_n}\right) \right|$$

$$\le \mathrm{E}\left[\frac{t^2 X_j^2}{s_n^2} I_{[|X_j| > \varepsilon s_n]} + \left| \frac{tX_j}{s_n} \right|^3 I_{[|X_j| \le \varepsilon s_n]} \right] + \frac{\sigma_j^4 t^4}{8 s_n^4}. \tag{5}$$

Thus, for $1 \le j \le n$, setting $S_0 = 0 = s_0$ and utilizing independence and (5),

$$\left| \mathrm{E} \exp\left\{ it\left(\frac{S_j}{s_n}\right) + \frac{s_j^2 t^2}{2 s_n^2} \right\} - \mathrm{E} \exp\left\{ it\left(\frac{S_{j-1}}{s_n}\right) + \frac{s_{j-1}^2 t^2}{2 s_n^2} \right\} \right|$$

$$= \left| \mathrm{E} \exp\left\{ it\left(\frac{S_{j-1}}{s_n}\right) + \frac{s_j^2 t^2}{2 s_n^2} \right\} \mathrm{E}\left(\exp\left\{ \frac{itX_j}{s_n} \right\} - \exp\left\{ -\frac{\sigma_j^2 t^2}{2 s_n^2} \right\} \right) \right|$$

$$\le e^{t^2/2} \left| \mathrm{E} \exp\left\{ \frac{itX_j}{s_n} \right\} - \exp\left\{ -\frac{\sigma_j^2 t^2}{2 s_n^2} \right\} \right|$$

$$\le e^{t^2/2}\, \mathrm{E}\left[\frac{t^2 X_j^2}{s_n^2} I_{[|X_j| > \varepsilon s_n]} + \varepsilon |t|^3 \frac{X_j^2}{s_n^2} I_{[|X_j| \le \varepsilon s_n]} + \frac{t^4 \sigma_j^2}{s_n^2} \max_{1 \le j \le n} \frac{\sigma_j^2}{s_n^2} \right] \tag{6}$$

Finally, via (6) and (3)

$$|\mathrm{E}\, e^{itS_n/s_n} - e^{-t^2/2}|$$

$$= \left| e^{-t^2/2} \sum_{j=1}^{n} \mathrm{E}\left(\exp\left\{ it\left(\frac{S_j}{s_n}\right) + \frac{s_j^2 t^2}{2 s_n^2} \right\} - \exp\left\{ it\left(\frac{S_{j-1}}{s_n}\right) + \frac{s_{j-1}^2 t^2}{2 s_n^2} \right\} \right) \right|$$

$$\le \frac{t^2}{s_n^2} \sum_{j=1}^{n} \mathrm{E}\, X_j^2 I_{[|X_j| > \varepsilon s_n]} + \varepsilon |t|^3 + o(1), \tag{7}$$

and, since ε is arbitrary, the conclusion follows from the hypothesis and Theorem 8.3.3.

Conversely, (4) entails (3) via

$$\max_{1 \le j \le n} \frac{\sigma_j}{s_n} \le \max_{1 \le j \le m} \frac{\sigma_j}{s_n} + \max_{m < j \le n} \frac{\sigma_j}{s_j} \to 0$$

as first n and then $m \to \infty$. This, in turn, yields as $n \to \infty$

$$\sum_{j=1}^{n}\left|\log \varphi_j\left(\frac{t}{s_n}\right) - \left(\varphi_j\left(\frac{t}{s_n}\right) - 1\right)\right| \le \sum_{j=1}^{n}\left|\varphi_j\left(\frac{t}{s_n}\right) - 1\right|^2$$

$$= o(1)\sum_{j=1}^{n}\left|\varphi_j\left(\frac{t}{s_n}\right) - 1\right| = o(1) \quad (8)$$

since Lemma 8.4.1 guarantees $|\varphi_j(t/s_n) - 1| \le t^2\sigma_j^2/2s_n^2$, which in conjunction with (3) implies

$$\max_{1\le j\le n}\left|\varphi_j\left(\frac{t}{s_n}\right) - 1\right| = o(1), \qquad \sum_{j=1}^{n}\left|\varphi_j\left(\frac{t}{s_n}\right) - 1\right| \le \frac{t^2}{2}.$$

The hypothesis ensures $\prod_1^n \varphi_j(t/s_n) \to e^{-t^2/2}$, whence $\sum_{j=1}^n \log \varphi_j(t/s_n) \to -t^2/2$ and, taking cognizance of (8), $-t^2/2 = \sum_{j=1}^n [\varphi_j(t/s_n) - 1] + o(1)$ or

$$\frac{t^2}{2} - \sum_{j=1}^{n}\int_{[|x|\le \varepsilon s_n]}\left(1 - \cos\frac{tx}{s_n}\right)dF_j(x)$$

$$= \sum_{j=1}^{n}\int_{[|x|> \varepsilon s_n]}\left(1 - \cos\frac{tx}{s_n}\right)dF_j(x) + o(1).$$

Since the integrand on the right $\le 2 \le 2x^2/\varepsilon^2 s_n^2$, while that on the left $\le t^2x^2/2s_n^2$, it follows that as $n \to \infty$

$$1 - \frac{1}{s_n^2}\sum_{j=1}^{n}\int_{[|x|\le \varepsilon s_n]} x^2\, dF_j(x) \le \frac{4}{t^2\varepsilon^2} + o(1),$$

$$\varliminf_{n\to\infty} s_n^{-2}\sum_{j=1}^{n}\int_{[|x|\le \varepsilon s_n]} x^2\, dF_j(x) \ge 1 - \frac{4}{t^2\varepsilon^2} \xrightarrow{t\to\infty} 1,$$

which is tantamount to (2′). □

Corollary 1 (Liapounov). *If* $\{X_n\}$ *are independent with* $\mathrm{E}\, X_n = 0$ *and* $\sum_{j=1}^n \mathrm{E}|X_j|^{2+\delta} = o(s_n^{2+\delta})$ *for some* $\delta > 0$, *then*

$$\lim_{n\to\infty} \mathrm{P}\left\{\frac{S_n}{s_n} < x\right\} = \frac{1}{\sqrt{2\pi}}\int_{-\infty}^{x} e^{-u^2/2}\, du. \quad (9)$$

Proof. Take $q = 2$ and $r = 2 + \delta$ in

$$0 \le \frac{1}{s_n^q}\sum_{j=1}^{n}\int_{[|x|>\varepsilon s_n]} |x|^q\, dF_j(x)$$

$$\le \frac{1}{\varepsilon^{r-q}s_n^r}\sum_{j=1}^{n}\int_{[|x|>\varepsilon s_n]} |x|^r\, dF_j = o(1), \qquad 0 < q < r. \quad □ \quad (10)$$

Corollary 2. *If* $\{X_n\}$ *are* i.i.d. *with* $\mathrm{E}\, X_n = \mu$, $\mathrm{E}(X_n - \mu)^2 = \sigma^2 \in (0, \infty)$, *then*

$$\lim_{n\to\infty} \mathrm{P}\left\{\frac{S_n - n\mu}{\sigma\sqrt{n}} < x\right\} = \frac{1}{\sqrt{2\pi}}\int_{-\infty}^{x} e^{-u^2/2}\, du. \tag{11}$$

If the $\{X_n\}$ have finite moments of order k, what additional assumptions will ensure moment convergence, i.e., that for any positive integer $k > 2$

$$\lim_{n\to\infty} \mathrm{E}\left(\frac{S_n}{s_n}\right)^k = \frac{1}{\sqrt{2\pi}}\int_{-\infty}^{\infty} x^k e^{-x^2/2}\, dx. \tag{12}$$

The answer is tied to the next

Definition. Random variables $\{X_n, n \geq 1\}$ with $\mathrm{E}\, X_n = 0$, $\mathrm{E}\, X_n^2 = \sigma_n^2$ are said to obey a Lindeberg condition of order $r > 0$ if

$$\sum_{j=1}^{n} \int_{[|x| > \varepsilon s_j]} |x|^r\, dF_j(x) = o(s_n^r), \quad \text{all } \varepsilon > 0. \tag{13}$$

For $r = 2$, this is just the ordinary Lindeberg condition. Surprisingly, for $r > 2$, (13) is equivalent to

$$\sum_{j=1}^{n} \mathrm{E}|X_j|^r = o(s_n^r) \tag{13'}$$

and also to (13″) (defined as (13) with s_j replaced by s_n). Clearly (13′) ⇒ (13) ⇒ (13″) and so to establish equivalence for $r > 2$ it suffices to verify that (13″) ⇒ (13′). The latter follows from the fact that for $r > 2$ and all $\varepsilon > 0$

$$\begin{aligned}\sum_{j=1}^{n} \mathrm{E}|X_j|^r &\leq \sum_{j=1}^{n} \mathrm{E}\{(\varepsilon s_n)^{r-2} X_j^2 I_{[|X_j| \leq \varepsilon s_n]} + |X_j|^r I_{[|X_j| > \varepsilon s_n]}\} \\ &\leq \varepsilon^{r-2} s_n^r + o(s_n^r).\end{aligned}$$

According to (10), a Lindeberg condition of order $r > 2$ implies that of order q, for all q in $[2, r]$; in particular, a Lindeberg condition of order $r > 2$ implies the central limit theorem (9).

Theorem 2. *Let* $\{X_n, n \geq 1\}$ *be independent random variables with* $\mathrm{E}\, X_n = 0$, $\mathrm{E}\, X_n^2 = \sigma_n^2$. *If* $\{X_n\}$ *satisfies a Lindeberg condition of order r for some integer* $r \geq 2$, *then* (12) *obtains for* $k = 1, 2, \ldots, r$.

Corollary 3. *If* $\{X_n, n \geq 1\}$ *are* i.i.d. *with* $\mathrm{E}\, X_1 = 0$, $\mathrm{E}\, X_1^{2k} \in (0, \infty)$ *for some positive integer k, then* $\lim_{n\to\infty} \mathrm{E}(S_n/s_n)^{2j} = (2j)!/2^j j!$, $\mathrm{E}(S_n/s_n)^{2j-1} = o(1)$, $1 \leq j \leq k$.

Proof of Theorem 2. Since $\mathrm{E}(S_n/s_n)^2 = 1$, $n \geq 1$, the theorem is clearly valid for $r = 2$. In the case $r > 2$, suppose inductively that the theorem holds for

$k = 2, 3, \ldots, r - 1$, whence, recalling that a Lindeberg condition of order r entails that of lower orders, (12) obtains for $2 \le k \le r - 1$. Thus,

$$\mathrm{E}\left|\frac{S_n}{s_n}\right|^{r-2} = O(1).$$

Let $\{Y_n, n \ge 0\}$ be independent, normally distributed random variables with means zero and variances σ_n^2, where $\sigma_0^2 = 1$ and, in addition, are independent of $\{X_n, n \ge 1\}$. Set

$$Q_{j,n} = \sum_{i=1}^{j-1} X_i + \sum_{i=j+1}^{n} Y_i, \qquad f(t) = t^r.$$

Then, $Q_{j,n} + X_j = Q_{j+1,n} + Y_{j+1}$, $1 \le j < n$, and $f^{(r)}(t) = r!$, whence

$$\begin{aligned} \mathrm{E} f(Y_0) - \mathrm{E} f\left(\frac{S_n}{s_n}\right) &= \mathrm{E}\sum_{j=1}^{n}\left[f\left(\frac{Q_{j,n} + Y_j}{s_n}\right) - f\left(\frac{Q_{j,n} + X_j}{s_n}\right)\right] \\ &= \sum_{j=1}^{n} \mathrm{E} \sum_{i=1}^{r} \frac{Y_j^i - X_j^i}{i!\, s_n^i} f^{(i)}\left(\frac{Q_{j,n}}{s_n}\right) \\ &= \sum_{i=3}^{r} \sum_{j=1}^{n} \mathrm{E} \frac{Y_j^i - X_j^i}{i!\, s_n^i} f^{(i)}\left(\frac{Q_{j,n}}{s_n}\right) \end{aligned} \tag{14}$$

noting that by independence, for all j

$$\mathrm{E}\frac{Y_j^i - X_j^i}{s_n^i} f^{(i)}\left(\frac{Q_{j,n}}{s_n}\right) = \mathrm{E}\left(\frac{Y_j^i - X_j^i}{s_n^i}\right)\mathrm{E} f^{(i)}\left(\frac{Q_{j,n}}{s_n}\right) = 0, \qquad i = 1, 2,$$

$$\begin{aligned} &= O(1)\mathrm{E}\left(\frac{|X_j|^i + |Y_j|^i}{s_n^i}\right)\mathrm{E}\left[\left|\frac{S_n}{s_n}\right|^{r-i} + |Y_0|^{r-i}\right] \\ &= O(1)\mathrm{E}\left(\frac{|X_j|^i + |Y_j|^i}{s_n^i}\right) \end{aligned}$$

for $i = 3, \ldots, r$, recalling that $\{|S_j|^\alpha, 1 \le j \le n\}$ is a submartingale, $\alpha \ge 1$. The latter together with the fact that $\{X_n\}$ obeys Lindeberg conditions of orders $3, 4, \ldots, r$ ensures that the last expression in (14) is $o(1)$, noting that for $i \ge 3$, $\mathrm{E}|Y_j|^i = C_i \sigma_j^i = C_i(\mathrm{E}\, X_j^2)^{i/2} \le C_i\, \mathrm{E}|X_j|^i$ for some constant C_i in $(0, \infty)$. □

A central limit theorem is greatly enhanced by knowledge of the rate of approach to normality and even more so by a precise bound on the error in approximating for fixed n. The latter, known as the Berry–Esseen theorem, has considerable practical and theoretical significance.

Lemma 1. *If F is a* d.f. *and G is a real differentiable function with $G(x) \to 0$ or 1 according as $x \to -\infty$ or $+\infty$ and $\sup_x |G'(x)| \le M > 0$, then there exists a real number c such that for all $T > 0$*

$$\left| \int_{-\infty}^{\infty} \left(\frac{\sin x}{x}\right)^2 H_c\left(\frac{2x}{T}\right) dx \right| \ge 2M\delta\left[\frac{\pi}{2} - 3\int_{T\delta/2}^{\infty} \frac{\sin^2 x}{x^2}\, dx\right] \ge 2M\delta\left[\frac{\pi}{2} - \frac{6}{T\delta}\right] \tag{15}$$

where $\delta = (1/2M)\sup_x |F(x) - G(x)|$ and $H_c(x) = F(x + c) - G(x + c)$.

PROOF. Since G is necessarily bounded, δ is finite and the integral on the left of (15) exists. In the nontrivial case $\delta > 0$, there exists a real sequence $\{x_n\}$ with $F(x_n) - G(x_n) \to \pm 2M\delta$. Since $F(x) - G(x) \to 0$ as $x \to \pm\infty$, $\{x_n\}$ has a finite limit point, say b, and the continuity of G ensures $F(b) - G(b) \le -2M\delta$ or $F(b+) - G(b) \ge 2M\delta$. Suppose the former for specificity and set $c = b - \delta$. Then if $|x| < \delta$, by the mean value theorem

$$\begin{aligned} H_c(x) &= F(x + c) - G(x + c) = F(b + x - \delta) - G(b + x - \delta) \\ &\le F(b) - [G(b) + (x - \delta)G'(\theta)] \\ &\le -2M\delta + (\delta - x)M = -M(\delta + x), \end{aligned}$$

whence

$$\begin{aligned} \int_{-\delta}^{\delta} \frac{1 - \cos Tx}{x^2} H_c(x)dx &\le -M \int_{-\delta}^{\delta} \frac{(x + \delta)(1 - \cos Tx)}{x^2}\, dx \\ &= -2M\delta \int_0^{\delta} \frac{1 - \cos Tx}{x^2}\, dx \\ &= -2M\delta T\left[\frac{\pi}{2} - \int_{\delta T/2}^{\infty} \frac{\sin^2 x}{x^2}\, dx\right]. \end{aligned} \tag{16}$$

Moreover,

$$\begin{aligned} \int_{[|x| > \delta]} \frac{1 - \cos Tx}{x^2} H_c(x)dx &\le 2M\delta \int_{[|x| > \delta]} \frac{1 - \cos Tx}{x^2}\, dx \\ &= 4M\delta T \int_{\delta T/2}^{\infty} \frac{\sin^2 x}{x^2}\, dx. \end{aligned} \tag{17}$$

Without loss of generality, suppose T large enough so that the middle of (15) is positive and hence the sum of the right sides of (16) and (17) negative. Then

$$\left| \int_{-\infty}^{\infty} \right| = -\left(\int_{[|x| \le \delta]} + \int_{[|x| > \delta]}\right) \ge 2M\delta T\left[\frac{\pi}{2} - 3\int_{\delta T/2}^{\infty} \frac{\sin^2 x}{x^2}\, dx\right],$$

which is tantamount to the first inequality in (15), and the second follows directly therefrom. □

Lemma 2. *If, in addition to the hypotheses of Lemma 1, G is of bounded variation on $(-\infty, \infty)$ and $F - G \in \mathscr{L}_1$, then for every $T > 0$*

$$\sup_x |F(x) - G(x)| \le \frac{2}{\pi}\int_0^T \left|\frac{\varphi_F(t) - \varphi_G(t)}{t}\right| dt + \frac{24M}{\pi T}, \tag{18}$$

where φ_F, φ_G are Fourier–Stieltjes transforms of F, G.

PROOF. In the nontrivial case where the integral is finite,

$$\varphi_F(t) - \varphi_G(t) = \int_{-\infty}^{\infty} e^{itx}\, d[F(x) - G(x)] = -it\int_{-\infty}^{\infty} [F(x) - G(x)]e^{itx}\, dx,$$

whence

$$\frac{\varphi_F(t) - \varphi_G(t)}{-it} e^{-itc} = \int_{-\infty}^{\infty} H_c(x)e^{itx}\, dx.$$

Since the above right side is bounded, via Fubini

$$\begin{aligned}\int_{-T}^{T} \frac{\varphi_F(t) - \varphi_G(t)}{-it} e^{-itc}[T - |t|]dt &= \int_{-T}^{T}\int_{-\infty}^{\infty} H_c(x)e^{itx}(T - |t|)dx\, dt \\ &= \int_{-\infty}^{\infty}\int_{-T}^{T} e^{itx}(T - |t|)H_c(x)dt\, dx \\ &= \int_{-\infty}^{\infty} \frac{2(1 - \cos Tx)}{x^2} H_c(x)dx \\ &= 2T\int_{-\infty}^{\infty} \frac{\sin^2 x}{x^2} H_c\left(\frac{2x}{T}\right)dx\end{aligned}$$

and so

$$\begin{aligned}\left|\int_{-\infty}^{\infty}\left(\frac{\sin x}{x}\right)^2 H_c\left(\frac{2x}{T}\right)dx\right| &= \frac{1}{2T}\left|\int_{-T}^{T}\frac{\varphi_F - \varphi_G}{-it} e^{-itc}(T - |t|)dt\right| \\ &\le \frac{1}{2}\int_{-T}^{T}\left|\frac{\varphi_F - \varphi_G}{t}\right| dt = \int_0^T \left|\frac{\varphi_F(t) - \varphi_G(t)}{t}\right| dt,\end{aligned}$$

whence by Lemma 1

$$\int_0^T \left|\frac{\varphi_F(t) - \varphi_G(t)}{t}\right| dt \ge 2M\delta\left(\frac{\pi}{2} - \frac{6}{T\delta}\right),$$

yielding (18). □

Lemma 3. *Let* $\varphi_n^*(t)$ *be the* c.f. *of* $S_n = \sum_1^n X_j$, *where* $\{X_n\}$ *are independent* r.v.s *with zero means and variances* σ_n^2. *If* $\Gamma_n^{2+\delta} = \sum_{j=1}^n \gamma_j^{2+\delta}$ *and* $s_n^2 = \sum_{j=1}^n \sigma_j^2$, *where* $\gamma_j^{2+\delta} = \mathrm{E}|X_j - \mathrm{E}\,X_j|^{2+\delta}$, *then for* $0 < \delta \le 1$

$$\left|\varphi_n^*\left(\frac{t}{s_n}\right) - e^{-t^2/2}\right| \le 3\left|\frac{\Gamma_n t}{s_n}\right|^{2+\delta} e^{-t^2/2} \quad \textit{for } |t| < \frac{s_n}{2\Gamma_n}. \tag{19}$$

PROOF. Let θ be a generic designation for a complex number with $|\theta| \le 1$. If φ_j is the c.f. of X_j, $1 \le j \le n$, by Theorem 8.4.1

$$\varphi_j\left(\frac{t}{s_n}\right) = 1 - \frac{t^2\sigma_j^2}{2s_n^2} + \theta\left(\frac{\gamma_j|t|}{s_n}\right)^{2+\delta}$$

For t as in (19), $|t\sigma_j/s_n| \le |t\gamma_j/s_n| \le |t\Gamma_n/s_n| < 1/2$ implies

$$\left| -\frac{t^2\sigma_j^2}{2s_n^2} + \theta\left(\frac{|t|\gamma_j}{s_n}\right)^{2+\delta}\right| \le \frac{1}{2}\cdot\frac{1}{4} + \frac{1}{4} = \frac{3}{8}.$$

Thus, since $\log(1+z) = z + (4\theta/5)|z|^2$ for $|z| \le 3/8$,

$$\log \varphi_j\left(\frac{t}{s_n}\right) = \frac{-t^2\sigma_j^2}{2s_n^2} + \theta\left(\frac{|t|\gamma_j}{s_n}\right)^{2+\delta} + \frac{8\theta}{5}\left(\frac{t^4\sigma_j^4}{4s_n^4} + \left|\frac{t\gamma_j}{s_n}\right|^{4+2\delta}\right)$$

$$= \frac{-t^2\sigma_j^2}{2s_n^2} + \theta\left|\frac{t\gamma_j}{s_n}\right|^{2+\delta}\left[1 + \tfrac{8}{5}\left(\tfrac{1}{8} + \tfrac{1}{4}\right)\right],$$

whence, summing on j,

$$\log \varphi^*\left(\frac{t}{s_n}\right) = \frac{-t^2}{2} + \frac{8\theta}{5}\left(\frac{\Gamma_n|t|}{s_n}\right)^{2+\delta}.$$

Hence, via $|e^z - 1| \le |z|e^{|z|}$

$$\left|\varphi_n^*\left(\frac{t}{s_n}\right) - e^{-t^2/2}\right| = e^{-t^2/2}\left|\exp\left\{\frac{8\theta}{5}\left(\frac{\Gamma_n|t|}{s_n}\right)^{2+\delta}\right\} - 1\right|$$

$$\le \tfrac{8}{5}e^{-t^2/2}\left(\frac{\Gamma_n|t|}{s_n}\right)^{2+\delta}\exp\left\{\frac{8}{5}\left(\frac{\Gamma_n|t|}{s_n}\right)^{2+\delta}\right\}$$

$$\le \frac{8}{5}\left(\frac{\Gamma_n|t|}{s_n}\right)^{2+\delta}\exp\left\{\frac{-t^2}{2} + \frac{2}{5}\right\} < 3\left(\frac{\Gamma_n|t|}{s_n}\right)^{2+\delta} e^{-t^2/2}$$

for $|t| < s_n/2\Gamma_n$. □

Lemma 4. *Under the conditions of Lemma* 3

$$\left|\varphi_n^*\left(\frac{t}{s_n}\right) - e^{-t^2/2}\right| \le 16\left(\frac{\Gamma_n|t|}{s_n}\right)^{2+\delta} e^{-t^2/3} \quad \textit{for } |t| \le \left[\frac{1}{36}\left(\frac{s_n}{\Gamma_n}\right)^{2+\delta}\right]^{1/\delta}$$

and $0 < \delta \le 1$.

PROOF. If $\{X_j', X_j\}$ are i.i.d., $1 \le j \le n$, employing Theorem 8.4.1 and Exercise 4.2.1,

$$\left|\varphi_j\left(\frac{t}{s_n}\right)\right|^2 = \mathrm{E}\exp\left(\frac{it(X_j - X_j')}{s_n}\right) \le 1 - \frac{t^2}{2s_n^2}\mathrm{E}(X_j - X_j')^2$$
$$+ \mathrm{E}\left|\frac{t}{s_n}(X_j - X_j')\right|^{2+\delta} \le 1 - \frac{t^2\sigma_j^2}{s_n^2}$$
$$+ 2^{2+\delta}\left|\frac{t}{s_n}\right|^{2+\delta}\mathrm{E}|X_j|^{2+\delta} \le 1 - \left(\frac{t\sigma_j}{s_n}\right)^2 + 8\left|\frac{\gamma_j t}{s_n}\right|^{2+\delta}$$
$$\le \exp\left(-\left(\frac{\sigma_j t}{s_n}\right)^2 + 8\left|\frac{\gamma_j t}{s_n}\right|^{2+\delta}\right),$$

implying since $36|t|^\delta \le (s_n/\Gamma_n)^{2+\delta}$ that

$$\left|\varphi_n^*\left(\frac{t}{s_n}\right)\right|^2 \le \exp\left(-t^2 + 8\left|\frac{t\Gamma_n}{s_n}\right|^{2+\delta}\right) \le e^{-t^2 + 2t^2/9} \le e^{-2t^2/3}$$

Hence, supposing (as is permissible by Lemma 3) that $|t| \ge s_n/(2\Gamma_n)$,

$$\left|\varphi_n^*\left(\frac{t}{s_n}\right) - e^{-t^2/2}\right| \le e^{-t^2/3} + e^{-t^2/2} \le 2e^{-t^2/3} \le 16\left|\frac{t\Gamma_n}{s_n}\right|^{2+\delta} e^{-t^2/3}. \quad \square$$

Theorem 3 (Berry–Esseen). *If* $\{X_n, n \ge 1\}$ *are independent random variables* $\mathrm{E}\,X_n = 0$, $\mathrm{E}\,X_n^2 = \sigma_n^2$, $s_n^2 = \sum_{i=1}^n \sigma_i^2 > 0$, $\Gamma_n^{2+\delta} = \sum_{i=1}^n \mathrm{E}\,|X_i|^{2+\delta} < \infty$, $n \ge 1$, *for some* δ *in* $(0, 1]$ *and* $S_n = \sum_1^n X_i$, *there exists a universal constant* C_δ *such that*

$$\sup_{-\infty < x < \infty}\left|\mathrm{P}\{S_n < xs_n\} - \frac{1}{\sqrt{2\pi}}\int_{-\infty}^x e^{-y^2/2}\,dy\right| \le C_\delta\left(\frac{\Gamma_n}{s_n}\right)^{2+\delta}. \tag{20}$$

Remark. The thrust of the theorem is for $\Gamma_n/s_n = o(1)$, in which case Esseen (1945) asserts that $C_\delta \le 7.5$ when $\delta = 1$.

PROOF. If $F(x) = \mathrm{P}\{S_n < xs_n\}$ and $\Phi(x) =$ standard normal d.f., $\Phi'(x) = (1/\sqrt{2\pi})e^{-x^2/2} \le 1/\sqrt{2\pi} = M$ and, since both F and Φ have mean zero and variance one, Tchebychev's inequality ensures $F(x) \le 1/x^2$, $x < 0$, and $1 - F(x) \le 1/x^2$, $x > 0$ (similarly for Φ). Thus, $F - \Phi \in \mathscr{L}_1$, whence by Lemma 2, taking $T^\delta = (1/36)\,(s_n/\Gamma_n)^{2+\delta}$ and recalling Lemma 4,

$$\sup_x|\mathrm{P}\{S_n < xs_n\} - \Phi(x)|$$
$$\le \frac{2}{\pi}\int_0^T\left|\frac{\varphi_n^*(t/s_n) - e^{-t^2/2}}{t}\right|dt + \frac{24M}{\pi T}$$
$$\le \frac{2}{\pi}\int_0^T 16\left(\frac{\Gamma_n}{s_n}\right)^{2+\delta} t^{1+\delta}e^{-t^2/3}\,dt + \frac{24M}{\pi}\left[\frac{1}{36}\left(\frac{s_n}{\Gamma_n}\right)^{2+\delta}\right]^{-1/\delta}$$
$$\le C_\delta \max\left\{\left(\frac{\Gamma_n}{s_n}\right)^{2+\delta}, \left(\frac{\Gamma_n}{s_n}\right)^{1+(2/\delta)}\right\}.$$

Consequently, (20) is valid for $\Gamma_n/s_n \le 1$ and holds trivially with $C_\delta = 1$ when $\Gamma_n/s_n > 1$. □

Corollary 4. *If* $\{X_n, n \ge 1\}$ *are* i.i.d. *random variables with* $\mathrm{E}\,X_n = 0$, $\mathrm{E}\,X_n^2 = \sigma^2$, $\mathrm{E}|X_n|^{2+\delta} = \gamma^{2+\delta} < \infty$ *for some* δ *in* $(0, 1]$ *and* Φ *is the standard normal* d.f., *there exists a universal constant* c_δ *such that*

$$\sup_{-\infty < x < \infty} |\mathrm{P}\{S_n < x\sigma n^{1/2}\} - \Phi(x)| \le \frac{c_\delta}{n^{\delta/2}} \left(\frac{\gamma}{\sigma}\right)^{2+\delta}. \tag{21}$$

Remark. It is asserted in Van Beek, 1972 that $c_\delta \le .7975$ when $\delta = 1$.

If $\{X_n, n \ge 1\}$ is a sequence of independent random variables with $\mathrm{E}\,X_n = 0$, $\mathrm{E}\,X_n^2 = 1$ obeying the Lindeberg condition, it seems extremely plausible that $(1/\sqrt{n}) \sum_{j=n+1}^{2n} X_j \xrightarrow{\mathrm{d}} N_{0,1}$. This would be encompassed by a central limit theorem involving a double sequence of **rowwise independent** random variables $\{X_{nj}, 1 \le j \le k_n \to \infty\}$, i.e., such that $\{X_{n1}, \ldots, X_{nk_n}\}$ were independent for each $n \ge 1$. For, if necessary and sufficient conditions for asymptotic normality of $\sum_{j=1}^{k_n} X_{nj}$ were available, it would suffice to set $X_{nj} = X_{n+j}/\sqrt{n}$, $1 \le j \le n$. Amazingly, the class of all possible limit distributions of (centered) row sums $\sum_{j=1}^{k_n} X_{nj}$ of rowwise independent $\{X_{nj}, 1 \le j \le k_n \to \infty\}$ can be characterized and coincides with the class of socalled infinitely divisible laws (under a minor additional proviso). But this will be the subject of Chapter 12.

A central limit theorem (C.L.T) has already been established for i.i.d. random variables with finite variance in Corollary 9.1.2, and the case of infinite variance is dealt with next. The latitude in choosing the normalizing constants A_n, B_n of Theorem 4 is governed by Corollary 8.2.2.

Theorem 4. *If* $\{X_n, n \ge 1\}$ *are* i.i.d. *random variables with non-degenerate* d.f. F *and* Φ *is the standard normal* d.f., *then*

$$\lim_{n \to \infty} \mathrm{P}\left\{\frac{1}{B_n} \sum_{i=1}^{n} X_i - A_n < x\right\} = \Phi(x) \tag{22}$$

for some $B_n > 0$ *and* A_n *iff*

$$\lim_{c \to \infty} \frac{\int_{[|x| > c]} dF(x)}{(1/c^2) \int_{[|x| < c]} x^2\, dF(x)} = 0; \tag{23}$$

moreover, B_n *may be chosen as in* (25) *while* A_n *may be taken as*

$$\frac{n}{B_n} \int_{[|x| < B_n]} x\, dF(x). \tag{24}$$

Proof. Note that either (22) or (23) implies that $\mathrm{E}\,X_1^2 > 0$. If $\int_{-\infty}^{\infty} x^2\, dF(x) < \infty$, then $c^2 \int_{[|x| > c]} dF \le \int_{[|x| > c]} x^2\, dF(x) = o(1)$ as $c \to \infty$, so that (23) obtains and in this case (22) has already been proved with

$$B_n = \sqrt{n}\sigma, \qquad A_n = \frac{\sqrt{n}}{\sigma} \int_{-\infty}^{\infty} x\, dF,$$

which differ from the choices of (25), (24) only in the sense of Corollary 8.2.2. Thus, it may and will be supposed that $E X^2 = \infty$.

By the dominated convergence theorem

$$c^{-2} \int_{[|x|<c]} x^2 \, dF(x) = o(1)$$

as $c \to \infty$ and so, defining

$$B_n = \sup\left\{c : c^{-2} \int_{[|x|<c]} x^2 \, dF(x) \geq \frac{1}{n}\right\}, \tag{25}$$

necessarily $B_n \uparrow \infty$ and via continuity

$$\frac{n}{B_n^2} \int_{[|x|<B_n]} x^2 \, dF(x) = 1. \tag{26}$$

Under the hypothesis of (23), if $0 < \varepsilon \leq 1$,

$$\frac{\int_{[\varepsilon B_n \leq |x| \leq B_n]} x^2 \, dF(x)}{\int_{[|x|<B_n]} x^2 \, dF(x)} \leq \frac{B_n^2 \int_{[|x| \geq \varepsilon B_n]} dF}{\int_{[|x|<\varepsilon B_n]} x^2 \, dF(x)} = o(1),$$

whence, recalling (26), for $0 < \varepsilon \leq 1$

$$\frac{n}{B_n^2} \int_{[|x|<\varepsilon B_n]} x^2 \, dF(x) = \frac{n}{B_n^2} \int_{[|x|<B_n]} x^2 \, dF(x) \left[1 - \frac{\int_{[\varepsilon B_n \leq |x| < B_n]} x^2 \, dF(x)}{\int_{[|x|<B_n]} x^2 \, dF(x)}\right] \to 1. \tag{27}$$

Thus, via (27) and (23)

$$\lim_{n \to \infty} n \, P\{|X_1| > \varepsilon B_n\} = \lim_{n \to \infty} \frac{n \int_{[|x| > \varepsilon B_n]} dF}{n B_n^{-2} \int_{[|x|<\varepsilon B_n]} x^2 \, dF} = 0 \tag{28}$$

for $0 < \varepsilon \leq 1$ and hence for all $\varepsilon > 0$. Likewise (27) holds for all $\varepsilon > 0$.

Define $X'_j = X_j I_{[|X_j|<B_n]}$, $1 \leq j \leq n$, and $S'_n = \sum_1^n X'_j$. Now (26) entails $n^{1/2} = o(B_n)$ so that if $Y_n = (n^{1/2}/B_n) X_1 I_{[|X_1|<\varepsilon B_n]}$, then $Y_n \xrightarrow{P} 0$ and, moreover, $E Y_n^2 \to 1$ by (27). Thus, $\{Y_n, n \geq 1\}$ is uniformly integrable, whence $(n^{1/2}/B_n) \int_{[|x|<\varepsilon B_n]} x \, dF = E Y_n = o(1)$ for all $\varepsilon > 0$, and so (27) implies

$$\frac{n}{B_n^2} \left[\int_{[|x|<\varepsilon B_n]} x^2 \, dF - \left(\int_{[|x|<\varepsilon B_n]} x \, dF\right)^2\right] \to 1, \qquad \varepsilon > 0. \tag{29}$$

Now for $\varepsilon = 1$, (29) asserts that $s_n^2 \equiv \sigma^2(S'_n) \sim B_n^2$ and, moreover, $E Y_n = o(1)$ implies $E X'_1 = o(B_n n^{-1/2})$. Thus, for $0 < \varepsilon \leq 1$ and all large n

$$\frac{1}{s_n^2} \sum_{j=1}^n E[X'_j - E X'_j]^2 I_{[|X'_j - E X'_j| > \varepsilon s_n]} = \frac{n}{s_n^2} E[X'_1 - E X'_1]^2 I_{[|X'_1 - E X'_1| > \varepsilon s_n]}$$

$$\leq \frac{2n}{s_n^2} \{E(X'_1)^2 I_{[|X'_1| > \varepsilon s_n/2]} + (E X'_1)^2\} = \frac{2n(1 + o(1))}{B_n^2} E X_1^2 I_{[(\varepsilon/2)s_n < |X_1| < B_n]} + o(1)$$

$$\leq 2n(1 + o(1)) P\left\{|X_1| > \frac{\varepsilon}{4} B_n\right\} + o(1) = o(1).$$

according to (28). Consequently, $(S_n' - \mathrm{E}\, S_n')/s_n \xrightarrow{d} N_{0,1}$ by Exercise 9.1.2 and,

$$\mathrm{P}\{S_n \neq S_n'\} \leq n\,\mathrm{P}\{|X_1| \geq B_n\} = o(1),$$

it follows that

$$\frac{S_n}{B_n} - A_n = \frac{S_n - \mathrm{E}\, S_n'}{B_n} \xrightarrow{d} N_{0,1}.$$

Conversely, if (22) prevails, then (28) and (29) hold for all $\varepsilon > 0$ by Corollary 12.2.3, whence

$$\frac{(\varepsilon B_n)^2 \int_{[|x| > B_n\varepsilon]} dF}{\int_{[|x| < B_n\varepsilon]} x^2\, dF(x)} \leq \frac{\varepsilon^2 n\, \mathrm{P}\{|X_1| > B_n \varepsilon\}}{nB_n^{-2}[\int_{[|x| < B_n\varepsilon]} x^2\, dF - (\int_{[|x| < \varepsilon B_n]} x\, dF)^2]} = o(1)$$

for all $\varepsilon > 0$, which is tantamount to (23). □

EXAMPLE 1 (Friedman, Katz, Koopmans). If $\{X, X_n, n \geq 1\}$ are i.i.d. r.v.s with $\mathrm{E}\, X = 0$, $\mathrm{E}\, X^2 = \sigma^2 \in (0, \infty)$, $\sigma_n^2 = \sigma^2(X_n')$ where $X_n' = (X \wedge n^{1/2}) \vee (-n^{1/2})$, $S_n = \sum_{j=1}^n X_j$, then

$$\sum_{n=1}^{\infty} n^{-1} \sup_x |\mathrm{P}\{S_n/n^{1/2} < x\} - \Phi(x/\sigma_n)| < \infty. \tag{30}$$

PROOF. Since $\sigma_n \to \sigma$, it may and will be supposed that $\sigma_n > 0$, $n \geq 1$ and that $\sigma = 1$. Let S_n' be the sum on n i.i.d. r.v.s having the common distribution of X_n' and set $\mu_n = \mathrm{E}\, X_n'$. In view of

$$|\mathrm{P}\{S_n < xn^{1/2}\} - \mathrm{P}\{S_n' < xn^{1/2}\}| \leq n\,\mathrm{P}\{|X| > n^{1/2}\}$$

it suffices to prove (30) with S_n replaced by S_n'. Now

$$\mathrm{P}\{S_n' < xn^{1/2}\} - \Phi(x/\sigma_n) = \left[\mathrm{P}\left\{\frac{S_n' - n\mu_n}{n^{1/2}\sigma_n} < \frac{xn^{1/2} - n\mu_n}{n^{1/2}\sigma_n}\right\} - \Phi\left(\frac{x - n^{1/2}\mu_n}{\sigma_n}\right)\right]$$
$$+ \left[\Phi\left(\frac{x - n^{1/2}\mu_n}{\sigma_n}\right) - \Phi\left(\frac{x}{\sigma_n}\right)\right] = A_n + B_n \text{ (say)}.$$

If C_j, $j \geq 1$ are positive constants, the Berry–Esseen theorem ensures that if $A_n' = \sup_x |A_n|$ and $B_n' = \sup_x |B_n|$

$$\sum_{n=1}^{\infty} n^{-1} A_n' \leq \sum_{n=1}^{\infty} C_1 n^{-3/2}\, \mathrm{E}|X_n' - \mu_n|^3/\sigma_n^3 \leq C_2 \sum_{n=1}^{\infty} n^{-3/2}\, \mathrm{E}|X_n'|^3 < \infty$$

since

$$\sum_{n=1}^{\infty} n^{-3/2}\, \mathrm{E}|X_n'|^3 = 3 \sum_{n=1}^{\infty} n^{-3/2} \int_0^{n^{1/2}} t^2\, \mathrm{P}\{|X| > t\} dt$$
$$= 3 \int_0^{\infty} t^2\, \mathrm{P}\{|X| > t\} \sum_{n \geq (t \vee 1)^2} n^{-3/2}\, dt$$
$$= C_3 \int_1^{\infty} t\, \mathrm{P}\{|X| > t\} dt < \infty.$$

On the other hand, since $\sigma_n \to \sigma$, via integration by parts, $\mathrm{E}\, X = 0$ and

Corollary 6.2.2 (30),

$$\sum_{n=1}^{\infty} n^{-1} B_n' \le \sum_{n=1}^{\infty} n^{-1} \left| \Phi\left(\frac{-n^{1/2}\mu_n}{\sigma_n}\right) - \Phi(0) \right| = C_4 \sum_{n=1}^{\infty} n^{-1} \int_0^{n^{1/2}|\mu_n|/\sigma_n} e^{-t^2/2}\,dt$$

$$\le C_4 \sum_{n=1}^{\infty} n^{-1/2} |\mu_n|/\sigma_n \le C_5 \sum_{n=1}^{\infty} n^{-1/2} \int_{n^{1/2}}^{\infty} \mathrm{P}\{|X| > t\}\,dt$$

$$= C_5 \int_1^{\infty} \mathrm{P}\{|X| > t\} \sum_{n=1}^{t^2} n^{-1/2}\,dt = C_6 \int_1^{\infty} t\,\mathrm{P}\{|X| > t\}\,dt < \infty. \quad \square$$

Dominated convergence in conjunction with (30) yields the result of Rosén that under the above conditions

$$\sum_{n=1}^{\infty} n^{-1} |\mathrm{P}\{S_n < 0\} - \tfrac{1}{2}| < \infty. \tag{31}$$

Theorem 5 (CLT for non-degenerate U-statistics). *Let $U_{k,n}(h)$, $n \ge k \ge 2$ be a sequence of U-statistics with $\mathrm{E}\,h = 0$ and $\sigma^2 = \mathrm{E}|\mathrm{E}\{h(X_1, \ldots, X_k)|X_1\}]^2 \in (0, \infty)$. If*

$$\mathrm{E}\{h(X_1, \ldots, X_k)|X_1, \ldots, X_j\} \in \mathscr{L}_{2j/(2j-1)}, \qquad 2 \le j \le k \tag{32}$$

(a fortiori, if $\mathrm{E}|h|^{4/3} < \infty$), then for any real x

$$\lim_{n\to\infty} \mathrm{P}\left\{\frac{n^{1/2}}{k\sigma} U_{k,n}(h) < x\right\} = \frac{1}{\sqrt{2\pi}} \int_{-\infty}^{x} e^{-y^2/2}\,dy. \tag{33}$$

PROOF. According to Lemma 7.5.5,

$$\frac{n^{1/2}}{k\sigma} U_{k,n}(h) = \frac{1}{\sigma n^{1/2}} \sum_{r=1}^{n} \mathrm{E}\{h(X_r, \ldots, X_{r+k-1})|X_r\} + o(1), \qquad \text{a.s.}$$

Since $Y_r = \mathrm{E}\{h(X_r, \ldots, X_{r+k-1})|X_r\}$, $r \ge 1$, are i.i.d. random variables with $\mathrm{E}\,Y_r = 0$, $\mathrm{E}\,Y_r^2 = \sigma^2 \in (0, \infty)$, the conclusion (33) follows from Corollary 2 and Theorem 8.1.1. $\square$

EXERCISES 9.1

1. Show that if $\{X_n, n \ge 1\}$ are independent r.v.s with $|X_n| \le C_n$, a.c., $n \ge 1$, and $C_n = o(s_n)$, where $s_n^2 = \sum_1^n \mathrm{E}(X_j - \mathrm{E}\,X_j)^2 \to \infty$, then $(S_n - \mathrm{E}\,S_n)/s_n \xrightarrow{d} N_{0,1}$.
2. If $\{X_{nj}, 1 \le j \le k_n \to \infty\}$ are rowwise independent r.v.s with $S_n = \sum_{j=1}^{k_n} X_{nj}$, $\mathrm{E}\,X_{nj} = 0$, $\mathrm{E}\,X_{nj}^2 = \sigma_{nj}^2$, $s_n^2 = \sum_{j=1}^{k_n} \sigma_{nj}^2 \to \infty$, then $S_n/s_n \xrightarrow{d} N_{0,1}$ if
$$\sum_{j=1}^{k_n} \mathrm{E}\,X_{nj}^2 I_{[|X_{nj}| > \varepsilon s_n]} = o(s_n^2), \ \varepsilon > 0.$$
3. If $\{X_n\}$ are independent with
$$\mathrm{P}\{X_n = \pm n^{\alpha}\} = \frac{1}{2n^{\beta}}, \qquad \mathrm{P}\{X_n = 0\} = 1 - \frac{1}{n^{\beta}}, \qquad 2\alpha > \beta - 1$$
the Lindeberg condition holds when and only when $0 \le \beta < 1$.

4. Failure of the Lindeberg condition does not preclude asymptotic normality. Let $\{Y_n\}$ be i.i.d. with $\mathrm{E}\, Y_n = 0$, $\mathrm{E}\, Y_n^2 = 1$; let $\{Z_n\}$ be independent with $\mathrm{P}\{Z_n = \pm n\} = 1/2n^2$, $\mathrm{P}\{Z_n = 0\} = 1 - (1/n^2)$ and $\{Z_n\}$ independent of $\{Y_n\}$. Prove that if $X_n = Y_n + Z_n$, $S_n = \sum_1^n X_i$, then $S_n/\sqrt{n} \xrightarrow{d} N_{0,1}$ and the Lindeberg condition cannot hold. Explain why this does not contravene Theorem 1.

5. Let $\{Y_n, n \geq 1\}$ be i.i.d. r.v.s with finite variance σ^2 (say $\sigma^2 = 1$) and let $\{\sigma_n^2, n \geq 1\}$ be nonzero constants with $s_n^2 = \sum_1^n \sigma_i^2 \to \infty$. Show that the **weighted i.i.d.** r.v.s $\{\sigma_n Y_n, n \geq 1\}$ obey the central limit theorem, i.e., $(1/s_n)\sum_1^n \sigma_j Y_j \xrightarrow{d} N_{0,1}$ if $\sigma_n = o(s_n)$ and $\mathrm{E}\, Y = 0$.

6. Let $\{X_n\}$ be independent with

$$\mathrm{P}\{X_n = \pm 1\} = \frac{1}{2a}, \ \mathrm{P}\{X_n = \pm n\} = \frac{1}{2}\left(1 - \frac{1}{a}\right)\frac{1}{n^2},$$

$$\mathrm{P}\{X_n = 0\} = \left(1 - \frac{1}{a}\right)\left(1 - \frac{1}{n^2}\right), n \geq 1, a > 1.$$

Again, $S_n/\sqrt{n}$ has a limiting normal distribution despite the Lindeberg condition being vitiated.

7. Prove that the Liapounov condition of Corollary 9.1.1 is more stringent the larger the value of δ.

8. (i) For what positive values of α, if any, does a C.L.T. hold for i.i.d. symmetric random variables with $F(x) = 1 - 1/(2x^\alpha)$, $x \geq 1$, $F(x) = \frac{1}{2}$, $0 \leq x \leq 1$.
(ii) Does a C.L.T. hold for independent $\{X_n\}$ with $\mathrm{P}\{X_n = \pm 1\} = \frac{1}{4}$, $\mathrm{P}\{X_n = \pm n\} = 1/(2n^3)$, $\mathrm{P}\{X_n = 0\} = (1/2) - (1/n^3)$? *Hint*: Apply Theorems 4 and 1.

9. C.L.T. for U-statistics: Let $\{X_n\}$ be i.i.d. and $\varphi(x_1, \ldots, x_m)$ a real, symmetric function of its m arguments. If $\mathrm{E}\, \varphi^2(X_1, \ldots, X_m) < \infty$ and $\mathrm{E}\, \varphi(X_1, \ldots, X_m) = \theta$, then $n^{1/2}(U_n - \theta) \xrightarrow{d} N_{0,\sigma}$, where $U_n = \binom{n}{m}^{-1} \sum_{1 \leq i_1 < \cdots < i_m \leq n} \varphi(X_{i_1}, \ldots, X_{i_m})$ and $\sigma^2 = m^2\, \mathrm{E}[\mathrm{E}^2\{\varphi(X_1, \ldots, X_m) | X_1\} - \theta^2]$. *Hint*: A C.L.T. applies to

$$V_n = \frac{m}{n^{1/2}} \sum_{i=1}^{n} [\mathrm{E}\{\varphi(X_i, X_2, \ldots, X_m) | X_i\} - \theta]$$

since the components of the sum are i.i.d. with mean 0 and variance σ^2/m^2, while $U_n - V_n \xrightarrow{P} 0$ via $\mathrm{E}(U_n - V_n)^2 = o(1)$.

10. Let $X_n = (X_{n1}, \ldots, X_{nk})$, $n \geq 1$, be i.i.d. random vectors with $\mathrm{E}\, X_{nj} = \mu_j$,

$$\mathrm{Cov}(X_{ni}, X_{nj}) = \sigma_{ij}, \ 1 \leq i \leq j \leq k.$$

If $S_{nj} = \sum_{i=1}^n X_{ij}$, $n \geq 1$, $1 \leq j \leq k$, prove that

$$(n^{-1/2}(S_{n1} - n\mu_1), \ldots, n^{-1/2}(S_{nk} - n\mu_k))$$

converges in distribution to the k-dimensional normal distribution with mean vector zero and covariance matrix $\{\sigma_{ij}\}$. *Hint*: Recall Exercise 8.5.6.

11. Let $\{X_n\}$ be independent r.v.s with $\mathrm{E}\, X_n \equiv 0$, $\mathrm{E}\, X_n^2 \equiv 1$ which obey the Lindeberg condition. If $n_j = [jn/k]$ = greatest integer $\leq jn/k$ for $j = 0, 1, \ldots, k$ and $S_n = \sum_{i=1}^n X_i$, prove that $((k/n)^{1/2} S_{n_1}, \ldots, (k/n)^{1/2} S_{n_k})$ converges in distribution to the k-dimensional normal with mean vector zero and covariance matrix $\{\sigma_{ij}\}$, where $\sigma_{ij} = \min(i, j)$. Conclude that

$$\lim_{n\to\infty} P\left\{\max_{1\le j\le k} S_{n_j} < xn^{1/2}\right\} = P\left\{\frac{1}{\sqrt{k}}\max_{1\le j\le k} Y_j < x\right\},$$

where $\{Y_j, 1 \le j \le k\}$ are i.i.d. r.v.s with the standard normal d.f.

12. Prove that if $\{X_n, n \ge 1\}$ are independent r.v.s with $E\,X_n = 0, E\,X_n^2 = \sigma_n^2$ which obey central limit theorem and $\lim_{n\to\infty} E(s_n^{-1/2}\sum_1^n X_i)^{2k} = (2k)!/2^k k!$, then $\{X_n\}$ obeys a Lindeberg condition of order $2k$.

13. Let $\{X_n, n \ge 1\}$ be i.i.d. with symmetric density $f(x) = |x|^{-3} I_{[|x|>1]}$. Show that $(n \log n)^{-1/2} \sum_1^n X_i \xrightarrow{d} N_{0,1}$.

9.2 Interchangeable Components

For an **infinite** sequence of interchangeable random variables $\{X_n\}$, Corollary 7.3.4 exhibits the joint distributions as mixtures of distributions corresponding to i.i.d. random variables and thereby provides a tool for proving central and other limit theorems.

Initially it will be assumed that $E\,X_1^2 < \infty$, whence no further generality is lost in supposing $E\,X_1 = 0$, $E\,X_1^2 = 1$.

Theorem 1. *Let $\{X_n, n \ge 1\}$ be interchangeable random values. If* $E\,X = 0$, $E\,X^2 = 1$, and

$$\mathrm{Cov}(X_1, X_2) = 0 = \mathrm{Cov}(X_1^2, X_2^2), \tag{1}$$

then $(1/\sqrt{n})\sum_1^n X_i \xrightarrow{d} N_{0,1}$. *Conversely, if* $E\,X^2 < \infty$ *and* $(1/\sqrt{n})\sum_1^n X_i \xrightarrow{d} N_{0,1}$, *then* (1) *holds.*

Proof. According to Theorem 7.3.2 the r.v.s $\{X_n\}$ are conditionally i.i.d. given the σ-algebra $\mathscr{G}$ of permutable events, and according to Corollary 7.3.5 there is a regular conditional distribution P^ω for X given $\mathscr{G}$ such that the coordinate r.v.s $\{\xi_n, n \ge 1\}$ are i.i.d. on the probability space $(R^\infty, \mathscr{B}^\infty, P^\omega)$. Moreover, by this same corollary, for $i \ne j$

$$\begin{aligned} E\,X_i X_j &= E[E\{X_i X_j \mid \mathscr{G}\}] = E[E^2\{X_i \mid \mathscr{G}\}], \\ E[X_i^2 - 1][X_j^2 - 1] &= E[E^2\{X_i^2 - 1) \mid \mathscr{G}\}] \end{aligned} \tag{2}$$

and so (1) is equivalent to

$$E\{X_i \mid \mathscr{G}\} = 0, \qquad E\{X_i^2 \mid \mathscr{G}\} = 1, \qquad i \ge 1, \quad \text{a.c.} \tag{1'}$$

Consequently, for all ω

$$E_\omega(\xi_1) \equiv \int_{-\infty}^{\infty} \xi_1\, dP^\omega = 0, \qquad \sigma_\omega^2(\xi_1) \equiv \int_{-\infty}^{\infty} \xi_1^2\, dP^\omega = 1,$$

and if $S_n = \sum_1^n X_i$, $T_n = \sum_1^n \xi_i$, then via Corollaries 7.2.2, 7.3.4, 7.3.5, and dominated convergence

$$\lim_{n\to\infty} P\left\{\frac{S_n}{\sqrt{n}} < x\right\} = \lim_{n\to\infty} \int P\left\{\frac{S_n}{\sqrt{n}} < x \,\middle|\, \mathscr{G}\right\} dP = \lim_{n\to\infty} \int P^{\omega}\left\{\frac{T_n}{\sqrt{n}} < x\right\} dP$$

$$= \int \lim_{n\to\infty} P^{\omega}\left\{\frac{T_n}{\sqrt{n}} < x\right\} dP, \tag{3}$$

and so sufficiency follows from Corollary 9.1.2.

Conversely, if $S_n/\sqrt{n} \xrightarrow{d} N_{0,1}$, then $S_n/n \xrightarrow{P} 0$. By Example 1 of 7.3.1, $S_n/n \xrightarrow{\text{a.c.}} E\{X|\mathscr{G}\}$ and the first half of (1)′ follows. Apropos of the second half,

$$e^{-t^2/2} = \lim_{n\to\infty} E[E\{e^{itS_n/n^{1/2}}|\mathscr{G}\}] = \lim_{n\to\infty} EE_{\omega} e^{itT_n/n^{1/2}}$$

$$= E\, e^{-\sigma_{\omega}^2(\xi)t^2/2}$$

so that

$$E\, e^{-[1-\sigma_{\omega}^2(\xi)]t^2/2} = 1, \qquad \text{all } t \in (-\infty, \infty),$$

implying ($t \to \infty$) that $\sigma_{\omega}^2(\xi) \le 1$, a.c. Consequently, $\sigma_{\omega}^2(\xi) = 1 = E\{X^2|\mathscr{G}\}$, a.c., whence (1)′ and hence (1) is obtained. □

If $S_n = \sum_1^n X_i$, $s_n^2 = E\, S_n^2$, it is quite possible that $S_n/s_n \xrightarrow{d} N_{0,1}$ despite violation of (1). The point is that if $\mathrm{Cov}(X_1, X_2) = \rho^2 > 0$, the order of magnitude of S_n is n rather than $n^{1/2}$. Recall that X_F denotes a fictitious r.v. with d.f. F.

Theorem 2. *If $\{X_n, n \ge 1\}$ are $\mathscr{L}_1$ interchangeable random variables with partial sums S_n, $n \ge 1$, then*

(i) $S_n/n \xrightarrow{d} S_G$, where G is the d.f. of $E\{X_1|\mathscr{G}\}$ and $\mathscr{G}$ is the σ-algebra of permutable events,

(ii) if F is any distribution function uniquely determined by its moments α_k, $k \ge 1$, and $E\, X_1 X_2 \cdots X_k$ exists, all $k \ge 1$, then $S_n/n \xrightarrow{d} S_F$ iff $E\, X_1 X_2 \cdots X_k = \alpha_k$, $k \ge 1$.

Corollary 1. *If $\{X_n, n \ge 1\}$ are $\mathscr{L}_1$ interchangeable random variables with $\mathrm{Cov}(X_1, X_2) = \rho^2 > 0$ and $E\, X_1 X_2 \cdots X_k$ exists for all $k \ge 1$ then $(1/\rho n) \sum_1^n X_i \xrightarrow{d} N_{0,1}$ iff $E\, X_1 X_2 \cdots X_k$ vanishes for odd integers and equals $1 \cdot 3 \cdots (k-1)\rho^k$ for even integers k.*

Proof of Theorem 2. (i) is an immediate consequence of the strong law of large numbers for interchangeable r.v.s (Example 7.3.1). To prove (ii) note via Corollary 7.3.5 that, for all $k \ge 1$,

$$E\, X_1 \cdots X_k = E[E\{X_1 \cdots X_k|\mathscr{G}\}] = E[E\{X_1|\mathscr{G}\}]^k. \tag{4}$$

Thus, if $S_n/n \xrightarrow{d} S_F$, where F has finite moments α_k, $k \ge 1$, part (i) ensures that the right and hence the left sides of (4) are α_k.

Conversely, if the left and hence the right sides of (4) equal α_k, then (i) and the Frechet–Shohat theorem 8.2.1 guarantee that $S_n/n \xrightarrow{d} S_F$. □

Unfortunately, interchangeable r.v.s encountered in practical situations of interest are likely to be finite in number and not embeddable in an infinite sequence, so that Corollary 7.3.4 and prior results are inapplicable. A case of this sort occurred in Chapter 3.1 with the random distribution of balls into cells.

Suppose that $N = N_n$ balls are distributed at random into n cells and set $X_{nj} = 1$ or 0 according as the jth cell, $1 \le j \le n$, is or is not empty. Then $\{X_{nj}, 1 \le j \le n, n \ge 1\}$ constitute a double sequence of r.v.s which, for each $n > 1$, form a finite collection of interchangeable r.v.s with

$$\operatorname{Cov}(X_{n1}, X_{n2}) = \left(1 - \frac{2}{n}\right)^N - \left(1 - \frac{1}{n}\right)^{2N} < 0. \tag{5}$$

Thus, recalling Exercises 7.3.6 and 6.3.10, $\{X_{nj}, 1 \le j \le n\}$ is not embeddable in an infinite sequence of interchangeable r.v.s. Nonetheless, asymptotic normality of the distribution of the number of empty cells, i.e., $\sum_{j=1}^n X_{nj}$, can be proved by more *ad hoc* methods.

By way of preliminaries, set $U = U_n = \sum_1^n X_{nj}$ and note that

$$\mathrm{E}\, U = \sum_{j=1}^{n} \mathrm{E}\, X_{nj} = n\left(1 - \frac{1}{n}\right)^N \tag{6}$$

and from (5)

$$\begin{aligned} \sigma_U^2 &= n\left[\left(1 - \frac{1}{n}\right)^N - \left(1 - \frac{1}{n}\right)^{2N}\right] + n(n-1)\left[\left(1 - \frac{2}{n}\right)^N - \left(1 - \frac{1}{n}\right)^{2N}\right] \\ &= n\left[\left(1 - \frac{1}{n}\right)^N + (n-1)\left(1 - \frac{2}{n}\right)^N - n\left(1 - \frac{1}{n}\right)^{2N}\right] \le n\left(1 - \frac{1}{n}\right)^N. \end{aligned} \tag{7}$$

Let $S_j = S_j^{(n)}$ denote the waiting time until the occupation of the jth new cell by a ball. Set $S_0 = 0$ and $Y_{n,j} = S_j - S_{j-1}$, $j \ge 1$. Clearly, $\{Y_{n,j}, 1 \le j \le n\}$ are independent with $Y_{n,1} = S_1 = 1$ and

$$\mathrm{P}\{Y_{n,j} = i\} = \left(\frac{j-1}{n}\right)^{i-1}\left(1 - \frac{j-1}{n}\right), \qquad j \ge 2, \qquad i \ge 1. \tag{8}$$

That is, $\{Y_{n,j} - 1, 2 \le j \le n\}$ are independent geometric r.v.s.

At most $n - k$ empty cells, i.e., at least k occupied cells in the random casting of N balls into n cells is tantamount to the kth new cell being entered by the Nth or a prior ball. Thus, for $2 \le k \le n$

$$\mathrm{P}\{S_k \le N\} = \mathrm{P}\{U_n \le n - k\}, \tag{9}$$

and the possibility arises of shunting the asymptotic normality of a sum U_n of interchangeable r.v.s to one S_k of independent r.v.s.

Theorem 3. *Let $U = U_n$ designate the number of empty cells engendered by distributing $N = N_n$ balls at random into n cells, and set $a = a_n = N/n$, $b = b_n = (e^a - 1 - a)^{1/2}$, and $\sigma = \sigma_{U_n}$.*

(i) *If $N \to \infty$ and $a/n \to 0$ as $n \to \infty$, then*

$$\sigma^2 = ne^{-2a}b^2(1 + o(1)). \tag{10}$$

(ii) *$\sigma \to \infty$ iff $ne^{-2a}b^2 \to \infty$, in which case $a = o(n)$, $N \to \infty$, (10) holds, and*

$$\frac{U_n - \mathrm{E}\,U_n}{\sigma} \xrightarrow{\mathrm{d}} N_{0,1}, \tag{11}$$

$$\frac{U_n - ne^{-a}}{\sqrt{ne^{-a}}\,b} \xrightarrow{\mathrm{d}} N_{0,1}. \tag{12}$$

PROOF. (i) Since for $-\infty < \alpha < \infty$

$$\left(1 - \frac{\alpha}{n}\right)^n = \exp\left\{n\left(-\frac{\alpha}{n} - \frac{\alpha^2}{2n^2} + O(n^{-3})\right)\right\} = \exp\left\{-\alpha - \frac{\alpha^2}{2n} + O(n^{-2})\right\}, \tag{13}$$

it follows via (7) that

$$\begin{aligned}
\sigma^2 &= n\left\{\left(1 - \frac{1}{n}\right)^{na} + (n-1)\left(1 - \frac{2}{n}\right)^{na} - n\left(1 - \frac{1}{n}\right)^{2na}\right\} \\
&= ne^{-2a}\left\{e^a \exp\left(-\frac{a}{2n}(1 + o(1))\right) + n\left[\exp\left(-\frac{2a}{n} + O(an^{-2})\right)\right.\right. \\
&\qquad \left.\left. - \exp\left(-\frac{a}{n} + O(an^{-2})\right)\right] - \exp\left(-\frac{2a}{n} + O(an^{-2})\right)\right\} \\
&= ne^{-2a}\left[e^a - a - 1 - \frac{ae^a}{2n}(1 + o(1)) + O(an^{-1}) + O(an^{-2})\right\} \\
&= ne^{-2a}b^2 - \frac{ae^{-a}}{2}(1 + o(1)) + O(ae^{-2a}) \\
&= ne^{-2a}b^2 + O(ae^{-a})
\end{aligned} \tag{14}$$

under the assumptions that $n \to \infty$ and $a = o(n)$. Moreover,

$$\frac{ae^{-a}}{ne^{-2a}b^2} = \frac{ae^a}{nb^2} = \frac{ae^a}{n(e^a - 1 - a)} = \begin{cases} \dfrac{a}{n}(1 + o(1)) = o(1) & \text{if } a \to \infty \\[2ex] \dfrac{2a(1 + o(1))}{na^2} = O\left(\dfrac{1}{N}\right) & \text{if } a \to 0 \\[2ex] O\left(\dfrac{1}{n}\right) & \text{if } a \to \alpha \in (0, \infty), \end{cases}$$

yielding (10) when $N \to \infty$ and $a = o(n)$.

(ii) From the definition of a and (14), if $\sigma^2 \to \infty$, then $N \to \infty$, and from (7)

$$\sigma^2 \le n\left(1 - \frac{1}{n}\right)^N \le ne^{-a},$$

implying $a = o(n)$ when $\sigma^2 \to \infty$. On the other hand, if $ne^{-2a}b^2 \to \infty$, then

$$\infty \leftarrow ne^{-2a}b^2 = \frac{N}{a}e^{-2a}(e^a - 1 - a) = O(N),$$

$$\frac{a}{n} \le \frac{ae^a}{nb^2} \le \frac{e^{2a}}{nb^2} = o(1),$$

and so again $N \to \infty$ and $a = o(n)$. Hence, if one of σ^2 and $ne^{-2a}b^2$ tends to ∞, then by (i), (10) holds and the other likewise approaches ∞. Now assume that

$$\sigma^2 = ne^{-2a}b^2(1 + o(1)) \to \infty. \tag{15}$$

From (15)

$$ne^{-a} \to \infty, \qquad \frac{e^a}{b\sqrt{n}} \to 0, \qquad \frac{b}{\sqrt{n}} \to 0, \qquad \frac{\sqrt{n}(e^a - 1)}{b} \to \infty. \tag{16}$$

In order to evaluate

$$P\left\{\frac{U_n - ne^{-a}}{\sqrt{n}be^{-a}} < x\right\} = P\{U_n < ne^{-a} + x\sqrt{n}be^{-a}\} \tag{17}$$

for $x \ne 0$, define $k = k_n$ so that $n - k + 1$ is the smallest integer $\ge ne^{-a} + x\sqrt{n}be^{-a}$. Then

$$n - k = ne^{-a} + x\sqrt{n}be^{-a} + O(1),$$

and from (15)

$$n - k = ne^{-a} + x(1 + o(1))\sqrt{n}be^{-a} = ne^{-a}\left[1 + \frac{xb(1 + o(1))}{\sqrt{n}}\right], \tag{18}$$

so that via (18)

$$\log\left(1 - \frac{k}{n}\right) = -a + \frac{xb(1 + o(1))}{\sqrt{n}}, \tag{19}$$

$$\frac{k}{n - k} = \frac{n}{n - k} - 1 = \frac{e^a}{1 + [xb(1 + o(1))/\sqrt{n}]} - 1 \tag{20}$$

$$= e^a - 1 + O\left(\frac{be^a}{\sqrt{n}}\right)$$

recalling (16).

From (9) and (17)

$$\mathrm{P}\left\{\frac{U_n - ne^{-a}}{\sqrt{n}be^{-a}} < x\right\} = \mathrm{P}\{U_n < n - k + 1\} = \mathrm{P}\{U_n \le n - k\}$$

$$= \mathrm{P}\{S_k \le N\} = \mathrm{P}\left\{\frac{S_k - \mathrm{E}\, S_k}{\sigma(S_k)} \le \frac{N - \mathrm{E}\, S_k}{\sigma(S_k)}\right\}. \quad (21)$$

Now by (8)

$$\mathrm{E}\, Y_{n,j} = \frac{n}{n - j + 1}, \qquad \sigma^2(Y_{n,j}) = \left(\frac{n}{n - j + 1}\right)^2 - \frac{n}{n - j + 1}$$

for $j \ge 1$, whence

$$\mathrm{E}\, S_k = \sum_{j=1}^{k} \frac{n}{n - j + 1} = n \sum_{n-k+1}^{n} \frac{1}{j},$$

$$\sigma^2(S_k) = \sum_{j=1}^{k}\left[\left(\frac{n}{n - j + 1}\right)^2 - \frac{n}{n - j + 1}\right] = \sum_{n-k+1}^{n}\left(\frac{n^2}{j^2} - \frac{n}{j}\right).$$

Since for $m = 1, 2$

$$0 \le \int_{j-1}^{j} \frac{dt}{t^m} - \frac{1}{j^m} \le \frac{1}{(j-1)^m} - \frac{1}{j^m}$$

it follows that

$$0 \le \log\frac{n}{n - k} - \sum_{n-k+1}^{n} \frac{1}{j} \le \frac{1}{n - k} - \frac{1}{n} = \frac{k}{n(n - k)},$$

$$0 \le \left(\frac{1}{n - k} - \frac{1}{n}\right) - \sum_{n-k+1}^{n} \frac{1}{j^2} \le \frac{1}{(n - k)^2} - \frac{1}{n^2} \le \frac{2nk}{n^2(n - k)^2},$$

whence, recalling (19) and (20),

$$N - \mathrm{E}\, S_k = N - n \log\frac{n}{n - k} + O\left(\frac{k}{n - k}\right)$$

$$= n\left[a + \log\left(1 - \frac{k}{n}\right)\right] + O(e^a)$$

$$= xb\sqrt{n}(1 + o(1)), \quad (22)$$

and via (20), (19), and (16)

$$\sigma^2(S_k) = \frac{nk}{n - k} - n \log\frac{n}{n - k} + O\left(\frac{nk}{(n - k)^2}\right)$$

$$= n\left\{\frac{k}{n-k} + \log\left(1 - \frac{k}{n}\right) + O\left(\frac{k}{(n-k)^2}\right)\right\}$$

$$= n\left\{e^a - 1 + O\left(\frac{be^a}{\sqrt{n}}\right) - a + O\left(\frac{e^a + O(be^a/\sqrt{n})}{ne^{-a}}\right)\right\}$$

$$= nb^2\left\{1 + O\left(\frac{e^a}{b\sqrt{n}}\right) + O\left(\frac{e^{2a}}{nb^2}\right)\right\}$$

$$= nb^2(1 + o(1)). \tag{23}$$

Moreover, by (22), (23)

$$\frac{N - \mathrm{E}\, S_k}{\sigma(S_k)} = x(1 + o(1)). \tag{24}$$

Now, from (6) and (13)

$$\mathrm{E}\, U_n = n\left(1 - \frac{1}{n}\right)^{na} = n \exp\left\{a\left[-1 - \frac{1}{2n} + O(n^{-2})\right]\right\} = ne^{-a}\left(1 + O\left(\frac{a}{n}\right)\right),$$

$$= ne^{-a} + O(1) \tag{25}$$

To complete the proof, by Exercise 9.1.2 it remains to verify that $\{Y_{n,j} - \mathrm{E}\, Y_{n,j}, 2 \le j \le k_n\}$ obey the Lindeberg condition.

Setting $Y_j' = Y_{n,j+1} - 1$ and $q_j = j/n$, $\mathrm{P}\{Y_j' = h\} = q_j^h(1 - q_j)$, $h = 0, 1, \ldots,$ $1 \le j \le k - 1$. Recalling Exercise 4.1.1 and (23),

$$\sigma_{S_k}^2 = \sum_{j=1}^{k-1}\left[\left(\frac{q_j}{1 - q_j}\right)^2 + \frac{q_j}{1 - q_j}\right] = nb^2(1 + o(1))$$

$$\sum_{j=1}^{k-1} \mathrm{E}\, Y_j'^3 = 6\sum_{j=1}^{k-1}\left[\left(\frac{q_j}{1 - q_j}\right)^3 + \left(\frac{q_j}{1 - q_j}\right)^2\right] + \sum_{j=1}^{k-1}\frac{q_j}{1 - q_j}.$$

For $1 \le j \le k$, noting (20) and (16),

$$\frac{q_j}{1 - q_j} \le \frac{k}{n - k} = e^a - 1 + O(be^a/n^{1/2}) = o(bn^{1/2}) = o(\sigma_{S_k})$$

whence

$$\sum_{j=1}^{k-1}\left(\frac{q_j}{1 - q_j}\right)^3 \le \frac{k}{n - k}\sum_{j=1}^{k-1}\left(\frac{q_j}{1 - q_j}\right)^2 = o(\sigma_{S_k}^3).$$

Clearly,

$$\sum_{j=1}^{k-1}\left[\left(\frac{q_j}{1 - q_j}\right)^2 + \left(\frac{q_j}{1 - q_j}\right)\right] = o(\sigma_{S_k}^3)$$

and so

$$\sum_{j=1}^{k-1} \mathrm{E}|Y_j' - \mathrm{E}\, Y_j'|^3 \le 8\sum_{j=1}^{k-1} \mathrm{E}\, Y_j'^3 = o(\sigma_{S_k}^3).$$

Thus, the Liapounov and hence also the Lindeberg condition of Exercise 9.1.2 holds. □

Remarks. (i) The denominator in (12) can be replaced by $\sqrt{n}e^{-\alpha}(e^{\alpha} - 1 - \alpha)^{1/2}$ if $a \to \alpha \in (0, \infty)$, by $a(n/2)^{1/2}$ if $a \to 0$, and by $(ne^{-a})^{1/2}$ if $a \to \infty$. (ii) If $a/n > \varepsilon > 0$, then $\sigma^2 \le ne^{-a} \to 0$ via (7); if $N = na \le C < \infty$, then $n^{1/2}a = o(1)$ and $\sigma^2 = (na^2/2)(1 + o(1)) + O(a) \to 0$ via (14). (iii) If $\sigma^2 \to \sigma_0^2 \in (0, \infty)$ and $a \to \infty$, then via (10) $a = \log n - \log \sigma_0^2 + o(1)$; if $\sigma^2 \to \sigma_0^2 \in (0, \infty)$ and $a \to 0$, then $a^2 = (2\sigma_0^2/n)(1 + o(1))$. Here, the possible limiting distributions can be completely determined (see Theorem 4, Theorem 3.1.4, and Exercise 6). (iv) If $\sigma^2 \to 0$, either $a \to 0$ or $a \to \infty$ and the limit distributions are degenerate (Exercise 6).

Theorem 4. *If U_n is the number of empty cells in distributing N_n balls at random into n cells and $N_n^2/n = N_n a_n \to \delta^2 \in [0, \infty)$, then*

$$\lim_{n\to\infty} P\{U_n - (n - N_n) = j\} = \frac{(\delta^2/2)^j e^{-\delta^2/2}}{j!}, \qquad j = 0, 1, \ldots. \tag{26}$$

PROOF. According to (9) and (8) with $N = N_{n,j}$

$$P\{U_n \le n - N + k\} = P\{S_{N-k} \le N\} = P\{S_{N-k} - (N - k) \le k\},$$

where $S_{N-k} - (N - k) = \sum_{j=1}^{N-k}(Y_j - 1)$ and $\{Y_j - 1, j \ge 2\}$ are independent, geometric r.v.s with success probabilities $1 - (j - 1)/n$. Hence, the characteristic function of $S_{N-k} - (N - k)$, say $\varphi(t)$, is given by

$$\begin{aligned}\varphi(t) &= \prod_{j=1}^{N-k}\left(1 - \frac{j-1}{n}\right)\left[1 - \frac{j-1}{n}e^{it}\right]^{-1}\\ &= \prod_{j=1}^{N-k}\exp\left\{-\frac{j-1}{n} + \frac{j-1}{n}e^{it} + O\left(\frac{j^2}{n^2}\right)\right\}\\ &= \exp\left\{\frac{N_n^2}{2n}(e^{it} - 1) + O\left(\frac{N_n^3}{n^2}\right)\right\}\\ &\to \exp\left\{\frac{\delta^2}{2}(e^{it} - 1)\right\},\end{aligned}$$

and Corollary 8.3.5 guarantees that the d.f. of $U_n - (n - N_n)$ tends to a Poisson d.f. with parameter $\delta^2/2$, yielding (26). □

EXERCISES 9.2

1. For an arbitrary d.f. G, concoct a sequence of interchangeable r.v.s for which $P\{S_n < x\sqrt{n}\} \to \int \Phi(x/y)dG(y)$.
2. If $\{X_n, n \ge 1\}$ are interchangeable r.v.s with $E X_1 = 0$, $E X_1^2 = 1$, $E|X_1|^3 < \infty$, $\text{Cov}(X_1, X_2) = 0 = \text{Cov}(X_1^2, X_2^2)$, then Corollary 9.1.4 holds.
3. Suppose that for each $n = 1, 2, \ldots, \{X_{ni}, i = 1, 2, \ldots\}$ constitute interchangeable random variables with $E X_{n1} = 0$, $E X_{n1}^2 = 1$, $E|X_{n1}|^3 < \infty$ and set

$$m_n(\omega) = \mathrm{E}\{X_{n1} \mid \mathscr{G}_n\}, \sigma_n^2(\omega) = \mathrm{E}[(X_{n1} - m_n)^2 \mid \mathscr{G}_n], \quad \alpha_n(\omega) = \mathrm{E}\{|X_{n1} - m_n|^3 \mid \mathscr{G}_n\}$$

when $\mathscr{G}_n$ is the σ-algebra relative to which $\{X_{ni}, i \geq 1\}$ are conditionally i.i.d. If $\sqrt{n}m_n(\omega) \xrightarrow{\mathrm{P}} 0$, $\sigma_n(\omega) \xrightarrow{\mathrm{P}} 1$, $\alpha_n(\omega)/\sqrt{n}\sigma_n^3(\omega) \xrightarrow{\mathrm{P}} 0$, and $S_n = \sum_{i=1}^n X_{ni}$, then $S_n/\sqrt{n} \xrightarrow{\mathrm{d}} N_{0,1}$.

4. If for each $n \geq 1$ $\{X_{ni}, i \geq 1\}$ are interchangeable r.v.s with $\mathrm{E}\, X_{n1} = 0$, $\mathrm{E}\, X_{n1}^2 = 1$, $\mathrm{E}|X_{n1}|^3 < \infty$ and $\mathrm{E}\, X_{n1}X_{n2} = o(n^{-1})$, $\mathrm{E}\, X_{n1}^2 X_{n2}^2 \to 1$, $\mathrm{E}|X_{n1}|^3 = o(n^{1/2})$, then $S_n/\sqrt{n} \xrightarrow{\mathrm{d}} N_{0,1}$.

5. If two of the last three conditions of Exercise 4 obtain but the third is violated, construct a sequence $\{X_{ni}, i \geq 1\}$ of interchangeable processes for which $S_n/\sqrt{n} \not\xrightarrow{\mathrm{d}} N_{0,1}$.

6. Apropos of Theorems 3 and 4, prove that (i) if $a_n - \log n \to \infty$, $\mathrm{P}\{\sum_1^n X_{ni} = 0\} \to 1$, (ii) if $Na_n \to 0$, $\mathrm{P}\{\sum X_{ni} = n - N\} \to 1$, (iii) if $a_n - \log n \to \delta$ finite and P_λ designates a Poisson r.v. with mean λ, $\sum_{i=1}^n X_{ni} \xrightarrow{\mathrm{d}} \mathrm{P}_{\exp\{-\delta\}}$ (see Theorem 3.1.2).

7. Let the i.i.d. r.v.s $\{Z_n, n \geq 1\}$ be independent of Y, where Y is uniformly distributed on $(0, 1)$ and $\mathrm{P}\{Z_n = \pm 1\} = \frac{1}{2}$. Then $X_n = Y^{-1/2}Z_n$, $n \geq 1$, are $\mathscr{L}_1$ interchangeable r.v.s with $(1/n)\sum_1^n X_i \xrightarrow{\text{a.c.}} 0$ but $\mathrm{E}\, X_1 X_2$ does not exist.

9.3 The Martingale Case

If the component r.v.s $\{X_n, n \geq 1\}$ are **martingale differences**, that is,

$$\mathrm{E}\{X_{n+1} \mid \mathscr{F}_n\} = 0, n \geq 1,$$

and, moreover, obey the Lindeberg condition, a central limit theorem will hold provided the conditional variances are sufficiently well behaved.

Theorem 1. *If $\{X_n, \mathscr{F}_n, n \geq 1\}$ is a stochastic sequence with $\mathrm{E}\{X_{n+1} \mid \mathscr{F}_n\} = 0$, $n \geq 0$ (with $\mathscr{F}_0 = \{\varnothing, \Omega\}$), such that $\{X_n\}$ obey the Lindeberg condition and, setting $\sigma_n^2 = \mathrm{E}\, X_n^2$, $s_n^2 = \sum_1^n \sigma_j^2$,*

$$\sum_{j=1}^n \mathrm{E}|\mathrm{E}\{X_j^2 \mid \mathscr{F}_{j-1}\} - \sigma_j^2| = o(s_n^2), \tag{1}$$

then $(1/s_n)\sum_{j=1}^n X_j \xrightarrow{\mathrm{d}} N_{0,1}$.

PROOF. The proof of Theorem 9.1.1 may easily be adapted to handle the current situation. In fact, let $Y_j(t)$, $a_j(t)$ be exactly as defined there and set

$$b_j(t) = \frac{t^2}{2}[\mathrm{E}\{X_j^2 \mid \mathscr{F}_{j-1}\} - \sigma_j^2].$$

Then 9.1(5) holds provided (i) all expectations E therein are changed to $\mathrm{E}\{\cdot \mid \mathscr{F}_{j-1}\}$, (ii) the term $b_j(t/s_n)$ is added to the expression within the absolute value signs on the right, and (iii) the term $|b_j(t/s_n)|$ is added within the brackets on the extreme right. Analogously, (6) obtains if the alteration (iii) is effected and E is replaced by $E\{\cdot|\mathscr{F}_{j-1}\}$ in the second expectation of the second expression (where independence was used) and in all succeeding expressions of (6). Finally, (7) holds if $E\sum_{j=1}^{n}|b_j(t/s_n)|$ is appended to the extreme right side and so, recalling (1), the theorem follows. □

A version of the Berry–Esseen theorem also carries over to martingales.

Theorem 2. *If $\{S_n = \sum_1^n X_i, n \geq 1\}$ is an $\mathscr{L}_3$ martingale with $E\,S_n = 0$, $E\,S_n^2 = s_n^2 = \sum_1^n \sigma_i^2$, $n \geq 1$, there exists an absolute constant c_0 such that*

$$\left|P\left\{\frac{S_n}{s_n} < x\right\} - \Phi(x)\right| \leq c_0\left[\frac{1}{s_n^3}\sum_1^n E|X_j|^3\right]^{1/4}$$

$$+\left[c_0 - \frac{1}{\sqrt{2\pi}}\right]\frac{\sum_1^n E|E\{X_j^2|X_1,\ldots,X_{j-1}\} - \sigma_j^2|}{s_n^{1/2}\sqrt{\sum_1^n E|X_j|^3}}.$$

Lemma 1. *If f is a real function with $f^{(n-1)}$ absolutely continuous on $[a, b]$, then, setting $f^{(0)} = f$,*

$$f(b) = \sum_{j=0}^{n-1}\frac{(b-a)^j}{j!}f^{(j)}(a) + \frac{1}{(n-1)!}\int_a^b (b-x)^{n-1}f^{(n)}(x)dx. \tag{2}$$

PROOF. Set

$$F(x) = f(b) - \sum_{j=0}^{n-1}\frac{(b-x)^j}{j!}f^{(j)}(x), \tag{3}$$

whence $F'(x) = [-(b-x)^{n-1}/(n-1)!]f^{(n)}(x)$, a.e. on $[a, b]$. Thus, F is absolutely continuous on $[a, b]$ and

$$F(a) = F(b) - \int_a^b F'(x)dx = \frac{1}{(n-1)!}\int_a^b (b-x)^{n-1}f^{(n)}(x)dx,$$

which is tantamount to (2) in view of (3). □

Lemma 2. *If $\{S_n = \sum_1^n X_i, n \geq 1\}$ is as in Theorem 2, then for any function f with absolutely continuous second derivative on every bounded interval and some absolute constant C in $(0, \frac{2}{3}]$*

$$\left|E\,f\left(\frac{S_n}{s_n}\right) - E\,f(Y_0)\right|$$

$$\leq C\left[\frac{\|f^{(3)}\|}{s_n^3}\sum_{j=1}^{n}E|X_j|^3 + \frac{\|f^{(2)}\|}{s_n^2}\sum_1^n E\left|E\{X_j^2|X_1,\ldots,X_{j-1}\} - \sigma_j^2\right|\right], \tag{4}$$

where Y_0 is normal with mean 0, variance 1, and $\|f^{(j)}\| = \inf[M: \mu\{|f^{(j)}(x)| \geq M\} = 0]$, $1 \leq j \leq 3$, with μ denoting Lebesgue measure on the real line.

PROOF. Let $\{Y_n, n \geq 0\}$ be independent normal r.v.s with means zero and variances σ_n^2 ($\sigma_0^2 = 1$) with $\{Y_n, n \geq 0\}$ independent of $\{X_n, n \geq 1\}$.

Set $Q_{j,n} = \sum_{i=1}^{j-1} X_i + \sum_{j+1}^{n} Y_i$, $1 \leq j \leq n$, and note that $Q_{j,n} + X_j = Q_{j+1,n} + Y_{j+1}$. By Lemma 1

$$\left| f(a+h) - f(a) - hf'(a) - \left(\frac{h^2}{2}\right) f^{(2)}(a) \right| \leq \frac{|h|^3}{6} \|f^{(3)}\|,$$

whence for some r.v.s θ_j with $|\theta_j| \leq 1$, $j = 1, 2$,

$$\mathrm{E}\, f\left(\frac{S_n}{s_n}\right) - \mathrm{E}\, f(Y_0) = \sum_{j=1}^{n} \mathrm{E}\left[f\left(\frac{Q_{j,n} + X_j}{s_n}\right) - f\left(\frac{Q_{j,n} + Y_j}{s_n}\right)\right]$$

$$= \sum_{j=1}^{n} \mathrm{E}\left[\frac{X_j - Y_j}{s_n} f'\left(\frac{Q_{j,n}}{s_n}\right) + \frac{X_j^2 - Y_j^2}{2s_n^2} f^{(2)}\left(\frac{Q_{j,n}}{s_n}\right) + \frac{\theta_j(|X_j|^3 + |Y_j|^3)}{6s_n^3} \|f^{(3)}\|\right]. \tag{5}$$

Since $\mathrm{E}|Y_j|^3 = c_1 \sigma_j^3 = c_1(\mathrm{E}\, X_j^2)^{3/2} \leq c_1\, \mathrm{E}|X_j|^3$ and

$$\mathrm{E}(X_j^i - Y_j^i) f^{(i)}\left(\frac{Q_{j,n}}{s_n}\right)$$

$$= \begin{cases} \mathrm{E}\left\{ f^{(i)}\left(\frac{Q_{j,n}}{s_n}\right) \mathrm{E}\{X_j^i - Y_j^i | X_1 \cdots X_{j-1}, Y_{j+1}, \ldots, Y_n\}\right\} = 0, & i = 1 \\ \mathrm{E}\, f^{(2)}\left(\frac{Q_{j,n}}{s_n}\right)[\mathrm{E}\{X_j^2 | X_1, \ldots, X_{j-1}\} - \sigma_j^2], & i = 2, \end{cases}$$

the conclusion of the lemma follows from (5). ☐

PROOF OF THEOREM 2. Let $\varepsilon > 0$ and define $h(-t) = -h(t)$, where

$$h(t) = \begin{cases} \frac{16t^3}{3\varepsilon^3} - \frac{2t}{\varepsilon}, & 0 \leq t \leq \frac{\varepsilon}{4} \\ \frac{16}{3\varepsilon^3}\left(\frac{\varepsilon}{2} - t\right)^3 - \frac{1}{2}, & \frac{\varepsilon}{4} \leq t \leq \frac{\varepsilon}{2} \\ -\frac{1}{2}, & t \geq \frac{\varepsilon}{2}. \end{cases} \tag{6}$$

Then h is a nonincreasing odd function with h'' absolutely continuous and $\|h''\| = 8\varepsilon^{-2}$, $\|h'''\| = 32\varepsilon^{-3}$, and consequently

$$f(t) = h\left(t - x - \frac{\varepsilon}{2}\right) + \frac{1}{2} \tag{7}$$

is a nonincreasing function on $(-\infty, \infty)$ whose second and third derivatives have the same norms as those of h and which vanishes on $(x + \varepsilon, \infty)$, while $f(t) = 1$ for $t \le x$. Since $f^{(2)}$ is absolutely continuous on $(-\infty, \infty)$, by Lemma 2, for any real x

$$P\left\{\frac{S_n}{s_n} < x\right\} \le E\, f\left(\frac{S_n}{s_n}\right) \le E\, f(Y_0) + I \le \Phi(x + \varepsilon) + I, \tag{8}$$

where

$$I = 32C\left[\frac{1}{\varepsilon^2 s_n^2}\sum_1^n E\left|E\{X_j^2 \mid X_1, \ldots, X_{j-1}\} - \sigma_j^2\right| + \frac{1}{\varepsilon^3 s_n^3}\sum_1^n E|X_j|^3\right].$$

By the mean value theorem $\Phi(x + \varepsilon) - \Phi(x) \le \varepsilon/\sqrt{2\pi}$, whence from (8)

$$P\left\{\frac{S_n}{s_n} < x\right\} \le \Phi(x) + \frac{\varepsilon}{\sqrt{2\pi}} + I. \tag{9}$$

Define

$$K_n = \frac{1}{s_n^3}\sum_{j=1}^n E|X_j|^3,$$

$$M_n = \frac{1}{s_n^2}\sum_{j=1}^n E\left|E\{X_j^2 \mid X_1, \ldots, X_{j-1}\} - \sigma_j^2\right|.$$

Then, choosing $\varepsilon = K_n^{1/4}$ in (9),

$$P\left\{\frac{S_n}{s_n} < x\right\} - \Phi(x) \le \frac{1}{\sqrt{2\pi}} K_n^{1/4} + 32C(K_n^{-1/2}M_n + K_n^{1/4}), \tag{10}$$

which is tantamount to the upper bound of the theorem. In analogous fashion a lower bound is obtained which combined with (10) yields the theorem. □

A discussion of martingale and other central limit theorems based on Dvoretzky (1970) and McLeish (1974) is given in Section 5.

Exercises 9.3

1. Utilize Theorem 9.3.1 to prove sufficiency in Theorem 9.2.1.
2. If $\{X_n\}$ are independent with $E\,X_i = 0$, $E\,X_i^2 = \sigma_i^2$, $E|X_i|^{2+\delta} < \infty$, $0 < \delta \le 1$, modify the right side of (4) of Lemma 2 to $(\|f^{(2)}\| + \|f^{(3)}\|)\sum_{j=1}^n E|X_i|^{2+\delta}$. *Hint*: Employ a 2- or 3-term Taylor expansion according as $|h| > 1$ or $|h| \le 1$.
3. If $\{X_n\}$ are i.i.d. with $E\,X_n = 0$, $E\,X_n^2 = 1$, $E|X_n|^{2+\delta} < \infty$, $0 < \delta \le 1$, prove the large deviation result that if $\delta \log n - 4\alpha_n^2 \to \infty$, $P\{S_n/\sqrt{n} \ge \alpha_n\} \sim 1 - \Phi(\alpha_n)$.

9.4 Miscellaneous Central Limit Theorems

The first central limit theorem concerns sums of random numbers, say t_n, of independent random variables $\{X_n, n \geq 1\}$ and permits t_n to be highly dependent upon the sequence $\{X_n\}$.

Theorem 1 (Doeblin–Anscombe). *Let $\{X_n, n \geq 1\}$ be independent random variables with $\mathrm{E}\,X_n = 0$, $\mathrm{E}\,X_n^2 = 1$, and $\{t_n, n \geq 1\}$ positive integer-valued random variables with $t_n/b_n \xrightarrow{\mathrm{P}} c$, where $\{c, b_n, n \geq 1\}$ are positive, finite constants such that $b_n \uparrow \infty$. If $n^{-1/2} \sum_{j=1}^{n} X_j \xrightarrow{\mathrm{d}} N_{0,1}$, then*

$$\frac{1}{t_n^{1/2}} \sum_{j=1}^{t_n} X_j \xrightarrow{\mathrm{d}} N_{0,1}.$$

PROOF. Set $S_n = \sum_1^n X_j$ and $k_n = [cb_n]$ = greatest integer $\leq cb_n$.

Now,

$$\frac{S_{t_n}}{t_n^{1/2}} = \left(\frac{k_n}{t_n}\right)^{1/2} \left[\frac{S_{k_n}}{k_n^{1/2}} + \frac{S_{t_n} - S_{k_n}}{k_n^{1/2}}\right] \tag{1}$$

and according to the hypothesis the first factor on the right side of (1) converges in probability to one. Moreover, $S_{k_n}/k_n^{1/2}$ as a subsequence of $S_n/n^{1/2}$ converges in distribution to $N_{0,1}$. Thus, to prove the theorem it suffices to establish that

$$\frac{S_{t_n} - S_{k_n}}{k_n^{1/2}} \xrightarrow{\mathrm{P}} 0. \tag{2}$$

To this end, note that for any positive numbers ε, δ

$$\begin{aligned} \mathrm{P}\{|S_{t_n} - S_{k_n}| > \varepsilon k_n^{1/2}\} \leq\ & \mathrm{P}\{|S_{t_n} - S_{k_n}| > \varepsilon k_n^{1/2}, |t_n - k_n| \leq \delta k_n\} \\ & + \mathrm{P}\{|t_n - k_n| > \delta k_n\}. \end{aligned} \tag{3}$$

Since the event corresponding to the first term on the right implies $A_n^+ \cup A_n^-$, where

$$A_n^+ = \left\{\max_{k_n \leq j \leq (1+\delta)k_n} |S_j - S_{k_n}| > \varepsilon k_n^{1/2}\right\},$$

$$A_n^- = \left\{\max_{k_n(1-\delta) \leq j \leq k_n} |S_j - S_{k_n}| > \varepsilon k_n^{1/2}\right\},$$

and since by Kolmogorov's inequality

$$\mathrm{P}(A_n^{\pm}) \leq (\varepsilon^2 k_n)^{-1}\, \mathrm{E}(S_{k_n \pm [\delta k_n]} - S_{k_n})^2 \leq (\varepsilon^2 k_n)^{-1} \delta k_n = \varepsilon^{-2}\delta,$$

it follows from (3) that

$$\mathrm{P}\{|S_{t_n} - S_{k_n}| > \varepsilon k_n^{1/2}\} \leq 2\delta\varepsilon^{-2} + o(1). \tag{4}$$

For arbitrary $\varepsilon > 0$, the first term on the right in (4) tends to zero with δ and the theorem follows. □

Actually, Theorem 1 holds even if t_n/n converges in probability to a positive random variable (Blum *et al.*, 1962–1963; Mogyorodi, 1962).

If r.v.s $\{X_n, n \geq 1\}$ obey the hypothesis of the ensuing theorem, then $N_c = \sup\{n \geq 1: \sum_{i=1}^n X_i \leq c\}$, is a bona fide r.v. In the special case of classical renewal theory (Section 5.4), where $X_n > 0, n \geq 1$, implies $N_c + 1 = T_c(0)$ [see (5)], asymptotic normality of N_c is an immediate consequence of Theorem 2 below. On the other hand, if merely $\mathrm{E}\, X_n \equiv \mu > 0$, then only the inequality $N_c \geq T_c(0) - 1$ may be inferred. Nonetheless, N_c is still asymptotically normal under the hypothesis of Corollary 2.

Theorem 2. *If* $\{X, X_n, n \geq 1\}$ *are* i.i.d. r.v.s *with* $\mathrm{E}\, X = \mu > 0$ *and* $\sigma^2 = \sigma_X^2 \in (0, \infty)$, *define* $S_n = \sum_1^n X_i$ *and*

$$T = T_c = T_c(\alpha) = \inf\{n \geq 1: S_n > cn^\alpha\}, \qquad -\infty < \alpha < 1, c > 0. \tag{5}$$

Then, as $c \to \infty$

$$\frac{\mu(1-\alpha)[T_c(\alpha) - (c/\mu)^{1/(1-\alpha)}]}{\sigma(c/\mu)^{1/2(1-\alpha)}} \xrightarrow{\mathrm{d}} N_{0,1}. \tag{6}$$

PROOF. For simplicity, take $\mu = 1$. Since via the strong law of large numbers $T_c/c^{1/(1-\alpha)} \xrightarrow{\text{a.c.}} 1$ as $c \to \infty$ (Exercise 5.4.6), by Theorems 1 and 8.1.1

$$\left(\frac{T}{c^{1/(1-\alpha)}}\right)^{1/2} \frac{(S_T - T)}{\sigma\sqrt{T}} = \frac{S_T - cT^\alpha}{\sigma c^{1/2(1-\alpha)}} + \frac{cT^\alpha - T}{\sigma c^{1/2(1-\alpha)}} \xrightarrow{\mathrm{d}} N_{0,1}. \tag{7}$$

Since $\mathrm{E}\, X_1^2 < \infty$, $X_n/\sqrt{n} \xrightarrow{\text{a.c.}} 0$, implying $X_T/\sqrt{T} \xrightarrow{\text{a.c.}} 0$ as $c \to \infty$, whence

$$0 \leq \frac{S_T - cT^\alpha}{\sigma c^{1/2(1-\alpha)}} \leq \frac{X_T}{\sqrt{T}} \frac{\sqrt{T}}{\sigma c^{1/2(1-\alpha)}} \xrightarrow{\text{a.c.}} 0,$$

and (7) ensures

$$\frac{cT^\alpha - T}{\sigma c^{1/2(1-\alpha)}} \xrightarrow{\mathrm{d}} N_{0,1}.$$

Now $Z_c = 1 + o(1)$ entails $Z_c^{1-\alpha} = 1 + (Z_c - 1)(1 - \alpha + o(1))$, so that

$$\begin{aligned} cT^\alpha - T &= T\{[T^{-1}c^{1/(1-\alpha)}]^{1-\alpha} - 1\} \\ &= T[T^{-1}c^{1/(1-\alpha)} - 1](1 - \alpha + o(1)) \\ &= -(1-\alpha)[T - c^{1/(1-\alpha)}](1 + o(1)), \end{aligned}$$

and the theorem follows. □

Corollary 1. *Under the hypothesis of Theorem 2, for* $-\infty < \alpha < 1$

$$\frac{\max_{1 \leq j \leq n}(S_j/j^\alpha) - \mu n^{1-\alpha}}{\sigma n^{(1/2)-\alpha}} \xrightarrow{\mathrm{d}} N_{0,1}. \tag{8}$$

PROOF. For $x \neq 0$, define

$$c = c_n = n^{1-\alpha}\mu - x\sigma n^{(1/2)-\alpha}. \tag{9}$$

Then, setting $q_c = [\sigma/\mu(1-\alpha)](c/\mu)^{1/2(1-\alpha)}$,

$$
\begin{aligned}
\mathrm{P}\left\{\max_{0\le j\le n}\frac{S_j}{j^\alpha} - \mu n^{1-\alpha} > -x\sigma n^{(1/2)-\alpha}\right\}
&= \mathrm{P}\left\{\max_{0\le j\le n}\frac{S_j}{j^\alpha} > c\right\} = \mathrm{P}\{T_c \le n\} \\
&= \mathrm{P}\left\{\frac{T_c - (c/\mu)^{1/(1-\alpha)}}{q_c} \le \frac{n-(c/\mu)^{1/(1-\alpha)}}{q_c}\right\} \to \Phi(x)
\end{aligned}
$$

by Theorem 2 via inversion of (9). □

Corollary 2. *If* $\{X, X_n, n \ge 1\}$ *are* i.i.d. *with* $\mathrm{E}\,X_1 = \mu > 0$, $\sigma^2 = \sigma_X^2 \in (0, \infty)$, *and* $N_c = \sup\{n \ge 1: \sum_{i=1}^n X_i \le c\}$, $c \ge 0$, *then*

$$\frac{N_c - c/\mu}{c^{1/2}\sigma\mu^{-3/2}} \xrightarrow{d} N_{0,1} \quad \text{as } c \to \infty.$$

PROOF. In view of Theorem 2, it suffices to show that $(N_c - T_c(0)/c^{1/2}$ converges to zero in probability or in $\mathscr{L}_1$, and, since $N_c - T_c(0) + 1 \ge 0$, it is enough to verify that $\mathrm{E}(N_c - T_c(0)) \le \mathrm{E}\,N_0 < \infty$. Since $\mathrm{E}\,T_c(0) < \infty$, by Corollary 5.4.1,

$$
\begin{aligned}
\mathrm{E}(N_c - T_c(0)) &= \sum_{n=1}^\infty (\mathrm{P}\{N_c \ge n\} - \mathrm{P}\{T_c(0) \ge n\}) \\
&\le \sum_{n=1}^\infty \mathrm{P}\{T_c(0) < n \le N_c\} \\
&= \sum_{n=1}^\infty \sum_{k=1}^{n-1} \mathrm{P}\{T_c(0) = k, S_j \le c \text{ for some } j \ge n\} \\
&\le \sum_{n=1}^\infty \sum_{k=1}^{n-1} \mathrm{P}\{T_c(0) = k\}\mathrm{P}\left\{\inf_{j\ge n}(S_j - S_k) < 0\right\} \\
&= \sum_{k=1}^\infty \sum_{n=k+1}^\infty \mathrm{P}\{T_c(0) = k\}\mathrm{P}\left\{\inf_{j\ge n-k} S_j < 0\right\} \\
&= \sum_{k=1}^\infty \mathrm{P}\{T_c(0) = k\}\sum_{n=1}^\infty \mathrm{P}\left\{\inf_{j\ge n} S_j < 0\right\} \\
&= \sum_{n=1}^\infty \mathrm{P}\left\{\inf_{j\ge n} S_j < 0\right\} \le \mathrm{E}\,N_0 < \infty
\end{aligned}
$$

via Corollary 10.4.5. □

The invariance principle, launched by Erdos and Kac (1946) consists in (i) showing that under some condition on r.v.s $\{X_n, n \ge 1\}$ (e.g., that of Lindeberg) the limit distribution of some functional (e.g., $\max_{1\le j\le n}\sum_{i=1}^j X_i$) of

$\{X_n\}$ is independent of the underlying distributions F_n of X_n, $n \geq 1$, (ii) evaluating this limit distribution for an expeditious choice of $\{F_n\}$.

By combining their technique with a sequence of stopping times, it is possible to eliminate step (ii) in

Theorem 3. *If* $\{X_n, n \geq 1\}$ *are independent* r.v.s *with* $\mathrm{E}\, X_n = 0$, $\mathrm{E}\, X_n^2 = \sigma^2 \in (0, \infty)$, $n \geq 1$, *which satisfy the Lindeberg condition,* $S_j = \sum_1^j X_i$, *and* $T_c^* = \inf\{j \geq 1 : S_j > c\}$, *then as* $c \to \infty$, $(\sigma^2/c^2)T_c^*$ *converges in distribution to the positive stable distribution of characteristic exponent* $\frac{1}{2}$ *or equivalently* $(1/\sigma n^{1/2})\max_{1 \leq j \leq n} S_j$ *converges to the positive normal distribution,* i.e., *for* $x > 0$, $y > 0$

$$\lim_{c\to\infty} \mathrm{P}\left\{T_{cx}^* > \frac{c^2 y}{\sigma^2}\right\} = \lim_{n\to\infty} \mathrm{P}\left\{\max_{1\leq j\leq ny} S_j \leq x\sigma n^{1/2}\right\} = 2\Phi\left(\frac{x}{y^{1/2}}\right) - 1. \tag{10}$$

Note. For $y = 1$, the right side of (10) is a d.f. in x, namely, the positive normal distribution, while for $x = 1$ it is one minus a d.f. in y; the latter, $2[1 - \Phi(y^{-1/2})]$, $y > 0$, is the so-called positive stable distribution of characteristic exponent $\frac{1}{2}$ (Chapter 12).

PROOF. Without loss of generality, take $\sigma = 1 = y$ and let $x > 0$. In view of the Lindeberg condition, for every $\delta > 0$

$$\sum_{j=1}^{n} \mathrm{P}\{|X_j| > \delta\sqrt{n}\} \leq \frac{1}{n\delta^2} \sum_{j=1}^{n} \mathrm{E}\, X_j^2 I_{[|X_j| > \delta\sqrt{n}]} = o(1). \tag{11}$$

For any positive integer k, set $n_j = [jn/k]$, $j = 0, 1, \ldots, k$ and $n = k, k + 1, \ldots$. If

$$Y_j = Y_{n,j} = X_{n_j+1} + \cdots + X_{n_{j+1}}, \qquad j = 0, 1, \ldots, k-1,$$

then $S_n = S_{n_k} = \sum_{i=0}^{k-1} Y_i$. Moreover, $Y_0, \ldots, Y_{k-1}$ are independent r.v.s and, furthermore, for fixed $j = 0, 1, \ldots, k - 1$ as $n \to \infty$

$$\frac{1}{\sigma^2(Y_j)} \sum_{i=n_j+1}^{n_{j+1}} \mathrm{E}\, X_i^2 I_{[|X_i| > \delta\sigma(Y_j)]} \leq \frac{2k}{n} \sum_{n_j+1}^{n_{j+1}} \mathrm{E}\, X_i^2 I_{[|X| > (\delta/2)\sqrt{n/k}]} = o(1). \tag{12}$$

Consequently, as $n \to \infty$ the r.v.s $Y_{n,j}/\sqrt{n} \xrightarrow{d} N_{0,1/k}$ for $j = 0, 1, \ldots, k - 1$.

Next, for each $i = 1, 2, \ldots, n$ let $m(i)$ be the integer for which $n_{m(i)-1} < i \leq n_{m(i)}$ and note that $0 < m(i) \leq k$. For any $\varepsilon > 0$, setting $A_i = A_{i,n}(\varepsilon) = \{|S_{n_{m(i)}} - S_i| \geq \varepsilon\sqrt{n}\}$, $1 \leq i \leq n$, and omitting the * in T_c^*

$$\begin{aligned} \mathrm{P}\{T_{x\sqrt{n}} \leq n\} - \mathrm{P}\{S_n > x\sqrt{n}\} &= \mathrm{P}\left\{S_n \leq x\sqrt{n},\ \max_{1\leq m<n} S_m > x\sqrt{n}\right\} \\ &\leq \sum_{i=1}^{n-1} \mathrm{P}\{T_{x\sqrt{n}} = i, A_i\} + \sum_{i=1}^{n-1} \mathrm{P}\{S_n \leq x\sqrt{n}, T_{x\sqrt{n}} = i, A_i^c\} \end{aligned} \tag{13}$$

$= D + B$ (say). Now, via Kolmogorov's inequality

$$D = \sum_{i=1}^{n-1} P\{T_{x\sqrt{n}} = i\}P\{A_i\}$$

$$\leq \sum_{i=1}^{n-1} \frac{1}{n\varepsilon^2}(n_{m(i)} - n_{m(i)-1})P\{T_{x\sqrt{n}} = i\} \leq \frac{1}{n\varepsilon^2}\frac{2n}{k} = \frac{2}{k\varepsilon^2}. \tag{14}$$

On the other hand, recalling that $n_k = n$ and $S_n/\sqrt{n} \xrightarrow{d} N_{0,1}$,

$$B \leq \sum_{i=1}^{n-1} [P\{S_n \leq (x - \varepsilon)\sqrt{n}, T_{x\sqrt{n}} = i, A_i^c\}$$

$$+ P\{(x - \varepsilon)\sqrt{n} < S_n \leq x\sqrt{n}, T_{x\sqrt{n}} = i\}]$$

$$\leq \sum_{i=1}^{n-1} P\{T_{x\sqrt{n}} = i, S_{n_{m(i)}} - S_n > 0\} + P\{(x - \varepsilon)\sqrt{n} < S_n \leq x\sqrt{n}\}$$

$$= \sum_{i=1}^{n-1} P\{T_{x\sqrt{n}} = i\}P\{S_{n_{m(i)}} - S_n > 0\} + O(\varepsilon). \tag{15}$$

By virtue of $n^{-1/2}Y_{n,j} \xrightarrow{d} N_{0,1/k}$, $j = 0, 1, \ldots, k - 1$, and independence, for fixed k as $n \to \infty$ necessarily

$$n^{-1/2}(S_{n_k} - S_{n_{m(i)}}) = n^{-1/2} \sum_{j=m(i)}^{k-1} Y_j \xrightarrow{d} N_{0,(k-m(i)/k} \quad \text{for } i = 1, 2, \ldots, n_{k-1}.$$

Noting that $1 \leq m(i) \leq k$ (despite $1 \leq i < n \to \infty$)

$$P\{S_{n_{m(i)}} - S_{n_k} > 0\} \to \tfrac{1}{2} \quad \text{uniformly for } 1 \leq i \leq n_{k-1}. \tag{16}$$

On the other hand, for $n_{k-1} < i < n$ necessarily $m(i) = k$, whence the left side of (16) is zero. Consequently, from (15) and (16)

$$\begin{aligned} B &\leq \tfrac{1}{2}\sum_{i=1}^{n-1} P\{T_{x\sqrt{n}} = i\} + o(1) + O(\varepsilon) \\ &\leq \tfrac{1}{2}P\{T_{x\sqrt{n}} \leq n\} + o(1) + O(\varepsilon), \end{aligned} \tag{17}$$

and so, combining (13), (14), and (17),

$$\tfrac{1}{2} \varlimsup_{n\to\infty} P\{T_{x\sqrt{n}} \leq n\} \leq 1 - \Phi(x) + \frac{2}{k\varepsilon^2} + O(\varepsilon). \tag{18}$$

To obtain the reverse inequality for the lower limit, observe that $A_n = \varnothing$, whence via (14)

$$P\{T_{x\sqrt{n}} \leq n\} - P\{S_n > (x + 2\varepsilon)\sqrt{n}\}$$

$$= P\{S_n \leq (x + 2\varepsilon)\sqrt{n}, \max S_j > x\sqrt{n}\}$$

$$\geq \sum_{i=1}^{n} P\{T_{x\sqrt{n}} = i, S_n < S_{n_{m(i)}}, |X_i| \leq \varepsilon\sqrt{n}, A_i^c\}$$

$$\geq \sum_{i=1}^{n} P\{T_{x\sqrt{n}} = i, S_n < S_{n_{m(i)}}, |X_i| \leq \varepsilon\sqrt{n}\} - \sum_{i=1}^{n} P\{T_{x\sqrt{n}} = i, A_i\}$$

$$\geq \sum_{i=1}^{n} \mathrm{P}\{T_{x\sqrt{n}} = i, S_n < S_{n_{m(i)}}\} - \sum_{i=1}^{n} \mathrm{P}\{|X_i| > \varepsilon\sqrt{n}\} - \frac{2}{k\varepsilon^2}$$

$$= \frac{1}{2}\sum_{i=1}^{n} \mathrm{P}\{T_{x\sqrt{n}} = i\} + o(1) - \frac{2}{k\varepsilon^2}$$

in view of (11) and the last equality of (15) and its aftermath. Thus,

$$\frac{1}{2}\varliminf_{n\to\infty} \mathrm{P}\{T_{x\sqrt{n}} \leq n\} \geq 1 - \Phi(x + 2\varepsilon) - \frac{2}{k\varepsilon^2}. \tag{19}$$

Letting $k \to \infty$ and then $\varepsilon \to 0$ in (18) and (19), it follows that as $n \to \infty$

$$\mathrm{P}\left\{\max_{1\leq j\leq n} S_j > x\sqrt{n}\right\} = \mathrm{P}\{T_{x\sqrt{n}} \leq n\} \to 2[1 - \Phi(x)], \qquad x > 0,$$

which is tantamount to (10) for $y = 1 = \sigma$. □

Exercises 9.4

1. Construct a sequence of independent r.v.s $\{X_n\}$ with identical means and variances for which the Lindeberg condition fails.
2. Let $\{X_n, n \geq 1\}$ be independent r.v.s obeying the central limit theorem with $\mathrm{E}\, X_n = \mu$, $\mathrm{E}\, X_n^2 \equiv \mu^2 + \sigma^2 < \infty$. If $N_c(\alpha) = \sup\{k \geq 1 : S_k \leq ck^\alpha\}$, $c > 0$, prove if $\mu > 0$, $0 \leq \alpha < 1$ that $N_c(\alpha)$ is a bona fide r.v.
3. Let $\{X_n, n \geq 1\}$ be independent r.v.s with $(1/\sqrt{n})\sum_1^n (\mathrm{E}\, X_j - \mu) \to 0$ for $\mu \in (0, \infty)$, $(1/n)\sum_1^n \sigma^2(X_j) \to \sigma^2 \in (0, \infty)$, and $(1/\sqrt{n}\sigma)(S_n - \sum_1^n \mathrm{E}\, X_j) \xrightarrow{d} N_{0,1}$. Show that the conclusions of Theorems 1 and 2 remain valid.
4. If $\{X_n, n \geq 1\}$ are i.i.d. r.v.s with $\mu = \mathrm{E}\, X > 0$ and $0 < b_n \to \infty$, then, if

$$\frac{S_n - n\mu}{b_n} \xrightarrow{d} S_F,$$

likewise $(1/b_n)(\max_{1\leq j\leq n} S_j - n\mu) \xrightarrow{d} S_F$. This yields an alternative proof of Corollary 9.4.1 in the special case $\alpha = 0$.

9.5 Central Limit Theorems for Double Arrays

The initial result employs a double sequence schema $\{X_{n,j}, 1 \leq j \leq k_n < \infty, n \geq 1\}$ (see Chapter 12 for the independent case) and furnishes conditions for the row sums $S_n = \sum_{j=1}^{k_n} X_{n,j}$ to converge in distribution to a mixture of normal distributions with means zero.

Since conditions (1), (2), and (5) of Theorem 1 can be interpreted as convergence in distribution (to the distribution with unit mass at zero), this theorem does not require the array of r.v.s $\{X_{ni}\}$ to be defined on a single probability space. In other words, for each $n \geq 1$, $\{X_{n,1}, \ldots, X_{n,k_n}\}$ may be r.v.s on some probability space $(\Omega_n, \mathscr{F}_n, \mathrm{P}_n)$ with the σ-algebras $\mathscr{F}_{n,1} \subset \mathscr{F}_{n,2} \subset \cdots \subset \mathscr{F}_{n,k_n} \subset \mathscr{F}_n$.

Theorem 1 (Hall, Heyde). *For each $n \ge 1$, let $\{S_{n,j} = \sum_{i=1}^{j} X_{n,i}, \mathscr{F}_{n,j}, 1 \le j \le k_n < \infty\}$ be an $\mathscr{L}_2$ stochastic sequence with $X_n^* = \max_{1 \le i \le k_n} |X_{n,i}|$, $U_{n,j}^2 = \sum_{i=1}^{j} X_{n,i}^2$, $1 \le j \le k_n$ such that for some $\mathscr{F}_{n,1}$-measurable* r.v. u_n^2

$$U_n^2 - u_n^2 \xrightarrow{P} 0 \qquad (\textit{where } U_n^2 = U_{n,k_n}^2) \tag{1}$$

$$X_n^* \xrightarrow{P} 0 \tag{2}$$

$$U_n^2 \xrightarrow{d} \eta_F^2 \tag{3}$$

$$\sup_{n \ge 1} \mathrm{E}\, X_n^{*2} < \infty \tag{4}$$

$$\sum_{j=1}^{k} \mathrm{E}\{X_{n,j} | \mathscr{F}_{n,j-1}\} \xrightarrow{P} 0, \qquad \sum_{j=1}^{k_n} \mathrm{E}^2\{X_{n,j} | \mathscr{F}_{n,j-1}\} \xrightarrow{P} 0 \tag{5}$$

with $\mathscr{F}_{n,0}$ the trivial σ-algebra, then $S_n = S_{n,k_n} \xrightarrow{d} S_G$ where $\mathrm{E}\, e^{itS_G} = \mathrm{E}\, e^{-(t^2/2)\eta_F^2}$.

PROOF. Set $X'_{n,j} = X_{n,j} - \mathrm{E}_{j-1} X_{n,j}$ where $\mathrm{E}_{j-1} X_{n,j} = \mathrm{E}\{X_{n,j} | \mathscr{F}_{n,j-1}\}$, and let $S'_{n,j}$, $U'_{n,j}$, $X_n'^*$ be the analogues of $S_{n,j}$, $U_{n,j}$, X_n^*. In view of (2) and (5), $X_n'^* \xrightarrow{P} 0$. By (5), (3), Schwarz's inequality and Slutsky's theorem,

$$U_n'^2 - U_n^2 = \sum_1^{k_n} \mathrm{E}_{j-1}^2 X_{n,j} - 2 \sum_1^{k_n} X_{n,j} \mathrm{E}_{j-1} X_{n,j} \xrightarrow{P} 0$$

whence

$$U_n'^2 - u_n^2 \xrightarrow{P} 0, \qquad U_n'^2 \xrightarrow{d} \eta_F^2$$

by (1) and (3). Moreover via Theorem 7.4.8

$$\begin{aligned} \mathrm{E} \max_{j \le k_n} (X_{n,j} - \mathrm{E}_{j-1} X_{n,j})^2 &\le 2\, \mathrm{E}(X_n^{*2} + \max_j \mathrm{E}_{j-1}^2 X_{n,j}) \\ &\le 2\, \mathrm{E}(X_n^{*2} + \max_{j \le k_n} \mathrm{E}_{j-1} X_n^{*2}) \le 10\, \mathrm{E}\, X_n^{*2} < \infty \end{aligned}$$

implying $\sup_{n \ge 1} \mathrm{E}(X_n'^*)^2 < \infty$. Hence (1), (2), (3) and (4) hold for the primed r.v.'s $\{X'_{nj}\}$.

For any $c > 0$, let $\eta''^2 = \eta^2 \wedge c$, $u_n''^2 = u_n^2 \wedge c$, $X''_{nj} = X'_{n,j} I_{[\alpha \ge j]}$ where $\alpha = \min\{1 \le i \le k_n : U'^2_{n,i} > c\}$ or k_n according as $U_n'^2 > c$ or not. Define $U''_{n,i}$, $S''_{n,j}$, $X_n''^*$ in an obvious fashion. Since either $U_n'^2 < c$ whence $U_n''^2 = U_n'^2$ or $\alpha \le k_n$ implying $c < U_n''^2 \le c + (X_n'^*)^2$,

$$U_n'^2 \wedge c \le U_n''^2 \le (U_n'^2 \wedge c) + (X_n'^*)^2$$

and so $U_n''^2 \xrightarrow{d} \eta^2$, $U_n''^2 - u_n''^2 \xrightarrow{P} 0$. Clearly $X_n''^* \xrightarrow{P} 0$ and $\sup \mathrm{E}(X_n''^*)^2 < \infty$. Hence the "double prime" analogues of (1), (2), (3), (4), say (1)″, (2)″, (3)″, (4)″ hold.

Next, setting $T_n = \prod_{i=1}^{k_n} (1 + iX''_{n,j})$ and noting that $X''_{n,j} = 0$ for $j > \alpha$

$$\mathrm{E} |T_n|^2 = \mathrm{E} \prod_{j=1}^{k_n} (1 + X_{nj}''^2) = \mathrm{E}(1 + X_{n,\alpha}''^2) \prod_{j < \alpha} (1 + X_{nj}''^2)$$

$$\leq E(1 + X''^2_{n,\alpha})e^{U''^2_{n,\alpha-1}} \leq e^c[1 + E(X''^*_n)^2]$$

implying via (4)″ that $\{T_n, n \geq 1\}$ is u.i.

Define $r(x)$ by $r(0) = 0$ and (see Lemma 12.1.1)

$$e^{ix} = (1 + ix)e^{-(x^2/2)+r(x)} \tag{6}$$

whence $1 = \exp\{r(x) + \sum_{j=3}^{\infty}(-ix)^j/j\}$ for $|x| < 1$ implying $r(x) = x^4a(x) + ix^3b(x)$, where

$$0 < a(x) = 1/4 - x^2/6 + x^4/8 \cdots < 1/4$$

$$0 < b(x) = 1/3 - x^2/5 + x^4/7 \cdots < 1/3$$

so that $|r(x)| < |x|^3$ for $|x| < 1$. Now (6) entails

$$e^{iS''_n} = T_n W''_n = (T_n - 1)W_n + (T_n - 1)(W''_n - W_n) + W''_n \tag{7}$$

where

$$W''_n = \exp\{-\tfrac{1}{2}U''^2_n + \sum_j r(X''_{nj})\}, \qquad W_n = \exp\{-\tfrac{1}{2}u''^2_n\}. \tag{8}$$

Since on $\{X''^*_n < 1\}$, $|\sum_{j=1}^{k_n} r(X''_{n,j})| < \sum_j |X''_{n,j}|^3 \leq X''^*_n U''^2_n \xrightarrow{P} 0$, (3)″ guarantees

$$W''_n \xrightarrow{d} e^{-\eta''^2/2} \tag{9}$$

while (1)″ and (8) ensure $W''_n/W_n \xrightarrow{P} 1$. Thus, $0 \leq W_n \leq 1$ entails $W''_n - W_n \xrightarrow{P} 0$ and so recalling that $\{T_n\}$ is u.i. (and a fortiori, tight)

$$(T_n - 1)(W''_n - W_n) \xrightarrow{P} 0. \tag{10}$$

Next, since $\{X''_{n,j}, \mathscr{F}_{n,j}, 1 \leq j \leq k_n\}$ are martingale differences,

$$E_1 T_n = E_1 \prod_{j=1}^{k_n-1} (1 + iX''_{n,j})E_{k_n-1}(1 + iX''_{n,k_n})$$

$$= E \prod_{j=1}^{k_n-1} (1 + iX''_{n,j}) = \cdots = 1 + iX''_{n,1}$$

and so, recalling that $0 \leq W_n \leq 1$

$$|E(T_n - 1)W_n| = |E\, W_n E_1(T_n - 1)| \leq E|E_1(T_n - 1)| \leq E|X''_{n,1}| \to 0 \tag{11}$$

via (2)″, (4)″ and dominated convergence.

Now in view of (7)

$$e^{iS''_n} - W_n(T_n - 1) = (T_n - 1)(W''_n - W_n) + W''_n \tag{12}$$

while (9) and (10) ensure that the right and hence left side of (12) $\xrightarrow{d} e^{-\eta''^2/2}$. Since the left side is u.i., Corollary 8.1.8 in conjunction with (11) guarantees that

$$E\, e^{iS''_n} \to E\, e^{-\eta''^2/2}. \tag{13}$$

Consequently,

$$|\mathrm{E}(e^{iS'_n} - e^{-\eta^2/2})| \le |\mathrm{E}(e^{iS'_n} - e^{iS''_n})| + |\mathrm{E}\, e^{iS''_n} - e^{\eta''^2/2}| + |\mathrm{E}(e^{-\eta''^2/2} - e^{-\eta^2/2})|$$
$$\le 2\mathrm{P}\{U_n'^2 > c\} + |\mathrm{E}(e^{iS''_n} - e^{-\eta''^2/2})| + \mathrm{P}\{\eta^2 > c\}$$

which, in conjunction with (13) implies that for any $\varepsilon > 0$

$$\overline{\lim}\, |\mathrm{E}(e^{iS'_n} - e^{-\eta^2/2})| \le 3\varepsilon$$

provided $c \ge C_\varepsilon$. Replacing S'_n by tS'_n for $t \ne 0$,

$$\lim_{n\to\infty} \mathrm{E}\, e^{itS'_n} = \mathrm{E}\, e^{-t^2\eta^2/2}.$$

Thus $S'_n \xrightarrow{d} S_G$ where the c.f. of S_G is $\mathrm{E}\, e^{-\eta^2 t^2/2}$. Finally, $S_n \xrightarrow{d} S_G$ via (5).

Corollary 1 (McLeish). *If for each* $n \ge 1$, $\{S_{n,j} = \sum_1^j X_{n,i}, \mathscr{F}_{n,j}, 1 \le j \le k_n < \infty\}$ *is an* $\mathscr{L}_2$ *stochastic sequence on* $(\Omega, \mathscr{F}, \mathrm{P})$ *satisfying* (2), (4), (5) *and*

$$U_n^2 \xrightarrow{\mathrm{P}} \eta^2 \tag{1'}$$

for some non-negative $\bigcap_{n=1}^\infty \mathscr{F}_{n,1}$*-measurable* r.v. η^2 *then the conclusion of Theorem* 1 *holds.*

In order to replace conditions (2), (4) by a conditional Lindeberg condition (22) and substitute $V_n^2 = \sum_{i=1}^{\eta} \mathrm{E}\{X_{nj}^2 | \mathscr{F}_{nj-1}\}$ for U_n^2, several lemmas will be needed.

Lemma 1 (Dvoretzky). *If the events* $A_j \in \mathscr{F}_j$, $1 \le j \le n$ *where* $\{\mathscr{F}_j, j \ge 0\}$ *is an increasing sequence of sub-σ-algebras of the basic σ-algebra* $\mathscr{F}$, *then*

$$\mathrm{P}\left\{\bigcup_{j=1}^n A_j\right\} \le \varepsilon + \mathrm{P}\left\{\sum_{j=1}^n \mathrm{P}\{A_j | \mathscr{F}_{j-1}\} > \varepsilon\right\}, \qquad \varepsilon > 0. \tag{14}$$

In particular, for any non-negative stochastic sequence $\{Y_{n,j}, \mathscr{F}_{n,j}, 1 \le j \le k_n\}$, *if* $\sum_{j=1}^{k_n} \mathrm{E}\{Y_{n,j} I_{[Y_{n,j} \ge \varepsilon]} | \mathscr{F}_{n,j-1}\} \xrightarrow{\mathrm{P}} 0$, *all* $\varepsilon > 0$, *then*

$$\max_{i \le j \le k_n} Y_{n,j} \xrightarrow{\mathrm{P}} 0. \tag{15}$$

PROOF. Setting $\mu_k = \sum_{j=1}^k \mathrm{P}\{A_j | \mathscr{F}_{j-1}\}$, $1 \le k \le n$

$$\mathrm{P}\left\{\bigcup_{j=1}^n A_j [\mu_n \le \varepsilon]\right\} \le \mathrm{P}\left\{\bigcup_{j=1}^n A_j [\mu_j \le \varepsilon]\right\} = \mathrm{E} \sum_{j=1}^n \mathrm{P}\{A_j | \mathscr{F}_{j-1}\} I_{[\mu_j \le \varepsilon]} \le \varepsilon$$

so that

$$\mathrm{P}\left\{\bigcup_{j=1}^n A_j\right\} \le \mathrm{P}\left\{\bigcup_{j=1}^n A_j [\mu_n \le \varepsilon]\right\} + \mathrm{P}\{\mu_n > \varepsilon\} \le \varepsilon + \mathrm{P}\{\mu_n > \varepsilon\}. \qquad \square$$

Lemma 2 (Dvoretzky). *Let* W *be any* $\mathscr{G}$*-measurable,* $\mathscr{L}_2$. r.v. *where* $\mathscr{G}$ *is a sub-σ-algebra of* $\mathscr{F}$. *Then for any* $\varepsilon > 0$ *and any* $\mathscr{L}_2$ r.v. Y *for which* $\mathrm{E}\{Y | \mathscr{G}\} \overset{\text{a.c.}}{=} 0$, *with probability one*

$$\mathrm{E}\{Y^2 I_{[Y^2 > \varepsilon]} | \mathscr{G}\} \le \mathrm{E}\{(2Y + W)^2 I_{[(2Y+W)^2 > \varepsilon]} | \mathscr{G}\}. \tag{16}$$

PROOF. Without loss of generality suppose $\varepsilon = 1$. Define

$$A = \{W^2 \geq 1\}, \qquad Z = (2Y + W)^2 - Y^2$$

$$Q(D) = \mathrm{E}(Y^2 + Z)I_{[Y^2+Z>1]D} - \mathrm{E}\, Y^2 I_{[Y^2>1]D}, \qquad D \in \mathscr{F}.$$

It suffices to prove for any $G \in \mathscr{G}$ that $Q(G) \geq 0$. Since $\mathrm{E}\, WYI_{AG} = \mathrm{E}\, WI_{AG}\mathrm{E}\{Y|\mathscr{G}\} = 0$ implies

$$\mathrm{E}\, ZI_{AG} = 3\,\mathrm{E}\, Y^2 I_{AG} + \mathrm{E}\, W^2 I_{AG}$$

and moreover

$$\mathrm{E}\, ZI_{[Y^2+Z\leq 1]AG} \leq \mathrm{E}\, ZI_{[0\leq Z\leq 1]AG} \leq \mathrm{P}\{AG\} \leq \mathrm{E}\, W^2 I_{AG}$$

necessarily

$$Q(AG) = \mathrm{E}\, ZI_{AG} - \mathrm{E}\, ZI_{[Y^2+Z\leq 1]AG} + \mathrm{E}\, Y^2 I_{[Y^2+Z>1]AG} - \mathrm{E}\, Y^2 I_{[Y^2>1]AG} \geq 0.$$

On A^c, $Y^2 + Z \leq 1$ iff $(2Y + W + 1)(2Y + W - 1) \leq 0$ iff $-(1 + W)/2 \leq Y \leq (1 - W)/2$. Hence $A^c[Y^2 + Z \leq 1] \subset [Y^2 \leq 1]$ so that $A^c[Y^2 + Z \leq 1]\cdot[Y^2 > 1] = \phi$. Furthermore, since $Z < 0$ iff $|2Y + W| < |Y|$ which, in turn, implies $|Y| < |W|$, necessarily $[Z < 0]A^c \subset [Y^2 < 1]$ so that $Z \geq 0$ on $A^c[Y^2 \geq 1]$. Thus

$$Q(A^cG) = \mathrm{E}(Y^2 + Z)I_{[Y^2+Z>1\geq Y^2]A^cG} - \mathrm{E}\, Y^2 I_{[Y^2+Z\leq 1<Y^2]A^cG}$$
$$+ \mathrm{E}\, ZI_{[Y^2+Z>1,\, Y^2>1]A^cG} \geq 0$$

whence

$$Q(G) = Q(AG) + Q(A^cG) \geq 0. \qquad \square$$

Corollary 2. *If* $\mathrm{E}|X|^2 < \infty$ *and* $\varepsilon > 0$, *with probability one*

$$\mathrm{E}\{|X - \mathrm{E}\{X|\mathscr{G}\}|^2 I_{[|X-\mathrm{E}\{X|\mathscr{G}\}|>2\varepsilon]}|\mathscr{G}\} \leq 4\,\mathrm{E}\{X^2 I_{[|X|>\varepsilon]}|\mathscr{G}\}. \tag{17}$$

PROOF. Take $W = 2\,\mathrm{E}\{X|\mathscr{G}\}$ and $Y = X - \mathrm{E}\{X|\mathscr{G}\}$ $\square$

Lemma 3 (McLeish). *For each* $n \geq 1$ *let* $\{Y_{n,j}, \mathscr{F}_{n,j}, 1 \leq j \leq k_n < \infty\}$ *be a non-negative* $\mathscr{L}_1$ *stochastic sequence and let*

$$S_{n,j} = \sum_{i=1}^{j} Y_{n,i}, \qquad \mu_{nj} = \sum_{i=1}^{j} \mathrm{E}\{Y_{nj}|\mathscr{F}_{n,j-1}\}, \qquad 1 \leq j \leq k_n,$$

$$S_n = S_{n,k_n}, \; \mu_n = \mu_{n,k_n}.$$

If

$$\sum_{i=1}^{k_n} \mathrm{E}\{Y_{n,j}I_{[Y_{n,j}>\varepsilon]}|\mathscr{F}_{n,j-1}\} \xrightarrow{\mathrm{P}} 0, \qquad \varepsilon > 0 \tag{18}$$

and

$$\{\mu_n, n \geq 1\} \text{ is tight} \tag{19}$$

then

$$\max_{1\leq j\leq k_n} |S_{n,j} - \mu_{n,j}| \xrightarrow{\mathrm{P}} 0. \tag{20}$$

PROOF. Via Lemma 1,

$$Y_n^* = \max_{j \le k_n} Y_{n,j} \xrightarrow{\mathrm{P}} 0.$$

Let $Y'_{n,j} = Y_{n,j} I_{[Y_{n,j} \le K^{-2}, \mu_{n,j} \le K]}$, $K > 0$.

Now

$$\left\{ \sum_{i=1}^{k_n} |Y_{n,j} - Y'_{n,j}| > 0 \right\} \subset \{\mu_n > K\} \cup \{Y_n^* > K^{-2}\}$$

implying

$$\overline{\lim_{n\to\infty}}\, \mathrm{P}\left\{ \sum_{1}^{k_n} |Y_{n,j} - Y'_{n,j}| > 0 \right\} \le \sup_{n \ge 1} \mathrm{P}\{\mu_n > K\} + \overline{\lim_{n\to\infty}}\, \mathrm{P}\{Y_n^* > K^{-2}\} \xrightarrow{K\to\infty} 0$$

whence defining $S'_{n,j}$, S'_n, $\mu'_{n,j}$, μ'_n analogously

$$\max_{1 \le j \le k_n} |S_{n,j} - S'_{n,j}| \le \sum_{j=1}^{k_n} |Y_{n,j} - Y'_{n,j}| \xrightarrow{\mathrm{P}} 0.$$

Moreover,

$$\begin{aligned}
\max_{j \le k_n} |\mu_{n,j} - \mu'_{n,j}| &\le \sum_{j=1}^{k_n} \mathrm{E}\{|Y_{n,j} - Y'_{n,j}| \,|\, \mathscr{F}_{n,j-1}\} \\
&\le \sum_{j=1}^{k_n} (\mathrm{E}\{Y_{n,j} I_{[Y_{n,j} > K^{-2}]} | \mathscr{F}_{n,j-1}\} + \mathrm{E}\{Y_{n,j} | \mathscr{F}_{n,j-1}\} I_{[\mu_{n,j} > K]}) \\
&\le \sum_{j=1}^{k_n} \mathrm{E}\{Y_{n,j} I_{[Y_{n,j} > K^{-2}]} | \mathscr{F}_{n,j-1}\} + \mu_n I_{[\mu_n > K]}
\end{aligned}$$

whence, in view of (18) and (19), for any $\delta > 0$

$$\begin{aligned}
\mathrm{P}\left\{ \max_{1 \le j \le k_n} |\mu_{n,j} - \mu'_{n,j}| > \delta \right\} &\le \mathrm{P}\left\{ \sum_{1}^{k_n} \mathrm{E}\{Y_{n,j} I_{[Y_{n,j} > K^{-2}]} | \mathscr{F}_{n,j-1}\} > \delta/2 \right\} \\
&\quad + \mathrm{P}\{\mu_n I_{[\mu_n > K]} > \delta/2\} \le \mathrm{P}\left\{ \sum_{1}^{k_n} \mathrm{E}\{Y_{n,j} I_{[Y_{n,j} > K^{-2}]} | \mathscr{F}_{n,j-1}\} > \delta/2 \right\} \\
&\quad + \sup_{n \ge 1} \mathrm{P}\{\mu_n > K\} \xrightarrow{n\to\infty} \sup_{n \ge 1} \mathrm{P}\{\mu_n > K\} \xrightarrow{K\to\infty} 0.
\end{aligned}$$

Furthermore, by Theorem 7.4.8 for any $\delta > 0$

$$\begin{aligned}
(\delta^2/4)\mathrm{P}\left\{ \max_j |S'_{n,j} - \mu'_{n,j}| > \delta \right\} &\le \frac{1}{4} \mathrm{E} \max_j (S'_{n,j} - \mu'_{n,j})^2 \le \mathrm{E}(S'_n - \mu'_n)^2 \\
&\le \sum_{j=1}^{k_n} \mathrm{E}\, Y'^2_{n,j} \le K^{-2} \mathrm{E} \sum_{1}^{k_n} Y'_{n,j} = K^{-2} \mathrm{E} \sum_{1}^{k_n} \mathrm{E}\{Y'_{n,j} | \mathscr{F}_{n,j-1}\} \\
&\le K^{-2} \mathrm{E} \sum_{j=1}^{k_n} \mathrm{E}\{Y_{n,j} | \mathscr{F}_{n,j-1}\} I_{[\mu_{n,j} \le K]} \le K^{-1} \xrightarrow{K\to\infty} 0.
\end{aligned}$$

Finally,

$$\max_j |S_{n,j} - \mu_{n,j}|$$
$$\leq \max_j |S_{n,j} - S'_{n,j}| + \max_j |S'_{n,j} - \mu'_{n,j}| + \max_j |\mu'_{n,j} - \mu_{n,j}| \xrightarrow{P} 0. \qquad \square$$

As earlier, $E_{j-1}\{\cdot\}$ abbreviates $E\{\cdot|\mathscr{F}_{n,j-1}\}$. Condition (22) is a conditional Lindeberg condition and hence less stringent than the classical version.

Theorem 2. *For each $n \geq 1$, let $\{S_{n,j} = \sum_{i=1}^{j} X_{ni}, \mathscr{F}_{n,j}, 1 \leq j \leq k_n < \infty\}$ be an $\mathscr{L}_2$ stochastic sequence such that for some $\mathscr{F}_{n,1}$-measurable* r.v. u_n^2

$$\sum_{j=1}^{k_n} [E_{j-1}\{X_{n,j}^2\} - E_{j-1}^2\{X_{n,j}\}] - u_n^2 \xrightarrow{P} 0 \tag{21}$$

$$\sum_{j=1}^{k_n} E_{j-1}\{X_{nj}^2 I_{[|X_{n,j}|>\varepsilon]}\} \xrightarrow{P} 0, \qquad \varepsilon > 0 \tag{22}$$

$$\sum_{j=1}^{k_n} [E_{j-1}\{X_{n,j}^2\} - E_{j-1}^2\{X_{n,j}\}] \xrightarrow{d} \eta_F^2 \tag{23}$$

$$\sum_{j=1}^{k_n} E_{j-1}\{X_{nj}\} \xrightarrow{P} 0, \qquad \sum_{j=1}^{k_n} E_{j-1}^2\{X_{nj}\} \xrightarrow{P} 0. \tag{24}$$

where $\mathscr{F}_{n,0}$ is the trivial σ-algebra. Then $S_n \equiv S_{n,k_n} \xrightarrow{d} S_G$ where $E\, e^{itS_G} = E\, e^{-\eta_F^2 t^2/2}$.

Proof. For $n \geq 1$, define $X'_{n,j} = X_{n,j} - E_{j-1}\{X_{n,j}\}$. By Corollary 2, $\{X'_{n,j}\}$ satisfies (22) and hence also condition (2) of Theorem 1, via Lemma 1. Set $V_n^2 = \sum_{j=1}^{k_n} E_{j-1}\{X_{n,j}^2\}$, $U_n^2 = \sum_1^{k_n} X_{nj}^2$ and define $V_n'^2$ and $U_n'^2$ analogously. Via (23)

$$V_n'^2 \xrightarrow{d} \eta_F^2$$

The latter ensures that $V_n'^2$ is tight which, in conjunction with (22) for $\{X'_{n,j}\}$ guarantees that

$$U_n'^2 - V_n'^2 \xrightarrow{P} 0$$

by Lemma 3. Consequently, (1) and (3) of Theorem 1 hold for $\{X'_{n,j}\}$.

Next, define the martingale difference sequence

$$X''_{n,j} = X'_{n,j} I_{[\sum_1^j E_{i-1}\{X'^2_{n,i} I_{[|X'_{n,i}|>\varepsilon]}\} \leq 1]}$$

and note that (2) and (5) hold trivially for $\{X''_{n,j}\}$. Likewise (1) and (3) obtain since $U_n'^2 - U_n''^2 \xrightarrow{P} 0$ in view of

$$U_n'^2 - U_n''^2 \leq U_n'^2 I_{[\sum_{n=1}^{k_n} E_{j-1}\{X'^2_{n,j} I_{[|X'_{n,j}|>\varepsilon]}\} > 1]}$$

and the conditional Lindeberg condition.

Moreover,

$$E \max_{j \leq k_n} X''^2_{n,j} \leq \varepsilon^2 + E \sum_{j=1}^{k_n} X'^2_{n,j} I_{[|X'_{nj}|>\varepsilon, \sum_{i=1}^j E_{i-1}\{X'^2_{n,i} I_{[|X'_{n,i}|>\varepsilon]}\} \leq 1]}$$

$$\leq \varepsilon^2 + E \sum_{j=1}^{k_n} E_{j-1}\{X'^2_{nj} I_{[|X'_{n,j}|>\varepsilon]}\} I_{[\sum_1^j E_{i-1}\{X'^2_{n,i} I_{[|X'_{n,i}|>\varepsilon]}\} \leq 1]}$$

$$\leq \varepsilon^2 + 1$$

so that (4) also holds for $\{X''_{n,j}\}$. Consequently, $S''_n = \sum_{j=1}^{k_n} X''_{n,j} \xrightarrow{d} S_G$ by Theorem 1. Again employing the conditional Lindeberg condition

$$\overline{\lim}\, P\{S'_n \neq S''_n\} \le \overline{\lim}\, P\left\{\sum_{i=1}^{k_n} E_{i-1}\{X'^2_{n,i} I_{[|X'_{n,i}|>\varepsilon]}\} > 1\right\} = 0$$

whence $S'_n \xrightarrow{d} S_G$. Finally, $S_n \xrightarrow{d} S_G$ via (24). □

Likewise in Theorem 2 the r.v.s $\{X_{nj}, 1 \le j \le k_n\}$ may be defined on different probability spaces $(\Omega_n, \mathscr{F}_n, P_n)$.

Corollary 3 (Dvoretzky). *If, for each* $n \ge 1$, $\{S_{n,j} = \sum_{i=1}^{j} X_{ni}, \mathscr{F}_{n,j}, 1 \le j \le k_n < \infty\}$ is an $\mathscr{L}_2$ stochastic sequence on $(\Omega, \mathscr{F}, P)$ satisfying (22), (24) and

$$V_n^2 = \sum_{j=1}^{k_n} [E\{X_{n,j}^2 | \mathscr{F}_{n,j-1}\} - E^2\{X_{n,j} | \mathscr{F}_{n,j-1}\}] \xrightarrow{P} \eta^2 \tag{23'}$$

for some non-negative $\bigcap_{n=1}^{\infty} \mathscr{F}_{n,1}$-measurable r.v. η^2, then the conclusion of Theorem 2 holds.

Exercises 9.5

1. (Helland) If for $n \ge 1$, $\{X_{nj}, \mathscr{F}_{nj}, 1 \le j \le k_n\}$ is a stochastic sequence on $(\Omega, \mathscr{F}, P)$ with $E\, X_n^* = E \max_j |X_{nj}| = o(1)$ as $n \to \infty$, then $\sum_{j=1}^{k_n} E_{j-1}\{|X_{nj}| I_{[|X_{nj}|>\varepsilon]}\} \xrightarrow{P} 0, \varepsilon > 0$. In particular, if events $A_{nj} \in \mathscr{F}_{nj}$ where $\mathscr{F}_{nl} \subset \cdots \subset \mathscr{F}_{nk_n} \subset \mathscr{F}$, then $P\{\bigcup_1^{k_n} A_{nj}\} = o(1)$ as $n \to \infty$ implies $\sum_{j=1}^{k_n} P\{A_{nj} | \mathscr{F}_{n,j-1}\} \xrightarrow{P} 0$.
   Hint: Let $T_n = \inf\{1 \le j \le k_n : |X_{nj}| > \varepsilon\}$ and $T_n = k_n$ otherwise. Then
   $$P\left\{\sum_1^k E_{j-1}\{|X_{nj}| I_{[|X_{nj}|>\varepsilon]}\} > \delta\right\} \le P\{X_n^* > \varepsilon\} + \delta^{-1} E\, X_n^*.$$
2. (Helland) (i) If for each $n \ge 1$, $\{X_{nj}, \mathscr{F}_{nj}, 1 \le j \le k_n\}$ is a stochastic sequence with $|X_{nj}| \le c, 1 \le j \le k_n$ and $U_n^2 = \sum_1^{k_n} X_{nj}^2 \xrightarrow{P} 1$, then $\{V_n^2, n \ge 1\}$ is tight.
   Hint: If $T_n = \inf\{1 \le j \le k_n, \sum_{i=1}^{j} X_{ni}^2 > 2\}$ and $T_n = k_n$ otherwise, then
   $$P\{V_n^2 > a\} \le P\{U_n^2 > 2\} + a^{-1} E \sum_1^{T_n} X_{nj}^2 \le o(1) + (c+2)a^{-1}.$$
   (ii) If, in addition, $X_n^* = \max_j |X_{nj}| \xrightarrow{P} 0$, then $V_n^2 \xrightarrow{P} 1$.
   Hint: The conditional Lindeberg condition (22) holds via Exercise 1.
3. (Gänssler–Häusler) If for each $n \ge 1$, $\{S_{nj} = \sum_{i=1}^{j} X_{ni}, \mathscr{F}_{nj}, 1 \le j \le k_n\}$ is a martingale with $E\, X_1 = 0, n \ge 1$ satisfying (i). $E\, X_n^* \to 0$ and (ii) $U_n^2 \xrightarrow{P} 1$, then $\sum_1^{k_n} X_{nj} \xrightarrow{d} N_{0,1}$.
   Hint: $X_n^* \xrightarrow{P} 0$ implies (iii) $\sum_{j=1}^{k_n} X_{nj}^r I_{[|X_{nj}|>1]} \xrightarrow{P} 0$, $r = 1, 2$ and via Exercise 1, (iv) $\sum_{j=1}^{k_n} E_{j-1}\{X_{nj} I_{[|X_{nj}| \le 1]}\} \xrightarrow{P} 0$. Then setting $X'_{nj} = X_{nj} I_{[|X_{nj}|\le 1]} - E_{j-1}\{X_{nj} I_{[|X_{nj}|\le 1]}\}$, Corollary 3 ensures $\sum_{j=1}^{k_n} X'_{nj} \xrightarrow{d} N_{0,1}$ since $\{X'_{nj}\}$ satisfies (22) and moreover via (ii), (iii) and (iv), $\sum_{j=1}^{k_n} (X'_{nj})^2 \xrightarrow{P} 1$ whence Exercise 2 (ii) yields $(V'_n)^2 \xrightarrow{P} 1$.

References

F. Anscombe, "Large sample theory of sequential estimation," *Proc. Cambr. Philos. Soc.* **48** (1952), 600–607.

S. Bernstein, "Several comments concerning the limit theorem of Liapounov," *Dokl. Akad. Nauk. SSSR* **24** (1939), 3–7.

A. C. Berry, "The accuracy of the Gaussian approximation to the sum of independent variates," *Trans. Amer. Math. Soc.* **49** (1941), 122–136.

J. Blum, D. Hanson, and J. Rosenblatt, "On the CLT for the sum of a random number of random variables," *Z. Wahr. Verw. Geb.* **1** (1962–1963), 389–393.

J. Blum, H. Chernoff, M. Rosenblatt, and H. Teicher, "Central limit theorems for interchangeable processes," *Can. Jour. Math.* **10** (1958), 222–229.

K. L. Chung, *A Course in Probability Theory*, Harcourt Brace, New York, 1968; 2nd ed., Academic Press, New York, 1974.

W. Doeblin, "Sur deux problèmes de M. Kolmogorov concernant les chaines denombrables," *Bull, Soc. Math. France* **66** (1938), 210–220.

J. L. Doob, *Stochastic Processes*, Wiley New York, 1953.

A. Dvoretzky, "Asymptotic normality for sums of dependent random variables," *Proc. Sixth Berkeley Symp. on Stat. and Prob.* 1970, 513–535.

P. Erdos and M. Kac, "On certain limit theorems of the theory of probability," *Bull. Amer. Math. Soc.* **52** (1946), 292–302.

C. Esseen, "Fourier analysis of distribution functions," *Acta Math.* **77** (1945), 1–125.

W. Feller, "Über den Zentralen Grenzwertsatz der wahrscheinlichkeitsrechnung," *Math, Zeit.* **40** (1935), 521–559.

N. Friedman, M. Katz, and L. Koopmans, "Convergence rates for the central limit theorem," *Proc. Nat. Acad. Sci.* **56** (1966), 1062–1065.

P. Hall and C. C. Heyde, *Martingale Limit Theory and its Application*, Academic Press, New York, 1980.

K. Knopp, *Theory and Application of Infinite Series*, Stechert–Hafner, New York, 1928.

P. Lévy, *Théorie de l'addition des variables aléatoiries*, Gauthier-Villars, Paris, 1937; 2nd ed., 1954.

J. Lindeberg, "Eine neue Herleitung des Exponentialgesetzes in der Wahrscheinlichkeitsrechnung," *Math. Zeit.* **15** (1922), 211–225.

D. L. McLeish, "Dependent Central Limit Theorems and invariance principles," *Ann. Prob.* **2** (1974), 620–628.

J. Mogyorodi, "A CLT for the sum of a random number of independent random variables," *Magyor. Tud. Akad. Mat. Kutato Int. Közl.* **7** (1962), 409–424.

A. Renyi, "Three new proofs and a generalization of a theorem of Irving Weiss," *Magyor. Tud. Akad. Mat. Kutato Int. Közl.* **7** (1962), 203–214.

A. Renyi, "On the CLT for the sum of a random number of independent random variables, *Acta Math. Acad. Sci. Hung.* **11** (1960), 97–102.

B. Rosen, "On the asymptotic distribution of sums of independent, identically distributed random variables," *Arkiv för Mat.* **4** (1962), 323–332.

D. Siegmund, "On the asymptotic normality of one-sided stopping rules," *Ann. Math. Stat.* **39** (1968), 1493–1497.

H. Teicher, "On interchangeable random variables," *Studi di Probabilita Statistica e Ricerca Operativa in Onore di Giuseppe Pompilj*, pp. 141–148, Gubbio, 1971.

H. Teicher, "A classical limit theorem without invariance or reflection, *Ann. Math. Stat.* **43** (1973), 702–704.

P. Van Beek, "An application of the Fourier method to the problem of sharpening the Berry–Esseen inequality," *Z. Wahr.* **23** (1972), 187–197.

I. Weiss, "Limit distributions in some occupancy problems," *Ann. Math. Stat.* **29** (1958), 878–884.

V. Zolotarev, "An absolute estimate of the remainder term in the C.L.T.," *Theor. Prob. and its Appl.* **11** (1966), 95–105.

10 Limit Theorems for Independent Random Variables

10.1 Laws of Large Numbers

Prior discussion of the strong and weak laws of large numbers centered around the i.i.d. case. Necessary and sufficient conditions for the weak law are available when the underlying random variables are merely independent and have recently been obtained for the strong law as well. Unfortunately, the practicality of the latter conditions leaves much to be desired.

A few words are in order on a method of considerable utility in probability theory, namely, that of symmetrization. In Chapter 6 it was pointed out that, given a sequence of r.v.s $\{X_n, n \geq 1\}$ on $(\Omega, \mathscr{F}, P)$, a symmetrized sequence of r.v.s $\{X_n^*, n \geq 1\}$ can be defined—if necessary by constructing a new probability space. The joint distributions of

$$X_n^* = X_n - X_n', \qquad n \geq 1,$$

are determined by the fact that $\{X_n', n \geq 1\}$ is independent of $\{X_n, n \geq 1\}$ and possesses the same joint d.f.s, that is, $\{X_n, n \geq 1\}$ and $\{X_n', n \geq 1\}$ are i.i.d. stochastic processes. In particular, if the initial r.v.s $X_n, n \geq 1$, are independent, so are the symmetrized X_n, $n \geq 1$, namely, the X_n^*, $n \geq 1$.

The salient point concerning symmetrization is that the X_n^* are symmetric about zero while the magnitude of the sum of the two tails of the corresponding d.f. F_n^* is roughly the same as that of F_n. The relation between the distributions and moments is stated explicitly in

Lemma 1. *If* $\{X_j, 1 \le j \le n\}$ *and* $\{X'_j, 1 \le j \le n\}$ *are* i.i.d. *stochastic processes with medians* m_j *and* $X^*_j = X_j - X'_j$, *then for any* $n \ge 1$, $\varepsilon > 0$, *and real* a

$$\tfrac{1}{2} P\left\{ \max_{1 \le j \le n} (X_j - m_j) \ge \varepsilon \right\} \le P\left\{ \max_{1 \le j \le n} X^*_j \ge \varepsilon \right\}, \tag{1}$$

$$\begin{aligned} \tfrac{1}{2} P\left\{ \max_{1 \le j \le n} |X_j - m_j| \ge \varepsilon \right\} &\le P\left\{ \max_{1 \le j \le n} |X^*_j| \ge \varepsilon \right\} \\ &\le 2\, P\left\{ \max_{1 \le j \le n} |X_j - a| \ge \frac{\varepsilon}{2} \right\}, \end{aligned} \tag{2}$$

$$\tfrac{1}{2} E\left(\max_{1 \le j \le n} |X_j - m_j| \right)^p \le E\left(\max_{1 \le j \le n} |X^*_j| \right)^p \le 2\, E\left(2 \max_{1 \le j \le n} |X_j - a| \right)^p, \tag{2'}$$

where $p > 0$. *Moreover, if* $E\, X_1 = 0$,

$$E|X_1|^p \le E|X^*_1|^p \le 2^p\, E|X_1|^p, \qquad p \ge 1. \tag{3}$$

PROOF. Set $A_j = \{X_j - m_j \ge \varepsilon\}$, $B_j = \{X'_j - m_j \le 0\}$, $C_j = \{X^*_j \ge \varepsilon\}$. Then $A_j \cdot B_j \subset C_j$ and by Lemma 3.3.3

$$\tfrac{1}{2} P\left\{ \bigcup_{j=1}^{n} A_j \right\} \le P\left\{ \bigcup_{j=1}^{n} A_j B_j \right\} \le P\left\{ \bigcup_{1}^{n} C_j \right\},$$

which is (1). To prove (2), set

$$T = \inf\{ j \ge 1 : |X_j - m_j| \ge \varepsilon \}.$$

Then via independence,

$$\begin{aligned} P\{T \le n\} &\le 2 \sum_{j=1}^{n} [P\{T = j, X_j - m_j \ge \varepsilon, X'_j \le m_j\} \\ &\quad + P\{T = j, X_j - m_j \le -\varepsilon, X'_j \ge m_j\}] \\ &\le 2 \sum_{j=1}^{n} P\{T = j, |X^*_j| \ge \varepsilon\} = 2 \sum_{j=1}^{n} P\{T = j, |X^*_T| \ge \varepsilon\} \\ &\le 2P\left\{ \max_{1 \le j \le n} |X^*_j| \ge \varepsilon \right\}. \end{aligned}$$

This yields the first inequality of (2) and the second follows from

$$\begin{aligned} P\left\{ \max_{1 \le j \le n} |X^*_j| \ge \varepsilon \right\} &\le P\left\{ \max_{1 \le j \le n} |X_j - a| \ge \frac{\varepsilon}{2} \text{ or } \max_{1 \ge j \ge n} |X'_j - a| \ge \frac{\varepsilon}{2} \right\} \\ &\le 2\, P\left\{ 2 \max_{1 \le j \le n} |X_j - a| \ge \varepsilon \right\}. \end{aligned}$$

Moreover, (2) in turn yields (2′) via Corollary 6.2.2.

Apropos of (3), since $\mathrm{E}\{X_1 - X'_1 | X_1\} = X_1$, the conditional Jensen inequality ensures $\mathrm{E}|X_1^*|^p \ge \mathrm{E}|X_1|^p$. The remaining portion follows trivially from Exercise 4.2.1 and does not require that $\mathrm{E}\,X_1 = 0$. □

Theorem 1 (Weak law). *For each* $n \ge 1$, *let* $\{X_{nj}, 1 \le j \le k_n \to \infty\}$ *be independent* r.v.s, $S_n = \sum_1^{k_n} X_{nj}$, *and let* m_{nj} *denote a median of* X_{nj}. *Then for some real numbers* A_n

$$S_n - A_n \xrightarrow{\mathrm{P}} 0, \tag{4}$$

$$\max_{1 \le j \le k_n} |m_{nj}| \to 0, \tag{5}$$

as $n \to \infty$ iff

$$\sum_{j=1}^{k_n} \mathrm{P}\{|X_{nj}| \ge \varepsilon\} \to 0, \qquad \varepsilon > 0, \tag{6}$$

$$\sum_{j=1}^{k_n} \sigma^2(X_{nj} I_{[|X_{nj}|<1]}) \to 0, \tag{7}$$

in which case

$$A_n - \sum_{j=1}^{k_n} \mathrm{E}\,X_{nj} I_{[|X_{nj}|<1]} \to 0. \tag{8}$$

PROOF. (6) implies (5) trivially. To prove (4), set

$$Y_{nj} = X_{nj} I_{[|X_{nj}|<1]}, \qquad V_n = \sum_{j=1}^{k_n} Y_{nj},$$

and note that via independence (7) ensures $V_n - \mathrm{E}\,V_n \xrightarrow{\mathrm{P}} 0$. Since by (6)

$$\mathrm{P}\{V_n \ne S_n\} \le \sum_{j=1}^{k_n} \mathrm{P}\{|X_{nj}| \ge 1\} = o(1), \tag{9}$$

$S_n - \mathrm{E}\,V_n \xrightarrow{\mathrm{P}} 0$, yielding (4) and (8).

Conversely, if (4) and (5) hold, let $(X'_{n1}, \ldots, X'_{nk_n})$ and $(X_{n1}, \ldots, X_{nk_n})$ be i.i.d. random vectors for each $n \ge 1$ and set $X_{nj}^* = X_{nj} - X'_{nj}$, $S_n^* = \sum_{j=1}^{k_n} X_{nj}^*$. Then (4) entails $S_n^* \xrightarrow{\mathrm{P}} 0$ and so by Levy's inequality (Corollary 3.3.5) for any $\varepsilon > 0$

$$\mathrm{P}\left\{\max_{j \le k_n} |X_{nj}^*| \ge \varepsilon\right\} \le \mathrm{P}\left\{\max_{k \le k_n} \left|\sum_1^k X_{nj}^*\right| \ge \frac{\varepsilon}{2}\right\} \le 2\,\mathrm{P}\left\{|S_n^*| \ge \frac{\varepsilon}{2}\right\} = o(1),$$

whence

$$\exp\left\{-\sum_1^{k_n} \mathrm{P}\{|X_{nj}^*| \ge \varepsilon\}\right\} \ge \prod_{j=1}^{k_n} \mathrm{P}\{|X_{nj}^*| < \varepsilon\} = \mathrm{P}\left\{\max_{j \le k_n} |X_{nj}^*| < \varepsilon\right\} \to 1$$

as $n \to \infty$, implying for all $\varepsilon > 0$ that

$$\sum_{j=1}^{k_n} \mathrm{P}\{|X_{nj}^*| \ge \varepsilon\} = o(1). \tag{10}$$

Since (5) ensures $|m_{nj}| < \varepsilon$, $1 \le j \le k_n$ for all large n, by Lemma 1

$$2\,P\{|X_{nj}^*| \ge \varepsilon\} \ge P\{|X_{nj} - m_{nj}| \ge \varepsilon\} \ge P\{|X_{nj}| \ge 2\varepsilon\}$$

for all large n, and (6) follows via (10).

To establish (7), set

$$Y'_{nj} = X'_{nj} I_{[|X'_{nj}| < 1]}, \qquad Y^*_{nj} = Y_{nj} - Y'_{nj}, \qquad V^*_n = \sum_{j=1}^{k_n} Y^*_{nj}.$$

By (6), (9), and (4), $V_n - A_n \xrightarrow{P} 0$ entailing $V_n^* \xrightarrow{P} 0$. Hence, if $V_{nk}^* = \sum_{j=1}^{k} Y_{nj}^*$, $1 \le k \le k_n$, by Lévy's inequality for all $\varepsilon > 0$

$$P\left\{\max_{k \le k_n} |V^*_{nk}| \ge \varepsilon\right\} \le 2\,P\{|V_n^*| \ge \varepsilon\} = o(1). \tag{11}$$

For fixed $n \ge 1$ and $\varepsilon > 0$, define $T = \inf\{j: 1 \le j \le k_n, |V_{nj}^*| \ge \varepsilon\}$ and $T = \infty$ if this set is empty. Then $T_n = \min(T, k_n)$ is a bounded stopping variable and since $|Y^*_{nj}| \le 2$, $1 \le j \le k_n$,

$$|V^*_{n, T_n}| \le \varepsilon I_{[T > k_n]} + (\varepsilon + 2) I_{[T \le k_n]},$$

whence it follows from the second moment analogue of Wald's equation (specifically, Corollary 7.4.7) that

$$\begin{aligned}\varepsilon^2 + 4(\varepsilon + 1)P\{T \le k_n\} &= E[V^*_{n, T_n}]^2 = E\sum_1^{T_n} \sigma^2(Y^*_{nj}) \\ &\ge 2\,P\{T \ge k_n\} \sum_1^{k_n} \sigma^2(Y_{nj}).\end{aligned}$$

As $n \to \infty$, (11) ensures that $P\{T \le k_n\} = o(1)$, yielding

$$\varepsilon^2 \ge 2 \varlimsup_{n\to\infty} \sum_1^{k_n} \sigma^2(Y_{nj}),$$

and since ε is arbitrary, (7) follows. □

Remark: Sufficiency only requires (6) for $\varepsilon = 1$.

Corollary 1. *If for each $n \ge 1$, $\{X_{nj}, 1 \le j \le k_n \to \infty\}$ are independent* r.v.s *with m_{nj} a median of X_{nj}, $1 \le j \le k_n$, and $S_n = \sum_{j=1}^{k_n} X_{nj}$, then*

$$S_n - A_n \xrightarrow{P} 0 \tag{12}$$

for some real numbers A_n iff

$$\sum_{j=1}^{k_n} P\{|X_{nj} - m_{nj}| \ge 1\} = o(1), \tag{13}$$

$$\sum_{j=1}^{k_n} \mathrm{E}(X_{nj} - m_{nj})^2 I_{[|X_{nj}-m_{nj}|<1]} = o(1), \tag{14}$$

in which case A_n may be chosen to be $\sum_{j=1}^{k_n} \{m_{nj} + \mathrm{E}(X_{nj} - m_{nj}) I_{[|X_{nj}-m_{nj}|<1]}\}$.

PROOF. Since $Z_{nj} = X_{nj} - m_{nj}$ has zero as a median, (12) follows from (13) and (14) by Theorem 1, noting (9). Conversely, under (12), setting $U_n = \sum_{j=1}^{k_n} Z_{nj}$, $B_n = A_n - \sum_{j=1}^{k_n} m_{nj}$,

$$U_n - B_n \xrightarrow{P} 0, \tag{15}$$

whence (13) holds by Theorem 1. To prove (14), set

$$Y_{nj} = Z_{nj} I_{[|Z_{nj}|<1]}$$

and let $(Y'_{n1}, \ldots, Y'_{nk_n})$ and $(Y_{n1}, \ldots, Y_{nk_n})$ be i.i.d. random vectors. It follows from (15) and (13) that

$$\sum_{j=1}^{k_n} \frac{(Y_{nj} - Y'_{nj})}{3} \xrightarrow{P} 0,$$

and since $|(Y_{nj} - Y'_{nj})/3| < 1$ and (Exercise 3.3.3) $m(Y_{nj}) = 0$, Theorem 1 and Lemma 1 ensure that

$$o(1) = \sum_{j=1}^{k_n} \sigma^2\left(\frac{Y_{nj} - Y'_{nj}}{3}\right) = \frac{1}{9}\sum_{j=1}^{k_n} \mathrm{E}(Y_{nj} - Y'_{nj})^2 \geq \frac{1}{18}\sum_{j=1}^{k_n} \mathrm{E}Y_{nj}^2,$$

which is tantamount to (14). □

Corollary 2. *If for each $n \geq 1$, $\{Y_{nj}, 1 \leq j \leq k_n \to \infty\}$ are independent, positive* r.v.s, *then $S_n = \sum_{j=1}^{k_n} Y_{nj} \xrightarrow{P} 1$ and $\max_{1 \leq j \leq k_n} m(Y_{nj}) \to 0$* iff *for all $\varepsilon > 0$*

$$\sum_{j=1}^{k_n} \mathrm{P}\{Y_{nj} \geq \varepsilon\} = o(1), \tag{16}$$

$$\sum_{j=1}^{k_n} \mathrm{E}\, Y_{nj} I_{[Y_{nj}<1]} \to 1. \tag{17}$$

Moreover, if $\sum_{j=1}^{k_n} \mathrm{E}\, Y_{nj} = 1$, then $S_n \xrightarrow{P} 1$ and $\max_{1 \leq j \leq k_n} m(Y_{nj}) \to 0$ iff

$$\sum_{j=1}^{k_n} \mathrm{E}\, Y_{nj} I_{[Y_{nj} \geq \varepsilon]} = o(1), \qquad \varepsilon > 0. \tag{18}$$

PROOF. Necessity of (16) and (17) is an immediate consequence of (6) and (8) of Theorem 1 with $A_n \equiv 1$. Sufficiency likewise follows from Theorem 1 once it is noted that for arbitrary ε in (0, 1)

$$0 \le \sum_{j=1}^{k_n} (\mathrm{E}\, Y_{nj}^2 I_{[Y_{nj}<1]} - \mathrm{E}^2\, Y_{nj} I_{[Y_{nj}<1]}) \le \sum_{j=1}^{k_n} \mathrm{E}\, Y_{nj}^2 I_{[Y_{nj}<1]}$$

$$\le \varepsilon \sum_{j=1}^{k_n} \mathrm{E}\, Y_{nj} I_{[Y_{nj}<\varepsilon]} + \sum_{j=1}^{k_n} \mathrm{P}\{Y_{nj} \ge \varepsilon\}.$$

The final remark is a direct consequence of

$$0 \le \varepsilon \sum_{j=1}^{k_n} \mathrm{P}\{Y_{nj} \ge \varepsilon\}$$

$$\le \sum_{j=1}^{k_n} \mathrm{E}\, Y_{nj} I_{[Y_{nj} \ge \varepsilon]} \le \sum_{j=1}^{k_n} (\mathrm{P}\{\varepsilon \le Y_{nj}\} + \mathrm{E}\, Y_{nj} I_{[Y_{nj} \ge 1]}). \qquad \square$$

Corollary 3. *If* $\{X_n, n \ge 1\}$ *are independent* r.v.s, $S_n = \sum_1^n X_j$, *and* $\{b_n, n \ge 1\}$ *are constants with* $0 < b_n \uparrow \infty$, *then* $S_n/b_n \xrightarrow{\mathrm{P}} 0$ iff

$$\sum_{j=1}^{n} \mathrm{P}\{|X_j| \ge b_n\} = o(1), \tag{19}$$

$$\frac{1}{b_n} \sum_{j=1}^{n} \mathrm{E}\, X_j I_{[|X_j|<b_n]} = o(1), \tag{20}$$

$$\frac{1}{b_n^2} \sum_{j=1}^{n} \sigma^2(X_j I_{[|X_j|<b_n]}) = o(1). \tag{21}$$

Proof. Apply Theorem 1 twice. Conditions (20), (21) imply

$$\frac{1}{b_n} \sum_{j=1}^{n} Z_{nj} \xrightarrow{\mathrm{P}} 0 \tag{22}$$

for $Z_{nj} = X_j I_{[|X_j|<b_n]} - \mathrm{E}\, X_j I_{[|X_j|<b_n]}$, and so (20) guarantees (22) with $Z_{nj} = X_j I_{[|X_j|<b_n]}$; moreover, (19) ensures (22) for $Z_{nj} = X_j I_{[|X_j| \ge b_n]}$, and these combine to yield $S_n/b_n \xrightarrow{\mathrm{P}} 0$. Conversely, $S_n/b_n \xrightarrow{\mathrm{P}} 0$ entails $|S_{n-1}/b_n| \le |S_{n-1}/b_{n-1}| \xrightarrow{\mathrm{P}} 0$, whence $X_n/b_n \xrightarrow{\mathrm{P}} 0$, and hence $m(X_n/b_n) = o(1)$. Since for $1 \le j \le n$

$$\left| m\left(\frac{X_j}{b_n}\right) \right| = \left| \frac{b_j}{b_n} m\left(\frac{X_j}{b_j}\right) \right| \le \left| m\left(\frac{X_j}{b_j}\right) \right|,$$

$$\max_{1 \le j \le n} \left| m\left(\frac{X_j}{b_n}\right) \right| \le \frac{1}{b_n} \max_{1 \le j \le n_0} |m(X_j)| + \max_{n_0 < j \le n} \left| m\left(\frac{X_j}{b_j}\right) \right|$$

$$\xrightarrow{n \to \infty} \sup_{j \ge n_0} \left| m\left(\frac{X_j}{b_j}\right) \right| \xrightarrow{n_0 \to \infty} 0,$$

conditions (4) and (5) of Theorem 1 hold with $A_n = 0$, so that (6), (7), (8) are tantamount to (19), (20), (21). □

The following lemmas will be useful in discussing the strong law.

Lemma 2. *If x_j, $1 \le j \le n$, are real numbers, $s_n = \sum_1^n x_j$, and*

$$Q_{k,n} = \sum_{1 \le i_1 < \cdots < i_k \le n} x_{i_1} x_{i_2} \cdots x_{i_k}, \qquad n \ge k \ge 1,$$

then for $n \ge k \ge 2$, $Q_{k,n} = \sum_{j=k}^n x_j Q_{k-1,j-1}$ and

$$s_n^k - k!\,Q_{k,n} = C_k, \tag{23}$$

where C_k is a generic designation for a finite linear combination (coefficients independent of n) of terms $\prod_{i=1}^m (\sum_{j=1}^n x_j^{h_i})$ of order k, that is, $\sum_{i=1}^m h_i = k$, $1 \le h_i \le k$, $1 \le m < k$.

PROOF. Suppose inductively that

$$s_n^h - h!\,Q_{h,n} = C_h, \qquad 1 \le h \le k \tag{24}$$

Since this implies $s_n^{k+1} - k!\,s_n Q_{k,n} = s_n C_k = C_{k+1}$, it suffices to verify that

$$s_n Q_{k,n} = (k+1)Q_{k+1,n} + C_{k+1}. \tag{25}$$

However, via the induction hypothesis and the identity below,

$$\begin{aligned} s_n Q_{k,n} &= (k+1)Q_{k+1,n} + \sum_{i=1}^{k-1} (-1)^{i+1} Q_{k-i,n} \sum_{j=1}^n x_j^{i+1} + (-1)^{k+1} \sum_{j=1}^n x_j^{k+1} \\ &= (k+1)Q_{k+1,n} + \sum_{i=1}^{k-1} (-1)^{i+1} \left[\frac{s_n^{k-i} + C_{k-i}}{(k-i)!}\right] \sum_{j=1}^n x_j^{i+1} + C_{k+1} \\ &= (k+1)Q_{k+1,n} + C_{k+1}. \end{aligned}$$

Since (24) clearly holds for $k = 1$ and 2, the lemma follows. □

Lemma 3. *If k is any positive integer, $\{y_n, n \ge 1\}$ is a sequence of positive numbers, $Q_{0n} \equiv 1$, and*

$$Q_{k,n} = \sum_{1 \le j_1 < \cdots < j_k \le n} y_{j_1} y_{j_2} \cdots y_{j_k} = \sum_{j=k}^n y_j Q_{k-1,j-1}, \qquad n \ge k, \tag{26}$$

then $y_j \le A$, $1 \le j \le n$, and $Q_{k,n} \le A^k$ imply $\sum_{j=1}^n y_j \le (2k-1)A$ for any positive integer n and any positive constant A.

PROOF. The lemma is trivially true for $k = 1$. Suppose inductively that it holds for $k - 1$ where $k \ge 2$. If the set $I = \{j : k \le j \le n, Q_{k-1,j-1} > A^{k-1}\}$ has an empty complement, then, since $Q_{k,n} \le A^k$, (26) yields $\sum_{j=k}^n y_j \le A$, implying $\sum_{j=1}^n y_j \le kA \le (2k-1)A$. Alternatively, there is a largest integer

m in I^c, whence by the induction hypothesis $\sum_{j=1}^{m-1} y_j \le (2k-3)A$, implying $\sum_{j=1}^{m} y_j \le (2k-2)A$. Consequently, noting that (26) entails $\sum_{j\in I} y_j \le A$,

$$\sum_1^n y_j \le \sum_1^m y_j + \sum_{j\in I} y_j \le (2k-1)A,$$

completing the induction. □

Let $0 < b_n \uparrow \infty$ and consider the series

$$\begin{aligned} \textstyle\sum_1 &= \sum_{j=1}^{\infty} b_j^{-2}\sigma_j^2, \\ \textstyle\sum_k &= \sum_{j_k=k}^{\infty} b_{j_k}^{-2k}\sigma_{j_k}^2 \sum_{j_{k-1}=k-1}^{j_k-1} \sigma_{j_{k-1}}^2 \cdots \sum_{j_1=1}^{j_2-1} \sigma_{j_1}^2, \qquad k \ge 2. \end{aligned} \tag{27}$$

Corollary 5.2.1 states that for independent r.v.s with $\mathrm{E}\, X_n = 0$, $\mathrm{E}\, X_n^2 = \sigma_n^2$, the convergence of $\sum_1$ is sufficient for the classical strong law (where $b_n = n$). The next theorem asserts that convergence of $\sum_k$ for some $k \ge 2$ in conjunction with the necessary condition $X_n = o(b_n)$, a.c., ensures the generalized strong law $S_n/b_n \xrightarrow{\text{a.c.}} 0$.

Theorem 2. *Let $\{X_n, n \ge 1\}$ be independent* r.v.s *with* $\mathrm{E}\, X_n = 0$, $\mathrm{E}\, X_n^2 = \sigma_n^2$ *and set $S_n = \sum_1^n X_i$. If the series Σ_k of (27) converges for some $k \ge 1$ and*

$$\sum_{n=1}^{\infty} \mathrm{P}\{|X_n| > \varepsilon b_n\} < \infty \quad \text{for all } \varepsilon > 0, \tag{28}$$

where $0 < b_n \uparrow \infty$, then $S_n/b_n \xrightarrow{\text{a.c.}} 0$.

PROOF. Define $W_{k,n} = \sum_{j=k}^n b_j^{-k} X_j U_{k-1,j-1}$, $n \ge k \ge 1$, where $U_{0,n} \equiv 1$ and

$$U_{k,n} = \sum_{j=k}^n X_j U_{k-1,j-1} = \sum_{1 \le i_1 < \cdots < i_k \le n} X_{i_1} X_{i_2} \cdots X_{i_k}, \qquad n \ge k \ge 1.$$

Then $\{W_{k,n}, \mathscr{F}_n, n \ge k\}$ is a martingale, where $\mathscr{F}_n = \sigma(X_j, 1 \le j \le n)$, and, moreover, $\mathrm{E}\, W_{k,n}^2$ is the series in (27) modified in that the first summation only goes up to n rather than ∞. The convergence of $\sum_k$ thus ensures that $\{W_{k,n}, \mathscr{F}_n, n \ge k\}$ is an $\mathscr{L}_2$ bounded martingale, hence convergent to some r.v. by Theorem 7.4.3. Consequently, by Kronecker's lemma

$$b_n^{-k} U_{k,n} = b_n^{-k} \sum_{j=k}^n X_j U_{k-1,j-1} \xrightarrow{\text{a.c.}} 0. \tag{29}$$

This proves the theorem for $k = 1$, in which case (28) is superfluous. Next, in view of (27)

$$Z_{k,n} = \sum_{j_k=k}^n b_{j_k}^{-2k} X_{j_k}^2 \sum_{j_{k-1}=k-1}^{j_k-1} X_{j_{k-1}}^2 \cdots \sum_{j_1=1}^{j_2-1} X_{j_1}^2, \qquad n \ge k,$$

is an $\mathscr{L}_1$ bounded submartingale and hence convergent a.c. Thus, by Kronecker's lemma

$$Q_{k,n} = \sum_{j_k=k}^{n} X_{j_k}^2 \sum_{j_{k-1}=k-1}^{j_k-1} X_{j_{k-1}}^2 \cdots \sum_{j_1=1}^{j_2-1} X_{j_1}^2 = o(b_n^{2k}), \qquad \text{a.c.}$$

Moreover, (28) ensures $X_n^2 = o(b_n^2)$, a.c. Applying Lemma 3 with $y_j = X_j^2$, it follows that with probability one

$$\left|\sum_{j=1}^{n} X_j^h\right| \le \left(\sum_{j=1}^{n} X_j^2\right)^{h/2} = o(b_n^h), \qquad h \ge 2. \tag{30}$$

By Lemma 2 there exist finite constants c_k; $c_1, \ldots, c_{k-2}$ (when $k = 2$, merely c_2) such that

$$(b_n^{-1}S_n)^k = \sum_{h=1}^{k-2} c_h(b_n^{-1}S_n)^h A_{k-h,n} + c_k A_{k,n} + k!\,b_n^{-k}U_{k,n}, \tag{31}$$

where for $0 \le h \le k-2$, $A_{k-h,n}$ is a finite linear combination of terms $\prod_{i=1}^{m} (b_n^{-h_i} \sum_{j=1}^{n} X_j^{h_i})$ satisfying $h_i \ge 2$ for $1 \le i < m < k$ and $\sum_{i=1}^{m} h_i = k - h$.

In view of (30), $A_{k-h,n} \xrightarrow{\text{a.c.}} 0$ for $0 \le h \le k-2$. Thus, according to (31) and recalling (29), S_n/b_n is a root of a kth degree polynomial in which the leading coefficient is unity and the remaining coefficients converge (a.c.) to zero. The conclusion of the theorem follows from the well-known relations between the roots and coefficients of a polynomial. □

The next corollary reveals, under the necessary (when $0 < b_n \uparrow \infty$) condition (28) with $b_n = s_n(\log s_n)^\alpha$, $\alpha > 0$, that independent r.v.s with zero means, variances σ_n^2 and $s_n^2 = \sum_{i=1}^{n} \sigma_i^2 \to \infty$, obey the strong law

$$s_n^{-1}(\log s_n)^{-\alpha} \sum_{j=1}^{n} X_j \xrightarrow{\text{a.c.}} 0,$$

thereby generalizing Exercise 5.2.10, where $\alpha > \frac{1}{2}$.

Corollary 4. *If* $\{X_n, n \ge 1\}$ *are independent* r.v.s *with* $\mathrm{E}\,X_n = 0$, $\mathrm{E}\,X_n^2 = \sigma_n^2$, *then* $S_n/b_n \xrightarrow{\text{a.c.}} 0$ *provided* (28) *and*

$$\left(\log \sum_{1}^{n} \sigma_j^2\right)^{\delta} \sum_{j=1}^{n} \sigma_j^2 = O(b_n^2) \quad \textit{for some } \delta > 0 \tag{32}$$

hold.

Proof. Setting $s_n^2 = \sum_1^n \sigma_j^2$, if $k\delta > 1$, the series Σ_k of (27) converges, being dominated by

$$\sum_n \left[b_n^{-2k}\sigma_n^2\left(\sum_1^n \sigma_j^2\right)^{k-1}\right] \le C \sum_{n=1}^{\infty} \frac{\sigma_n^2}{s_n^2(\log s_n^2)^{k\delta}} < \infty. \qquad \square$$

If higher-order moments are assumed finite, the next result asserts that convergence of a single series suffices for the classical strong law.

Theorem 3 (Brunk–Chung). *If* $\{X_n, n \geq 1\}$ *are independent* r.v.s *with* $\mathrm{E}\,X_n = 0$, $n \geq 1$, *and for some* $r \geq 1$

$$\sum_{n=1}^{\infty} \frac{\mathrm{E}|X_n|^{2r}}{n^{r+1}} < \infty, \tag{33}$$

then $(1/n)\sum_{i=1}^{n} X_i \xrightarrow[\mathscr{L}_{2r}]{\text{a.c.}} 0$.

Proof. Setting $S_n = \sum_1^n X_i$, the submartingale inequality of Theorem 7.4.8 (33) yields

$$\begin{aligned} \varepsilon^{2r}\,\mathrm{P}\left\{\sup_{j\geq n}\frac{|S_j|}{j} \geq \varepsilon\right\} &= \varepsilon^{2r}\lim_{m\to\infty}\mathrm{P}\left\{\max_{n\leq j\leq m}\frac{|S_j|^{2r}}{j^{2r}} \geq \varepsilon^{2r}\right\} \\ &\leq \frac{1}{n^{2r}}\,\mathrm{E}|S_n|^{2r} + \sum_{n+1}^{\infty}\frac{1}{j^{2r}}\,\mathrm{E}(|S_j|^{2r} - |S_{j-1}|^{2r}), \end{aligned} \tag{34}$$

and so, in view of Lemma 3.3.1 it suffices to show that the right side of (34) is $o(1)$. Now,

$$\begin{aligned} &\sum_{j=2}^{n}\frac{1}{j^{2r}}\,\mathrm{E}(|S_j|^{2r} - \mathrm{E}|S_{j-1}|^{2r}) \\ &\quad\leq \sum_{j=3}^{n}\left(\frac{1}{(j-1)^{2r}} - \frac{1}{j^{2r}}\right)\mathrm{E}|S_{j-1}|^{2r} + \frac{\mathrm{E}|S_n|^{2r}}{n^{2r}}, \end{aligned} \tag{35}$$

and by the Marcinkiewicz–Zygmund (Section 3) and Hölder inequalities

$$E|S_j|^{2r} \leq \mathrm{E}\left(\sum_1^j X_i^2\right)^r \leq \mathrm{E}\,j^{r-1}\sum_1^j |X_i|^{2r}.$$

It follows via (33) and Kronecker's lemma that $\mathrm{E}|S_n|^{2r} = o(n^{2r})$ and, moreover, that the series on the right of (35) is bounded by

$$\begin{aligned} \sum_{j=2}^{n-1}\left(\frac{1}{j^{2r}} - \frac{1}{(j+1)^{2r}}\right)j^{r-1}\sum_1^j \mathrm{E}|X_i|^{2r} &\leq C\sum_{j=2}^{n-1}\frac{1}{j^{r+2}}\sum_{i=1}^{j}\mathrm{E}|X_i|^{2r} \\ &\leq C'\sum_{j=2}^{n}\frac{\mathrm{E}|X_j|^{2r}}{j^{r+1}} + O(1), \end{aligned}$$

which converges as $n \to \infty$ by hypothesis. □

Theorem 4. *Let* $\{S_n, n \geq 1\}$ *be the partial sums of independent random variables* $\{X_n, n \geq 1\}$ *with* $\mathrm{E}\,X_n = 0$, $\mathrm{E}|X_n|^\alpha \leq a_{n,\alpha}$, $A_n \equiv A_{n,\alpha} = (\sum_{i=1}^n a_{i,\alpha})^{1/\alpha} \to \infty$

where $1 < \alpha \le 2$ *and* $A_{n+1,\alpha}/A_{n,\alpha} \le \gamma < \infty$, *all* $n \ge 1$. *If, for some* β *in* $[0, 1/\alpha)$ *and positive* δ, c,

$$\sum_{n=1}^{\infty} P\{|X_n| > \delta A_n(\log_2 A_n)^{1-\beta}\} < \infty \tag{36}$$

$$\sum_{n=1}^{\infty} \frac{1}{A_n^2(\log_2 A_n)^{2(1-\beta)}} E\, X_n^2 I_{[cA_n(\log_2 A_n)^{-\beta} < |X_n| \le \delta A_n(\log_2 A_n)^{1-\beta}]} < \infty \tag{37}$$

then

$$\frac{S_n}{A_n(\log_2 A_n)^{1-\beta}} \xrightarrow{\text{a.c.}} 0. \tag{38}$$

PROOF. Set $Y_n = X_n I_{[|X_n| \le cA_n(\log_2 A_n)^{-\beta}]}$, $W_n = X_n I_{[|X_n| > \delta A_n(\log_2 A_n)^{1-\beta}]}$, $V_n = X_n - Y_n - W_n$. Now since $\beta < 1/\alpha$,

$$\begin{aligned}\left|\sum_{i=1}^{n} E\, W_i\right| &\le \sum_{i=1}^{n} (E|X_i| I_{[\delta A_i(\log_2 A_i)^{1-\beta} < |X_i| \le A_n(\log_2 A_n)^{-\beta}} \\ &\quad + E|X_i| I_{[|X_i| > A_n(\log_2 A_n)^{-\beta}]}) \\ &\le A_n(\log_2 A_n)^{-\beta} \sum_{i=1}^{n} P\{|X_i| > \delta A_i(\log_2 A_i)^{1-\beta}\} \\ &\quad + \frac{(\log_2 A_n)^{\beta(\alpha-1)}}{A_n^{\alpha-1}} \sum_{i=1}^{n} E|X_i|^{\alpha} I_{[|X_i| > A_n(\log_2 A_n)^{-\beta}]} \\ &= o(A_n(\log_2 A_n)^{1-\beta})\end{aligned}$$

and so, in view of (36),

$$\frac{1}{A_n(\log_2 A_n)^{1-\beta}} \sum_{i=1}^{n} (W_i - E\, W_i) \xrightarrow{\text{a.c.}} 0.$$

Secondly, (37) and Kronecker's lemma guarantee

$$\frac{1}{A_n(\log_2 A_n)^{1-\beta}} \sum_{i=1}^{n} (V_i - E\, V_i) \xrightarrow{\text{a.c.}} 0.$$

Thus, since $E\, X_n = 0$, it suffices to verify that

$$\frac{1}{A_n(\log_2 A_n)^{1-\beta}} \sum_{i=1}^{n} (Y_i - E\, Y_i) \xrightarrow{\text{a.c.}} 0. \tag{39}$$

To this end, note that if $n_k = \inf\{n \ge 1: A_n \ge \gamma^k\}$,

$$A_{n_k} \le \gamma A_{n_k - 1} < \gamma^{k+1} \le A_{n_{k+1}},$$

and so $\{n_k, k \ge 1\}$ is strictly increasing. Moreover, for all $k \ge 1$

$$\frac{A_{n_k}}{A_{n_{k-1}}} < \frac{\gamma^{k+1}}{\gamma^{k-1}} = \gamma^2 > 1.$$

Therefore, setting $U_n = \sum_{i=1}^n (Y_i - \mathrm{E}\, Y_i)$, for all $\varepsilon > 0$

$$\begin{aligned} &\mathrm{P}\{U_n > 2\gamma^2 \varepsilon A_n (\log_2 A_n)^{1-\beta}, \text{ i.o.}\} \\ &\quad \le \mathrm{P}\left\{ \max_{n_{k-1} < n \le n_k} U_n > 2\gamma^2 \varepsilon A_{n_{k-1}} (\log_2 A_{n_{k-1}})^{1-\beta}, \text{ i.o.}\right\} \\ &\quad \le \mathrm{P}\left\{ \max_{1 \le n \le n_k} U_n > \varepsilon A_{n_k} (\log_2 A_{n_k})^{1-\beta}, \text{ i.o.}\right\}. \end{aligned} \tag{40}$$

Now, $|Y_n - \mathrm{E}\, Y_n| \le c_n t_n$, where $c_n = 2cA_n(\log_2 A_n)^{-\beta}/t_n$ and

$$t_n^2 = \mathrm{E}\, U_n^2 \le \sum_{i=1}^n \mathrm{E}\, X_i^2 I_{[|X_i| \le cA_i(\log_2 A_i)^{-\beta}]}$$

$$\le \frac{c^{2-\alpha} A_n^{2-\alpha}}{(\log_2 A_n)^{\beta(2-\alpha)}} \sum_{i=1}^n \mathrm{E}|X_i|^\alpha = \frac{c^{2-\alpha} A_n^2}{(\log_2 A_n)^{\beta(2-\alpha)}}. \tag{41}$$

Consequently, setting

$$\lambda_n = \frac{\varepsilon A_n^2 (\log_2 A_n)^{1-2\beta}}{t_n^2}, \qquad x_n = \frac{t_n (\log_2 A_n)^\beta}{A_n}$$

it follows that $c_n \cdot x_n = 2c$. Since (41) ensures that $2c\lambda_n x_n \ge 2c^{\alpha-1}\varepsilon(\log_2 A_n)^{1-\beta\alpha} \to \infty$ and $h(x) \ge (x/2)\log(1+x)$ as $x \to \infty$, for all large n (see 10.2(1))

$$x_n^2 h(2c\lambda_n) \ge c\lambda_n x_n^2 \log(1 + 2c\lambda_n) = \varepsilon(\log_2 A_n)\log(1 + 2c\lambda_n) > 8c^2 \log_2 A_n.$$

Thus, via (3) of Lemma 10.2.1

$$\mathrm{P}\left\{ \max_{1 \le n \le n_k} U_n > \varepsilon A_{n_k} (\log_2 A_{n_k})^{1-\beta} \right\} = \mathrm{P}\left\{ \max_{1 \le n \le n_k} U_n > \lambda_{n_k} x_{n_k} t_{n_k} \right\}$$

$$\le \exp\{-\tfrac{1}{2} x_{n_k}^2 h(2c\lambda_{n_k})/4c^2\} \le \exp\{-2\log_2 A_{n_k}\} \le \frac{1}{(k \log \gamma)^2}$$

and so (40) and the Borel–Cantelli lemma ensure

$$\varlimsup_{n\to\infty} \frac{U_n}{A_n(\log_2 A_n)^{1-\beta}} \le 0, \text{ a.c.} \tag{42}$$

Since $\{-(Y_n - \mathrm{E}\, Y_n), n \ge 1\}$ have the same bounds and variances as $\{Y_n - \mathrm{E}\, Y_n, n \ge 1\}$, (42) likewise obtains with $-U_n$ replacing U_n thereby proving (39) and the theorem. □

Corollary 5. *Let* $\{X_n, n \geq 1\}$ *be independent random variables with* $\mathrm{E}\, X_n = 0$, $\mathrm{E}\, X_n^2 = \sigma_n^2$, $s_n^2 = \sum_1^n \sigma_i^2 \to \infty$ and $s_{n+1}/s_n \leq \gamma < \infty$, $n \geq 1$. If for some β in $[0, \frac{1}{2})$ *and some positive* c, δ

$$\sum_{n=1}^{\infty} \mathrm{P}\{|X_n| > \delta s_n(\log_2 s_n)^{1-\beta}\} < \infty \tag{43}$$

$$\sum_{n=1}^{\infty} \frac{1}{s_n^2(\log_2 s_n)^{2(1-\beta)}} \mathrm{E}\, X_n^2 I_{[cs_n(\log_2 s_n)^{-\beta} < |X_n| \leq \delta s_n(\log_2 s_n)^{1-\beta}]} < \infty \tag{44}$$

then

$$\frac{S_n}{s_n(\log_2 s_n)^{1-\beta}} \xrightarrow{\text{a.c.}} o. \tag{45}$$

Note that Corollary 5 precludes $\beta = \frac{1}{2}$. In fact, (43) and (44) when $\beta = \frac{1}{2}$ comprise two of the three conditions for the Law of the Iterated Logarithm in Theorem 10.2.3.

Corollary 6. *If* $\{X_n, n \geq 1\}$ *are independent random variables with* $\mathrm{E}\, X_n = 0$, $\mathrm{E}\, X_n^2 = \sigma_n^2$, $s_n^2 = \sum_1^n \sigma_i^2 \to \infty$ *and* $|X_n| \leq c_\beta s_n(\log_2 s_n)^{-\beta}$, a.s., $n \geq 1$ *where* $0 \leq \beta < \frac{1}{2}, c_\beta > 0$ *then* (45) *obtains provided in the case* $\beta = 0$ *that* $s_{n+1} = O(s_n)$.

Prohorov (1950) has shown for $b_n = n$ and $n_k = 2^k$ that convergence of the series

$$\sum_{k=1}^{\infty} \exp\{-\varepsilon b_{n_{k-1}}^2/(s_{n_k}^2 - s_{n_{k-1}}^2)\}, \ \varepsilon > 0$$

is necessary and sufficient for $S_n/n \xrightarrow{\text{a.c.}} 0$ when $|X_n| < Kn/\log_2 n$, a.c. for $n \geq 1$. Unrestricted necessary and sufficient conditions depending upon solutions of equations involving truncated moment generating functions have been given by Nagaev (1972).

Exercises 10.1

1. Verify via independent $\{X_n\}$ with

$$\mathrm{P}\{X_n = (-1)^{n+1} n^{1/2}\} = \frac{n^2}{n^2 + n^{1/2}} = 1 - \mathrm{P}\{X_n = (-1)^n n^2\}$$

that (21) of Corollary 3 ($b_n = n$) cannot be replaced by an analogous condition with a truncated second moment replacing the variance.

2. If $\{X_n\}$ are i.i.d. with $p_j = \mathrm{P}\{X_1 = 2^j\} = 1/[2^j(j+1)j]$, $j \geq 1$, and $\mathrm{P}\{X_1 = 0\} = 1 - \sum_1^\infty p_j$, prove that $(\mathrm{Log}\, n/n)(S_n - n) \xrightarrow{\mathrm{P}} -1$, where Log denotes logarithm to the base 2. *Hint*: Consider $Y_{nj} = X_j I_{[X_j \leq n/\mathrm{Log}\, n]}$.

3. Let $\{X_n, n \geq 1\}$ be independent r.v.s with $\mathrm{E}\,X_n = 0$, $\mathrm{E}\,X_n^2 = \sigma_n^2 < \infty$, $s_n^2 = \sum_1^n \sigma_i^2 \to \infty$, and $\sigma_n = o(s_n)$. Prove the result of Raikov that

$$\frac{1}{s_n}\sum_{j=1}^n X_j \xrightarrow{\text{d}} N_{0,1} \quad \text{iff} \quad \frac{1}{s_n^2}\sum_{j=1}^n X_j^2 \xrightarrow{\text{P}} 1.$$

Hint: Apply Corollary 2 to $Y_{nj} = X_j^2/s_n^2$, $1 \leq j \leq n$,

4. Necessary and sufficient conditions for the strong law for independent $\{X_n\}$ with $\mathrm{E}\,X_n = 0$, $\mathrm{E}\,X_n^2 = \sigma_n^2 < \infty$ cannot be framed solely in terms of variances. *Hint*: Consider

$$\mathrm{P}\{X_n = \pm cn\} = \frac{1}{2c^2 n \log n} = \tfrac{1}{2}[1 - \mathrm{P}\{X_n = 0\}], \qquad n \geq 2,$$

and

$$\mathrm{P}\left\{X_n = \pm\frac{n}{\log n}\right\} = \frac{\log n}{2n} = \tfrac{1}{2}[1 - \mathrm{P}\{X_n = 0\}], \qquad n \geq 2.$$

5. Define a sequence of r.v.s for which $(1/n)\sum_1^n X_i \xrightarrow{\text{a.c.}} 0$ but $\sum (X_n/n)$ diverges a.c., revealing limitations to the "Kronecker lemma approach."

6. Let $\{X_n, n \geq 1\}$ be independent r.v.s with $\mathrm{E}\,X_n = 0$, $\mathrm{E}\,X_n^2 = \sigma_n^2$, satisfying (28).
   (i) Show that if $\sigma_n^2 \sim n/(\log n)^\delta$, $\delta > 0$, then (*) $(1/n)\sum_1^n X_i \xrightarrow{\text{a.c.}} 0$.
   (ii) If, rather, $\sigma_n^2 \sim n/(\log\log n)^\delta$, gives necessary and sufficient conditions for (*) in terms of δ when $|X_n| = O(n/\log\log n)$, a.c.

7. Let $\{X_n, n \geq 1\}$ be independent r.v.s with $\mathrm{E}\,X_n = 0$, $\mathrm{E}\,X_n^2 = \sigma_n^2$, $s_n^2 = \sum_1^n \sigma_i^2$, and $\sigma_n^2 s_{n-1}^2 \leq \mathrm{E}\,X_n^4$, where (*) $\sum_{n=1}^\infty n^{-4}\,\mathrm{E}\,X_n^4 < \infty$. Prove that the classical SLLN holds. Compare (*) with (33) when $r = 2$.

8. Show that Theorem 2 extends to the case where $\{X_n, n \geq 1\}$ are martingale differences with constant conditional variances σ_n^2.

9. If $\{X_n\}$ are independent r.v.s with variances σ_n^2, $s_n^2 = \sum_1^n \sigma_i^2 = o(b_n^2)$, and $\sum_1^\infty (\mathrm{E}\,X_n^4/b_n^4) < \infty$, where $b_n \uparrow \infty$, then $(1/b_n^2)\sum_{i=1}^n X_i^2 \xrightarrow{\text{a.c.}} 0$.

10. Let $X_n = b^n Y_n$, $n \geq 1$, $b > 1$, where $\{Y_n\}$ are bounded i.i.d. random variables. Prove that $(1/b_n)\sum_{i=1}^n X_i \xrightarrow{\text{a.c.}} 0$ provided $b_n/b^n \to \infty$. Compare with Exercise 5.2.8.

11. Let $\{X_n, n \geq 1\}$ be independent, symmetric r.v.s and $X_n' = X_n I_{[|X_n| < b_n]}$, where $0 < b_n < \infty$. If $X_n^* = X_n I_{[|X_n| \leq b_n]} - X_n I_{[|X_n| > b_n]}$, then $\{X_n^*, n \geq 1\}$ are independent, symmetric r.v.s with the same joint d.f.s as $\{X_n, n \geq 1\}$. Let S_n, S_n', S_n^* be the corresponding partial sums and $\varepsilon_n > 0$. Then

$$[S_n' > \varepsilon_n] \subset [S_n > \varepsilon_n] \cup [S_n^* > \varepsilon_n]$$

and

$$\mathrm{P}\{S_n' > \varepsilon_n, \text{i.o.}\} \leq 2\,\mathrm{P}\{S_n > \varepsilon_n, \text{i.o.}\}.$$

12. Let $X_n = \sigma_n Y_n$, $n > 1$, where $\{Y_n\}$ are bounded i.i.d. r.v.s and

$$\sigma_n^2 = (\log n)^{-1}\exp\left\{\frac{2\lambda n}{\log n}\right\}, \qquad n > 1,\ \lambda > 0.$$

Show that $s_n^2 = \sum_1^n \sigma_i^2 \sim (2\lambda)^{-1}\exp\{2\lambda n/\log n\}$ and that $1/b_n \sum_{i=1}^n X_i \xrightarrow{\text{a.c.}} 0$ whenever $s_n^2 \log\log s_n^2 = o(b_n^2)$.

10.2 Law of the Iterated Logarithm

One of the most beautiful and profound discoveries of probability theory is the celebrated iterated logarithm law. This law, due in the case of suitably bounded independent random variables to Kolmogorov, was the culmination of a series of strides by mathematicians of the caliber of Hausdorff, Hardy, Littlewood, and Khintchine. The crucial instruments in the proof are sharp exponential bounds for the probability of large deviations of sums of independent r.v.s with vanishing means.

The next lemmas are generalized analogues of Kolmogorov's exponential bounds. The probabilistic inequality (3) is of especial interest in the cases (i) $\lambda_n \equiv \lambda$, $x_n \to \infty$, $c_n \to 0$, and (ii) $\lambda_n \to \infty$, $c_n x_n = a > 0$.

Define

$$h(x) = (1 + x)\log(1 + x) - x, \quad x \geq 0$$
$$g(x) = x^{-2}(e^x - 1 - x), \quad -\infty < x < \infty \tag{1}$$

Lemma 1. *Let* $S_n = \sum_{j=1}^n X_j$ *where* $\{X_j, 1 \leq j \leq n\}$ *are independent* r.v.s *with* $\mathrm{E}\, X_j = 0$, $\mathrm{E}\, X_j^2 = \sigma_j^2$, $s_n^2 = \sum_{j=1}^n \sigma_j^2 > 0$.

(i) *If* $\mathrm{P}\{X_j \leq c_n s_n\} = 1$, $1 \leq j \leq n$, *then*

$$\mathrm{E}\, e^{tS_n/s_n} \leq e^{t^2 g(c_n t)}, \qquad t > 0, \tag{2}$$

and if, in addition, $c_n x_n \leq a_n$, *then for all* λ_n *and* x_n *such that* $a_n \cdot x_n > 0$, $a_n \lambda_n > 0$

$$\mathrm{P}\left\{\max_{1 \leq j \leq n} S_j \geq \lambda_n x_n s_n\right\} \leq e^{-h(a_n \lambda_n)x_n^2/a_n^2} \tag{3}$$

(ii) *If* $\mathrm{P}\{X_j \geq -c_n s_n\} = 1$, $1 \leq j \leq n$, *then*

$$\mathrm{E}\, e^{tS_n/s_n} \geq \exp\left\{t^2 g(-c_n t)\left[1 - \frac{t^2 g(-tc_n)}{s_n^4} \sum_{j=1}^n \sigma_j^4\right]\right\}, \qquad t > 0, \tag{4}$$

and if, in addition, $\sigma_j \leq c_n s_n$, $1 \leq j \leq n$,

$$\begin{aligned} \mathrm{E}\, e^{tS_n/s_n} &\geq \exp\{t^2 g(-tc_n)[1 - t^2 c_n^2 g(-tc_n)]\} \\ &\geq \exp\{t^2 g(-tc_n)[1 - (t^2 c_n^2/2)]\}, \qquad t > 0. \end{aligned} \tag{5}$$

Proof. The representation $g(x) = \int_0^1 \int_0^u e^{xy}\, dy\, du$ shows that g is non-negative, increasing and convex. The point of departure for (2), (3), (4) is the simple observation

$$\mathrm{E} e^{tX_j/s_n} = 1 + \mathrm{E}\left[e^{tX_j/s_n} - 1 - \frac{tX_j}{s_n}\right] = 1 + t^2 \mathrm{E}\, \frac{X_j^2}{s_n^2}\, g\left(\frac{tX_j}{s_n}\right). \tag{6}$$

Hence, monotonicity ensures under (i) that if $t > 0$,

$$\mathrm{E}\, e^{tX_j/s_n} \leq 1 + \frac{t^2 \sigma_j^2}{s_n^2}\, g(tc_n) \leq \exp\{t^2 g(tc_n)\sigma_j^2/s_n^2\} \tag{7}$$

and (1) follows via independence.

If, rather, (ii) obtains, then (6) in conjunction with the elementary inequality $(1 + u)e^{u^2} > e^u$, $u > 0$, yields for $t > 0$

$$\mathrm{E}\, e^{tX_j/s_n} \geq 1 + \frac{t^2\sigma_j^2}{s_n^2} g(-tc_n) \geq \exp\left\{\frac{t^2\sigma_j^2}{s_n^2} g(-tc_n) - \frac{t^4\sigma_j^4}{s_n^4} g^2(-tc_n)\right\},$$

whence (4) is immediate. Under the additional hypothesis $\sigma_j \leq c_n s_n$, $1 \leq j \leq n$, necessarily $s_n^{-4} \sum_{j=1}^n \sigma_j^4 \leq c_n^2$, and (5) follows from (4) in view of $g(0) = \frac{1}{2}$.

To establish (3), note that via Example 7.4.9 and (1), for $t > 0$

$$\mathrm{P}\left\{\max_{1 \leq j \leq n} S_j \geq \lambda x_n s_n\right\} \leq e^{-\lambda t x_n s_n}\, \mathrm{E}\, e^{tS_n} \leq \exp\{-\lambda t x_n s_n + t^2 s_n^2 g(c_n s_n t)\}, \quad (8)$$

and so, setting $a = a_n$ and $t = bx_n/s_n$ where $bx_n > 0$,

$$\begin{aligned}\mathrm{P}\left\{\max_{1 \leq j \leq n} S_j \geq \lambda x_n s_n\right\} &\leq \exp\{-x_n^2[\lambda b - b^2 g(c_n x_n b)]\} \\ &\leq \exp\{-x_n^2[\lambda b - b^2 g(ab)]\}\end{aligned}$$

Employing the definition (1) of g, the prior exponent is minimized at $b = a^{-1}\log(1 + a\lambda)$ with a value of

$$\frac{-x_n^2}{a^2}[(1 + a\lambda)\log(1 + a\lambda) - a\lambda] = \frac{-x_n^2}{a^2} h(a\lambda)$$

Clearly, nothing precludes λ or a from being a function of n. □

The simple inequality of (3) yields a generalization of the easier half of the law of the iterated logarithm. In what follows $\log_2 n$ will abbreviate $\log\log n$.

Corollary 1. *Let* $\{X_n, n \geq 1\}$ *be independent* r.v.s *with* $\mathrm{E}\, X_n = 0$, $\mathrm{E}\, X_n^2 = \sigma_n^2$, $s_n^2 = \sum_1^n \sigma_i^2 \to \infty$.

(i) *If* $\mathrm{P}\{X_n \leq d_n\} = 1$, $n \geq 1$ *where* $d_n > 0$, *then with probability one*

$$\overline{\lim_{n\to\infty}} \frac{S_n}{s_n(\log_2 s_n^2)^{1/2}} \leq \sqrt{2} \quad \text{or} \quad \overline{\lim_{n\to\infty}} \frac{S_n}{s_n(\log_2 s_n^2)^{1/2}} \leq \frac{1}{a} h^{-1}(a^2) \quad (9)$$

according as $(\log_2 s_n^2)^{1/2} d_n/s_n \to 0$ *or* $(\log_2 s_n^2)^{1/2} d_n/s_n \to a > 0$.

(ii) *If* $\mathrm{P}\{|X_n| \leq d_n\} = 1$, $n \geq 1$ *where* $(\log_2 s_n^2)^{1/2} d_n/s_n \to a$, *then with probability one,*

$$\overline{\lim_{n\to\infty}} \frac{|S_n|}{s_n(\log_2 s_n^2)^{1/2}} \leq \frac{1}{a} h^{-1}(a^2). \quad (10)$$

PROOF. Let $b_n = s_n(\log_2 s_n^2)^{-1/2}$ and suppose $d_n/b_n \to a \geq 0$. Since $0 < b_n \uparrow \infty$, it follows that $b_n^{-1}(\max_{1 \leq i \leq n} d_i) \to a$ and so d_n may be taken to be increasing.

For any $\alpha > 1$, define $n_0 = \inf\{n \geq 1: s_n \geq \alpha\}$ and $n_k = \inf\{n > n_{k-1}: s_n \geq \alpha s_{n_{k-1}}\}$, $k \geq 1$. Then $s_{n_{k-1}} \leq s_{n_k - 1} < \alpha s_{n_{k-1}}$ so that $s_{n_k - 1} \geq \alpha^k$ and $\log_2 s_{n_{k-1}}^2 \sim \log_2 s_{n_k - 1}^2$. Hence,

$$\mathrm{P}\{S_n > \lambda\alpha^2 s_n(\log_2 s_n^2)^{1/2}, \text{i.o.}\} \le \mathrm{P}\left\{\max_{n_{k-1} \le n < n_k} S_n > \lambda\alpha^2 s_{n_{k-1}}(\log_2 s_{n_{k-1}}^2)^{1/2}, \text{i.o.}\right\}$$

$$\le \mathrm{P}\left\{\max_{1 \le j < n_k} S_j > \lambda s_{n_k-1}(\log_2 s_{n_k-1}^2)^{1/2}, \text{i.o.}\right\} \quad (11)$$

For any $\gamma > 0$, setting $\lambda = (1/u)h^{-1}(\gamma)$, $x_n = (\log_2 s_n^2)^{1/2}$, $n = n_k - 1$ and noting for any $u > a$ and all large n that $c_n(\log_2 s_n^2)^{1/2} = (\log_2 s_n^2)^{1/2} d_n/s_n < u$, Lemma 1 ensures that for all large k

$$\mathrm{P}\left\{\max_{1 \le j < n_k} S_j > \frac{1}{u}h^{-1}(\gamma)s_{n_k-1}(\log_2 s_{n_k-1}^2)^{1/2}\right\} \le \exp\left\{-\frac{1}{u^2}h(h^{-1}(\gamma))\log_2 s_{n_k-1}^2\right\}$$

$$\le (2k\log\alpha)^{-\gamma/u^2}$$

Hence, via (11) and the Borel–Cantelli lemma, for all $\gamma > u^2$

$$\mathrm{P}\left\{S_n > \frac{\alpha^2}{u}h^{-1}(\gamma)s_n(\log_2 s_n^2)^{1/2}, \text{i.o.}\right\} = 0$$

whence with probability one for $a > 0$,

$$\varlimsup_{n\to\infty} \frac{S_n}{s_n(\log_2 s_n^2)^{1/2}} \le \frac{\alpha^2}{a} \inf_{\gamma > u^2} h^{-1}(\gamma) \downarrow \frac{h^{-1}(a^2)}{a} \quad (12)$$

proving the second part of (i). If rather, $a = 0$ then (12) holds for arbitrarily small a. Since $h(a) \sim a^2/2$ as $a \to 0$, necessarily $h^{-1}(a) \sim (2a)^{1/2}$ whence $(1/a)h^{-1}(a^2) \to \sqrt{2}$ as $a \to 0$ yielding the first portion of (i). Finally, under (ii), (12) holds for both S_n and $-S_n$ so that (10) obtains. □

For any positive integer n, let $\{X_{n,j}, 1 \le j \le n\}$ constitute independent r.v.s with d.f.s $F_{n,j}$ and finite moment generating functions $\varphi_{n,j}(t) = \exp\{\psi_{n,j}(t)\}$ for $0 \le t < t_0$. Suppose that $S_{n,n} = \sum_{j=1}^n X_{n,j}$ has d.f. F_n. For any t in $[0, t_0)$, define **associated d.f.s** $F_{n,j}^{(t)}$ by

$$F_{n,j}^{(t)}(x) = \frac{1}{\varphi_{n,j}(t)} \int_{-\infty}^{x} e^{ty}\, dF_{n,j}(y)$$

and let $\{X_{n,j}(t), 1 \le j \le n\}$ be (fictitious) independent r.v.s with d.f.s $\{F_{n,j}^{(t)}, 1 \le j \le n\}$. Since the c.f. of $X_{n,j}(t)$ is $\varphi_{n,j}(t + iu)/\varphi_{n,j}(t)$, setting $\psi_n(t) = \sum_{j=1}^n \psi_{n,j}(t)$, the c.f. of $S_n(t) = \sum_{j=1}^n X_{n,j}(t)$ is given by

$$\mathrm{E}\, e^{iuS_n(t)} = \prod_{j=1}^{n} \frac{\varphi_{n,j}(t + iu)}{\varphi_{n,j}(t)} = \exp\{\psi_n(t + iu) - \psi_n(t)\}.$$

Thus, the mean and variance of $S_n(t)$ are $\psi_n'(t)$ and $\psi_n''(t)$ respectively and, moreover, the d.f. of $S_n(t)$ is

$$F_n^{(t)}(x) = e^{-\psi_n(t)} \int_{-\infty}^{x} e^{ty}\, dF_n(y),$$

whence for any t in $[0, t_0)$ and real u

$$P\{S_{n,n} > u\}$$
$$= \exp\{\psi_n(t) - t\psi_n'(t)\} \int_{[u-\psi_n'(t)]/\sqrt{\psi_n''(t)}}^{\infty} \exp\{-ty\sqrt{\psi_n''(t)}\} dF_n^{(t)}(y\sqrt{\psi_n''(t)} + \psi_n'(t)). \tag{13}$$

If ψ_n and its derivatives can be approximated with sufficient accuracy, (13) holds forth the possibility of obtaining a lower bound for the probability that a sum of independent r.v.s with zero means exceeds a multiple of its standard deviation.

Lemma 2. *Let $\{X_j, 1 \le j \le n\}$ be independent* r.v.s *with* $E\, X_j = 0$, $E\, X_j^2 = \sigma_j^2$, $s_n^2 = \sum_1^n \sigma_j^2 > 0$, *and* $P\{|X_j| \le d_n\} = 1$, $1 \le j \le n$. *If* $S_n = \sum_{j=1}^n X_j$ *and* $\lim_{n\to\infty} d_n x_n/s_n = 0$, *where* $x_n > x_0 > 0$, *then for every* γ *in* $(0, 1)$, *some* C_γ *in* $(0, \frac{1}{2})$, *and all large* n

$$P\{S_n > (1-\gamma)^2 s_n x_n\} \ge C_\gamma \exp\{-x_n^2(1-\gamma)(1-\gamma^2)/2\}. \tag{14}$$

PROOF. Let $\varphi_j(t)$ denote the m.g.f. of X_j and set $S_{n,n} = S_n/s_n = \sum_{j=1}^n X_j/s_n$ and $c_n = d_n/s_n$. Since, in the notation leading to (13), $\varphi_{n,j}(t) = \varphi_j(t/s_n)$, $1 \le j \le n$, and $g_1(x) = x^{-1}(e^x - 1)\uparrow$, it follows for $t > 0$ and $1 \le j \le n$ that

$$\varphi_{n,j}'(t) = \frac{d}{dt} E\, e^{tX_j/s_n} = E\, \frac{X_j}{s_n}(e^{tX_j/s_n} - 1) \lesseqgtr tg_1(\pm tc_n)\frac{\sigma_j^2}{s_n^2},$$

$$\varphi_{n,j}''(t) = E\, \frac{X_j^2}{s_n^2} e^{tX_j/s_n} \lesseqgtr e^{\pm tc_n}\frac{\sigma_j^2}{s_n^2},$$

and so, noting $\varphi_{n,j}(t) \ge 1$, (7), and $\sigma_j^2 \le c_n^2 s_n^2$, $1 \le j \le n$,

$$\psi_{n,j}'(t) = \frac{\varphi_{n,j}'(t)}{\varphi_{n,j}(t)} \begin{cases} \le tg_1(tc_n)\dfrac{\sigma_j^2}{s_n^2} \\[2ex] \ge \dfrac{tg_1(-tc_n)\sigma_j^2/s_n^2}{1 + t^2c_n^2 g(tc_n)}, \end{cases}$$

$$\psi_{n,j}''(t) = \frac{\varphi_{n,j}''(t)}{\varphi_{n,j}(t)} - [\psi_{n,j}'(t)]^2 \begin{cases} \le e^{tc_n}\dfrac{\sigma_j^2}{s_n^2} \\[2ex] \ge \dfrac{\sigma_j^2}{s_n^2}[\exp\{-tc_n - t^2c_n^2 g(tc_n)\} - t^2c_n^2 g_1^2(tc_n)], \end{cases}$$

where g is as in Lemma 1. Hence if $\psi_n(t) = \sum_{j=1}^n \psi_{n,j}(t)$,

$$\psi_n'(t) = \sum_{j=1}^n \psi_{n,j}'(t) \begin{cases} \le tg_1(tc_n) \\ \ge tg_1(-tc_n)/[1 + t^2c_n^2 g(tc_n)]. \end{cases} \tag{15}$$

Moreover, since $|g_1(x) - 1| < (|x|/2)[1 - (|x|/3)]^{-1}$ for $0 < |x| < 3$ and

$|g(x) - \frac{1}{2}| < (|x|/6)(1 - (|x|/4))^{-1}$ for $0 < |x| < 4$, if $\lim t_n c_n = 0$,

$$\psi_n''(t_n) = \sum_{j=1}^{n} \psi_{n,j}''(t_n) = 1 + O(t_n c_n). \tag{16}$$

Thus, via (5), (15), (16) and $g(0) = \frac{1}{2}$, $g_1(0) = 1$, for any γ in (0, 1) and all large n

$$\psi_n(t_n) - t_n\psi_n'(t_n) \geq t_n^2[g(-t_n c_n) - t_n^2 c_n^2 g^2(-t_n c_n) - g_1(t_n c_n)] \geq \frac{-t_n^2}{2}(1 + \gamma),$$

$$v_n \equiv \frac{(1-\gamma)t_n - \psi_n'(t_n)}{\sqrt{\psi_n''(t_n)}} = -\gamma t_n(1 + o(1)) \leq \frac{-\gamma t_n}{2}.$$

Consequently, taking $u = (1-\gamma)t_n$ in (13),

$$\begin{aligned} P\{S_n > (1-\gamma)s_n t_n\} &\geq \exp\{\psi_n(t_n) - t_n\psi_n'(t_n)\} \\ &\quad \times \int_{v_n}^{0} \exp\{-ty\sqrt{\psi_n''(t_n)}\}dF_n^{(t_n)}(y\sqrt{\psi_n''(t_n)} + \psi_n'(t_n)) \\ &\geq \exp\{-t_n^2/2)(1+\gamma)\} \int_{-\gamma t_n/2}^{0} dF_n^{(t_n)}(y\sqrt{\psi_n''(t_n)} + \psi_n'(t_n)) \\ &\geq C_\gamma \exp\{(-t_n^2/2)(1+\gamma)\} \end{aligned} \tag{17}$$

since

$$\frac{S_n(t_n) - \psi_n'(t_n)}{\sqrt{\psi_n''(t_n)}} = \sum_{j=1}^{n} Z_{nj} \xrightarrow{d} N_{0,1}$$

by Exercise 9.1.2 or Corollary 12.2.2 in view of

$$\text{E } Z_{n,j} = 0, \quad \sum_{j=1}^{n} \text{E } Z_{nj}^2 = 1, \quad \text{and} \quad |Z_{nj}| = \left| \frac{X_{n,j}(t_n) - \text{E } X_{n,j}(t_n)}{\sqrt{\psi_n''(t_n)}} \right|$$

$$\leq \frac{2c_n}{\sqrt{\psi_n''(t_n)}} = o(1).$$

Finally, set $t_n = (1-\gamma)x_n$ in (17) to obtain (14). □

Remark. If $x_n \to \infty$, then for every γ in (0, 1) the constant $C_\gamma > \frac{1}{2} - \varepsilon$ for all $\varepsilon > 0$ provided $n \geq$ some integer N_ε.

The strong law asserts under certain conditions that with probability one sums S_n of independent r.v.s with zero means are $o(n)$. In the symmetric Bernoulli case, Hausdorff proved in 1913 that $S_n \overset{\text{a.c.}}{=} O(n^{(1/2)+\varepsilon})$, $\varepsilon > 0$. The order of magnitude was improved to $O(\sqrt{n \log n})$ by Hardy and Littlewood in 1914 and to $O(\sqrt{n \log_2 n})$ by Khintchine in 1923. (Here, as elsewhere, $\log_2 n$ denotes $\log \log n$ and $\log_{k+1} n = \log \log_k n$, $k \geq 1$). One year later Khintchine obtained the iterated logarithm law for the special case in question and in 1929 Kolmogorov proved

Theorem 1 (Law of the Iterated Logarithm). *Let* $\{X_n, n \geq 1\}$ *be independent* r.v.s *with* $\mathrm{E}\, X_n = 0$, $\mathrm{E}\, X_n^2 = \sigma_n^2$, $s_n^2 = \sum_1^n \sigma_i^2 \to \infty$. *If* $|X_n| \leq d_n$, a.c., *where the constant* $d_n = o(s_n/(\log_2 s_n)^{1/2})$ *as* $n \to \infty$, *then, setting* $S_n = \sum_{i=1}^n X_i$,

$$\mathrm{P}\left\{\overline{\lim_{n\to\infty}} \frac{S_n}{s_n(\log_2 s_n)^{1/2}} = \sqrt{2}\right\} = 1. \tag{18}$$

PROOF. Choose the integers n_k, $k \geq 1$ such that $s_{n_k} \leq \alpha^k < s_{n_{k+1}}$ and note that $\sigma_n^2/s_n^2 = o(1)$ whence $s_{n_k} \sim \alpha^k$. According to Corollary 1,

$$\overline{\lim_{n\to\infty}} \frac{S_n}{s_n\sqrt{\log_2 s_n}} \leq \sqrt{2}, \qquad \text{a.c.} \tag{19}$$

To establish the reverse inequality, choose γ in $(0, 1)$ and define independent events

$$A_k = \{S_{n_k} - S_{n_{k-1}} > (1-\gamma)^2 g_k h_k\}, \qquad k \geq 1,$$

where, since $s_{n_k} \sim \alpha^k$, $\alpha > 1$,

$$g_k^2 \equiv s_{n_k}^2 - s_{n_{k-1}}^2 \sim s_{n_k}^2\left(1 - \frac{1}{\alpha^2}\right)$$
$$h_k^2 \equiv 2\log_2 g_k \sim 2\log_2 s_{n_k} < (1+\gamma)(2\log k) \tag{20}$$

for all large k. Thus, taking $x_{n_k} = h_k$ in Lemma 2, noting (20) and that $d_{n_k} h_k/g_k = o(1)$,

$$\mathrm{P}\{A_k\} \geq C_\gamma \exp\{-h_k^2(1-\gamma)(1-\gamma^2)/2\} \geq C_\gamma \exp\{-(1-\gamma^2)^2 \log k\}$$
$$= \frac{C_\gamma}{k^{(1-\gamma^2)^2}}$$

for all large k, whence by the Borel–Cantelli theorem

$$\mathrm{P}\{A_k, \text{i.o.}\} = 1. \tag{21}$$

Next, choose α so large that $(1-\gamma)^2(1-\alpha^{-2})^{1/2} - (2/\alpha) > (1-\gamma)^3$ and set $t_n = \sqrt{2\log_2 s_n}$, implying for all large k that

$$(1-\gamma)^2 g_k h_k - 2s_{n_{k-1}}t_{n_{k-1}} \sim [(1-\gamma)^2(1-\alpha^{-2})^{1/2} - 2\alpha^{-1}]s_{n_k}t_{n_k}$$
$$> (1-\gamma)^3 s_{n_k}t_{n_k}.$$

Hence, setting $B_k = \{|S_{n_{k-1}}| \leq 2s_{n_{k-1}}t_{n_{k-1}}\}$,

$$A_k B_k \subset \{S_{n_k} > (1-\gamma)^2 g_k h_k - 2s_{n_{k-1}}t_{n_{k-1}}\} \subset \{S_{n_k} > (1-\gamma)^3 s_{n_k}t_{n_k}\}$$

again for all large k. However, (ii) of Corollary 1 guarantees $P\{B_k^c, \text{i.o.}\} = 0$, which, in conjunction with (21), entails

$$\mathrm{P}\{S_{n_k} > (1-\gamma)^3 s_{n_k}t_{n_k}, \text{i.o.}\} \geq \mathrm{P}\{A_k \cdot B_k, \text{i.o.}\} = 1.$$

Thus, with probability one

$$\overline{\lim_{n\to\infty}} \frac{S_n}{s_n t_n} \geq \overline{\lim_{k\to\infty}} \frac{S_{n_k}}{s_{n_k} t_{n_k}} \geq (1-\gamma)^3,$$

and letting $\gamma \downarrow 0$ the reverse inequality of (19) is proved. □

Corollary 2. *Under the conditions of Theorem* 1

$$P\left\{\underline{\lim_{n\to\infty}} \frac{S_n}{s_n\sqrt{\log_2 s_n}} = -\sqrt{2}\right\} = 1 = P\left\{\overline{\lim_{n\to\infty}} \frac{|S_n|}{s_n\sqrt{\log_2 s_n}} = \sqrt{2}\right\}.$$

To extend the law of the iterated logarithm (LIL) from bounded to unbounded r.v.s without losing ground, a refined truncation is necessary. This means that the truncation constants, far from being universal, should (as first realized by Hartman and Wintner in the i.i.d. case) depend upon the tails of the distributions of the r.v.s involved.

Let $\{X_n, n \geq 1\}$ denote independent random variables with $\mathrm{E}\,X_n = 0$, $\mathrm{E}\,X_n^2 = \sigma_n^2$, $s_n^2 = \sum_{i=1}^n \sigma_i^2 \to \infty$. Then $\{X_n, n \geq 1\}$ **obeys the LIL** if (17) obtains.

Theorem 2. *If* $\{X_n, n \geq 1\}$ *are independent* r.v.s *with* $\mathrm{E}\,X_n = 0$, $\mathrm{E}\,X_n^2 = \sigma_n^2$, $s_n^2 = \sum_1^n \sigma_i^2 \to \infty$, *in order that* $\{X_n, n \geq 1\}$ *obey the* LIL *it is necessary that*

$$\sum_{n=1}^{\infty} P\{X_n > \delta s_n(\log_2 s_n^2)^{1/2}\} < \infty, \qquad \delta > \sqrt{2}. \tag{22}$$

PROOF. If $b_n^2 = 2s_n^2 \log_2 s_n^2$ and $S_n = \sum_1^n X_i$, $S_0 = 0$, then $\overline{\lim}_{n\to\infty} S_n/b_n \overset{\text{a.c.}}{=} 1$. Now $S_{n-1}/b_n \xrightarrow{P} 0$ since $\sigma^2(S_{n-1}) = s_{n-1}^2 = o(b_n^2)$, and clearly S_{n-1} is independent of $(X_n, X_{n+1}, \ldots)$ for all $n \geq 1$. Hence, by Lemma 3.3.4(ii) $\overline{\lim}\, X_n/b_n \leq 1 + \varepsilon$, a.c. and (22) follows by the Borel–Cantelli theorem. □

Corollary 3. *Under the hypothesis of Theorem* 2, *in order that both* $\{X_n\}$ *and* $\{-X_n\}$ *obey the* LIL, *it is necessary that*

$$\sum_{n=1}^{\infty} P\{|X_n| > \delta s_n(\log_2 s_n^2)^{1/2}\} < \infty, \qquad \delta > \sqrt{2}. \tag{23}$$

The next result stipulates two conditions which, conjoined with (23) for a fixed δ, are sufficient for the LIL. One of these, (25), clearly implies the Lindeberg criterion and hence the asymptotic normality of $\sum_1^n X_j/s_n$.

Theorem 3. *If* $\{X_n, n \geq 1\}$ *are independent* r.v.s *with* $\mathrm{E}\,X_n = 0$, $\mathrm{E}\,X_n^2 = \sigma_n^2$, $s_n^2 = \sum_1^n \sigma_i^2 \to \infty$, *and* d.f.s $\{F_n, n \geq 1\}$ *satisfying for* **some** $\delta > 0$

$$\sum_{n=1}^{\infty} P\{|X_n| > \delta s_n(\log_2 s_n^2)^{1/2}\} < \infty, \tag{24}$$

$$\frac{1}{s_n^2}\sum_{j=1}^{n}\int_{[|x|>\varepsilon s_j(\log_2 s_j^2)^{-1/2}]} x^2\,dF_j(x) = o(1) \quad \textit{for all } \varepsilon > 0, \tag{25}$$

$$\sum_{n=1}^{\infty}\frac{1}{s_n^2(\log_2 s_n^2)}\int_{[\varepsilon s_n(\log_2 s_n^2)^{-1/2}<|x|\le \delta s_n(\log_2 s_n^2)^{1/2}]} x^2\,dF_n(x) < \infty \quad \textit{for all } \varepsilon > 0, \tag{26}$$

then the law of the iterated logarithm (17) *holds for* $\{X_n\}$ *and* $-\{X_n\}$. *Alternatively, if* (24) *is valid for* **all** $\delta > 0$, (25) *obtains and* (26) *is replaced by*

$$\sum_{j_k=k}^{\infty}(s_{j_k}^2 \log_2 s_{jk}^2)^{-k}\gamma_{j_k}\sum_{j_{k-1}=k-1}^{j_k-1}\gamma_{j_{k-1}}\cdots\sum_{j_1=1}^{j_2-1}\gamma_{j_1} < \infty, \tag{27}$$

for some $k \ge 2$ *and all* $\varepsilon > 0$, *where*

$$\gamma_n = \gamma_n(\varepsilon) = \int_{[\varepsilon s_n(\log_2 s_n^2)^{-1/2}<|x|\le s_n(\log_2 s_n^2)^{1/2}]} x^2\,dF_n(x),$$

then the LIL *likewise holds for* $\{X_n\}$ *and* $\{-X_n\}$.

PROOF. Condition (25) implies

$$\varphi_n(\varepsilon) = \max_{m\ge n} s_m^{-2}\sum_{j=1}^{m}\int_{[x^2>\varepsilon^2 s_j^2(\log_2 s_j^2)^{-1}]} x^2\,dF_j(x) = o(1), \qquad \varepsilon > 0,$$

and hence permits the choice of integers $n_{k+1} > n_k$ such that $\varphi_n(k^{-2}) < k^{-2}$ for $n \ge n_k$, $k \ge 1$. Define $\varepsilon_n' = k^{-2}$, $n_k \le n < n_{k+1}$, $k \ge 1$. Then $\varepsilon_n' \downarrow 0$ and for $n_k \le n < n_{k+1}$

$$\frac{1}{s_n^2}\sum_{j=1}^{n}\int_{[x^2>\varepsilon_j^2 s_j^2/\log_2 s_j^2]} x^2\,dF_j(x) \le \frac{1}{s_n^2}\sum_{j=1}^{n}\int_{[x^2>\varepsilon_n^2 s_j^2/\log_2 s_j^2]} x^2\,dF_j(x)$$

$$\le \varphi_n(\varepsilon_n) \le \varphi_{n_k}(\varepsilon_n) < k^{-2} = o(1) \tag{28}$$

as $n \to \infty$ provided $\varepsilon_n = \varepsilon_n'$. Proceeding in a similar spirit with the tail of the series of (26), there is a sequence $\varepsilon_n'' = o(1)$ such that

$$\sum_{n=1}^{\infty}(s_n^2 \log_2 s_n^2)^{-1}\int_{[\varepsilon_n s_n(\log_2 s_n^2)^{-1/2}<|x|\le \delta s_n(\log^2 s_n^2)^{1/2}]} x^2\,dF_n(x)$$

$$= \sum_{k=1}^{\infty}\sum_{n>n_k}^{n_{k+1}}\cdots \le \sum_{k=1}^{\infty}k^{-2} < \infty, \tag{29}$$

where $\varepsilon_n = \varepsilon_n''$.

Consequently, $\varepsilon_n = \max(\varepsilon_n', \varepsilon_n'') = o(1)$ and both (25) and (26) hold with ε replaced by ε_j and ε_n respectively.

Define truncation constants $\{b_n, n \ge 1\}$ by

$$b_n = \frac{\varepsilon_n s_n}{(\log_2 s_n^2)^{1/2}}$$

and set

$$X'_n = X_n I_{[|X_n| \le b_n]}, \qquad X'''_n = X_n I_{[|X_n| > \delta s_n (\log_2 s_n^2)^{1/2}]},$$
$$X''_n = X_n - X'_n - X'''_n, \tag{30}$$
$$S'_n = \sum_1^n X'_j, \qquad S''_n = \sum_1^n X''_j, \qquad S'''_n = \sum_1^n X'''_j.$$

Now

$$\sigma_n^2 - \sigma_{X'_n}^2 = \mathrm{E}\, X_n^2 I_{[|X_n| > b_n]} + \mathrm{E}^2\, X_n I_{[|X_n| \le b_n]} \le 2\, \mathrm{E}\, X_n^2 I_{[|X_n| > b_n]},$$

recalling that $\mathrm{E}\, X_n = 0$, and so (28) ensures $\sigma_{S'_n}^2 \sim \sigma_{S_n}^2$. Thus Theorem 1 yields

$$\varlimsup_{n \to \infty} \frac{S'_n - \mathrm{E}\, S'_n}{s_n (2 \log_2 s_n^2)^{1/2}} \stackrel{\text{a.c.}}{=\!=} 1. \tag{31}$$

Secondly, Kronecker's lemma and (29) guarantee that

$$\frac{S''_n - \mathrm{E}\, S''_n}{s_n (\log_2 s_n^2)^{1/2}} \xrightarrow{\text{a.c.}} 0. \tag{32}$$

Thirdly, (24) implies that $S'''_n = O(1)$ with probability one, and, furthermore,

$$\begin{aligned}
|\mathrm{E}\, S'''_n| &\le \sum_{j=1}^n \int_{[|x| > \delta s_j (\log_2 s_j^2)^{1/2}]} |x|\, dF_j(x) \\
&\le \sum_{j=1}^n \int_{[\delta s_j (\log_2 s_j^2)^{1/2} < |x| \le s_n (\log_2 s_n^2)^{-1/2}]} |x|\, dF_j(x) \\
&\quad + \sum_{j=1}^n \int_{[|x| > s_n (\log_2 s_n^2)^{-1/2}]} |x|\, dF_j(x) \\
&\le \frac{s_n}{(\log_2 s_n^2)^{1/2}} \sum_{j=1}^n \mathrm{P}\{|X_j| > \delta s_j (\log_2 s_j^2)^{1/2}\} \\
&\quad + \frac{(\log_2 s_n^2)^{1/2}}{s_n} \sum_{j=1}^n \int_{[|x| > s_n (\log^2 s_n^2)^{-1/2}]} x^2\, dF_j(x) = o(s_n (\log_2 s_n^2)^{1/2})
\end{aligned}$$

via (24) and (25).

The first portion of the theorem is an immediate consequence of (30), (31), (32), and the assertion just after (32).

In the alternative case, note that since $\gamma_n(\varepsilon)$ and hence the series of (27) is decreasing in ε, there exists, as earlier, a sequence $\varepsilon_n = o(1)$ such that (25) and (27) hold with ε replaced by ε_j and ε_n respectively. Define

$$b_n = \varepsilon_n s_n (\log_2 s_n^2)^{-1/2}$$

and X'_n, X''_n, X'''_n as in (30), but now with $\delta = 1$ and the new choice of b_n. The only link in the prior chain of argument requiring modification is that used to establish (32).

Now

$$s_n^{-1}|\mathrm{E}\,X_n''| \le \left(s_n^{-2}\int_{[x^2>\varepsilon_n^2 s_n^2(\log_2 s_n^2)^{-1}]} x^2\,dF_n(x)\right)^{1/2} = o(1)$$

in view of the strengthened version of (25), and so for any $\delta > 0$

$$\begin{aligned}\sum_{n=1}^{\infty} &\mathrm{P}\{|X_n'' - \mathrm{E}\,X_n''| > \delta s_n(\log_2 s_n^2)^{1/2}\}\\ &\le O(1) + \sum_{n=1}^{\infty}\mathrm{P}\left\{|X_n''| > \frac{\delta}{2}s_n(\log_2 s_n^2)^{1/2}\right\}\\ &\le O(1) + \sum_{n=1}^{\infty}\mathrm{P}\left\{|X_n| > \frac{\delta}{2}s_n(\log_2 s_n^2)^{1/2}\right\} < \infty \qquad (33)\end{aligned}$$

for all $\delta > 0$ as hypothesized.

Since the variance of X_n'' is dominated by $\gamma_n(\varepsilon_n)$, it follows from the strengthened (or ε_n) version of (27) and (33) that Theorem 10.1.2 applies to $X_n'' - \mathrm{E}\,X_n''$ with $b_n = s_n(\log_2 s_n^2)^{1/2}$. Thus (32) and the final portion of the theorem follow. □

The first corollary reduces the number of conditions of the theorem while the second circumvents the unwieldy series of (27).

Corollary 4. *If* $\{X_n\}$ *are independent random variables with* $\mathrm{E}\,X_n = 0$, $\mathrm{E}\,X_n^2 = \sigma_n^2$, $s_n^2 = \sum_1^n \sigma_i^2 \to \infty$, *satisfying* (25), *and for some* α *in* (0, 2],

$$\sum_{n=1}^{\infty}\frac{1}{(s_n^2\log_2 s_n^2)^{\alpha/2}}\int_{[|x|>\varepsilon s_n(\log_2 s_n^2)^{-1/2}]}|x|^\alpha\,dF_n(x) < \infty \quad \text{for all } \varepsilon > 0 \qquad (34)$$

then the LIL *holds for* $\{X_n\}$ *and* $\{-X_n\}$.

Proof. Clearly, the series of (34) exceeds the series obtained from (34) by restricting the range of integration (i) to $(\varepsilon s_n(\log_2 s_n^2)^{-1/2}, \delta s_n(\log_2 s_n^2)^{1/2}]$ or (ii) to $(\varepsilon s_n(\log_2 s_n^2)^{1/2}, \infty)$. But the series corresponding to (i) dominates the series of (26) multiplied by $\delta^{\alpha-2}$, while the series corresponding to (ii) (with $\varepsilon < \delta$) majorizes the series of (24) multiplied by δ^α. □

Corollary 5. *Let* $\{X_n\}$ *be independent random variables with* $\mathrm{E}\,X_n = 0$, $\mathrm{E}\,X_n^2 = \sigma_n^2$, $s_n^2 = \sum_1^n \sigma_i^2 \to \infty$, *satisfying* (24) *for all* $\delta > 0$, *and* (25). *If for some* $\rho > 0$,

$$\left[\log\sum_1^n \gamma_j(\varepsilon)\right]^\rho \sum_{j=1}^n \gamma_j(\varepsilon) = O(s_n^2\log_2 s_n^2) \quad \text{for all } \varepsilon > 0 \qquad (35)$$

where

$$\gamma_j = \gamma_j(\varepsilon) = \int_{[\varepsilon s_j(\log_2 s_j^2)^{-1/2} < |x| \le s_j(\log_2 s_j^2)^{1/2}]} x^2\,dF_j(x),$$

then the LIL *holds for* $\{X_n\}$ *and* $\{-X_n\}$.

PROOF. For all $\varepsilon > 0$,

$$\sum_{n=k}^{\infty} \frac{\gamma_n}{(s_n^2 \log_2 s_n^2)^k} \sum_{j_{k-1}=k-1}^{n-1} \gamma_{j_{k-1}} \cdots \sum_{1}^{j_2-1} \gamma_{j_1} \le \sum_{n=k}^{\infty} \frac{\gamma_n}{\sum_1^n \gamma_j} \left(\frac{\sum_1^n \gamma_j}{s_n^2 \log_2 s_n^2} \right)^k$$

$$\le C \sum_{n=k}^{\infty} \frac{\gamma_n}{(\sum_1^n \gamma_j)(\log \sum_1^n \gamma_j)^{\rho k}} < \infty$$

for $k > 1/\rho$. □

All the prior conditions for the LIL simplify greatly in the special case of weighted i.i.d. random variables. Define such a class Q by

$$Q = \{\sigma_n Y_n, n \ge 1: Y_n, n \ge 1, \text{are i.i.d. random variables with mean 0, variance } \sigma_Y^2 \text{ in } (0, \infty] \text{ and } \sigma_n, n \ge 1 \text{ are nonzero constants satisfying } s_n^2 = \textstyle\sum_1^n \sigma_i^2 \to \infty\} \tag{36}$$

and let F denote the common d.f. of $\{Y_n\}$.

To obtain the classical Hartman–Wintner theorem governing the i.i.d. case only part (i) of the following theorem is needed.

Theorem 4. *Let $\{\sigma_n Y_n\} \in Q$ with $\sigma_n^2 = o(s_n^2/\log_2 s_n^2)$ and $\sigma_Y^2 = 1$. If either* (i) *for some α in* $(0, 2]$

$$\sum_{n=1}^{\infty} \left(\frac{\sigma_n^2}{s_n^2 \log_2 s_n^2} \right)^{\alpha/2} \int_{[y^2 \ge \varepsilon s_n^2/\sigma_n^2 \log_2 s_n^2]} |y|^\alpha \, dF(y) < \infty \quad \textit{for all } \varepsilon > 0 \tag{37}$$

or (ii)

$$\sum_{n=1}^{\infty} P\left\{ Y_1^2 > \frac{\delta s_n^2 \log_2 s_n^2}{\sigma_n^2} \right\} < \infty \quad \textit{for all } \delta > 0 \tag{38}$$

and for some $\rho > 0$,

$$q_n(\varepsilon) \equiv \int_{[\varepsilon s_n^2/\sigma_n^2 \log_2 s_n^2 < y^2 \le (s_n^2 \log_2 s_n^2)/\sigma_n^2]} y^2 \, dF(y) = O\left(\frac{\log_2 s_n^2}{(\log s_n^2)^\rho} \right) \quad \textit{for all } \varepsilon > 0 \tag{39}$$

then the LIL *holds for $\{\sigma_n Y_n\}$, that is,*

$$P\left\{ \overline{\lim_{n\to\infty}} \frac{\sum_{j=1}^n \sigma_j Y_j}{s_n (2 \log_2 s_n^2)^{1/2}} = 1 \right\} = 1. \tag{40}$$

PROOF. In the weighted i.i.d. case, condition (25) becomes

$$\frac{1}{s_n^2} \sum_{j=1}^n \sigma_j^2 \int_{[y^2 > \varepsilon s_j^2/\sigma_j^2 \log_2 s_j^2]} y^2 \, dF(y) = o(1), \qquad \varepsilon > 0, \tag{41}$$

and is automatic whenever the integral therein is $o(1)$, that is, whenever

$\sigma_n^2 = o(s_n^2/\log_2 s_n^2)$. The first part of the theorem thus follows from Corollary 4 since (37) is just a transcription of (34).

Likewise, (38) is a transliteration of (24), and so the second portion will follow from Corollary 5 once (35) is established. To this end, note that $\sigma_n^2 = o(s_n^2/\log_2 s_n^2)$ entails $\lim \sigma_n^2/s_{n-1}^2 = 0$, whence

$$a_n \equiv \frac{\log(1 + (\sigma_n^2/s_{n-1}^2))}{\log s_{n-1}^2} = \frac{O(\sigma_n^2/s_{n-1}^2)}{\log s_{n-1}^2} = \frac{o(1)}{\log s_{n-1}^2}$$

and

$$\log s_n^2 = (1 + a_n)\log s_{n-1}^2, \tag{42}$$

so that,

$$\begin{aligned}
&s_n^2(\log s_{n-1}^2)^\rho \log_2 s_n^2 \\
&= (s_{n-1}^2 + \sigma_n^2)[(1 + a_n)^{-1} \log s_n^2]^\rho[\log_2 s_{n-1}^2 + \log(1 + a_n)] \\
&= \left(1 + \frac{\sigma_n^2}{s_{n-1}^2}\right)\left[1 + \rho \frac{o(1)}{\log s_{n-1}^2}\right]\left[1 + \frac{o(1)}{(\log s_{n-1}^2)\log s_n^2}\right] \\
&\quad \times s_{n-1}^2(\log s_n^2)^\rho \log_2 s_{n-1}^2 \\
&= s_{n-1}^2(\log s_n^2)^\rho \log_2 s_{n-1}^2 + (1 + o(1))\sigma_n^2(\log s_n^2)^\rho \log_2 s_{n-1}^2.
\end{aligned}$$

Hence, if $q_n \equiv q_n(\varepsilon) = C_\varepsilon(\log_2 s_n^2/(\log s_n^2)^\rho)$, noting that

$$\log_i s_{n-1}^2 = (1 + o(1)\log_i s_n^2, \ i = 1, 2,$$

$$\frac{s_n^2 \log_2 s_n^2}{(\log s_n^2)^\rho} - \frac{s_{n-1}^2 \log_2 s_{n-1}^2}{(\log s_{n-1}^2)^\rho} = \frac{(1 + o(1))\sigma_n^2 \log_2 s_{n-1}^2}{(\log s_{n-1}^2)^\rho} = C_\varepsilon^{-1}(1 + o(1))\sigma_n^2 q_n,$$

implying for all large n that

$$\sum_{j=1}^n \sigma_j^2 q_j \leq 2C_\varepsilon \frac{s_n^2 \log_2 s_n^2}{(\log s_n^2)^\rho} \leq \frac{2C_\varepsilon s_n^2 \log_2 s_n^2}{(\log \sum_1^n \sigma_j^2 q_j)^\rho} \tag{43}$$

Consequently, $q_n(\varepsilon) = C_\varepsilon(\log_2 s_n^2/(\log s_n^2)^\rho)$ (and *a fortiori* $q_n(\varepsilon) = O(\log_2 s_n^2/(\log s_n^2)^\rho)$) entails (43). But (43) is precisely (35) in the weighted i.i.d. case since then $\gamma_n(\varepsilon) = \sigma_n^2 q_n(\varepsilon)$. □

The status of the LIL in Q is conveniently described in terms of

$$\gamma_n = n \frac{\sigma_n^2}{s_n^2}, \qquad n \geq 1. \tag{44}$$

Note that $\gamma_1 = 1, 0 < \gamma_n < n, n > 1$, and

$$\frac{s_n^2}{s_1^2} = \prod_{j=2}^n \left(1 - \frac{\gamma_j}{j}\right)^{-1}. \tag{45}$$

Consequently, under the hypothesis

$$\frac{\gamma_n}{n} = \frac{\sigma_n^2}{s_n^2} \le 1 - \frac{1}{\delta} \quad \text{for some } \delta > 1, \tag{46}$$

it follows from (45) that for some $c > 0$

$$s_n^2 \le c\delta^n, \qquad \log_2 s_n^2 \le (1 + o(1))\log n. \tag{47}$$

Lemma 3. *If* $\{\sigma_n, n \ge 1\}$ *satisfies* (46), $s_n^2 \to \infty$ *and*

$$\gamma_n = o((\log s_n^2)\log_2 s_n^2), \tag{48}$$

then for every $\mu_1 > 0$ *and real* μ_2 *necessarily* $n^{\mu_1}/(\log_2 s_n^2)^{\mu_2} \uparrow \infty$ *(all large n) and*

$$\sum_{j=1}^{n} \frac{1}{j^{1-\mu_1}(\log_2 s_j^2)^{\mu_2}} = O\left(\frac{n^{\mu_1}}{(\log_2 s_n^2)^{\mu_2}}\right). \tag{49}$$

PROOF. Under (46), recalling (45) and employing

$$\left(1 - \frac{\gamma_n}{n}\right)^{-1} = 1 + \frac{\gamma_n}{n}\left(1 - \frac{\gamma_n}{n}\right)^{-1} \le \exp\{\delta\gamma_n/n\},$$

there follows

$$\log s_n^2 = \log s_{n-1}^2 - \log\left(1 - \frac{\gamma_n}{n}\right) \le \left(1 + \frac{\delta\gamma_n}{n \log s_{n-1}^2}\right)\log s_{n-1}^2,$$

implying

$$\log_2 s_n^2 \le \left(1 + \frac{\delta\gamma_n}{n(\log s_{n-1}^2)\log_2 s_{n-1}^2}\right)\log_2 s_{n-1}^2.$$

Therefore, noting that these entail $\log_i s_n^2 = (1 + o(1))\log_i s_{n-1}^2, \quad i = 1, 2,$

$$\begin{aligned} 1 - \left(\frac{n-1}{n}\right)^{\mu_1}\left(\frac{\log_2 s_n^2}{\log_2 s_{n-1}^2}\right)^{\mu_2} &\ge 1 - \left(1 - \frac{1}{n}\right)^{\mu_1}\left[1 + \frac{\delta\gamma_n}{n(\log s_{n-1}^2)\log_2 s_{n-1}^2}\right]^{\mu_2} \\ &\ge 1 - \left[1 - \frac{\mu_1}{n} + O\left(\frac{1}{n^2}\right)\right]\left[1 + o\left(\frac{1}{n}\right)\right] \\ &= \frac{\mu_1 + o(1)}{n} \end{aligned}$$

for $\mu_2 \ge 0$; the same conclusion is obvious when $\mu_2 < 0$, so that for all μ_2

$$\frac{n^{\mu_1}}{(\log_2 s_n^2)^{\mu_2}} - \frac{(n-1)^{\mu_1}}{(\log_2 s_{n-1}^2)^{\mu_2}} \ge \frac{\mu_1(1 + o(1))}{n^{1-\mu_1}(\log_2 s_n^2)^{\mu_2}}, \tag{50}$$

whence for all large n

$$\sum_{j=1}^{n} \frac{1}{j^{1-\mu_1}(\log_2 s_j^2)^{\mu_2}} \le \frac{2}{\mu_1} \frac{n^{\mu_1}}{(\log_2 s_n^2)^{\mu_2}},$$

which is tantamount to (49). Moreover, (50) ensures that $n^{\mu_1}(\log_2 s_n^2)^{-\mu_2}$ is increasing for all large n. When $\mu_2 > 0$, (47) guarantees that it tends to ∞ as $n \to \infty$, whereas this is obvious if $\mu_2 \le 0$. □

Theorem 5. *If $\{\sigma_n Y_n\} \in Q$, where $\sigma_n^2 = o(s_n^2/\log_2 s_n^2)$* and *$\gamma_n = \mathcal{O}((\log_2 s_n^2)^\beta)$ for some $\beta < 1$, then the* LIL *holds for $\{\sigma_n Y_n, n \ge 1\}$ provided* $\mathrm{E}\, Y^2 < \infty$.

PROOF. According to Theorem 4 it suffices to verify (37) for some α in $(0, 2]$. Now the hypotheses entail $\gamma_n = o(n)$, thus *a fortiori* (46) and also $\gamma_n \le K(\log_2 s_n^2)^\beta$, whence Lemma 3 is applicable. Setting

$$r_j \equiv \frac{\varepsilon j}{\gamma_j \log_2 s_j^2} \ge \frac{\varepsilon j}{K(\log_2 s_j^2)^{1+\beta}} \equiv q_j,$$

this lemma guarantees $q_j \uparrow \infty$ all large j (and for convenience this will be supposed for all $j \ge 1$); the lemma also certifies for any α in $[0, 2)$ that

$$\sum_{j=1}^{n} \left(\frac{\gamma_j}{j \log_2 s_j^2}\right)^{\alpha/2} \le K \sum_{j=1}^{n} j^{-\alpha/2}(\log_2 s_j^2)^{(\alpha/2)(\beta-1)} \le K_1 n^{1-(\alpha/2)}(\log_2 s_n^2)^{(\alpha/2)(\beta-1)}.$$

Consequently, for any $\varepsilon > 0$ and some constant K_ε in $(0, \infty)$

$$\sum_{j=1}^{\infty} \left(\frac{\sigma_j^2}{s_j^2 \log_2 s_j^2}\right)^{\alpha/2} \int_{[y^2 \ge r_j]} |y|^\alpha \, dF(y) \le \sum_{j=1}^{\infty} \left(\frac{\gamma_j}{j \log_2 s_j^2}\right)^{\alpha/2} \int_{[y^2 \ge q_j]} |y|^\alpha \, dF(y)$$

$$= \sum_{j=1}^{\infty} \left(\frac{\gamma_j}{j \log_2 s_j^2}\right)^{\alpha/2} \sum_{n=j}^{\infty} \int_{[q_n \le y^2 < q_{n+1}]} |y|^\alpha \, dF(y)$$

$$\le \sum_{n=1}^{\infty} \frac{K_1 n^{1-(\alpha/2)}}{(\log_2 s_n^2)^{(1-\beta)\alpha/2}} \int_{[q_n \le y^2 < q_{n+1}]} |y|^\alpha \, dF(y)$$

$$\le K_\varepsilon \sum_{n=1}^{\infty} (\log_2 s_n^2)^{1+\beta-\alpha} \int_{[q_n \le y^2 < q_{n+1}]} y^2 \, dF(y) < \infty$$

provided $1 + \beta \le \alpha < 2$. Thus, (37) obtains and the theorem is proved. □

Corollary 6. *If $s_n^2 \to \infty$, $\gamma_n = O(1)$, and $\{Y, Y_n, n \ge 1\}$ are* i.i.d. *random variables with* $\mathrm{E}\, Y = 0$, *the* LIL *holds for $\{\sigma_n Y_n\}$ and $\{-\sigma_n Y_n\}$* iff $\mathrm{E}\, Y^2 < \infty$.

PROOF. The hypothesis implies (46), whence (47) ensures

$$\frac{\sigma_n^2 \log_2 s_n^2}{s_n^2} = \frac{\gamma_n \log_2 s_n^2}{n} \le \frac{C \log n}{n} = o(1),$$

and so the conclusions follow from Theorems 5 and 6. □

In the special case $\sigma_n = 1, n \geq 1$, necessarily $\gamma_n = 1, n \geq 1$, and Corollary 6 reduces to

Corollary 7 (Hartman–Wintner). *If* $\{Y_n\}$ *are* i.i.d. *random variables with* $E\,Y_1 = 0$, *the* LIL *holds for* $\{Y_n\}$ *and* $\{-Y_n\}$ iff $E\,Y_1^2 < \infty$.

In Q, the necessary condition (23) for the two-sided LIL becomes

$$\sum_{n=1}^{\infty} P\left\{Y_1^2 > \delta \frac{n \log_2 s_n^2 \sigma_Y^2}{\gamma_n}\right\} < \infty, \qquad \delta > 2\sigma_Y^2. \tag{51}$$

If γ_n increases faster than $C \log_2 s_n^2$, (51) asserts that something beyond a finite second moment for Y_1 is required for a two-sided LIL. On the other hand, if $\gamma_n = O(1)$, (51) does not even stipulate that the variance be finite. Nonetheless, this is necessary for a two-sided LIL according to

Theorem 6. *Let* $\{\sigma_n, n \geq 1\}$ *be nonzero constants satisfying* $s_n^2 = \sum_1^n \sigma_i^2 \to \infty$, $\sigma_n^2 = o(s_n^2/\log_2 s_n^2)$. *If* $\{Y, Y_n, n \geq 1\}$ *are* i.i.d. *with* $E\,Y = 0$, $E\,Y^2 = \infty$, *then*

$$P\left\{\overline{\lim_{n\to\infty}} \frac{|\sum_{j=1}^n \sigma_j Y_j|}{s_n(\log_2 s_n)^{1/2}} = \infty\right\} = 1. \tag{52}$$

PROOF. Let $\{Y_n^*, n \geq 1\}$ denote the symmetrized $\{Y_n\}$ and for $c > 0$ set $Y_n' = Y_n^* I_{[|Y_n^*| \leq c]}$, $X_n' = \sigma_n Y_n'$, and $\sigma_c^2 = E\,Y_n'^2$. Then $s_n'^2 \equiv \sum_{j=1}^n \sigma_{X_j'}^2 = \sigma_c^2 s_n^2$, whence $\{X_n'\}$ obey the conditions of Theorem 1, implying

$$P\left\{\overline{\lim_{n\to\infty}} \frac{\sum_{j=1}^n \sigma_j Y_j'}{s_n(\log_2 s_n)^{1/2}} > \sigma_c\right\} = 1.$$

By Lemma 4.2.6

$$P\left\{\overline{\lim_{n\to\infty}} \frac{\sum_{j=1}^n \sigma_j Y_j^*}{s_n(\log_2 s_n)^{1/2}} \geq \sigma_c\right\} = 1, \tag{53}$$

and since $\sigma_c \to \infty$ as $c \to \infty$, (53) holds with σ_c replaced by $+\infty$, which, in turn, yields (52). □

Theorem 7 (LIL for non-degenerate U-statistics). *Let* $U_{k,n}(h)$, $n \geq k \geq 2$ *be a sequence of U-statistics with* $E\,h = 0$ *and* $\sigma^2 = E[E\{h(X_1, \ldots, X_k)|X_1\}]^2 \in (0, \infty)$. *If*

$$E\{h(X_1, \ldots, X_k)|X_1, \ldots, X_j\} \in \mathcal{L}_{2j/(2j-1)}, \qquad 2 \leq j \leq k$$

(a fortiori, if $E|h|^{4/3} < \infty$*), then with probability one*

$$-k\sigma = \underline{\lim}_{n\to\infty} \left(\frac{n}{2 \log \log n}\right)^{1/2} U_{k,n}(h)$$

$$\leq \overline{\lim}_{n\to\infty} \left(\frac{n}{2 \log \log n}\right)^{1/2} U_{k,n}(h) = k\sigma.$$

PROOF. Via Lemma 7.5.5,

$$\left(\frac{n}{2\log\log n}\right)^{1/2} U_{k,n}(h) = \frac{k}{(2n\log\log n)^{1/2}}\sum_{r=1}^{n} \mathrm{E}\{h(X_r,\ldots,X_{r+k-1})|X_r\} + o((\log\log n)^{-1/2}), \quad \text{a.s.}$$

Since $Y_r = \mathrm{E}\{h(X_r,\ldots,X_{r+k-1})|X_r\}$, $r \geq 1$ are i.i.d. random variables with $\mathrm{E}\,Y_r = 0$, $\mathrm{E}\,Y_r^2 = \sigma^2 \in (0, \infty)$, the conclusion follows from Corollary 7. □

EXERCISES 10.2

1. Show under the conditions of Theorem 1 or Corollary 7 that

$$\overline{\lim_{n\to\infty}} \frac{S_n}{(\sum_{j=1}^n X_j^2 \log_2 \sum_{j=1}^n X_j^2)^{1/2}} = \sqrt{2}, \quad \text{a.c.}$$

2. If $a_n\lambda_n \geq e - 1$, the upper bound in (3) of Lemma 1 may be replaced by $\exp\{-(\lambda_n \cdot x_n^2)/a_n(e-1)\}$.

3. Verify that the LIL holds for independent r.v.s X_n distributed as $N(0, \sigma_n^2)$ provided $s_n^2 = \sum_1^n \sigma_i^2 \to \infty$, $\sigma_n = o(s_n)$. *Hint*: Use a sharp estimate of the normal tail for the d.f. of S_n/s_n.

4. If $\{X_j, 1 \leq j \leq n\}$ are independent r.v.s with $\mathrm{P}\{X_j \leq \mu + d_n\} = 1$, $d_n > 0$ where $\mathrm{E}\,X_j = \mu$, $\sigma_{X_j}^2 = \sigma^2 \in (0, \infty)$, $1 \leq j \leq n$ then for h as in (1),

$$\mathrm{P}\{S_n \geq n(1+\varepsilon)\mu\} \leq \exp\left\{\frac{-n\sigma^2}{d_n^2} h(\varepsilon\mu d_n/\sigma^2)\right\} \quad \text{or} \quad \exp\{-n\mu\varepsilon/d_n(e-1)\}$$

according as $\varepsilon\mu > 0$ or $\varepsilon\mu \geq (e-1)\sigma^2/d_n$. *Hint*: Apply Lemma 1 with $\lambda_n = \varepsilon\mu/\sigma^2$, $x_n = \sigma n^{1/2}$.

5. When $\{X_n\}$ are independent with $\mathrm{E}\,X_n = 0$, $\mathrm{E}\,X_n^2 = \sigma_n^2$, $s_n^2 = \sum_1^n \sigma_j^2$, $X_n \leq c_n s_n \uparrow$, a.c., $\lim c_n x_n = 0$, check via (2) that for all $\gamma > 0$, $r > 0$, and all large n

$$\mathrm{P}\left\{\max_{1\leq j\leq n} S_j > (1+\gamma)^r x_n s_n\right\} \leq \exp\{-\tfrac{1}{2}x_n^2(1+\gamma)^{2r-1}\}.$$

6. Under the conditions of Theorem 1, show that with probability one that every point of $[-1, 1]$ is a limit point of $S_n/s_n(2\log_2 s_n)^{1/2}$. *Hint*: For $d \neq 0$, $0 < d < 1$, $\gamma > 0$, setting $a_k = (1-\gamma)dh_k$, $b_k = (1+\gamma)^2 dh_k$, and $T_k = S_{n_k} - S_{n_{k-1}}$, for all large k

$$\mathrm{P}\{g_h a_k < T_k < b_k g_k\} = \mathrm{P}\{T_k > a_k g_k\} - \mathrm{P}\{T_k > b_k g_k\}$$

$$\geq C_\gamma \exp\left\{\frac{-(1+\gamma)a_k^2}{2}\right\} - \exp\left\{\frac{-(1+\gamma)b_k^2}{2}\right\} > \frac{1}{2}C_\gamma \exp\left\{\frac{-(1-\gamma)d^2h_k^2}{2}\right\}$$

via Exercise 5.

7. Let $\{Y_n\}$ be i.i.d. with $\mathrm{E}\,Y_1 = 0$ and let $s_n^2 = \sum_1^n \sigma_i^2 = \exp\{\prod_1^3 (\log_i n)^{\alpha_i}\}$, where $\alpha_i \geq 0$, $i = 1, 2, 3$. Note that if $0 < \alpha_1 < 1$, or $\alpha_1 = 1$, $0 < \alpha_2 < 1$, Theorem 5 applies. Show that if $\alpha_1 > 1$, the two-sided LIL holds for $\{\sigma_n Y_n\}$ iff

$$\mathrm{E}\,Y^2(\log|Y|)^{\alpha_1-1}(\log_2|Y|)^{\alpha_2-1}(\log_3|Y|)^{\alpha_3} < \infty.$$

8. If $\{X_n, n \geq 1\}$ are independent r.v.s with $\mathrm{P}\{X_n = \pm n^\alpha\} = \frac{1}{2}n^{-\beta}$, $\mathrm{P}\{X_n = 0\} = 1 - n^{-\beta}$, then $\{X_n\}$ obeys the LIL if $1 - \beta > \max(0, -2\alpha)$, $\beta > 0$.

9 Let $S_n = \sum_{i=1}^n X_i$ where $\{X, X_n, n \geq 1\}$ are i.i.d. with $\mathrm{E}\, X = 0$, $\mathrm{E}\, X^2 = 1$ and let T be an $\{X_n\}$-time with $\mathrm{E}\, T < \infty$. If $T_n = \sum_{j=1}^n T^{(j)}$ where $T^{(j)}, j \geq 1$ are copies of T, then

$$\overline{\lim_{n\to\infty}} \frac{S_{T_n}}{(2T_n \log_2 T_n)^{1/2}} \overset{\text{a.c.}}{=} 1.$$

10. If $\{X_n, n \geq 1\}$ are i.i.d. with $\mathrm{E}\, e^{itX_1} = e^{-|t|^\alpha}$, $0 < \alpha \leq 1$, prove that

$$\mathrm{P}\{\overline{\lim} |n^{-1/\alpha} S_n|^{1/\log_2 n} = e^{1/\alpha}\} = 1,$$

that is, $\mathrm{P}\{|S_n| > n^{1/\alpha}(\log n)^{(1+\varepsilon)/\alpha}, \text{ i.o.}\} = 0$ or 1 according as $\varepsilon > 0$ or $\varepsilon < 0$. *Hint*: Show that $\mathrm{P}\{n^{-1/\alpha}|S_n| > x\} = \mathrm{P}\{|X_1| > x\}$ and use the known (Chapter 12) fact that $\mathrm{P}\{X_1| > x\} \sim Cx^{-\alpha}$ as $x \to \infty$.

11. If $\{X_n, n \geq 1\}$ are interchangeable r.v.s with $\mathrm{E}\, X_1 = 0$, $\mathrm{E}\, X_1^2 = 1$, $\mathrm{Cov}(X_1, X_2) = 0 = \mathrm{Cov}(X_1^2, X_2^2)$, then $\overline{\lim}(2n \log_2 n)^{-1/2} \sum_1^n X_i = 1$, a.c.

12. For $\{X_n, n \geq 1\}$ as in Lemma 2 except that (*) $\lim d_n x_n / s_n = a > 0$, $x_n \to \infty$, prove that for all γ in $(0, 1)$ and all u in $(0, u_0)$

$$\mathrm{P}\left\{S_n > \frac{1-\gamma}{a}\left(\frac{1-e^{-u}}{e^u - u}\right) s_n x_n\right\} \geq (\tfrac{1}{2} + o(1))\exp\left\{\frac{-x_n^2}{a^2}[h(u) + o(1)]\right\},$$

where u_0 is the root of the equation $e^{-u} = (e^u - u)(e^u - 1)^2$ and

$$h(u) = u^2[g_1(u) - g(-u) + u^2 g^2(-u)]$$

with g and g_1 as in Lemma 2. Utilize this to conclude under these conditions with $x_n = (\log_2 s_n)^{1/2}$ that $\overline{\lim}\, S_n(s_n^2 \log_2 s_n)^{-1/2} = C \in (0, \infty)$, where C depends upon a and perhaps also the underlying d.f.s.

10.3 Marcinkiewicz–Zygmund Inequality, Dominated Ergodic Theorems

The first theorem, an inequality due to Khintchine concerning symmetric Bernoulli trials, will play a vital role in establishing an analogous inequality due to Marcinkiewicz and Zymund applicable to sums of independent r.v.s with vanishing expectations.

Theorem 1 (Khintchine Inequality). *If* $\{X_n, n \geq 1\}$ *are* i.i.d. r.v.*s with* $\mathrm{P}\{X_1 = 1\} = \mathrm{P}\{X_1 = -1\} = \frac{1}{2}$ *and* $\{c_n\}$ *are any real numbers, then for every* p *in* $(0, \infty)$ *there exist positive, finite constants* $\bar{A}_p$, $\bar{B}_p$ *such that*

$$\bar{A}_p\left(\sum_1^n c_j^2\right)^{1/2} \leq \left\|\sum_{i=1}^n c_i X_i\right\|_p \leq \bar{B}_p\left(\sum_1^n c_j^2\right)^{1/2}. \tag{1}$$

PROOF. Suppose initially that $p = 2k$, where k is a positive integer. Then, setting $S_n = \sum_{i=1}^n c_i X_i$,

$$\mathrm{E}\, S_n^{2k} = \sum A_{\alpha_1, \ldots, \alpha_j} c_{i_1}^{\alpha_1} \cdots c_{i_j}^{\alpha_j} \mathrm{E}\, X_{i_1}^{\alpha_1} \cdots X_{i_j}^{\alpha_j},$$

where $\alpha_1, \ldots, \alpha_j$ are positive integers with $\sum_{i=1}^{j} \alpha_i = 2k$, $A_{\alpha_1, \ldots, \alpha_j} = (\alpha_1 + \cdots + \alpha_j)!/\alpha_1! \cdots \alpha_j!$, and $i_1, \ldots, i_j$ are distinct integers in $[1, n]$. Since $\mathrm{E}\, X_{i_1}^{\alpha_1} \cdots X_{i_1}^{\alpha_j} = 1$ when $\alpha_1, \ldots, \alpha_j$ are all even and zero otherwise,

$$\mathrm{E}\, S_n^{2k} = \sum A_{2\beta_1, \ldots, 2\beta_j} c_{i_1}^{2\beta_1} \cdots c_{i_j}^{2\beta_j},$$

$\beta_1, \ldots, \beta_j$ being positive integers with $\sum_{i=1}^{j} \beta_i = k$. Hence

$$\mathrm{E}\, S_n^{2k} = \sum \frac{A_{2\beta_1, \ldots, 2\beta_j}}{A_{\beta_1, \ldots, \beta_j}} \cdot A_{\beta_1, \ldots, \beta_j} c_{i_1}^{2\beta_1} \cdots c_{i_j}^{2\beta_j}$$

$$\leq B_{2k}^{2k} s_n^{2k},$$

where $s_n^2 = \sum_{i=1}^{n} c_i^2$ and

$$B_{2k}^{2k} = \sup \frac{A_{2\beta_1, \ldots, 2\beta_j}}{A_{\beta_1, \ldots, \beta_j}} = \sup \frac{(2k)!}{(2\beta_1)! \cdots (2\beta_j)!} \frac{\beta_1! \cdots \beta_j!}{k!}$$

$$\leq \sup \frac{2k(2k-1) \cdots (k+1)}{\prod_{i=1}^{j} 2\beta_i(2\beta_i - 1) \cdots (\beta_i + 1)} \leq \frac{2k(2k-1) \cdots (k+1)}{2^{\beta_1 + \cdots + \beta_j}}$$

$$= \frac{2k(2k-1) \cdots (k+1)}{2^k} \leq k^k.$$

Thus, when $p = 2k$ the upper inequality of (1) holds with $\bar{B}_{2k} \leq k^{1/2}$. Since $\|S_n\|_p$ is increasing in p, $\|S_n\|_p \leq \|S_n\|_{2k} \leq \bar{B}_{2k} s_n$ for $p \leq 2k$, whence the upper inequality of (1) obtains with $\bar{B}_p \leq k^{1/2}$, where k is the smallest integer $\geq p/2$.

It suffices to establish the lower inequality for $0 < p < 2$ since $\|S_n\|_p \geq \|S_n\|_2 = s_n$ for $p \geq 2$. Recalling the logarithmic convexity of the $\mathscr{L}_p$ norm established in Section 4.3 and choosing $r_1, r_2 > 0$ such that $r_1 + r_2 = 1$, $pr_1 + 4r_2 = 2$,

$$s_n^2 = \|S_n\|_2^2 \leq \|S_n\|_p^{pr_1} \|S_n\|_4^{4r_2} \leq \|S_n\|_p^{pr_1} (2^{1/2} s_n)^{4r_2},$$

whence

$$\|S_n\|_p^{pr_1} \geq 4^{-r_2} s_n^{2-4r_2} = 4^{-r_2} s_n^{pr_1},$$

$$\|S_n\|_p \geq 4^{-r_2/pr_1} s_n$$

Hence, the lower inequality holds for $0 < p < 2$ with $\bar{A}_p \geq 4^{-r_2/pr_1} = 2^{-(2-p)/p}$ and for $p \geq 2$ with $\bar{A}_p \geq 1$. □

Corollary 1. *Under the hypothesis of Theorem* 1, *if* $s^2 = \sum_1^\infty c_j^2 < \infty$, *then* (i) $S_n = \sum_1^n c_i X_i \xrightarrow{\text{a.c.}} S$, (ii) $\|S\|_p \leq k^{1/2} s$, *where k is the smallest integer* $\geq p/2$, (iii) $\mathrm{E}\, e^{tS^2} < \infty$ *for all* $t > 0$.

Proof. Theorem 5.1.2 guarantees (i) while (ii) follows from Khintchine's inequality and Fatou's lemma. Apropos of (iii),

$$\mathrm{E}\, e^{tS^2} = \sum_{j=0}^{\infty} \frac{t^j}{j!} \mathrm{E}\, S^{2j} \leq \sum_{j=0}^{\infty} \frac{t^j}{j!} (j^{1/2} s)^{2j} = \sum_{j=0}^{\infty} \frac{j^j}{j!} (ts^2)^j < \sum_{j=0}^{\infty} (ts^2 e)^j$$

since $j^j/j! < \sum_{n=0}^{\infty} j^n/n! = e^j$. Thus, $\mathrm{E}\, e^{tS^2} < \infty$ for $ts^2 e < 1$. Finally, since

$s_n \to s$, for any $t > 0$ the integer n may be chosen so that $2te(s^2 - s_n^2) < 1$. Then

$$\mathrm{E}\, e^{tS^2} = \mathrm{E}\, e^{t(S - S_n + S_n)^2} \le \mathrm{E}[e^{2tS_n^2} \cdot e^{2t(S - S_n)^2}] < \infty$$

since S_n^2 is a bounded r.v. for fixed n. □

Theorem 2 (Marcinkiewicz–Zygmund Inequality). *If* $\{X_n, n \ge 1\}$ *are independent* r.v.s *with* $\mathrm{E}\, X_n = 0$, *then for every* $p \ge 1$ *there exist positive constants* A_p, B_p *depending only upon* p *for which*

$$A_p \left\| \left(\sum_1^n X_j^2 \right)^{1/2} \right\|_p \le \left\| \sum_1^n X_j \right\|_p \le B_p \left\| \left(\sum_{j=1}^n X_j^2 \right)^{1/2} \right\|_p. \tag{2}$$

PROOF. Clearly (Exercise 4.2.4), $\sum_1^n X_j \in \mathscr{L}_p$ iff $X_j \in \mathscr{L}_p$, $1 \le j \le n$, iff $(\sum_1^n X_j^2)^{1/2} \in \mathscr{L}_p$, whence the latter may be supposed. Let $\{X_n^*, n \ge 1\}$ be the symmetrized $\{X_n, n \ge 1\}$, that is, $X_n^* = X_n - X_n'$, $n \ge 1$, where $\{X_n', n \ge 1\}$ is independent of and identically distributed with $\{X_n, n \ge 1\}$. Moreover, let $\{V_n, n \ge 1\}$ constitute a sequence of i.i.d. r.v.s independent of $\{X_n, X_n', n \ge 1\}$ with $\mathrm{P}\{V_1 = 1\} = \mathrm{P}\{V_1 = -1\} = \frac{1}{2}$. Since

$$\mathrm{E}\left\{ \sum_1^n V_i(X_i - X_i') \,\middle|\, V_1, \ldots, V_n, X_1, \ldots, X_n \right\} = \sum_1^n V_i X_i,$$

it follows that for any integer $n > 0$, $\{\sum_1^n V_i X_i, \sum_1^n V_i(X_i - X_i')\}$ is a two-term martingale, leading to the first inequality of

$$\begin{aligned} \mathrm{E}\left| \sum_1^n V_i X_i \right|^p \le \mathrm{E}\left| \sum_1^n V_i X_i^* \right|^p &\le 2^{p-1}\, \mathrm{E}\left\{ \left| \sum_1^n V_i X_i \right|^p + \left| \sum_1^n V_i X_i' \right|^p \right\} \\ &= 2^p\, \mathrm{E}\left| \sum_1^n V_i X_i \right|^p. \end{aligned} \tag{3}$$

Since Khintchine's inequality (1) is applicable to $\mathrm{E}\{|\sum_1^n V_i X_i|^p \mid X_1, X_2, \ldots\}$, necessarily

$$\bar{A}_p^p\, \mathrm{E}\left(\sum_1^n X_i^2 \right)^{p/2} \le \mathrm{E}\left| \sum_1^n V_i X_i \right|^p \le \bar{B}_p^p\, \mathrm{E}\left(\sum_1^n X_i^2 \right)^{p/2},$$

which, in conjunction with (3), yields

$$\bar{A}_p^p\, \mathrm{E}\left(\sum_1^n X_i^2 \right)^{p/2} \le \mathrm{E}\left| \sum_1^n V_i X_i^* \right|^p \le 2^p \bar{B}_p^p\, \mathrm{E}\left(\sum_1^n X_i^2 \right)^{p/2}. \tag{4}$$

However, in view of the symmetry of $\{X_j^*, 1 \le j \le n\}$

$$\mathrm{E}\left| \sum_1^n V_j X_j^* \right|^p = \mathrm{E}\left| \sum_1^n X_j^* \right|^p \le 2^p\, \mathrm{E}\left| \sum_1^n X_j \right|^p$$

whence, recalling Lemma 10.1.1 (or repeating the earlier two-term martingale argument),

$$\mathrm{E}\left|\sum_1^n X_j\right|^p \le \mathrm{E}\left|\sum_1^n X_j^*\right|^p = \mathrm{E}\left|\sum_1^n V_j X_j^*\right|^p \le 2^p \,\mathrm{E}\left|\sum_1^n X_j\right|^p, \tag{5}$$

and so (2) follows from (4) and (5), the upper and lower constants B_p and A_p being twice and one half respectively those of the Khintchine inequality. □

Corollary 2. *If* $\{X_n, n \ge 1\}$ *are* i.i.d. *with* $\mathrm{E}\, X_1 = 0$, $\mathrm{E}|X_1|^p < \infty$, $p \ge 2$, *and* $S_n = \sum_1^n X_i$, *then* $\mathrm{E}|S_n|^p = O(n^{p/2})$. (6)

PROOF. If $p > 2$, by Holder's inequality $\sum_1^n X_j^2 \le n^{(p-2)/p}(\sum_1^n |X_i|^p)^{2/p}$, and the conclusion follows from (2).

Corollary 3. *If* $\{X_n\}$ *are independent* r.v.s *with* $\mathrm{E}\, X_n = 0$, $n \ge 1$, *and both* $\sum_1^n X_i$ *and* $\sum_1^n X_i^2$ *converge* a.c. *as* $n \to \infty$, *then, denoting the limits by* $\sum_1^\infty X_i$, $\sum_1^\infty X_i^2$ *respectively, for* $p \ge 1$

$$A_p \left\| \left(\sum_1^\infty X_i^2\right)^{1/2} \right\|_p \le \left\| \sum_1^\infty X_i \right\|_p \le B_p \left\| \left(\sum_1^\infty X_i^2\right)^{1/2} \right\|_p. \tag{7}$$

If T is a stopping time relative to sums S_n of independent r.v.s and $\{c_n\}$ is a sequence of positive constants, the finiteness of $\mathrm{E}\, c_T |S_T|$, which is of interest in problems of optimal stopping (see, e.g., Theorem 5.4.6), is guaranteed by that of $\mathrm{E}\sup_{n\ge 1} c_n|S_n|$. Questions such as the latter have long been of interest in ergodic theory in a framework far more general than will be considered here. In fact, the classical dominated ergodic theorem of Wiener encompasses the sufficiency part of

Theorem 3 (Marcinkiewicz–Zygmund). *For* $r \ge 1$, *independent, identically distributed* r.v.s $\{X, X_n, n \ge 1\}$ *satisfy*

$$\mathrm{E}\sup_{n\ge 1} n^{-r}\left|\sum_1^n X_i\right|^r < \infty \tag{8}$$

iff

$$\mathrm{E}|X|^r < \infty, \quad r > 1, \quad \text{and} \quad \mathrm{E}|X|\log^+|X| < \infty, \quad r = 1. \tag{9}$$

PROOF. Since $\{X_n^+, n \ge 1\}$ and $\{X_n^-, n \ge 1\}$ are each i.i.d. with moments of the same order as $\{X_n\}$, Example 7.4.3 stipulates that $\{(1/n)\sum_1^n X_i^+, \mathscr{F}_n, n \ge 1\}$ and $\{(1/n)\sum_1^n X_i^-, \mathscr{F}_n, n \ge 1\}$ are positive (reversed) martingales, whence (9) and Theorem 7.4.8 (35) ensure that (8) holds with X replaced by X^+ or X^-. The exact conclusion (8) then follows by Lemma 4.2.3.

Conversely, if (8) obtains for $r \ge 1$, $\mathrm{E}|X_1|^r \le \mathrm{E}\sup_{n\ge 1} n^{-r}|\sum_1^n X_i|^r < \infty$, so that only the case $r = 1$ needs further attention. Now

$$\mathrm{E}\sup n^{-1}|X_n| = \mathrm{E}\sup n^{-1}\left|\sum_1^n X_i - \sum_1^{n-1} X_i\right| \le 2\,\mathrm{E}\sup n^{-1}\left|\sum_1^n X_i\right| < \infty,$$

and thus, choosing $M > 1$ so that $P\{|X| < M\} > 0$,

$$\infty > \int_0^\infty P\left\{\sup_{n\geq 1} n^{-1}|X_n| \geq t\right\}dt \geq \int_M^\infty P\left\{\sup_{n\geq 1} n^{-1}|X_n| \geq t\right\}dt$$

$$= \int_M^\infty \sum_{n=1}^\infty P\{|X| \geq nt\} \prod_{j=1}^{n-1} P\{|X| < jt\}dt$$

$$\geq \prod_{j=1}^\infty P\{|X| < jM\} \int_M^\infty \sum_{n=1}^\infty P\{|X| \geq nt\}dt.$$

Now $E|X| < \infty$ entails positivity of the infinite product, whence

$$\infty > \int_M^\infty \sum_{n=1}^\infty P\{|X| \geq nt\}dt = \int_{[|X|\geq M]} \int_M^\infty \sum_{n=1}^\infty I_{[n\leq t^{-1}|X|]}\,dt\,dP$$

$$> \int_{[|X|\geq M]} \int_M^{|X|} \left(\frac{|X|}{t} - 1\right)dt\,dP > \int_{[|X|\geq M]} |X|(\log|X| - \log M - 1)\,dP$$

$$= E|X|\log^+|X| + O(1),$$

establishing (9) for $r = 1$. □

If the i.i.d. random variables of the prior theorem have mean zero, the order of magnitude n^{-r} appearing therein can almost be halved provided $r \geq 2$.

A useful preliminary step in proving this is furnished by

Lemma 1. *If $\{Y_n, n \geq 1\}$ are independent, nonnegative* r.v.s *then* $E(\sum_1^\infty Y_n)^r < \infty$ *for some $r \geq 1$ provided*

$$\sum_{n=1}^\infty E\,Y_n^r < \infty, \qquad \sum_{n=1}^\infty E\,Y_n^\alpha < \infty, \tag{10}$$

where $\alpha = 1$ if r is an integer and $\alpha = r - [r]$ = fractional part of r otherwise.

PROOF. Since

$$\left(\sum_1^\infty Y_n\right)^r = \left(\sum_1^\infty Y_n\right)^\alpha \left(\sum_1^\infty Y_n\right)^{r-\alpha} \leq \sum_1^\infty Y_n^\alpha\left[Y_n + \sum_{j\neq n} Y_j\right]^{r-\alpha}$$

$$\leq 2^{r-\alpha}\sum_1^\infty Y_n^\alpha\left[Y_n^{r-\alpha} + \left(\sum_{j\neq n} Y_j\right)^{r-\alpha}\right],$$

independence guarantees that

$$E\left(\sum_1^\infty Y_n\right)^r \leq 2^{r-\alpha}\sum_{n=1}^\infty\left[E\,Y_n^r + E\,Y_n^\alpha E\left(\sum_{j\neq n} Y_j\right)^{r-\alpha}\right]$$

$$\leq 2^{r-\alpha}\left[\sum_1^\infty E\,Y_n^r + \left(\sum_1^\infty E\,Y_n^\alpha\right)E\left(\sum_1^\infty Y_n\right)^{r-\alpha}\right]. \tag{11}$$

Since (*) $Y_n^{r-\alpha} \le Y_n^r + Y_n^\alpha$ (or via Exercise 4.3.8), (10) ensures

$$\sum_1^\infty \mathrm{E}\, Y_n^{r-\alpha} < \infty. \tag{12}$$

The lemma is trivial for $r = 1$. If r is an integer ≥ 2, the lemma follows inductively via (11) and (12). If r is not an integer, (10) and (*) entail $\sum_1^\infty \mathrm{E}\, Y_n < \infty$ and (12), whence the conclusion follows from (11) and the already established integral case. □

Theorem 4. *For $r \ge 2$, independent, identically distributed* r.v.s $\{X, X_n, n \ge 1\}$ *with* $\mathrm{E}\, X = 0$ *satisfy*

$$\mathrm{E} \sup_{n \ge e^e} \frac{|\sum_{i=1}^n X_i|^r}{(n \log_2 n)^{r/2}} < \infty \tag{13}$$

iff

$$\mathrm{E}|X|^r < \infty, \qquad r > 2, \quad \text{and} \quad \mathrm{E}\frac{X^2 \log|X|}{\log_2|X|} I_{[|X| > e^e]} < \infty, \qquad r = 2. \tag{14}$$

Proof. Let $\mathrm{E}\, X^2 = 1$, $S_n = \sum_{i=1}^n X_i$, $c_n = (n \log_2 n)^{-1/2}$ or 1 according as $n > e^e$ or not, and set $b_n = n^{1/r}$ or $(n/\log_2 n)^{1/2}$ according as $r > 2$ or $r = 2$. Assume initially that X_n is symmetric; in proving (13), r will be supposed >2, the case of equality requiring only minor emendations. Define

$$X_n' = X_n I_{[|X_n| \le b_n]}, \qquad X_n'' = X_n - X_n',$$

$$S_n' = \sum_i^n X_i', \qquad S_n'' = \sum_1^n X_i''.$$

Now for $h = \alpha = r - [r] > 0$ or $h = 1$ and positive constants n_0, K_1, K

$$\begin{aligned}\sum_{n=n_0}^\infty \mathrm{E}|c_n X_n''|^h &= \sum_{n=n_0}^\infty \sum_{j=n}^\infty \frac{1}{(n \log_2 n)^{h/2}} \int_{[b_j < |X| \le b_{j+1}]} |X|^h \\ &\le K_1 \sum_{j=n_0}^\infty \frac{j^{1-(h/2)}}{(\log_2 j)^{h/2}} \int_{[b_j < |X| \le b_{j+1}]} |X|^h \\ &\le K_1 \sum_{j=n_0}^\infty \frac{j^{(r^{-1}-(1/2))h}}{(\log_2 j)^{h/2}} \int_{[b_j \le |X| < b_{j+1}]} |X|^r \\ &< K\, \mathrm{E}|X|^r < \infty,\end{aligned}$$

and the same conclusion follows analogously when $h = r$. Hence, by Lemma 1

$$\mathrm{E} \sup_{n \ge e^e} \frac{|S_n''|^r}{(n \log_2 n)^{r/2}} \le \mathrm{E}\left(\sup_{n \ge 1} c_n \sum_{j=1}^n |X_j''|\right)^r \le \mathrm{E}\left(\sum_1^\infty c_n |X_n''|\right)^r < \infty.$$

Thus, to complete the proof of sufficiency for symmetric $\{X_n\}$ it suffices via Lemma 4.2.3 to prove that $\mathrm{E}(\sup c_n|S'_n|)^r < \infty$, and this will be done for all $r > 0$, supposing only that $\mathrm{E}\,X = 0$, $\mathrm{E}\,X^2 = 1$. To this end, set $n_k = [3^k]$, whence by Levy's inequality and symmetry

$$\mathrm{P}\left\{\sup_{n > e^e} c_n|S'_n| \geq u\right\} \leq \sum_{k=1}^{\infty} \mathrm{P}\left\{c_{n_k} \sup_{n_k \leq n < n_{k+1}} |S'_n| \geq u\right\}$$

$$\leq 4 \sum_{k=1}^{\infty} \mathrm{P}\{c_{n_k} S'_{n_{k+1}} \geq u\}. \tag{15}$$

It follows from Example 7.4.9 and (1) of Lemma 10.2.1, noting $g(1) < 1$, that for $0 < tb_n \leq 1$

$$\mathrm{P}\{S'_n \geq x\} \leq e^{-tx} \prod_{j=1}^{n} \mathrm{E}\, e^{tX'_j} \leq \exp\left(-tx + t^2 \sum_{1}^{n} \mathrm{E}\, X'^2_j\right) \leq e^{-tx + nt^2}.$$

Hence, taking $t = ((\log_2 n_{k+1})/n_{k+1})^{1/2}$, $c = b_{n_{k+1}}$,

$$\log \mathrm{P}\{S'_{n_{k+1}} \geq c_{n_k}^{-1} u\} \leq -u(n_k \log_2 n_k)^{1/2}\left(\frac{\log_2 n_{k+1}}{n_{k+1}}\right)^{1/2} + \log_2 n_{k+1}$$

$$\leq -(\alpha u - 1)\log_2 n_{k+1}$$

for some positive constant α. Therefore, choosing u_0 such that $\alpha u_0 > 5$, it follows via (15) that

$$\int_{u_0}^{\infty} u^{r-1}\, \mathrm{P}\{\sup c_n|S'_n| \geq u\}du \leq 4 \sum_{k=1}^{\infty} \int_{u_0}^{\infty} u^{r-1} \exp\{-(\alpha u - 1)\log(k+1)\}du$$

$$< C \sum_{1}^{\infty} \frac{1}{k^2} < \infty,$$

so that $\mathrm{E}(\sup c_n|S'_n|)^r < \infty$ by Corollary 6.2.2.

In the case of general $\{X_n\}$, if $X_n^* = X_n - X'_n$, $n \geq 1$, are the symmetrized r.v.s,

$$\mathrm{E} \sup c_n^r |S_n|^r = \mathrm{E} \sup c_n^r |\mathrm{E}\{S_n^* | X_1, X_2, \ldots\}|^r$$

$$\leq \mathrm{E} \sup c_n^r\, \mathrm{E}\{|S_n^*|^r | X_1, X_2, \ldots\}$$

$$\leq \mathrm{E} \sup c_n^r |S_n^*|^r < \infty.$$

For $r > 2$ the converse follows from the necessity part of Theorem 3. When $r = 2$, Theorem 3 merely yields $\mathrm{E}\,X_1^2 < \infty$. However,

$$\mathrm{E} \sup_{n \geq e^e} \frac{X_n^2}{n \log_2 n} \leq 2\,\mathrm{E} \sup_{n \geq e^e} \frac{S_n^2 + S_{n-1}^2}{n \log_2 n} < \infty,$$

and so, choosing $M > e^e$ such that $\mathrm{P}\{|X| < M\} > 0$,

$$\infty > \int_0^\infty \mathrm{P}\left\{\sup \frac{X_n^2}{n \log_2 n} \geq t\right\} dt \geq \int_M^\infty \mathrm{P}\left\{\sup \frac{X_n^2}{n \log_2 n} \geq t\right\} dt$$

$$\geq \int_M^\infty \sum_n \mathrm{P}\{X_n^2 \geq tn \log_2 n\} \prod_{j=1}^{n-1} \mathrm{P}\{X_j^2 < tj \log_2 j\} dt$$

$$\geq \prod_{j=1}^\infty \mathrm{P}\{X^2 < Mj \log_2 j\} \int_M^\infty \sum_n \mathrm{P}\{X_n^2 \geq tn \log_2 n\} dt.$$

Now $\mathrm{E}\, X^2 < \infty$ entails positivity of the infinite product, whence for some $C > 0$

$$\infty > \int_M^\infty \sum_n \mathrm{P}\{X^2 \geq tn \log_2 n\} dt = \int_{[|X| \geq C]} \int_M^\infty \sum_n I_{[n \log_2 n \leq X^2 t^{-1}]} \, dt \, d\mathrm{P}$$

$$\geq \int_{[|X| \geq C]} \int_M^{X^2} \left(\frac{X^2}{t \log_2 X^2 t^{-1}} - e^e\right) dt \, d\mathrm{P}$$

$$\geq \int_{[|X| \geq C]} \int_M^{X^2} \frac{X^2}{t \log_2 X^2} \, dt \, d\mathrm{P} + O(1)$$

$$= 2\, \mathrm{E} \frac{X^2 \log|X|}{\log_2 |X|} I_{[|X| > e^e]} + O(1). \qquad \square$$

Exercises 10.3

1. Show via examples that Theorem 3 is false for $0 < r < 1$ and Theorem 4 fails for $1 \leq r < 2$.

2. Verify under the hypothesis of Theorem 4 that for any finite $\{X_n\}$-time T
$$\mathrm{E}[(T \log_2 T)^{-1/2} |S_T|]^r < \infty.$$

3. If $X^{(1)}, X^{(2)}$ are r.v.s. with respective **symmetric** d.f.s F and G, where $F(x) \geq G(x)$ for all $x > 0$, then $F_n(x) \geq G_n(x)$, $x > 0$, where F_n (resp. G_n) is the n-fold convolution of F (resp. G) with itself. Hence, if $S_n^{(j)} = \sum_{i=1}^n X_i^{(j)}$, $j = 1, 2$, where $X_i^{(j)}$, $1 \leq i \leq n$, are are i.i.d., then $\mathrm{E}|S_n^{(1)}|^p \leq \mathrm{E}|S_n^{(2)}|^p$, $n \geq 1$, $p > 0$.

4. Show that in Theorem 3 sufficiency and necessity for $r > 1$ extend to interchangeable r.v.s. Is (9) necessary for $r = 1$?

5. If $\{X_n, n \geq 1\}$ are independent r.v.s with $\mathrm{E}\, X_n = 0$, $n \geq 1$, and $p \in (1, 2]$, then $\mathrm{E} \max_{1 \leq j \leq n} |\sum_{i=1}^j X_i|^p \leq A_p \sum_{j=1}^n \mathrm{E}|X_j|^p$ for some finite constant A_p. *Hint*: Use the Doob and Marcinkiewicz–Zygmund inequalities.

6. If $S_n = \sum_{i=1}^n X_i$ where $\{X, X_n, n \geq 1\}$ are i.i.d. r.v.s, prove that $\mathrm{E} \exp\{t \sup c_n |S_n|\} < \infty$ for some $t > 0$ iff $\mathrm{E}\{\exp t|X|\} < \infty$ for some $t > 0$ where $c_n = (n \log_2 n)^{-1/2}$, $n \geq e^e$.

10.4 Maxima of Random Walks

In Section 5.2 moments of i.i.d. r.v.s $\{X_n, n \geq 1\}$ and hence behavior of the corresponding random walk $\{S_n = \sum_1^n X_i, n \geq 1\}$ were linked to convergence of series involving $|S_n|$ and $S_n^* = \max_{1 \leq i \leq n}|S_i|$. Here, analogous results for the one-sided case involving explicit bounds for the series will be obtained. For any positive v, set

$$X_0 = 0, \qquad \bar{X}_v = \max_{0 \leq j \leq v} X_j, \qquad S_v = \sum_{i=0}^{[v]} X_i, \qquad \bar{S}_v = \max_{0 \leq j \leq v} S_j,$$
$$X_v^* = \max_{0 \leq j \leq v} |X_j|, \qquad S_v^* = \max_{0 \leq j \leq v} |S_j|. \tag{1}$$

Theorem 1. *If* $\{X, X_n, n \geq 1\}$ *are* i.i.d. r.v.s *with* $\mathrm{E}\,X = 0$ *and* p, α, γ *are constants satisfying* $1 \leq \gamma \leq 2$, $\alpha\gamma > 1$, $\alpha p > 1$, *there exists a constant* $C = C_{p,\alpha,\gamma} \in (0, \infty)$ *such that*

$$\sum_{n=1}^{\infty} n^{\alpha p - 2}\,\mathrm{P}\{\bar{S}_n \geq n^\alpha\} \leq C[\mathrm{E}(X^+)^p + (\mathrm{E}|X|^\gamma)^{(\alpha p - 1)/(\alpha\gamma - 1)}]. \tag{2}$$

PROOF. Suppose without loss of generality that $\mathrm{E}|X| > 0$, $\mathrm{E}(X^+)^p < \infty$, and $\mathrm{E}|X|^\gamma < \infty$. Clearly, for any $k > 0$, by Lemma 5.3.5,

$$\mathrm{P}\{\bar{S}_n \geq n^\alpha\} \leq n\mathrm{P}\{X > n^\alpha/2k\} + \mathrm{P}^k\{\bar{S}_n \geq n^\alpha/2k\}. \tag{3}$$

Now via Theorem 7.4.8 and the Marcinkiewicz–Zygmund inequality

$$\mathrm{P}\left\{\bar{S}_n \geq \frac{n^\alpha}{2k}\right\} \leq \mathrm{P}\left\{S_n^* \geq \frac{n^\alpha}{2k}\right\} \leq C_\gamma'\left(\frac{n^\alpha}{2k}\right)^{-\gamma}\mathrm{E}|S_n|^\gamma$$
$$\leq C_\gamma\left(\frac{n^\alpha}{2k}\right)^{-\gamma}\mathrm{E}\left(\sum_1^n X_j^2\right)^{\gamma/2} \leq C_{\gamma,k}\,n^{1-\alpha\gamma}\,\mathrm{E}|X|^\gamma \tag{4}$$

Set $k = 1 + [(\alpha p - 1)/(\alpha\gamma - 1)]$. Then $\lambda = k - (\alpha p - 2)/(\alpha\gamma - 1) > 1/(\alpha\gamma - 1)$, and if $\mathrm{E}|X|^\gamma \geq 1$, from (4)

$$I \equiv \sum_1^{\infty} n^{\alpha p - 2}\,\mathrm{P}^k\left\{\bar{S} \geq \frac{n^\alpha}{2k}\right\} = \sum_{n^{\alpha\gamma - 1} > \mathrm{E}|X|^\gamma} + \sum_{n^{\alpha\gamma - 1} \leq \mathrm{E}|X|^\gamma}$$
$$\leq C_1\left[\sum n^{\alpha p - 2 + k(1 - \alpha\gamma)}\,\mathrm{E}^k|X|^\gamma + \sum n^{\alpha p - 2}\right]$$
$$= C_2\left[(\mathrm{E}|X|^\gamma)^{\{[1 - \lambda(\alpha\gamma - 1)]/(\alpha\gamma - 1)\} + k} + (\mathrm{E}|X|^\gamma)^{(\alpha p - 1)/(\alpha\gamma - 1)}\right]$$
$$= C(\mathrm{E}|X|^\gamma)^{(\alpha p - 1)/(\alpha\gamma - 1)}. \tag{5}$$

On the other hand, if $\mathrm{E}|X|^\gamma < 1$, again via (4)

$$I \leq C_1 \sum_{n=1}^{\infty} n^{-\lambda(\alpha\gamma - 1)}\,\mathrm{E}^k|X|^\gamma \leq C_2\,\mathrm{E}^k|X|^\gamma \leq C(\mathrm{E}|X|^\gamma)^{(\alpha p - 1)/(\alpha\gamma - 1)}. \tag{6}$$

Hence, from (2), (5), and (6)

$$\sum_{n=1}^{\infty} n^{\alpha p-2}\,\mathrm{P}\{\bar{S}_n \ge n^{\alpha}\} \le \sum_{1}^{\infty}\left[n^{\alpha p-1}\,\mathrm{P}\left\{X > \frac{n^{\alpha}}{2k}\right\} + n^{\alpha p-2}\,\mathrm{P}^k\left\{\bar{S}_n \ge \frac{n^{\alpha}}{2k}\right\}\right]$$
$$\le C[\mathrm{E}(X^+)^p + (\mathrm{E}|X|^{\gamma})^{(\alpha p-1)/(\alpha\gamma-1)}]. \qquad \square$$

Taking $\alpha = 1$, $\gamma = p$ in Theorem 1 yields

Corollary 1. *If* $\{X, X_n, n \ge 1\}$ *are* i.i.d. r.v.s *with* $\mathrm{E}\,X = 0$ *and* $1 < p \le 2$, *then for some constant* $C_p \in (0, \infty)$

$$\sum_{n=1}^{\infty} n^{p-2}\,\mathrm{P}\{\bar{S}_n \ge n\} \le C_p\,\mathrm{E}|X|^p,$$
$$\sum_{n=1}^{\infty} n^{p-2}\,\mathrm{P}\{S_n^* \ge n\} \le 2C_p\,\mathrm{E}|X|^p. \tag{7}$$

Corollary 2 (Hsu–Robbins). *If* $\{X, X_n, n \ge 1\}$ *are* i.i.d. r.v.s, *then*

$$\sum_{n=1}^{\infty} \mathrm{P}\{|S_n| \ge n\varepsilon\} < \infty, \qquad \varepsilon > 0, \tag{8}$$

iff $\mathrm{E}\,X = 0$, $\mathrm{E}\,X^2 < \infty$.

PROOF. (8) follows immediately from Corollary 1 applied to X/ε while the converse is the special case $\alpha = 1$, $p = 2$ of Theorem 6.4.5. $\square$

EXAMPLE 1. **Strong Laws for Arrays.** Let $\{X_{ni}, 1 \le i \le n, n \ge 1\}$ be an array of random variables that are identically distributed and *rowwise independent*, i.e., $\{X_{n1}, \ldots, X_{nn}\}$ are independent for $n \ge 2$. If $\mathrm{E}|X_{11}|^q < \infty, 0 < q < 4$, and $\mathrm{E}\,X_{11} = 0$ whenever $1 \le q < 4$, then $n^{-2/q}\sum_{i=1}^{n} X_{ni} \xrightarrow{\text{a.s.}} 0$.

PROOF. The case $0 < q < 2$ is covered in Exercise 5.2.17. Set $S_{nn} = \sum_{i=1}^{n} X_{ni}$, $n \ge 1$. When $1 \le q < 4$, replacing p by q and choosing $\gamma = 2$, $\alpha = 2/q$ in Theorem 1, it follows that

$$\sum_{n=1}^{\infty} \mathrm{P}\{S_{nn} > \varepsilon n^{2/q}\} < \infty, \qquad \varepsilon > 0. \tag{9}$$

Since X_i may be replaced by $-X_i$ in Theorem 1 (when $\mathrm{E}|X|^p < \infty$), (9) holds with S_{nn} replaced by $-S_{nn}$, and so

$$\sum_{n=1}^{\infty} \mathrm{P}\{|S_{nn}| > \varepsilon n^{2/q}\} < \infty, \qquad \varepsilon > 0, \tag{10}$$

whence the Borel–Cantelli lemma implies that $n^{-2/q}S_{nn} \xrightarrow{\text{a.s.}} 0$. $\square$

For any r.v.s $\{Y_n, n \ge 1\}$ and nonnegative v, set $Y_0 = 0$,

$$Y_v = Y_{[v]}, \qquad \bar{Y}_v = \max_{0 \le j \le v} Y_j, \tag{11}$$

and for $p > 0$, $\alpha > 0$ define (as usual, $\sup \varnothing = 0$)

$$M(\varepsilon) = M(\varepsilon, \alpha) = \sup_{n\geq 0}(Y_n - \varepsilon n^\alpha) = \sup_{n\geq 0}(\overline{Y}_n - \varepsilon n^\alpha),$$

$$\overline{L}(\varepsilon) = \overline{L}(\varepsilon, \alpha) = \sup\{n \geq 1 : \overline{Y}_n \geq \varepsilon n^\alpha\},$$

$$I(\varepsilon) = I(\varepsilon, p, \alpha) = \int_0^\infty v^{\alpha p-2}\, \mathrm{P}\left\{\sup_{j\geq v} j^{-\alpha}\overline{Y}_j \geq \varepsilon\right\}dv,$$

$$J(\varepsilon) = J(\varepsilon, p, \alpha) = \int_0^\infty v^{\alpha p-2}\, \mathrm{P}\{\overline{Y}_v \geq \varepsilon v^\alpha\}dv. \tag{12}$$

Clearly, $J(\varepsilon) \leq I(\varepsilon)$.

Lemma 1. *For* $\alpha p > 1$,

$$I(2\varepsilon) \leq \frac{1}{\alpha p - 1}\,\varepsilon^{(1-\alpha p)/\alpha}\,\mathrm{E}[M(\varepsilon)]^{(\alpha p - 1)/\alpha} \leq (2^{(\alpha p - 1)/\alpha} - 1)^{-1} J\left(\frac{\varepsilon}{2}\right), \tag{13}$$

$$J(\varepsilon) \leq \max(2^{\alpha p - 2}, 1)\sum_{n=1}^\infty n^{\alpha p - 2}\,\mathrm{P}\{\overline{Y}_n \geq \varepsilon n^\alpha\}, \tag{14}$$

$$\mathrm{E}[\overline{L}(\varepsilon)]^{\alpha p - 1} \leq (\alpha p - 1)I(\varepsilon). \tag{15}$$

PROOF. (14) is trivial. For (13), set $\beta = (\alpha p - 1)/\alpha$. Then

$$\begin{aligned}
\varepsilon^{-\beta}\,\mathrm{E}[M(\varepsilon)]^\beta &= \int_0^\infty \mathrm{P}\left\{\sup_{n\geq 1}(Y_n - \varepsilon n^\alpha) \geq \varepsilon v^{1/\beta}\right\}dv \\
&\leq \sum_{k=1}^\infty \int_0^\infty \mathrm{P}\left\{\max_{(2^{k-1}-1)v^{1/\beta} < n^\alpha \leq 2^k v^{1/\beta}} (Y_n - \varepsilon n^\alpha) \geq \varepsilon v^{1/\beta}\right\}dv \\
&\leq \sum_{k=1}^\infty \int_0^\infty \mathrm{P}\{\overline{Y}_{(2^k v^{1/\beta})^{1/\alpha}} \geq 2^{k-1}\varepsilon v^{1/\beta}\}\,dv \\
&= \sum_1^\infty 2^{-k\beta}(\alpha p - 1)\int_0^\infty u^{\alpha p - 2}\,\mathrm{P}\{\overline{Y}_u \geq \varepsilon u^\alpha/2\}du \\
&= (\alpha p - 1)(2^\beta - 1)^{-1} J\left(\frac{\varepsilon}{2}\right),
\end{aligned}$$

establishing the second half of (13). Moreover,

$$\begin{aligned}
I(2\varepsilon) &= \int_0^\infty v^{\alpha p - 2}\,\mathrm{P}\left\{\sup_{j\geq v} j^{-\alpha}\overline{Y}_j \geq 2\varepsilon\right\}dv \\
&\leq \int_0^\infty v^{\alpha p - 2}\,\mathrm{P}\left\{\sup_{n\geq 0}(\overline{Y}_n - \varepsilon n^\alpha) \geq \varepsilon v^\alpha\right\}dv
\end{aligned}$$

$$= \int_0^\infty v^{\alpha p-2} \, \mathrm{P}\left\{\left(\frac{M(\varepsilon)}{\varepsilon}\right)^{1/\alpha} \geq v\right\} dv = \frac{1}{\alpha p - 1} \, \mathrm{E}\left[\frac{M(\varepsilon)}{\varepsilon}\right]^{(\alpha p-1)/\alpha},$$

which is the first half of (13). Apropos of (15),

$$\mathrm{P}\{\bar{L}(\varepsilon) \geq v\} = \mathrm{P}\{\bar{Y}_n \geq \varepsilon n^\alpha \text{ for some } n \geq v\}$$
$$\leq \mathrm{P}\left\{\sup_{n \geq v} n^{-\alpha} \bar{Y}_n \geq \varepsilon\right\},$$

so that

$$\mathrm{E}[\bar{L}(\varepsilon)]^{\alpha p-1} = (\alpha p - 1) \int_0^\infty v^{\alpha p-2} \, \mathrm{P}\{\bar{L}(\varepsilon) \geq v\} dv \leq (\alpha p - 1) I(\varepsilon). \qquad \square$$

The combination of Lemma 1 and Theorem 1 yields

Theorem 2. *If $\{X, X_n, n \geq 1\}$ are* i.i.d. r.v.s *with* $\mathrm{E}\,X = 0$, $S_n = \sum_1^n X_i$, $S_0 = 0$, *and* $1 \leq \gamma \leq 2$, $\alpha\gamma > 1$, $\alpha p > 1$, *then for some constant* $C = C_{p,\alpha,\gamma} \in (0, \infty)$

$$\mathrm{E}\left[\sup_{n \geq 0}(S_n - n^\alpha)\right]^{(\alpha p-1)/\alpha} \leq C\left[\mathrm{E}(X^+)^p + (\mathrm{E}|X|^\gamma)^{(\alpha p-1)/(\alpha\gamma-1)}\right]. \tag{16}$$

Lemma 2. *Let $\{X, X_n, n \geq 1\}$ be* i.i.d. r.v.*s with* $\mathrm{E}|X|^{1/\alpha} < \infty$, *where* $\alpha > \frac{1}{2}$ *and* $\mathrm{E}\,X = 0$ *if* $\alpha \leq 1$. *If* $S_n = \sum_1^n X_i$ *and*

$$L(\varepsilon) = L(\varepsilon, \alpha) = \sup\{n \geq 1: S_n \geq \varepsilon n^\alpha\}, \tag{17}$$
$$L_1(\varepsilon) = L_1(\varepsilon, \alpha) = \sup\{n \geq 1: X_n \geq \varepsilon n^\alpha\}, \tag{18}$$

where $\varepsilon > 0$ *and* $\sup \varnothing = 0$, *then* $\mathrm{E}\,L^\gamma(\varepsilon) < \infty$ *implies* $\mathrm{E}\,L_1^\gamma(2\varepsilon) < \infty$, $\gamma > 0$.

PROOF. Set $A_j = \{X_j \geq 2\varepsilon j^\alpha\}$, $B_j = \{|S_{j-1}| \leq \varepsilon j^\alpha\}$. Now $n^{-\alpha}S_n \xrightarrow{\text{a.c.}} 0$ by Theorem 5.2.2, and so $\mathrm{P}\{B_n\} \to 1$ as $n \to \infty$. Since for $n \geq 1$ the classes $\{B_n\}$ and $\{A_n, A_n A_{n+1}^c, \ldots\}$ are independent, by Lemma 3.3.3 for $n \geq n_0$

$$\mathrm{P}\{L(\varepsilon) \geq n\} \geq \mathrm{P}\left\{\bigcup_{j=n}^\infty A_j B_j\right\} \geq \mathrm{P}\left\{\bigcup_{j=n}^\infty A_j\right\} \inf_{j \geq n} \mathrm{P}\{B_j\}$$
$$\geq \tfrac{1}{2} \mathrm{P}\{L_1(2\varepsilon) \geq n\},$$

and the lemma follows. $\square$

Lemma 3. *Let $\{X, X_n, n \geq 1\}$ be* i.i.d., $\alpha > 0$, $\alpha p > 1$, $\varepsilon > 0$, *and*

$$J_1(\varepsilon) = \int_0^\infty v^{\alpha p-2} \, \mathrm{P}\{\bar{X}_v \geq \varepsilon v^\alpha\} dv. \tag{19}$$

Then, if $L_1(\varepsilon)$ is as in (18),

$$(\alpha p - 1) J_1(\varepsilon) \leq \mathrm{E}[L_1(2^{-\alpha}\varepsilon)]^{\alpha p-1}, \qquad \alpha p \varepsilon^p J_1(\varepsilon) \leq \mathrm{E}(X^+)^p, \tag{20}$$

$$J_1(1) < \infty \Rightarrow \mathrm{E}(X^+)^p < \infty \quad and \quad \frac{\mathrm{E}(X^+)^p - 1}{2\alpha p[1 + \mathrm{E}(X^+)^{1/\alpha}]} \le J_1(1). \tag{21}$$

PROOF. The second inequality of (20) follows immediately from

$$\mathrm{P}\{\bar{X}_v \ge \varepsilon v^\alpha\} \le v\,\mathrm{P}\{X \ge \varepsilon v^\alpha\}.$$

As for the first,

$$\begin{aligned}(\alpha p - 1)J_1(\varepsilon) &= (\alpha p - 1)\int_0^\infty v^{\alpha p - 2}\,\mathrm{P}\Big\{\max_{v < n \le 2v - 1} X_n \ge \varepsilon v^\alpha\Big\}dv \\ &\le (\alpha p - 1)\int_0^\infty v^{\alpha p - 2}\,\mathrm{P}\Big\{\bigcup_{n=v}^{\infty}\Big[X_n \ge \varepsilon\Big(\frac{n}{2}\Big)^\alpha\Big]\Big\}dv \\ &= (\alpha p - 1)\int_0^\infty v^{\alpha p - 2}\,\mathrm{P}\{L_1(2^{-\alpha}\varepsilon) \ge v\}dv = \mathrm{E}[L_1(2^{-\alpha}\varepsilon)]^{\alpha p - 1}.\end{aligned}$$

To prove (21), note that for $v \ge 1$

$$\begin{aligned}\mathrm{P}\{\bar{X}_v \ge v^\alpha\} &= \sum_{j=1}^{[v]} \mathrm{P}\{X_j \ge v^\alpha\}\mathrm{P}^{j-1}\{X < v^\alpha\} \\ &\ge [v]\mathrm{P}^{[v]}\{X < v^\alpha\}\mathrm{P}\{X \ge v^\alpha\} \\ &= [v]P\{X \ge v^\alpha\}[1 - \mathrm{P}\{\bar{X}_v \ge v^\alpha\}],\end{aligned}$$

whence for $v \ge 1$

$$\begin{aligned}[v]\mathrm{P}\{X \ge v^\alpha\} &\le (1 + [v]\mathrm{P}\{X \ge v^\alpha\})\mathrm{P}\{\bar{X}_v \ge v^\alpha\} \\ &\le [1 + \mathrm{E}(X^+)^{1/\alpha}]\mathrm{P}\{\bar{X}_v \ge v^\alpha\}\end{aligned}$$

by the Markov inequality. Hence

$$\begin{aligned}\mathrm{E}(X^+)^p &= \alpha p\int_0^\infty v^{\alpha p - 1}\,\mathrm{P}\{X \ge v^\alpha\}dv \\ &\le 1 + 2\alpha p\int_1^\infty v^{\alpha p - 2}\,\mathrm{P}\{\bar{X}_v \ge v^\alpha\}dv\cdot[1 + \mathrm{E}(X^+)^{1/\alpha}]. \end{aligned}\tag{22}$$

Finally, if $X' = \max(X, c)$, $c > 0$, then (22) holds with X^+ replaced by X'. If $\mathrm{E}(X^+)^p = \infty$, then, since $\alpha p > 1$, $\mathrm{E}(X')^{1/\alpha} = o(\mathrm{E}(X')^p)$ as $c \to \infty$, implying $J_1(1) = \infty$. Thus, (21) obtains. □

Theorem 3. *Let* $\{X, X_n, n \ge 1\}$ *be* i.i.d. r.v.s, $S_n = \sum_1^n X_i$, $S_0 = X_0 = 0$, *and* $p > 1/\alpha > 0$, $\varepsilon > 0$. *Set*

$$M(\varepsilon) = \sup_{n \ge 0}(S_n - \varepsilon n^\alpha), \qquad M_1(\varepsilon) = \sup_{n \ge 0}(X_n - \varepsilon n^\alpha),$$

$$L(\varepsilon) = \sup\{n \ge 1: S_n \ge \varepsilon n^\alpha\}, \qquad L_1(\varepsilon) = \sup\{n \ge 1: X_n \ge \varepsilon n^\alpha\}.$$

Then, (i)

$$E(X^+)^p < \infty \quad \text{iff } J_1(1) = \int_0^\infty v^{\alpha p - 2} P\{\bar{X}_v \ge v^\alpha\} dv < \infty$$

$$\text{iff } E[M_1(\varepsilon)]^{(\alpha p - 1)/\alpha} < \infty; \qquad \varepsilon > 0,$$

$$\text{iff } E[L_1(\varepsilon)]^{\alpha p - 1} < \infty, \qquad \varepsilon > 0.$$

(ii) Suppose $E(X^+)^p < \infty$ for $p \ge 1$, $E X = 0$, and $E|X|^\gamma < \infty$ for some $\gamma \in (1/\alpha, 2]$ when $\frac{1}{2} < \alpha < 1$. Then for any $\alpha > \frac{1}{2}$ and all $\varepsilon > 0$

$$E[M(\varepsilon)]^{(\alpha p - 1)/\alpha} < \infty, \qquad E[L(\varepsilon)]^{\alpha p - 1} < \infty, \tag{23}$$

and $J(\varepsilon) \le I(\varepsilon) < \infty$, where the latter are as in (12) but with $\bar{S}_v$ replacing $\bar{Y}_v$.

(iii) Let $\alpha > \frac{1}{2}$, $E|X|^{1/\alpha} < \infty$, and $E X = 0$ if $\alpha \le 1$. If either of the conditions of (23) holds or $I(\varepsilon) < \infty$ or $J(\varepsilon) < \infty$, then $E(X^+)^p < \infty$.

PROOF. By Lemma 3, $E(X^+)^p < \infty$ iff $J_1(1) < \infty$ iff $J_1(\varepsilon) < \infty$ for all $\varepsilon > 0$. Then Lemma 1 ensures $E[M_1(\varepsilon)]^{(\alpha p - 1)/\alpha} < \infty$, $I(2\varepsilon) < \infty$, $\varepsilon > 0$, and hence $E[L_1(\varepsilon)]^{\alpha p - 1} < \infty$ since $L_1(\varepsilon) \le \bar{L}_1(\varepsilon)$. Conversely, by Lemma 1

$$E[M_1(\varepsilon)]^{(\alpha p - 1)/\alpha} < \infty$$

implies $I_1(\varepsilon) < \infty$, $\varepsilon > 0$, and hence $J_1(\varepsilon) < \infty$, $\varepsilon > 0$. Moreover, if $E[L_1(\varepsilon)]^{\alpha p - 1} < \infty$, $\varepsilon > 0$, then by Lemma 3, $J_1(\varepsilon) < \infty$, $\varepsilon > 0$.

Apropos of (ii), if $\frac{1}{2} < \alpha < 1$, the first half of (23) follows from Theorem 2. Then by Lemma 1, $J(\varepsilon) \le I(\varepsilon) < \infty$ and $E[L(\varepsilon)]^{\alpha p - 1} \le E[\bar{L}(\varepsilon)]^{\alpha p - 1} < \infty$. If, rather, $\alpha \ge 1$, define $X'_n = X_n I_{[X_n \ge -C]}$, $C > 0$. Then $S'_n = \sum_1^n X'_j \ge S_n$. Since $E(X^+)^p < \infty$ for some $p \ge 1$ ($p > 1$ if $\alpha = 1$), necessarily $\gamma \equiv \min(p, 2) \in [1, 2]$ and $E|X'_1|^\gamma < \infty$. Hence, by Theorem 2

$$E\left[\sup_{n \ge 0}(S_n - n^\alpha(\varepsilon + E X'_1))\right]^{(\alpha p - 1)/\alpha}$$

$$\le E\left[\sup_{n \ge 0}\left(\sum_{j=1}^n (X'_j - E X'_1) - \varepsilon n^\alpha\right)\right]^{(\alpha p - 1)/\alpha} < \infty$$

by Theorem 1, and since $E X'_1 = o(1)$ as $C \to \infty$, the first half of (23) is established. The remainder of (23) and (ii) follow from $L'(\varepsilon) \ge L(\varepsilon)$, $I'(\varepsilon) \ge I(\varepsilon)$.

To prove (iii), note that by Lemma 1

$$J(\varepsilon) < \infty \Rightarrow E[M(\varepsilon)]^{(\alpha p - 1)/\alpha} < \infty \Rightarrow I(\varepsilon) < \infty \Rightarrow E[\bar{L}(\varepsilon)]^{\alpha p - 1} < \infty,$$

whence $E[L(\varepsilon)]^{\alpha p - 1} < \infty$. Then by Lemma 2, $E[L_1(\varepsilon)]^{\alpha p - 1} < \infty$, implying $E(X^+)^p < \infty$ by part (i). □

Corollary 3 (Kiefer–Wolfowitz). *If* $\{X, X_n, n \ge 1\}$ *are* i.i.d. *with* $E X < 0$ *and* $S_n = \sum_1^n X_i$, $S_0 = 0$, $p > 1$, *then* $E(\sup_{n \ge 0} S_n)^{p-1} < \infty$ iff $E(X^+)^p < \infty$.

PROOF. Apply Theorem 3 with $\alpha = 1$ and $\varepsilon = -\mathrm{E}\, X$ to $\{X_n - \mathrm{E}\, X, n \geq 1\}$.

Corollary 4. *Let* $\{X, X_n, n \geq 1\}$ *be* i.i.d. r.v.s *with* $\mathrm{E}\, X = 0$, $S_n = \sum_1^n X_i$, *and define* $\hat{L}(\varepsilon) = \sup\{n \geq 1 : |S_n| \geq n\varepsilon\}$, $\varepsilon > 0$. *Then for* $p > 1$, $\mathrm{E}[\hat{L}(\varepsilon)]^{p-1} < \infty$ *for all* $\varepsilon > 0$ iff $\mathrm{E}|X|^p < \infty$.

PROOF. Clearly, $\hat{L}(\varepsilon) = \max[L^+(\varepsilon), L^-(\varepsilon)] \leq L^+(\varepsilon) + L^-(\varepsilon)$, where $L^-(\varepsilon) = \sup\{n \geq 1 : S_n \leq -n\varepsilon\}$ and $L^+(\varepsilon) = \sup\{n \geq 1 : S_n \geq \varepsilon n\}$. □

Corollary 5. *If* $\{X, X_n, n \geq 1\}$ *are* i.i.d. *with* $\mathrm{E}\, X = \mu > 0$, $\mathrm{E}(X^-)^2 < \infty$ *and* $N_0 = \sup\{n \geq 0 : S_n \leq 0\}$, *where* $S_n = \sum_1^n X_i$, *then* $\mathrm{E}\, N_0 < \infty$ *and*

$$\sum_{n=1}^{\infty} \mathrm{P}\left\{\inf_{j \geq n} S_j \leq 0\right\} < \infty.$$

PROOF. Set $Y_n = \mu - X_n$, whence

$$\begin{aligned}\sum_{n=1}^{\infty} \mathrm{P}\left\{\inf_{j \geq n} S_j \leq 0\right\} &= \sum_{n=1}^{\infty} \mathrm{P}\left\{\sup_{j \geq n} \sum_{i=1}^{j} (Y_i - \mu) \geq 0\right\} \\ &\leq \sum_{n=1}^{\infty} \mathrm{P}\left\{\sup_{j \geq n} \sum_{i=1}^{j} \left(Y_i - \frac{\mu}{2}\right) \geq \frac{n\mu}{2}\right\} \\ &\leq \sum_{n=1}^{\infty} \mathrm{P}\left\{\sup_{j \geq 1} \sum_{i=1}^{j} \left(Y_i - \frac{\mu}{2}\right) > \frac{n\mu}{2}\right\} < \infty\end{aligned}$$

by Corollary 3. Since $\{N_0 \geq n\} = \bigcup_{j=n}^{\infty} \{S_j \leq 0\} \subset \{\inf_{j \geq n} S_j \leq 0\}$, necessarily $\mathrm{E}\, N_0 < \infty$. □

For any r.v.s Y, Z let the relation $Y \sim Z$ signify that Y and Z have identical distributions.

Theorem 4. *Let* $\{X, X_n, n \geq 1\}$ *be* i.i.d. r.v.s *with* $S_n = \sum_1^n X_i$, $n \geq 1$, $S_0 = 0$, $\bar{S}_n = \max_{0 \leq j \leq n} S_j$, $\bar{S} = \sup_{n \geq 0} S_n$. *Then*

i. $\bar{S}_n \sim (\bar{S}_{n-1} + X)^+$, $n \geq 1$,
ii. if $S_n \xrightarrow{\text{a.c.}} -\infty$, then $\bar{S} \sim (\bar{S} + X)^+$,
iii. If $\mathrm{E}\, X \in (-\infty, 0)$ and $\mathrm{E}\, X^2 < \infty$, then $\mathrm{E}\, \bar{S} = [\sigma_X^2 - \sigma_{(\bar{S}+X)^-}^2]/(-2\, \mathrm{E}\, X)$.

PROOF. If $W_0 = 0$, $W_n = (W_{n-1} + X_n)^+$, $n \geq 1$, then

$$W_n = \max[S_n, S_n - S_1, \ldots, S_n - S_{n-1}, 0], \qquad n \geq 1. \tag{24}$$

Clearly, (24) obtains for $n = 1$ and, assuming its validity for $n - 1 \geq 1$,

$$\begin{aligned} W_n &= (\max[S_{n-1}, S_{n-1} - S_1, \ldots, S_{n-1} - S_{n-2}, 0] + X_n)^+ \\ &= (\max[S_n, S_n - S_1, \ldots, S_n - S_{n-1}])^+ \\ &= \max[S_n, S_n - S_1, \ldots, S_n - S_{n-1}, 0], \end{aligned}$$

whence (24) holds for $n \geq 1$. Thus,

$$W_n \sim \max[S_n, S_{n-1}, \ldots, S_1, 0] = \bar{S}_n, \qquad n \geq 1,$$

and (i) follows.

Next, if $S_n \xrightarrow{\text{a.c.}} -\infty$, then $\bar{S}_n \xrightarrow{\text{a.c.}} \bar{S}$ (finite), whence (ii) is an immediate consequence of (i).

Under the hypothesis of (iii), $\mathrm{E}\,\bar{S} < \infty$ by Corollary 3. Moreover, assuming temporarily that $\mathrm{E}(X^+)^3 < \infty$, this same corollary ensures $\mathrm{E}\,\bar{S}^2 < \infty$. By (ii),

$$\mathrm{E}\,\bar{S} = \mathrm{E}(\bar{S} + X)^+ = \mathrm{E}(\bar{S} + X) + \mathrm{E}(\bar{S} + X)^-,$$

implying

$$\mathrm{E}\,X = -\mathrm{E}(\bar{S} + X)^-. \tag{25}$$

Similarly,

$$\begin{aligned} \mathrm{E}[(\bar{S} + X)^-]^2 - \mathrm{E}\,X^2 &= \mathrm{E}[(\bar{S} + X)^2 - 2(\bar{S} + X)(\bar{S} + X)^+ \\ &\quad + (\bar{S} + X)^{+2}] - \mathrm{E}\,X^2 \\ &= \mathrm{E}\,\bar{S}^2 + 2\,\mathrm{E}\,\bar{S}X - \mathrm{E}[(\bar{S} + X)^+]^2 \\ &= 2\,\mathrm{E}\,\bar{S}\,\mathrm{E}\,X, \end{aligned}$$

and combining this with (25) yields (iii) when $\mathrm{E}(X^+)^3 < \infty$. In general, set $X_n' = \min(X_n, C)$ and define $\bar{S}'$, X' analogously via X_n', whence

$$-2\,\mathrm{E}\,\bar{S}'\,\mathrm{E}\,X' = \sigma_{X'}^2 - \sigma_{(\bar{S}' + X')^-}^2. \tag{26}$$

Since $(\bar{S}' + X')^- \leq (X')^- = X^-$, $\bar{S}' \leq \bar{S}$, and $\lim_{C\to\infty} \bar{S}' = S$, by Lebesgue's dominated convergence theorem $\mathrm{E}\,\bar{S}' \longrightarrow \mathrm{E}\,S$ and $\sigma_{(\bar{S}'+X')^-}^2 \to \sigma_{(S+X)^-}^2$ and (iii) follows from (26). □

Corollary 6. *If* $\{X, X_n, n \geq 1\}$ *are* i.i.d. r.v.s *with* $\mathrm{E}\,X = 0$, $\sigma^2 = \mathrm{E}\,X^2 < \infty$, $S_n = \sum_1^n X_i$, $S_0 = 0$, *and*

$$M = M(\varepsilon) = \sup_{n\geq 0}(S_n - n\varepsilon) = \sup_{n\geq 0} \sum_{i=1}^{n} (X_i - \varepsilon), \tag{27}$$

where $\varepsilon > 0$, *then*

$$\lim_{\varepsilon\to 0} \varepsilon\,\mathrm{E}\,M(\varepsilon) = \frac{\sigma^2}{2}. \tag{28}$$

Proof. To avoid triviality, suppose $\sigma^2 > 0$. By Theorem 4(iii)

$$2\varepsilon\,\mathrm{E}\,M(\varepsilon) = \sigma_{X-\varepsilon}^2 - \sigma_{(M+X-\varepsilon)^-}^2. \tag{29}$$

Since $(M + X - \varepsilon)^- \leq X^- + \varepsilon$ and

$$\lim_{\varepsilon\to 0} M(\varepsilon) \geq \lim_{\varepsilon\to 0} \sup_{0<j\leq n} (S_j - j\varepsilon) = \sup_{0\leq j<n} S_j \xrightarrow[n\to\infty]{\text{a.c.}} \infty,$$

$\sigma_{(M+X-\varepsilon)^-}^2 \to 0$ by dominated convergence and (28) follows from (29). □

Theorem 5. *Let* $\{X, X_n, n \ge 1\}$ *be* i.i.d. r.v.s *with* $\mathrm{E}\,X = 0$, $\mathrm{E}|X| > 0$, $S_n = \sum_{i=1}^n X_i$, $S_0 = 0$ *and set* $M(\varepsilon) = \sup_{n \ge 0}(S_n - n\varepsilon)$. *If*

$$U_- = U_-(\varepsilon) = \inf\{n \ge 1: S_n \le n\varepsilon\}, \qquad T_- = \inf\{n \ge 1: S_n \le 0\},$$

$$U_+ = U_+(\varepsilon) = \inf\{n \ge 1: S_n > n\varepsilon\}, \qquad T_+ = \inf\{n \ge 1: S_n > 0\},$$

then $\lim_{\varepsilon \to 0} U_\pm(\varepsilon) = T_\pm < \infty$, a.c., *and*

i. $\mathrm{E}\,M(\varepsilon) = (1/\mathrm{P}\{U_+ = \infty\})\int_{[U_+ < \infty]}(S_{U_+} - \varepsilon U_+)$
$\quad = (\mathrm{E}\,U_-)\int_{[U_+ < \infty]}(S_{U_+} - \varepsilon U_+)$.

ii. if $\sigma^2 = \mathrm{E}\,X^2 < \infty$, then $\mathrm{E}\,S_{T_+} < \infty$, $\mathrm{E}\,S_{T_-} < \infty$, and

$$\mathrm{E}(\varepsilon U_- - S_{U_-})\int_{[U_+ < \infty]}(S_{U_+} - \varepsilon U_+) \to \frac{\sigma^2}{2},$$

$$\mathrm{E}(-S_{T_-})\cdot \mathrm{E}\,S_{T_+} \le \frac{\sigma^2}{2},$$

iii. $\int_{[U_+ < \infty]}(S_{U_+} - \varepsilon U_+) < \infty$ iff $\mathrm{E}\,M(\varepsilon) < \infty$ iff $\mathrm{E}(X^+)^2 < \infty$.

PROOF. By Corollaries 5.4.1, 5.4.2, T_+ and T_- are finite stopping times and, clearly, $\lim_{\varepsilon \to 0} U_\pm(\varepsilon) = T_\pm$. The second equality of (i) follows immediately from Theorem 5.4.2. Apropos of the first, define $U^{(1)} = U_+$, $U^{(i+1)} = \infty$ on $\{U^{(i)} = \infty\}$ and otherwise $U_n = \sum_{i=1}^n U^{(i)}$ where

$$U^{(i+1)} = \inf\left\{n \ge 1: \sum_{j=U_i+1}^{U_i+n}(X_j - \varepsilon) > 0\right\}, \qquad i \ge 1.$$

Thus, $\{U^{(i)}, i \ge 1\}$ are copies of U_+ which, however, is not a finite stopping time. Nonetheless, analogously to Corollary 5.3.2, for any positive integers m_i, $1 \le i \le n$,

$$\mathrm{P}\left\{\bigcap_{i=1}^n [U^{(i)} = m_i]\right\} = \prod_{i=1}^n \mathrm{P}\{U^{(1)} = m_i\}. \tag{30}$$

Now, setting

$$W_n = S_n - n\varepsilon, \qquad n \ge 1, \tag{31}$$

necessarily

$$M(\varepsilon) = W_{U_1} I_{[U_1 < \infty]} + \sum_{n=2}^\infty (W_{U_n} - W_{U_{n-1}}) I_{[U^{(n)} < \infty]}. \tag{32}$$

Since

$$\mathrm{E}(W_{U_2} - W_{U_1}) I_{[U^{(2)} < \infty]} = \sum_{j=1}^\infty \mathrm{E}(W_{U^{(2)}+j} - W_j) I_{[U^{(1)} = j,\, U^{(2)} < \infty]}$$

$$= \sum_{j=1}^{\infty} P\{U^{(1)} = j\} E\, W_{U_+} I_{[U_+ < \infty]}$$

$$= P\{U_+ < \infty\} E\, W_{U_+} I_{[U_+ < \infty]}$$

and similarly for $n > 2$, it follows from (32) that

$$E\, M(\varepsilon) = \sum_{n=1}^{\infty} P^{n-1}\{U_+ < \infty\} E\, W_{U_+} I_{[U_+ < \infty]}$$

$$= \frac{1}{P\{U_+ = \infty\}} E\, W_{U_+} I_{[U_+ < \infty]}.$$

To prove (ii), note that via Wald's equation, the initial notation of (31), and part (i),

$$E(-W_{U_-}) \cdot E\, W_{U_+} I_{[U_+ < \infty]} = \varepsilon\, E\, U_- \cdot E\, W_{U_+} I_{[U_+ < \infty]}$$

$$= \varepsilon\, E\, M(\varepsilon) \to \frac{\sigma^2}{2}$$

as $\varepsilon \to 0$ by Corollary 6. Hence, from Fatou's lemma

$$E(-S_{T_-}) \cdot E\, S_{T_+} \le \varliminf_{\varepsilon \to 0} E(-W_{U_-}) \cdot \varliminf_{\varepsilon \to 0} E\, W_{U_+} I_{[U_+ < \infty]}$$

$$\le \varliminf_{\varepsilon \to 0} E(-W_{U_-}) E\, W_{U_+} I_{[U_+ < \infty]} = \frac{\sigma^2}{2}.$$

Since both $E(-S_{T_-})$ and $E\, S_{T_+}$ are positive, each is therefore finite and (ii) is proved.

Finally, as already observed, $E\, U_- = 1/P\{U_+ = \infty\} < \infty$, whence (iii) follows from (i) and Corollary 3.

References

L. E. Baum and M. Katz, "Convergence rates in the law of large numbers," *Trans. Amer. Math. Soc.* **120** (1965), 108–123.

H. D. Brunk, "The strong law of large numbers," *Duke Math. J.* **15** (1948), 181–195.

Y. S. Chow, "A martingale inequality and the law of large numbers," *Proc. Amer. Math. Soc.* **11** (1960), 107–111.

Y. S. Chow, "On a strong law of large numbers for martingales," *Ann. Math. Stat.* **38** (1967), 610.

Y. S. Chow, "Delayed sums and Borel summability of independent, identically distributed random variables, *Bull. Inst. Math., Academia Sinica* **1** (1973), 207–220.

Y. S. Chow and T. L. Lai, "Some one-sided theorems on the tail distribution of sample sums with applications to the last time and largest excess of boundary crossings, *Trans. Amer. Math. Soc.* (1975).

Y. S. Chow, H. Robbins, and D. Siegmund, *Great Expectations: The Theory of Optimal Stopping*, Houghton Mifflin, Boston, 1972.

K. L. Chung, "Note on some strong laws of large numbers," *Amer. J. Math.* **69** (1947), 189–192.

K. L. Chung, "The strong law of large numbers," *Proc. 2nd Berkeley Symp. Stat. and Prob.* (1951), 341–352.

J. L. Doob, *Stochastic Processes*, Wiley, New York, 1953.

V. A. Egorov, "On the strong law of large numbers and the law of the iterated logarithm for sequences of independent random variables." *Theor. Prob. Appl.* **15** (1970), 509–514.

W. Feller, *An Introduction to Probability Theory and Its Applications*, Vol. 2, Wiley, New York, 1966.

W. Feller, "An extension of the law of the iterated logarithm to variables without variance," *Journ. Math. and Mech.* **18** (1968), 343–356.

B. V. Gnedenko and A. N. Kolmogorov, *Limit Distributions for Sums of Independent Random Variables* (K. L. Chung, translator), Addison-Wesley, Reading, Mass., 1954.

P. Hartman and A. Wintner, "On the law of the iterated logarithm," *Amer. Jour. Math.* **63** (1941), 169–176.

P. L. Hsu and H. Robbins, "Complete convergence and the law of large numbers, *Proc. Nat. Acad. Sci. U.S.A.* **33** (1947), 25–31.

A. Khintchine, "Über dyadische Bruche," *Math. Zeit.* **18** (1923), 109–116.

A. Khintchine, "Über einen Satz der Wahrscheinlichkeitsrechnung," *Fund. Math.* **6** (1924), 9–20.

J. Kiefer and J. Wolfowitz, "On the characteristics of the general queueing process with applications to random walk," *Ann. Math. Stat.* **27** (1956), 147–161.

J. F. C. Kingman, "Some inequalities for the queue G1/G/1," *Biometrika* **49** (1962), 315–324.

A. Kolmogorov, "Über der Gesetz des Iterierten Logarithmus," *Math. Annalen* **101** (1929), 126–135.

M. Loève, *Probability Theory*, 3rd ed., Van Nostrand, Princeton, 1963; 4th ed., Springer-Verlag, Berlin and New York, 1977–1978.

J. Marcinkievicz and A. Zygmund, "Sur les fonctions indépendantes," *Fund. Math.* **29** (1937), 60–90.

J. Marcinkiewicz and A. Zygmund, "Quelques théorèmes sur les fonctions indépendantes," *Studia Math.* **7** (1938), 104–120.

S. V. Nagaev, "On necessary and sufficient conditions for the strong law of large numbers," *Theor. Prob. Appl.* **17** (1972), 573–581.

Y. V. Prohorov. "The strong law of large numbers," *Izv. Akad. Nauk. Ser. Mat.* **14** (1950), 523–536 [in Russian].

Y. V. Prohorov, "Some remarks on the strong law of large numbers," *Theor. Prob. Appl.* **4** (1959), 201–208.

D. A. Raikov, "On a connection between the central limit theorem in the theory of probability and the law of large numbers," *Izv. Nauk USSR, Sov. Math.* (1938), 323–338.

D. Siegmund, "On moments of the maximum of normed sums," *Ann. Math. Stat.* **40** (1969), 527–531.

F. Spitzer, "A combinatorial lemma and its applications to probability theory," *Trans. Amer. Math. Soc.* **82** (1956), 323–339.

F. Spitzer, "The Wiener–Hopf equation whose kernal is a probability density," *Duke Math. Jour.* **24** (1957), 327–343.

V. Strassen, "A converse to law of the iterated logarithm," *Z. Wahr.* **4** (1966), 265–268.
H. M. Taylor, "Bounds for stopped partial sums," *Ann. Math. Stat.* **43** (1972), 733–747.
H. Teicher, "A dominated ergodic type theorem," *Z. Wahr.* **8** (1967), 113–116.
H. Teicher, "Some new conditions for the strong law," *Proc. Nat. Acad. Sci. U.S.A.* **59** (1968), 705–707.
H. Teicher, "Completion of a dominated ergodic theorem," *Ann. Math. Stat.* **42** (1971), 2156–2158.
H. Teicher, "On interchangeable random variables," *Studi di Probabilità, Statistica e Ricerca Operative in Onore di Giuseppe Pompilj*, pp. 141–148, Gubbio, 1971.
H. Teicher, "On the law of the iterated logarithm, "*Ann. Prob.* **2** (1974), 714–728.
H. Teicher, "A necessary condition for the iterated logarithm law," *Z. Wahr. Verw. Geb.* (1975), 343–349.
H. Teicher, "Generalized exponential bounds, iterated logarithm and strong laws," *Z. Wahr. Verw. Geb.* (1979), 293–307.
N. Wiener, "The ergodic theorem," *Duke Math. Jour.* **5** (1938) 1–18.
A. Zygmund, *Trigonometric Series*, Vol. I, Cambridge, 1959.

11 Martingales

An introduction to martingales appeared in Section 7.4, where convergence theorems for submartingales $\{S_n, \mathscr{F}_n, n \geq 1\}$ (relating to differentiation theory) were discussed. Here, emphasis will fall upon convergence theorems for martingales $\{S_{-n}, \mathscr{F}_{-n}, n \leq -1\}$ (relating to ergodic theorems). In demarcating the two cases, it is natural to refer to a martingale $\{S_n, \mathscr{F}_n, n \geq 1\}$ as an upward martingale and to allude to a martingale $\{S_{-n}, \mathscr{F}_{-n}, n \leq -1\}$ when written $\{S_n, \mathscr{F}_n, n \geq 1\}$ as a downward or reverse martingale. Martingale and stochastic inequalities will also be dealt with.

11.1 Upcrossing Inequality and Convergence

The convergence approach of Section 7.4 does not carry over to downward martingales since the formal analogue of a "first time" in the former case is a "last time" in the latter, whereas a genuine first time is not well defined. The ensuing upcrossing inequality provides an avenue of approach to downward martingales.

Definition. If $r_1 < r_2$ and $a_1, \ldots, a_n$ are finite, real numbers, the number of upcrossings of the interval $[r_1, r_2]$ by the sequence $a_1, a_2, \ldots, a_n$ is defined as the number of times the elements a_j pass from "on or below r_1" to "on or above r_2." More precisely, let α_1 be the smallest integer (if any) for which $a_{\alpha_1} \leq r_1$, and in general for $j \geq 2$ let α_j be the smallest integer (if any) exceeding α_{j-1} for which

$$a_{\alpha_j} \begin{cases} \geq r_2 & \text{if } j \text{ is even} \\ \leq r_1 & \text{if } j \text{ is odd.} \end{cases}$$

If $2u$ is the largest even integer j for which α_j is defined then u is called the number of upcrossings (if α_2 is undefined, then $u = 0$).

Lemma 1 (Panzone). *If $\{S_j, \mathscr{F}_j \equiv \sigma(S_i, 1 \le i \le j), 1 \le j \le n\}$ is a nonnegative submartingale, r is a positive $\mathscr{F}_1$-measurable random variable and U is the number of upcrossings of $[0, r]$ by $(S_1, \ldots, S_n)$, then*

$$\mathrm{E}\, rU + \mathrm{E}\, S_1 \le \int_{[S_{n-1} > 0]} S_n + \int_{[S_{n-1} = 0,\, S_n \ge r]} S_n \le \mathrm{E}\, S_n. \tag{1}$$

PROOF. For $n = 2$, (1) is confirmed by

$$\mathrm{E}\, rU + \mathrm{E}\, S_1 = \int_{[S_1 = 0,\, S_2 \ge r]} r + \int_{[S_1 > 0]} S_1 \le \int_{[S_1 = 0,\, S_2 \ge r]} S_2$$

$$+ \int_{[S_1 > 0]} S_2 \le \mathrm{E}\, S_2.$$

Suppose inductively that (1) holds with n replaced by $n - 1$ for all submartingales, and set

$$T_j = S_j, \qquad 1 \le j \le n - 2$$

$$T_{n-1} = \begin{cases} S_n & \text{if } 0 < S_{n-1} < r \\ S_{n-1} & \text{otherwise.} \end{cases}$$

For $A \in \sigma(T_j, 1 \le j \le n - 2) = \sigma(S_j, 1 \le j \le n - 2)$

$$\int_A T_{n-2} = \int_A S_{n-2} \le \int_A S_{n-1}$$

$$\le \int_{A[S_{n-1} \ge r]} S_{n-1} + \int_{A[0 < S_{n-1} < r]} S_n = \int_A T_{n-1}.$$

Hence, $\mathrm{E}\{T_{n-1} \mid T_j, 1 \le j \le n - 2\} \ge T_{n-2}$, a.c. Clearly, for $2 \le m < n - 1$ $\mathrm{E}\{T_m \mid T_j, 1 \le j \le m - 1\} \ge T_{m-1}$, a.c., and so $\{T_j, \mathscr{F}_j, 1 \le j \le n - 1\}$ is a nonnegative submartingale.

Let V be the number of upcrossings of $[0, r]$ by $(T_1, \ldots, T_{n-1})$. Then

$$U = V + I_{[S_{n-1} = 0,\, S_n \ge r]},$$

whence by the induction hypothesis

$$\begin{aligned} \mathrm{E}\, rU + \mathrm{E}\, S_1 &= \mathrm{E}\, rV + \mathrm{E}\, rI_{[S_{n-1} = 0,\, S_n \ge r]} + \mathrm{E}\, S_1 \\ &\le \mathrm{E}\, T_{n-1} + \mathrm{E}\, rI_{[S_{n-1} = 0,\, S_n \ge r]} \\ &= \int_{[0 < S_{n-1} < r]} S_n + \int_{[S_{n-1} \ge r]} S_{n-1} + \mathrm{E}\, rI_{[S_{n-1} = 0,\, S_n \ge r]} \\ &\le \int_{[0 < S_{n-1} < r]} S_n + \int_{[S_{n-1} \ge r]} S_n + \int_{[S_{n-1} = 0,\, S_n \ge r]} S_n \\ &\le \int_{[S_{n-1} > 0]} S_n + \int_{[S_{n-1} = 0,\, S_n \ge r]} S_n \le \mathrm{E}\, S_n. \end{aligned}$$

□

Corollary 1. *Let $\{S_j, \mathcal{F}_j = \sigma(X_i, 1 \le i \le j), 1 \le j \le n\}$ be a submartingale and let $r_1 < r_2$ be finite real numbers. If U is the number of upcrossings of $[r_1, r_2]$ by $(S_1, \ldots, S_n)$, then*

$$\mathrm{E}\, U \le (r_2 - r_1)^{-1} \mathrm{E}(S_n - r_1)^+ \le (r_2 - r_1)^{-1}[\mathrm{E}\, S_n^+ + |r_1|].$$

PROOF. Since $\{(S_j - r_1)^+, \mathcal{F}_j, 1 \le j \le n\}$ is a nonnegative submartingale and U is the number of upcrossings of $[0, r_2 - r_1]$ by $((S_1 - r_1)^+, \ldots, (S_n - r_1)^+)$, the lemma ensures

$$(r_2 - r_1)\mathrm{E}\, U \le \mathrm{E}(S_n - r_1)^+ - \mathrm{E}(S_1 - r_1)^+ \le \mathrm{E}(S_n - r_1)^+. \qquad \square$$

Theorem 1. *Let $\{S_n, \mathcal{F}_n, -\infty < n \le 0\}$ be a submartingale and $\mathcal{F}_{-\infty} = \bigcap_{-\infty}^{0} \mathcal{F}_n$.*

(i) *If* $\mathrm{E}\, S_0^+ < \infty$, *then* $S_{-\infty} = \lim_{n \to -\infty} S_n$ *exists* a.c. *with* $\mathrm{E}\, S_{-\infty}^+ < \infty$. *Moreover, if* $\lim_{n \to -\infty} \mathrm{E}\, S_n = K > -\infty$, *then* $\sup_{n \le 0} \mathrm{E}|S_n| < \infty$, $S_n \xrightarrow{\mathcal{L}_1} S_{-\infty}$ *and* $\{S_n, \mathcal{F}_n, -\infty \le n \le 0\}$ *is a submartingale.*

(ii) *If* $\{S_n, \mathcal{F}_n, -\infty < n \le 0\}$ *is a nonnegative submartingale with* $\mathrm{E}\, S_0^p < \infty$ *for some* $p \ge 1$, *then* $S_n \xrightarrow{\mathcal{L}_p} S_{-\infty}$ *as* $n \to -\infty$.

PROOF. Set $S = \overline{\lim}_{n \to -\infty} S_n$, $S_{-\infty} = \underline{\lim}_{n \to -\infty} S_n$ and suppose that for some choice of $r_1 < r_2$ the set $A = [S > r_2 > r_1 > S_{-\infty}]$ has positive probability. Then if U_n is the number of upcrossings of $[r_1, r_2]$ by $(S_{-n}, \ldots, S_0)$, the prior corollary yields

$$\infty > (r_2 - r_1)^{-1}[\mathrm{E}\, S_0^+ + |r_1|] \ge \mathrm{E}\, U_n \ge \int_A U_n \to \infty$$

as $n \to \infty$, a contradiction. Hence, $\mathrm{P}\{A\} = 0$ for all choices of $r_1 < r_2$, implying $S_{-\infty} = S$, a.c., and $S_n \xrightarrow{\text{a.c.}} S_{-\infty}$. By Fatou's lemma

$$\mathrm{E}\, S_{-\infty}^+ = \mathrm{E} \varliminf_{n \to -\infty} S_n^+ \le \varliminf_{n \to -\infty} \mathrm{E}\, S_n^+ \le \mathrm{E}\, S_0^+,$$

yielding the initial portion of (i). Moreover, if $\lim_{n \to -\infty} \mathrm{E}\, S_n = K > -\infty$, for $-\infty < n \le 0$

$$\mathrm{E}|S_n| = 2\,\mathrm{E}\, S_n^+ - \mathrm{E}\, S_n \le 2\,\mathrm{E}\, S_0^+ - K,$$

implying $\sup_{n \le 0} \mathrm{E}|S_n| < \infty$ and via Fatou that $\mathrm{E}|S_{-\infty}| < \infty$. To prove $S_n \xrightarrow{\mathcal{L}_1} S_{-\infty}$, it suffices to verify uniform integrability of $\{S_n\}$. To this end, choose $\varepsilon > 0$ and fix $m \le 0$ so that $\mathrm{E}\, S_m < K + \varepsilon$. Then for $n \le m$ and $a > 0$,

$$\begin{aligned}\int_{[|S_n| \ge a]} |S_n| &= \int_{[S_n \ge a]} S_n + \left(\int_{[S_n > -a]} S_n\right) - \mathrm{E}\, S_n \\ &\le \int_{[S_n \ge a]} S_m + \left(\int_{[S_n > -a]} S_m\right) - K \\ &\le \int_{[|S_n| \ge a]} |S_m| + \mathrm{E}\, S_m - K < \int_{[|S_n| \ge a]} |S_m| + \varepsilon < 2\varepsilon \qquad (2)\end{aligned}$$

for large a since $P\{|S_n| \geq a\} \leq a^{-1} E|S_n| \to 0$ as $a \to \infty$. The conclusion of (2) clearly also holds for the finitely many integers n in $[m, 0]$, and so uniform integrability is established.

Next, for $A \in \mathscr{F}_{-\infty}$ and every $m \leq 0$

$$\int_A S_{-\infty} = \lim_{n \to -\infty} \int_A S_n \leq \int_A S_m,$$

whence $E\{S_m | \mathscr{F}_{-\infty}\} \geq S_{-\infty}$, a.c. for all m, concluding the proof of (i).

Apropos of (ii), $\{S_n^p, \mathscr{F}_n, -\infty < n \leq 0\}$ is a nonnegative submartingale which is easily seen to be u.i., whence the conclusion of (ii) follows. □

Since U-statistics

$$U_{m,n} = \binom{n}{m}^{-1} \sum_{1 \leq i_1 < \cdots < i_m \leq n} \varphi(X_{i_1}, \ldots, X_{i_m}), \qquad n \geq m,$$

constitute a downward martingale (Example 3, Section 7.4), Theorem 1 yields directly a.c. and $\mathscr{L}_1$ convergence of $U_{m,n}$ as $n \to \infty$. Moreover, when the $\{X_n\}$ are i.i.d., the Hewitt–Savage zero–one law (Corollary 7.3.7) ensures that the limit is a constant which clearly must coincide with $E\,\varphi(X_1, \ldots, X_m)$.

Corollary 2. *If $\{S_n, \mathscr{F}_n, -\infty < n < \infty\}$ is an $\mathscr{L}_1$ submartingale with $\sup_{-\infty<n<\infty} E|S_n| < \infty$ and $X_n = S_n - S_{n-1}$, then*

$$\sum_{-\infty}^{\infty} X_i = \lim_{m, n \to \infty} \sum_{-m}^{n} X_i = S_\infty - S_{-\infty}, \qquad \text{a.c.}$$

where $S_{\pm\infty} = \lim_{n \to \pm\infty} S_n$ and, moreover, $E|S_\infty - S_{-\infty}| < \infty$.

Proof. By Theorem 1, $\sum_{-m}^{0} X_j = S_0 - S_{-m-1} \xrightarrow{\text{a.c.}} S_0 - S_{-\infty}$ with $E|S_{-\infty}| < \infty$ by Fatou. Moreover, by Theorem 7.4.2, $\sum_1^m X_j = S_m - S_0 \xrightarrow{\text{a.c.}} S_\infty - S_0$ with $E|S_\infty| < \infty$. □

Corollary 3. *If $\{S_n, \mathscr{F}_n, -\infty \leq n \leq 0\}$ is a submartingale with $E\,S_0^+ < \infty$ and $\lim_{n \to -\infty} E\,S_n = K > -\infty$, then $S_n \xrightarrow[\mathscr{L}_1]{\text{a.c.}} S$ and $E\{S | \mathscr{F}_{-\infty}\} \geq S_{-\infty}$, a.c.*

Proof. By Theorem 1, $S_n \xrightarrow[\mathscr{L}_1]{\text{a.c.}} S$ as $n \to \infty$. For $A \in \mathscr{F}_{-\infty}$

$$\int_A S_{-\infty} \leq \int_A S_n \to \int_A S$$

by uniform integrability and the corollary follows. □

Corollary 4. *Let $\{Y_n, -\infty < n < \infty\}$ be a sequence of r.v.s on a probability space $\{\Omega, \mathscr{F}, P\}$ and $\{\mathscr{F}_n, -\infty < n < \infty\}$ a sequence of increasing sub-σ-algebras of $\mathscr{F}$ with $\mathscr{F}_\infty = \sigma(\bigcup_0^\infty \mathscr{F}_n)$ and $\mathscr{F}_{-\infty} = \bigcap_{-\infty}^{-1} \mathscr{F}_n$. If $\lim_{n \to \infty} Y_n = Y_\infty$, a.c., $\lim_{n \to -\infty} Y_n = Y_{-\infty}$, a.c., and $E \sup_{-\infty<n<\infty} |Y_n| < \infty$, then*

$$\lim_{n \to \infty} E\{Y_n | \mathscr{F}_n\} = E\{Y_\infty | \mathscr{F}_\infty\}, \qquad \text{a.c.,}$$
$$\lim_{n \to -\infty} E\{Y_n | \mathscr{F}_n\} = E\{Y_{-\infty} | \mathscr{F}_{-\infty}\}, \qquad \text{a.c.,} \tag{3}$$

In particular, for any $\mathscr{L}_1$ r.v. Y

$$\begin{aligned}\lim_{n\to\infty} \mathrm{E}\{Y|\mathscr{F}_n\} &= \mathrm{E}\{Y|\mathscr{F}_\infty\}, \quad \text{a.c.}\\ \lim_{n\to-\infty} \mathrm{E}\{Y|\mathscr{F}_n\} &= \mathrm{E}\{Y|\mathscr{F}_{-\infty}\}, \quad \text{a.c.}\end{aligned} \tag{4}$$

PROOF. The initial portion of (3) is Corollary 7.4.1. Apropos of the second, note that by Fatou, $\mathrm{E}|Y_{-\infty}| < \infty$. If $D_n = \mathrm{E}\{Y_n|\mathscr{F}_n\} - \mathrm{E}\{Y_{-\infty}|\mathscr{F}_{-\infty}\}$, then for $n \le m$

$$\begin{aligned}|D_n| &\le |\mathrm{E}\{Y_n|\mathscr{F}_n\} - \mathrm{E}\{Y_{-\infty}|\mathscr{F}_n\}| + |\mathrm{E}\{Y_{-\infty}|\mathscr{F}_n\} - \mathrm{E}\{Y_{-\infty}|\mathscr{F}_{-\infty}\}|\\ &\le \mathrm{E}\left\{\sup_{j\le m}|Y_j - Y_{-\infty}|\,\middle|\,\mathscr{F}_n\right\} + |\mathrm{E}\{Y_{-\infty}|\mathscr{F}_n\} - \mathrm{E}\{Y_{-\infty}|\mathscr{F}_{-\infty}\}|.\end{aligned}$$

By Corollary 3, $\lim_{n\to-\infty} \mathrm{E}\{Y_{-\infty}|\mathscr{F}_n\} = \mathrm{E}\{Y_{-\infty}|\mathscr{F}_{-\infty}\}$ and

$$\begin{aligned}\overline{\lim_{n\to-\infty}}\, |D_n| &\le \lim_{n\to-\infty} \mathrm{E}\left\{\sup_{j\le m}|Y_j - Y_{-\infty}|\,\middle|\,\mathscr{F}_n\right\}\\ &= \mathrm{E}\left\{\sup_{j\le m}|Y_j - Y_{-\infty}|\,\middle|\,\mathscr{F}_{-\infty}\right\} \xrightarrow[m\to-\infty]{\text{a.c.}} 0.\end{aligned}$$ □

Theorem 2 (Austin). *If* $\{S_n, \mathscr{F}_n, -\infty < n < \infty\}$ *is a martingale with* $\sup_n \mathrm{E}|S_n| = K < \infty$, *then* $\sum_{-\infty}^{\infty} (S_n - S_{n-1})^2 < \infty$, a.c.

PROOF. By Theorem 1, $S_n \xrightarrow{\text{a.c.}}$ an $\mathscr{F}_{-\infty}$-measurable r.v. $S_{-\infty}$ as $n \to -\infty$. Since $S_{-\infty} \in \mathscr{L}_1$, it may and will be supposed that $S_{-\infty} = 0$, a.c. Set $X_n = S_n - S_{n-1}$, whence by Corollary 2, $\sum_{j=-\infty}^{n} X_j = S_n$, a.c. For any $C > 0$ define $T = \inf\{n > -\infty: |S_n| > C\}$, where $\inf\{\varnothing\} = \infty$. Then, noting that $|S_j| \le C$ on $\{T > j\}$,

$$\begin{aligned}\int_{[T=\infty]} \sum_{-\infty}^{n} X_j^2 &\le \sum_{j=-\infty}^{n} \int_{[T>j]} (S_j - S_{j-1})^2\\ &= \sum_{-\infty}^{n}\left\{\int_{[T>j]} (S_j^2 - S_{j-1}^2) - 2\int_{[T>j]} S_{j-1}(S_j - S_{j-1})\right\}.\end{aligned} \tag{5}$$

Now

$$\begin{aligned}\sum_{-\infty}^{n} \int_{[T>j]} (S_j^2 - S_{j-1}^2) &= \sum_{-\infty}^{n}\left\{\left(\int_{[T>j]} S_j^2 - \int_{[T>j-1]} S_{j-1}^2\right) + \int_{[T=j]} S_{j-1}^2\right\}\\ &= \int_{[T>n]} S_n^2 + \int_{[T\le n]} S_{T-1}^2 \le C^2 + \int_{[T\le n]} S_{T-1}^2.\end{aligned}$$

Moreover, since $\int_{[T \geq j]} S_{j-1} X_j = 0$,

$$-\sum_{j=-\infty}^{n} \int_{[T>j]} S_{j-1}(S_j - S_{j-1}) = \sum_{-\infty}^{n} \int_{[T=j]} (S_{j-1}S_j - S_{j-1}^2)$$
$$\leq \int_{[T \leq n]} (C|S_T| - S_{T-1}^2)$$
$$\leq CK - \int_{[T \leq n]} S_{T-1}^2,$$

recalling Exercise 7.4.4. Consequently, (5) ensures that for all $n > -\infty$

$$\int_{[T=\infty]} \sum_{-\infty}^{n} X_j^2 \leq C^2 + 2CK.$$

Thus, $\sum_{-\infty}^{\infty} X_j^2 < \infty$, a.c. on the set $\{T = \infty\}$. By Corollary 2, $P\{T = \infty\} = P\{\sup_{n \geq -\infty} |S_n| \leq C\} \to 1$ as $C \to \infty$, and so $\sum_{-\infty}^{\infty} X_j^2$ converges a.c. □

Corollary 5. *If $\{S_n = \sum_1^n X_j, \mathscr{F}_n, 1 \leq n < \infty\}$ is an $\mathscr{L}_1$ bounded martingale, then for every integer $k \geq 2$, $U_{k,n} = \sum_{1 \leq i_1 < \cdots < i_k \leq n} X_{i_1} X_{i_2} \cdots X_{i_k}$ converges* a.c. *as $n \to \infty$.*

PROOF. Since for $k \geq 2$, $\sum_{j=1}^n |X_j|^k \leq (\sum_{j=1}^n X_j^2)^{k/2}$, $n > 1$, Theorem 2 ensures that $\sum_{j=1}^n |X_j|^k$ converges a.c. as $n \to \infty$, $k \geq 2$, while Theorem 7.4.3 guarantees that S_n converges a.c. The corollary now follows from the identity of Lemma 10.1.2. □

In view of the identity $U_{k,n} - U_{k,n-1} = X_n U_{k-1,n-1}$, $n \geq k > 1$, the prior corollary asserts that if $S_n = \sum_1^n X_j$, $n \geq 1$, is an $\mathscr{L}_1$ bounded martingale, then $U_{k,n} = \sum_{j=k}^n X_j U_{k-1,j-1}$ converges a.c. Note that in the formation of $U_{k,n}$ the martingale differences X_j have been multiplied by an $\mathscr{F}_{j-1}$-measurable function, viz., $U_{k-1,j-1}$. A more general result of this nature appears in part (ii) of the ensuing theorem.

For any sequence of r.v.s $\{Y_n\}$, the generic notation

$$Y^* = \sup_n |Y_n| \tag{6}$$

will be employed for the **maximal function** $\sup_n |Y_n|$.

Lemma 2. *Let $\{S_n = \sum_1^n X_j, \mathscr{F}_n, n \geq 1\}$ be an $\mathscr{L}_1$ martingale and $V_r = (\sum_1^\infty |X_n|^r)^{1/r}$, $r \geq 1$. Then for every $K > 0$*

$$X_n = X_n^{(1)} + X_n^{(2)} + X_n^{(3)}, \tag{7}$$

where for $j = 1, 2, 3$, $\{X_n^{(j)}, \mathscr{F}_n, n \geq 1\}$ are martingale difference sequences satisfying

$$\mathrm{E} \sum_1^\infty |X_n^{(1)}|^r \leq c_r \, \mathrm{E} \min(V_r, K)^r \tag{8}$$

with $c_r = 2^r, r \neq 2, c_2 = 1$,

$$\mathrm{E}\sum_1^\infty |X_n^{(2)}| \leq 2\,\mathrm{E}|X_T I_{[T<\infty]}| \leq 2\,\mathrm{E}\,X^*, \tag{9}$$

where

$$T = \inf\left\{n \geq 1: V_{n,r} \equiv \left(\sum_1^n |X_j|^r\right)^{1/r} > K\right\},$$

$$\mathrm{E}\left(\sum_1^\infty |X_n^{(3)}|^r\right)^{1/r} \leq \mathrm{E}\,V_r, \qquad \{X^{(3)*} > 0\} \subset \{V_r > K\}, \tag{10}$$

$$\mathrm{P}\left\{X^{(3)*} > 0\right\} \leq \frac{EV_r}{K}.$$

PROOF. For $K > 0$ and T, $V_{n,r}$ as in (9) note that $V_{n,r} \xrightarrow{\text{a.c.}} V_r$. Moreover, for $n \geq 1$ define

$$X_n^{(1)} = X_n I_{[T>n]} - \mathrm{E}\{X_n I_{[T>n]} | \mathscr{F}_{n-1}\},$$
$$X_n^{(2)} = X_n I_{[T=n]} - \mathrm{E}\{X_n I_{[T=n]} | \mathscr{F}_{n-1}\},$$
$$X_n^{(3)} = X_n I_{[T<n]},$$

where $\mathscr{F}_0$ is any sub-σ-algebra of $\mathscr{F}_1$. For $r \geq 1$

$$\begin{aligned}
\mathrm{E}\sum_1^\infty |X_n^{(1)}|^r &= \sum_1^\infty \mathrm{E}|X_n I_{[T>n]} - \mathrm{E}\{X_n I_{[T>n]}|\mathscr{F}_{n-1}\}|^r \\
&\leq 2^{r-1}\sum_1^\infty \mathrm{E}(|X_n I_{[T>n]}|^r + |E\{X_n I_{[T>n]}|\mathscr{F}_{n-1}\}|^r) \\
&\leq 2^r\,\mathrm{E}\sum_1^\infty |X_n|^r I_{[T>n]} = 2^r\,\mathrm{E}\sum_1^{T-1} |X_n|^r \\
&\leq 2^r\,\mathrm{E}\min(V_r, K)^r,
\end{aligned}$$

while for $r = 2$

$$\mathrm{E}\sum_1^\infty |X_n^{(1)}|^2 \leq \mathrm{E}\sum_1^\infty X_n^2 I_{[T>n]} \leq \mathrm{E}\sum_1^{T-1} X_n^2 \leq \mathrm{E}\min(V_2, K)^2,$$

so that (8) obtains. Furthermore,

$$\mathrm{E}\sum_1^\infty |X_n^{(2)}| \leq 2\,\mathrm{E}\sum_1^\infty |X_n I_{[T=n]}| \leq 2\,\mathrm{E}\,|X_T| I_{[T<\infty]} \leq 2\,\mathrm{E}\,X^*,$$

$$\mathrm{E}\left(\sum_1^\infty |X_n^{(3)}|^r\right)^{1/r} \leq \mathrm{E}\left(\sum_1^\infty |X_n|^r\right)^{1/r} = \mathrm{E}\,V_r,$$

and, since $\{X^{(3)*} > 0\} \subset \{V_r > K\}$,

$$\mathrm{P}\{X^{(3)*} > 0\} \leq \mathrm{P}\{V_r > K\} \leq K^{-1}\,\mathrm{E}\,V_r.$$

Clearly, $\{X_n^{(j)}, \mathscr{F}_n, n \geq 1\}$, $j = 1, 2, 3$, are martingale difference sequences satisfying (7). □

Theorem 3. *Let $\{S_n, \mathscr{F}_n, n \geq 1\}$ be a martingale and $\{Y_n, \mathscr{F}_{n-1}, n \geq 1\}$ a stochastic sequence with $Y^* < \infty$, a.c., where $\mathscr{F}_0$ is any sub-σ-algebra of $\mathscr{F}$.*

(i) *If $\mathrm{E}\,X^* < \infty$, where $X_n = S_n - S_{n-1}$, then S_n converges a.c. on $\{\sum_1^\infty X_n^2 < \infty\}$. In particular, if $\mathrm{E}(\sum_1^\infty X_n^2)^{1/2} < \infty$, S_n converges a.c.*

(ii) *If $\sup_{n\geq 1} \mathrm{E}|S_n| < \infty$, then $\sum_1^\infty X_n Y_n$ converges a.c.*

PROOF. Under the hypothesis of (i), taking $r = 2$ in the prior lemma, for every $K > 0$ there exists a decomposition as in (7) with $S_n^{(1)} = \sum_1^n X_k^{(1)}$, $n \geq 1$, an $\mathscr{L}_2$ bounded martingale, $\sum_1^\infty |X_n^{(2)}| < \infty$, a.c., and

$$\{X^{(3)*} > 0\} \subset \left\{\sum_1^\infty X_i^2 > K^2\right\}.$$

Thus, in view of the arbitrariness of K, the martingale S_n converges a.c. on $\{\sum_1^\infty X_n^2 < \infty\}$.

If, as in (ii), $\{S_n, \mathscr{F}_n, n \geq 1\}$ is an $\mathscr{L}_1$ bounded martingale, Theorem 2 guarantees that $\sum_1^\infty X_n^2 < \infty$, a.c. Then, since $Y^* < \infty$ a.c., clearly

$$\sum_1^\infty X_n^2 Y_n^2 < \infty, \qquad \text{a.c.}$$

Define $T = \inf\{n \geq 1 : |S_n| \geq K \text{ or } |Y_{n+1}| \geq K\}$, $X_n' = X_n Y_n I_{[T \geq n]}$, and $S_n' = \sum_{j=1}^n X_j'$, $n \geq 1$. Now $\{S_n', \mathscr{F}_n, n \geq 1\}$ is a martingale and $\sum_1^\infty (X_n')^2 < \infty$, a.c. Moreover,

$$|X_n'| \leq K(|S_n| + |S_{n-1}|)I_{[T \geq n]} \leq K[2K + |S_n| I_{[T=n]}],$$

$$\sup_n |X_n'| \leq K(2K + |S_T| I_{[T<\infty]}),$$

implying $\mathrm{E}\,X'^* < \infty$. By (i), S_n' converges a.c., whence $\sum_1^\infty X_n Y_n$ converges a.c. on $\{T = \infty\}$. Since S_n converges a.c. and $Y^* < \infty$ a.c., $\mathrm{P}\{T = \infty\} \to 1$ as $K \to \infty$, whence $\sum_1^\infty X_n Y_n$ converges a.c. □

Corollary 6. *Let $\{S_n = \sum_1^n X_j, \mathscr{F}_n, n \geq 1\}$ be an $\mathscr{L}_1$ martingale such that*

$$\sup\{\mathrm{E}|X_T I_{[T<\infty]}| : \textit{stopping times } T\} < \infty.$$

Then S_n converges a.c. on $[\sum_1^\infty X_n^2 < \infty]$.

EXERCISES 11.1

1. If U_n is the number of upcrossings of $[r_1, r_2]$ by the submartingale $\{S_j, \mathscr{F}_j = \sigma(S_i, 1 \leq i \leq j), 1 \leq j \leq n\}$ prove that

$$\mathrm{P}\{U_n \geq K\} \leq (r_2 - r_1)^{-1}\, \mathrm{E}(S_n - r_1)^+ I_{[U_n = k]}, \; k \geq 1.$$

2. If $\{X_n, n \geq 1\}$ are $\mathscr{L}_1$ interchangeable r.v.s, prove that $1/n \sum_1^n X_i \xrightarrow{\text{a.c.}}$ some r.v. Y (also in $\mathscr{L}_1$). Verify that $Y = \mathrm{E}\{X_1 | \bigcap_1^\infty \mathscr{G}_n\}$, where $\mathscr{G}_n = \sigma(\sum_1^j X_i, j \geq n)$. *Hint*: $n^{-1} \sum_1^n X_i = \mathrm{E}\{n^{-1} \sum_1^n X_i | \mathscr{G}_n\} = \mathrm{E}\{X_1 | \mathscr{G}_n\}$.

3. Let Y be a r.v. with $\mathrm{E}\, Y = 0$ and $\mathrm{E}|Y|\log(1 + |Y|) < \infty$. If $\{\mathscr{G}_n, n \geq 1\}$ is a sequence of independent σ-algebras of events, then $\mathrm{E}\{Y | \mathscr{G}_n\} \xrightarrow{\text{a.c.}} 0$ as $n \to \infty$. *Hint*: Apply Corollary 4 and Theorem 7.4.8.

11.2 Martingale Extension of Marcinkiewicz–Zygmund Inequalities

Let $(\Omega, \mathscr{F}, \mathrm{P})$ be a probability space and let $\{\emptyset, \Omega\} = \mathscr{F}_0 \subset \mathscr{F}_1 \subset \cdots \subset \mathscr{F}_\infty \subset \mathscr{F}$ be a sequence of σ-algebras with $\mathscr{F}_\infty = \sigma(\bigcup_1^\infty \mathscr{F}_n)$. For any stochastic sequence $\{f_n, \mathscr{F}_n, n \geq 1\}$ define $f_0 = 0$ and

$$f = \{f_n, n \geq 1\}, \qquad f^* = \sup_{n \geq 1} |f_n|, \qquad \|f\|_p = \sup_{n \geq 1} \|f_n\|_p, \qquad f_\infty = \overline{\lim_{n \to \infty}}\, f_n,$$

$$f_n^* = \max_{1 \leq j \leq n} |f_j|, \qquad d_n = f_n - f_{n-1}, \qquad n \geq 1, \qquad d = \{d_n, n \geq 1\},$$

$$S(f) = S_\infty(f) = \left(\sum_1^\infty d_j^2\right)^{1/2}, \qquad S_n(f) = \left(\sum_1^n d_j^2\right)^{1/2}.$$

Recall that f is $\mathscr{L}_p$ bounded if $\|f\|_p < \infty$. Moreover, f will be called a submartingale or martingale whenever $\{f_n, \mathscr{F}_n, n \geq 1\}$ is such. For any real numbers a, b let $a \wedge b = \min(a, b)$ and $a \vee b = \max(a, b)$.

Lemma 1. *If f is an $\mathscr{L}_1$ bounded martingale or nonnegative $\mathscr{L}_1$ bounded submartingale and $T = \inf\{n \geq 1 : |f_n| > \lambda\}$, $\lambda > 0$, then*

$$\mathrm{E}\, S_{T-1}^2(f) + \mathrm{E}\, f_{T-1}^2 \leq 2\, \mathrm{E} f_T f_{T-1} \leq 2\lambda \|f\|. \tag{1}$$

Proof. Since $|f_{T-1}| \leq \lambda$, and $f_\infty = \overline{\lim}\, f_n = \underline{\lim}\, f_n$, a.c., by Fatou

$$\mathrm{E}|f_T f_{T-1}| \leq \lambda\, \mathrm{E}|f_T| \leq \lambda \lim_{n \to \infty} \mathrm{E}|f_n| = \lambda \|f\|,$$

and so it remains to prove the first inequality in (1). Setting $T_n = T \wedge n$ for $n \geq 1$,

$$S_{n-1}^2(f) + f_{n-1}^2 = 2 \sum_{1 \leq j \leq k \leq n-1} d_j d_k = 2 \sum_{j=1}^{n-1} d_j (f_{n-1} - f_{j-1})$$

$$= 2 f_{n-1}^2 - 2 \sum_1^{n-1} f_{j-1} d_j = 2\left[f_n f_{n-1} - \sum_1^n f_{j-1} d_j\right],$$

whence (employing Corollary 7.4.5 in the martingale case)

$$\mathrm{E}[S_{T_n-1}^2(f) + f_{T_n-1}^2] \leq 2\, \mathrm{E}\, f_{T_n} f_{T_n-1}.$$

Since $|f_{T_n} f_{T_n-1}| \le \lambda(\lambda + |f_T|)$ and $\mathrm{E}|f_T| < \infty$ by Exercise 7.4.4, the first inequality of (1) follows as $n \to \infty$. □

Lemma 2. *If f is a martingale or nonnegative submartingale, for every $\lambda > 0$*

$$\lambda \, \mathrm{P}\{S(f) > \lambda, f^* \le \lambda\} < 2\|f\|, \tag{2}$$

$$\lambda \, \mathrm{P}\{S(f) > \lambda\} \le 3\|f\|. \tag{3}$$

Proof. Theorem 7.4.8 (34) implies $\lambda \, \mathrm{P}\{f^* > \lambda\} \le \|f\|$, whence (3) is an immediate consequence of (2). To prove the latter, suppose without loss of generality that f is $\mathscr{L}_1$ bounded and define $T = \inf\{n \ge 1 : |f_n| > \lambda\}$. Now $S_{T-1}(f) = S(f)$ on the set $\{T = \infty\} = \{f^* \le \lambda\}$ and, utilizing Lemma 1,

$$\begin{aligned} \lambda \, \mathrm{P}\{S(f) > \lambda, f^* \le \lambda\} &\le \lambda \, \mathrm{P}\{S_{T-1}(f) > \lambda\} \\ &\le \lambda^{-1} \, \mathrm{E}\, S_{T-1}^2(f) \le 2\|f\|. \end{aligned}$$ □

Lemma 3. *Let f be a nonnegative submartingale, $0 < \theta < \infty$, $Y_n = S_n(\theta f) \vee f_n^*$, $n \ge 1$. Then, for $\lambda > 0$, $\beta = (1 + 2\theta^2)^{1/2}$, and $p \in (1, \infty)$*

$$\lambda \, \mathrm{P}\{Y_n > \beta\lambda\} \le 3 \int_{[Y_n > \lambda]} f_n, \tag{4}$$

$$\|S_n(f)\|_p \le \frac{9p^{3/2}}{p-1} \|f_n\|_p, \tag{5}$$

$$\|S(f)\|_p \le \frac{9p^{3/2}}{p-1} \|f\|_p. \tag{6}$$

Proof. Define $I_j = I_{[S_j(\theta f) > \lambda]}$ and $g_j = I_j f_j$, $j \ge 1$. Since $I_{j+1} \ge I_j$, necessarily $g = \{g_n, n \ge 1\}$ is a nonnegative submartingale. Let $T = \inf\{n \ge 1: S_n(\theta f) > \lambda\}$. On the set $\{S_n(\theta f) > \beta\lambda, f_n^* \le \lambda\}$, note that $T \le n$, $g_n^* \le \lambda$, and $|d_T| = |f_T - f_{T-1}| \le f_T \vee f_{T-1} \le f_n^* \le \lambda$, so that, recalling the definition of β,

$$\begin{aligned} (1 + 2\theta^2)\lambda^2 < S_n^2(\theta f) &= S_{T-1}^2(\theta f) + \theta^2 d_T^2 + \theta^2 \sum_{T<j\le n} d_j^2 \\ &\le \lambda^2 + \theta^2\lambda^2 + \theta^2 \sum_{T<j\le n} (g_j - g_{j-1})^2 \\ &\le (1 + \theta^2)\lambda^2 + \theta^2 S_n^2(g), \end{aligned}$$

implying that $S_n(g) > \lambda$ on the set in question. Hence, applying Lemma 2 to $\{g_1, \ldots, g_n, g_n, \ldots\}$,

$$\lambda \, \mathrm{P}\{S_n(\theta f) > \beta\lambda, f_n^* \le \lambda\} \le \lambda \, \mathrm{P}\{S_n(g) > \lambda, g_n^* \le \lambda\} \le 2\|g_n\| = 2\|I_n f_n\|.$$

On the other hand, by Theorem 7.4.8 (34)

$$\lambda \, \mathrm{P}\{f_n^* > \lambda\} \le \int_{[f_n^* > \lambda]} f_n \le \int_{[Y_n > \lambda]} f_n,$$

and so, combining these estimates,

$$\lambda \, P\{Y_n > \beta\lambda\} \leq \lambda \, P\{f_n^* > \lambda\} + \lambda \, P\{S_n(\theta f) > \beta\lambda, f_n^* \leq \lambda\}$$

$$\leq 3 \int_{[Y_n > \lambda]} f_n$$

and (4) is proved. To obtain (5), note that via (4)

$$\beta^{-p} \, E \, Y_n^p = p \int_0^\infty \lambda^{p-1} P\{Y_n > \beta\lambda\} d\lambda \leq 3p \int_0^\infty \lambda^{p-2} \int_{[Y_n > \lambda]} f_n \, dP \, d\lambda$$

$$= 3p \, E\left(f_n \int_0^{Y_n} \lambda^{p-2} \, d\lambda \right) = \frac{3p}{p-1} E f_n Y_n^{p-1}$$

$$\leq \frac{3p}{p-1} \|f_n\|_p \|Y_n\|_p^{p-1},$$

implying

$$\theta \|S_n(f)\|_p \leq \|Y_n\|_p \leq \frac{3p\beta^p}{p-1} \|f_n\|_p.$$

Choose $\theta = p^{-1/2}$, whence $\beta^p = (1 + (2/p))^{p/2} < 3$ and (5) follows. Finally, let $n \to \infty$ in (5) to obtain (6). □

Theorem 1 (Burkholder). *If* $f = \{f_n, n \geq 1\}$ *is an* $\mathscr{L}_1$ *martingale and* $p \in (1, \infty)$, *there exist constants* $A_p = [18p^{3/2}/(p-1)]^{-1}$ *and* $B_p = 18p^{3/2}/(p-1)^{1/2}$ *such that*

$$A_p \|S_n(f)\|_p \leq \|f_n\|_p \leq B_p \|S_n(f)\|_p, \tag{7}$$

$$A_p \|S(f)\|_p \leq \|f\|_p \leq B_p \|S(f)\|_p. \tag{8}$$

Proof. It suffices to verify (7) since (8) then follows as $n \to \infty$. To this end, set $g_j = E\{f_n^+ | \mathscr{F}_j\}$, $h_j = E\{f_n^- | \mathscr{F}_j\}$. Then $g_n = f_n^+$, $h_n = f_n^-$ and $f_j = E\{f_n | \mathscr{F}_j\} = g_j - h_j$ for $1 \leq j \leq n$. Since $S_n(f) \leq S_n(g) + S_n(h)$, by Minkowski's inequality and Lemma 3

$$\|S_n(f)\|_p \leq \|S_n(g)\|_p + \|S_n(h)\|_p$$

$$\leq \frac{9p^{3/2}}{p-1} (\|g_n\|_p + \|h_n\|_p) \leq \frac{18p^{3/2}}{p-1} \|f_n\|_p,$$

yielding the first inequality of (7). Apropos of the second, suppose without loss of generality that $\|f_n\|_p > 0$ and $\|S_n(f)\|_p < \infty$. Then $f_j \in \mathscr{L}_p$, $1 \leq j \leq n$, whence, if

$$g_n = \frac{(\text{sgn } f_n)|f_n|^{p-1}}{\|f_n\|_p^{p-1}}, \qquad g_j = E\{g_n | \mathscr{F}_j\}, \qquad 1 \leq j \leq n,$$

and $(1/p) + (1/q) = 1$, it follows that $\{g_j, 1 \le j \le n\}$ is an $\mathscr{L}_q$ martingale with $\|g_n\|_q = 1$ and $\mathrm{E}\, f_n g_n = \|f_n\|_p$. Consequently, if $e_1 = g_1$, $e_j = g_j - g_{j-1}$ for $2 \le j \le n$, then via the Schwarz and Hölder inequalities

$$\|f_n\|_p = \mathrm{E} f_n g_n = \mathrm{E}(f_{n-1} + d_n)(g_{n-1} + e_n) = \mathrm{E}(f_{n-1} g_{n-1} + d_n e_n)$$

$$= \mathrm{E} \sum_1^n d_j e_j \le \mathrm{E}\, S_n(f) S_n(g) \le \|S_n(f)\|_p \|S_n(g)\|_q$$

$$\le \|S_n(f)\|_p \cdot A_q^{-1} \|g_n\|_q = \frac{18p^{3/2}}{(p-1)^{1/2}} \|S_n(f)\|_p,$$

utilizing the portion of (7) already proved. □

Theorem 1 in conjunction with Theorem 7.4.8 yields

Corollary 1. *If* $\{f_n, n \ge 1\}$ *is an* $\mathscr{L}_1$ *martingale and* $p \in (1, \infty)$, *there exist constants* $A_p = [18p^{3/2}/(p-1)]^{-1}$ *and* $B'_p = 18p^{5/2}/(p-1)^{3/2}$ *such that*

$$A_p \|S_n(f)\|_p \le \|f_n^*\|_p \le B'_p \|S_n(f)\|_p, \tag{9}$$

$$A_p \|S(f)\|_p \le \|f^*\|_p \le B'_p \|S(f)\|_p. \tag{10}$$

The usefulness of Theorem 1 will be demonstrated in obtaining a martingale strong law of large numbers and an extension of Wald's equation that does not require integrability of the stopping rule.

Corollary 2. *If* $f = \{f_n, \mathscr{F}_n, n \ge 1\}$ *is an* $\mathscr{L}_{2r}$ *martingale such that for some* $r \ge 1$

$$\sum_{n=1}^{\infty} \frac{\mathrm{E}|f_n - f_{n-1}|^{2r}}{n^{r+1}} < \infty, \tag{11}$$

then $f_n/n \xrightarrow[\mathscr{L}_{2r}]{\text{a.c.}} 0$.

PROOF. The argument of Theorem 10.1.3 carries over verbatim. □

Corollary 3. *If* $f = \{f_n, \mathscr{F}_n, n \ge 1\}$ *is an* $\mathscr{L}_r$ *martingale such that for some* r *in* $(1, 2]$ *and* $B \in (0, \infty)$

$$\sup_{n \ge 1} n^{-1} \sum_{j=1}^n \mathrm{E}\{|f_j - f_{j-1}|^r | \mathscr{F}_{j-1}\} \le B, \qquad \text{a.c.}, \tag{12}$$

then $\mathrm{E}\, f_T = \mathrm{E}\, f_1$ *for every stopping time* T *with* $\mathrm{E}\, T^{1/r} < \infty$.

PROOF. If $T_n = T \wedge n$, then $\mathrm{E}\, f_{T_n} = \mathrm{E}\, f_1$ for $n \ge 1$ by Corollary 7.4.4. Hence, it suffices via dominated convergence to prove that $Z = \sup_{n \ge 1} |f_{T_n}|$ is integrable. To this end, set $m = [v^r]$ for $v > 0$, whence, employing Theorem 7.4.8,

$$\begin{aligned}P\{Z \geq v\} &\leq P\{T \geq v^r\} + P\{T \leq m, Z \geq v\} \\ &\leq P\{T \geq v^r\} + P\Big\{\max_{1 \leq j \leq m} |f_{T_j}| \geq v\Big\} \\ &\leq P\{T \geq v^r\} + v^{-r} E|f_{T_m}|^r. \end{aligned} \tag{13}$$

By Theorem 1 there exists a constant $C \in (0, \infty)$ such that

$$\begin{aligned} E|f_{T_m}|^r &\leq C\, E\left(\sum_1^{T_m} d_j^2\right)^{r/2} \leq C\, E \sum_1^{T_m} |d_j|^r \\ &\leq C\, E \sum_1^{T_m} E\{|d_j|^r | \mathscr{F}_{j-1}\} \leq CB\, E\, T_m \\ &= CB\left(m\, P\{T \geq v^r\} + \int_{[T \leq v^r]} T\right) \end{aligned} \tag{14}$$

in view of (29) of Corollary 7.4.7 (which holds in general) and (12). Consequently, from (13) and (14)

$$P\{Z \geq v\} \leq (1 + CB)P\{T \geq v^r\} + CBv^{-r} \int_{[T \leq v^r]} T,$$

whence

$$\begin{aligned} E\, Z = \int_0^\infty P\{Z \geq v\}dv &\leq (1 + CB)E\, T^{1/r} + CB \int_0^\infty v^{-r} \int_{[T \leq v^r]} T\, dP\, dv \\ &= (1 + CB)E\, T^{1/r} + CB \int_\Omega T\, dP \int_{T^{1/r}}^\infty v^{-r}\, dv \\ &= \left(1 + CB + \frac{CB}{r-1}\right)E\, T^{1/r} < \infty. \end{aligned}$$

□

Thus, if $\{X_n, n \geq 1\}$ are independent r.v.s with common mean zero and common variance $\sigma^2 < \infty$, $E \sum_1^T X_i = 0$ whenever $E\, T^{1/2} < \infty$.

In contrast to the Marcinkiewicz–Zygmund theorem (Theorem 10.3.2), Theorem 1 does not hold for $p = 1$. For example, if $\{X_n, n \geq 1\}$ are i.i.d. r.v.s with $P\{X_n = \pm 1\} = \frac{1}{2}$, $W_n = \sum_1^n X_j$ and $T = \inf\{n \geq 1: W_n = 1\}$, then $\{f_n \equiv W_{T \wedge n}, n \geq 1\}$ is an $\mathscr{L}_1$ martingale with

$$\|f_n\| = E|f_n| = E f_n^+ + E f_n^- = 2\, E f_n^+ \to 2$$

as $n \to \infty$, noting that $E f_n = 0$. However,

$$\|S_n(f)\| = E\, S_n(f) = E\left(\sum_{j=1}^{T \wedge n} 1\right)^{1/2} = E(T \wedge n)^{1/2} \to \infty,$$

so that the first inequality of (7) fails for $p = 1$. However, the second inequality does hold. More precisely, Corollary 1 obtains when $p = 1$, as will be shown in the next theorem.

Lemma 4. *Let $f = \{f_n \equiv \sum_1^n d_j, n \geq 1\}$ be an $\mathscr{L}_1$ martingale with $|d_j| \leq V_j$, where $\{V_n, \mathscr{F}_{n-1}, n \geq 1\}$ is a stochastic sequence. If $\lambda > 0, \beta > 1$ and $0 < \delta < \beta - 1$ then*

$$P\{f^* > \beta\lambda, S(f) \vee V^* \leq \delta\lambda\} \leq \frac{2\delta^2}{(\beta - \delta - 1)^2} P\{f^* > \lambda\}, \tag{15}$$

$$P\{S(f) > \beta\lambda, f^* \vee V^* \leq \delta\lambda\} \leq \frac{9\delta^2}{\beta^2 - \delta^2 - 1} P\{S(f) > \lambda\}. \tag{16}$$

PROOF. Set $S_0(f) = 0$ and define

$$\mu = \inf\{n \geq 1: |f_n| > \lambda\}, \qquad \nu = \inf\{n \geq 1: |f_n| > \beta\lambda\},$$
$$\sigma = \inf\{n \geq 0: S_n(f) \vee V_{n+1} > \delta\lambda\}.$$

If

$$h_n = \sum_{j=1}^n d_j I_{[\mu < j \leq \nu \wedge \sigma]}$$

then $h = \{h_n, n \geq 1\}$ is a martingale with $S(h) = 0$ on $\{\mu = \infty\} = \{f^* \leq \lambda\}$. Moreover, recalling that $|d_j| \leq V_j$ by hypothesis,

$$\begin{aligned} S^2(h) &\leq S_\sigma^2(f) = [S_{\sigma-1}^2(f) + d_\sigma^2] I_{[\sigma < \infty]} + S_\sigma^2(f) I_{[\sigma = \infty]} \\ &\leq 2\delta^2\lambda^2. \end{aligned}$$

Hence,

$$\|h\|_2^2 = E\, S^2(h) I_{[f^* > \lambda]} \leq 2\delta^2\lambda^2\, P\{f^* > \lambda\},$$

implying via the Kolmogorov inequality (Theorem 7.4.8) that

$$\begin{aligned} &P\{f^* > \beta\lambda, S(f) \vee V^* \leq \delta\lambda\} \\ &= P\{\sigma = \infty, \nu < \infty, h_n = f_{\nu \wedge n} - f_{\mu \wedge n}, n \geq 1\} \\ &\leq P\{h^* > \beta\lambda - (1 + \delta)\lambda\} \leq [(\beta - 1 - \delta)\lambda]^{-2} \|h\|_2^2 \\ &\leq \frac{2\delta^2}{(\beta - \delta - 1)^2} P\{f^* > \lambda\}. \end{aligned}$$

To obtain (16), define analogously

$$\mu' = \inf\{n \geq 1: S_n(f) > \lambda\}, \qquad \nu' = \inf\{n \geq 1: S_n(f) > \beta\lambda\}$$
$$\sigma' = \inf\{n \geq 0: f_n^* \vee V_{n+1} > \delta\lambda\}.$$

If

$$g_n = \sum_1^n d_j I_{[\mu' < j \leq \nu' \wedge \sigma']},$$

then $g = \{g_n, n \geq 1\}$ is a martingale with $g^* = 0$ on $\{\mu' \geq \sigma'\} \supset \{S(f) \leq \lambda\}$, whence

$$g^* \leq (f^*_{\sigma'} + f^*_{\mu'})I_{[\mu' < \sigma' < \infty]} + 2f^* I_{[\mu' < \sigma' = \infty]} \\ \leq 3\delta\lambda,$$

implying $E(g^*)^2 \leq 9\delta^2\lambda^2 P\{S(f) > \lambda\}$. Thus,

$$\begin{aligned} &P\{S(f) > \beta\lambda, f^* \vee V^* \leq \delta\lambda\} \\ &= P\{v' < \infty = \sigma', S_n^2(g) = S^2_{v' \wedge n}(f) - S^2_{\mu' \wedge n}(f), n \geq 1\} \\ &\leq P\{S^2(g) > [\beta^2 - (1 + \delta^2)]\lambda^2\} \leq [(\beta^2 - 1 - \delta^2)\lambda^2]^{-1} E\, S^2(g) \\ &\leq [(\beta^2 - 1 - \delta^2)\lambda^2]^{-1} E\,(g^*)^2 \leq \frac{9\delta^2}{\beta^2 - \delta^2 - 1} P\{S(f) > \lambda\}. \end{aligned}$$ □

Lemma 5. *If* $f = \{f_n = \sum_1^n d_i, n \geq 1\}$ *is an* $\mathscr{L}_1$ *martingale and*

$$g_n = \sum_1^n a_j, \qquad a_j = d_j I_{[|d_j| \leq 2d^*_{j-1}]} - E\{d_j I_{[|d_j| \leq 2d^*_{j-1}]} | \mathscr{F}_{j-1}\}, \tag{17}$$

$$h_n = \sum_1^n b_j, \qquad b_j = d_j I_{[|d_j| > 2d^*_{j-1}]} - E\{d_j I_{[|d_j| > 2d^*_{j-1}]} | \mathscr{F}_{j-1}\}, \tag{18}$$

then $g = \{g_n, n \geq 1\}$ *and* $h = \{h_n, n \geq 1\}$ *are* $\mathscr{L}_1$ *martingales with* $f_n = g_n + h_n, n \geq 1$ *and*

$$|a_n| \leq 4d^*_{n-1}, \tag{19}$$

$$\sum_{n=1}^{\infty} |d_n I_{[|d_n| > 2d^*_{n-1}]}| \leq 2d^*, \tag{20}$$

$$\sum_{n=1}^{\infty} E|b_n| \leq 4\, E\, d^*. \tag{21}$$

Proof. The validity of (19) is clear. On the set $\{|d_j| > 2d^*_{j-1}\}, |d_j| + 2d^*_{j-1} \leq 2|d_j| \leq 2d^*_j$, implying

$$\sum_1^{\infty} |d_j I_{[|d_j| > 2d^*_{j-1}]}| \leq 2 \sum_1^{\infty} (d^*_j - d^*_{j-1}) = 2d^*,$$

which is (20). This, in turn, yields (21) via

$$\sum_1^{\infty} E|b_j| \leq E \sum_1^{\infty} |d_j I_{[|d_j| > 2d^*_{j-1}]}| + \sum_1^{\infty} E|E\{d_j I_{[|d_j| > 2d^*_{j-1}]} | \mathscr{F}_{j-1}\}| \\ \leq 2\, E\, d^* + 2\, E\, d^* = 4\, E\, d^*.$$ □

Theorem 2 (Davis). *There exist constants* $0 < A < B < \infty$ *such that, for any* $\mathscr{L}_1$ *martingale* $f = \{f_n, n \geq 1\}$,

$$A\, E\, S(f) \leq E f^* \leq B\, E\, S(f). \tag{22}$$

PROOF. Writing $f_n = g_n + h_n$ as in Lemma 5, it follows therefrom that

$$\mathrm{E}\, f^* \le \mathrm{E}(g^* + h^*) \le \mathrm{E}\, g^* + \sum_1^\infty \mathrm{E}|b_j| \le \mathrm{E}\, g^* + 4\,\mathrm{E}\, d^*, \tag{23}$$

$$\mathrm{E}\, S(f) \le \mathrm{E}[S(g) + S(h)] \le \mathrm{E}\, S(g) + \sum_1^\infty \mathrm{E}|b_j| \le \mathrm{E}\, S(g) + 4\,\mathrm{E}\, d^*. \tag{24}$$

Since $g_n = \sum_1^n a_j$ is a martingale with $|a_n| \le 4d_{n-1}^*$, Lemma 4 ensures that for $\lambda > 0$, $\beta > \delta$, and $0 < \delta < 1$

$$P\{g^* > \beta\lambda, S(g) \vee 4d^* \le \delta\lambda\} \le \frac{2\delta^2}{(\beta - \delta - 1)^2} P\{g^* > \lambda\},$$

$$P\{S(g) > \beta\lambda, g^* \vee 4d^* \le \delta\lambda\} \le \frac{9\delta^2}{\beta^2 - \delta^2 - 1} P\{S(g) > \lambda\}.$$

Hence,

$$P\{g^* > \beta\lambda\} \le P\{S(g) > \delta\lambda\} + P\{4d^* > \delta\lambda\} + \frac{2\delta^2}{(\beta - \delta - 1)^2} P\{g^* > \lambda\},$$

$$P\{S(g) > \beta\lambda\} \le P\{g^* > \delta\lambda\} + P\{4d^* > \delta\lambda\} + \frac{9\delta^2}{\beta^2 - \delta^2 - 1} P\{S(g) > \lambda\},$$

implying

$$\beta^{-1}\,\mathrm{E}\, g^* \le \delta^{-1}\,\mathrm{E}\, S(g) + 4\delta^{-1}\,\mathrm{E}\, d^* + \frac{2\delta^2}{(\beta - \delta - 1)^2}\,\mathrm{E}\, g^*, \tag{25}$$

$$\beta^{-1}\,\mathrm{E}\, S(g) \le \delta^{-1}\,\mathrm{E}\, g^* + 4\delta^{-1}\,\mathrm{E}\, d^* + \frac{9\delta^2}{\beta^2 - \delta^2 - 1}\,\mathrm{E}\, S(g). \tag{26}$$

Now, recalling Lemma 5,

$$\mathrm{E}\, g^* \le \mathrm{E}(f^* + h^*) \le \mathrm{E}\, f^* + \sum_1^\infty \mathrm{E}|b_j| \le \mathrm{E}\, f^* + 4\,\mathrm{E}\, d^*, \tag{27}$$

$$\mathrm{E}\, S(g) \le \mathrm{E}[S(f) + S(h)] \le \mathrm{E}\, S(f) + \sum_1^\infty \mathrm{E}|b_j| \le \mathrm{E}\, S(f) + 4\,\mathrm{E}\, d^*. \tag{28}$$

In order to subtract in (25) and (26), rewrite these for the martingale $\{g_1, g_2, \ldots, g_n, g_n, g_n, \ldots\}$, obtaining via (27) and (28)

$$\left[\beta^{-1} - \frac{2\delta^2}{(\beta - \delta - 1)^2}\right]\mathrm{E}\, g_n^* \le \delta^{-1}\,\mathrm{E}\, S_n(f) + 8\delta^{-1}\,\mathrm{E}\, d_n^* \le 9\delta^{-1}\,\mathrm{E}\, S(f), \tag{29}$$

$$\left[\beta^{-1} - \frac{9\delta^2}{\beta^2 - \delta^2 - 1}\right]\mathrm{E}\, S_n(g) < \delta^{-1}\,\mathrm{E}\, f_n^* + 8\delta^{-1}\,\mathrm{E}\, d_n^* \le 17\delta^{-1}\,\mathrm{E}\, f^*. \tag{30}$$

For small δ, the coefficients on the left in (29), (30) are positive, and so, letting $n \to \infty$ and recalling (23), (24), there exist constants B_1, B, A_1, A for which

$$\mathrm{E}\, f^* \le B_1 \,\mathrm{E}\, S(f) + 4\,\mathrm{E}\, d^* \le B\,\mathrm{E}\, S(f),$$
$$\mathrm{E}\, S(f) \le A_1 \,\mathrm{E}\, f^* + 4\,\mathrm{E}\, d^* \le A^{-1}\,\mathrm{E}\, f^*. \qquad \square$$

Corollary 4. *If the stochastic sequence $\{|X_n|^p, \mathscr{F}_n, n \ge 1\}$ is uniformly integrable for some $p \in (0, 2)$ and $S_n = \sum_1^n X_i$, then*

$$\lim_{n\to\infty} \frac{1}{n} \mathrm{E}|S_n - a_n|^p = 0, \tag{31}$$

where $a_n \equiv 0$ for $0 < p < 1$ and $a_n = \sum_{j=1}^n E\{X_j | \mathscr{F}_{j-1}\}$ if $1 \le p < 2$.

Proof. Define $Y_n = X_n$ if $0 < p < 1$ and $Y_n = X_n - \mathrm{E}\{X_n | \mathscr{F}_{n-1}\}$ when $1 \le p < 2$. Then $\{|Y_n|^p, n \ge 1\}$ is u.i., whence for any $\varepsilon > 0$ there exists a constant $M > 1$ with

$$\sup_{n\ge 1} \int_{[|Y_n| > M]} |Y_n|^p < \varepsilon.$$

Set

$$Y_n' = Y_n I_{[|Y_n| \le M]}, \qquad Y_n'' = Y_n - Y_n'.$$

Then, if $0 < p < 1$,

$$\mathrm{E}|S_n|^p = \mathrm{E}\left|\sum_1^n (Y_j' + Y_j'')\right|^p \le \mathrm{E}\left(\sum_1^n |Y_j'|\right)^p + \mathrm{E}\left(\sum_1^n |Y_j''|\right)^p$$
$$\le (nM)^p + \sum_1^n \mathrm{E}|Y_j''|^p \le (nM)^p + n\varepsilon$$

and (31) follows. If, rather, $1 \le p < 2$, then Theorems 1 and 2 guarantee the existence of a constant C with

$$\mathrm{E}|S_n - a_n|^p = \mathrm{E}\left|\sum_1^n Y_j\right|^p \le C\,\mathrm{E}\left(\sum_1^n Y_j^2\right)^{p/2}$$
$$\le C\,\mathrm{E}\left[\left(\sum_1^n Y_j'^2\right)^{p/2} + \left(\sum_1^n Y_j''^2\right)^{p/2}\right]$$
$$\le C\left[(nM^2)^{p/2} + \mathrm{E}\sum_1^n |Y_j''|^p\right] \le C[(nM^2)^{p/2} + n\varepsilon],$$

again yielding (31). $\square$

Exercises 11.2

1. Let $\{X_n, n \ge 1\}$ be independent r.v.s with $\mathrm{E}\, X_n = 0$, $\mathrm{E}\, X_n^2 = 1$ for $n \ge 1$. If $T = \inf\{n \ge 1: \sum_1^n X_j > 0\}$, then $\mathrm{P}\{T < \infty\} = 1$ and $\mathrm{E}\, T^{1/2} = \infty$.
2. Let $\{X_n, n \ge 1\}$ be interchangeable r.v.s with $\mathrm{E}\, X_1 = 0 = \mathrm{E}\, X_1 X_2$, $\mathrm{E}\, X_1^2 = 1$. If $T = \inf\{n \ge 1: \sum_1^n X_i > 0\}$, then $\mathrm{P}\{T < \infty\} = 1$ and $\mathrm{E}\, T^{1/2} = \infty$.

3. If $\{X, X_n, n \geq 1\}$ are martingale differences with $\mathrm{E}\, X = 0$, $\mathrm{E}|X|^p < \infty$ for some $p \geq 2$, and $\{S_n, n \geq 1\}$ are the partial sums, then $\{|S_n/n^{1/2}|^p, n \geq 1\}$ are u.i. *Hint*: Consider $\mathrm{E}|S_n'|^{p+1}$ and $\mathrm{E}|S_n''|^p$ where $X_n' = X_n I_{[|X_n| \leq M]} - \mathrm{E}\{X_n I_{[|X_n| \leq M]} | X_1 \cdots X_{n-1}\}$ and $X_n'' = X_n - X_n'$.

4. If $\{X_n, n \geq 1\}$ are martingale differences and $p \in (1, 2]$ then $\mathrm{E} \sup_{n \geq 1} |\sum_1^n X_j|^p \leq A_p \sum_1^\infty \mathrm{E}|X_n|^p$ for some finite constant A_p.

11.3 Convex Function Inequalities for Martingales

In the prior section it was shown that for $p \geq 1$ and any martingale f there exist constants $0 < A_p < B_p < \infty$ such that

$$A_p \, \mathrm{E}[S(f)]^p \leq \mathrm{E}(f^*)^p \leq B_p \, \mathrm{E}[S(f)]^p, \tag{1}$$

where f^* and $S(f)$ are defined in Section 2. Here, it will be shown for any convex, nondecreasing function Φ on $[0, \infty)$ with $\Phi(0) = 0$ and

$$\Phi(2\lambda) \leq c\Phi(\lambda) \quad \text{for all } \lambda \geq 0 \text{ and some } c > 0$$

that there exist constants $0 < A_c < B_c < \infty$ such that for any martingale f

$$A_c \, \mathrm{E}\, \Phi(S(f)) \leq \mathrm{E}\, \Phi(f^*) \leq B_c \, \mathrm{E}\, \Phi(S(f)). \tag{2}$$

In what follows, let $(\Omega, \mathcal{F}, \mathrm{P})$ be a probability space, $\{\mathcal{F}_n, n \geq 1\}$ an increasing sequence of sub-σ-algebras of $\mathcal{F}$, and $\mathcal{F}_0 = \{\emptyset, \Omega\}$.

Lemma 1. *Let Φ be a nondecreasing function on $[0, \infty]$, continuous on $[0, \infty)$ with $\Phi(0) = 0$, $\Phi(\infty) = \Phi(\infty -)$, and*

$$\Phi(2\lambda) \leq c\Phi(\lambda) \quad \text{for all } \lambda \in [0, \infty) \text{ and some } c > 0. \tag{3}$$

If f and g are nonnegative, measurable functions on $(\Omega, \mathcal{F}, \mathrm{P})$ and δ, ε, β are positive constants with $\beta > 1$ satisfying

$$\varepsilon c^{1 + \log \beta} < 1 \quad \text{where } 2^{\log \beta} = \beta, \tag{4}$$

$$\mathrm{P}\{g > \beta\lambda, f \leq \delta\lambda\} \leq \varepsilon \mathrm{P}\{g > \lambda\} \quad \text{for all } \lambda \in (0, \infty), \tag{5}$$

there exists $A = A_{c,\beta,\delta,\varepsilon} \in (0, \infty)$ such that

$$\mathrm{E}\, \Phi(g) \leq A \, \mathrm{E}\, \Phi(f). \tag{6}$$

PROOF. Suppose without loss of generality that $\mathrm{E}\, \Phi(f) < \infty$. From (3) and (4) there exists $\gamma = \gamma_{c,\beta} \in (0, \infty)$ and $\eta = \eta_{c,\delta} \in (0, \infty)$ with $\gamma\varepsilon < 1$ such that for all $\lambda > 0$

$$\Phi(\beta\lambda) \leq \gamma\Phi(\lambda), \quad \Phi\left(\frac{\lambda}{\delta}\right) \leq \eta\Phi(\lambda). \tag{7}$$

Since $\Phi(t) = \int_0^\infty I_{[0,t)}(\lambda)d\Phi(\lambda)$, for any nonnegative measurable function h on $(\Omega, \mathscr{F}, P)$

$$E\,\Phi(h) = E\int_0^\infty I_{[0,h)}(\lambda)d\Phi(\lambda) = E\int_0^\infty I_{(\lambda,\infty]}(h)d\Phi(\lambda)$$
$$= \int_0^\infty P\{h > \lambda\}d\Phi(\lambda). \tag{8}$$

Now (5) ensures that

$$P\{g > \beta\lambda\} \le P\{f > \delta\lambda\} + \varepsilon\,P\{g > \lambda\},$$

and so via (8) and (7)

$$E\,\Phi(\beta^{-1}g) = \int_0^\infty P\{g > \beta\lambda\}d\Phi(\lambda)$$
$$\le \int_0^\infty P\{f > \delta\lambda\}d\Phi(\lambda) + \varepsilon\int_0^\infty P\{g > \lambda\}d\Phi(\lambda)$$
$$= E\,\Phi(\delta^{-1}f) + \varepsilon\,E\,\Phi(g) \le \eta\,E\,\Phi(f) + \varepsilon\,E\,\Phi(g). \tag{9}$$

Let $g_n = g \wedge n$ for $n \ge 1$. Then g_n satisfies (5), so that from (6) and (9)

$$E\,\Phi(g_n) = E\,\Phi(\beta\beta^{-1}g_n) \le \gamma[\eta\,E\,\Phi(f) + \varepsilon\,E\,\Phi(g_n)],$$

implying

$$(1 - \gamma\varepsilon)E\,\Phi(g_n) \le \gamma\eta\,E\,\Phi(f),$$

and hence (6) and $A = \gamma\eta/(1 - \gamma\varepsilon)$. □

Lemma 2 (Zygmund). *If Φ is a convex function on an interval $[a, b)$, where a is finite, there exists a nondecreasing, integrable φ on $[a, c)$ for every $c \in (a, b)$ such that*

$$\Phi(t) = \Phi(a) + \int_a^t \varphi(u)du, \qquad t \in [a, b). \tag{10}$$

PROOF. Since Φ is convex, $\varphi(t) = \lim_{0<h\to 0}[\Phi(t + h) - \Phi(t)]/h$, $t \in [a, b)$ exists, is a finite, nondecreasing function on $[a, b)$, and coincides with $d\Phi(t)/dt$, a.e. Now Φ is absolutely continuous on $[a, c)$ for every $c \in (a, b)$ and so (10) follows. □

Lemma 3 (Burkholder–Davis–Gundy). *Let Φ be a nondecreasing function on $[0, \infty]$, finite and convex on $[0, \infty)$ with $\Phi(0) = 0$, $\Phi(\infty) = \Phi(\infty -) = \infty$, and $\Phi(2\lambda) \le c\Phi(\lambda)$ for all $\lambda \ge 0$ and some $c \in (0, \infty)$. Then there exists a constant $B = B_c \in (0, \infty)$ such that for every sequence $\{z_n, n \ge 1\}$ of nonnegative measurable functions on $(\Omega, \mathscr{F}, P)$*

$$E\,\Phi\left(\sum_1^\infty E\{z_j \mid \mathscr{F}_{j-1}\}\right) \le B\,E\,\Phi\left(\sum_1^\infty z_j\right). \tag{11}$$

PROOF. Set $W_0 = Z_0 = 0$ and define for $n \geq 1$

$$Z_n = \sum_1^n z_j, \qquad Z = Z_\infty = \sum_1^\infty z_j,$$

$$W_n = \sum_1^n \mathrm{E}\{z_j | \mathscr{F}_{j-1}\}, \qquad W = W_\infty = \sum_1^\infty \mathrm{E}\{z_j | \mathscr{F}_{j-1}\}.$$

In verifying (11) it may be supposed that $\mathrm{E}\, Z > 0$ and $\mathrm{E}\, \Phi(Z) < \infty$. Then convexity and $\Phi(\infty-) = \infty$ entail $\mathrm{E}\, Z < \infty$, so that $\mathrm{E}\, W < \infty$. For $\lambda \geq 0$ define the $\mathscr{F}_n$-time $T = \inf\{n \geq 0: W_{n+1} > \lambda\}$, whence

$$\begin{aligned} \int_{[0 \leq T \leq n]} (W - Z - \lambda) &= \sum_{j=0}^n \int_{[T=j]} (W_j - \lambda + W - W_j - Z) \\ &\leq \sum_{j=0}^n \int_{[T=j]} \mathrm{E}\{W - W_j - Z | \mathscr{F}_j\} \leq 0 \end{aligned}$$

or

$$\int_{[0 \leq T \leq n]} (W - \lambda) \leq \int_{[0 \leq T \leq n]} Z,$$

and so, letting $n \to \infty$,

$$\mathrm{E}(W - \lambda)^+ = \int_{[0 \leq T < \infty]} (W - \lambda) \leq \int_{[0 \leq T < \infty]} Z = \int_{[W > \lambda]} Z. \tag{12}$$

Next, setting $d = c - 1$ and applying Lemma 2,

$$t\varphi(t) \leq \int_t^{2t} \varphi(u)du = \Phi(2t) - \Phi(t) \leq d\Phi(t). \tag{13}$$

Since φ is nonnegative, nondecreasing on $[0, \infty)$, and bounded on every finite interval $[0, M]$, necessarily φ is a function of bounded variation. Hence, via integration by parts (for Riemann–Stieltjes integrals), for $t \in (0, \infty)$

$$\begin{aligned} \int_0^t \varphi(u)du &= t\varphi(t) - \int_0^t u\, d\varphi(u) \\ &= \int_0^t (t - u)d\varphi(u) + t\varphi(0) = \int_0^\infty (t - u)^+\, d\varphi(u) + t\varphi(0), \end{aligned}$$

and so by Theorem 6.2.4 and Lemma 2,

$$\Phi(t) = \int_0^\infty (t - u)^+\, d\varphi(u) + t\varphi(0), \tag{14}$$

where the integral is in the Lebesgue–Stieltjes sense. From (14), recalling (12),

$$\begin{aligned} \mathrm{E}\,\Phi(W_n) &= \mathrm{E}\left[\int_0^\infty (W_n - u)^+ \, d\varphi(u) + W_n \varphi(0)\right] \\ &= \int_0^\infty \mathrm{E}(W_n - u)^+ \, d\varphi(u) + \varphi(0)\mathrm{E}\,W_n \\ &\le \int_0^\infty \int_{[W_n > u]} Z_n \, d\mathrm{P} \, d\varphi(u) + \varphi(0)\mathrm{E}\,Z_n \\ &= \mathrm{E}\,Z_n \int_{[0,\, W_n)} d\varphi(u) + \varphi(0)\mathrm{E}\,Z_n = \mathrm{E}\,Z_n \varphi(W_n), \end{aligned}$$

implying

$$a\,\mathrm{E}\,\Phi\left(\frac{W_n}{a}\right) \le \mathrm{E}\,Z_n \varphi\left(\frac{W_n}{a}\right) \qquad \text{for } a \in (0, \infty). \tag{15}$$

Let ψ on $[0, \infty)$ be an inverse function of φ, that is, $\psi(\varphi(u)) = u = \varphi(\psi(u))$ for $u \in (0, \infty)$, and set

$$\Psi(t) = \int_0^t \psi(u)du.$$

Then, by Young's inequality (Exercise 6.2.15) $uv \le \Phi(u) + \Psi(v)$ for u, $v \in (0, \infty)$ with equality holding for $v = \varphi(u)$. Consequently,

$$\frac{Z_n}{b}\,\varphi\left(\frac{W_n}{a}\right) \le \Phi\left(\frac{Z_n}{b}\right) + \Psi\left(\varphi\left(\frac{W_n}{a}\right)\right),$$

and so from (15)

$$a\,\mathrm{E}\,\Phi\left(\frac{W_n}{a}\right) \le b\left[\mathrm{E}\,\Phi\left(\frac{Z_n}{b}\right) + \mathrm{E}\,\Psi\left(\varphi\left(\frac{W_n}{a}\right)\right)\right]. \tag{16}$$

Next, take $v = \varphi(u)$ in Young's inequality, yielding via (13)

$$\Psi(\varphi(t)) = t\varphi(t) - \Phi(t) \le (d-1)\Phi(t), \qquad d \ge 1,$$

and, combining this with (16),

$$[a - b(d-1)]E\,\Phi\left(\frac{W_n}{a}\right) \le b\,\mathrm{E}\,\Phi\left(\frac{Z_n}{b}\right). \tag{17}$$

From (13), $\varphi(u) \le (d/u)\Phi(u)$, whence for $r \in [1, \infty)$

$$\log \frac{\Phi(ru)}{\Phi(u)} = \int_u^{ru} \frac{\varphi(t)}{\Phi(t)}\,dt \le d \int_u^{ru} \frac{dt}{t} = \log r^d$$

or

$$\Phi(ru) \le r^d \Phi(u), \qquad r \ge 1. \tag{18}$$

Taking $a = d$ and $b = 1$ in (17),

$$\mathrm{E}\,\Phi\left(\frac{W_n}{d}\right) \le \mathrm{E}\,\Phi(Z_n),$$

whence, recalling (18),

$$\mathrm{E}\,\Phi(W_n) \le \mathrm{E}\,\Phi(dZ_n) \le d^d\,\mathrm{E}\,\Phi(Z_n), \tag{19}$$

and (11) follows with $B = d^d$ by monotone convergence. □

Corollary 1 (Garsia). *For every sequence $\{z_n, n \ge 1\}$ of nonnegative measurable functions on $(\Omega, \mathscr{F}, \mathrm{P})$ and any $p > 1$*

$$\mathrm{E}\left[\sum_1^\infty \mathrm{E}\{z_j|\mathscr{F}_{j-1}\}\right]^p \le p^p\,\mathrm{E}\left(\sum_1^\infty z_j\right)^p. \tag{20}$$

PROOF. When $\Phi(t) = t^p$, (13) yields $d = p$, whence (20) follows directly from (19).

Theorem 1 (Burkholder–Davis–Gundy). *Let Φ be a nondecreasing function on $[0, \infty]$, finite and convex on $[0, \infty)$ with $\Phi(0) = 0$, $\Phi(\infty) = \Phi(\infty-)$, and $\Phi(2\lambda) \le c\Phi(\lambda)$ for all $\lambda > 0$ and some c in $(0, \infty)$. Then there exist constants $0 < A_c < B_c < \infty$ such that for any martingale f*

$$A_c\,\mathrm{E}\,\Phi(S(f)) \le \mathrm{E}\,\Phi(f^*) \le B_c\,\mathrm{E}\,\Phi(S(f)), \tag{21}$$

where $S(f)$ and f^* are as in Section 2.

PROOF. By Lemma 11.2.5, $f = g + h$, where the martingales g, h are defined by $g_n = \sum_{j=1}^n a_j$, $h_n = \sum_{j=1}^n b_j$ with

$$a_n = d_n I_{[|d_n| \le 2d_{n-1}^*]} - \mathrm{E}\{d_n I_{[|d_n| \le 2d_{n-1}^*]}|\mathscr{F}_{n-1}\},$$

$$b_n = d_n I_{[|d_n| > 2d_{n-1}^*]} + \mathrm{E}\{d_n I_{[|d_n| \le 2d_{n-1}^*]}|\mathscr{F}_{n-1}\},$$

$$Z \equiv \sum_1^\infty |d_j I_{[|d_j| > 2d_{j-1}^*]}| \le 2d^*, \qquad |a_n| \le 4d_{n-1}^*,$$

$$W \equiv \sum_1^\infty |\mathrm{E}\{d_j I_{[d_j| \le 2d_{j-1}^*]}|\mathscr{F}_{j-1}\}| \le \sum_1^\infty \mathrm{E}\{|d_j|I_{[|d_j| > 2d_{j-1}^*]}|\mathscr{F}_{j-1}\}. \tag{22}$$

Then

$$\begin{aligned} f^* &\le g^* + h^* \le g^* + \sum_1^\infty |b_j| \le g^* + Z + W, \\ S(g) &\le S(f) + S(h) \le S(f) + \sum_1^\infty |b_j| \le S(f) + Z + W, \end{aligned} \tag{23}$$

and

$$\begin{aligned} S(f) &\le S(g) + S(h) \le S(g) + Z + W, \qquad d^* \le S(f) \\ g^* &\le f^* + h^* \le f^* + Z + W, \qquad d^* \le 2f^*. \end{aligned} \tag{24}$$

By Lemmas 11.2.4 and 1 there are finite, positive constants $B_j = B_j(c)$, $j = 1, 2$ such that

$$\begin{aligned} \mathrm{E}\,\Phi(g^*) &\le B_1\,\mathrm{E}\,\Phi(S(g) \vee 4d^*) \le B_1\,\mathrm{E}[\Phi(S(g)) + \Phi(4d^*)], \\ \mathrm{E}\,\Phi(S(g)) &\le B_2\,\mathrm{E}\,\Phi(g^* \vee 4d^*) \le B_2\,\mathrm{E}[\Phi(g^*) + \Phi(4d^*)]. \end{aligned} \tag{25}$$

Moreover, by Lemma 3, for some $B_3 = B_3(c) \in (0, \infty)$

$$\mathrm{E}\,\Phi(W) \le B_3\,\mathrm{E}\,\Phi(Z), \tag{26}$$

whence via (23)

$$\begin{aligned} \mathrm{E}\,\Phi(f^*) &\le \mathrm{E}\,\Phi(g^* + Z + W) \le \mathrm{E}[\Phi(3g^*) + \Phi(3Z) + \Phi(3W)] \\ &\le c^2\,\mathrm{E}[\Phi(g^*) + \Phi(Z) + \Phi(W)] \\ &\le c^2 B_1\,\mathrm{E}[\Phi(S(g)) + \Phi(4d^*)] + c^2\,\mathrm{E}[\Phi(Z) + \Phi(W)] && \text{by (25)} \\ &\le c^2 B_1\,\mathrm{E}\,\Phi(S(f) + Z + W) + c^2 B_1\,\mathrm{E}\,\Phi(4d^*) \\ &\quad + c^2\,\mathrm{E}[\Phi(Z) + \Phi(W)] && \text{by (23)} \\ &\le c^4 B_1\,\mathrm{E}\,\Phi(S(f)) + c^2 B_1\,\mathrm{E}\,\Phi(4d^*) \\ &\quad + (c^4 B_1 + c^2)\mathrm{E}[\Phi(Z) + \Phi(W)] \\ &\le c^4 B_1\,\mathrm{E}\,\Phi(S(f)) + c^2 B_1\,\mathrm{E}\,\Phi(4d^*) \\ &\quad + (c^4 B_1 + c^2)(1 + B_3)\mathrm{E}\,\Phi(Z) && \text{by (26)} \\ &\le c^4 B_1\,\mathrm{E}\,\Phi(S(f)) + c^4 B_1\,\mathrm{E}\,\Phi(d^*) \\ &\quad + (c^5 B_1 + c^3)(1 + B_3)\mathrm{E}\,\Phi(d^*) && \text{by (22)} \\ &\le [2c^4 B_1 + (1 + B_3)(c^5 B_1 + c^3)]\mathrm{E}\,\Phi(S(f)) && \text{by (24),} \end{aligned}$$

yielding the upper inequality in (21). Similarly, for some finite, positive constants $A_j = A_j(c)$, $1 \le j \le 7$,

$$\begin{aligned} \mathrm{E}\,\Phi(S(f)) &\le \mathrm{E}\,\Phi(S(g) + Z + W) && \text{by (24)} \\ &\le A_2\,\mathrm{E}[\Phi(g^*) + \Phi(4d^*) + \Phi(Z) + \Phi(W)] && \text{by (25)} \\ &\le A_3\,\mathrm{E}[\Phi(f^* + Z + W) + \Phi(4d^*) + \Phi(Z) + \Phi(W)] && \text{by (24)} \\ &\le A_4\,\mathrm{E}[\Phi(f^*) + \Phi(d^*) + \Phi(Z) + \Phi(W)] \\ &\le A_5\,\mathrm{E}[\Phi(f^*) + \Phi(d^*) + \Phi(Z)] && \text{by (26)} \\ &\le A_6\,\mathrm{E}[\Phi(f^*) + \Phi(d^*)] && \text{by (22)} \\ &\le A_7\,\mathrm{E}\,\Phi(f^*), \end{aligned}$$

completing the proof of (21).

As will be demonstrated shortly, it is useful to have a version of Theorem 1 with $S(f) = (\sum_1^\infty d_j^2)^{1/2}$ replaced by $s(f) = (\sum_{j=1}^\infty \mathrm{E}\{d_j^2 \mid \mathscr{F}_{j-1}\})^{1/2}$, and for this an analogue of Lemma 11.2.4 is needed.

Lemma 4. *If f is a martingale, $1 \le \alpha \le 2$, and*

$$s(f) = s_\alpha(f) = \left(\sum_1^\infty \mathrm{E}\{|d_j|^\alpha \mid \mathscr{F}_{j-1}\}\right)^{1/\alpha}, \tag{27}$$

then for any $\lambda > 0$, $\beta > 1$, and $\delta \in (0, \beta - 1)$ there exists a finite, positive constant B_α such that

$$P\{f^* > \beta\lambda, s(f) \vee d^* \le \delta\lambda\} \le \frac{B_\alpha \delta^\alpha}{(\beta - 1 - \delta)^\alpha} P\{f^* > \lambda\}. \tag{28}$$

PROOF. Define $s_n(f) = (\sum_{j=1}^n E\{|d_j|^\alpha | \mathscr{F}_{j-1}\})^{1/\alpha}$ and

$$\mu = \inf\{n \ge 1: |f_n| > \lambda\},$$
$$\nu = \inf\{n \ge 1: |f_n| > \beta\lambda\},$$
$$\sigma = \inf\{n \ge 0: |d_n| \vee s_{n+1}(f) > \delta\lambda\}$$

and

$$h_n = \sum_{j=1}^n d_j I_{[\mu < j \le \nu \wedge \sigma]} \equiv \sum_{j=1}^n a_j \quad \text{(say)}.$$

Then $\{h_n, \mathscr{F}_n, n \ge 1\}$ is a martingale with $S(h) = 0$ on $\{\mu = \infty\}$ and

$$s^\alpha(h) = s_\sigma^\alpha(h) I_{[\mu < \infty]} \le \delta^\alpha \lambda^\alpha I_{[f^* > \lambda]}.$$

Hence, by Theorems 11.2.1 and 11.2.2, for some $B = B_\alpha \in (0, \infty)$

$$\|h\|_\alpha^\alpha = \sup_{n \ge 1} E|h_n|^\alpha \le E\, h^{*\alpha} \le B_\alpha\, E\left(\sum_1^\infty a_j^2\right)^{\alpha/2} \le B_\alpha\, E \sum_1^\infty |a_j|^\alpha$$
$$= B_\alpha\, E\, s^\alpha(h) \le B_\alpha \delta^\alpha \lambda^\alpha\, P\{f^* > \lambda\},$$

whence

$$P\{f^* > \beta\lambda, s(f) \vee d^* \le \delta\lambda\}$$
$$= P\{\nu < \infty, \sigma = \infty, h_n = f_{\nu \wedge n} - f_{\mu \wedge n} \quad \text{for all } n \ge 1\}$$
$$\le P\{h^* > \beta\lambda - \lambda - \delta\lambda\}$$
$$\le B_\alpha \left(\frac{\delta}{\beta - 1 - \delta}\right)^\alpha P\{f^* > \lambda\}. \qquad \square$$

Theorem 2. *Let Φ be a nondecreasing function on $[0, \infty]$, continuous on $[0, \infty)$ with $\Phi(0) = 0$, $\Phi(\infty) = \Phi(\infty -)$, and $\Phi(2\lambda) \le c\Phi(\lambda)$ for all $\lambda > 0$ and some $c \in (0, \infty)$. Then, for every α in $[1, 2]$ there exists a finite positive constant $B = B_{c,\alpha}$ such that for any martingale f*

$$E\,\Phi(f^*) \le B\, E\,\Phi(s(f)) + B\, E\,\Phi(d^*), \tag{29}$$

where $s(f) = (\sum_{j=1}^\infty E\{|d_j|^\alpha | \mathscr{F}_{j-1}\})^{1/\alpha}$.

PROOF. Choose $\beta = 3$ and $\delta \in (0, 1)$ such that $B_\alpha \delta^\alpha c^3 < 1$, where B is the constant in (28). Then by Lemmas 1 and 4 there exists $B = B_{c,\alpha} \in (0, \infty)$ such that

$$E\,\Phi(f^*) \le B\, E\,\Phi(s(f) \vee d^*) \le B\, E[\Phi(s(f)) + \Phi(d^*)]. \qquad \square$$

Theorem 2 will be applied in generalizing Theorem 9.1.2 so as to encompass convergence of absolute moments.

Corollary 2 (Brown). *Let* $\{X_n, n \geq 1\}$ *be independent* r.v.s *with* $\mathrm{E}\, X_n = 0$, $\mathrm{E}\, X_n^2 = \sigma_n^2$, $s_n^2 = \sum_1^n \sigma_i^2$. *If* $\{X_n\}$ *obey a Lindeberg condition of order* $r \geq 2$, *that is,*

$$\sum_{j=1}^{n} \int_{[|X_j| > \varepsilon s_j]} |X_j|^r = o(s_n^r), \qquad \text{all } \varepsilon > 0, \tag{30}$$

then

$$\lim_{n\to\infty} \mathrm{E}\left|\frac{1}{s_n}\sum_1^n X_j\right|^r = \frac{1}{\sqrt{2\pi}}\int_{-\infty}^{\infty} |t|^r e^{-t^2/2}\, dt. \tag{31}$$

PROOF. Set

$$X_n' = X_n I_{[|X_n| \leq s_n]} - \mathrm{E}\, X_n I_{[|X_n| \leq s_n]},$$
$$X_n'' = X_n I_{[|X_n| > s_n]} - \mathrm{E}\, X_n I_{[|X_n| > s_n]},$$

and $S_n = \sum_1^n X_j$, $S_n' = \sum_1^n X_j'$, $S_n'' = \sum_1^n X_j''$. Then $S_n = S_n' + S_n''$ and, by Theorem 2, for some constant $C \in (0, \infty)$

$$\begin{aligned}
\mathrm{E}|S_n'|^{r+1} &\leq C\left[\sum_1^n E(X_j')^2\right]^{(r+1)/2} + C\,\mathrm{E}\max_{1\leq j\leq n}|X_j'|^{r+1} \\
&\leq C(s_n^{r+1} + 2^{r+1}s_n^{r+1}).
\end{aligned}$$

Thus, $\mathrm{E}|S_n'/s_n|^{r+1} \leq C(1 + 2^{r+1})$, implying $\{|S_n'/s_n|^r, n \geq 1\}$ is uniformly integrable.

Again via Theorem 2, for some B in $(0, \infty)$

$$\begin{aligned}
\mathrm{E}|S_n''|^r &\leq B\left[\sum_1^n \mathrm{E}(X_j'')^2\right]^{r/2} + B\,\mathrm{E}\max_{1\leq j\leq n}|X_j''|^r \\
&\leq B\left(\sum_1^n \int_{[|X_j|>s_j]} X_j^2\right]^{r/2} + 2^r B\sum_{j=1}^n \int_{[|X_j|>s_j]} |X_j|^r \\
&= o(s_n^2)^{r/2} + o(s_n^r) = o(s_n^r)
\end{aligned}$$

since as noted in Section 9.1 (see, e.g., (10) therein), a Lindeberg condition of order $r > 2$ ensures that of lower orders. Thus, $\mathrm{E}|S_n''/s_n|^r = o(1)$, implying $\{|S_n''/s_n|^r, n \geq 1\}$ is uniformly integrable. Consequently, the same is true of $\{|S_n/s_n|^r, n \geq 1\}$, and so, in view of the central limit theorem (Theorem 9.1.1) and Corollary 8.1.8, the conclusion (31) follows.

In order to complement Theorem 2 a counterpart to (14) of Lemma 3 is needed for concave functions. A function Φ is **concave** if $-\Phi$ is convex.

Lemma 5. *If* Φ *is a nondecreasing function on* $[0, \infty]$, *finite and concave on* $(0, \infty)$ *with* $\Phi(0) = 0$, *then*

$$\Phi(t) = \int_0^t \varphi(u)du, \qquad t \in [0, \infty), \tag{32}$$

for some finite nonnegative, nonincreasing function φ *on* $[0, \infty)$ *and, moreover,*

$$\Phi(t) = -\int_0^\infty (t \wedge u)d\varphi(u) + t\varphi(\infty) \tag{33}$$

for $t \in [0, \infty)$, *where* $\varphi(\infty) = \lim_{t\to\infty} \varphi(t)$.

PROOF. Lemma 2 ensures (32) and, since Φ is nondecreasing, $\varphi \geq 0$. Via integration by parts in (32) for Riemann–Stieltjes integrals,

$$\begin{aligned}\Phi(t) &= t\varphi(t) - \int_0^t u\, d\varphi(u) \\ &= -\int_0^t u\, d\varphi(u) - \int_t^\infty t\, d\varphi(u) + t\varphi(\infty) \\ &= -\int_0^\infty (t \wedge u)d\varphi(u) + t\varphi(\infty).\end{aligned}$$ □

The following is a direct analogue of Lemma 3.

Lemma 6. *If* Φ *is a nondecreasing function on* $[0, \infty]$, *finite and concave on* $(0, \infty)$ *with* $\Phi(0) = 0$ *and* $\Phi(\infty) = \Phi(\infty -)$, *then for every sequence* $\{z_n, n \geq 1\}$ *of nonnegative measurable functions on* $(\Omega, \mathscr{F}, P)$

$$E\,\Phi\left(\sum_1^\infty z_j\right) \leq 2\,E\,\Phi\left(\sum_1^\infty E\{z_j | \mathscr{F}_{j-1}\}\right). \tag{34}$$

PROOF. Define $W_n, Z_n, W = W_\infty$, and $Z = Z_\infty$ as in the proof of Lemma 3, and for $\lambda > 0$ set $T = \inf\{n \geq 0: W_{n+1} > \lambda\}$. Then $W_T \leq W \wedge \lambda$ and

$$\begin{aligned}E\,Z_T &= E \sum_1^\infty z_j I_{[T \geq j]} = E \sum_1^\infty E\{z_j | \mathscr{F}_{j-1}\} I_{[T \geq j]} \\ &= E\,W_T \leq E(W \wedge \lambda),\end{aligned}$$

whence

$$\begin{aligned}E(Z \wedge \lambda) &\leq E(Z_T + \lambda I_{[T<\infty]}) \\ &\leq E(W \wedge \lambda) + \lambda\,P\{W > \lambda\} \leq 2\,E(W \wedge \lambda).\end{aligned}$$

Consequently, employing Lemma 5,

$$\begin{aligned}E\,\Phi(Z) &= E\left[-\int_0^\infty (Z \wedge u)d\varphi(u) + Z\varphi(\infty)\right] \\ &= -\int_0^\infty E(Z \wedge u)d\varphi(u) + \varphi(\infty)E\,Z \\ &\leq -2\int_0^\infty E(W \wedge u)d\varphi(u) + \varphi(\infty)E\,W = 2E\,\Phi(W).\end{aligned}$$ □

Theorem 3. *If Φ is a nondecreasing function on $[0, \infty]$, finite and concave on $(0, \infty)$ with $\Phi(0) = 0$, $\Phi(\infty) = \Phi(\infty -)$ and $\alpha \in [1, 2]$, there exists a constant $A = A_\alpha \in (0, \infty)$ such that for any martingale f*

$$\mathrm{E}\,\Phi(f^{*\alpha}) \le A\,\mathrm{E}\,\Phi(s^\alpha(f)), \tag{35}$$

where $s^\alpha(f) = \sum_1^\infty \mathrm{E}\{|d_j|^\alpha|\mathscr{F}_{j-1}\}$.

PROOF. By Theorems 11.2.1 and 11.2.2, for some constant $C = C_\alpha \in (0, \infty)$

$$\mathrm{E}(f^*)^\alpha \le C\,\mathrm{E}\left(\sum_1^\infty d_j^2\right)^{\alpha/2} \le C\,\mathrm{E}\sum_1^\infty |d_j|^\alpha = C\,\mathrm{E}\,s^\alpha(f). \tag{36}$$

For $u > 0$, define

$$T = \inf\left\{n \ge 0 : s_{n+1}^\alpha(f) \equiv \sum_1^{n+1} \mathrm{E}\{|d_j|^\alpha|\mathscr{F}_{j-1}\} > u\right\}.$$

Then $s_T^\alpha(f) \le s^\alpha(f) \wedge u$ and

$$(f^*)^\alpha \wedge u \le u I_{[T<\infty]} + \sup_{n\ge 1}|f_{T\wedge n}|^\alpha.$$

Hence, via (36),

$$\begin{aligned}\mathrm{E}[(f^*)^\alpha \wedge u] &\le u\,\mathrm{P}\{s^\alpha(f) > u\} + C\,\mathrm{E}\,s^\alpha(f)\\ &\le (C+1)\mathrm{E}[s^\alpha(f) \wedge u].\end{aligned}$$

Consequently, from Lemma 5 and (36)

$$\begin{aligned}\mathrm{E}\,\Phi(f^{*\alpha}) &= -\int_0^\infty \mathrm{E}(f^{*\alpha} \wedge u)d\varphi(u) + \varphi(\infty)\mathrm{E}\,f^{*\alpha}\\ &\le -(C+1)\int_0^\infty \mathrm{E}[s^\alpha(f) \wedge u]d\varphi(u) + C\varphi(\infty)\mathrm{E}\,s^\alpha(f)\\ &\le (C+1)\left[-\int_0^\infty \mathrm{E}[s^\alpha(f) \wedge u]d\varphi(u) + \varphi(\infty)\mathrm{E}\,s^\alpha(f)\right]\\ &= (C+1)\mathrm{E}\,\Phi(s^\alpha(f)).\end{aligned}$$

□

Corollary 3. *If* $0 < p < \alpha$ *and* $1 \le \alpha \le 2$, *there exists a constant* $A = A_\alpha \in (0, \infty)$ *such that, for any martingale f,*

$$\mathrm{E}(f^*)^p \le A\,\mathrm{E}\left(\sum_1^\infty \mathrm{E}\{|d_j|^\alpha|\mathscr{F}_{j-1}\}\right)^{p/\alpha}. \tag{37}$$

PROOF. $\Phi(t) = t^{p/\alpha}$, $t \ge 0$, satisfies the requirements of Theorem 3, whence

$$\mathrm{E}(f^*)^p = \mathrm{E}\,\Phi(f^{*\alpha}) \le A\,\mathrm{E}\,\Phi(s^\alpha(f)) = A\,\mathrm{E}\,s^p(f).$$

□

Lemma 7. *If $\{X_n, \mathscr{F}_n, n \ge 1\}$ is a stochastic sequence and T is an $\mathscr{F}_n$-time satisfying*

$$\mathrm{E}\sum_{i=1}^{T}\mathrm{E}\{|X_i|^\gamma|\mathscr{F}_{i-1}\}<\infty,\qquad \mathrm{E}\left(\sum_1^T \mathrm{E}\{|X_i||\mathscr{F}_{i-1}\}\right)^\gamma<\infty \tag{38}$$

for some $\gamma \geq 1$, *then*

$$\mathrm{E}\left(\sum_1^T |X_i|\right)^\gamma<\infty. \tag{39}$$

PROOF. When $\gamma = 1$, the first two lines of the proof of Theorem 7.4.6 (with X_n^+ replaced by $|X_n|$) yield (39). For $\gamma > 1$, set $V_{\alpha,n} = \sum_{i=1}^n \mathrm{E}\{|X_n|^\alpha|\mathscr{F}_{i-1}\}$, $\alpha > 0$, $n \geq 1$, and $Y_i = |X_i| - \mathrm{E}\{|X_i||\mathscr{F}_{i-1}\}$, $i \geq 1$, and consider the martingale $f_n = \sum_{i=1}^n Y_i I_{[T\geq i]}$, $n \geq 1$. Since $\sum_{i=1}^\infty \mathrm{E}\{|Y_i|I_{[T\geq i]}|\mathscr{F}_{i-1}\} = \sum_1^T \mathrm{E}\{|Y_i||\mathscr{F}_{i-1}\}$ $\leq 2V_{1,T}$, it follows, taking $\alpha = 1$, $\Phi(x) = |x|^\gamma$ in Theorem 11.3.3, that

$$\begin{aligned}
\mathrm{E}\left|\sum_1^{T\wedge n} Y_i\right|^\gamma &\leq \mathrm{E}\sup_{n\geq 1}\left|\sum_1^n Y_i I_{[T\geq i]}\right|^\gamma \\
&\leq B\cdot\mathrm{E}\left[(2V_{1,T})^\gamma + \left(\sup_{i\geq 1}|Y_i|I_{[T\geq i]}\right)^\gamma\right] \\
&\leq B\cdot\mathrm{E}\left[(2V_{1,T})^\gamma + \sum_{i=1}^\infty |Y_i|^\gamma I_{[T\geq i]}\right] \\
&= B\cdot\left[\mathrm{E}(2V_{1,T})^\gamma + \mathrm{E}\sum_{i=1}^\infty I_{[T\geq i]}\mathrm{E}\{|Y_i|^\gamma|\mathscr{F}_{i-1}\}\right] \\
&\leq 2^{\gamma+1}B[\mathrm{E}\,V_{1,T}^\gamma + \mathrm{E}\,V_{\gamma,T}].
\end{aligned}$$

Consequently,

$$\mathrm{E}\left(\sum_1^{T\wedge n}|X_i|\right)^\gamma \leq 2^\gamma\left[\mathrm{E}\left|\sum_1^{T\wedge n} Y_i\right|^\gamma + \mathrm{E}\,V_{1,T}^\gamma\right] = O(1),$$

and so monotone convergence yields (39). □

Theorem 4. *If* $\{S_n = \sum_{j=1}^n X_i, \mathscr{F}_n, n \geq 1\}$ is a martingale and T is an $\mathscr{F}_n$-time with

$$\mathrm{E}\sum_1^T \mathrm{E}\{|X_i|^\gamma|\mathscr{F}_{i-1}\}<\infty,\qquad \mathrm{E}\left(\sum_1^T \mathrm{E}\{X_i^2|\mathscr{F}_{i-1}\right)^{\gamma/2}<\infty \tag{40}$$

for some $\gamma \geq 2$, *then*

$$\mathrm{E}|S_T|^\gamma<\infty. \tag{41}$$

PROOF. By Theorem 11.2.1,

$$\mathrm{E}|S_{T\wedge n}|^\gamma \leq O(1)\cdot\mathrm{E}\left(\sum_1^{T\wedge n}X_i^2\right)^{\gamma/2} \leq O(1)\mathrm{E}\left(\sum_1^T X_i^2\right)^{\gamma/2} = O(1)$$

via Lemma 7 with $|X|$ replaced by X^2, whence the conclusion follows by Fatou's lemma. □

Corollary 4. *If* $S_n = \sum_{j=1}^n X_j$, $n \geq 1$, *where* $\{X_n, n \geq 1\}$ *are independent random variables with* $\mathrm{E}\,X_n = 0$, $n \geq 1$, *and* $\sup_{n\geq 1} \mathrm{E}|X_n|^\gamma < \infty$ *for some* $\gamma \geq 2$, *then* $\mathrm{E}|S_T|^\gamma < \infty$ *for any* $\{X_n\}$*-time* T *with* $\mathrm{E}\,T^{\gamma/2} < \infty$.

EXERCISE 11.3

1. Prove Corollary 11.2.3 by applying Corollary 11.3.3.

11.4 Stochastic Inequalities

Throughout this section the generic notation

$$\bar{a} = 1 - a \tag{1}$$

will be adopted and the conditional expectation $\mathrm{E}\{U \mid \mathcal{G}\}$ and conditional variance will be denoted $\mathrm{E}_{\mathcal{G}}\,U$ and $\sigma_{\mathcal{G}}^2(U)$ respectively. Integrability requirements in many of the lemmas of this section can clearly be weakened to a.c. finiteness of the conditional expectations therein.

Lemma 1. *If* $\mathcal{G}$ *is a* σ*-algebra of measurable sets,* β *is a* $\mathcal{G}$*-measurable* r.v. *at most equal to one* a.c., *and* U *is an* $\mathcal{L}_2$ *random variable satisfying*

$$\mathrm{E}_{\mathcal{G}}\,U \leq -\alpha\sigma_{\mathcal{G}}^2(U), \qquad \mathrm{P}\{\beta + U \leq 1\} = 1, \tag{2}$$

for some $\alpha > 0$, *then*

$$\mathrm{E}_{\mathcal{G}}\,\frac{\beta + U}{1 + \alpha\overline{\beta + U}} \leq \frac{\beta}{1 + \alpha\bar{\beta}}.$$

PROOF.

$$\frac{1}{1 + \alpha\overline{\beta + U}} - \frac{1}{1 + \alpha\bar{\beta}} = \frac{1}{1 + \alpha\overline{\beta + U}}\,\frac{\alpha U}{1 + \alpha\bar{\beta}}$$

$$= \frac{\alpha U}{1 + \alpha\bar{\beta}}\left(\frac{1}{1 + \alpha\bar{\beta}} + \frac{1}{1 + \alpha\overline{\beta + U}}\,\frac{\alpha U}{1 + \alpha\bar{\beta}}\right) \leq \frac{\alpha U + \alpha^2 U^2}{(1 + \alpha\bar{\beta})^2}, \tag{3}$$

and so via (2)

$$\frac{1}{1 + \alpha\overline{\beta + \mathrm{E}_{\mathcal{G}}\,U}} - \frac{1}{1 + \alpha\bar{\beta}} = \frac{1}{1 + \alpha\overline{\beta + \mathrm{E}_{\mathcal{G}}\,U}}\,\frac{\alpha\,\mathrm{E}_{\mathcal{G}}\,U}{1 + \alpha\bar{\beta}}$$

$$\leq \frac{-\alpha^2\sigma_{\mathcal{G}}^2(U)}{(1 + \alpha\overline{\beta + \mathrm{E}_{\mathcal{G}}\,U})^2}. \tag{4}$$

Hence, from (3) and then (4)

$$\mathrm{E}_{\mathscr{G}} \frac{1}{1+\alpha\beta+U} = \mathrm{E}_{\mathscr{G}} \frac{1}{1+\alpha(\beta+\mathrm{E}_{\mathscr{G}}\, U)+(U-\mathrm{E}_{\mathscr{G}}\, U)}$$

$$\leq \frac{1}{1+\alpha\beta+\mathrm{E}_{\mathscr{G}}\, U} + \frac{\alpha^2\sigma_{\mathscr{G}}^2(U)}{(1+\alpha\beta+\mathrm{E}_{\mathscr{G}}\, U)^2} \leq \frac{1}{1+\alpha\beta}.$$

Consequently,

$$\mathrm{E}_{\mathscr{G}} \frac{\beta+U}{1+\alpha\beta+U} = \mathrm{E}_{\mathscr{G}} \frac{1}{\alpha}\left[-1+\frac{1+\alpha}{1+\alpha\beta+U}\right] \leq \frac{1}{\alpha}\left[-1+\frac{1+\alpha}{1+\alpha\beta}\right]$$

$$= \frac{\beta}{1+\alpha\beta}. \qquad \square$$

Lemma 2. *If $\{S_n = \sum_1^n X_i, n \geq 1\}$ is an $\mathscr{L}_2$ stochastic sequence satisfying*

$$\mathrm{E}_{\mathscr{F}_n} X_{n+1} \leq -\alpha\sigma_{\mathscr{F}_n}^2(X_{n+1}), \qquad \mathrm{P}\{S_n \leq 1\} = 1, \qquad n \geq 1, \tag{5}$$

for some $\alpha > 0$, then $\{S_n/[1+\alpha(1-S_n)], \mathscr{F}_n, n \geq 1\}$ is a supermartingale.

PROOF. By Lemma 1, for $n \geq 1$

$$\mathrm{E}_{\mathscr{F}_n} \frac{S_{n+1}}{1+\alpha\bar{S}_{n+1}} = \mathrm{E}_{\mathscr{F}_n} \frac{S_n + X_{n+1}}{1+\alpha\bar{S}_n+\bar{X}_{n+1}} \leq \frac{S_n}{1+\alpha\bar{S}_n}. \qquad \square$$

According to the ensuing theorem, it is uncertain that a nonnegative supermartingale S_n starting in (0, 1) will ever exceed one if the conditional coefficients of variation of the differences are bounded away from zero.

Theorem 1 (Dubins–Savage). *If $\mathrm{P}\{0 \leq X_1 \leq 1\} = 1$ and $\{S_n = \sum_1^n X_i, \mathscr{F}_n, 1 \leq n < \infty\}$ is a nonnegative $\mathscr{L}_2$ supermartingale satisfying*

$$\mathrm{E}\{X_n \mid \mathscr{F}_{n-1}\} \leq -\alpha\sigma_{\mathscr{F}_{n-1}}^2(X_n), \qquad n > 1, \tag{6}$$

for some $\alpha > 0$, then

$$\mathrm{P}\{S_n \geq 1, \textit{some } n \geq 1\} \leq \mathrm{E}\frac{X_1}{1+\alpha(1-X_1)}. \tag{7}$$

PROOF. Let $T = \inf\{n \geq 1 : S_n \geq 1\}$ and set $V_n = \sum_{i=1}^n Y_i$, where

$$Y_n = X_n I_{[T>n]} + (1-S_{n-1})I_{[T=n]}, \qquad S_0 = 0. \tag{8}$$

On $[T = n]$, $0 < Y_n = 1 - S_{n-1} \leq X_n$, and so $Y_n^2 \leq X_n^2$ on Ω, whence $\mathrm{E}_{\mathscr{F}_{n-1}} Y_n^2 \leq \mathrm{E}_{\mathscr{F}_{n-1}} X_n^2$, $n > 1$. On the set $[T \geq n]$

$$\mathrm{E}_{\mathscr{F}_{n-1}} Y_n \leq \mathrm{E}_{\mathscr{F}_{n-1}} X_n \leq -\alpha\sigma_{\mathscr{F}_{n-1}}^2(X_n),$$

so that $\mathrm{E}_{\mathscr{F}_{n-1}}^2 Y_n \geq \mathrm{E}_{\mathscr{F}_{n-1}}^2 X_n$, implying $\sigma_{\mathscr{F}_{n-1}}^2(Y_n) \leq \alpha_{\mathscr{F}_{n-1}}^2(X_n)$ on $[T \geq n]$. Therefore, (6) obtains for $\{Y_n\}$ on $[T \geq n]$, and on the complement $[T < n]$

it is trivially true since $Y_n = 0$. Hence, Lemma 2 ensures that $\{V_n/(1 + \alpha\bar{V}_n), \mathscr{F}_n, n \geq 1\}$ is a supermartingale. Consequently, noting that $V_T = 1$, setting $T(n) = \min[T, n]$, and employing Theorem 7.4.5.

$$P\{S_n \geq 1, \text{ some } n \geq 1\} = P\{V_n = 1, \text{ some } n \geq 1\} = \int_{[T<\infty]} \frac{V_T}{1 + \alpha\bar{V}_T}$$

$$\leq \varliminf_{n\to\infty} E \frac{V_{T(n)}}{1 + \alpha\bar{V}_{T(n)}} \leq E \frac{Y_1}{1 + \alpha\bar{Y}_1}.$$

Since $X_1 = Y_1$, (7) follows. □

Theorem 2. *If $\{S_n = \sum_1^n X_i, \mathscr{F}_n, n \geq 1\}$ is an $\mathscr{L}_2$ martingale with* $E\, S_1 = 0$ *and $\mathscr{F}_0 = (\varnothing, \Omega)$, then for any positive constants a, b*

$$P\left\{S_n \geq a \sum_1^n E\{X_j^2 | \mathscr{F}_{j-1}\} + b, \text{ some } n \geq 1\right\} \leq (1 + ab)^{-1}. \tag{9}$$

PROOF. It suffices to prove the theorem when $a = 1$. Define $S_0 = X_0 = 0$, $\mathscr{F}_0 = \mathscr{F}_{-1}$, and for $K > 0$ set

$$U_n = \frac{S_n + K - \sum_{j=0}^n E_{\mathscr{F}_{j-1}} X_j^2}{K + b}, \qquad n \geq 0,$$

whence $u_n = U_n - U_{n-1} = (K + b)^{-1}[X_n - E_{\mathscr{F}_{n-1}} X_n^2]$, $n \geq 1$, and $u_0 = K(K + b)^{-1} = U_0$. Note that for $n \geq 1$

$$E_{\mathscr{F}_{n-1}} u_n = \frac{-1}{K + b} E_{\mathscr{F}_{n-1}} X_n^2$$

and

$$\sigma^2_{\mathscr{F}_{n-1}}(u_n) = (K + b)^{-2} E_{\mathscr{F}_{n-1}} X_n^2 = -(K + b)^{-1} E_{\mathscr{F}_{n-1}} u_n, \qquad n \geq 1,$$

so that for $\alpha = (K + b)$

$$E_{\mathscr{F}_{n-1}} u_n = -\alpha\sigma^2_{\mathscr{F}_{n-1}}(u_n). \tag{10}$$

Then, via Theorem 1

$$P\left\{S_n \geq \sum_1^n E_{j-1} X_j^2 + b, \text{ some } n \geq 1\right\} = P\{U_n \geq 1, \text{ some } n \geq 0\}$$

$$\leq \frac{U_0}{1 + \alpha\bar{U}_0} = \frac{K}{K + b}\frac{1}{1 + b} \xrightarrow{K\to\infty} \frac{1}{1 + b}.$$ □

Corollary 1. *If $\{X_n, n \geq 1\}$ are independent* r.v.s *with* $E\, X_n = 0$, $E\, X_n^2 = 1$, *$n \geq 1$, and $S_n = \sum_1^n X_i$, then for positive a,b*

$$P\{S_n \geq an + b, \text{ some } n \geq 1\} \leq \frac{1}{1 + ab}.$$

Notation. In the remainder of this section, the generic notation

$$a' = e^a - 1 - a \tag{11}$$

will be employed.

Next, analogues of Lemmas 1 and 2 will yield a counterpart of Theorem 1.

Lemma 3. *If U, V, Y are r.v.s with $Y \geq 0$, $|U| \leq 1$, $\mathrm{E}\,U = 0$, and both Y and V are $\mathscr{G}$-measurable for some σ-algebra $\mathscr{G}$ of measurable sets, then*

$$\mathrm{E}_{\mathscr{G}}[\cosh Y(U + V)] \leq \exp(Y'\,\mathrm{E}_{\mathscr{G}}\,U^2)\cosh YV.$$

PROOF. $(YU)' = \sum_{j=2}^{\infty} Y^j U^j/j! \leq U^2 Y'$, whence

$$\mathrm{E}_{\mathscr{G}}\,e^{YU} \leq 1 + Y'\,\mathrm{E}_{\mathscr{G}}\,U^2 \leq \exp(Y'\,\mathrm{E}_{\mathscr{G}}\,U^2). \tag{12}$$

Applying (12) to $-U$ also,

$$\mathrm{E}_{\mathscr{G}} \cosh Y(V + U) = \tfrac{1}{2}\,\mathrm{E}_{\mathscr{G}}(e^{Y(V+U)} + e^{-Y(V+U)}) \leq \exp(Y'\,\mathrm{E}_{\mathscr{G}}\,U^2)\cosh YV.$$

□

Lemma 4. *Let $\{S_n = \sum_1^n X_j, \mathscr{F}_n, n \geq 1\}$ be a martingale with $|X_n| \leq 1$, $n \geq 1$, and set $u_n = \mathrm{E}_{\mathscr{F}_{n-1}} X_n^2$, $n \geq 1$, where $\mathscr{F}_0 = (\varnothing, \Omega)$. For any real y and positive numbers λ, u*

$$\left\{\exp\left\{-\sum_{j=1}^n u_j \left(\frac{\lambda}{u + \sum_1^j u_i}\right)'\right\}\cosh \lambda\left(\frac{y + S_n}{u + \sum_1^n u_j}\right), \mathscr{F}_n, n \geq 1\right\}$$

is a supermartingale.

PROOF. Designate the putative supermartingale by $\{V_n, \mathscr{F}_n, n \geq 1\}$ and set $u_0 = u$. By Lemma 3, for $n \geq 2$

$$\mathrm{E}_{\mathscr{F}_{n-1}} V_n = \exp\left\{-\sum_1^n u_j\left(\frac{\lambda}{u + \sum_1^j u_i}\right)'\right\}\mathrm{E}_{\mathscr{F}_{n-1}} \cosh \frac{\lambda(y + S_{n-1} + X_n)}{\sum_0^n u_j}$$

$$\leq \exp\left\{-\sum_{j=1}^n u_j\left(\frac{\lambda}{\sum_0^j u_i}\right)' + u_n\left(\frac{\lambda}{\sum_0^n u_j}\right)'\right\}\cosh\frac{\lambda(y + S_{n-1})}{\sum_0^n u_j} \leq v_{n-1}.$$

Setting $V_0 = \cosh(\lambda y/u)$, the same argument shows that if $\mathrm{E}\,X_1 = 0$, $\{V_n, \mathscr{F}_n, n \geq 0\}$ is likewise a supermartingale. □

Theorem 3. *If $\{S_n = \sum_1^n X_j, \mathscr{F}_n, n \geq 1\}$ is an $\mathscr{L}_2$ martingale with $\mathrm{E}\,S_1 = 0$ and T is a stopping variable for which $X_n^2 \leq 1$ on $\{T \geq n\}$, $n \geq 1$, then for any real y and positive numbers λ, u*

$$\mathrm{E} \cosh \frac{\lambda(y + S_T)}{u + \sum_1^T \mathrm{E}_{\mathscr{F}_{j-1}} X_j^2} \leq e^{u\left(\frac{\lambda}{u}\right)'} \cosh\frac{\lambda y}{u}. \tag{13}$$

PROOF. Set $u_0 = u$ and $u_n = E_{\mathscr{F}_{n-1}} X_n^2$, $n \geq 1$. If $Y_n = X_n I_{[T \geq n]}$, $\tilde{u}_n = E_{\mathscr{F}_{n-1}} Y_n^2$, then $E_{\mathscr{F}_{n-1}} Y_n = 0$ and $\tilde{u}_n = u_n I_{[T \geq n]}$. Since $Y_n^2 \leq 1$ and the indices in the sums below are at most T,

$$\frac{y + \sum_1^T X_j}{u + \sum_1^T u_j} = \frac{u + \sum_1^T Y_j}{u + \sum_1^T \tilde{u}_j},$$

whence no generality is lost in supposing $X_n^2 \leq 1$, a.c. Now, for $a > 0$

$$\left(\frac{\lambda}{u+a}\right)' = \sum_{j=2}^{\infty} \frac{1}{j!}\left(\frac{\lambda}{u+a}\right)^j \leq \sum_{j=1}^{\infty} \frac{1}{j!}\left(\frac{\lambda}{u}\right)^j \left(\frac{u}{u+a}\right)^2$$
$$= \left(\frac{u}{u+a}\right)^2 \left(\frac{\lambda}{u}\right)',$$

implying

$$\sum_{j=1}^{n} u_j \left(\frac{\lambda}{\sum_0^j u_i}\right)' \leq \left(\frac{\lambda}{u}\right)' \sum_{j=1}^{n} u_j \left(\frac{u}{\sum_{i=0}^j u_i}\right)^2 \leq u^2 \left(\frac{\lambda}{u}\right)' \sum_1^n \left(\frac{1}{\sum_0^{j-1} u_i} - \frac{1}{\sum_{i=0}^j u_i}\right)$$
$$= u^2 \left(\frac{\lambda}{u}\right)' \frac{1}{u_0} = u \left(\frac{\lambda}{u}\right)'. \tag{14}$$

Employing the notation and final remark of Lemma 4, setting $T(n) = \min[T, n]$, and invoking Theorem 7.4.5,

$$\text{E} \cosh \frac{\lambda(y + S_T)}{u + \sum_1^T u_j} = \text{E}\, V_T \exp\left\{\sum_1^T u_j \left(\frac{\lambda}{\sum_0^j u_i}\right)'\right\}$$
$$\leq \underline{\lim}_n \text{E}\, V_{T(n)} \exp\left\{\sum_1^{T(n)} u_j \left(\frac{\lambda}{\sum_0^j u_i}\right)'\right\}$$
$$\leq \exp\left\{u\left(\frac{\lambda}{u}\right)'\right\} \underline{\lim}\, \text{E}\, V_{T(n)}$$
$$\leq \exp\left\{u\left(\frac{\lambda}{u}\right)'\right\} \text{E}\, V_0 = \exp\left\{u\left(\frac{\lambda}{u}\right)'\right\} \cosh \frac{\lambda y}{u}$$

recalling (14). □

Corollary 2. *If* $\{S_n = \sum_1^n X_j, \mathscr{F}_n, n \geq 1\}$ *is an* $\mathscr{L}_2$ *martingale with* $\text{E}\, X_1 = 0$, $u_1 = \text{E}\, X_1^2 > 0$, *and* T *is a stopping variable with* $X_n^2 \leq 1$ *on* $[T \geq n]$, $n \geq 1$, *then for any* $\lambda > 0$

$$\text{E} \exp\left\{\frac{\lambda S_T}{\sum_1^T \text{E}_{\mathscr{F}_{j-1}} X_j^2}\right\} \leq 2 \exp\left\{2u_1 \left(\frac{\lambda}{u_1}\right)'\right\}.$$

PROOF. Via Theorem 3 and then Lemma 3 with $V = 0$, $Y = \lambda/u_1$, $U = X_1$, $\mathscr{G} = (\varnothing, \Omega)$,

$$\mathrm{E}\exp\left\{\frac{\lambda S_T}{\sum_1^T u_i}\right\} \le 2\,\mathrm{E}\cosh\frac{\lambda S_T}{\sum_1^T u_i} = 2\,\mathrm{E}\,\mathrm{E}\left\{\cosh\frac{\lambda S_T}{\sum_1^T u_i}\,\middle|\,\mathscr{F}_1\right\}$$

$$\le 2\,\mathrm{E}\exp\left\{u_1\left(\frac{\lambda}{u_1}\right)'\right\}\cosh\frac{\lambda X_1}{u_1} \le 2\exp\left\{2u_1\left(\frac{\lambda}{u_1}\right)'\right\}. \quad \square$$

Corollary 3. *Under the conditions of Corollary 2*

$$\mathrm{E}\exp\left\{\lambda S_T - \lambda' \sum_1^T \mathrm{E}_{\mathscr{F}_{j-1}} X_j^2\right\} \le 1. \tag{15}$$

PROOF. As in the theorem, suppose $X_n^2 \le 1$, a.c. By (12) with $Y = \lambda$, $\mathscr{G} = \mathscr{F}_{n-1}$,

$$\begin{aligned}
\mathrm{E}_{\mathscr{F}_{n-1}}\exp\left\{\lambda S_n - \lambda'\sum_1^n u_j\right\}
&= \exp\left(\lambda S_{n-1} - \lambda'\sum_1^n \mu_j\right)\mathrm{E}_{\mathscr{F}_{n-1}} e^{\lambda X_n} \\
&\le \exp\left\{\lambda S_{n-1} - \lambda'\sum_1^n u_j + \lambda' u_n\right\} \\
&= \exp\left\{\lambda S_{n-1} - \lambda'\sum_1^{n-1} u_j\right\},
\end{aligned}$$

whence $\{\exp\{\lambda S_n - \lambda' \sum_1^n u_j\}, \mathscr{F}_n, n \ge 1\}$ is a nonnegative supermartingale. Again via (12)

$$\begin{aligned}
\mathrm{E}\exp\left\{\lambda S_T - \lambda'\sum_1^T u_j\right\} &\le \underline{\lim}\,\mathrm{E}\exp\left\{\lambda S_{T(n)} - \lambda'\sum_1^{T(n)} u_j\right\} \\
&\le \mathrm{E}\, e^{\lambda X_1 - \lambda' u_1} \le 1. \quad \square
\end{aligned}$$

Theorem 4 (Blackwell). *If $\{S_n = \sum_1^n X_i, \mathscr{F}_n, n \ge 0\}$ is a supermartingale with $S_0 = 0$, $|X_n| \le 1$, and $\mathrm{E}_{\mathscr{F}_{n-1}} X_n \le -\alpha$ for some α in $(0, 1)$ and all $n \ge 1$, then*

$$P\{S_n \ge \lambda > 0, \textit{some } n \ge 1\} \le \left(\frac{1-\alpha}{1+\alpha}\right)^{\lambda}. \tag{16}$$

PROOF. If F_n is the conditional distribution of X_n given $\mathscr{F}_{n-1}$, by convexity of the exponential function for $\theta > 0$ and $n \ge 1$

$$\mathrm{E}_{\mathscr{F}_{n-1}} e^{\theta X_n} = \int_{-1}^{1} e^{\theta x}\, dF_n(x) \le \int_{-1}^{1}\left[\frac{1}{2}(e^{\theta} + e^{-\theta}) + x\frac{e^{\theta} - e^{-\theta}}{2}\right] dF_n(x)$$

$$\leq \frac{1-\alpha}{2} e^{\theta} + \frac{1+\alpha}{2} e^{-\theta} \leq 1 \quad \text{for } 0 \leq \theta \leq \log \frac{1+\alpha}{1-\alpha}$$

$= \theta_0$ (say). Consequently, for $n \geq 1$

$$E_{\mathscr{F}_{n-1}} e^{\theta_0 S_n} = e^{\theta_0 S_{n-1}} E_{\mathscr{F}_{n-1}} e^{\theta_0 X_n} \leq e^{\theta_0 S_{n-1}},$$

whence $\{e^{\theta_0 S_n}, \mathscr{F}_n, n \geq 0\}$ is a supermartingale. Hence, for any stopping rule T, setting $T(n) = \min[T, n]$,

$$E\, e^{\theta_0 S_T} \leq \underline{\lim}_n E\, e^{\theta_0 S_{T(n)}} \leq E\, e^{\theta_0 S_0} = 1.$$

Finally, if $T = \inf\{n \geq 1 : S_n \geq \lambda > 0\}$,

$$e^{\theta_0 \lambda} P\{S_n \geq \lambda \text{ for some } n \geq 1\} = e^{\theta_0 \lambda} P\{T < \infty\} \leq E\, e^{\theta_0 S_T} \leq 1,$$

which is tantamount to (16). □

Suppose that a gambler with initial capital $f_0 \in (0, 1)$ wins or loses (at each play) whatever amount he stakes with probabilities p and q respectively, where $0 < p \leq \frac{1}{2}$, the successive plays being independent. What is the best gambling scheme for boosting his capital to one dollar? It will be shown that a bold strategy is optimal, namely, whenever his capital $f_n \leq \frac{1}{2}$, the entire capital should be staked, whereas whenever $f_n > \frac{1}{2}$, only the amount $1 - f_n$ (needed to obtain his goal) should be bet.

Let $\{X_n, n \geq 1\}$ be i.i.d. random variables with

$$P\{X_n = 1\} = 1 - P\{X_n = -1\} = p \leq \tfrac{1}{2}$$

and let $\bar{a} = \min(a, 1 - a)$ whenever $0 \leq a \leq 1$.

For any constant $0 \leq f = f_0 \leq 1$, define for $n \geq 1$,

$$f_n = f_{n-1} + \bar{f}_{n-1} X_n = \begin{cases} f_{n-1}(1 + X_n) & \text{if } 0 \leq f_{n-1} \leq \frac{1}{2} \\ f_{n-1}(1 - X_n) + X_n & \text{if } \frac{1}{2} \leq f_{n-1} \leq 1, \end{cases} \tag{17}$$

and note that $\{f_n, n \geq 1\}$ represent the successive (random) fortunes associated with the bold strategy just described. Set

$$p_n(f) = P\left\{\max_{0 \leq j \leq n} f_j \geq 1\right\}, \qquad 0 \leq f \leq 1, n = 0, 1, \ldots,$$

$$p_n(f) = 1 \quad \text{if } f \geq 1, \qquad p_n(f) = 0 \quad \text{if } f \leq 0. \tag{18}$$

Lemma 5. *For any constants* $0 \leq f - s \leq f$, *setting* $q = 1 - p$,

$$p_{n+1}(f) \geq p p_n(f + s) + q p_n(f - s), \qquad n = 0, 1, \ldots. \tag{19}$$

PROOF. For every n, by (17) $p_n(f) \uparrow$ in f. If $A_n(f) = \{\max_{0 \leq j \leq n} f_j \geq 1\}$, then

$$\begin{aligned} p_{n+1}(\tfrac{1}{2}f) &= p\, P\{A_{n+1}(\tfrac{1}{2}f) | X_1 = 1\} + q\, P\{A_{n+1}(\tfrac{1}{2}f) | X_1 = -1\} \\ &= p\, P\{A_n(f)\} = p p_n(f) \end{aligned}$$

and

$$p_{n+1}\left(\frac{1+f}{2}\right) = p\,\mathrm{P}\left\{A_{n+1}\left(\frac{1+f}{2}\right)\middle| X_1 = 1\right\}$$
$$+ q\,\mathrm{P}\left\{A_{n+1}\left(\frac{1+f}{2}\right)\middle| X_1 = -1\right\}$$
$$= p + q\,\mathrm{P}\{A_n(f)\} = p + qp_n(f).$$

Hence, for $0 \le f \le 1$ and $n \ge 0$

$$p_{n+1}(f) = pp_n(2f) + qp_n(2f - 1), \tag{20}$$

and obviously (20) holds for $f \le 0$ or $f \ge 1$. Consequently, (20) obtains for all $f \in (-\infty, \infty)$ and all $n \ge 0$.

To prove (19), define for $0 \le f - s \le f$ and $n \ge 0$

$$\Delta_n(f, s) = p_{n+1}(f) - pp_n(f + s) - qp_n(f - s). \tag{21}$$

If $f + s < 1$, then $\Delta_0(f, s) = p_1(f) \ge 0$, while if $f + s \ge 1$, monotonicity of p_n and (20) ensure $\Delta_0(f, s) \ge 0$. Hence, for $0 \le f - s \le f$ and $n = 0$

$$\Delta_n(f, s) \ge 0. \tag{22}$$

Now from (20) and (21)

$$\Delta_n(f, s) = p[p_n(2f) - p_n(f + s)] + q[p_n(2f - 1) - p_n(f - s)],$$

and employing (20) once more

$$\begin{aligned}\Delta_{n+1}(f, s) = {} & p^2[p_n(4f) - p_n(2f + 2s)] + q^2[p_n(4f - 3) - p_n(2f - 2s - 1)] \\ & + pq[p_n(4f - 1) - p_n(2f + 2s - 1) \\ & + p_n(4f - 2) - p_n(2f - 2s)],\end{aligned} \tag{23}$$

and so

$$\begin{aligned}\Delta_{n+1}(f, s) = {} & p\Delta_n(2f, 2s) + q\{p[p_n(4f - 2) - p_n(2f + 2s - 1)] \\ & + q[p_n(4f - 3) - p_n(2f - 2s - 1)]\}.\end{aligned} \tag{24}$$

To verify (22) for $n \ge 1$, suppose inductively that it holds for a fixed but arbitrary integer n. If (i) $f - s \ge \frac{1}{2}$, then $2f - 1 \ge 2s$, whence by (24) $\Delta_{n+1}(f, s) = p\Delta_n(2f, 2s) + q\Delta_n(2f - 1, 2s) \ge 0$; if (ii) $f + s \le \frac{1}{2}$, then $2f - 2s \le 2f + 2s \le 1$ and again via (24), $\Delta_{n+1}(f, s) \ge p\Delta_n(2f, 2s) \ge 0$; finally, if (iii) $0 \le f - s < \frac{1}{2} < f + s$, then $\frac{1}{4} < f$, and by (23)

$$\Delta_{n+1}(f, s) \ge pq[p_n(4f - 1) - p_n(2f + 2s - 1) + p_n(4f - 2) - p_n(2f - 2s)]. \tag{25}$$

Since $q \ge p$, if $s > \frac{1}{4}$, then from (25)

$$\Delta_{n+1}(f, s) \ge p\{[p[p_n(4f - 1) - p_n(2f + 2s - 1)]$$

$$+ q[p_n(4f - 2) - p_n(2f - 2s)]\}$$
$$= p\Delta_n(2f - \tfrac{1}{2}, 2s - \tfrac{1}{2}) \geq 0,$$

and if $s \leq \frac{1}{4}$, again via (25)

$$\begin{aligned}\Delta_{n+1}(f, s) &\geq p\{p[p_n(4f - 1) - p_n(2f - 2s)] \\ &\quad + q[p_n(4f - 2) - p_n(2f + 2s - 1)] \\ &= p\Delta_n(2f - \tfrac{1}{2}, \tfrac{1}{2} - 2s) \geq 0,\end{aligned}$$

completing the induction. □

Lemma 6. *Let $\{X_n, \mathscr{F}_n, 1 \leq n \leq N < \infty\}$ be an $\mathscr{L}_1$ stochastic sequence, let C_N be the class of all stopping rules T with $\mathrm{P}\{T \leq N\} = 1$ and define*

$$\begin{aligned}\gamma_N &= \gamma_N^N = X_N, \\ \gamma_n &= \gamma_n^N = \max[X_n, \mathrm{E}\{\gamma_{n+1} | \mathscr{F}_n\}], \qquad 1 \leq n < N.\end{aligned} \tag{26}$$

Then, if $\sigma = \inf\{n \geq 1 : X_n = \gamma_n\}$,

$$\sup_{T \in C_N} \mathrm{E}\, X_T = \mathrm{E}\, \gamma_1 = \mathrm{E}\, X_\sigma. \tag{27}$$

Proof. Clearly, $\{\gamma_n, \mathscr{F}_n, 1 \leq n \leq N\}$ is a supermartingale. Thus, for any $T \in C_N$,

$$\mathrm{E}\, X_T = \sum_{j=1}^{N} \int_{[T=j]} X_j \leq \sum_{1}^{N} \int_{[T=j]} \gamma_j = \mathrm{E}\, \gamma_T \leq \mathrm{E}\, \gamma_1.$$

Moreover,

$$\begin{aligned}\mathrm{E}\, \gamma_1 &= \int_{[\sigma=1]} \gamma_1 + \int_{[\sigma>1]} \gamma_1 = \int_{[\sigma=1]} \gamma_1 + \int_{[\sigma>1]} \mathrm{E}\{\gamma_2 | \mathscr{F}_1\} \\ &= \int_{[\sigma=1]} \gamma_1 + \int_{[\sigma>1]} \gamma_2 = \int_{[\sigma=1]} \gamma_1 + \int_{[\sigma=2]} \gamma_2 + \int_{[\sigma>2]} \gamma_2 \\ &= \cdots = \sum_{j=1}^{N} \int_{[\sigma=j]} \gamma_j = \mathrm{E}\, \gamma_\sigma = \mathrm{E}\, X_\sigma,\end{aligned}$$

and (27) follows. □

Lemma 7. *For any random variables $Y_1, \ldots, Y_N$ setting $X_n = I_{[Y_n \geq 1]}$ and $\mathscr{F}_n = \sigma(Y_1, \ldots, Y_n)$, $1 \leq n \leq N$,*

$$\mathrm{P}\left\{\max_{1 \leq n \leq N} Y_n \geq 1\right\} = \mathrm{E}\, \gamma_1 \tag{28}$$

where γ_1 is as in (26).

PROOF. Set $T = \inf\{1 \le n \le N: X_n = 1\}$ where $\inf \emptyset = N$. Then $T \in C_N$ and

$$\begin{aligned} P\left\{\max_{1 \le j \le N} Y_j \ge 1\right\} &= P\left\{\max_{1 \le j \le N} X_j = 1\right\} \\ &= P\{T < N\} + P\{T = N, X_N = 1\} \\ &= E\, X_T = E\, \gamma_1 \end{aligned}$$

by Lemma 6. □

Theorem 5 (Dvoretzky). *Let* $\{X_n, n \ge 1\}$ *be* i.i.d. *with* $P\{X_1 = 1\} = p = 1 - P\{X_1 = -1\}$ *and* $q = 1 - p \ge \frac{1}{2}$. *Set* $\mathscr{F}_n = \sigma(X_1, \ldots, X_n)$, $n \ge 1$, *and* $\mathscr{F}_0 = \{\emptyset, \Omega\}$. *For any constant* $0 < f = f_0 \le 1$, *designate the fortunes associated with the bold strategy by*

$$f_n = f_{n-1} + \bar{f}_{n-1} X_n \quad \textit{where } \bar{a} = \min(a, 1 - a). \tag{29}$$

If g_n, $n \ge 1$, *are any* $\mathscr{F}_n$*-measurable functions with* $0 \le g_n \le 1$ *and* h_n, $n \ge 1$, *are the fortunes associated with this alternative betting strategy, that is,* $h_0 = f$ *and*

$$h_n = h_{n-1}(1 + g_{n-1} X_n), \qquad n \ge 1, \tag{30}$$

then for all $n = 0, 1, \ldots,$ *and all* $f \in (0, 1)$

$$p_N(f) = P\left\{\max_{0 \le j \le N} f_j \ge 1\right\} \ge P\left\{\max_{0 \le j \le N} h_j \ge 1\right\}. \tag{31}$$

PROOF. Without loss of generality suppose $0 < f < 1$. Set

$$d_N = I_{[h_N \ge 1]}, \qquad d_n = \max[I_{[h_n \ge 1]}, E\{d_{n+1} | \mathscr{F}_n\}] \tag{32}$$

for $0 \le n < N$. By Lemma 7

$$P\left\{\max_{0 \le j \le N} h_j \ge 1\right\} = E\, d_0, \tag{33}$$

and obviously $d_N = p_0(h_N)$, a.c.

Suppose inductively for some n in $[1, N)$ that

$$d_{n+1} \le p_{N-n-1}(h_{n+1}), \text{ a.c.}$$

Now (32) entails

$$d_n = 1 = p_{N-n}(h_n) \quad \text{if } h_n \ge 1,$$

while if $h_n < 1$, via the induction hypothesis with probability one,

$$\begin{aligned} d_n = E\{d_{n+1} | \mathscr{F}_n\} &\le E\{p_{N-n-1}(h_{n+1}) | \mathscr{F}_n\} \\ &= E\{p_{N-n-1}(h_n + h_n g_n X_{n+1}) | \mathscr{F}_n\} \end{aligned}$$

$$= p \cdot p_{N-n-1}(h_n + h_n g_n) + q \cdot p_{N-n-1}(h_n - h_n g_n)$$
$$\leq p_{N-n}(h_n),$$

recalling Lemma 5. This completes the (backward) induction, whence for $0 \leq n \leq N$

$$d_n \leq p_{N-n}(h_n),$$

and, in particular, recalling (33),

$$\mathrm{P}\left\{\max_{0 \leq j \leq N} h_j \geq 1\right\} = \mathrm{E}\, d_0 \leq p_N(h_0) = p_N(f). \qquad \square$$

References

D. G. Austin, "A sample property of martingales," *Ann. Math. Stat.* **37** (1966), 1396–1397.

D. Blackwell, "On optimal systems," *Ann. Math. Stat.* **25** (1954), 394–397.

B. M. Brown, "A note on convergence of moments," *Ann. Math. Stat.* **42** (1971), 777–779.

D. L. Burkholder, "Martingale transforms," *Ann. Math. Stat.* **37** (1966), 1494–1504.

D. L. Burkholder, "Distribution function inequalities for martingales," *Ann. Probability* **1** (1973), 19–42.

D. L. Burkholder and R. F. Gundy, "Extrapolation and interpolation of quasi-linear operators on martingales," *Acta Math.* **124** (1970), 249–304.

D. L. Burkholder, B. J. Davis, and R. F. Gundy, "Inequalities for convex functions of operators on martingales," *Proc. Sixth Berkeley Symp. Math. Stat. Prob.* **2** (1972), 223–240.

Y.S. Chow, "On a strong law of large numbers for martingales," *Ann. Math. Stat.* **38** (1967), 610.

Y. S. Chow, "Convergence of sums of squares of martingale differences," *Ann. Math. Stat.* **39** (1968), 123–133.

Y. S. Chow, "On the L_p-convergence for $n^{-1/p}S_n$, $0 < p < 2$," *Ann. Math. Stat.* **42** (1971), 393–394.

K. L. Chung, *A Course in Probability Theory*, Harcourt Brace, New York, 1968; 2nd ed., Academic Press, New York, 1974.

B. Davis, "A comparison test for martingale inequalities," *Ann. Math. Stat.* **40** (1969), 505–508.

J. L. Doob, *Stochastic Processes*, Wiley, New York, 1953.

L. E. Dubins and D. A. Freedman, "A sharper form of the Borel–Cantelli lemma and the strong law," *Ann. Math. Stat.* **36** (1965), 800–807.

L. E. Dubins and L. J. Savage, *How to Gamble If You Must*, McGraw–Hill, New York, 1965.

A. M. Garsia, "On a convex function inequality for martingales," *Ann. Probability* **1** (1973), 171–174.

R. F. Gundy, "A decomposition for $\mathscr{L}_1$-bounded martingales," *Ann. Math. Stat.* **39** (1968), 134–138.

J. Neveu, *Martingales à temps discrets*, Masson, Paris, 1972.

R. Panzone, "Alternative proofs for certain upcrossing inequalities," *Ann. Math. Stat.* **38** (1967), 735–741.

E. M. Stein, *Topics in Harmonic Analysis Related to the Littlewood–Paley Theory*, Princeton Univ. Press, Princeton, 1970.

H. Teicher, "Moments of randomly stopped sums-revisited," *Jour. Theor. Prob.* **9** (1995), 779–793.

A. Zygmund, *Trigonometric Series*, Vol. I, Cambridge, 1959.

12 Infinitely Divisible Laws

Row sums $\sum_{i=1}^{k_n} X_{ni}$ of arrays of random variables $\{X_{ni}, 1 \le i \le k_n \to \infty, n \ge 1\}$ that are rowwise independent have been considered briefly with respect to the Marcinkiewicz–Zygmund type strong laws of large numbers (Example 10.4.1). In this same context, non-Iterated Logarithm laws and generalizations thereof have been dealt with by H. Cramér and C. Esseen (see references at the end of this chapter). Here, limit distributions of row sums of the variables in such an array will be treated.

It is a remarkable fact that the class of limit distributions of normed sums of i.i.d. random variables is severely circumscribed. If the underlying r.v.s, say $\{X_n, n \ge 1\}$ have merely absolute moments of order r, then for $r \ge 2$ only the normal distribution can arise as a limit, while if $0 < r \le 2$, the limit law belongs to a class called stable distributions. If the basic r.v.s are merely independent (and infinitesimal when normed cf. (1) of Section 2), a larger class of limit laws, the so-called class $\mathscr{L}$ emerges. But even the class $\mathscr{L}$ does not contain a distribution of such crucial importance as the Poisson. A perusal of the derivation (Chapter 2) of the Poisson law as a limit of binomial laws B_n reveals that the success probability associated with B_n is a function of n. Thus, if B_{n-1} is envisaged as the distribution of the sum of i.i.d. random variables $Y_1, \ldots, Y_{n-1}$, then B_n must be the distribution of the sum of n **different** i.i.d. random variables which, therefore, may as well be labeled $X_{n,1}, X_{n,2}, \ldots, X_{n,n}$. In other words, to obtain the Poisson law as a limit of distributions of sums of i.i.d. (or even independent) random variables, a double sequence schema $\{X_{nj}, j = 1, \ldots, k_{n \to \infty}\}$ must be employed (with $X_{n,1}, \ldots, X_{n,k_n}$ independent within rows for each $n = 1, 2, \ldots$). Under one further proviso, the class of limit laws of (row) sums of such r.v.s coincides with the class of infinitely divisible laws.

12.1 Infinitely Divisible Characteristic Functions

It should be borne in mind that the notion of infinite divisibility as presented below is a distribution concept requiring no mention or consideration of r.v.s. In fact, an attempt to define it via r.v.s can lead to unnecessary difficulty and complication (see Gnedenko and Kolmogorov (1954)).

Definition. A d.f. F is called **infinitely divisible** (i.d.) if for every integer $n \geq 1$ there exists a d.f. F_n such that $F = F_n * F_n * \cdots * F_n = (F_n)^{n*}$ or equivalently if its c.f. φ (also called i.d.) is the nth power of a c.f. φ_n for every integer $n \geq 1$.

Clearly, the normal, Poisson, and degenerate distributions are i.d. Moreover, if $F(x)$ is i.d., all distributions of the same type $F(ax + b)$ are i.d.

A number of useful facts about i.d. distributions and c.f.s will be amassed in the propositions which follow.

Proposition 1. *A* d.f. F *with bounded support is* i.d. iff *it is degenerate.*

PROOF. Although the proof can be couched solely in terms of d.f.s, it is more easily intuited, and hence will be presented, in terms of r.v.s. Thus, if X is a r.v. with d.f. F, the hypothesis implies that with probability one, $|X| \leq C < \infty$. Without loss of generality suppose $\mathrm{E}\, X = 0$. Then, since $F = (F_n)^{n*}$, $n \geq 1$, if $\{X_{ni}, 1 \leq i \leq n\}$ are (fictitious) i.i.d. random variables with d.f. F_n, necessarily $\mathrm{E}\, X_{n1} = 0$ and $|X_{n1}| \leq C/n$ with probability one. Consequently, if σ^2 denotes the variance of X, $0 \leq \sigma^2 = \sum_{i=1}^n \mathrm{E}(X_{ni} - \mathrm{E}\, X_{ni})^2 = n\, \mathrm{E}\, X_{n1}^2 \leq n(C/n)^2 = o(1)$, whence $\sigma^2 = 0$, implying $\mathrm{P}\{X = \mathrm{E}\, X\} = 1$, i.e., F degenerate. □

Proposition 2. *An* i.d. c.f. $\varphi(t)$ *does not vanish* (*for real* t).

PROOF. By hypothesis, $\varphi = \varphi_n^n$, $n \geq 1$, with φ_n a c.f. Then $\psi = |\varphi|^2$ and $\psi_n = |\varphi_n|^2$ are real-valued c.f.s and the positive real function ψ has a unique real, positive nth root, say $\psi^{1/n}$, $n \geq 1$. Since necessarily $\psi = \psi_n^n$, $n \geq 1$, the positive real function ψ_n must coincide with $\psi^{1/n}$. Thus, $0 \leq \psi \leq 1$ implies $\Gamma(t) = \lim_{n\to\infty} \psi_n(t) = 1$ or 0 according as $\psi(t) > 0$ or $\psi(t) = 0$. Then, $\psi(0) = 1$ and continuity of ψ imply $\Gamma(t) = 1$ throughout some neighborhood of $t = 0$. Theorem 8.3.3 ensures that $\Gamma(t)$ is a c.f., whence continuity dictates that Γ is nonvanishing. Hence, ψ and therefore also φ is nonvanishing. □

If φ is an i.d. c.f. so that $\varphi = \varphi_n^n$, $n \geq 1$, it seems plausible that $\varphi_n = \varphi^{1/n}$. But how can one choose a continuous version of $\varphi^{1/n}$? The following lemma asserts that a continuous logarithm (hence also nth root) of a continuous, nonvanishing complex function on $(-\infty, \infty)$ can be defined.

Lemma 1. *If* $f(t)$ *is a continuous, nonvanishing complex function on* $[-T, T]$ *with* $f(0) = 1$, *there is a unique* (*single-valued*) *continuous function* $\lambda(t)$ *on*

$[-T, T]$ *with* $\lambda(0) = 0$ *and* $f(t) = e^{\lambda(t)}$. *Moreover,* $[-T, T]$ *is replaceable by* $(-\infty, \infty)$.

PROOF. If $\rho_T = \inf_{[-T,T]}|f(t)|$, then $0 < \rho_T \le 1$. Since f is uniformly continuous on $[-T, T]$, there exists δ_T in $(0, \rho_T)$ such that $|t' - t| \le \delta_T$ implies $|f(t') - f(t)| < \rho_T/2 \le \frac{1}{2}$. Choose points $\{t_j\}$ with $t_0 = 0$ such that $-T = t_{-m} < \cdots < t_{-1} < t_0 < t_1 < \cdots < t_m = T$ and $t_{j+1} - t_j = t_1 - t_0 \le \delta_T$. Define

$$L(z) = \sum_{j=1}^{\infty} \frac{(-1)^{j-1}}{j}(z-1)^j, \qquad |z - 1| < 1.$$

Then $L(z)$ is the unique determination (principal value) of $\log z$ vanishing at $z = 1$. For $t \in [t_{-1}, t_1]$, $|f(t) - 1| = |f(t) - f(t_0)| \le \frac{1}{2}$, and so $L(f(t))$ is defined. Set

$$\lambda(t) = L(f(t)), \qquad t \in [t_{-1}, t_1].$$

Then $\lambda(0) = L(1) = 0$ and $\lambda(t)$ is continuous with $\exp\{\lambda(t)\} = f(t)$ in $[t_{-1}, t_1]$. Since for $t \in [t_k, t_{k+1}]$, $|(f(t)/f(t_k)) - 1| \le (\rho_T/2)/\rho_T = \frac{1}{2}$, for any k the definition of λ may be extended from $[t_{-k}, t_k]$ to $[t_k, t_{k+1}]$ by $\lambda(t) = \lambda(t_k) + L((f(t)/f(t_k)))$; analogously, replacing t_k by t_{-k}, the definition extends to $[t_{-k-1}, t_{-k}]$. Then $\lambda(t)$ is defined and continuous in $[-T, T]$, and for $t \in [t_k, t_{k+1}]$, $k \ge 1$,

$$e^{\lambda(t)} = \exp\left(L\left(\frac{f(t)}{f(t_k)}\right) + \lambda(t_k)\right) = \exp\left(L\left(\frac{f(t)}{f(t_k)}\right) + \sum_{j=0}^{k-1} L\left(\frac{f(t_{j+1})}{f(t_j)}\right)\right) = f(t).$$

A similar statement holds in $[t_{-k-1}, t_{-k}]$. Next, given λ in $[-T, T]$, it may be extended by the prior method to $[-T-1, T+1]$, and hence by induction to $(-\infty, \infty)$. Finally, if two such functions λ and λ' exist, $e^{\lambda(t)} = e^{\lambda'(t)}$, whence $\lambda(t) - \lambda'(t) = 2\pi i k(t)$ with $k(t)$ an integer. Since $k(t)$ is continuous with $k(0) = 0$, necessarily $k(t) = 0$ and λ is unique.

Definition. The function $\lambda(t)$ defined by Lemma 1 is called the **distinguished logarithm** of $f(t)$ and is denoted by $\operatorname{Log} f(t)$. Also $\exp\{(1/n)\lambda(t)\}$ is called the **distinguished *n*th root** of $f(t)$ and is denoted by $f^{1/n}(t)$.

Note. Clearly, if $\psi(t)$ is a continuous complex function on $(-\infty, \infty)$ with $\psi(0) = 0$, then $\operatorname{Log} e^{\psi(t)} = \psi(t)$. Moreover, for f, g as in Lemma 1, $\operatorname{Log} f \cdot g = \operatorname{Log} f + \operatorname{Log} g$, $\operatorname{Log}(f/g) = \operatorname{Log} f - \operatorname{Log} g$, and $\operatorname{Log} f = L(f)$ for $|t| \le T$ whenever $\sup_{|t| \le T}|f(t) - 1| < 1$. Thus, for k an integer, $\operatorname{Log}(e^{ait + 2k\pi i}) = ait$ and

$$(e^{ait + 2k\pi i})^{1/n} = e^{ait/n}.$$

Lemma 2. *Let* f, f_k, $k \ge 1$ *be as in Lemma* 1. *If* $f_k \to f$ *uniformly in* $[-T, T]$, *then* $\operatorname{Log} f_k - \operatorname{Log} f \to 0$ *uniformly in* $[-T, T]$.

PROOF. Since $\min_{|t|\le T}|f(t)| > 0$ and $f_k \to f$ uniformly in $[-T, T]$, $\sup_{|t|\le T}|(f_k(t)/f(t)) - 1| \le \frac{1}{2}$ for $k \ge K_0$. Then

$$\operatorname{Log} f_k - \operatorname{Log} f = \operatorname{Log}\frac{f_k}{f} = L\left(\frac{f_k(t)}{f(t)}\right) \to L(1) = 0, \quad \text{uniformly in } [-T, T].$$

□

Proposition 3. *A* c.f. $\varphi(t)$ *is* i.d. iff *its distinguished nth root* $\varphi^{1/n}(t) = e^{(1/n)\operatorname{Log}\varphi(t)}$ *is a* c.f. *for every positive integer.*

PROOF. If φ is i.d., $\varphi = \varphi_n^n$, $n \ge 1$, and so by Proposition 2, φ and hence also φ_n is nonvanishing, whence their distinguished nth roots and logarithms are well defined by Lemma 1. Moreover, $e^{\operatorname{Log}\varphi} = \varphi = \varphi_n^n = e^{n\operatorname{Log}\varphi_n}$, so that $\operatorname{Log}\varphi(t) = n\operatorname{Log}\varphi_n(t) + 2\pi i k(t)$ with $k(t)$ an integer. Since $\operatorname{Log}\varphi$ and $\operatorname{Log}\varphi_n$ are continuous and vanish at zero, $k(t) = 0$, whence $\operatorname{Log}\varphi_n = (1/n)\operatorname{Log}\varphi$, which is tantamount to $\varphi_n = \varphi^{1/n}$.

Conversely if the distinguished nth root of φ exists and is a c.f. for every $n \ge 1$, $\varphi = e^{\operatorname{Log}\varphi} = (e^{(1/n)\operatorname{Log}\varphi})^n$ shows that φ is i.d. □

Proposition 4. *A finite product of* i.d. c.f.s *is* i.d. *Moreover, if* i.d. c.f.s $\varphi_k \to \varphi$, *a* c.f., *then this limit* c.f. φ *is* i.d.

PROOF. Clearly, if $\varphi = \varphi_n^n$, $\psi = \psi_n^n$, $n \ge 1$, then $\varphi \cdot \psi = [\varphi_n \cdot \psi_n]^n$, $n \ge 1$, shows that a product of two and hence any finite number of i.d. c.f.s is i.d. Suppose next that the i.d. c.f.s $\varphi_k \to \varphi$, a c.f. Then, the i.d. c.f.s $\psi_k(t) = |\varphi_k(t)|^2 = \varphi_k(t)\cdot\varphi_k(-t) \to$ the c.f. $\psi(t) = |\varphi(t)|^2$. Consequently, $\psi_k^{1/n}$ as the positive nth root of the positive function ψ_k tends as $k \to \infty$ to the nonnegative nth root $\psi^{1/n}$ of the nonnegative function ψ. Since for $n \ge 1$, $\psi_k^{1/n}$ is a sequence of c.f.s whose limit $\psi^{1/n}$ is continuous at $t = 0$, $\psi^{1/n}$ is a c.f. for $n \ge 1$. Thus, ψ is i.d. and hence nonvanishing. Consequently, φ is nonvanishing, whence $\operatorname{Log}\varphi$ is defined by Lemma 1. By Lemma 2,

$$\varphi_k^{1/n} - \varphi^{1/n} = \varphi^{1/n}[e^{(1/n)(\operatorname{Log}\varphi_k - \operatorname{Log}\varphi)} - 1] \to 0, n \ge 1,$$

as $k \to \infty$, and since $\varphi^{1/n}$ is continuous at $t = 0$, it is a c.f. for all $n \ge 1$, so that φ is i.d. by Proposition 3. □

Since c.f.s $\exp\{\lambda(e^{itu} - 1) + it\theta\}$, $\lambda > 0$, of Poisson type are i.d., it follows from Proposition 4 that $\exp\{\sum_{j=1}^{k}[\lambda_j(e^{itu_j} - 1) + it\theta_j]\}$ and hence also $\exp\{it\theta + \int_{-\infty}^{\infty}(e^{itu} - 1)dG(u)\}$ with G a bounded, increasing function is i.d. The latter comes close but does not quite exhaust the class of i.d. c.f.s.

Proposition 5. *The class of* i.d. *laws coincides with the class of distribution limits of finite convolutions of distributions of Poisson type.*

PROOF. That every such limit is i.d. follows directly from Proposition 4. Conversely, if φ is i.d., so that $\varphi = \varphi_n^n$, $n \ge 1$, then

$$n[\varphi_n(t) - 1] = n[e^{(1/n)\operatorname{Log}\varphi} - 1] \to \operatorname{Log}\varphi,$$

that is

$$\lim_{n\to\infty} \exp\{n[\varphi_n(t) - 1]\} = \varphi(t).$$

Now,

$$n[\varphi_n(t) - 1] = \int_{-\infty}^{\infty} n(e^{itu} - 1)dF_n(u), \qquad n \geq 1,$$

and a net $-\infty < -M_n = u_{n,1} < u_{n,2} < \cdots < u_{n,k_n+1} = M_n < \infty$ may be chosen whose points are continuity points of F_n and such that

$$\max_j(u_{n,j+1} - u_{n,j}) \leq \frac{1}{2n^3}$$

and

$$\int_{[|u| \geq M_n]} dF_n(u) \leq \frac{1}{4n^2}.$$

Then for $|t| \leq n$, choosing $\lambda_{n,j} = n[F_n(u_{n,j+1}) - F_n(u_{n,j})]$,

$$\left| \int_{u_{n,j}}^{u_{n,j+1}} (e^{itu} - 1)n\,dF_n(u) - \lambda_{n,j}(e^{itu_{n,j}} - 1) \right|$$

$$= \left| \int_{u_{n,j}}^{u_{n,j+1}} (e^{itu} - e^{itu_{n,j}})n\,dF_n(u) \right|$$

$$\leq \int_{u_{n,j}}^{u_{n,j+1}} |1 - e^{it(u_{n,j} - u)}|n\,dF_n(u) \leq \int_{u_{n,j}}^{u_{n,j+1}} |t(u_{n,j} - u)|n\,dF_n(u)$$

$$\leq \frac{1}{2n} \int_{u_{n,j}}^{u_{n,j+1}} dF_n(u).$$

Hence, for $|t| \leq n$, summing over $1 \leq j \leq k_n$,

$$\left| \int_{-\infty}^{\infty} (e^{itu} - 1)n\,dF_n(u) - \sum_{j=1}^{k_n} \lambda_{n,j}(e^{itu_{n,j}} - 1) \right|$$

$$\leq 2n \int_{[|u| \geq M_n]} dF_n(u) + \sum_{1}^{k_n} \frac{1}{2n} \int_{u_{n,j}}^{u_{n,j+1}} dF_n(u) \leq \frac{1}{n}.$$

Consequently, for $|t| \leq n$ and sufficiently large n

$$\left| e^{n(\varphi_n(t) - 1)} - \prod_{j=1}^{k_n} \exp(\lambda_{n,j}(e^{itu_{n,j}} - 1)) \right|$$

$$= \left| \exp\left(\sum_{1}^{k_n} \lambda_{n,j}(e^{itu_{n,j}} - 1) \right) - \exp(n(\varphi_n(t) - 1)) \right|$$

$$= |e^{n(\varphi_n(t) - 1)}| \left| \exp\left(\sum_{1}^{k_n} \lambda_{n,j}(e^{itu_{n,j}} - 1) - \int_{-\infty}^{\infty} (e^{itu} - 1)n\,dF_n(u) \right) - 1 \right|$$

$$\leq 2|\varphi(t)|2(e^{1/n} - 1) = o(1),$$

recalling Corollary 8.4.1. Consequently, for all real t

$$\varphi(t) = \lim_{n\to\infty} e^{n[\varphi_n(t)-1]} = \lim_{n\to\infty} \prod_{j=1}^{k_n} \exp(\lambda_{n,j}(e^{itu_{n,j}} - 1)). \qquad \square$$

For any real γ and nondecreasing, left-continuous $G(u)$ with $G(-\infty) = 0$, $G(\infty) < \infty$, set

$$\psi(t) = \psi(t; \gamma, G) = i\gamma t + \int_{-\infty}^{\infty} \left(e^{itu} - 1 - \frac{itu}{1+u^2}\right)\left(\frac{1+u^2}{u^2}\right) dG(u), \quad (1)$$

where the integrand, say $h(t, u)$, is defined at $u = 0$ by continuity (whence $h(t, 0) = -t^2/2$). Since $e^{itu} - 1 = O(1)$ as $|u| \to \infty$ and $e^{itu} - 1 = itu + O(u^2)$ as $u \to 0$, dominated convergence ensures that $\psi(t)$ is continuous, and, clearly, $\psi(0) = 0$ and $\psi(t) = \text{Log } e^{\psi(t)}$.

Theorem 1. $\varphi(t) = \exp\{\psi(t; \gamma, G)\}$ *as defined by* (1) *is an* i.d. c.f. *for every real* γ *and* G *as stipulated. Moreover,* φ *uniquely determines* γ *and* G.

PROOF. The integrand $h(t, u)$ of ψ satisfies $|h(t, u)| \le C < \infty$ for $|t| \le T$ and all real u. Choose $-M_n = u_{n,1} < u_{n,2} < \cdots < u_{n,k_n+1} = M_n$ to be nonzero continuity points of G for which

$$\int_{[|u|\ge M_n]} dG(u) \le \frac{1}{2Cn}, \quad \max_{\substack{u_{n,j}\le u\le u_{n,j+1}\\ 1\le j\le k_n,\, |t|\le T}} |h(t, u_{n,j}) - h(t, u)| \le \frac{1}{2nG(\infty)}.$$

Then, for $|t| \le T$

$$\left|\int_{-\infty}^{\infty} h(t, u)dG(u) - \sum_{j=1}^{k_n} h(t, u_{n,j})[G(u_{n,j+1}) - G(u_{n,j})]\right|$$

$$\le C \int_{[|u|\ge M_n]} dG(u) + \sum_{j=1}^{k_n} \frac{G(u_{n,j+1}) - G(u_{n,j})}{2nG(\infty)} \le \frac{1}{n}.$$

Thus, setting $\lambda_{n,j} = [(1 + u_{n,j}^2)/u_{n,j}^2][G(u_{n,j+1}) - G(u_{n,j})]$ and

$$a_{n,j} = \frac{\gamma}{k_n} - \frac{G(u_{n,j+1}) - G(u_{n,j})}{u_{n,j}},$$

$$e^{\psi(t)} = \lim_{n\to\infty} \exp\left\{i\gamma t + \sum_{j=1}^{k_n} h(t, u_{n,j})[G(u_{n,j+1}) - G(u_{n,j})]\right\}$$

$$= \lim_{n\to\infty} \prod_{j=1}^{k_n} \exp(ita_{n,j} + \lambda_{n,j}(e^{itu_{n,j}} - 1)),$$

and so $\varphi(t)$ is a c.f. and i.d. by Proposition 4.

Apropos of uniqueness, define

$$\begin{aligned}
-V(t) &= \int_{t-1}^{t+1} \psi(w)dw - 2\psi(t) = \int_{t-1}^{t+1}\int_{-\infty}^{\infty} h(w, u)dG(u)dw + 2i\gamma t - 2\psi(t) \\
&= \int_{-\infty}^{\infty}\left[e^{itu}\left(\frac{e^{iu} - e^{-iu}}{iu}\right) - 2 - \frac{2itu}{1+u^2}\right]\left(\frac{1+u^2}{u^2}\right)dG(u) \\
&\quad - 2\int_{-\infty}^{\infty} h(t, u)dG(u) \\
&= -2\int_{-\infty}^{\infty} e^{itu}\left(1 - \frac{\sin u}{u}\right)\left(\frac{1+u^2}{u^2}\right) dG(u) = -\int_{-\infty}^{\infty} e^{itu}\, dH(u),
\end{aligned}$$

where

$$H(u) = 2\int_{-\infty}^{u}\left(1 - \frac{\sin x}{x}\right)\left(\frac{1+x^2}{x^2}\right)dG(x).$$

Clearly, H is nondecreasing and left continuous with $H(-\infty) = 0$, $H(\infty) = C_1 < \infty$. Thus, ψ determines V, which, in turn, determines H (by Theorem 8.3.1), which by Theorem 6.5.2 determines G (hence, also γ). □

Let $\{G, G_n, n \geq 1\}$ be nondecreasing, left-continuous functions of bounded variation with $G(-\infty) = G_n(-\infty) = 0$, $n \geq 1$. Recall as in Section 8.1 that $G_n \xrightarrow{c} G$ iff $G_n \xrightarrow{w} G$ and $G_n(\infty) \to G(\infty)$, $G_n(-\infty) \to G(-\infty)$.

Theorem 2. *Let $\{\gamma, \gamma_n, n \geq 1\}$ be finite real numbers and $\{G, G_n, n \geq 1\}$ nondecreasing left-continuous functions of bounded variation which vanish at $-\infty$. If $\gamma_n \to \gamma$ and $G_n \xrightarrow{c} G$, then $\psi(t, \gamma_n, G_n) \to \psi(t; \gamma, G)$ for all real t, where ψ is as in (1). Conversely, if $\psi(t; \gamma_n, G_n)$ tends to a continuous function $g(t)$ as $n \to \infty$, then necessarily $g(t) = \psi(t, \gamma, G)$ and $\gamma_n \to \gamma$, $G_n \xrightarrow{c} G$.*

Proof. If $G(\infty) = 0$, then $G_n(\infty) \to 0$ by complete convergence, whence $\psi(t; \gamma_n, G_n) \to \psi(t; \gamma, G) = i\gamma t$, recalling that the integrand $h(t, u)$ of (1) is bounded in modulus for fixed t. If $G(\infty) > 0$, then $G_n(\infty) > 0$ for all large n by Lemma 8.1.1, whence $(1/G_n(\infty))\psi(t; \gamma_n, G_n) \to (1/G(\infty))\, \psi(t; \gamma, G)$ by the Helly–Bray theorem, and so $\psi(t; \gamma_n, G_n) \to \psi(t; \gamma, G)$.

Apropos of the final assertion, Theorem 1 ensures that the i.d. c.f.s $e^{\psi_n(t)} = e^{\psi(t;\, \gamma_n,\, G_n)} \to e^{g(t)}$, continuous. Thus, $e^{g(t)}$ is a c.f. and i.d. by Theorem 1 and Proposition 4. Define $\alpha(t) = \operatorname{Log} e^{g(t)}$ and $\alpha_n(t) = \operatorname{Log} e^{\psi_n(t)} = \psi_n(t)$. By Theorem 8.3.3 $e^{\psi_n(t)} \to e^{g(t)}$ uniformly in $|t| \leq T$ for all $T \in (0, \infty)$, whence by Lemma 2, $\psi_n(t) \to \alpha(t)$ uniformly in $|t| \leq T$ and $\alpha(t)$ is continuous. Hence, recalling the proof and notation of the last part of Theorem 1 and defining

$$V_n(t) = 2\psi_n(t) - \int_{t-1}^{t+1} \psi_n(y)dy = \int_{-\infty}^{\infty} e^{ity}\, dH_n(y),$$

$$V(t) = 2\alpha(t) - \int_{t-1}^{t+1} \alpha(y)dy,$$

it follows that $V_n(t) \to V(t)$, continuous, and, in particular, $H_n(\infty) = V_n(0) \to V(0)$, whence $V(0) \geq 0$. If $V(0) = 0$, then

$$H_n(\infty) = 2\int_{-\infty}^{\infty}\left(1 - \frac{\sin x}{x}\right)\left(\frac{1 + x^2}{x^2}\right)dG_n(x) \to 0,$$

implying $G_n(\infty) \to 0$, whence $G(u) \equiv 0$ and necessarily γ_n tends to a finite limit γ. If $V(0) > 0$, the d.f.s $H_n(u)/H_n(\infty)$ (whose c.f.s $V_n(t)/V_n(0) \to V(t)/V(0)$) converge to a limit d.f., say $H(u)/V(0)$. Thus, $H_n \xrightarrow{c} H$, and by the Helly–Bray theorem for any continuity point u of H, recalling Theorem 6.5.2,

$$\begin{aligned} 2G_n(u) &= \int_{-\infty}^{u}\left(1 - \frac{\sin y}{y}\right)^{-1}\frac{y^2}{1 + y^2}\,dH_n(y) \\ &\to \int_{-\infty}^{u}\left(1 - \frac{\sin y}{y}\right)^{-1}\frac{y^2}{1 + y^2}\,dH(y). \end{aligned} \tag{2}$$

Define $G(u)$ to be the integral on the right side of (2). Since the continuity points of G and H are identical, $G_n \xrightarrow{c} G$. Hence, γ_n tends to a finite limit γ. Clearly, $\psi(t, \gamma, G) = g(t)$. □

From the preceding, a canonical form for i.d. c.f.s known as the Levy–Khintchine representation follows readily.

Theorem 3 (Lévy–Khintchine representation). *A c.f. $\varphi(t)$ is i.d. iff*

$$\varphi(t) = \exp\left\{i\gamma t + \int_{-\infty}^{\infty}\left(e^{itu} - 1 - \frac{itu}{1 + u^2}\right)\left(\frac{1 + u^2}{u^2}\right)dG(u)\right\}, \tag{3}$$

where γ, G are as stipulated in (1).

PROOF. Theorem 1 asserts that $\varphi(t) = e^{\psi(t;\,\gamma,\,G)}$ as above is i.d., and so it suffices to prove the converse. If $\varphi = \varphi_n^n$, $n \geq 1$, as in the proof of Proposition 5, $n[\varphi_n(t) - 1] \to \text{Log}\,\varphi(t)$. Now,

$$\begin{aligned} n[\varphi_n(t) - 1] &= \int_{-\infty}^{\infty} n(e^{itu} - 1)dF_n(u) \\ &= it\int_{-\infty}^{\infty}\frac{u}{1 + u^2}\,n\,dF_n(u) \\ &\quad + \int_{-\infty}^{\infty}\left(e^{itu} - 1 - \frac{itu}{1 + u^2}\right)\left(\frac{1 + u^2}{u^2}\right)\frac{u^2}{1 + u^2}\,n\,dF_n(u). \end{aligned}$$

Set

$$\gamma_n = \int_{-\infty}^{\infty}\frac{nu}{1 + u^2}\,dF_n(u), \qquad G_n(u) = \int_{-\infty}^{u}\frac{ny^2}{1 + y^2}\,dF_n(y),$$

$$\psi_n = \psi(t; \gamma_n, G_n).$$

As noted above, $\psi_n(t) \to \operatorname{Log} \varphi(t)$, which is continuous. Thus, by Theorem 2 $\gamma_n \to \gamma$, $G_n \xrightarrow{c} G$, and

$$\varphi(t) = \lim_{n\to\infty} e^{\psi_n(t)} = e^{\psi(t;\,\gamma,\,G)}. \qquad \square$$

In the case of distributions with finite variance, the canonical form in (3) admits considerable simplification.

Theorem 4 (Kolmogorov). *A function $\varphi(t)$ is the* c.f. *of an* i.d. *distribution with finite variance* iff *for some real γ^* and nondecreasing, left-continuous G^* with $G^*(-\infty) = 0$, $G^*(\infty) < \infty$*

$$\varphi(t) = \exp\left\{i\gamma^* t + \int_{-\infty}^{\infty} (e^{itu} - 1 - itu)\frac{1}{u^2}\, dG^*(u)\right\} \tag{4}$$

and, moreover, φ uniquely determines γ^* and G^*.

PROOF. If the d.f. corresponding to the i.d. c.f. $\varphi(t) = e^{\psi(t)}$ has a finite second moment, ψ has a finite second derivative at zero and *a fortiori* a generalized second derivative at zero. Now, via (3)

$$\frac{\psi(2t) - 2\psi(0) + \psi(-2t)}{(2t)^2} = \int_{-\infty}^{\infty} \frac{e^{2itu} - 2 + e^{-2itu}}{-(2it)^2}\frac{1+u^2}{u^2}\, dG(u)$$

$$= -\int_{-\infty}^{\infty}\left(\frac{e^{itu} - e^{-itu}}{2it}\right)^2 \frac{1+u^2}{u^2}\, dG(u) = -\int_{-\infty}^{\infty}\left(\frac{\sin tu}{tu}\right)^2 (1+u^2)\, dG(u),$$

whence

$$-\psi''(0) \geq \lim_{t\to 0}\int_{[|u|\leq 1/t]}\left(\frac{\sin tu}{tu}\right)^2 (1+u^2) dG(u) \geq (\sin 1)^2 \int_{-\infty}^{\infty} (1+u^2) dG(u).$$

Thus, $G^*(u) = \int_{-\infty}^{u} (1+y^2) dG(y)$ has the asserted properties i.e., $G^*(\infty) < \infty$ and

$$\int\left(e^{itu} - 1 - \frac{itu}{1+u^2}\right)\left(\frac{1+u^2}{u^2}\right) dG(u) = \int\left(e^{itu} - 1 - \frac{itu}{1+u^2}\right)\frac{1}{u^2}\, dG^*(u)$$

$$= \int (e^{itu} - 1 - itu)\frac{1}{u^2}\, dG^*(u) + it\int\left(u - \frac{u}{1+u^2}\right)\frac{1}{u^2}\, dG^*(u),$$

whence (4) holds with $\gamma^* = \gamma + \int u\, dG$.

Conversely, if $\varphi(t) = \exp\{\Gamma(t)\}$ is as in (4), then, since

$$\frac{1}{u^2}\left(u - \frac{u}{1+u^2}\right) = O(1),$$

necessarily $\Gamma(t) = \psi(t; \gamma, G) + itc$ for some constant c, where

$$G(u) = \int_{-\infty}^{u} (1+y^2)^{-1}\, dG^*(y),$$

whence $\varphi(t)$ is an i.d. c.f. by Theorem 1. Moreover, as $t' \to t$

$$\frac{\Gamma(t') - \Gamma(t)}{t' - t} = i\gamma^* + \int \left(\frac{e^{it'u} - e^{itu}}{t' - t} - iu \right) \frac{dG^*(u)}{u^2}$$

$$\to i\gamma^* + i \int (e^{itu} - 1) \frac{dG^*(u)}{u} = \Gamma'(t) \tag{5}$$

since for $|t' - t| \le 1$, recalling Lemma 8.4.1,

$$\left| \frac{e^{it'u} - e^{itu}}{t' - t} - iu \right| = \left| \frac{e^{i(t'-t)u} - 1}{t' - t} - iue^{-itu} \right|$$

$$\le \left| \frac{e^{i(t'-t)u} - 1 - i(t' - t)u}{t' - t} \right| + |u(1 - e^{-itu})| \le u^2(\tfrac{1}{2} + |t|),$$

which is integrable with respect to $u^{-2}\, dG^*(u)$. Analogously, as $t' \to t$

$$\frac{\Gamma'(t') - \Gamma'(t)}{t' - t} = i \int \frac{e^{it'u} - e^{itu}}{u(t' - t)}\, dG^*(u) \to - \int e^{itu}\, dG^*(u).$$

Thus, Γ and hence φ has a finite second derivative, whence the transform of φ has a finite second moment. Moreover, Γ'' and hence φ'' uniquely determines G^* and therefore also γ^*. From (5), $\gamma^* = -i\Gamma'(0)$ is the mean of the underlying distribution and it is readily verified that $G^*(\infty)$ is the variance. □

Exercises 12.1

1. Prove that if φ is an i.d. c.f., then φ^λ is a c.f. for every $\lambda \ge 0$.
2. Prove Proposition 1 without mentioning r.v.s.
3. Verify that the function G of (3) has a finite moment of even order $2k$ iff the same is true for the underlying i.d. distribution.
4. Let φ be a non-i.d. c.f. having the representation (3) with G not nondecreasing (but otherwise as in (1)). Prove that γ and G are still uniquely determined by φ.
5. If

$$P\{X = -1\} = \frac{\alpha(1 - \beta)}{1 + \alpha}, \qquad P\{X = k\} = \frac{1 - \beta}{1 + \alpha}(1 + \alpha\beta)\beta^k,\ k = 0, 1, \ldots,$$

where $0 < \alpha < \beta < 1$, show that X has an i.d. c.f. $\varphi(t)$ iff $\alpha = 0$ and, further, that $|\varphi(t)|^2$ is i.d. even when $\alpha \ne 0$.

6. Show for an i.d. c.f. $\varphi(t) = \exp\{\psi(t; \gamma, G)\}$ that if the support of its d.f. F is bounded from below, so is that of G. Is the converse true?
7. Prove that if $\varphi_0 = \varphi_k^{n_k}$, $k \ge 1$, where φ_k is a c.f. for $k \ge 0$ and n_k is a sequence of positive integers $\to \infty$, then φ is i.d.
8. Prove that if $\{X_n, n \ge 1\}$ are i.i.d. r.v.s with d.f. G and N is a Poisson r.v. (parameter λ) independent of $\{X_n, n \ge 1\}$, then the c.f. of $\sum_1^N X_i$ is $\exp\{\lambda \int (e^{itu} - 1) dG(u)\}$.
9. Show that an i.d. mixing G of an additively closed family $\mathscr{F} = \{F(x; \lambda), \lambda \in \Lambda \subset R^m\}$ yields an i.d. mixture H. *Hint*: Recall Exercise 8.3.14.

12.2 Infinitely Divisible Laws as Limits

On several occasions double sequences of random variables (independent within rows) have been briefly encountered. Such a general schema comprises an array of r.v.s

$$\{X_{n,k}, 1 \le k \le k_n < \infty, n \ge 1, k_n \to \infty\} \tag{1}$$

with corresponding d.f.s $\{F_{n,k}\}$ and c.f.s $\{\varphi_{n,k}\}$ such that within each row, i.e., for each $n \ge 1$, the r.v.s $X_{n,1}, X_{n,2}, \ldots, X_{n,k_n}$ are independent. The r.v.s of an array such as in (1) will be called **infinitesimal** if

(i) $$\max_{1 \le k \le k_n} P\{|X_{n,k}| > \varepsilon\} = o(1), \quad \text{all } \varepsilon > 0,$$

that is, if the row elements become uniformly small in the sense of (i).

Exactly as in the proof of the weak law of large numbers (Section 10.1), this implies

(ii) $$\max_{1 \le k \le k_n} |m(X_{n,k})| = o(1),$$

where, as usual, $m(X)$ is a median of X. Moreover, since

$$\max_k \int_{[|x| < \tau]} |x|^r \, dF_{nk} \le \varepsilon^r + \max_k \int_{[\varepsilon < |x| < \tau]} |x|^r \, dF_{nk}$$
$$\le \varepsilon^r + \tau^r \max_k P\{|X_{nk}| > \varepsilon\},$$

infinitesimality also entails

(iii) $$\max_{1 \le k \le k_n} \int_{[|x| < \tau]} |x|^r \, dF_{nk}(x) = o(1) \quad \text{for all } r > 0, \tau > 0.$$

Lemma 1. *The infinitesimality condition* (i) *is equivalent to either*

(i′) $$\max_{1 \le k \le k_n} \int_{-\infty}^{\infty} \frac{x^2}{1 + x^2} \, dF_{nk}(x) = o(1)$$

or

(i″) $$\max_{1 \le k \le k_n} |1 - \varphi_{n,k}(t)| = o(1) \quad \textit{uniformly in } |t| \le T \quad \textit{for all } T > 0.$$

Proof. $\max_k \int [x^2/(1 + x^2)] dF_{nk}(x) \le \varepsilon^2 + \max_k \int_{[|x| \ge \varepsilon]} dF_{nk} \to 0$ as $n \to \infty$, and then $\varepsilon \to 0$ under (i). Conversely, under (i′), for all $\varepsilon > 0$

$$\frac{\varepsilon^2}{1 + \varepsilon^2} \max_k P\{|X_{n,k}| > \varepsilon\} \le \max_k \int_{[|x| \ge \varepsilon]} \frac{x^2}{1 + x^2} \, dF_{nk}(x) = o(1),$$

and so (i) obtains. Next, for $|t| \le T$, recalling Lemma 8.4.1,

$$\begin{aligned}\max_k |1 - \varphi_{nk}(t)| &= \max_k \left| \int (e^{itx} - 1) dF_{nk} \right| \\ &\le \max_k \left[\int_{[|x| \le \varepsilon]} |tx| dF_{nk} + 2 \int_{[|x| > \varepsilon]} dF_{nk} \right] \\ &\le \varepsilon T + 2 \max_k \mathrm{P}\{|X_{nk}| > \varepsilon\} \to 0 \quad .\end{aligned}$$

as $n \to \infty$, and then $\varepsilon \to 0$, so that (i) implies (i″). Finally, since $1 - \mathscr{R}\{\varphi\} \le |1 - \varphi|$, Lemma 8.3.1 stipulates a positive constant $a(c, \delta)$ such that

$$\max_k \mathrm{P}\{|X_{nk}| > c\} \le a(c, \delta) \int_0^\delta \max_k |1 - \varphi_{n,k}(t)| dt = o(1)$$

for all $c > 0$, whence (i″) ensures (i). □

For fixed but arbitrary τ, $0 < \tau < \infty$, define

$$\begin{aligned} a_{n,k} = a_{n,k}(\tau) = \int_{[|x| < \tau]} x\, dF_{nk}, \qquad \tilde{X}_{n,k} = X_{n,k} - a_{n,k}, \\ \tilde{F}_{n,k}(x) = F_{n,k}(x + a_{n,k}), \qquad \tilde{\varphi}_{n,k}(t) = e^{-ita_{n,k}} \varphi_{n,k}(t). \end{aligned} \tag{2}$$

Since (i) entails (iii), $\max_{1 \le k \le k_n} |a_{n,k}| = o(1)$ for all $\tau > 0$, and so $\{X_{n,k}\}$ infinitesimal implies $\{\tilde{X}_{n,k}\}$ infinitesimal and hence also, via Lemma 1,

(iv) $\max_{1 \le k \le k_n} |1 - \tilde{\varphi}_{n,k}(t)| = o(1)$ uniformly in $|t| \le T$ for all $T > 0$.

Lemma 2. *If $\{X_{n,k}\}$ are infinitesimal and $\{\tilde{F}_{n,k}, \tilde{\varphi}_{n,k}\}$ are defined by (2), then for any τ, $T > 0$ and $n \ge N_\tau$, there exist positive constants $c_i = c_i(T, \tau)$, $i = 1, 2$, such that for $1 \le k \le k_n$*

$$c_1 \sup_{|t| \le T} |1 - \tilde{\varphi}_{n,k}(t)| \le \int \frac{x^2}{1 + x^2} d\tilde{F}_{n,k}(x) \le -c_2 \int_0^T \log|\varphi_{n,k}(t)| dt.$$

PROOF. For $|t| \le T$, omitting subscripts,

$$\begin{aligned} |1 - \tilde{\varphi}(t)| &\le \left| \int_{[|x| < \tau]} (e^{it(x-a)} - 1) dF \right| + 2 \int_{[|x| \ge \tau]} dF \\ &\le \left| \int_{[|x| < \tau]} (e^{it(x-a)} - 1 - it(x-a)) dF \right| \\ &\quad + \left| \int_{[|x| < \tau]} it(x-a) dF(x) \right| + 2P\{|X| \ge \tau\} \\ &\le \frac{T^2}{2} \int_{[|x| < \tau]} (x-a)^2\, dF + T \left| \int_{[|x| < \tau]} (x-a) dF \right| \\ &\quad + 2\mathrm{P}\{|X| \ge \tau\}. \end{aligned} \tag{3}$$

Now, noting via (2) that $|a| < \tau$,

$$\int_{[|x|<\tau]} (x-a)^2\, dF \le [1 + (\tau + |a|)^2] \int_{[|x|<\tau]} \frac{(x-a)^2}{1+(x-a)^2}\, dF,$$

$$\left|\int_{[|x|<\tau]} (x-a)\, dF\right| = |a| \int_{[|x|\ge\tau]} dF$$

$$\le \frac{|a|[1 + (\tau + |a|)^2]}{(\tau - |a|)^2} \int_{[|x|\ge\tau]} \frac{(x-a)^2}{1+(x-a)^2}\, dF, \tag{4}$$

and so it follows from (3) that

$$\sup_{|t|\le T} |1 - \tilde{\varphi}_{n,k}(t)| \le [1 + (\tau + |a|)^2]\left(\frac{T^2}{2} + \frac{2 + |a|T}{(\tau - |a|)^2}\right) \int \frac{x^2}{1+x^2}\, d\tilde{F}_{n,k}$$

$$= \frac{1}{d_1} \int \frac{x^2}{1+x^2}\, d\tilde{F}_{n,k},$$

where, noting that (i) entails $\max_k |a_{n,k}| < \tau/2$ for all large n,

$$\frac{1}{d_1} = [1 + (\tau + |a|)^2]\left[\frac{2 + T|a|}{(\tau - |a|)^2} + \frac{T^2}{2}\right]$$

$$\le \left(1 + \frac{9\tau^2}{4}\right)\left[\frac{2 + \tau T}{\tau^2/4} + \frac{T^2}{2}\right] = \frac{1}{c_1},$$

yielding the lower bound. Next, if F^* denotes the d.f. of a r.v. X^* with c.f. $|\varphi(t)|^2$, from the elementary inequality

$$\left(1 - \frac{\sin Tx}{Tx}\right)\left(\frac{1+x^2}{x^2}\right) \ge c(T) > 0,$$

$$\int_0^T (1 - |\varphi(u)|^2)du = \int_{-\infty}^{\infty} \int_0^T (1 - \cos ux)du\, dF^*$$

$$= T \int_{-\infty}^{\infty} \left(1 - \frac{\sin Tx}{Tx}\right) dF^*(x)$$

$$\ge Tc(T) \int_{-\infty}^{\infty} \frac{x^2}{1+x^2}\, dF^*(x). \tag{5}$$

For any r.v. X with d.f. F and median m, let X^* denote the symmetrized X and define

$$F^m(x) = P\{X - m < x\}, \qquad q^m(x) = P\{|X - m| \ge x\},$$

$$q^*(x) = P\{|X^*| \ge x\}.$$

By Lemmas 6.2.1 (iii) and 10.1.1

$$\int_{-\infty}^{\infty} \frac{x^2}{1+x^2}\, dF^m = \int_0^{\infty} q^m(x)\, d\left(\frac{x^2}{1+x^2}\right) \le 2 \int_0^{\infty} q^*(x)\, d\left(\frac{x^2}{1+x^2}\right)$$

$$= 2 \int_{-\infty}^{\infty} \frac{x^2}{1+x^2}\, dF^*(x). \tag{6}$$

Moreover, from the elementary inequality $(x-a)^2 \le (x-m)^2 + 2(m-a)(x-a)$ and the first equality in (4),

$$\int_{[|x|<\tau]} (x-a)^2\, dF \le \int_{[|x|<\tau]} (x-m)^2\, dF + 2(\tau + |m|) \left| \int_{[x|<\tau]} (x-a) dF \right|$$

$$\le \int_{[|x|<\tau]} (x-m)^2\, dF + 2\tau(\tau + |m|) \int_{[|x|\ge\tau]} dF,$$

whence

$$\int \frac{x^2}{1+x^2}\, d\tilde{F} = \int \frac{(x-a)^2}{1+(x-a)^2}\, dF \le \int_{[|x|<\tau]} (x-a)^2\, dF + \int_{[|x|\ge\tau]} dF$$

$$\le \int_{[|x|<\tau]} (x-m)^2\, dF + [1 + 2\tau(\tau + |m|)] \int_{[|x|\ge\tau]} dF. \tag{7}$$

But

$$\int_{[|x|<\tau]} (x-m)^2\, dF \le [1 + (\tau + |m|)^2] \int_{[|x|<\tau]} \frac{(x-m)^2}{1+(x-m)^2}\, dF$$

$$\le [1 + (\tau + |m|)^2] \int_{-\infty}^{\infty} \frac{x^2}{1+x^2}\, dF^m$$

and as in (4)

$$\int_{[|x|\ge\tau]} dF \le \frac{1 + (\tau + |m|)^2}{(\tau - |m|)^2} \int_{[|x|\ge\tau]} \frac{(x-m)^2}{1+(x-m)^2}\, dF$$

$$\le \frac{1 + (\tau + |m|)^2}{(\tau - |m|)^2} \int_{-\infty}^{\infty} \frac{x^2}{1+x^2}\, dF^m,$$

so that, combining these with (7) and recalling (6) and (5),

$$\int_{-\infty}^{\infty} \frac{x^2}{1+x^2}\, d\tilde{F} \le d_2 \int \frac{x^2}{1+x^2}\, dF^m \le 2d_2 \int \frac{x^2}{1+x^2}\, dF^*$$

$$\le \frac{2d_2}{Tc(T)} \int_0^T (1 - |\varphi(u)|^2) du. \tag{8}$$

In view of $1 - |\varphi|^2 \le -\log|\varphi|^2 = -2\log|\varphi|$, the upper bound follows from (8) with $c_2 = 2d_2(\tau)/Tc(T)$, noting that for sufficiently large n

$$d_2 = [1 + (\tau + |m|)^2]\left(1 + \frac{1 + 2\tau(\tau + |m|)}{(\tau - |m|)^2}\right)$$

$$\le (1 + 4\tau^2)\left(1 + \frac{1 + 4\tau^2}{(\tau/2)^2}\right) = d_2(\tau). \qquad \square$$

Lemma 3. *If $\{X_{n,k}\}$ are infinitesimal* r.v.s *with* c.f.s *$\varphi_{n,k}$ for which*

$$\lim_{n\to\infty} \prod_{k=1}^{k_n} |\varphi_{n,k}(t)| = f(t)$$

exists and is continuous at $t = 0$, then for any τ in $(0, \infty)$ there exists a constant C depending on τ and $\{\varphi_{n,k}\}$ such that

$$\sum_{k=1}^{k_n} \int \frac{x^2}{1 + x^2}\, d\tilde{F}_{n,k}(x) \le C.$$

PROOF. Since $\prod_{k=1}^{k_n} |\varphi_{n,k}(t)|^2 \to f^2(t)$, continuous at zero, f^2 is a c.f., whence T may be chosen so that $f^2 > \frac{3}{4}$ for $|t| \le T$. Then, by uniform convergence $\prod_1^{k_n} |\varphi_{n,k}(t)|^2 > \frac{1}{2}$ for $n \ge N_T$ and $|t| \le T$ and $\sum_{k=1}^{k_n} \log|\varphi_{n,k}(t)| \to \log f(t)$ uniformly in $|t| \le T$. Hence from Lemma 2

$$\sum_{k=1}^{k_n} \int_{-\infty}^{\infty} \frac{x^2}{1 + x^2}\, d\tilde{F}_{n,k}(x) \le -c_2 \sum_{k=1}^{k_n} \int_0^T \log|\varphi_{n,k}(t)|\,dt$$

$$\to -c_2 \int_0^T \log f(t)\,dt. \qquad \square$$

The next lemma indicates that the c.f. of a sum of infinitesimal rowwise independent random variables behaves like that of a related i.d. c.f.

Lemma 4. *If $\{X_{n,k}\}$ are infinitesimal* r.v.s *such that for some τ in $(0, \infty)$ there exists a C in $(0, \infty)$ with $\sum_{k=1}^{k_n} \int [x^2/(1 + x^2)]\,d\tilde{F}_{n,k}(x) \le C$, $n \ge 1$, then $\sum_{k=1}^{k_n} [\operatorname{Log} \tilde{\varphi}_{n,k}(t) - (\tilde{\varphi}_{n,k}(t) - 1)] = o(1)$ for all real t. Moreover, for any constants A_n and all real t*

$$\operatorname{Log} e^{-itA_n} \prod_{k=1}^{k_n} \varphi_{n,k}(t) - \psi(t; \gamma_n, G_n) = o(1), \tag{9}$$

where

$$\gamma_n = -A_n + \sum_{k=1}^{k_n}\left(a_{n,k} + \int \frac{x}{1 + x^2}\, d\tilde{F}_{n,k}(x)\right),$$

$$G_n(u) = \sum_{k=1}^{k_n} \int_{-\infty}^{u} \frac{x^2}{1 + x^2}\, d\tilde{F}_{n,k}(x)$$

and ψ is as in (1) of Section 1.

PROOF. By hypothesis and Lemma 2, for $|t| \le T$ and $n \ge N_\tau$

$$\sum_{k=1}^{k_n} |\tilde{\varphi}_{n,k}(t) - 1| \le \frac{1}{c_1} \sum_{k=1}^{k_n} \int \frac{x^2}{1+x^2} d\tilde{F}_{n,k}(x) \le \frac{C}{c_1}.$$

Furthermore, infinitesimality implies (iv), whence $\operatorname{Log} \tilde{\varphi}_{n,k}$ is well defined for $|t| \le T$ and $1 \le k \le k_n$ provided $n \ge n(T, \tau)$, whence under these circumstances

$$|\operatorname{Log} \tilde{\varphi}_{n,k}(t) - (\tilde{\varphi}_{n,k}(t) - 1)| \le |\tilde{\varphi}_{n,k}(t) - 1|^2.$$

Thus, since T is arbitrary, for all real t if $n \ge n(T, |\tau|)$,

$$\left| \sum_{k=1}^{k_n} [\operatorname{Log} \tilde{\varphi}_{n,k}(t) - (\tilde{\varphi}_{n,k}(t) - 1)] \right| \le \sum_{k=1}^{k_n} |\tilde{\varphi}_{n,k}(t) - 1|^2$$

$$\le \frac{C}{c_1} \max_{1 \le k \le k_n} |\tilde{\varphi}_{n,k}(t) - 1| = o(1). \quad (10)$$

Next,

$$\operatorname{Log} \tilde{\varphi}_{n,k}(t) - (\tilde{\varphi}_{n,k}(t) - 1) = \operatorname{Log} \varphi_{n,k}(t) - ita_{n,k} - \int (e^{itu} - 1) d\tilde{F}_{n,k}(u)$$

$$= \operatorname{Log} \varphi_{n,k}(t) - \left[ita_{n,k} + it \int \frac{x}{1+x^2} d\tilde{F}_{n,k} \right.$$

$$\left. + \int \left(e^{itx} - 1 - \frac{itx}{1+x^2} \right) \left(\frac{1+x^2}{x^2} \right) \frac{x^2}{1+x^2} d\tilde{F}_{n,k} \right],$$

and so, upon summing and setting

$$\gamma_n = -A_n + \sum_{k=1}^{k_n} \left(a_{n,k} + \int \frac{x}{1+x^2} d\tilde{F}_{n,k} \right),$$

$$G_n(u) = \sum_1^{k_n} \int_{-\infty}^{u} \frac{x^2}{1+x^2} d\tilde{F}_{n,k},$$

(9) follows from (10). □

The connections between i.d. laws and the array of (1) is unfolded in Theorem 1 below.

Theorem 1. *If $\{X_{n,k}, 1 \le k \le k_n \to \infty, n \ge 1\}$ are infinitesimal, rowwise independent* r.v.s, *the class of limit distributions of centered sums $\sum_{k=1}^{k_n} X_{n,k} - A_n$ coincides with the class of* i.d. *laws. Moreover, $\sum_{k=1}^{k_n} X_{n,k} - A_n \xrightarrow{d}$* i.d. *distribution characterized by (γ, G) iff $\gamma_n \to \gamma$, $G_n \xrightarrow{c} G$, where*

$$\gamma_n = -A_n + \sum_{k=1}^{k_n} \left(a_{n,k} + \int \frac{x}{1+x^2} d\tilde{F}_{n,k}(x) \right),$$

$$G_n(u) = \sum_{k=1}^{k_n} \int_{-\infty}^{u} \frac{x^2}{1+x^2} d\tilde{F}_{n,k}(x),$$

and τ is an arbitrary but fixed constant in $(0, \infty)$.

PROOF. Any i.d. law characterized by (γ, G) is obtainable as a limit of distributions of row sums of independent, infinitesimal r.v.s $X_{n,k}$. It suffices to choose $k_n = n$ and take as the c.f. of $X_{n,k}$ the i.d. c.f. characterized by $(\gamma/k_n, (1/k_n)G)$ since such $X_{n,k}$ are clearly infinitesimal. Next, if for some constants A_n, $e^{-itA_n}\prod_{k=1}^{k_n}\varphi_{n,k}(t) \to g(t)$, a c.f., then Lemma 3 applies with $f = |g|$, and hence also Lemma 4, so that by Theorem 12.1.2, $\gamma_n \to \gamma$, $G_n \xrightarrow{c} G$, and $g = \exp\{\psi(t; \gamma, G)\}$. Finally, if $\gamma_n \to \gamma$, $G_n \xrightarrow{c} G$, Theorem 12.1.2 ensures $\psi(t, \gamma_n, G_n) \to \psi(t; \gamma, G)$, and $\sum_{k=1}^{k} \int [x^2/(1 + x^2)]\, d\tilde{F}_{n,k} = G_n(\infty) \to G(\infty) < \infty$ whence Lemma 4 guarantees that $e^{-itA_n}\prod_{k=1}^{k_n}\varphi_{n,k}(t) \to e^{\psi(t;\gamma,G)}$. □

Corollary 1. *The only admissible choices of the constants* A_n *are*

$$A_n = \sum_{k=1}^{k_n}\left(a_{n,k} + \int \frac{x}{1 + x^2}\, d\tilde{F}_{n,k}\right) - \gamma + o(1)$$

for some constants γ *and* $\tau > 0$.

The next question that poses itself is under what conditions on $F_{n,k}$ a particular i.d. limit is obtained.

Theorem 2. *If* $\{X_{n,k}, 1 \le k \le k_n \to \infty, n \ge 1\}$ *are rowwise independent, infinitesimal* r.v.s, *then for any constants* A_n, $e^{-iA_nt}\prod_{k=1}^{k_n}\varphi_{n,k}(t) \to$ *the* i.d. c.f. $\exp\{\psi(t; \gamma, G)\}$ iff

$$\sum_1^{k_n} F_{nk}(u) \to \int_{-\infty}^{u} \frac{1 + x^2}{x^2}\, dG(x), \qquad 0 > u \in C(G)$$

$$\sum_1^{k_n} [1 - F_{nk}(u)] \to \int_u^{\infty} \frac{1 + x^2}{x^2}\, dG(x), \qquad 0 < u \in C(G) \tag{11}$$

$$\lim_{\varepsilon \to 0} \varliminf_{n \to \infty} \sum_{k=1}^{k_n}\left[\int_{|x|<\varepsilon]} x^2\, dF_{nk}(x) - \left(\int_{|x|<\varepsilon]} x\, dF_{nk}(x)\right)^2\right] = G(0+) - G(0-), \tag{12}$$

$$-A_n + \sum_1^{k_n} \int_{[|x|<\tau]} x\, dF_{nk}(x) \to \gamma + \int_{[|x|<\tau]} x\, dG(x) - \int_{[|x|\ge\tau]} \frac{1}{x}\, dG(x) \tag{13}$$

for some fixed $\tau > 0$ *for which* $\pm\tau$ *are continuity points of* G.

PROOF. By Theorem 1 it suffices to show that (11), (12), (13) are equivalent to

$$\sum_{k=1}^{k_n} \int_{-\infty}^{u} \frac{x^2}{1 + x^2}\, d\tilde{F}_{nk}(x) = G_n(u) \xrightarrow{c} G(u) \tag{14}$$

and

$$-A_n + \sum_1^{k_n}\left(a_{nk} + \int \frac{x}{1 + x^2}\, d\tilde{F}_{nk}(x)\right) = \gamma_n \to \gamma. \tag{15}$$

This will be effected by proving (14) ⇔ (14′) ⇔ (11′), (12′) ⇔ (11), (12), and, moreover, that when (14) obtains (13) ⇔ (15). In this schema, (11′) is (11) with $F_{n,k}$ replaced by $\tilde{F}_{n,k}$, and (12′) and (14′) are defined by

$$\lim_{\varepsilon\to 0+} \overline{\lim_n}\, [G_n(\varepsilon) - G_n(-\varepsilon)] = G(0+) - G(0-) \tag{12′}$$

(where G_n is as in (14)), and

$$\begin{aligned} G_n(x) &\to G(x), & 0 > x \in C(G), \\ G_n(\infty) - G_n(x) &\to G(\infty) - G(x), & 0 < x \in C(G), \end{aligned} \tag{14′}$$

$$\lim_{\varepsilon\to 0} \overline{\lim_{n\to\infty}}\, [G_n(\varepsilon) - G_n(-\varepsilon)] = G(0+) - G(0-).$$

Since (14) implies $\lim_{\varepsilon\to 0} \overline{\lim}[G_n(\varepsilon) - G_n(-\varepsilon)] = \lim_{\varepsilon\to 0}[G(\varepsilon) - G(-\varepsilon)] = G(0+) - G(0-)$, it is apparent that (14) $\Rightarrow$ (14′). Conversely, under (14′), taking $\pm x \in C(G)$,

$$\begin{aligned} G(0+) - G(0-) &= \lim_{x\to 0+} \overline{\lim_n}[G_n(x) - G_n(-x)] \\ &= \lim_x \overline{\lim_n}[G_n(x) - G_n(\infty) - G_n(-x) + G_n(\infty)] \\ &= \lim_{x\to 0+} [G(x) - G(\infty) - G(-x) + \overline{\lim}\, G_n(\infty)] \\ &= G(0+) - G(0-) - G(\infty) + \overline{\lim_n}\, G_n(\infty), \end{aligned}$$

so that $\overline{\lim}_n G_n(\infty) = G(\infty)$, whence (14) holds. Of course, (14) $\Rightarrow$ (12′) trivially and (14′) $\Rightarrow$ (11′) since for continuity points x, by the Helly–Bray theorem,

$$\begin{aligned} \sum_1^{k_n} \tilde{F}_{nk}(x) &= \sum_1^{k_n} \int_{-\infty}^x \frac{1+u^2}{u^2}\frac{u^2}{1+u^2}\, d\tilde{F}_{nk}(u) \\ &= \int_{-\infty}^x \frac{1+u^2}{u^2}\, dG_n(u) \to \int_{-\infty}^x \frac{1+u^2}{u^2}\, dG(u), \qquad x < 0, \end{aligned}$$

$$\sum_1^{k_n} [1 - \tilde{F}_{nk}(x)] = \int_x^\infty \frac{1+u^2}{u^2}\, dG_n(u) \to \int_x^\infty \frac{1+u^2}{u^2}\, dG(u), \qquad x > 0.$$

Conversely, (11′), (12′) $\Rightarrow$ (14′) since for continuity points u, by the Helly–Bray theorem,

$$\begin{aligned} G_n(u) &= \int_{-\infty}^u dG_n(x) = \int_{-\infty}^u \frac{x^2}{1+x^2}\, d\left(\sum_1^{k_n} \tilde{F}_{nk}(x)\right) \\ &\to \int_{-\infty}^u \frac{x^2}{1+x^2}\frac{1+x^2}{x^2}\, dG(x) = G(u), \qquad u < 0, \end{aligned}$$

$$\begin{aligned} G_n(\infty) - G_n(u) &= \int_u^\infty \frac{x^2}{1+x^2}\, d\left(\sum_1^{k_n} \tilde{F}_{nk}(x)\right) \\ &= -\int_u^\infty \frac{x^2}{1+x^2}\, d\left(\sum_1^{k_n} [1 - \tilde{F}_{nk}(x)]\right) \\ &\to \int_u^\infty dG(x) = G(\infty) - G(u), \qquad u > 0 \end{aligned}$$

Thus, (14)$\Leftrightarrow$(14′)$\Leftrightarrow$(11′), (12′). Next, if $a_n = \max_{1 \le k \le k_n} \int_{[|x|<\tau]} |x|\, dF_{nk}$, then $\max_k |a_{nk}| \le a_n = o(1)$ by infinitesimality. Now,

$$\sum_{k=1}^{k_n} F_{nk}(x - a_n) \le \sum_k \tilde{F}_{nk}(x) \le \sum_{k=1}^{k_n} F_{nk}(x + a_n),$$

whence

$$\sum_k \tilde{F}_{nk}(x - a_n) \le \sum_k F_{nk}(x) \le \sum_k \tilde{F}_{nk}(x + a_n).$$

Thus, if $u < 0$, for any $\varepsilon > 0$ for which $u, u \pm \varepsilon$ are negative continuity points of G, if (11) obtains,

$$\overline{\lim} \sum_k \tilde{F}_{nk}(u) \le \overline{\lim} \sum_k F_{nk}(u + \varepsilon) = \int_{-\infty}^{u+\varepsilon} \frac{1 + x^2}{x^2}\, dG(x),$$

$$\underline{\lim} \sum_k \tilde{F}_{nk}(u) \ge \underline{\lim} \sum_k F_{nk}(u - \varepsilon) = \int_{-\infty}^{u-\varepsilon} \frac{1 + x^2}{x^2}\, dG(x),$$

and so, letting $\varepsilon \to 0$, $\lim \sum_{k=1}^{k_n} \tilde{F}_{nk}(u) = \int_{-\infty}^{u} [(1 + x^2)/x^2] dG(x)$. An analogous statement holds for continuity points $u > 0$, whence (11) $\Rightarrow$ (11′). The same argument works in reverse, yielding (11′) $\Rightarrow$ (11). Consequently, (11′) $\Leftrightarrow$ (11). Next, since for any $\varepsilon > 0$

$$\frac{1}{1 + \varepsilon^2} \sum_{k=1}^{k_n} \int_{[|x|<\varepsilon]} x^2\, d\tilde{F}_{nk}(x) \le \sum_k \int_{[|x|<\varepsilon]} \frac{x^2}{1 + x^2}\, d\tilde{F}_{nk}(x) \le \sum_k \int_{[|x|<\varepsilon]} x^2\, d\tilde{F}_{nk},$$

(12′) $\Leftrightarrow \lim_{\varepsilon \to 0+} \overline{\lim}_n \sum_{k=1}^{k_n} \int_{[|x|<\varepsilon]} x^2\, d\tilde{F}_{nk}(x) = G(0+) - G(0-)$, and therefore, to complete the verification that (11′), (12′) $\Leftrightarrow$ (11), (12) it suffices to show that for all $\varepsilon > 0$

$$J_n(\varepsilon) = \sum_{k=1}^{k_n} \left| \int_{[|x|<\varepsilon]} x^2\, d\tilde{F}_{nk}(x) - \left[\int_{[|x|<\varepsilon]} x^2\, dF_{nk}(x) - \left(\int_{[|x|<\varepsilon]} x\, dF_{nk}(x) \right)^2 \right] \right|$$
$$= o(1).$$

Recalling that $|a_{nk}| \le a_n = o(1)$, for $0 < \varepsilon < \tau$ (omitting subscripts temporarily)

$$\left| \int_{[|x|<\varepsilon]} x^2\, d\tilde{F}_{nk} - \int_{[|x|<\varepsilon]} (x - a_{nk})^2\, dF_{nk} \right|$$
$$= \left| \int_{[|x-a|<\varepsilon]} (x - a)^2\, dF - \int_{[|x|<\varepsilon]} (x - a)^2\, dF \right|$$
$$= \left| \int_{[|x-a|<\varepsilon\le|x|]} - \int_{[|x|<\varepsilon\le|x-a|]} \right|$$
$$\le \varepsilon^2 \int_{[\varepsilon\le|x|\le\varepsilon+|a|]} dF + \int_{[\varepsilon-|a|\le|x|<\varepsilon]} (\varepsilon + |a|)^2\, dF$$
$$\le (\varepsilon + a_n)^2 \int_{[\varepsilon-a_n\le|x|\le\varepsilon+a_n]} dF_{nk}$$

and

$$\left| \int_{[|x|<\varepsilon]} (x - a_{nk})^2 \, dF_{nk} - \int_{[|x|<\varepsilon]} x^2 \, dF_{nk} + \left(\int_{[|x|<\varepsilon]} x \, dF_{nk} \right)^2 \right|$$

$$= \left| \int_{[|x|<\varepsilon]} (-2ax + a^2) dF + \left(\int_{[|x|<\varepsilon]} x \, dF \right)^2 \right|$$

$$= \left| \left(\int_{[|x|<\varepsilon]} x \, dF - a \right)^2 - a^2 \int_{[|x|\geq\varepsilon]} dF \right|$$

$$= \left| \left(\int_{[\varepsilon\leq|x|<\tau]} x \, dF \right)^2 - a^2 \int_{[|x|\geq\varepsilon]} dF \right|$$

$$\leq a_n \int_{[\varepsilon\leq|x|<\tau]} \tau \, dF + a_n^2 \int_{[|x|\geq\varepsilon]} dF$$

$$\leq a_n(a_n + \tau) \int_{[|x|\geq\varepsilon]} dF_{nk},$$

implying

$$J_n(\varepsilon) \leq \sum_{k=1}^{k_n} \int_{[\varepsilon - a_n \leq |x| \leq \varepsilon + a_n]} (\varepsilon + a_n)^2 \, dF_{nk} + a_n(a_n + \tau) \sum_{k=1}^{k_n} \int_{[|x|\geq\varepsilon]} dF_{nk} = o(1)$$

since, choosing η in $(0, \varepsilon)$, where $\pm\varepsilon$, $\varepsilon \pm \eta$ are continuity points of G, and noting that $a_n < \eta$ for $n \geq N_\eta$ (and recalling that (11) $\Leftrightarrow$ (11′))

$$\sum_{k=1}^{k_n} \int_{[|x|\geq\varepsilon]} dF_{nk} = \sum_k F_{nk}(-\varepsilon) + \sum_k [1 - F_{nk}(\varepsilon)] \to \int_{-\infty}^{-\varepsilon} \frac{1 + x^2}{x^2} \, dG(x)$$

$$+ \int_{\varepsilon}^{\infty} \frac{1 + x^2}{x^2} \, dG(x) < \infty,$$

$$\sum_{k=1}^{k_n} \int_{[\varepsilon - a_n \leq |x| \leq \varepsilon + a_n]} dF_{nk}(x) \leq \sum_k \int_{[\varepsilon - \eta \leq |x| \leq \varepsilon + \eta]} dF_{nk}(x)$$

$$\to \int_{[\varepsilon - \eta \leq |x| \leq \varepsilon + \eta]} \frac{1 + x^2}{x^2} \, dG(x) \tag{16}$$

$$= o(1) \quad \text{as } \eta \to 0.$$

To complete the proof of the theorem it remains to show that (13) $\Leftrightarrow$ (15) under (14). Since for all k

$$\begin{aligned}
\left|\int_{[|x|<\tau]} x\, d\tilde{F}_{nk}(x)\right| &= \left|\int_{[|x-a|<\tau]} (x-a)dF\right| \\
&= \left|\int_{[|x-a|<\tau]} (x-a)dF - \int_{[|x|<\tau]} (x-a)dF + a\int_{[|x|\geq\tau]} dF\right| \\
&\leq \left|\int_{[|x-a|<\tau\leq|x|]} (x-a)dF - \int_{[|x|<\tau\leq|x-a|]} (x-a)dF\right| \\
&\quad + |a|\int_{[|x|\geq\tau]} dF \\
&\leq \tau\int_{[\tau\leq|x|<\tau+|a|]} dF \\
&\quad + (\tau+|a|)\int_{[\tau-|a|\leq|x|<\tau]} dF + |a|\int_{[|x|\geq\tau]} dF \\
&\leq \int_{[\tau-a_n\leq|x|\leq\tau+a_n]} (2\tau+a_n)dF + a_n\int_{[|x|\geq\tau]} dF,
\end{aligned}$$

$$\begin{aligned}
\left|\sum_{k=1}^{k_n}\int_{[|x|<\tau]} x\, d\tilde{F}_{nk}(x)\right| &\leq \sum_{k=1}^{k_n}\int_{[\tau-a_n\leq|x|\leq\tau+a_n]} (2\tau+a_n)dF_{nk} \\
&\quad + a_n\sum_{k=1}^{k_n}\int_{[|x|\geq\tau]} dF_{nk} = o(1)
\end{aligned}$$

via (16), and so, recalling that $\pm\tau$ are continuity points of G, it follows from

$$\begin{aligned}
\sum_{k=1}^{k_n}\int \frac{x}{1+x^2}\, d\tilde{F}_{nk}(x) &= \sum_{k=1}^{k_n}\left(\int_{[|x|<\tau]} x\, d\tilde{F}_{nk} - \int_{[|x|<\tau]} \frac{x^3}{1+x^2}\, d\tilde{F}_{nk}(x)\right. \\
&\quad \left. + \int_{[|x|\geq\tau]} \frac{x}{1+x^2}\, d\tilde{F}_{nk}(x)\right) \\
&= o(1) - \int_{[|x|<\tau]} x\, dG_n(x) + \int_{[|x|\geq\tau]} \frac{1}{x}\, dG_n(x)
\end{aligned}$$

that (13) $\Leftrightarrow$ (15) under (14). □

Theorem 3. *Let* $\{X_{n,k}, 1 \leq k \leq k_n \to \infty, n \geq 1\}$ *be rowwise independent* r.v.s.

(i) *If* $\{X_{n,k}\}$ *are infinitesimal and* $\sum_{k=1}^{k_n} X_{n,k} - A_n$ *converges in distribution for some choice of constants* A_n, *the limit is normal* iff

$$\max_{1\leq k\leq k_n} |X_{n,k}| \xrightarrow{P} 0 \tag{17}$$

or equivalently

$$\sum_{k=1}^{k_n} \int_{[|x| \geq \varepsilon]} dF_{nk} = o(1), \qquad \varepsilon > 0. \tag{18}$$

(ii) $\{X_{n,k}\}$ *are infinitesimal and* $\sum_{k=1}^{k_n} (X_{n,k} - a_{n,k})$ *has a limiting normal distribution necessarily* $N(0, \sigma^2)$ *iff for all* $\varepsilon > 0$, (18) *holds and*

$$\lim_n \sum_{k=1}^{k_n} \left[\int_{[|x| < \varepsilon]} x^2 \, dF_{nk}(x) - \left(\int_{[|x| < \varepsilon]} x \, dF_{nk}(x) \right)^2 \right] = \sigma^2. \tag{19}$$

PROOF. (i) If $\sum X_{n,k} - A_n$ has a limiting distribution, then (11), (12), and (13) hold. Moreover, if the limit is normal, since $\exp\{\psi(t, \gamma, G)\} = e^{-(\sigma^2 t^2/2) + it\theta}$ iff $\gamma = 0$, $G(u) = 0$ for $u < 0$ and $G(u) = \sigma^2$, $u > 0$, (11) of Theorem 2 requires $\sum_1^{k_n} F_{nk}(-\varepsilon) = o(1) = \sum_1^{k_n} [1 - F_{nk}(\varepsilon)]$, $\varepsilon > 0$, which is tantamount to (18) and also to $\prod_1^{k_n} (1 - \int_{[|x| > \varepsilon]} dF_{nk}(x)) \to 1$, $\varepsilon > 0$, in view of $1 - \sum p_i \leq \prod (1 - p_i) \leq e^{-\sum p_i} \leq 1$ (where $0 \leq p_i \leq 1$). It is therefore also equivalent to (17) via $P\{\max_{1 \leq k \leq k_n} |X_{nk}| \geq \varepsilon\} = 1 - \prod_{k=1}^{k_n} (1 - P\{|X_{n,k}| \geq \varepsilon\})$.

Apropos of (ii), (18) and (19) ensure (11) and (12) with G as above and $\sigma^2 = G(0+) - G(0-)$, while (13) is satisfied with $A_n = \sum_{k=1}^{k_n} a_{n,k}$, $\gamma = 0$. Thus, $\sum_{k=1}^{k_n} (X_{n,k} - a_{n,k})$ has a limiting normal d.f. $N(0, \sigma^2)$ by Theorem 2. As for the converse, (18) holds via (i), and therefore it suffices to verify that (12) and (18) entail (19). Now, for $0 < \varepsilon' < \varepsilon$

$$\begin{aligned}
&\sum_{k=1}^{k_n} \left\{ \int_{[|x| < \varepsilon]} x^2 \, dF_{nk} - \left(\int_{[|x| < \varepsilon]} x \, dF_{nk} \right)^2 \right\} \\
&= \sum_{k=1}^{k_n} \left\{ \int_{[|x| < \varepsilon']} x^2 \, dF_{nk} - \left(\int_{[|x| < \varepsilon']} x \, dF_{nk} \right)^2 \right\} \\
&\quad + \sum_{k=1}^{k_n} \left\{ \int_{[\varepsilon' \leq |x| < \varepsilon]} x^2 \, dF_{nk} - \left(\int_{[\varepsilon' \leq |x| < \varepsilon]} x \, dF_{nk} \right)^2 \right\} \\
&\quad - 2 \sum_{k=1}^{k_n} \left(\int_{[|x| < \varepsilon']} x \, dF_{nk} \right) \left(\int_{[\varepsilon' \leq |x| < \varepsilon]} x \, dF_{nk} \right),
\end{aligned} \tag{20}$$

and in view of (18),

$$\begin{aligned}
0 &\leq \sum_{k=1}^{k_n} \left\{ \int_{[\varepsilon' < |x| < \varepsilon]} x^2 \, dF_{nk} - \left(\int_{[\varepsilon' \leq |x| < \varepsilon]} x \, dF_{nk} \right)^2 \right\} \\
&\leq \sum_{k=1}^{k_n} \int_{[\varepsilon' \leq |x| < \varepsilon]} x^2 \, dF_{nk} \leq \varepsilon^2 \sum_{k=1}^{k_n} \int_{[|x| \geq \varepsilon']} dF_{nk} = o(1), \\
0 &\leq 2 \sum_{k=1}^{k_n} \left| \int_{[|x| < \varepsilon']} x \, dF_{nk} \right| \left| \int_{[\varepsilon' \leq |x| < \varepsilon]} x \, dF_{nk} \right| \\
&\leq 2\varepsilon\varepsilon' \sum_{k=1}^{k_n} \int_{[|x| \geq \varepsilon']} dF_{nk} = o(1).
\end{aligned}$$

Consequently, from (20), for all positive ε, ε′

$$\overline{\lim_{n\to\infty}} \sum_{k=1}^{k_n} \left\{ \int_{[|x|<\varepsilon]} x^2 \, dF_{nk} - \left(\int_{[|x|<\varepsilon]} x \, dF_{nk} \right)^2 \right\}$$

$$= \overline{\lim_{n\to\infty}} \sum_{k=1}^{k_n} \left\{ \int_{[|x|<\varepsilon']} x^2 \, dF_{nk} - \left(\int_{[|x|<\varepsilon']} x \, dF_k \right)^2 \right\}.$$

Thus, the prior upper and lower limits are independent of ε, whence (12) ensures (19). □

Corollary 2. *If* $\{X_{nk}, 1 \le k \le k_n \to \infty, n \ge 1\}$ *are infinitesimal rowwise independent* r.v.s *with zero means and variances* σ_{nk}^2 *satisfying* $\sum_{k=1}^{k_n} \sigma_{nk}^2 = 1$, $n \ge 1$, *then* $\sum_{k=1}^{k_n} X_{nk}$ *has a limiting standard normal distribution* iff *for all* $\varepsilon > 0$

$$\sum_{k=1}^{k_n} \int_{[|x|\ge\varepsilon]} x^2 \, dF_{nk}(x) = o(1). \tag{21}$$

PROOF. In view of

$$\sum_{k=1}^{k_n} \int_{[|x|\ge\varepsilon]} x^2 \, dF_{nk}(x)$$

$$\ge \max\left[\sum_{k=1}^{k_n} \left(\int_{[|x|\ge\varepsilon]} x \, dF_{nk}(x) \right)^2, \varepsilon \sum_{k=1}^{k_n} \int_{[|x|\ge\varepsilon]} |x| \, dF_{nk}(x), \varepsilon^2 \sum_{k=1}^{k_n} \int_{[|x|\ge\varepsilon]} dF_{nk}(x) \right], \tag{22}$$

(21) implies (18) and also $\sum_{k=1}^{n} (\int_{|x|<\varepsilon} x \, dF_{nk})^2 = o(1)$ for all $\varepsilon > 0$, noting that $\mathrm{E}\, X_{nk} = 0$. Consequently (21), which is equivalent to $\sum_{k=1}^{k_n} \int_{[|x|<\varepsilon]} x^2 \, dF_{nk}(x) \to 1$ via $\sum_{k=1}^{k_n} \mathrm{E}\, X_{nk}^2 = 1$, also implies (19). Then, Theorem 3 guarantees a limiting standard normal d.f. for $\sum_{k=1}^{k_n} X_{n,k}$ since $\sum_{k=1}^{k_n} a_{nk} = o(1)$ via (22). Conversely, if $\sum_{k=1}^{k_n} X_{n,k}$ is asymptotically $N(0, 1)$, then taking $\gamma = 0 = A_n$ in Theorem 2, (11) and (13) ensure $\sum_1^{k_n} a_{nk} = o(1)$ so that for all $\varepsilon > 0$,

$$1 = \lim_{n\to\infty} \sum_{k=1}^{k_n} \mathrm{E}\, X_{nk}^2 \ge \lim_{n} \sum_{k=1}^{k_n} \int_{[|x|<\varepsilon]} x^2 \, dF_{nk}(x) \ge 1$$

via (19), implying $\sum_{k=1}^{k_n} \int_{[|x|<\varepsilon]} x^2 \, dF_{nk}(x) \to 1$ for all $\varepsilon > 0$, and, as already noted, this is tantamount to (21). □

Corollary 3. *If* $\{X_n, n \ge 1\}$ *are independent* r.v.s *with* d.f.s $\{F_n, n \ge 1\}$, *then* $(1/B_n) \sum_1^n X_i - A_n$ *has a limiting standard normal* d.f. *and* $\{X_k/B_n, 1 \le k \le n\}$ *are infinitesimal for some sequence* $\{B_n\}$ *of positive constants tending to* ∞ iff *for all* $\varepsilon > 0$

$$\sum_{k=1}^{n} \int_{[|x|\ge\varepsilon B_n]} dF_k(x) = o(1), \tag{23}$$

$$\frac{1}{B_n^2}\sum_{k=1}^{n}\left\{\int_{[|x|<\varepsilon B_n]} x^2\,dF_k(x) - \left(\int_{[|x|<\varepsilon B_n]} x\,dF_k(x)\right)^2\right\} \to 1. \tag{24}$$

where

$$A_n = \sum_{1}^{k_n}\int_{[|x|<B_n]} x\,dF_k(x) + o(1).$$

Proof. Define $X_{n,k} = X_k/B_n$, $1 \le k \le n$, $n \ge 1$. Then $\{X_{n,k}, 1 \le k \le n, n \ge 1\}$ are rowwise independent r.v.s with $F_{nk}(x) = F_k(B_n x)$, and (23), (24) are simply transcriptions of (18), (19). □

Exercises 12.2

1. If $\{X_{n,k}, 1 \le k \le k_n \to \infty, n \ge 1\}$ are rowwise independent r.v.s, prove that $\sum_1^{k_n} X_{nk} \xrightarrow{P}$ some constant γ and that $\{X_{n,k}\}$ are infinitesimal iff for every $\varepsilon > 0$

$$\text{(i)}\ \sum_1^{k_n}\int_{[|x|\ge\varepsilon]} dF_{nk}(x) = o(1), \quad \text{(ii)}\ \sum_1^{k_n}\int_{[|x|<\varepsilon]} x\,dF_{nk}(x) \to \gamma,$$

$$\text{(iii)}\ \sum_{k=1}^{k_n}\left\{\int_{[|x|<\varepsilon]} x^2\,dF_{nk}(x) - \left(\int_{|x|<\varepsilon} x\,dF_{n,k}(x)\right)^2\right\} = o(1).$$

2. Let $\{X_{n,k}, 1 \le k \le k_n \to \infty, n \ge 1\}$ be rowwise independent and positive r.v.s. Prove that $\sum_{k=1}^{k_n} X_{n,k} \xrightarrow{P} 1$ and $\{X_{n,k}\}$ are infinitesimal iff for every $\varepsilon > 0$

$$\text{(iv)}\ \sum_{k=1}^{k_n}\int_{\varepsilon}^{\infty} dF_{n,k}(x) = o(1), \quad \text{(v)}\ \sum_{k=1}^{k_n}\int_0^{\varepsilon} x\,dF_{nk}(x) \to 1.$$

3. If $\{X_{n,k}, 1 \le k \le k_n \to \infty\}$ are rowwise independent positive r.v.s with finite expectations satisfying $\sum_{k=1}^{k_n} \mathrm{E}\, X_{nk} = 1$, then $\sum_{k=1}^{k_n} X_{n,k} \xrightarrow{P} 1$ and $\{X_{n,k}\}$ are infinitesimal iff for every $\varepsilon > 0$, (v) holds.
4. (Raikov) Let $\{X_{n,k}\}$ have finite variances and $\{X_{n,k} - \mathrm{E}\, X_{n,k}, 1 \le k \le k_n, n \ge 1\}$ be rowwise independent, infinitesimal r.v.s satisfying $\sum_{k=1}^{k_n} \mathrm{E}\,(X_{n,k} - \mathrm{E}\, X_{nk})^2 = 1$, $n \ge 1$. Then $\sum_{k=1}^{k_n} (X_{n,k} - \mathrm{E}\, X_{n,k})$ has a limiting standard normal d.f. iff $u_n = \sum_{k=1}^{k_n} (X_{n,k} - \mathrm{E}\, X_{n,k})^2 \xrightarrow{P} 1$.
5. Construct rowwise independent, infinitesimal r.v.s $\{X_{n,k}\}$ which do not satisfy (i).
6. Since the uniform distribution is not i.d., (prove) why does Exercise 8.3.7 not contradict Theorem 1?
7. Give necessary and sufficient conditions for sums $\sum_1^{k_n} X_{nk}$ of rowwise independent, infinitesimal r.v.s $\{X_{n,k}\}$ to have limiting Poisson distributions.
8. Prove that if $\{X_n, n \ge 1\}$ are independent r.v.s, there exist constants A_n, $B_n > 0$ such that $B_n^{-1}\sum_1^n X_i - A_n$ has a limiting standard normal d.f. and $\{X_k/B_n, 1 \le k \le n\}$ are infinitesimal iff there exists constants $C_n \to \infty$ with

$$\text{(vi)}\ \sum_1^n \int_{[|x|>C_n]} dF_k = o(1), \quad \text{(vii)}\ \frac{1}{C_n^2}\sum_1^n\left\{\int_{[|x|<C_n]} x^2\,dF_k - \left(\int_{[|x|<C_n]} x\,dF_k\right)^2\right\} \to \infty.$$

Hint: Under (23) and (24), choose $\varepsilon_n \to 0$ such that $\varepsilon_n B_n \to \infty$ and then determine n_j such that for $n \geq n_j$ the left side of (23) (resp. (24)) with $\varepsilon = \varepsilon_j$ is $< 1/j$ (resp. $> 1 - 1/j$). Then take $C_n = \varepsilon_j B_n$ for $n_j \leq n < n_{j+1}$. Conversely, under (vi), (vii) choose B_n^2 to be C_n^2 multiplied by the left side of (vii), whence $C_n = o(B_n)$ and (23), (24) hold.

9. If $\{X_n, n \geq 1\}$ are independent r.v.s with $P\{X_k = \pm k\} = 1/2k$, $P\{X_k = 0\} = 1 - 1/k$, does $B_n^{-1} \sum_1^n X_i - A_n$ have a limiting standard normal d.f. for some A_n, $B_n > 0$?
10. If $\{X, X_n, n \geq 1\}$ are i.i.d. with $E\,X = 0$, $E\,X^2 = 1$ and $\{a_{n,i}, 1 \leq i \leq n\}$ are constants with $\max_i |a_{n,i}| = o(1)$ and $\sum_{i=1}^n a_{n,i}^2 = 1$ then $\sum_{i=1}^n a_{n,i} X_i \xrightarrow{d} N_{0,1}$.
11. The subclass of infinitely divisible distributions which are limit laws of normed sums $(1/B_n) \sum_1^n X_i - A_n$ of independent r.v.s $\{X_n, n \geq 1\}$ $(0 < B_n \to \infty)$ is known as the class $\mathscr{L}$ (Lévy). Employ characteristic functions to prove that $F \in \mathscr{L}$ iff for every α in (0, 1) there exists a d.f. G_α such that $F(x) = F(x/\alpha) * G_\alpha$. (If (α, G) characterizes an i.d. c.f. whose distribution $\in \mathscr{L}$, then its left and right derivatives, denoted $G'(x)$, exist on $(-\infty, 0)$ and $(0, \infty)$ and $[(1 + x^2)/x]G'(x)$ is nonincreasing.)

12.3 Stable Laws

As indicated at the outset of this chapter, the class of limit laws of normed sums of i.i.d. random variables is a narrow subclass of the infinitely divisible laws consisting of stable distributions.

Definition. A d.f. F or its c.f. φ is called **stable** if for every pair of positive constants b_1, b_2 and real constants a_1, a_2 there exists $b > 0$ and real a such that

$$F(b_1 x + a_1) * F(b_2 x + a_2) = F(bx + a). \tag{1}$$

Clearly, if $F(x)$ is stable, so is $F(cx + d)$, $c > 0$, so that one may speak of "stable types." Patently, degenerate distributions and normal distributions are stable, and in fact these are the only stable d.f.s with finite variance. The class of stable c.f.s will be completely characterized but, unfortunately, explicit expressions for stable d.f.s are known in only a handful of cases.

Theorem 1. *The class of limit distributions of normed sums* $(1/B_n) \sum_1^n X_i - A_n$ *of* i.i.d. *random variables* $\{X_n, n \geq 1\}$ *coincides with the class of stable laws.*

Proof. If F is a stable d.f. and $\{X_n, n \geq 1\}$ are i.i.d. with distribution F, then via (1), $P\{\sum_1^n X_i < x\} = [F(x)]^{n*} = F(bx + a)$, where the parameters depend on n, say $b = 1/B_n > 0$, and $a = -A_n$. Then

$$P\left\{\frac{1}{B_n}\sum_{i=1}^{n} X_i - A_n < x\right\} = P\left\{\sum_{1}^{n} X_i < B_n x + A_n B_n\right\}$$

$$= F\left[\frac{1}{B_n}(B_n x + A_n B_n) - A_n\right] = F(x)$$

for all $n \geq 1$ and, *a fortiori*, in the limit.

Conversely, suppose that F is a limit distribution of normed sums $(1/B_n)\sum_1^n X_i - A_n$ of i.i.d. $\{X_n, n \geq 1\}$. If F is improper, it is certainly stable, while otherwise by Theorem 8.4.2, (i) $B_n \to \infty$ and (ii) $B_n/B_{n-1} \to 1$. For any constants $0 < b_1 < b_2 < \infty$, define $m = m_n = \inf\{m > n: B_m/B_n > b_2/b_1\}$, whence $B_m/B_n \to b_2/b_1$ via (i) and (ii). For any real a_1, a_2, define constants $A_{m,n}$ so that

$$\left[b_1\left(\frac{1}{B_n}\sum_{1}^{n} X_i - A_n\right) - a_1 b_1\right] + \left[b_1\frac{B_m}{B_n}\left(\frac{1}{B_m}\sum_{n+1}^{n+m} X_i - A_m\right) - a_2 b_2\right]$$
$$= \frac{b_1}{B_n}\sum_{1}^{m+n} X_i - A_{m,n}. \tag{2}$$

By hypothesis, the left and hence right side of (2) converges in distribution to $F(b_1^{-1}x + a_1) * F(b_2^{-1}x + a_2)$. On the other hand, $(1/B_{m+n})\sum_1^{m+n} X_i - A_{m+n}$ converges in distribution. According to Corollary 8.2.2, the two limit distributions must be of the same type, that is, (1) obtains for some $b > 0$ and a. □

It follows immediately from Theorem 1 that the stable distributions form a subclass of the infinitely divisible laws and hence (1) may be used in conjunction with the representation of i.d. c.f.s to glean further information.

Theorem 2. *A function φ is a stable* c.f. iff

$$\varphi(t) = \varphi_\alpha(t; \gamma, \beta, c) = \exp\left\{i\gamma t - c|t|^\alpha\left[1 + i\beta\frac{t}{|t|}w(t, \alpha)\right]\right\}, \tag{3}$$

where $0 < \alpha \leq 2$, $|\beta| \leq 1$, $c \geq 0$, *and*

$$w(t, \alpha) = \begin{cases} \tan \pi\alpha/2, & \alpha \neq 1 \\ (2/\pi)\log|t|, & \alpha = 1. \end{cases}$$

Note. The subclass with $\beta = 0 = \gamma$ comprises the symmetric stable distributions. The parameter α is called the **characteristic exponent**. If $\alpha = 2$, necessarily $\beta = 0$, yielding the normal c.f. When $\alpha < 2$, absolute moments of order r are finite iff $r < \alpha$.

PROOF. If φ is a stable c.f. it is i.d. by Theorems 1 and 12.2.1, whence $\varphi(t) = \exp\{\psi(t)\}$, where, according to the representation theorem (Theorem 12.1.3),

$$\begin{aligned}
\psi(bt) &= itb\gamma + \int_{-\infty}^{\infty}\left(e^{itbx} - 1 - \frac{itbx}{1+x^2}\right)\frac{1+x^2}{x^2}\,dG(x) \\
&= itb\gamma + \int_{-\infty}^{\infty}\left(e^{ity} - 1 - \frac{ity}{1+y^2/b^2}\right)\frac{b^2+y^2}{y^2}\,dG\left(\frac{y}{b}\right) \\
&= it\left[b\gamma + (1-b^2)\int_{-\infty}^{\infty}\frac{y}{1+y^2}\,dG\left(\frac{y}{b}\right)\right] \\
&\quad + \int_{-\infty}^{\infty}\left(e^{ity} - 1 - \frac{ity}{1+y^2}\right)\frac{b^2+y^2}{y^2}\,dG\left(\frac{y}{b}\right) \\
&= itb' + \int_{-\infty}^{\infty}\left(e^{ity} - 1 - \frac{ity}{1+y^2}\right)\left(\frac{1+y^2}{y^2}\right)\frac{b^2+y^2}{1+y^2}\,dG\left(\frac{y}{b}\right). \qquad (4)
\end{aligned}$$

Since φ is stable (taking $a_1 = a_2 = 0$) for any positive pair b_1, b_2, there exists $b > 0$ and real a with $\psi(b_1 t) + \psi(b_2 t) = ita + \psi(bt)$. Hence, from (4) and uniqueness of the i.d. representation, for all x

$$\int_{-\infty}^{x}\frac{b_1^2+y^2}{1+y^2}\,dG\left(\frac{y}{b_1}\right) + \int_{-\infty}^{x}\frac{b_2^2+y^2}{1+y^2}\,dG\left(\frac{y}{b_2}\right) = \int_{-\infty}^{x}\frac{b^2+y^2}{1+y^2}\,dG\left(\frac{y}{b}\right),$$

$$(b_1^2+y^2)dG\left(\frac{y}{b_1}\right) + (b_2^2+y^2)dG\left(\frac{y}{b_2}\right) = (b^2+y^2)dG\left(\frac{y}{b}\right), \qquad (5)$$

$$(b_1^2+b_2^2-b^2)[G(0+) - G(0-)] = 0. \qquad (6)$$

Set

$$J(x) = \int_{e^x}^{\infty}\frac{1+y^2}{y^2}\,dG(y), \qquad J^-(x) = \int_{-\infty}^{-e^x}\frac{1+y^2}{y^2}\,dG(y)$$

for real x. If $b = e^{-h}$, $b_i = e^{-h_i}$, $i = 1, 2$,

$$J(x+h) = \int_{(1/b)e^x}^{\infty}\frac{1+y^2}{y^2}\,dG(y) = \int_{e^x}^{\infty}\frac{b^2+u^2}{u^2}\,dG\left(\frac{u}{b}\right),$$

$$J^-(x+h) = \int_{-\infty}^{-e^x}\frac{b^2+u^2}{u^2}\,dG\left(\frac{u}{b}\right).$$

Thus, from (5), for all x and arbitrary h_1, h_2 there exists h such that

$$J(x+h_1) + J(x+h_2) = J(x+h),$$

$$J^-(x+h_1) + J^-(x+h_2) = J^-(x+h). \qquad (7)$$

Taking $h_1 = h_2 = 0$, there exists δ_2 such that $2J(x) = J(x+\delta_2)$, and inductively $nJ(x) = J(x+\delta_{n-1}) + J(x) = J(x+\delta_n)$ for some $\delta_n \in (-\infty, \infty)$. Hence, $(m/n)J(x) = (1/n)J(x+\delta_m) = J(x+\delta_m-\delta_n) = J(x+\delta_{(m/n)})$ say, for any positive integers m, n, whence $rJ(x) = J(x+\delta_r)$ for all real x, every positive rational r, and some function δ_r.

If $J(x_0)$ is positive, $J(x_0+\delta_{(1/2)}) = (1/2)J(x_0) > 0$, implying $\delta_{(1/2)} > 0$

since J is nonincreasing. In similar fashion,

$$J(x_0 + n\delta_{(1/2)}) = (1/2)J(x_0 + (n-1)\delta_{(1/2)}) > 0$$

for every positive integer n, implying $J(x) > 0$ for all x. Thus, either $J \equiv 0$ or (as will be supposed) J is nonvanishing.

Since $0 < J(x)\downarrow$ as $x\uparrow$, it follows from $rJ(x) = J(x + \delta_r)$ that for rational numbers $r' > r > 0$, $\delta_r > \delta_{r'}$, whence as $r \uparrow 1$ (through rationals), $0 \le \delta_r \downarrow$ some number δ'. Thus, $J(x) = \lim_{r\uparrow 1} rJ(x) = \lim_{r\uparrow 1} J(x + \delta_r) \le J(x + \delta') \le J(x)$, implying $J(x) = J(x+)$ and $J(x + n\delta') = J(x)$ for $n \ge 1$. Since $J(\infty -) = 0$, $\delta' = 0$. Analogously, for rational r $\downarrow 1$, $0 \ge \delta_r \uparrow \delta^*$ and $J(x) = J(x-)$ and $\delta^* = 0$. Consequently, J is continuous and if rational $r\uparrow$ any positive r_0, then $\delta_r \downarrow$ some δ_{r_0}, whence

$$r_0 J(x) = \lim_{r\uparrow r_0} rJ(x) = \lim_{r\uparrow r_0} J(x + \delta_r) = J(x + \delta_{r_0}).$$

Thus, δ_r is defined and strictly decreasing (the same functional equation obtains) for all real, positive r and $\delta_1 = 0$.

(i) Note that by definition $J(x) < \infty$ for all $x > -\infty$, and so $\infty = \lim_{r\uparrow\infty} rJ(x) = \lim_{r\uparrow\infty} J(x + \delta_r)$ implies $\delta_r \downarrow -\infty$ as $r\uparrow\infty$ and $J(-\infty) = \infty$. As $r\downarrow 0$, $\delta_r \uparrow \infty$ since $0 = \lim_{r\to 0} rJ(x) = \lim_{r\to\infty} J(x + \delta_r)$, implying $\delta_r \to \infty$ as $r \to 0$.

(ii) $J(x + \varepsilon) < J(x)$ for $\varepsilon > 0$ and all $x \in (-\infty, \infty)$. Suppose contrariwise that $J(x_0 + \varepsilon) = J(x_0)$ for some x_0 and $\varepsilon > 0$. Since $\delta_{1-} = 0$, the quantity r may be chosen so that $0 < \delta_r < \varepsilon$, implying $rJ(x_0) = J(x_0 + \delta_r) = J(x_0)$, a contradiction since $J(x_0) > 0$. Thus J is strictly decreasing implying δ_r continuous.

(iii) For all positive r_1, r_2

$$J(x + \delta_{r_1 r_2}) = r_1 r_2 J(x) = r_1 J(x + \delta_{r_2}) = J(x + \delta_{r_2} + \delta_{r_1}),$$

and so by strict monotonicity $\delta_{r_1 r_2} = \delta_{r_1} + \delta_{r_2}$ for all $r_i > 0$, $i = 1, 2$. This is the multiplicative form of Cauchy's functional equation, and since δ_r is continuous, necessarily $\delta_r = -(1/\alpha)\log r$ for some constant α. As r increases from 0 to ∞, δ_r decreases from ∞ to $-\infty$, necessitating $\alpha > 0$. Moreover $rJ(0) = J(\delta_r) = J(-(1/\alpha)\log r)$, implying for $x \in (-\infty, \infty)$ that

$$J(x) = J(0)\cdot e^{-\alpha x} = \frac{c_1}{\alpha} e^{-\alpha x},$$

where $c_1 = \alpha J(0) \ge 0$. Note that, $c_1 > 0$ if $G(\infty) - G(0+) > 0$. Hence,

$$\int_x^\infty \frac{1 + y^2}{y^2}\, dG(y) = J(\log x) = \frac{c_1}{\alpha} x^{-\alpha}, \qquad x > 0,$$

or

$$dG(x) = c_1 \frac{x^{1-\alpha}}{1 + x^2}\, dx, \qquad x > 0 \qquad c_1 = \alpha J(0). \tag{8}$$

Since $G(\infty) - G(0+) < \infty$, necessarily $0 < \alpha < 2$ and moreover, from (7) $e^{-\alpha(x+h_1)} + e^{-\alpha(x+h_2)} = e^{-\alpha(x+h)}$ or

$$b_1^\alpha + b_2^\alpha = b^\alpha. \tag{9}$$

Proceeding in similar fashion with $J^-(x)$ if $G(0-) > 0$, it follows that

$$dG(x) = \frac{c_2|x|^{1-\alpha_0}}{1+x^2}\,dx, \qquad x < 0, \qquad c_2 = \alpha_0 J^-(0) > 0, \tag{10}$$

and again via (7)

$$b_1^{\alpha_0} + b_2^{\alpha_0} = b^{\alpha_0}. \tag{11}$$

Setting $b_1 = b_2 = 1$ in (9) and (11) reveals that $\alpha_0 = \alpha$.

Summarizing, if $G \not\equiv 0$, either $G(\infty) - G(0+) = 0 = c_1$ and $G(0-) = 0 = c_2$, whence $G(0+) - G(0-) = \sigma^2 > 0$, entailing $b^2 = b_1^2 + b_2^2$ via (6) and $\varphi(t)$ normal, i.e., $\alpha = 2$ or alternatively $0 < \alpha < 2$ and $G(0+) = G(0-)$ via (6), (9) and (11) with $b_1 = b_2 = 1$. In the latter case, from (4), (8), (10)

$$\psi(t) = i\gamma t + c_1 \int_{0+}^{\infty}\left(e^{itx} - 1 - \frac{itx}{1+x^2}\right)\frac{dx}{x^{1+\alpha}}$$

$$+ c_2 \int_{-\infty}^{0-}\left(e^{itx} - 1 - \frac{itx}{1+x^2}\right)\frac{dx}{|x|^{1+\alpha}}. \tag{12}$$

Next, (12) will be evaluated in terms of elementary functions.

(i) $0 < \alpha < 1$.

$$\psi(t) = it\left[\gamma - c_1 \int_0^\infty \frac{dx}{x^\alpha(1+x^2)} + c_2 \int_{-\infty}^0 \frac{dx}{|x|^\alpha(1+x^2)}\right]$$

$$+ c_1 \int_{0+}^{\infty} (e^{itx} - 1)\frac{dx}{x^{1+\alpha}} + c_2 \int_{-\infty}^{0-} (e^{itx} - 1)\frac{dx}{|x|^{1+\alpha}}. \tag{13}$$

By contour integration

$$0 = \int_Q (e^{iz} - 1)\frac{dz}{z^{1+\alpha}} = \int_{R_1}^{R_2} (e^{iv} - 1)\frac{dv}{v^{1+\alpha}} + \int_{R_2}^{R_1} (e^{-u} - 1)\frac{du}{i^\alpha u^{1+\alpha}}$$

$$+ \sum_{j=1}^{2} (-1)^j \int_0^{\pi/2} (\exp(iR_j e^{i\theta}) - 1)\frac{R_j e^{i\theta} i\,d\theta}{R_j^{1+\alpha} e^{i\theta(1+\alpha)}}.$$

Now, since $v \equiv |\exp(iRe^{i\theta}) - 1| \le 2R$ for $0 < R < 1$ and $v \le e^{-R\sin\theta} + 1 \le 2$ for $0 \le \theta \le \pi/2$,

$$\left|\int_0^{\pi/2} (\exp(iRe^{i\theta}) - 1)\frac{i\,d\theta}{R^\alpha e^{i\theta\alpha}}\right| \le \begin{cases} 2\displaystyle\int_0^{\pi/2} R^{1-\alpha}\,d\theta = o(1) & \text{as } R \to 0 \\ 2\displaystyle\int_0^{\pi/2} \frac{d\theta}{R^\alpha} = o(1) & \text{as } R \to \infty, \end{cases}$$

it follows that

$$\int_0^\infty (e^{iv} - 1)\frac{dv}{v^{1+\alpha}} = -i^{-\alpha} \lim_{\substack{R_1\to 0\\ R_2\to\infty}} \int_{R_2}^{R_1} (e^{-u} - 1)\frac{du}{u^{1+\alpha}} = i^{-\alpha}\int_0^\infty (e^{-u} - 1)\frac{du}{u^{1+\alpha}}$$

$$= \frac{-i^{-\alpha}}{\alpha}\int_0^\infty u^{-\alpha}e^{-u}\,du = \frac{-i^{-\alpha}}{\alpha}\Gamma(1-\alpha) = i^{-\alpha}\Gamma(-\alpha)$$

via integration by parts and the recursion formula for the Γ function. Thus, if $t > 0$,

$$\int_0^\infty (e^{itx} - 1)\frac{dx}{x^{1+\alpha}} = t^\alpha \int_0^\infty (e^{iv} - 1)\frac{dv}{v^{1+\alpha}} = t^\alpha i^{-\alpha}\Gamma(-\alpha) = t^\alpha e^{-(i\pi\alpha/2)}\Gamma(-\alpha).$$

Since

$$\int_{-\infty}^0 (e^{itx} - 1)\frac{dx}{|x|^{1+\alpha}} = \int_0^\infty (e^{-itx} - 1)\frac{dx}{x^{1+\alpha}} = \overline{\int_{0+}^\infty (e^{itx} - 1)\frac{dx}{x^{1+\alpha}}}$$

$$= t^\alpha e^{i\pi\alpha/2}\Gamma(-\alpha),$$

setting

$$\gamma' = \gamma - c_1\int_0^\infty \frac{dx}{x^\alpha(1+x^2)} + c_2\int_{-\infty}^0 \frac{dx}{|x|^\alpha(1+x^2)}$$

and

$$c = -(c_1 + c_2)\cdot\Gamma(-\alpha)\cos\frac{\alpha\pi}{2} \ge 0,$$

for $t > 0$, from (13),

$$\psi(t) = it\gamma' + \Gamma(-\alpha)t^\alpha\left[(c_1 + c_2)\cos\frac{\pi}{2}\alpha + i(c_2 - c_1)\sin\frac{\pi}{2}\alpha\right]$$

$$= it\gamma' + \alpha\Gamma(-\alpha)t^\alpha\left([J(0) + J^-(0)]\cos\frac{\pi}{2}\alpha + [J^-(0) - J(0)]\sin\frac{\pi}{2}\alpha\right)$$

$$= it\gamma' - ct^\alpha\left\{1 + i\beta\tan\left(\frac{\pi}{2}\right)\alpha\right\},$$

where

$$\beta = \frac{c_2 - c_1}{c_2 + c_1} = \frac{J^-(0) - J(0)}{J^-(0) + J(0)} \in [-1, 1]. \tag{14}$$

For $t < 0$,

$$\psi(t) = \overline{\psi(-t)} = -i\gamma'(-t) - c(-t)^{\alpha}\left\{1 - i\beta \tan\left(\frac{\pi}{2}\right)\alpha\right\}$$

$$= i\gamma' t - c|t|^{\alpha}\left\{1 + i\beta \frac{t}{|t|} \tan\left(\frac{\pi}{2}\right)\alpha\right\},$$

which dovetails with (3).

(ii) $1 < \alpha < 2$. Since $x/(1 + x^2) = x - [x^3/(1 + x^2)]$, it follows from (12) that for suitable γ''

$$\psi(t) = it\gamma'' + c_1 \int_0^{\infty} (e^{itx} - 1 - itx) \frac{dx}{x^{1+\alpha}}$$

$$+ c_2 \int_{-\infty}^{0} (e^{itx} - 1 - itx) \frac{dx}{|x|^{1+\alpha}}. \tag{15}$$

By the same contour integration

$$\int_0^{\infty} (e^{iv} - 1 - iv) \frac{dv}{v^{1+\alpha}} = i^{-\alpha} \int_0^{\infty} (e^{-y} - 1 + y) \frac{dy}{y^{1+\alpha}}$$

$$= i^{-\alpha}\alpha^{-1} \int_0^{\infty} (1 - e^{-y}) y^{-\alpha}\, dy$$

$= i^{-\alpha} M(\alpha)$ say, where $0 < M(\alpha) < \infty$, and so, setting

$$c = -M(\alpha)(c_1 + c_2)\cos \frac{\pi}{2} \alpha \geq 0,$$

for $t > 0$, from (15)

$$\psi(t) = it\gamma'' - ct^{\alpha}\left\{1 + i\beta \tan\left(\frac{\pi}{2}\right)\alpha\right\}$$

$$= it\gamma'' - c|t|^{\alpha}\left\{1 + i \frac{t}{|t|} \beta \tan\left(\frac{\pi}{2}\right)\alpha\right\},$$

and exactly as in case (i) the above also holds for $t < 0$.

(iii) $\alpha = 1$. Since

$$\int_0^{\infty} \frac{1 - \cos u}{u^2}\, du = \int_0^{\infty} \frac{\sin u}{u}\, du = \frac{\pi}{2}$$

and

$$\lim_{\varepsilon \to 0+} \int_{\varepsilon}^{\varepsilon u} \frac{\sin v}{v^2}\, dv = \lim_{\varepsilon \to 0+} \int_{\varepsilon}^{\varepsilon u} \frac{v + o(v^2)}{v^2}\, dv = \log u, \qquad u > 0,$$

if $t > 0$,

$$\int_0^{\infty}\left(e^{itx} - 1 - \frac{itx}{1+x^2}\right)\frac{dx}{x^2}$$

$$= \int_0^{\infty} \frac{\cos tx - 1}{x^2}\,dx + i\int_0^{\infty}\left(\sin tx - \frac{tx}{1+x^2}\right)\frac{dx}{x^2}$$

$$= -\frac{\pi}{2}t + i\lim_{\varepsilon\to 0+}\left[-t\int_{\varepsilon}^{\varepsilon t}\frac{\sin v}{v^2}\,dv + t\int_{\varepsilon}^{\infty}\left(\frac{\sin v}{v^2} - \frac{1}{v(1+v^2)}\right)dv\right]$$

$$= -\frac{\pi}{2}t - it\log t + it\gamma_0,$$

noting

$$\frac{\sin v}{v^2} - \frac{1}{v(1+v^2)} = \frac{v + O(v^3)}{v^2} - \frac{1}{v} + \frac{v}{1+v^2} = O(v) \quad \text{as } v \to 0.$$

Thus, setting $c = (\pi/2)(c_1 + c_2) \geq 0$, for $t > 0$, from (12)

$$\psi(t) = it\tilde{\gamma} + c_1\left[-\frac{\pi}{2}t - it\log t\right] + c_2\left[-\frac{\pi}{2}t + it\log t\right]$$

$$= it\tilde{\gamma} - ct\left[1 + i\beta\frac{2}{\pi}\log t\right] = it\tilde{\gamma} - c|t|\left[1 + i\beta\frac{t}{|t|}\frac{2}{\pi}\log|t|\right],$$

which coincides with (3) for $t > 0$ and also for $t < 0$. Clearly, $|\beta| \leq 1$ from (14).

Conversely, suppose that $\varphi(t)$ is defined by (3). If $\alpha = 2$, then $\varphi(t) = \exp\{i\gamma t - ct^2/2\}$, so that φ is a normal c.f. and hence a stable c.f. If, rather, $0 < \alpha < 2$, let $J^-(0)$ and $J^+(0)$ be determined by

$$\frac{J^-(0) - J(0)}{J^-(0) + J(0)} = \beta, \tag{16}$$

$$c = \begin{cases} -\alpha\Gamma(-\alpha)[J^-(0) + J(0)]\cos\dfrac{\pi}{2}\alpha & \text{if } \alpha \neq 1 \\[2ex] \dfrac{\pi\alpha}{2}[J^-(0) + J(0)] & \text{if } \alpha = 1. \end{cases} \tag{17}$$

Set

$$G(x) = \begin{cases} \alpha J^-(0)\displaystyle\int_{-\infty}^{x}\frac{dy}{|y|^{\alpha-1}(1+y^2)}, & x < 0 \\[2ex] \alpha J(0)\displaystyle\int_0^x\frac{dy}{y^{\alpha-1}(1+y^2)} + G(0), & x > 0 \end{cases} \tag{18}$$

and define for arbitrary $\gamma' \in (-\infty, \infty)$

$$\psi(t) = it\gamma' + \int_{-\infty}^{\infty} \left(e^{itx} - 1 - \frac{itx}{1 + x^2}\right) \frac{1 + x^2}{x^2}\, dG(x). \tag{19}$$

Then

$$\psi(t) = it\gamma' + \alpha J(0) \int_0^{\infty} \left(e^{itx} - 1 - \frac{itx}{1 + x^2}\right) \frac{dx}{x^{1+\alpha}}$$

$$+ \alpha J^-(0) \int_{-\infty}^{0} \left(e^{itx} - 1 - \frac{itx}{1 + x^2}\right) \frac{dx}{|x|^{1+\alpha}}$$

and, from the computations following (12), for some $\gamma'' \in (-\infty, \infty)$

$$\psi(t) = it(\gamma'' + \gamma') - c|t|^{\alpha}\left\{1 + i\beta \frac{t}{|t|} w(t, \alpha)\right\}. \tag{20}$$

Hence, from (3), (19), and (20), choosing $\gamma' = \gamma - \gamma''$,

$$\varphi(t) = \exp\{\psi(t)\},$$

whence φ is an i.d. c.f. by Theorem 12.1.3.

Moreover, $\varphi(t)$ is stable, since setting

$$s(t) = it\gamma_{\cdot} - c|t|^{\alpha}\left[1 + i\beta \frac{t}{|t|} w(t, \alpha)\right],$$

for $\alpha \neq 1, 2$ and positive b_i, $i = 1, 2$,

$$s(b_1 t) + s(b_2 t) = s(bt) + ita,$$

where $b = b_1 + b_2 - a > 0$ and $b^{\alpha} = b_1^{\alpha} + b_2^{\alpha}$. If, rather, $\alpha = 1$ then (21) obtains with $b = b_1 + b_2$ and

$$a = \frac{2}{\pi} c\beta[b \log b - b_1 \log b_1 - b_2 \log b_2\}.$$

Thus, φ is a stable c.f. □

Exercises 12.3

1. Show that the mass of a proper stable d.f. is confined to $(0, \infty)$ (resp. $(-\infty, 0)$) iff $\beta = -1$ (resp. $\beta = 1$).

2. Prove that if X has a stable distribution with characteristic exponent $\alpha \in (0, 2)$, then $\mathrm{E}|X|^{\beta} < \infty$ for $\beta < \alpha$ and $\mathrm{E}|X|^{\beta} = \infty$ for $\beta > \alpha$. *Hint*: If $X = X_1 + X_2$, where X_1, X_2 are i.i.d., then $\mathrm{E}|X|^{\beta} < \infty$ for $\beta < \alpha$ by Exercise 8.4.11. If $\beta > \alpha$, then $\mathrm{E}|X|^{\beta} = \infty$ by Theorem 8.4.1 (4).

3. Prove that all proper stable distributions are continuous and infinitely differentiable.

4. A d.f. F (with c.f. φ) is said to be in the **domain of attraction** (resp. domain of normal attraction) of a (stable) distribution G (with c.f. ψ) if for suitable constants A_n, B_n (resp. for some A_n and $B_n = bn^{1/\alpha}$), $\lim e^{itA_n}\varphi^n(t/B_n) = \psi(t)$. Show that every stable distribution belongs to its own domain of (normal) attraction.

5. In coin tossing, prove that the probability that the mth return to equilibrium (equal number of heads and tails) occurs before m^2x tends to $2[1 - \Phi(x^{-1/2})]$ as $m \to \infty$. As usual, Φ is the standard normal d.f.

6. The limit distribution of Exercise 5 has density $f(x) = (2\pi x^3)^{-1/2}e^{-1/2x}$, $x > 0$. This is actually the stable density function corresponding to $\alpha = \frac{1}{2}$, $\beta = -1$, $\gamma = 0$, $c = 1$.

7. If $S_n = \sum_1^n X_i$, $n \geq 1$ is a random walk with X_1 having a symmetric stable distribution of characteristic exponent α, show that $S_n/n^{1/\alpha}$ also has this distribution. Hence, if $1 \leq r < \alpha < 2$, $\mathrm{E}|S_n| = Cn^{1/\alpha}$ for some C in $(0, \infty)$ whence the conclusion of Theorem 10.3.4 fails for $1 \leq r < 2$.

References

K. L. Chung, *A Course in Probability Theory*, Harcourt Brace, New York, 1968; 2nd ed., Academic Press, New York, 1974.

H. Cramér, "Su un teorema relativo alla legge uniforme dei grande numeri," *Giornale dell' Istituto degli Attuari* **5** (1934), 1–13.

C. Esseen, "Fourier analysis of distribution functions," *Acta Math* **77** (1945), 79.

B. V. Gnedenko and A. N. Kolmogorov, *Limit Distributions for Sums of Independent Random Variables*, Addison–Wesley, Reading, Mass., 1954.

P. Lévy, *Théorie de l'addition des variables aléatoires*, Garthier–Villars, Paris, 1937; 2nd ed., 1954.

M. Loève, *Probability Theory*, 3rd ed., Van Nostrand, Princeton, 1963; 4th ed., Springer-Verlag, Berlin and New York, 1977–1978.

Index

Springer Texts in Statistics *(continued from page ii)*

Madansky: Prescriptions for Working Statisticians
McPherson: Applying and Interpreting Statistics: A Comprehensive Guide, Second Edition
Mueller: Basic Principles of Structural Equation Modeling: An Introduction to LISREL and EQS
Nguyen and Rogers: Fundamentals of Mathematical Statistics: Volume I: Probability for Statistics
Nguyen and Rogers: Fundamentals of Mathematical Statistics: Volume II: Statistical Inference
Noether: Introduction to Statistics: The Nonparametric Way
Nolan and Speed: Stat Labs: Mathematical Statistics Through Applications
Peters: Counting for Something: Statistical Principles and Personalities
Pfeiffer: Probability for Applications
Pitman: Probability
Rawlings, Pantula and Dickey: Applied Regression Analysis
Robert: The Bayesian Choice: From Decision-Theoretic Foundations to Computational Implementation, Second Edition
Robert and Casella: Monte Carlo Statistical Methods
Rose and Smith: Mathematical Statistics with *Mathematica*
Santner and Duffy: The Statistical Analysis of Discrete Data
Saville and Wood: Statistical Methods: The Geometric Approach
Sen and Srivastava: Regression Analysis: Theory, Methods, and Applications
Shao: Mathematical Statistics, Second Edition
Shorack: Probability for Statisticians
Shumway and Stoffer: Time Series Analysis and Its Applications
Simonoff: Analyzing Categorical Data
Terrell: Mathematical Statistics: A Unified Introduction
Timm: Applied Multivariate Analysis
Toutenburg: Statistical Analysis of Designed Experiments, Second Edition
Whittle: Probability via Expectation, Fourth Edition
Zacks: Introduction to Reliability Analysis: Probability Models and Statistical Methods

Ruppert: Statistics and Finance: An Introduction
